Ulrich Hohenester

Nano and Quantum Optics

An Introduction to Basic Principles and Theory

Springer

Ulrich Hohenester
Institut für Physik, Theoretische Physik
Karl-Franzens-Universität Graz
Graz, Austria

ISSN 1868-4513 ISSN 1868-4521 (electronic)
Graduate Texts in Physics
ISBN 978-3-030-30506-2 ISBN 978-3-030-30504-8 (eBook)
https://doi.org/10.1007/978-3-030-30504-8

This Springer imprint is published by the registered company Springer Nature Switzerland AG.
The registered company address is: Gewerbestrasse 11, 6330 Cham, Switzerland

Preface

Nano optics combines the research areas of optics and nanoscience. Through light we acquire information about the world around us, and the controlled manipulation of light forms the backbone of numerous optics applications, such as fiber-based communication or light harvesting. Nanoscience, on the other hand, deals with the controlled manufacturing and manipulation of matter at the atomic scale, and has driven the digital revolution that has irrevocably shaped our everyday life, for instance, in the form of computers or mobile phones. A combination of these areas is expected to bring together the best of two worlds. Yet, optics and nanoscience don't come together easily. The diffraction limit dictates that light cannot be squeezed into volumes with dimensions smaller than the wavelength, which are on the order of micrometers rather than nanometers, and conversely in optical microscopy only objects further apart than approximately the light wavelength can be spatially distinguished.

Nano optics deals with the manipulation of light at length scales comparable or smaller than the light wavelength, ideally down to the nanometer scale. In the last decades, scientists have succeeded in devising schemes to let optics go nano. For instance, localization microscopy and optical tweezers have been awarded the Nobel Prizes 2014 and 2018 for light manipulations in the threshold region of the diffraction limit. To overcome the limit, one can collect light at the nanoscale, for instance, in scanning nearfield optical microscopy, or bind within the field of plasmonics light to electron charge oscillations at the surface of metallic nanostructures, hereby squashing light into extreme subwavelength volumes.

This book provides an introduction to nano optics and plasmonics. It is based on a lecture series I have taught over several years at the University of Graz and other places. My main focus is on the basic principles and the theoretical tools needed in nano optics, whereas applications are discussed only exemplary. In this respect, the book is expected to differ from most related textbooks. I have kept references to the current research literature somewhat sparse, but have tried to cite the many excellent review articles for further information, whenever possible. I have also tried to keep the discussion self-contained, and have refrained from using the phrase "it can be shown" or equivalent whenever possible. This has made the presentation considerably longer than I initially thought, but it will hopefully facilitate reading the book.

The book is separated into two parts. The first one deals with classical nano optics, where classical can be understood both in terms of classical electrodynamics and in terms of the classical, canonical presentation of the subject. The second part brings nano optics to the quantum realm. I have tried hard to make the presentation as entertaining as possible, but reading through the final version I realize that it has become somewhat technical and busy—apologies for that. As always, the natural way of reading a book is from the beginning to the end, and in principle there is nothing wrong with this traditional approach. However, since I rarely stick to this order myself, I will not provide any particular advice for using the book: start reading wherever it looks interesting, and go back to the basics if needed.

Many colleagues and students have helped me to bring the book to its present form; they are acknowledged separately below. I sincerely hope that this book will be helpful to both experienced researchers seeking for selected information and to beginners who are interested in a first glance of the topic. For the field of nano optics as a whole, I hope that its future will be as bright and shiny as its past has been.

Graz, Austria
June 2019

Ulrich Hohenester

Acknowledgements

This book presents my personal account of nano optics and plasmonics, but my way of seeing the field has been strongly influenced by many colleagues and predecessors. I thank all of them. My first encounter with the topic has been through the NANOOPTICS group headed by Joachim Krenn at the University of Graz who has worked in the field of plasmonics long before it has become fashionable and has been given this name. I have strongly benefited from their deep insights as well as their serenity in judging novel developments in the light of the long history of the field. I am indebted to my long-term collaborator Andi Trügler who has accompanied me for more than a decade on our joint nano optics and plasmonics activities. Many thanks also to the numerous experimental and theoretical collaborators for sharing their results and opinions, as well as for making science such an exciting and pleasant undertaking.

Teaching the subject has been always important to me, unfortunately, it has never come easy. In 2011, Jussi Toppari invited me to teach a course at a summer school in Jyväskylä, Finland, where I enjoyed both the interactions with the students in the class room and the traditional Finnish sauna. However, I had to realize that I should probably spend more time with the basic things, which one often erroneously calls "simple" after having employed them for a sufficiently long time. For several years, I have used the loose collection of slides compiled for this summer school as lecture notes for a course I have taught at Graz University. The kind invitation of Guido Goldoni and Elisa Molinari for teaching a course on "Nano and Quantum Optics" at the University of Modena and Reggio Emilia in the spring of 2018 finally triggered my (surprisingly spontaneous) decision to start writing a textbook on the subject. The conveniences of the superb Modenese food, the beautiful bike tours to the Apennin, as well as a class of extremely bright students helped me to make the start as enjoyable as possible. Of course, I have completely underestimated writing a textbook, and after having worked on it for about one and a half years my emotions towards the project have remained as mixed and diverse as they have been from the beginning.

Special thanks go to my wife Olga Flor, among many other things for organizing during my stay at Modena, a memorable trip to the eroding castle of Canossa, which has been given up by the Italian state but is kept alive by a few brave volunteers, as well as a visit together with Elisa to the Osteria Francescana, and for showing me how to write real books. I am indebted to numerous colleagues who have read through specific parts of the book and have given most valuable feedback. In alphabetic order, I wish to thank Javier Aizpurua, Stefano Corni, Hari Ditlbacher, Hans Gerd Evertz, Antonio Fernández-Domínguez, Christian Hill, Mathieu Kociak, Joachim Krenn, Olivier Martin, Walter Pötz, Stefan Scheel, and Gerhard Unger. They have helped me to detect the most obvious errors and mistakes in the manuscript. I will provide an updated list of errata on my homepage at the University of Graz, and I invite all readers to inform me about possible errors and to provide feedback on how the presentation could be made even more clear.

Contents

Chapter 1
What Is Nano Optics?

In this chapter we introduce the concepts of propagating and evanescent waves. The removal of the latter waves in conventional optics is responsible for the diffraction limit of light, which we will explain in terms of the scalar wave equation. A discussion within the framework of Maxwell's equations will be given in later parts of this book. We start by introducing the one-dimensional scalar wave equation, and then ponder on the generalization to higher dimensions. Many of these concepts will be familiar to most readers, but are repeated here for clarity. Once we have set up the stage, we will focus on the role of evanescent waves and how to live with or without them in the field of nano optics. We conclude the chapter with a brief summary of Chaps. 2–11 forming the first part of this book.

1.1 Wave Equation

1.1.1 One-Dimensional Waves

What are waves? I encourage the reader to reflect a while about this question and to come up with a meaningful answer. After all, waves are abundant in physics, ranging from water and sound waves to electromagnetic ones, which are the central objects of this book. However, it seems rather difficult to explain what a wave really is. In the book "Introduction to Electrodynamics" Griffiths comes up with the following definition [1]:

> A wave is a disturbance of a continuous medium that propagates with a fixed shape at a constant velocity.

Electronic Supplementary Material The online version of this chapter (https://doi.org/10.1007/978-3-030-30504-8_1) contains supplementary material, which is available to authorized users.

U. Hohenester, *Nano and Quantum Optics*, Graduate Texts in Physics,
https://doi.org/10.1007/978-3-030-30504-8_1

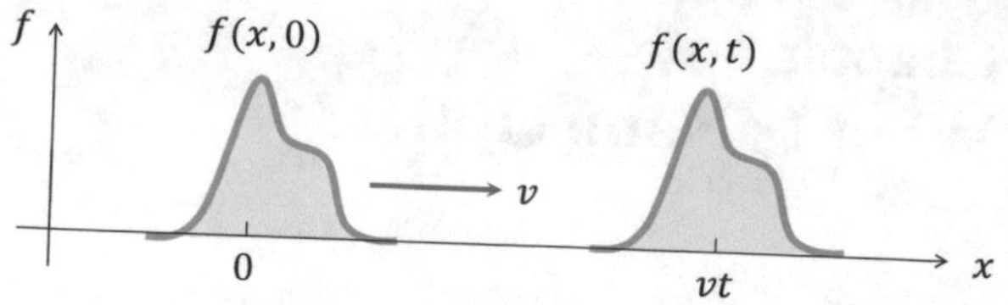

Fig. 1.1 Wave in one spatial dimension. A wave $f(x,t)$ propagates with a fixed speed v without changing its shape. After a time t it has propagated over a distance vt

This definition leaves a number of open questions (what is the continuous medium in case of electromagnetic waves? what about dispersive media?), and I will propose later a modified, albeit more technical definition. To get started, let us take Griffiths' description and consider waves in one spatial dimension. We denote the wave disturbance propagating along x with $f(x,t)$, where t is the time. Figure 1.1 shows a schematic sketch of such a wave propagation. After a time t the initial wave has been displaced by a distance vt. We can thus write

$$f(x,0)=g(x)\,,\quad f(x,t)=g(x-vt)\,,$$

which shows that f is a function of one combined variable $u=x-vt$ rather than of two independent variables x, t. The same analysis applies to a wave that moves to the left, and the general solution is a superposition of left- and right-moving waves

$$f(x,t)=g_-(x-vt)+g_+(x+vt)=g_-(u_-)+g_+(u_+)\,,\qquad u_\pm=x\pm vt\,.$$

It is now easy to show that

$$\left(\frac{\partial}{\partial x}+\frac{1}{v}\frac{\partial}{\partial t}\right)g_-(u_-)=\left(\frac{\partial u_-}{\partial x}+\frac{1}{v}\frac{\partial u_-}{\partial t}\right)\frac{dg(u_-)}{du_-}=\left(1-\frac{v}{v}\right)\frac{dg(u_-)}{du_-}=0\,.$$

Thus, the operator on the left-hand side equates all right-moving waves to zero. Similarly, we find for the left-moving waves

$$\left(\frac{\partial}{\partial x}-\frac{1}{v}\frac{\partial}{\partial t}\right)g_+(u_+)=\left(\frac{\partial u_-}{\partial x}-\frac{1}{v}\frac{\partial u_+}{\partial t}\right)\frac{dg(u_+)}{du_+}=\left(1-\frac{v}{v}\right)\frac{dg(u_+)}{du_+}=0\,.$$

If we apply both operators on the wavefunction $f(x,t)$, we equate the left- and right-moving waves to zero, and we arrive at the scalar wave equation in one spatial dimension

Scalar Wave Equation for One Spatial Dimension

$$\left(\frac{\partial}{\partial x}+\frac{1}{v}\frac{\partial}{\partial t}\right)\left(\frac{\partial}{\partial x}-\frac{1}{v}\frac{\partial}{\partial t}\right)f(x,t)=\left(\frac{\partial^2}{\partial x^2}-\frac{1}{v^2}\frac{\partial^2}{\partial t^2}\right)f(x,t)=0\,. \tag{1.1}$$

In what follows, we consider the most simple wave form, a sinusoidal wave, which can be written as

$$f(x,t) = A\cos(kx - \omega t + \delta)\,. \tag{1.2}$$

Here A is the amplitude, k is the wavenumber, ω is the angular frequency, and δ is a phase factor. The wavenumber and angular frequency are related to the wavelength λ and the oscillation period T via

$$k = \frac{2\pi}{\lambda}\,, \qquad \omega = \frac{2\pi}{T}\,.$$

With this, we find that the sinusoidal wave is periodic in λ and T,

$$f(x+\lambda,t) = A\cos\left[2\pi\left(\frac{x}{\lambda}+1\right) - \omega t + \delta\right] = f(x,t)$$

$$f(x,t+T) = A\cos\left[kx - 2\pi\left(\frac{t}{T}+1\right) + \delta\right] = f(x,t)\,.$$

It turns out to be convenient to expand our definition of sinusoidal waves to the complex plane. Starting from Euler's formula

$$e^{i\phi} = \cos\phi + i\sin\phi\,, \tag{1.3}$$

where ϕ is some phase argument, we express the sinusoidal wave in the form

$$f(x,t) = \mathrm{Re}\left[A\,e^{i(kx-\omega t)}e^{i\delta}\right] = \mathrm{Re}\left[\tilde{A}\,e^{i(kx-\omega t)}\right]\,, \quad \tilde{A} = Ae^{i\delta}\,.$$

In the last expression we have introduced a complex amplitude $\tilde{A}$, which is the product of A and the phase factor $e^{i\delta}$. It turns out that the use of complex waves and of the real part operation (in order to get the physically meaningful part) is so successful that throughout this book we will no longer explicitly indicate the real part operation. Then, a sinusoidal wave is of the form

Sinusoidal Wave in One Spatial Dimension

$$f(x,t) = A\,e^{i(kx-\omega t)} \tag{1.4}$$

and take real part operation $\mathrm{Re}\{\dots\}$ at end.

We have dropped the tilde from the amplitude, which from now on is always understood as a complex quantity. To make this point clear again: physical waves are always real, the complex notation is only adopted for simplicity and we assume implicitly that the "real" wave (that can be compared with experiment) is obtained by taking the real part of the complex expressions.

So far we have described the sinusoidal wave in terms of the wavenumber k and angular frequency ω. In the same way as $f(x,t)$ is not a function that depends independently on x and t, but on a single variable $x \mp vt$, ω and k are related to each other through the so-called dispersion relation. This relation can be obtained by inserting the sinusoidal wave ansatz into the scalar wave equation, Eq. (1.1),

$$\left(\frac{\partial^2}{\partial x^2}-\frac{1}{v^2}\frac{\partial^2}{\partial t^2}\right)A\,e^{i(kx-\omega t)}=-\left(k^2-\frac{\omega^2}{v^2}\right)A\,e^{i(kx-\omega t)}=0\,.$$

To fulfill the wave equation for arbitrary x, t values the term in parentheses must be zero. We thus find for the dispersion relation of the scalar wave equation

Dispersion Relation for Scalar Wave Equation (1D)

$$\omega(k)=vk\,. \tag{1.5}$$

Here $\omega(k)$ is the angular frequency, which is a function of the wavenumber k, and v is the velocity of the wave propagation. In what follows, we show that any wave can be decomposed into sinusoidal waves, and that the dispersion relation determines the propagation properties of waves.

Fourier Transform. An important theorem in mathematics states that any function (which is sufficiently well behaved) can be decomposed into sinusoidal-like waves through

$$f(x)=\int_{-\infty}^{\infty}e^{+ikx}\tilde{f}(k)\,\frac{dk}{2\pi} \tag{1.6a}$$

$$\tilde{f}(k)=\int_{-\infty}^{\infty}e^{-ikx}f(x)\,dx\,. \tag{1.6b}$$

Here $\tilde{f}(k)$ is called the Fourier transform of $f(x)$. The magic of Eq. (1.6) is that both $f(x)$ and $\tilde{f}(k)$ contain the same information. Thus, if we know $f(x)$ we can immediately compute $\tilde{f}(k)$, and vice versa. Note that the factor of $1/(2\pi)$ in the integration over k could be also shifted to the integration over x, or both integrations could acquire a $1/\sqrt{2\pi}$ prefactor in a symmetric fashion. In this book we will usually adopt the above definition, but will occasionally deviate from it.

Wave Propagation. Suppose that we have given a wave $f(x,0)$ at time zero, and would like to compute its shape at later time. This can be done easily when using sinusoidal waves and the Fourier transform. At time zero we decompose $f(x,0)$ into sinusoidal waves via Eq. (1.6a). As time goes on, each sinusoidal wave evolves according to Eq. (1.4), and we get

$$f(x,t)=\int_{-\infty}^{\infty} e^{i[kx-\omega(k)t]}\,\tilde f(k)\,\frac{dk}{2\pi}\,. \tag{1.7}$$

With the dispersion relation of Eq. (1.5) for the scalar wave equation we can work out the integral explicitly, and obtain

$$f(x,t)=\int_{-\infty}^{\infty} e^{ik(x-vt)}\,\tilde f(k)\,\frac{dk}{2\pi}=f(x-vt,0)\,,$$

in agreement to our initial discussion that waves propagate without changing their shapes.

Dispersion. Let us pause for a moment and consider a more complicated dispersion relation. We will encounter such modifications in later parts of this book when discussing dispersive media. In principle we can still use Eq. (1.7) for the wave propagation, but the evaluation of the integral is now more complicated because of the modified $\omega(k)$ function. As a representative example we consider for $\tilde f(k)$ a Gaussian centered around k_0 with a width of σ_0^{-1}. For small widths the function is strongly peaked around k_0, and we can approximate $\omega(k)$ through a Taylor series around k_0,

$$\omega(k)\approx\omega_0+v_g\big(k-k_0\big)+\frac{\beta}{2}\big(k-k_0\big)^2\,,\quad v_g=\left[\frac{d\omega}{dk}\right]_{k_0}\,,\ \beta=\left[\frac{d^2\omega}{dk^2}\right]_{k_0}\,.$$

Here v_g is the group velocity and β a dispersion parameter. As explicitly worked out in Exercise 1.5, the integral of Eq. (1.7) can be solved analytically for the Gaussian and the approximated dispersion relation, and we obtain

$$f(x,t)=\sigma^{-\frac{1}{2}}(t)\,e^{i(k_0x-\omega_0t)}\exp\left[-\frac{(x-v_gt)^2}{2\sigma^2(t)}\right]\,,\qquad \sigma(t)=\sqrt{\sigma_0^2+i\beta t}\,. \tag{1.8}$$

Thus, the Gaussian wavepacket propagates with the group velocity v_g, but owing to β it does not conserve its shape but broadens while propagating, as described by $\sigma(t)$. In the remainder of this chapter we will not be overly concerned with dispersive media. However, we have added this brief discussion here to emphasize that most of our analysis not only applies to wave propagation in free space and non-dispersive media, but can be easily extended to more complicated situations.

1.1.2 Three-Dimensional Waves

So how do things change when we go from one to three spatial dimensions? Formally not that much. Instead of Eq. (1.1) we get

Scalar Wave Equation for Three Spatial Dimensions

$$\left(\nabla^2 - \frac{1}{v^2}\frac{\partial^2}{\partial t^2}\right) f(\boldsymbol{r},t) = 0\,, \tag{1.9}$$

where $f(\boldsymbol{r},t)$ is the scalar wavefunction depending on $\boldsymbol{r} = (x,y,z)$, and ∇^2 is the usual Laplace operator

$$\nabla^2 = \frac{\partial^2}{\partial x^2} + \frac{\partial^2}{\partial y^2} + \frac{\partial^2}{\partial z^2}\,.$$

Similarly to the decomposition into sinusoidal waves of Eq. (1.4) we introduce plane waves

Plane Wave in Three Spatial Dimensions

$$f(x,t) = A\, e^{i(\boldsymbol{k}\cdot\boldsymbol{r}-\omega t)}\,, \tag{1.10}$$

where A is the amplitude and $\boldsymbol{k} = k\hat{\boldsymbol{n}}$ is the *wavevector* that has the length $k = 2\pi/\lambda$ determined by the wavelength λ and points in the direction of the wave propagation, see Fig. 1.2. With these plane waves we can define in complete analogy to Eq. (1.6) the three-dimensional Fourier transform

$$f(\boldsymbol{r}) = \int_{-\infty}^{\infty} e^{+i\boldsymbol{k}\cdot\boldsymbol{r}}\,\tilde{f}(\boldsymbol{k})\,\frac{d^3k}{(2\pi)^3} \tag{1.11a}$$

$$\tilde{f}(\boldsymbol{k}) = \int_{-\infty}^{\infty} e^{-i\boldsymbol{k}\cdot\boldsymbol{r}}\,f(\boldsymbol{r})\,d^3r\,. \tag{1.11b}$$

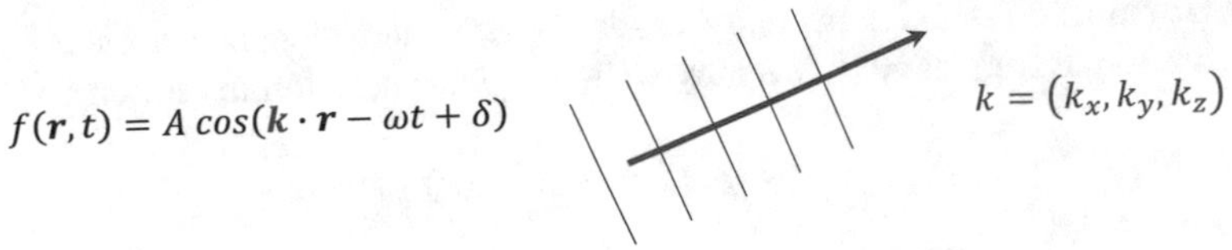

Fig. 1.2 Plane wave in three dimensions. The *wavevector* $\boldsymbol{k} = k\hat{\boldsymbol{n}}$ has the length of the wavenumber $k = 2\pi/\lambda$, and points in the wave propagation direction $\hat{\boldsymbol{n}}$. The lines perpendicular to $\boldsymbol{k}$ indicate planes of constant phase

Finally, upon insertion of plane waves into the scalar wave equation of Eq. (1.9) we get

$$k_x^2 + k_y^2 + k_z^2 - \frac{\omega^2}{v^2} = 0\,.$$

In principle, we obtain positive and negative solutions for ω. However, we only keep the positive ones for reasons to become more clear in Chap. 5 when discussing the solutions of the wave equation in terms of in- and out-going spherical waves. Only the latter ones, which correspond to positive frequencies, fulfill the requirement of causality and must be kept. We thus find for the dispersion relation of waves in three dimensions

Dispersion Relation for Scalar Wave Equation (3D)

$$\omega(k) = v\big|\boldsymbol{k}\big| = v\sqrt{k_x^2 + k_y^2 + k_z^2}\,. \tag{1.12}$$

We emphasize that this dispersion relation has been directly derived from the scalar wave equation, and no approximations have been adopted. For this reason, the dispersion relation must be always fulfilled—it is strict and not negotiable. We will come back to this in a moment when discussing evanescent waves. A few further comments might be in place.

Linearity. First, the wave equation is linear which has the consequence that if f_1 and f_2 are two solutions of the wave equation, then also the sum $f_1 + f_2$ is a solution of the wave equation. The Fourier transform of Eq. (1.11) is a special case of this, where we have decomposed the wave into particularly simple plane waves. If we know how a single plane wave evolves in time, we can describe the time evolution of a more complicated wave by decomposing it into such simple plane waves. This always works because plane waves form a complete basis, as stated by Fourier's theorem.

Time Harmonic Fields. In many cases we are interested in waves oscillating with a single frequency ω. We may write the wave solution in the form[1]

$$f(\boldsymbol{r},t)=e^{-i\omega t}f(\boldsymbol{r})\,, \tag{1.13}$$

where for notational simplicity we keep the same symbol for $f(\boldsymbol{r},t)$ and $f(\boldsymbol{r})$. The above form is less restrictive than it seems because we can always decompose a wave into harmonic components (using a Fourier transformation), and it thus suffices to investigate time-harmonic fields of the form given by Eq. (1.13) solely. Any more complicated wave form can then be constructed from the superposition of these simple waves.

Wave Equation. If we insert Eq. (1.13) into the wave equation of Eq. (1.9) we get

$$(\nabla^2+k^2)f(\boldsymbol{r})=0\,,$$

where we have cancelled the common term $e^{-i\omega t}$. At the beginning of this chapter I have promised to come up with a more general definition for the wave equation. Indeed, we can define a wave as a solution of the generalized wave equation

$$\left(\nabla^2+n^2(\omega)k^2\right)f(\boldsymbol{r})=0\,, \tag{1.14a}$$

where $n(\omega)$ is some frequency-dependent refractive index. The above form then also applies to dispersive media. We could go even a step further and define waves as solutions of the inhomogeneous wave equation

$$\left(\nabla^2+n^2(\boldsymbol{r},\omega)k^2\right)f(\boldsymbol{r})=0\,, \tag{1.14b}$$

where the refractive index depends on the spatial coordinate and on frequency. These definitions are not as instructive as the one given by Griffiths, but we shall find it convenient to refer to "waves" or "wave-like" solutions when dealing with solutions of Eq. (1.14).

1.2 Evanescent Waves

We are now ready to get real. The situation we have in mind is depicted in Fig. 1.3. Suppose that we know a scalar field $f(x,y,0)$ at position $z=0$, here the "nano"

---

[1]Note that in physics one usually introduces the time-harmonic form $e^{-i\omega t}$. In engineering one usually writes $e^{j\omega t}$, where j is the imaginary unit and the sign in the exponential is reversed in comparison to the physics convention.

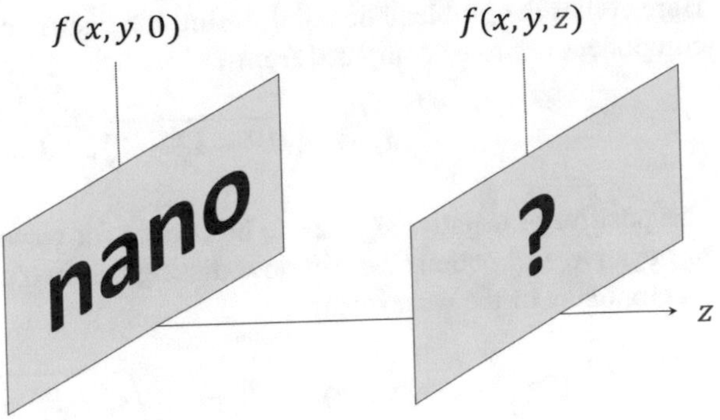

Fig. 1.3 Suppose that we know the field $f(x, y, 0)$ at position $z = 0$, here the "nano" letters. How does the field evolve when propagating over a distance z?

word, and suppose that the field is oscillating with a single frequency ω. In the following we ask the questions:

- How does the field evolve when propagating over a distance z?
- And how can we compute the field $f(x, y, z)$ at position z?

In principle, with the tools developed in the previous section we can analyze the situation quite easily.

Plane Wave Decomposition. We start by decomposing the initial field $f(x, y, 0)$ in a plane wave basis using

$$f(x, y, 0) = (2\pi)^{-2} \int_{-\infty}^{\infty} e^{i(k_x x + k_y y)} \tilde{f}(k_x, k_y)\, dk_x dk_y \,.$$

Wave Propagation. When moving away from $z = 0$, we can make the general ansatz

$$f(x, y, z) = (2\pi)^{-3} \int_{-\infty}^{\infty} e^{i(k_x x + k_y y + k_z z)} \tilde{f}(k_x, k_y, k_z)\, dk_x dk_y dk_z \,.$$

However, we must additionally fulfill the constraint of the wave equation and therefore cannot chose ω and k_x, k_y, k_z independently. We can thus express one variable, for instance, k_z, in terms of the others and are led to

$$f(x, y, z) = (2\pi)^{-2} \int_{-\infty}^{\infty} \exp\Big\{i\Big[k_x x + k_y y + k_z(k_x, k_y) z\Big]\Big\} \tilde{f}(k_x, k_y)\, dk_x dk_y \,, \tag{1.15}$$

where we have explicitly indicated the dependence of k_z on k_x, k_y. From this expression one observes that for $z > 0$ each plane wave acquires an additional phase

$$\tilde{f}(k_x, k_y) \underset{z>0}{\longrightarrow} e^{ik_z z} \tilde{f}(k_x, k_y) \,.$$

Here comes the problematic point. Using the dispersion relation of Eq. (1.3), the k_z component has to be computed from

$$k_z = \pm\sqrt{k^2 - k_x^2 - k_y^2}\,, \quad k = \frac{\omega}{v}\,. \tag{1.16}$$

The positive or negative sign has to be chosen for waves propagating in the positive or negative z direction. We can now distinguish two cases. For $k_x^2 + k_y^2 \le k^2$ the z-component of the wavevector

$$\text{(Propagating wave)} \qquad k_z = \pm\sqrt{k^2 - k_x^2 - k_y^2} \tag{1.17}$$

is a real number, corresponding to a normal wave propagation. However, for $k_x^2 + k_y^2 \ge k^2$ we get

$$\text{(Evanescent wave)} \qquad k_z = \pm\sqrt{k^2 - k_x^2 - k_y^2} = \pm i\sqrt{k_x^2 + k_y^2 - k^2} \equiv \pm i\kappa\,, \tag{1.18}$$

which corresponds to an imaginary wavenumber! Readers familiar with the concept of evanescent waves will not be overly surprised by this finding. However, those unfamiliar with this concept should take their time to check carefully whether we have done everything properly up to this point, or whether we have missed something important. However, the only two ingredients in the derivation of evanescent waves are the plane wave decomposition, which is based on Fourier's theorem (and which we better should not question), and the dispersion relation, which is deeply rooted in the wave equation itself (which forms the basis of our whole analysis). So obviously there is nothing wrong with evanescent waves, and they are here to stay.

In order to understand how these evanescent waves propagate, we insert the imaginary wavenumber into the plane wave ansatz to get

$$\exp\Big[i\big(k_x + k_y y \pm i\kappa z\big)\Big] = \exp\Big[i\big(k_x + k_y y\big) \mp \kappa z\Big]\,.$$

Thus, evanescent waves grow or decay exponentially when moving away from z. To be physically meaningful, we only keep the decaying waves, this is $e^{-\kappa z}$ for $z > 0$ and $e^{\kappa z}$ for $z < 0$. Evanescent waves are better known in quantum mechanics. Figure 1.4a shows a quantum mechanical particle that impinges on a potential barrier. If the kinetic energy of the particle is smaller than the height of the barrier, it is reflected. However, part of the wave penetrates into the barrier where its amplitude decays exponentially. This is the analog to evanescent waves for the scalar wave equation. If the width of the potential barrier is reduced, see panel (c), the particle can tunnel through the barrier, where it becomes converted again into a propagating wave.

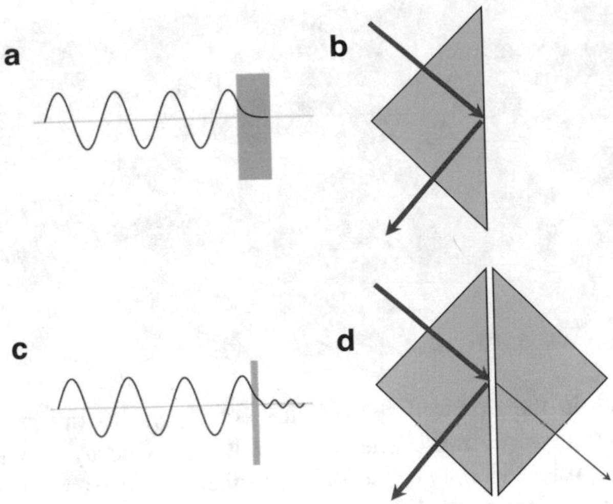

Fig. 1.4 (**a**) Transmission of a quantum mechanical particle at a potential barrier. The particle tunnels into the barrier and becomes reflected. (**c**) When the width of the barrier is reduced, the wavefunction penetrates through the barrier and the quantum mechanical particle can tunnel through a classically forbidden region. (**b**) Classical wave analog to quantum tunneling. A light wave propagates through a prism and becomes reflected under conditions of total internal reflection. The wave "tunnels" into the classically forbidden region at the prism-air side, and the wave acquires a small phase shift usually known as the Goos–Hänchen shift. (**d**) When a second prism is brought close to the first one, with a distance comparable to the light wavelength, light can "tunnel" through the air gap, and becomes converted on the other side of the prism into a propagating wave

Tunneling is a general wave phenomenon, and can not only be observed in quantum mechanics but also in electrodynamics. Panel (c) shows a prism where an incoming light beam becomes reflected under the condition of total internal reflection. Similarly to the situation shown in panel (a), the reflection is not abrupt but part of the light field penetrates to the air side of the prism where it decays exponentially (evanescent wave). This penetration can be observed as the so-called Goos–Hänchen phase shift a reflected wave suffers in comparison to an abruptly reflected wave [2]. When a second prism is brought close to the first one, panel (d), the exponentially decaying field amplitude of the first prism can "tunnel" to the second prism, where it becomes converted again into a propagating light field.

While the above examples are of somewhat limited use, evanescent waves play a more important role in the understanding of the resolution limit of light. We return to our previous plane wave decomposition of Eq. (1.15), and express the fields at larger z values in the form

Scalar Wave Propagation ($z > 0$)

$$f(x,y,z) = (2\pi)^{-2}\int_{k^2>k_x^2+k_y^2} e^{i\left(k_xx+k_yy+\sqrt{k^2-k_x^2-k_y^2}z\right)}\tilde{f}(k_x,k_y)\,dk_xdk_y$$
$$+\,(2\pi)^{-2}\int_{k^2<k_x^2+k_y^2} e^{i(k_xx+k_yy)-\sqrt{k_x^2+k_y^2-k^2}z}\tilde{f}(k_x,k_y)\,dk_xdk_y\,. \tag{1.19}$$

Here the terms in the first and second line correspond to the propagating and evanescent waves, respectively. The evanescent waves decay exponentially and contribute only in close vicinity to $z = 0$. The further we move away from this plane, the more the evanescent waves become exponentially damped. In later parts of the book we will show that evanescent waves are always bound to matter. Figure 1.5 demonstrates the impact of the decay of evanescent waves. Panel (a) shows the Fourier transform of the image formed by the "nano" letters shown in the inset. When part of the k-space is removed, for instance, through the decay of evanescent waves, the inverse Fourier transform gives a modified function. Panel (b) shows the function reconstructed from the Fourier components located inside the circle of panel (a), and panels (c,d) report images with a further reduced k-space content. From these images it is clear that the large wavenumber components of $\tilde{f}(k_x, k_y)$ carry the high-resolution information. The more these waves become removed from the propagated wave $f(x, y, z)$, the more the function blurs and all fine details are washed out.

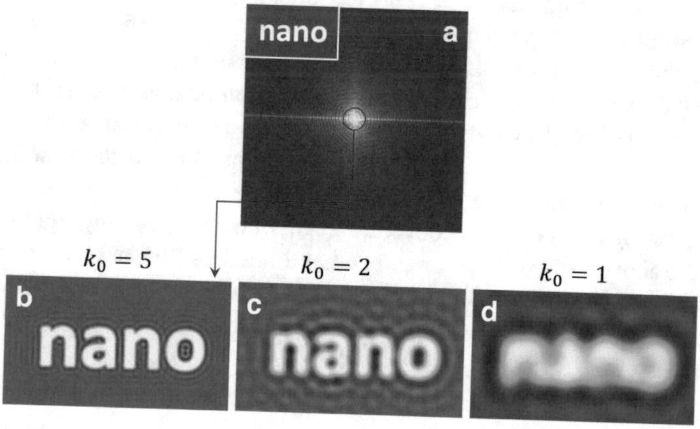

Fig. 1.5 Dependence of resolution on cutoff parameter k_0. (**a**) Fourier transform of "nano" letters (inset). The red circle reports wavenumbers with $k_0 = 5$. (**b–d**) Inverse Fourier transform for different cutoff parameters, with $k_x^2 + k_y^2 \leq k_0^2$ in arbitrary units

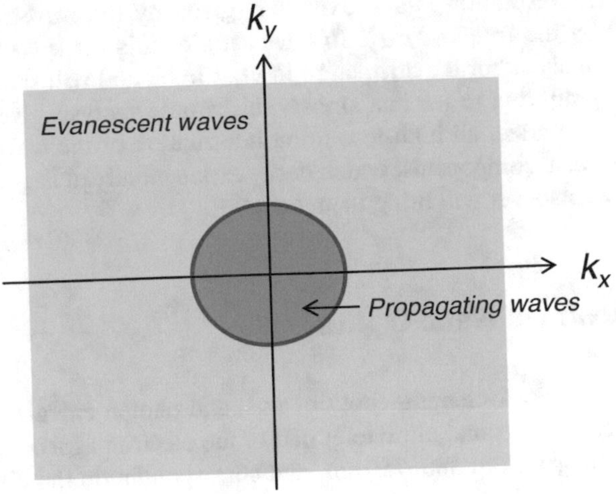

Fig. 1.6 Diffraction limit of light. Only the inner part of the wavevector space corresponds to propagating waves, the larger wavenumber components, which carry information about the fine spatial details of $f(x, y)$, correspond to evanescent waves which decay exponentially when moving away from $z = 0$. Far away from this plane only the propagating waves survive, and in an optical image reconstruction all high-resolution information is lost

We are now in the position to qualitatively discuss the diffraction limit of light. As shown in Fig. 1.6, only wavevectors with a sufficiently small modulus correspond to propagating waves, whereas components with a larger wavenumber, which carry the high spatial resolution, correspond to evanescent waves which decay exponentially when moving away from the plane $z = 0$. If we assume that the spatial resolution Δ is approximately given by the largest wavenumber k_{max} available, we get

$$\Delta \approx \frac{2\pi}{k_{\text{max}}}\,.$$

The largest wavenumber associated with a propagating mode can be found from the dispersion relation

$$\sup\left(k_x^2 + k_y^2\right) = k_{\text{max}} = \frac{\omega}{v}\,.$$

Combining these two equations, we then obtain for the best resolution one can obtain from the propagating modes

Diffraction Limit of Scalar Wave Equation (approximate)

$$\Delta \approx \frac{2\pi v}{\omega} = \lambda\,. \tag{1.20}$$

Thus, the spatial resolution is approximately given by the wavelength λ (which is determined by the frequency ω). In later parts of this book we will provide a more rigorous analysis for the diffraction limit of light, and will show that $\Delta \approx \frac{\lambda}{2}$. However, it is gratifying to see that already this simple analysis gives a reasonable estimate. To summarize, all high-resolution information of the wave is carried by the evanescent wave components, which decay exponentially at larger distances and no procedure whatsoever will bring them back.

1.3 The Realm of Nano Optics

Figure 1.7 shows the wavelengths (bottom axis) and photon energies (top axis) for the near-infrared, visible, and ultraviolet part of the electromagnetic spectrum. The visible regime ranges from 380–750 nm, and correspondingly the diffraction limit is in the micrometer rather than nanometer regime. Thus, optics and nanoscience do not come naturally together! Nano optics is the science that tries to push optics to the nanoscale despite these limitations.

First, and most importantly, we have to realize that the diffraction limit is based on fundamental laws of physics, most importantly the dispersion relation which is deeply rooted in the fundamental wave equation. From the dispersion relation we find that there exist two types of waves, propagating and evanescent ones, and the decay of the latter waves is responsible for the loss of resolution. Using conventional optics it is not possible to resolve objects that are closer to each other than the wavelength of light λ, and conversely we cannot focus light to spots that are smaller in dimension than λ. In order to overcome the diffraction limit of light we can hardly compete with the fundamental laws of physics, thus we have to change the rules of the game. Nano optics has come up with a number of successful solutions, which will be discussed in detail in this book. Figure 1.8 shows three representative examples.

Nearfield Optics. In scanning nearfield optical microscopy (SNOM) an optical fiber is brought into close vicinity of a nano object, see panel (a). Through the fiber tip, the evanescent nearfields of the nano object can be converted into propagating photons, which are detected at the other end of the fiber. By raster-

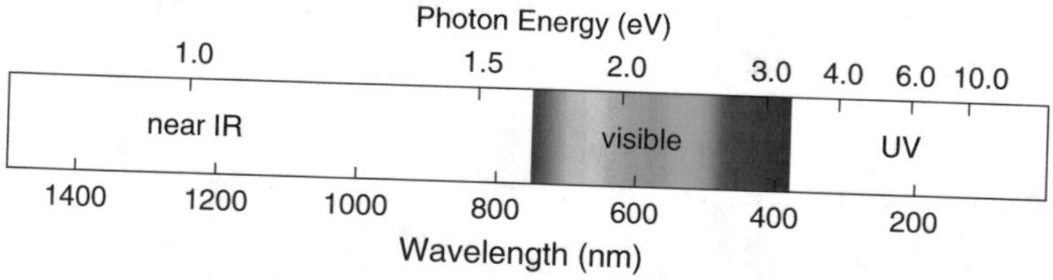

Fig. 1.7 Wavelengths (bottom) and photon energies (top) for near-infrared (770–3000 nm), visible (380–750 nm), and ultraviolet (10–380 nm) regime

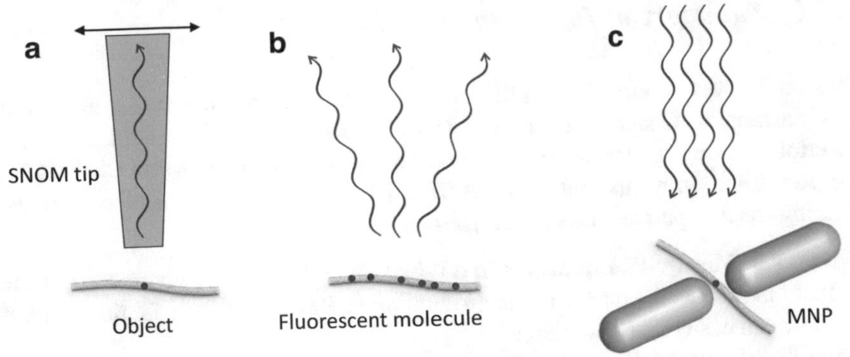

Fig. 1.8 Three examples of how to bring optics to the nanoscale. (**a**) In scanning nearfield optical microscopy (SNOM) an optical fiber is brought in close vicinity to some nano object. Through coupling to the fiber the evanescent nearfields of the object are converted to propagating photons, which can be detected at the end of the fiber. By raster-scanning the fiber over the specimen, one obtains spatial information about the nearfields. (**b**) In localization microscopy fluorescent molecules are attached to the nano object to be investigated. In the far-field one can measure the location of these molecules with nanometer spatial resolution, and thus obtains indirectly information with high spatial resolution about the nano object under investigation. (**c**) In plasmonics one or several metallic nanoparticles (MNPs) act as nano antennas which convert far-field radiation to evanescent fields, which allows delivering light to the nanoscale

scanning the fiber over the specimen, one obtains information about the optical nearfields with nanometer resolution.

Localization Microscopy. While in conventional optics there is a priori no information available about the system to be studied, in localization microscopy the objects under study (typically biological systems such as cells) are decorated with fluorescent molecules. In conventional far-field optics the positions of the fluorescing molecules can be measured with nanometer resolution, provided that the optically active molecules are further apart than the light wavelength, which allows to obtain optical images of nano objects with nanometer resolution. Localization microscopy has been awarded the Nobel Prize in Chemistry 2014.

Plasmonics. In plasmonics one exploits coherent electron charge oscillations at the interface between a metal and a dielectric, so-called surface plasmons or particle plasmons, to convert far-field radiation to plasmons. These plasmons come along with strongly localized evanescent waves, which allows delivering light at the nanoscale. By a similar token, a quantum emitter located close to a plasmonic nanoparticle can use the particle as a nano antenna in order to emit light much more efficiently. This principle form the basis for techniques such as surface-enhanced fluorescence or surface-enhanced Raman scattering (SERS).

1.3.1 Summary of Book Chaps. 2–11

This book gives an introduction to the theoretical concepts underlying nano optics and plasmonics. Chapters 2–11 form the first part of the book and deal with the description using the framework of classical electrodynamics, while Chaps. 12–18 are concerned with quantum aspects. In all brevity, the contents of the chapters forming the first part can be summarized as follows.

Chapter 2: Maxwell's Equations in a Nutshell. We start by giving a short summary of Maxwell's theory of electrodynamics. Readers already familiar with the topic can easily skip this chapter.

Chapter 3: Angular Spectrum Representation. This chapter provides the tools needed for the theoretical description of optical imaging. We show how to compute the optical far-fields and how to theoretically describe the operation of collection and imaging lenses.

Chapter 4: Symmetry and Forces. Here we investigate the symmetries underlying Maxwell's equations, in particular momentum conservation and optical forces, energy conservation and optical cross sections, and orbital angular momentum. We also discuss the working principles of optical tweezers, for which one half of the Nobel Prize in Physics 2018 has been awarded.

Chapter 5: Green's Functions. This chapter gives an introduction to the concept of Green's functions, which are of paramount importance in the field of nano optics. We derive the Green's functions for the Helmholtz and vector wave equations, and obtain the respective representation formulas.

Chapter 6: Diffraction Limit and Beyond. Here we provide a thorough discussion of the diffraction limit of light, and show how to overcome this limit using scanning nearfield optical microscopy (SNOM) and localization microscopy, for which the Nobel Prize in Chemistry 2014 has been awarded.

Chapter 7: Material Properties. This chapter marks the end of the part concerned with optical fields only. We introduce generic models for the material response, including the Drude–Lorentz and Drude models, and show how to obtain through suitable averaging over the microscopic properties the macroscopic Maxwell's equations.

Chapter 8: Stratified Media. The most simple hybrid system combining Maxwell's equations with materials are stratified media, consisting of stacked planar layers of different materials. We show that a novel type of excitation exists at the interface between a metal and a dielectric, so-called surface plasmons, and discuss how these and related excitations can be tailored and exploited for numerous applications.

Chapter 9: Particle Plasmons. Surface plasmons in restricted geometries ("particles") give rise to particle plasmons, which exhibit resonances and come along with strongly localized evanescent nearfields. We investigate simple geometries for which analytic models are available, and develop general description schemes for more complicated particle shapes.

Chapter 10: Photonic Local Density of States. Through the combination of plasmonic nanoparticles with quantum emitters it becomes possible to completely alter the optical properties of the emitters. We introduce to surface-enhanced fluorescence and Raman scatterings (SERS), and discuss the basic principles underlying electron energy loss spectroscopy (EELS).

Chapter 11: Computational Methods in Nano Optics. The final chapter of the first part deals with computational methods in nano optics and photonics, including finite difference time domain (FDTD), boundary element method (BEM), and finite element method (FEM) simulation approaches.

Exercises

Exercise 1.1 Consider a complex number $z = x + iy$ and its complex conjugate $z^* = x - iy$. Express the real part $\text{Re}(z)$ and imaginary part $\text{Im}(z)$ in terms of z and z^*.

Exercise 1.2 Show that a standing wave $f(x,t) = A\sin(kz)\cos(kvt)$ satisfies the wave equation, and express it as a sum of a waves traveling to the left and right.

Exercise 1.3 Obtain the solution of the one-dimensional wave equation, Eq. (1.1), through separation of variables, $f(x,t) = \phi(x)\psi(t)$.

Exercise 1.4 Consider a Gaussian wavepacket of the form

$$f(x,0) = A\exp\left[ik_0x + \frac{x^2}{2\sigma^2}\right].$$

Compute the Fourier transform, and use Eq. (1.7) to get the wave at a later time t. The following Gaussian integral might be useful:

$$\int_{-\infty}^{\infty} e^{-ax^2+bx+c}\,dx = \sqrt{\frac{\pi}{a}}e^{\frac{b^2}{4a}+c}.$$

Exercise 1.5 Same as Exercise 1.4 but for a dispersion relation of the form

$$\omega(k) = \omega_0 + v_g\left(k - k_0\right) + \frac{\beta}{2}\left(k - k_0\right)^2.$$

Interpret the final result:

(a) How does the wave propagate in time?
(b) How does the width of the wavepacket change as a function of time?

Exercise 1.6 The position $x = 0$ separates two media:

- for $x < 0$ the wave velocity is v_1 and
- for $x > 0$ the wave velocity is v_2.

A wave impinges from the left-hand side on the interface. Part of the wave becomes reflected and part transmitted,

$$\begin{aligned} x < 0: \quad & f(x,t) = \quad e^{i(k_1 x - \omega t)} + R\, e^{i(-k_1 x - \omega t)} \\ x > 0: \quad & f(x,t) = T\, e^{i(k_2 x - \omega t)}\,, \end{aligned}$$

where R and T are reflection and transmission coefficients.

(a) Use the dispersion relations to compute k_1 and k_2.
(b) Assume that the function and its derivative are continuous at $x = 0$ in order to compute R, T.

Exercise 1.7 Same as Exercise 1.6 but for two dimensions. Consider a wave ansatz with a given wavenumber k_y (with $k_y < \omega/v_1$)

$$\begin{aligned} x < 0: \quad & f(x,y,t) = \quad e^{i(k_{1x} x + k_y y - \omega t)} + R\, e^{i(-k_{1x} x + k_y y - \omega t)} \\ x > 0: \quad & f(x,y,t) = T\, e^{i(k_{2x} x + k_y y - \omega t)}\,. \end{aligned}$$

(a) What kind of solution does this function describe?
(b) Use the dispersion relations to compute k_{1x} and k_{2x}.
(c) Assume that the function and its derivative in x direction are continuous at $x = 0$ in order to compute R, T.
(d) Discuss under which conditions the solutions for $x > 0$ have an evanescent character.

Exercise 1.8 In the fabrication of computer chips one nowadays uses photons with an energy of about 10 eV. Use the diffraction limit of light to estimate the smallest structure sizes achievable, and compare with actual gate lengths of about 15 nm.

Chapter 2
Maxwell's Equations in a Nutshell

In this book I will assume that readers are already familiar with the basic concepts of electrodynamics. Many textbooks can be found on the topic and in general they will all suffice as a suitable starting point. Jackson's "Classical Electrodynamics" [2] is certainly among the most comprehensive accounts of the topic. My favorite book for teaching is the "Introduction to Electrodynamics" by Griffiths [1], and readers familiar with this book will probably recognize some of his notations here (albeit not the famous $\mathscr{r}$). In the following I briefly summarize the basic ideas of electrodynamics, however, without going too much into details.

2.1 The Concept of Fields

Electrostatics can be briefly summarized through Coulomb's law that describes how a particle with charge q_1 situated at position $\boldsymbol{r}_1$ becomes attracted or repelled by a second particle with charge q_2 situated at position $\boldsymbol{r}_2$,

$$\boldsymbol{F}_{12} = \frac{1}{4\pi\varepsilon_0}\frac{q_1 q_2}{r_{12}^2}\,\hat{\boldsymbol{r}}_{12}\,. \tag{2.1}$$

Here ε_0 is the vacuum permittivity, which appears because of the SI unit system under use, $\boldsymbol{r}_{12} = \boldsymbol{r}_1 - \boldsymbol{r}_2$ is the distance vector between the two charges, and $\hat{\boldsymbol{r}}_{12}$ is the unit vector pointing in the direction of $\boldsymbol{r}_{12}$. Let me emphasize a few important points about Coulomb's law of Eq. (2.1).

Symmetry. Coulomb's law only depends on the *relative* distance vector $\boldsymbol{r}_{12}$. For this reason, it respects the homogeneity of space (no point of space is distinguished with respect to any other one) and the isotropy of space (no direction in space is distinguished with respect to another one). We will come back to this point in Chap. 4 when discussing the symmetries of the electromagnetic fields.

U. Hohenester, *Nano and Quantum Optics*, Graduate Texts in Physics,
https://doi.org/10.1007/978-3-030-30504-8_2

We also note in passing that the $1/r^2$ dependence of Coulomb's law is the only distance dependence compatible with massless photons as force carriers of the field [2].

Superposition. When two or more charged particles are present, the total force can be simply computed by adding the respective forces together,

$$\boldsymbol{F}_1 = \boldsymbol{F}_{12} + \boldsymbol{F}_{13} + \cdots + \boldsymbol{F}_{1n} = \frac{1}{4\pi\varepsilon_0}\sum_{j=2}^{n}\frac{q_1 q_j}{r_{1j}^2}\,\hat{\boldsymbol{r}}_{1j}\,. \tag{2.2}$$

This is the essence of the so-called *superposition principle* that has been tested experimentally to the highest degree of precision [2], and which plays an important role in the theory of electromagnetism.

Charge Distribution. In many situations we do not want to deal with point-like particles but with a continuous charge distribution $\rho(\boldsymbol{r})$. Suppose that many particles are present within a small volume element ΔV_i and we are only interested in the fields on sufficiently larger length scales. We may then group together the particles in small bunches Δq_i and relate them to the charge distribution $\rho(\boldsymbol{r})$ via

$$\Delta q_i \approx \rho(\boldsymbol{r}_i)\Delta V_i\,.$$

Although the limit $\Delta V \to 0$ is not meaningful for point-like particles, we can still introduce a continuous charge distribution $\rho(\boldsymbol{r})$, which is expected to vary smoothly as a function of $\boldsymbol{r}$ (see Chap. 7 for a more thorough discussion of such an averaging procedure), and obtain instead of Eq. (2.2) the expression

$$\boldsymbol{F}_1 = \frac{1}{4\pi\varepsilon_0}\int\frac{q_1\rho(\boldsymbol{r}')}{|\boldsymbol{r}_1-\boldsymbol{r}'|^2}\,\hat{\boldsymbol{R}}_1\,d^3r'\,, \tag{2.3}$$

where $\hat{\boldsymbol{R}}_1$ is the unit vector pointing in the direction of $\boldsymbol{r}_1 - \boldsymbol{r}'$.

From Eqs. (2.2) and (2.3) one observes that one can pull out the charge q_1 from the sum or integral. This can be done because of the superposition principle that allows separating a many-body force into its mutual two-body forces. In the field of electrostatics it is then convenient to introduce an electric field $\boldsymbol{E}(\boldsymbol{r})$ defined through

Electric Field of Given Charge Distribution

$$\boldsymbol{E}(\boldsymbol{r}) = \frac{1}{4\pi\varepsilon_0}\int\frac{\rho(\boldsymbol{r}')}{|\boldsymbol{r}-\boldsymbol{r}'|^2}\,\hat{\boldsymbol{R}}\,d^3r'\,, \tag{2.4}$$

where $\hat{\boldsymbol{R}}$ is the unit vector pointing in the direction of $\boldsymbol{r} - \boldsymbol{r}'$. $\boldsymbol{E}(\boldsymbol{r})$ is a so-called vector function which assigns to each space point $\boldsymbol{r}$ a vector. The force acting on a charge q_1 located at position $\boldsymbol{r}_1$ can then be computed from

$$\boldsymbol{F}_1 = q_1\, \boldsymbol{E}(\boldsymbol{r}_1)\,. \tag{2.5}$$

Thus, the electric field $\boldsymbol{E}(\boldsymbol{r})$ gives the force acting on a unit charge located at position $\boldsymbol{r}$.

Up to this point the electric field has appeared as a completely auxiliary device. It provides us with the answer to the question of how the force on a charged particle *would be* if a particle was there. However, nothing hinders us to go one step further and to give physical sense to $\boldsymbol{E}(\boldsymbol{r})$. In fact, the tremendous success of the theory of electromagnetism is the identification of the electromagnetic fields as *the* central objects of the theory. As we will discuss in the remainder of this chapter, the electromagnetic fields acquire a dynamics that is determined not only by the charge and current distributions, which act as sources, but also by the electromagnetic fields themselves. In this way fields can become decoupled from their sources and propagate through space independently.

Before we will present the full set of Maxwell's equations in Sect. 2.2, we should take a step back and ponder on the theory of vector fields. In fact, the central objects in electrodynamics are the electric $\boldsymbol{E}(\boldsymbol{r}, t)$ and magnetic $\boldsymbol{B}(\boldsymbol{r}, t)$ vector fields which depend on both space and time coordinates. Maxwell's equations state how these fields change in space and time, and the solutions have to be supplemented by appropriate boundary conditions. The spatial variations are best described in terms of the so-called "nabla" operator, which we will introduce next.

Faraday's Lines of Force

The concept of fields was originally brought up by Michael Faraday, shown in Fig. 2.1, an ingenious experimentalist with only limited mathematical skills. In an article by Basil Mahon [3], which appeared at the occasion of the 150'th anniversary of Maxwell's equation in Nature Photonics and which is absolutely worth reading, the author explains how Faraday's "lines of force," as he called them, were ahead of their time.

> Faraday's thoughts were running along lines quite different from those of everyone else. The general scientific opinion was still that electric and magnetic forces resulted from material bodies acting on one another at a distance with the intervening space playing only a passive role. The Astronomer Royal, Sir George Biddell Airy, spoke for many when he described Faraday's lines of force as "vague and varying". One can understand this view. Action-at-a-distance gave exact formulae, whereas Faraday supplied none. While they respected Faraday as a superb experimenter, most scientists thought him ill-equipped to theorize as he knew no mathematics.

(continued)

Conscious of such views, Faraday was circumspect when publishing his thoughts on lines of force. Only once, in 1846, did he venture into speculation. A colleague, Charles Wheatstone, was due to speak at the Royal Institution about one of his inventions but took fright at the last minute. Faraday decided to give the talk himself but ran out of things to say on the advertised topic well before the allotted hour was up. Caught offguard he let his private thoughts escape and gave the audience an astonishingly prescient outline of the electromagnetic theory of light. All space, he surmised, was filled with electric and magnetic lines of force which vibrated laterally when disturbed, and sent waves of energy along their lengths at a rapid but finite speed. Light, he said, was probably one manifestation of these 'ray-vibrations'. We now know he was close to the truth, but to most of Faraday's fellow scientists the ray-vibrations seemed like an absurd fantasy. Even his supporters were embarrassed, and Faraday regretted letting his guard slip. He had left his contemporaries behind, and it would take someone forty years his junior, a man of equal stature and complementary talents, to reveal Faraday's true greatness. The man was James Clerk Maxwell.

2.1.1 *The Nabla Operator*

Gradient. Consider a scalar function $f(x, y, z)$ that depends on all three spatial coordinates. The total derivative of f becomes

$$df = \frac{\partial f}{\partial x}dx + \frac{\partial f}{\partial y}dy + \frac{\partial f}{\partial z}dz = \nabla f \cdot d\boldsymbol{\ell}\,. \tag{2.6}$$

We have introduced the infinitesimal position change

$$d\boldsymbol{\ell} = \hat{\boldsymbol{x}}\,dx + \hat{\boldsymbol{y}}\,dy + \hat{\boldsymbol{z}}\,dz\,,$$

and the nabla operator

$$\nabla = \hat{\boldsymbol{x}}\frac{\partial}{\partial x} + \hat{\boldsymbol{y}}\frac{\partial}{\partial y} + \hat{\boldsymbol{z}}\frac{\partial}{\partial z}\,. \tag{2.7}$$

To be meaningful, ∇ must act on some function such as $f(\boldsymbol{r})$ in Eq. (2.6). If we rewrite Eq. (2.6) in the form

$$df = |\nabla f|\,|d\boldsymbol{\ell}|\,\cos\theta\,,$$

where θ is the angle between ∇f and $d\boldsymbol{\ell}$, we observe that df becomes largest when both vectors are parallel, this is for $\theta = 0$. In other words, if we move in the same direction as ∇f the total change df is maximized. Thus ∇f, which is usually called the "gradient" of f, points into the direction where $f(\boldsymbol{r})$ changes most.

Fig. 2.1 Two of the grounding fathers of electrodynamics. Michael Faraday (left, 1791–1867) was the first to envision the "lines of force" as the fundamental objects in electrodynamics, but he did not have the mathematical skills to develop a rigid mathematical theory. This was accomplished by James Clerk Maxwell (right, 1831–1879) who first wrote down the equations that are now named after him

Divergence. The nabla operator can also act on a vector function $\boldsymbol{F}(\boldsymbol{r})$ either as an inner product $\nabla \cdot \boldsymbol{F}(\boldsymbol{r})$, which is called the divergence, or as an outer product $\nabla \times \boldsymbol{F}(\boldsymbol{r})$, which is called the curl. Let us discuss first the divergence

$$\nabla \cdot \boldsymbol{F}(\boldsymbol{r}) = \frac{\partial F_x(\boldsymbol{r})}{\partial x} + \frac{\partial F_y(\boldsymbol{r})}{\partial y} + \frac{\partial F_z(\boldsymbol{r})}{\partial z} . \tag{2.8}$$

To get some insight into the physical meaning of the divergence, we approximate the derivatives by finite differences and assume for simplicity that $\boldsymbol{F}$ has no dependence on z. We then get

$$\nabla \cdot \boldsymbol{F} \approx \frac{F_x\left(x + \frac{\Delta x}{2}, y\right) - F_x\left(x - \frac{\Delta x}{2}, y\right)}{\Delta x} + \frac{F_y\left(x, y + \frac{\Delta y}{2}\right) - F_y\left(x, y - \frac{\Delta y}{2}\right)}{\Delta y} . \tag{2.9}$$

We can represent this equation symbolically through

$$\nabla \cdot \boldsymbol{F} \approx (\Delta x)^{-1}\left\{ \boxed{\Rightarrow} + \boxed{\Leftarrow} \right\} + (\Delta y)^{-1}\left\{ \boxed{\Uparrow} + \boxed{\Downarrow} \right\} \equiv \boxed{\Leftrightarrow\!\Updownarrow} .$$

The arrow to the right gives the vector component F_x at position $x + \Delta/2$, and the arrow to the left the component $-F_x$ at position $x - \Delta/2$, with a similar interpretation for the up and down directions. By applying the plaquette $\boxed{\Leftrightarrow\!\Updownarrow}$ to a given vector field one obtains its divergence. When we think about $\boldsymbol{F}$ as a fluid, the divergence gives us information about the sources and sinks of the fluid.

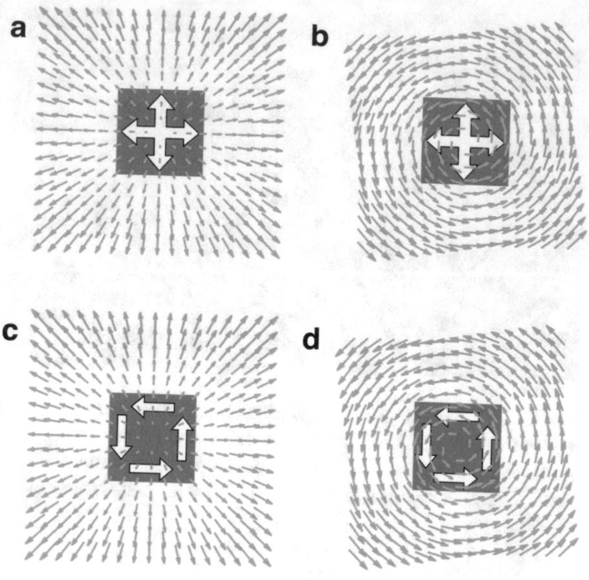

Fig. 2.2 Application of (**a**, **b**) divergence and (**c**, **d**) curl plaquettes to vector fields. (**a**) $\nabla\cdot\boldsymbol{F}\neq 0$ for point source, (**b**) $\nabla\cdot\boldsymbol{F}=0$ for curl field, (**c**) $\nabla\times\boldsymbol{F}=0$ for point source, and (**d**) $\nabla\times\boldsymbol{F}\neq 0$ for curl field

$\nabla\cdot\boldsymbol{F}=0$ means that the fluid simply flows through a given square element, but the in-flow equals the out-flow, whereas $\nabla\cdot\boldsymbol{F}>0$ means that more flows out from the element than flows in, in accordance to a source within the square. A similar interpretation in terms of a sink holds for $\nabla\cdot\boldsymbol{F}<0$. In Fig. 2.2 we show the examples of point-like source fields and curl fields.

Curl. We finally discuss the outer product $\nabla\times\boldsymbol{F}(\boldsymbol{r})$ which becomes

$$\nabla\times\boldsymbol{F}(\boldsymbol{r})=\hat{\boldsymbol{x}}\left(\frac{\partial F_z}{\partial y}-\frac{\partial F_y}{\partial z}\right)+\hat{\boldsymbol{y}}\left(\frac{\partial F_x}{\partial z}-\frac{\partial F_z}{\partial x}\right)+\hat{\boldsymbol{z}}\left(\frac{\partial F_y}{\partial x}-\frac{\partial F_x}{\partial y}\right). \tag{2.10}$$

Using finite differences, we find for the z-component of the curl

$$(\nabla\times\boldsymbol{F})_z\approx\frac{F_y\left(x+\frac{\Delta x}{2},y\right)-F_y\left(x-\frac{\Delta x}{2},y\right)}{\Delta x}-\frac{F_x\left(x,y+\frac{\Delta y}{2}\right)-F_x\left(x,y-\frac{\Delta y}{2}\right)}{\Delta y}. \tag{2.11}$$

By a similar token as for the divergence we define the plaquette

$$(\nabla\times F)_z\approx(\Delta x)^{-1}\Big\{\boxed{\Uparrow}+\boxed{\Downarrow}\Big\}+(\Delta y)^{-1}\Big\{\boxed{\Leftarrow}+\boxed{\Rightarrow}\Big\}\equiv\boxed{\circlearrowleft},$$

which symbolically describes how to sum up the field contributions in order to get the curl of a vector field. From this graphical representation one observes that this operation measures how much a vector function "curls" around a given point of interest. In terms of fluids, a curl can be interpreted as a vortex where one gains or loses energy by making a round trip. Figure 2.2c, d shows examples of source and curl fields. We will return to such finite difference approximations of the curl in Chap. 11 when discussing finite difference time domain (FDTD) simulations.

2.1.2 *Helmholtz Theorem*

Usually we are dealing with problems where we are seeking for a vector field $\boldsymbol{F}(\boldsymbol{r})$ within a given volume Ω. Throughout this book we shall denote the boundary of a volume with $\partial\Omega$. The boundary $\partial\Omega$ can consist of several unconnected parts, and often the volume Ω fills the entire space such that $\partial\Omega$ is pushed towards infinity. See also Fig. 2.3. Nevertheless, a boundary is always present and one has to be careful about the behavior of $\boldsymbol{F}(\boldsymbol{r})$ at the boundaries. An important theorem in vector analysis states:

Helmholtz Theorem
A vector function $\boldsymbol{F}(\boldsymbol{r})$ is uniquely determined upon knowledge of the following quantities:

$$\nabla\cdot\boldsymbol{F},\quad \nabla\times\boldsymbol{F},\quad \text{and } \boldsymbol{F}(\boldsymbol{r}\in\partial\Omega) \text{ at the boundary.}$$

Below we will interpret Maxwell's equations in light of the Helmholtz theorem. In many cases the boundary conditions will not be explicitly stated but will be somehow built into the solutions, such as outgoing waves at infinity or electric fields that become zero at infinity

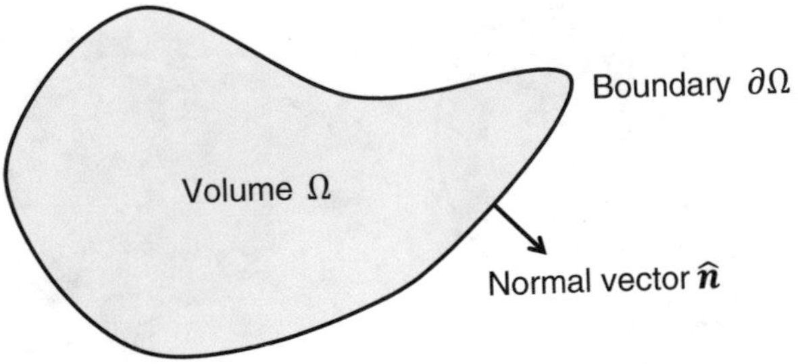

Fig. 2.3 Volumes and boundaries. In solving Maxwell's equation we often deal with dielectric or metallic bodies. We denote a given regions of space (volume) with Ω, the boundary of this volume with $\partial\Omega$, and the vector perpendicular to the boundary (and pointing outwards of the volume) with $\hat{\boldsymbol{n}}$

2.1.3 Gauss' and Stokes' Theorem

We will often employ two integral theorems. The first one is **Gauss' theorem**

$$\int_{\Omega} \nabla \cdot \boldsymbol{F}(\boldsymbol{r})\, d^3r = \oint_{\partial\Omega} \boldsymbol{F}(\boldsymbol{r}) \cdot d\boldsymbol{S}\,, \tag{2.12}$$

which states that the integral of $\nabla \cdot \boldsymbol{F}$ over a volume Ω equals the (directed) flow of the vector field through the boundary $\partial\Omega$ of the volume. Here $d\boldsymbol{S} = \hat{\boldsymbol{n}}\, dS$, where $\hat{\boldsymbol{n}}$ is the outer surface normal and dS denotes an infinitesimal surface element. Figure 2.4a gives a graphical interpretation of this theorem in terms of the previously introduced plaquettes. As the divergence measures the net difference between in- and out-flow in a given square element, the in and out fluxes 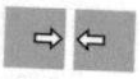 of two neighbor elements precisely cancel each other, and the only non-vanishing contributions are located at the boundary.

The second theorem is **Stokes' theorem**

$$\int_{S} \nabla \times \boldsymbol{F}(\boldsymbol{r}) \cdot d\boldsymbol{S} = \oint_{\partial S} \boldsymbol{F}(\boldsymbol{r}) \cdot d\boldsymbol{\ell}\,, \tag{2.13}$$

which states that the integration of the curl of a vector function over an open surface S equals the line integral of $\boldsymbol{F}(\boldsymbol{r})$ along the boundary ∂S of the surface. Figure 2.4b gives a graphical interpretation of this theorem in terms of the previously introduced plaquettes. The curl contributions cancel each other at the edges of neighbor elements, such as 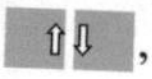, and the only non-vanishing contributions are located at the surface boundary.

In which direction does the outer surface normal of $d\boldsymbol{S} = \hat{\boldsymbol{n}}\, dS$ point? And in which direction goes $d\boldsymbol{\ell}$? In case of Gauss' theorem $\hat{\boldsymbol{n}}$ points to the outside of the volume. If one wants to define the boundary differently, and we will do so in later parts of the book, one has to be careful about this point. Similarly, the direction of

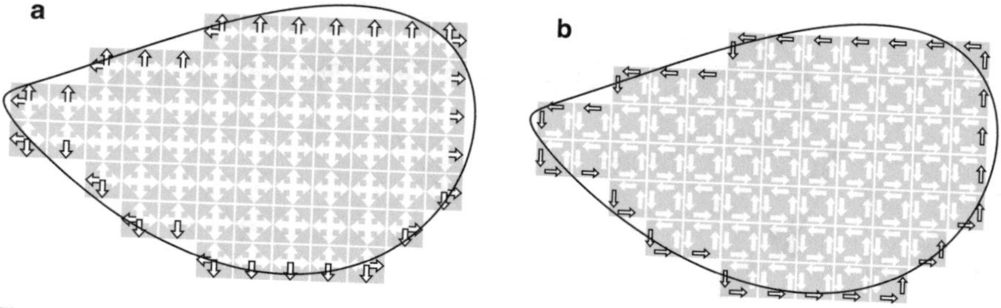

Fig. 2.4 Graphical representation of (**a**) Gauss' and (**b**) Stokes' theorem. In the Gauss' theorem $\nabla \cdot \boldsymbol{F}$ is integrated over a volume, where $\boldsymbol{F}$ flows from one volume element to a neighbor one, and the in and out fluxes compensate each other with the exception of boundary terms. Similarly, when integrating the curl over an open boundary, the curl terms in neighbor elements cancel each other with the exception of boundary elements with no neighbors

dS dictates the circulation of $d\ell$ according to the right-hand rule, which means that if one points with the thumb of the right hand upwards (pointing in the direction of $\hat{n}$) the other fingers point in the direction of $d\ell$.

2.2 Maxwell's Equations

In electrodynamics we deal with time-dependent electric $E(r, t)$ and magnetic $B(r, t)$ fields. The force acting on a particle with charge q at position r and moving with velocity v can be computed from the **Lorentz force**

$$F = q\left[E(r, t) + v \times B(r, t)\right] . \tag{2.14}$$

The electromagnetic fields themselves are created by charge and current distributions $\rho(r, t)$ and $J(r, t)$, respectively. The dynamics of these fields is determined by Maxwell's equations which form the basis of the theory of electrodynamics,

Maxwell's Equations

Gauss' law $$\nabla \cdot E = \frac{\rho}{\varepsilon_0} \tag{2.15a}$$

no name $$\nabla \cdot B = 0 \tag{2.15b}$$

Faraday's law $$\nabla \times E = -\frac{\partial B}{\partial t} \tag{2.15c}$$

Ampere's law $$\nabla \times B = \mu_0 J + \mu_0 \varepsilon_0 \frac{\partial E}{\partial t} . \tag{2.15d}$$

μ_0 is the vacuum permeability. There are various ways of "reading" Maxwell's equations. The first one is in terms of the Helmholtz theorem stating that a vector function is determined once its divergence and curl together with the boundary conditions are known. In this way, Gauss' and Faradays's law determine the electric field $E(r, t)$, whereas the second equation (no name) and Ampere's law determine the magnetic field $B(r, t)$. Additionally we have to specify appropriate boundary conditions. Maxwell's equations can be interpreted as follows:

Gauss' Law. The sources and sinks of the electric fields are given by the charge distribution.

No Name. There exist no magnetic charges.

Faraday's Law. A time-dependent magnetic field induces an electric curl field. This equation is also called the induction law and describes in electric circuits the creation of electromotive forces, which are of paramount importance for electric motors and generators.

Ampere's Law. A magnetic curl field is either produced through an electric current or a time-dependent electric field. The latter term was first introduced by Maxwell and is usually referred to as "Maxwell's displacement current."

The Breakthrough of Maxwell's Equations

Maxwell did not write down "his" equations in the nowadays well-known form of Eq. (2.15), but used more complicated equations that kept both the electromagnetic fields and potentials as basic quantities. For this reason, they were originally not well accepted by the community and it took another twenty years before Heaviside, who is usually known for "his" step function, brought it to the form nowadays familiar to the physics community [3]:

> Up to the time Maxwell died, in 1879, and for several years afterwards, no one else really understood his theory. It sat like an exhibit in a glass case, admired by some but out of reach. The man who made it accessible was a self-taught former telegraph operator called Oliver Heaviside. In 1885 he summed up the theory in what we now call the four Maxwell's equations. [...]
>
> Heaviside had greatly simplified things by using his new system of vector analysis, in which three-dimensional vectors were represented by single letters, and by pushing the electric and magnetic potentials to the background. When Heinrich Hertz's discovery of electromagnetic waves in 1888 prompted a surge of interest in electromagnetism, people turned not to Maxwell's original expression of his theory but to Heaviside's compact version.

In order to understand how electromagnetic waves emerge from Maxwell's theory we consider

- a time-dependent current distribution $\boldsymbol{J}(\boldsymbol{r}, t)$ which, according to Ampere's law, Eq. (2.15d), gives rise to
- a time-dependent magnetic field $\boldsymbol{B}(\boldsymbol{r}, t)$. According to Faraday's law, Eq. (2.15c), this leads to
- a time-dependent electric field $\boldsymbol{E}(\boldsymbol{r}, t)$ which acts as a displacement current in Eq. (2.15d) (last term on right-hand side), and gives rise to
- a time-dependent magnetic field $\boldsymbol{B}(\boldsymbol{r}, t)$, and so on.

Through the coupled curl equations the fields can thus propagate away from the sources and become decoupled from them. A more formal account of electromagnetic waves as fundamental solutions of Maxwell's equations in free space $\rho = \boldsymbol{J} = 0$ can be obtained by taking the curl on both sides of Faraday's law

$$\nabla \times \nabla \times \boldsymbol{E} = -\nabla \times \frac{\partial \boldsymbol{B}}{\partial t} \implies \nabla(\nabla \cdot \boldsymbol{E}) - \nabla^2 \boldsymbol{E} = -\frac{\partial}{\partial t} \nabla \times \boldsymbol{B} \,,$$

where we have employed the usual identities for vector functions, see Exercise 2.6, and have exchanged spatial and time derivatives on the right-hand side. In absence of any sources we have $\nabla \cdot \boldsymbol{E} = 0$ and $\boldsymbol{J} = 0$, and we are thus led together with Ampere's law to the wave equation

$$\left(\nabla^2 - \mu_0\varepsilon_0\frac{\partial^2}{\partial t^2}\right)\boldsymbol{E}(\boldsymbol{r},t) = 0\,. \tag{2.16}$$

A similar equation can be obtained for the magnetic fields, see Exercise 2.8. From the above equation we also observe that the speed of light with which electromagnetic waves propagate can be expressed as

$$c = \frac{1}{\sqrt{\mu_0\varepsilon_0}}\,.$$

We will return in more detail to electromagnetic waves in Sect. 2.4

Cosmic Background Radiation My favorite example for the predictive power of Maxwell's equation is the cosmic background radiation that was created about 400 000 years after the big bang, when the temperature of the initial plasma had sufficiently cooled down such that atoms could form and radiation was released from the plasma. Since then this radiation has propagated freely, according to the laws of Maxwell's equations, although the radiation frequency has decreased because of the expansion of the universe and the distribution maximum is nowadays in the microwave regime. I find it fascinating how well the laws of physics are fulfilled even over billions of years and how well Maxwell's equations describe the dynamics of electromagnetic waves. Although we usually take this success for granted, I think that we should sometimes be more thankful for these powerful tools we have at hand to describe processes in nature.

2.2.1 *Electromagnetic Potentials*

Electromagnetic potentials play an important role in classical electrodynamics [1, 2], but are of significantly less importance in the field of nano optics. As we will discuss in Chap. 5, in nano optics one usually introduces different objects, the so-called Green's functions, which take over many of the advantages of electromagnetic potentials. Nevertheless, at several places, most noteworthy certainly in quantum optics, we will rely on the concept of electromagnetic potentials.

We start by "reading" Maxwell's equations in a way that will become even more important in the next section, namely in terms of homogeneous and inhomogeneous equations. The inhomogeneities in Maxwell's equations are the external charge and current distributions. The idea behind electrodynamic potentials is to introduce new quantities, the scalar and vector potentials $V(\boldsymbol{r},t)$ and $\boldsymbol{A}(\boldsymbol{r},t)$, which are chosen such that the homogeneous Maxwell's equations are automatically fulfilled. The equations that determine these potentials are then provided by the inhomogeneous Maxwell equations.

Let us start with $\nabla \cdot \boldsymbol{B} = 0$. We may now relate $\boldsymbol{B}$ to the **vector potential** $\boldsymbol{A}$ through

$$\boldsymbol{B} = \nabla \times \boldsymbol{A}\,. \tag{2.17}$$

With this choice the magnetic field is guaranteed to have no sources or sinks, because the divergence of a curl field $\nabla \cdot \nabla \times \boldsymbol{A}$ is always zero. As for the other homogeneous equation, Faraday's law, we start with

$$\nabla \times \left(\boldsymbol{E} + \frac{\partial \boldsymbol{A}}{\partial t}\right) = -\nabla \times \nabla V = 0\,.$$

Here we have replaced the expression in parentheses by $-\nabla V$ because the curl of a gradient field is automatically zero. V is the **scalar potential**. The negative sign is a convention adopted from electrostatics where the potential can be related to the work done by a charge against the electric field [2]. Thus, we can express the electric field in terms of the scalar and vector potentials via

$$\boldsymbol{E} = -\nabla V - \frac{\partial \boldsymbol{A}}{\partial t}\,. \tag{2.18}$$

With these relations between $\boldsymbol{E}$, $\boldsymbol{B}$ and the electromagnetic potentials V, $\boldsymbol{A}$ the homogeneous Maxwell equations are automatically fulfilled.

The electromagnetic potentials are not uniquely defined but we can add certain contributions to them without causing any changes to $\boldsymbol{E}$ and $\boldsymbol{B}$. Let $\lambda(\boldsymbol{r}, t)$ be an arbitrary scalar function. We can then modify V and $\boldsymbol{A}$ according to

$$\boldsymbol{A}' = \boldsymbol{A} + \nabla\lambda\,, \quad V' = V - \frac{\partial \lambda}{\partial t}\,. \tag{2.19}$$

As can be easily proven through direct calculation, both V, $\boldsymbol{A}$ as well as V', $\boldsymbol{A}'$ lead to the same electromagnetic fields $\boldsymbol{E}$, $\boldsymbol{B}$. We will make use of these so-called **gauge transformations** of Eq. (2.19) in a moment.

The defining equations for $\boldsymbol{V}$ and $\boldsymbol{A}$ care obtained by inserting $\boldsymbol{E}$, $\boldsymbol{B}$ expressed in terms of the electromagnetic potentials into the inhomogeneous Maxwell equations, namely Gauss' law of Eq. (2.15a) and Ampere's law of Eq. (2.15d). We then obtain

$$\nabla \cdot \left(-\nabla V - \frac{\partial \boldsymbol{A}}{\partial t}\right) = -\nabla^2 V - \nabla \cdot \frac{\partial \boldsymbol{A}}{\partial t} = \frac{\rho}{\varepsilon_0} \tag{2.20}$$

$$\nabla \times \nabla \times \boldsymbol{A} = \nabla(\nabla \cdot \boldsymbol{A}) - \nabla^2 \boldsymbol{A} = \mu_0 \boldsymbol{J} + \mu_0\varepsilon_0 \frac{\partial}{\partial t}\left(-\nabla V - \frac{\partial \boldsymbol{A}}{\partial t}\right).$$

As properly noted by Griffiths [1], this is a rather ugly set of equations and one might wonder about the reason for introducing electromagnetic potentials. Fortunately, there is a simple way out and we can introduce the so-called Lorenz gauge condition to decouple the two equations for V and $\boldsymbol{A}$. There is often some confusion in the literature whether the transformation should be attributed to the Loren(t)z with or without a "t," but there recently seems to be some consensus that it should be properly called "Lorenz condition" after his inventor Ludvig Lorenz. We can impose on the electromagnetic potentials the additional Lorenz gauge condition (see also Exercise 2.9)

$$\nabla \cdot \boldsymbol{A} + \mu_0 \varepsilon_0 \frac{\partial V}{\partial t} = 0 \tag{2.21}$$

in order to decouple the defining equations for V and $\boldsymbol{A}$, and we obtain the final set of equations

$$\begin{aligned} \left(\nabla^2 - \mu_0 \varepsilon_0 \frac{\partial^2}{\partial t^2}\right) V(\boldsymbol{r}, t) &= -\frac{\rho(\boldsymbol{r}, t)}{\varepsilon_0} \\ \left(\nabla^2 - \mu_0 \varepsilon_0 \frac{\partial^2}{\partial t^2}\right) \boldsymbol{A}(\boldsymbol{r}, t) &= -\mu_0 \boldsymbol{J}(\boldsymbol{r}, t)\,. \end{aligned} \tag{2.22}$$

Thus, in the Lorenz gauge both V and $\boldsymbol{A}$ obey wave equations. We will come back to the solution of these equations in Chap. 5. There exist other popular gauges, such as the Coulomb or transverse gauge, which will be adopted in later parts of this book in the context of the quantization of Maxwell's equations.

2.3 Maxwell's Equations in Matter

We continue with our "reading" of Maxwell's equations in terms of homogeneous and inhomogeneous equations. Apparently, the inhomogeneities of charge and current distributions ρ, $\boldsymbol{J}$ are the means of how the material world "communicates" with the electromagnetic fields. In fact, one of the main reasons why optics and nano optics have seen such a tremendous boost in recent years is the progress in nano and material science. This has brought up numerous novel charge and current sources that allow for an unprecedented control of light–matter interaction. Was it just for the electromagnetic part of Maxwell's equations, the field of electrodynamics would have probably turned into a completely boring discipline by now.

When dealing with Maxwell's equations in matter, it is convenient to separate the charge and current distributions into *external* parts, which can be controlled from the outside, and *induced* contributions associated with polarizations and magnetizations. The latter can usually not be easily controlled, yet, in presence of matter microscopic polarizations and magnetizations will be induced and will inevitably act back on the fields. Figure 2.5 gives a brief sketch of the principle underlying this separation. The separation into external and induced contributions

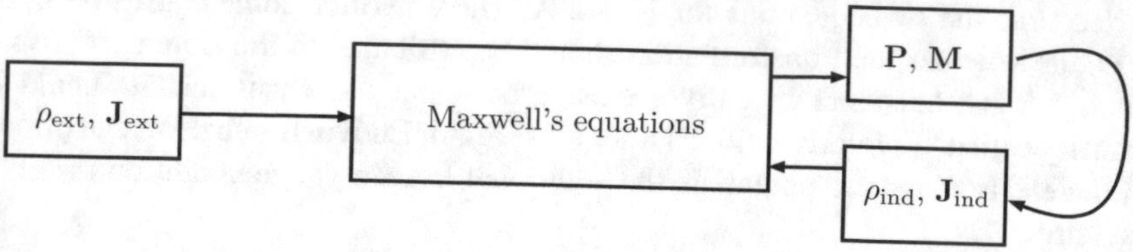

Fig. 2.5 Schematics of Maxwell's equations in matter. The charge and current sources are separated into external and induced contributions. $\rho_{\rm ext}$, $\boldsymbol{J}_{\rm ext}$ can be controlled externally and act as sources for Maxwell's equations. In addition, the electromagnetic fields $\boldsymbol{E}$, $\boldsymbol{B}$ induce a polarization $\boldsymbol{P}$ and magnetization $\boldsymbol{M}$, which act back through the induced contributions $\rho_{\rm ind}$, $\boldsymbol{J}_{\rm ind}$ on Maxwell's equations

is not always completely clear and there is sometimes some freedom of choice what is "external" and what is "induced."

We next introduce, in close analogy to the charge and current distributions, the **polarization** $\boldsymbol{P}(\boldsymbol{r},t)$ as an electric dipole moment per unit volume, and the **magnetization** $\boldsymbol{M}(\boldsymbol{r},t)$ as a magnetic dipole moment per unit volume. These quantities account for the material response in presence of electromagnetic fields, and we have to provide a prescription of how they are related to the electromagnetic fields. With $\boldsymbol{P}$ and $\boldsymbol{M}$ we can separate the charge and current distributions into free and bound contributions according to [1, 2]

External and Induced Charge and Current Distributions

$$\begin{aligned}\rho &= \rho_{\rm ext} + \rho_{\rm ind} = \rho_{\rm ext} - \nabla\cdot\boldsymbol{P}\\ \boldsymbol{J} &= \boldsymbol{J}_{\rm ext} + \boldsymbol{J}_{\rm ind} = \boldsymbol{J}_{\rm ext} + \nabla\times\boldsymbol{M} + \frac{\partial \boldsymbol{P}}{\partial t}\,. \end{aligned} \tag{2.23}$$

This separation process will be discussed in more detail in Chap. 7. The strategy to be pursued next is to develop a theory that depends on the external sources only, and to eliminate the bound sources which cannot be controlled externally. We first insert the charge distribution of Eq. (2.23) into Gauss' law and obtain

$$\varepsilon_0\nabla\cdot\boldsymbol{E} = \rho_{\rm ext} - \nabla\cdot\boldsymbol{P} \implies \nabla\cdot(\varepsilon_0\boldsymbol{E}+\boldsymbol{P}) = \rho_{\rm ext}\,.$$

The expression in parentheses is usually called the **dielectric displacement**

$$\boldsymbol{D} = \varepsilon_0\boldsymbol{E} + \boldsymbol{P}\,. \tag{2.24}$$

We next insert the current distribution of Eq. (2.23) into Ampere's law and get

$$\frac{1}{\mu_0}\nabla\times\boldsymbol{B}=\boldsymbol{J}_{\text{ext}}+\nabla\times\boldsymbol{M}+\frac{\partial\boldsymbol{P}}{\partial t}+\varepsilon_0\frac{\partial\boldsymbol{E}}{\partial t}\implies\nabla\times\left(\frac{1}{\mu_0}\boldsymbol{B}-\boldsymbol{M}\right)=\boldsymbol{J}_{\text{ext}}+\frac{\partial\boldsymbol{D}}{\partial t}.$$

The expression in parentheses is usually called the **magnetic field**

$$\boldsymbol{H}=\frac{1}{\mu_0}\boldsymbol{B}-\boldsymbol{M}. \tag{2.25}$$

Magnetic Field The alert reader may have noticed that I have previously denoted also $\boldsymbol{B}$ as the "magnetic field", and there is indeed some confusion in the literature about this point. Table 2.1 summarizes the most convenient notations for the various quantities appearing for electrodynamics in matter, with "magnetic induction" being used for $\boldsymbol{B}$ and "magnetic field" for $\boldsymbol{H}$. Throughout this book I will be somewhat sloppy about the proper terminology and will often call both $\boldsymbol{B}, \boldsymbol{H}$ the magnetic field, buth the precise meaning should be always clear from the context.

With these new definitions, we can now express Maxwell's equations in matter in the form

Maxwell's Equations in Matter

Gauss' law	$\nabla\cdot\boldsymbol{D}=\rho_{\text{ext}}$	(2.26a)
no name	$\nabla\cdot\boldsymbol{B}=0$	(2.26b)
Faraday's law	$\nabla\times\boldsymbol{E}=-\dfrac{\partial\boldsymbol{B}}{\partial t}$	(2.26c)
Ampere's law	$\nabla\times\boldsymbol{H}=\boldsymbol{J}_{\text{ext}}+\dfrac{\partial\boldsymbol{D}}{\partial t}.$	(2.26d)

Note that the homogeneous equations remain unchanged, and only the inhomogeneous ones become modified because of our decomposition of the source terms into external and induced contributions.

Constitutive Relations. Eqs. (2.26a–d) do not form a closed set of equations but must be additionally complemented by constitutive equations

$$\boldsymbol{D}=\boldsymbol{D}(\boldsymbol{E},\boldsymbol{B}),\quad \boldsymbol{H}=\boldsymbol{H}(\boldsymbol{E},\boldsymbol{B}),$$

which relate $\boldsymbol{D}, \boldsymbol{H}$ to the true fields $\boldsymbol{E}, \boldsymbol{B}$. This can be done by either using some microscopic or phenomenological description of the polarization and magnetization, or by considering linear materials.

True Versus Auxiliary Fields. The "true" fields in electrodynamics are $\boldsymbol{E}, \boldsymbol{B}$. The dielectric displacement $\boldsymbol{D}$ and the magnetic field $\boldsymbol{H}$ are auxiliary quantities

Table 2.1 List of different quantities involved in Maxwell's equations in matter

Maxwell' equations	Symbol	Name
Free space	$\boldsymbol{E}(\boldsymbol{r},t)$	Electric field
	$\boldsymbol{B}(\boldsymbol{r},t)$	Magnetix induction
	ε_0	Vacuum permittivity
	μ_0	Vacuum permeability
Matter	$\boldsymbol{P}(\boldsymbol{r},t)$	Polarization
	$\boldsymbol{M}(\boldsymbol{r},t)$	Magnetization
	$\boldsymbol{D}=\varepsilon_0\boldsymbol{E}+\boldsymbol{P}$	Dielectric displacement
	$\boldsymbol{H}=\frac{1}{\mu_0}\boldsymbol{B}-\boldsymbol{M}$	Magnetic field
Linear materials	$\boldsymbol{P}=\varepsilon_0\chi_e\boldsymbol{E}$	Electric susceptibility χ_e
	$\boldsymbol{M}=\chi_m\boldsymbol{H}$	Magnetic susceptibility χ_m
	$\boldsymbol{D}=\varepsilon\boldsymbol{E}$	Permittivity ε
	$\boldsymbol{B}=\mu\boldsymbol{M}$	Permeability μ

which are helpful when dealing with Maxwell's equations in matter. Unfortunately, there is some historical confusion about the role of $\boldsymbol{B}$ and $\boldsymbol{H}$, which will become more clear in the next section. As a consequence, it is often more convenient to work with the electromagnetic fields $\boldsymbol{E}$, $\boldsymbol{H}$ because expressions become more symmetric then. In this book we will use both descriptions in terms of $\boldsymbol{E}$, $\boldsymbol{B}$ and $\boldsymbol{E}$, $\boldsymbol{H}$, however, the reader should keep in the back of his or her mind that the true fields felt by charged particles are *always* $\boldsymbol{E}$, $\boldsymbol{B}$.

2.3.1 Linear Materials

For a wide class of materials we can assume a linear relation between the material response and the external fields. More specifically, we get

Linear Materials

$$\boldsymbol{P}=\varepsilon_0\chi_e\boldsymbol{E}\,,\quad \boldsymbol{M}=\chi_m\boldsymbol{H}\,. \tag{2.27}$$

Here χ_e and χ_h are the electric and magnetic susceptibilities, respectively.

Polarization. Let me first discuss the polarization expression. As I would like to argue, there is a strong physical motivation for relating $\boldsymbol{P}$ to the electric field $\boldsymbol{E}$. We first recall that $\boldsymbol{D}$ is an auxiliary field that is solely created by the external charge distribution ρ_{ext}. If we had erroneously assumed $\boldsymbol{P} = \chi_e\boldsymbol{D}$ (wrong relation), the polarization at a given position would be only due to the external fields. In reality, however, the true field $\boldsymbol{E}$ is the sum of the external field and of the polarization field, which is produced by the entire polarized body under investigation, in agreement to the choice made in Eq. (2.27).

Magnetization. Things are different for the magnetization, which is only induced by the free currents governing $\boldsymbol{H}$. This directly points to the previously mentioned confusion regarding the proper role of $\boldsymbol{B}$ and $\boldsymbol{H}$. Fortunately, things are not as bad as they seem. For practically all materials under study the magnetization is very small, and for this reason the error made through $\boldsymbol{M} = \chi_m \boldsymbol{H}$ is usually negligible in comparison to the arguably more correct choice $\boldsymbol{M} = \chi_m \boldsymbol{B}$ (wrong relation).

We can now continue to establish a relation between $\boldsymbol{D}$ and $\boldsymbol{E}$,

$$\boldsymbol{D} = \varepsilon_0 (1 + \chi_e)\, \boldsymbol{E} = \varepsilon \boldsymbol{E}\,, \tag{2.28}$$

where we have introduced the permittivity $\varepsilon = \varepsilon_0 (1 + \chi_e)$. Similarly, we get

$$\boldsymbol{B} = \mu_0 (1 + \chi_m)\, \boldsymbol{H} = \mu \boldsymbol{H}\,, \tag{2.29}$$

where we have introduced the permeability $\mu = \mu_0 (1 + \chi_m)$. In case of anisotropic materials both ε and μ become tensorial quantities, but we will not consider such materials unless stated differently. Finally, Maxwell's equations for a linear medium can be expressed in the form

$$\begin{aligned} \nabla \cdot \varepsilon \boldsymbol{E} &= \rho_{\text{ext}}\,, & \nabla \times \boldsymbol{E} &= -\frac{\partial \boldsymbol{B}}{\partial t} \\ \nabla \cdot \boldsymbol{B} &= 0\,, & \nabla \times \frac{1}{\mu} \boldsymbol{B} &= \boldsymbol{J}_{\text{ext}} + \varepsilon \frac{\partial \boldsymbol{E}}{\partial t}\,. \end{aligned} \tag{2.30}$$

2.3.2 *Boundary Conditions*

In electrodynamics and also nano optics we often deal with problems where bodies consisting of different materials are separated by sharp boundaries. From Maxwell's equations in matter, Eq. (2.26), we can obtain boundary conditions that connect the fields directly above and below the interface. In the following we adopt the notation that $\hat{\boldsymbol{n}}$ is a unit vector that is perpendicular to the interface and points from the lower to the upper medium. In our analysis we have to be careful about possible singular layers of surface charges or currents located directly at the interfaces. We will next use the integral theorems of Sect. 2.1.3 in the following ways:

Gauss' Theorem. For the Maxwell's equations that define the divergence of the electromagnetic fields we integrate over a very thin cylinder $\delta\Omega_h$ with height $h \to 0$. The top cover of the disk is assumed to be just above the interface and

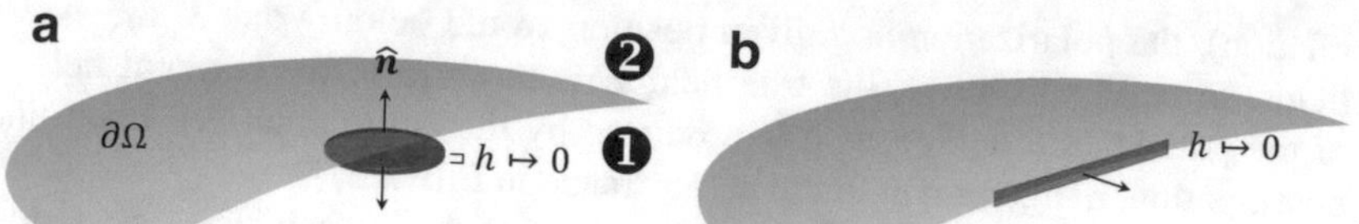

Fig. 2.6 Integration volumes and areas for deriving the boundary conditions of Maxwell's equations. (**a**) The divergence equations are integrated over a small box with height $h \to 0$, and the integrals are converted through Gauss' theorem into boundary integrals. $\hat{\boldsymbol{n}}$ is the outer surface normal of the boundary $\partial\Omega$ which separates the upper medium 2 from the lower medium 1. (**b**) The curl equations are integrated over a small area with height $h \to 0$, and the integrals are converted through Stokes' theorem into line integrals

the bottom cover just below, see also Fig. 2.6. Using Gauss' theorem we convert the volume integral into a boundary integral, where the contributions from the side ribbon of the cylinder becomes zero in the limit $h \to 0$.

Stokes' Theorem. For the Maxwell's equations that define the curl of the electromagnetic fields we integrate over a very thin rectangle δS_h with height $h \to 0$. We use Stokes' theorem to convert the boundary integral into a line integral, where only the contributions parallel to the interface contribute in the limit $h \to 0$. For both the cylinder and the rectangle we assume that the sizes are so small that we can well approximate the fields and sources within the integration regions by constant values.

Let us start with Gauss' law of Eq. (2.26a). Employing the scheme described above we are led to

$$\lim_{h\to 0}\int_{\delta\Omega_h} \nabla\cdot\boldsymbol{D}\,d^3r = \lim_{h\to 0}\oint_{\partial\,\delta\Omega_h} \boldsymbol{D}\cdot d\boldsymbol{S}$$

$$= (\boldsymbol{D}_2 - \boldsymbol{D}_1)\cdot\hat{\boldsymbol{n}}\,\delta S = \lim_{h\to 0}\int_{\delta\Omega_h} \rho_{\text{ext}}\,d^3r = \sigma_{\text{ext}}\,\delta S\,.$$

Here σ_{ext} denotes a possible surface charge distribution located at the interface, and the subscripts 1, 2 denote the regions below and above the interface. Note that the sign of $\boldsymbol{D}_1$ is negative because the surface normal of the lower cap points in the $-\hat{\boldsymbol{n}}$ direction. From this expression and using a similar analysis for $\nabla\cdot\boldsymbol{B}$, we are led to the first set of boundary conditions

First Set of Boundary Conditions

$$\hat{\boldsymbol{n}}\cdot(\boldsymbol{D}_2 - \boldsymbol{D}_1) = \sigma_{\text{ext}}\,,\quad \hat{\boldsymbol{n}}\cdot(\boldsymbol{B}_2 - \boldsymbol{B}_1) = 0\,. \tag{2.31}$$

We next take Faraday's law and integrate over a small rectangle,

$$\begin{aligned}\lim_{h\to 0}\int_{\delta S_h}\nabla\times\boldsymbol{E}\cdot d\boldsymbol{S} &= \lim_{h\to 0}\oint_{\partial\,\delta S_h}\boldsymbol{E}\cdot d\boldsymbol{\ell} = (\boldsymbol{E}_2-\boldsymbol{E}_1)\cdot\delta\boldsymbol{\ell}\\ &= -\lim_{h\to 0}\int_{\delta S_h}\frac{\partial\boldsymbol{B}}{\partial t}\cdot d\boldsymbol{S} = -\lim_{h\to 0}\frac{d}{dt}\int_{\delta S_h}\boldsymbol{B}\cdot d\boldsymbol{S} = 0\,.\end{aligned}$$

In the last step we have used that the flux of the magnetic field through a rectangle becomes zero in the limit $h \to 0$. A similar analysis holds for Ampere's law, Eq. (2.26a), and we are led to

$$(\boldsymbol{H}_2-\boldsymbol{H}_1)\cdot\delta\boldsymbol{\ell} = \lim_{h\to 0}\boldsymbol{J}_{\rm ext}\cdot\delta\boldsymbol{S}_h\,.$$

Here we have to be careful about a possible surface current distribution $\boldsymbol{K}_{\rm ext}$ that may give a non-vanishing contribution in the limit $h \to 0$. This finally provides us with the second set of boundary conditions

Second Set of Boundary Conditions

$$\hat{\boldsymbol{n}}\times(\boldsymbol{E}_2-\boldsymbol{E}_1)=0\,,\quad \hat{\boldsymbol{n}}\times(\boldsymbol{H}_2-\boldsymbol{H}_1)=\boldsymbol{K}_{\rm ext}\,. \tag{2.32}$$

In most cases of interest, in this book we will not be dealing with free surface charge or current contributions. The boundary conditions of Maxwell's equations at an interface between two media can then be summarized as follows:

– the normal components of $\boldsymbol{D}$, $\boldsymbol{B}$ are continuous and
– the parallel components of $\boldsymbol{E}$, $\boldsymbol{H}$ are continuous.

2.4 Time-Harmonic Fields

In particular for linear materials one often deals with light excitations oscillating with a single frequency ω, and because of the linearity of the material response the system oscillates after an initial transient with the same frequency ω. We here follow the notation of the introduction chapter and assume electromagnetic fields of the form

$$\boldsymbol{E}(\boldsymbol{r},t)=e^{-i\omega t}\boldsymbol{E}(\boldsymbol{r})\,,\quad \boldsymbol{B}(\boldsymbol{r},t)=e^{-i\omega t}\boldsymbol{B}(\boldsymbol{r})\,. \tag{2.33}$$

A few words of caution are at place. First, the true electromagnetic fields are of course real quantities. Whenever we are interested in these real fields we have to take the real parts of the corresponding expressions. Second, we use the same symbols for $\boldsymbol{E}(\boldsymbol{r},t)$ and $\boldsymbol{E}(\boldsymbol{r})$. In general, this should not cause too much confusions because we will always clearly state whether we work in the time domain or use time-harmonic fields. For time-harmonic fields and a linear material response, Maxwell's equations can be written in the form

Maxwell's Equations for Time-Harmonic Fields

$$\nabla\cdot\varepsilon\boldsymbol{E}=\rho\,,\qquad \nabla\times\boldsymbol{E}=i\omega\boldsymbol{B}$$
$$\nabla\cdot\boldsymbol{B}=0\,,\qquad \nabla\times\frac{1}{\mu}\boldsymbol{B}=\boldsymbol{J}-i\omega\,\varepsilon\boldsymbol{E}\,. \tag{2.34}$$

We have not explicitly indicated that ρ, $\boldsymbol{J}$ correspond to external charge and current distributions, and will suppress this dependence from here on whenever it is clear from the context.

Products of Time-Harmonic Fields Occasionally we will need the time-averaged product of two expressions oscillating with the *same* frequency ω,

$$\langle fg\rangle=\frac{1}{T}\int_0^T \mathrm{Re}\left(fe^{-i\omega t}\right)\,\mathrm{Re}\left(ge^{-i\omega t}\right)\,dt\,.$$

Here $T=2\pi/\omega$. A typical example is the Poynting vector accounting for the averaged energy flow, which will be discussed in more detail in Chap. 4. Expanding the real parts in the form

$$\langle fg\rangle=\frac{1}{4T}\int_0^T\left(fe^{-i\omega t}+f^*e^{i\omega t}\right)\left(ge^{-i\omega t}+g^*e^{i\omega t}\right)\,dt\,,$$

we next note that all expressions oscillating with $e^{\pm 2i\omega t}$ become zero when integrated over an entire oscillation period. The only non-vanishing terms are then

$$\langle fg\rangle=\frac{1}{4}\left(fg^*+f^*g\right)=\frac{1}{2}\mathrm{Re}\left(fg^*\right)\,. \tag{2.35}$$

2.4.1 *Sinusoidal Waves*

Any electric field can be decomposed into plane waves by means of a Fourier transformation

$$\boldsymbol{E}(\boldsymbol{r})=\int e^{i\boldsymbol{k}\cdot\boldsymbol{r}}\,\boldsymbol{E}_{\boldsymbol{k}}\,\frac{d^3k}{(2\pi)^3}\,. \tag{2.36}$$

Here $\boldsymbol{k}$ denotes a wavevector and $\boldsymbol{E}_{\boldsymbol{k}}$ is a Fourier component. A corresponding decomposition can be obtained for the magnetic field. Consider next a single sinusoidal wave

$$\boldsymbol{E}(\boldsymbol{r}) = \boldsymbol{E}_{\boldsymbol{k}} e^{i\boldsymbol{k}\cdot\boldsymbol{r}}\,, \quad \boldsymbol{B}(\boldsymbol{r}) = \boldsymbol{B}_{\boldsymbol{k}} e^{i\boldsymbol{k}\cdot\boldsymbol{r}}\,. \tag{2.37}$$

If we are interested in the propagation of a more complicated field, we can always decompose the field into plane waves, using Eq. (2.36), propagate the sinusoidal components separately, and put together the final result at the end. The action of the nabla operator on a plane wave becomes $\nabla \to i\boldsymbol{k}$, see Exercise 2.10. Thus, in absence of sources $\rho = \boldsymbol{J} = 0$ Maxwell's equations take the following form:

Maxwell's Equations for a Plane Wave

$$\begin{aligned} &\varepsilon\, \boldsymbol{k}\cdot\boldsymbol{E}_{\boldsymbol{k}} = 0\,, && \boldsymbol{k}\times\boldsymbol{E}_{\boldsymbol{k}} = \omega\boldsymbol{B}_{\boldsymbol{k}} \\ &\boldsymbol{k}\cdot\boldsymbol{B}_{\boldsymbol{k}} = 0\,, && \boldsymbol{k}\times\boldsymbol{B}_{\boldsymbol{k}} = -\omega\,\mu\varepsilon\boldsymbol{E}_{\boldsymbol{k}}\,. \end{aligned} \tag{2.38}$$

A few important things can be directly inferred from these equations. First, electromagnetic waves are transverse waves where $\boldsymbol{E}_{\boldsymbol{k}}$, $\boldsymbol{B}_{\boldsymbol{k}}$ are perpendicular to the wavenumber $\boldsymbol{k}$, as can be seen from $\boldsymbol{k}\cdot\boldsymbol{E}_{\boldsymbol{k}} = \boldsymbol{k}\cdot\boldsymbol{B}_{\boldsymbol{k}} = 0$. Second, $\boldsymbol{E}_{\boldsymbol{k}}$, $\boldsymbol{B}_{\boldsymbol{k}}$, and $\boldsymbol{k}$ form a right-handed orthogonal basis. Finally from

$$\boldsymbol{k}\times\boldsymbol{k}\times\boldsymbol{E}_{\boldsymbol{k}} = -k^2\boldsymbol{E}_{\boldsymbol{k}} = \omega\boldsymbol{k}\times\boldsymbol{B}_{\boldsymbol{k}} = -\omega^2\,\mu\varepsilon\boldsymbol{E}_{\boldsymbol{k}}$$

we recover the wave equation for a plane wave together with the dispersion relation

$$k = \sqrt{\mu\varepsilon}\,\omega\,. \tag{2.39}$$

We can use this expression together with Faraday's law

$$\boldsymbol{k}\times\boldsymbol{E}_{\boldsymbol{k}} = \sqrt{\mu\varepsilon}\,\omega\,\hat{\boldsymbol{k}}\times\boldsymbol{E}_{\boldsymbol{k}} = \omega\mu\,\boldsymbol{H}_{\boldsymbol{k}}$$

to establish a relation between $\boldsymbol{E}_{\boldsymbol{k}}$ and $\boldsymbol{H}_{\boldsymbol{k}}$,

$$Z\boldsymbol{H}_{\boldsymbol{k}} = \hat{\boldsymbol{k}}\times\boldsymbol{E}_{\boldsymbol{k}}\,. \tag{2.40}$$

Here we have introduced the impedance

$$Z = \sqrt{\frac{\mu}{\varepsilon}} \tag{2.41}$$

that allows to express via Eq. (2.41) the electric field $\boldsymbol{E}_k$ through the magnetic field $\boldsymbol{H}_k$, and vice versa.

Transverse Electric and Magnetic Fields We finish this section with a short discussion of transverse electric (TE) and transverse magnetic (TM) fields, which will be of importance in later parts of this book. Suppose that $\hat{\boldsymbol{n}}$ is a vector that is normal to an interface. We can then decompose the electromagnetic fields into components perpendicular (transverse) and parallel to $\hat{\boldsymbol{n}}$,

$$\begin{aligned} \boldsymbol{E}_k &= \left[\boldsymbol{E}_k - \hat{\boldsymbol{n}}(\hat{\boldsymbol{n}}\cdot\boldsymbol{E}_k)\right] + \hat{\boldsymbol{n}}(\hat{\boldsymbol{n}}\cdot\boldsymbol{E}_k) = \boldsymbol{E}_k^{TE} + \hat{\boldsymbol{n}}(\hat{\boldsymbol{n}}\cdot\boldsymbol{E}_k) \\ \boldsymbol{H}_k &= \left[\boldsymbol{H}_k - \hat{\boldsymbol{n}}(\hat{\boldsymbol{n}}\cdot\boldsymbol{H}_k)\right] + \hat{\boldsymbol{n}}(\hat{\boldsymbol{n}}\cdot\boldsymbol{H}_k) = \boldsymbol{H}_k^{TM} + \hat{\boldsymbol{n}}(\hat{\boldsymbol{n}}\cdot\boldsymbol{H}_k)\,. \end{aligned} \tag{2.42}$$

Next, we use Eq. (2.40) to express the normal component of $\boldsymbol{H}_k$ in terms of the parallel component of $\boldsymbol{E}_k$,

$$Z\hat{\boldsymbol{n}}\cdot\boldsymbol{H}_k = \hat{\boldsymbol{n}}\cdot\hat{\boldsymbol{k}}\times\boldsymbol{E}_k \underset{\text{c.p.}}{=} \hat{\boldsymbol{k}}\cdot\boldsymbol{E}_k\times\hat{\boldsymbol{n}} = \hat{\boldsymbol{k}}\cdot\boldsymbol{E}_k^{TE}\times\hat{\boldsymbol{n}} = \hat{\boldsymbol{n}}\cdot\hat{\boldsymbol{k}}\times\boldsymbol{E}_k^{TE}\,.$$

In the above expression we have performed a cyclic permutation (c.p.) for the triple product. We are thus led to the decomposition of electromagnetic fields in the form

$$\begin{aligned} \boldsymbol{E}_k &= \boldsymbol{E}_k^{TE} - \hat{\boldsymbol{n}}\left(\hat{\boldsymbol{n}}\cdot\hat{\boldsymbol{k}}\times\boldsymbol{H}_k^{TM}\right)Z \\ \boldsymbol{H}_k &= \boldsymbol{H}_k^{TM} + \hat{\boldsymbol{n}}\left(\hat{\boldsymbol{n}}\cdot\hat{\boldsymbol{k}}\times\boldsymbol{E}_k^{TE}\right)Z^{-1}\,. \end{aligned} \tag{2.43}$$

The nice thing about this decomposition is that it is often easier to deal with the transverse electric $\boldsymbol{E}_k^{TE}$ and magnetic $\boldsymbol{H}_k^{TM}$ fields, for instance, when employing the boundary conditions at an interface. Equation (2.43) shows that any field can be uniquely decomposed into TE and TM components.

2.5 Longitudinal and Transverse Fields

We conclude this chapter with a brief discussion of longitudinal and transverse fields. Our starting point is the Helmholtz theorem introduced in Sect. 2.1.2 which states that any vector field is specified upon knowledge of its divergence and curl, together with the appropriate boundary conditions. In the spirit of this theorem, we can decompose a vector function $\boldsymbol{F}(\boldsymbol{r})$ into a longitudinal and transverse part

Decomposition of $\boldsymbol{F}$ Into Longitudinal and Transverse Parts

$$\boldsymbol{F}(\boldsymbol{r}) = \boldsymbol{F}^L(\boldsymbol{r}) + \boldsymbol{F}^\perp(\boldsymbol{r})\,, \tag{2.44}$$

where F^L and $F^{\perp}$ are specified according to

$$\begin{aligned} \nabla \cdot \boldsymbol{F}^L(\boldsymbol{r}) &= f(\boldsymbol{r}), & \nabla \times \boldsymbol{F}^L(\boldsymbol{r}) &= 0 \\ \nabla \cdot \boldsymbol{F}^{\perp}(\boldsymbol{r}) &= 0, & \nabla \times \boldsymbol{F}^{\perp}(\boldsymbol{r}) &= \boldsymbol{g}(\boldsymbol{r}). \end{aligned} \tag{2.45}$$

Here $f(\boldsymbol{r})$ and $\boldsymbol{g}(\boldsymbol{r})$ are arbitrary scalar and vector functions. The longitudinal part has a non-zero divergence and the transverse part a non-zero curl. The above decomposition becomes particularly transparent for a sinusoidal wave of the form of Eq. (2.37). The longitudinal part (with respect to the wavevector $\boldsymbol{k}$) can be written as

$$\boldsymbol{F}_k^L = \hat{\boldsymbol{k}} F_k^L = \hat{\boldsymbol{k}}\left(\hat{\boldsymbol{k}} \cdot \boldsymbol{F}\right) = \bar{\bar{\mathcal{P}}}_k^L \cdot \boldsymbol{F}_k,$$

where in the last expression we have introduced the projection matrix

$$\left(\bar{\bar{\mathcal{P}}}_k^L\right)_{ij} = \hat{k}_i \hat{k}_j .$$

A vector function $\boldsymbol{F}_k$ can then be decomposed into its longitudinal and transverse parts through

$$\boldsymbol{F}_k = \left(\bar{\bar{\mathcal{P}}}_k^L\right) \cdot \boldsymbol{F}_k + \left(\mathbb{1} - \bar{\bar{\mathcal{P}}}_k^L\right) \cdot \boldsymbol{F}_k = \left(\bar{\bar{\mathcal{P}}}_k^L\right) \cdot \boldsymbol{F}_k + \left(\bar{\bar{\mathcal{P}}}_k^{\perp}\right) \cdot \boldsymbol{F}_k . \tag{2.46}$$

Here $\mathbb{1}$ is the unit matrix and in the last expression we have introduced the projection matrix on the transverse part. From Eq. (2.45) we then get for the longitudinal part

$$ikF_k^L = f_k . \tag{2.47}$$

Similarly, we find for the transverse part

$$i\boldsymbol{k} \times \boldsymbol{F}_k^{\perp} = \boldsymbol{g}_k \implies ik\boldsymbol{F}_k^{\perp} = -\hat{\boldsymbol{k}} \times \boldsymbol{g}_k . \tag{2.48}$$

Thus, a possible strategy for splitting in Eq. (2.44) a vector function into its longitudinal and transverse parts is to decompose it into its sinusoidal components using a Fourier transform, apply the projection matrices, and to finally perform the inverse Fourier transform to arrive at the desired splitting.

Maxwell's Equations

In the spirit of the above discussion, we can write Maxwell's equations of Eq. (2.34) in the form

$$\begin{aligned} i\boldsymbol{k} \cdot \varepsilon \boldsymbol{E}_k^L &= \rho_k, & \boldsymbol{k} \times \boldsymbol{E}_k^{\perp} &= \omega \boldsymbol{B}_k \\ \boldsymbol{k} \cdot \boldsymbol{B}_k^L &= 0, & \boldsymbol{k} \times \boldsymbol{B}_k^{\perp} &= -i\mu \boldsymbol{J}_k - \omega\,\mu\varepsilon \boldsymbol{E}_k . \end{aligned} \tag{2.49}$$

We have indicated the longitudinal and transverse character whenever it is clear from the context. ρ_k and $\boldsymbol{J}_k$ are the Fourier transforms of the charge and current distributions, respectively. It is obvious that the longitudinal component B_k^L is zero and the magnetic field is purely transverse. Using Eq. (2.47) we obtain from Gauss' law

$$E_k^L = -i\frac{\rho_k}{\varepsilon k}\,. \tag{2.50}$$

The continuity equation relates the charge distribution to the longitudinal component of the current distribution

$$\omega\rho_k = kJ_k^L \implies J_k^L = \frac{\omega\rho_k}{k} = i\omega\varepsilon\, E_k^L\,.$$

Thus, Ampere's law in Eq. (2.49) can be rewritten in the form

$$\boldsymbol{k}\times\boldsymbol{B}_k^\perp = -i\mu\boldsymbol{J}_k^\perp - \omega\,\mu\varepsilon\boldsymbol{E}_k^\perp\,, \tag{2.51}$$

because the longitudinal components of J_k^L and E_k^L cancel each other. We thus observe that in Eq. (2.49) the two divergence equations specify the longitudinal components of $\boldsymbol{E}_k$, $\boldsymbol{B}_k$, whereas the curl equations relate the transverse components of the electromagnetic fields.

Electromagnetic Potentials

The concept of longitudinal and transverse fields can be also applied to the electromagnetic potentials V, $\boldsymbol{A}$ introduced in Sect. 2.2.1. The magnetic field is obviously determined by the transverse component of $\boldsymbol{A}_k$

$$\boldsymbol{B}_k^\perp = i\boldsymbol{k}\times\boldsymbol{A}_k^\perp\,. \tag{2.52}$$

The electric field has both longitudinal and transverse components

$$E_k^L = -ikV_k + i\omega A_k^L\,,\quad \boldsymbol{E}_k^\perp = i\omega\boldsymbol{A}_k^\perp\,. \tag{2.53}$$

Gauge transformations of the form given by Eq. (2.19) then leave the transverse components of $\boldsymbol{A}_k^\perp$ unaffected and only modify

$$\boldsymbol{A}_k^{L\prime} = \boldsymbol{A}_k^L + i\boldsymbol{k}\,\lambda_k\,,\quad V_k' = V_k + i\omega\,\lambda_k\,.$$

Exercises

Before starting with the exercises, I would like to add a personal advertisement about component notations and the Levi-Civita tensor. Both concepts are overly useful when it comes to the evaluation of more complicated expressions.

Dot Product. The dot product can be written in the form $\boldsymbol{a}\cdot\boldsymbol{b} = a_i b_i$.

Einstein's Sum Convention. Einstein's sum convention states that one has to sum over indices that appear twice in an expression, such as i in the above dot product.

Cross Product. The cross product can be expressed as $(\boldsymbol{a}\times\boldsymbol{b})_i = \epsilon_{ijk}\, a_j b_k$. Here ϵ_{ijk} is the Levi-Civita tensor, or ϵ tensor in short.

Levi-Civita Tensor. The Levi-Civita tensor, or ϵ tensor in short, is a totally antisymmetric tensor defined as

$$\epsilon_{ijk} = \begin{cases} 1 & \text{for even permutations } \epsilon_{123}, \epsilon_{312}, \epsilon_{231} \\ -1 & \text{for odd permutations } \epsilon_{213}, \epsilon_{321}, \epsilon_{132} \\ 0 & \text{else}. \end{cases} \tag{2.54}$$

From the above relation it is clear that one can cyclically shuffle the indices around, $\epsilon_{ijk} = \epsilon_{kij} = \epsilon_{jki}$, and that it changes sign when one exchanges two indices, for instance, $\epsilon_{ijk} = -\epsilon_{jik}$.

The by far most important relation for the Levi-Civita tensor reads

$$\epsilon_{ijk}\epsilon_{mnk} = \delta_{im}\delta_{jn} - \delta_{in}\delta_{jm}\,. \tag{2.55}$$

I really urge you to remember this relation as it can considerably help you for working out most expressions we will encounter in this book.

Exercise 2.1 Use the Levi-Civita tensor to compute the following expressions:

(a) $\boldsymbol{a}\times(\boldsymbol{b}\times\boldsymbol{c})$.
(b) $(\boldsymbol{a}\times\boldsymbol{b})\times\boldsymbol{c}$. Show that the result is different from a.

Exercise 2.2 Use the Levi-Civita tensor to prove the following relation:

$$(\boldsymbol{a}\times\boldsymbol{b})\cdot(\boldsymbol{c}\times\boldsymbol{d}) = (\boldsymbol{a}\cdot\boldsymbol{c})(\boldsymbol{b}\cdot\boldsymbol{d}) - (\boldsymbol{a}\cdot\boldsymbol{d})(\boldsymbol{b}\cdot\boldsymbol{c})\,.$$

Exercise 2.3 Use the definition of Eq. (2.54) to prove $\epsilon_{imn}\epsilon_{jmn} = 2\delta_{ij}$.

Exercise 2.4 Prove for two vector functions $\boldsymbol{F}(\boldsymbol{r}), \boldsymbol{G}(\boldsymbol{r})$ that the following relation holds:

$$\nabla\cdot\boldsymbol{F}\times\boldsymbol{G} = (\nabla\times\boldsymbol{F})\cdot\boldsymbol{G} - \boldsymbol{F}\cdot(\nabla\times\boldsymbol{G})\,.$$

Hint: Use the product rule for the ∇ operator acting on the two vector functions on its right. In the above expression and in many parts of this book we will use parentheses to indicate that ∇ acts on only one expression.

Exercise 2.5 Use the definition of Eq. (2.54) to argue why $\nabla\cdot\nabla\times\boldsymbol{F}(\boldsymbol{r}) = 0$.

Exercise 2.6 Show that $\nabla\times\nabla\times\boldsymbol{F} = \nabla(\nabla\cdot\boldsymbol{F}) - \nabla^2\boldsymbol{F}$ holds for any vector function $\boldsymbol{F}(\boldsymbol{r})$.

Exercise 2.7 Apply the divergence $\nabla\cdot$ on both sides of Ampere's law, Eq. (2.15d), to derive the continuity equation $\frac{\partial\rho}{\partial t} = -\nabla\cdot\boldsymbol{J}$. You will need another Maxwell equation to arrive at this result. Which one?

Exercise 2.8 Apply the curl $\nabla\times$ on both sides of Ampere's law, Eq. (2.15d), to derive the wave equation for $\boldsymbol{B}$ using $\boldsymbol{J} = 0$. You will additionally need Faraday's law to arrive at the final expression.

Exercise 2.9 Show that with the Lorenz gauge condition of Eq. (2.21) one obtains the wave equations for the electromagnetic potentials, see Eq. (2.22).

Exercise 2.10 Evaluate for a plane wave $\boldsymbol{E}(\boldsymbol{r}) = \boldsymbol{E}_0\, e^{i\boldsymbol{k}\cdot\boldsymbol{r}}$ with constant vectors $\boldsymbol{E}_0, \boldsymbol{k}$ the expressions $\nabla\cdot\boldsymbol{E}$ and $\nabla\times\boldsymbol{E}$. Argue why for such a plane wave one can make the replacement $\nabla \to i\boldsymbol{k}$.

Exercise 2.11 At a given instant of time, say at $t = 0$, the electric field of a wave propagating along z can be written in the form

$$\hat{\boldsymbol{x}}\, E_x(z) + \hat{\boldsymbol{y}}\, E_y(z) = \mathrm{Re}\left[E_0 \hat{\boldsymbol{e}}\, e^{ikz}\right],$$

where E_0 is an amplitude (assumed to be real) and $\hat{\boldsymbol{e}}$ the polarization vector of the wave. Compute $E_x(z)$, $E_y(z)$ for the following polarization vectors:

(a) $\hat{\boldsymbol{e}} = \hat{\boldsymbol{x}}$.
(b) $\hat{\boldsymbol{e}} = \cos\theta\, \hat{\boldsymbol{x}} + \sin\theta\, \hat{\boldsymbol{y}}$, where θ is a real angle.
(c) $\hat{\boldsymbol{e}} = \hat{\boldsymbol{x}} + i\,\hat{\boldsymbol{y}}$.
(d) $\hat{\boldsymbol{e}} = \cos\theta\, \hat{\boldsymbol{x}} + i\,\sin\theta\, \hat{\boldsymbol{y}}$, where θ is a real angle.

Exercise 2.12 Use the wave of Exercise 2.11 to compute the corresponding magnetic field components $B_x(z)$, $B_y(z)$. Use Faraday's law to relate $\boldsymbol{E}$ and $\boldsymbol{B}$. How does $\boldsymbol{B}(z)$ look for the different polarization vectors?

Exercise 2.13 Consider a wave $\boldsymbol{E}(\boldsymbol{r}) = E_0 \hat{\boldsymbol{y}}\, e^{i\boldsymbol{k}\cdot\boldsymbol{r}}$ with a complex wavevector $\boldsymbol{k} = k_x\,\hat{\boldsymbol{x}} + ik_z\,\hat{\boldsymbol{z}}$. Such a wave is called an evanescent wave which propagates along x and is exponentially attenuated for $z > 0$. Compute the real part of $\boldsymbol{E}(\boldsymbol{r})$ and discuss the polarization properties.

Chapter 3
Angular Spectrum Representation

The situation we will consider in this chapter is depicted in Fig. 3.1 and can be summarized as follows. Suppose that we know the electric field at a given plane $z = 0$,

$$\boldsymbol{E}(x, y, 0) = \boldsymbol{E}_0(x, y) .$$

We are now asking the following questions. How does the field $\boldsymbol{E}(x, y, z)$ evolve when we move away from the plane? How does $\boldsymbol{E}(\boldsymbol{r})$ behave in the far-field zone? And what happens when we focus the far-fields through a lens or a system of lenses? To answer these questions, we will introduce a few novel concepts that play an important role in the field of nano optics.

Angular Spectrum Representation. In the angular spectrum representation we decompose $\boldsymbol{E}_0(x, y)$ into its Fourier components $\tilde{\boldsymbol{E}}_0(k_x, k_y)$. These components can then be easily propagated away from $z = 0$ by adding appropriate phase factors.

Far-Field Representation. To compute the electromagnetic fields far away from the plane $z = 0$, we introduce the concept of the stationary phase approximation. We will find that in the far-field the propagating fields are closely related to $\tilde{\boldsymbol{E}}_0(k_x, k_y)$.

Focusing of Fields. In experiments the focusing of fields in order to retrieve an image of the original field distribution $\boldsymbol{E}_0(x, y)$ is achieved by optical lenses. In theoretical nano optics one can introduce a transformation for the far-fields in order to mimic the lens performance.

U. Hohenester, *Nano and Quantum Optics*, Graduate Texts in Physics,
https://doi.org/10.1007/978-3-030-30504-8_3

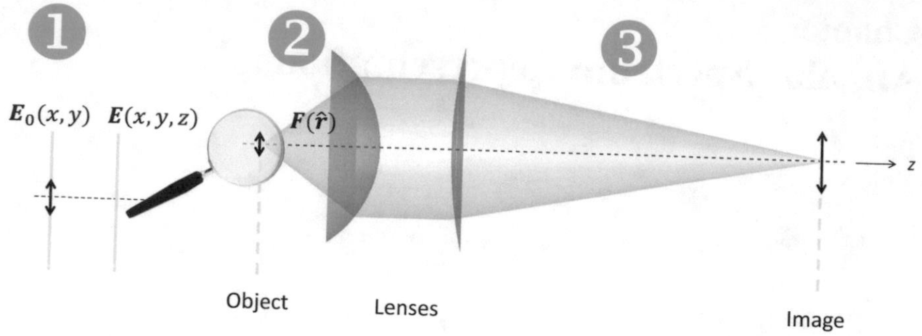

Fig. 3.1 Sketch of angular spectrum representation and imaging for an object located in the focus of a lens at $z = 0$. In this chapter we show (1) how the fields $\boldsymbol{E}(x, y, z)$ can be propagated away from the object using the angular spectrum representation, (2) how to compute the far-fields $\boldsymbol{F}(\hat{\boldsymbol{r}})$ far away from the object, and (3) how to get an image of the object using a system of lenses described in terms of Gaussian reference spheres. In the transition from the near to the far-fields all evanescent waves are removed from the field distribution, which is responsible for the diffraction limit of light to be discussed in Chap. 6

3.1 Fourier Transform of Fields

Suppose that we know the electric field distribution $\boldsymbol{E}_0(x, y)$ at a given plane $z = 0$. We can introduce a Fourier transformation

$$\tilde{\boldsymbol{E}}_0(k_x, k_y) = (2\pi)^{-2} \int e^{-i(k_x x + k_y y)} \boldsymbol{E}_0(x, y)\, dx dy \tag{3.1}$$

to decompose the fields into their wavevector components. Obviously, through the inverse Fourier transformation

$$\boldsymbol{E}_0(x, y) = \int e^{i(k_x x + k_y y)} \tilde{\boldsymbol{E}}_0(k_x, k_y)\, dk_x dk_y \tag{3.2}$$

we get back the original fields. When we move away from $\boldsymbol{E}(x, y, 0) = \boldsymbol{E}_0(x, y)$ each Fourier component acquires a phase factor

$$e^{ik_z z} \tilde{\boldsymbol{E}}_0(k_x, k_y)\,,$$

where k_z can be determined from the k value given by the light frequency and the dispersion relation

$$k_z = \pm\sqrt{k^2 - k_x^2 - k_y^2}\,. \tag{3.3}$$

The sign has to be determined such that we obtain waves that propagate either in the positive or negative z direction. For $k^2 < k_x^2 + k_y^2$ we get an imaginary wavenumber corresponding to an evanescent field, and the sign has to be chosen such that the wave decays away from $z = 0$. Thus, for the half-space with $z > 0$

we then obtain, in accordance to Eq. (1.19) for the scalar wave equation, the angular spectrum representation

Angular Spectrum Representation ($z > 0$)

$$\begin{aligned} \boldsymbol{E}(x,y,z) &= \int_{k^2>k_x^2+k_y^2} e^{i(k_x x+k_y y+\sqrt{k^2-k_x^2-k_y^2}z)}\,\tilde{\boldsymbol{E}}_0(k_x,k_y)\,dk_x dk_y \\ &+ \int_{k^2<k_x^2+k_y^2} e^{i(k_x x+k_y y)-\sqrt{k_x^2+k_y^2-k^2}z}\,\tilde{\boldsymbol{E}}_0(k_x,k_y)\,dk_x dk_y\,. \end{aligned} \tag{3.4}$$

The expressions in the first and second line correspond to propagating and evanescent waves, respectively. From Eq. (3.4) one observes that in order to compute fields away from a given plane we must first decompose $\boldsymbol{E}(x,y,0)$ into its Fourier components, and then each Fourier component simply acquires a phase or decays exponentially when moving away from $z = 0$. Obviously, the further we move away from $z = 0$ the more the evanescent fields decay and only the propagating ones survive for sufficiently large propagation lengths z.

As a representative example, we investigate the situation shown in Fig. 3.2 consisting of an ensemble of densely packed dipoles oriented along the z direction, which form the letters "micron." As shown in Fig. 3.3, the image blurs when moving away from the sheet of dipoles as an effect of both loss of evanescent waves and defocusing.

3.2 Far-Field Representation

For large propagation distances the electric fields of Eq. (3.4) are outgoing waves

$$\boldsymbol{E}(\boldsymbol{r}) \underset{kr\gg 1}{\longrightarrow} \frac{e^{ikr}}{r}\,\boldsymbol{F}(\hat{\boldsymbol{r}})\,, \tag{3.5}$$

modulated by the far-field amplitude

$$\boldsymbol{F}(\hat{\boldsymbol{r}}) = \lim_{r\to\infty} r e^{-ikr} \int_{k_x^2+k_y^2\le k^2} e^{i\boldsymbol{k}\cdot\boldsymbol{r}}\,\tilde{\boldsymbol{E}}_0(k_x,k_y)\,dk_x dk_y\,. \tag{3.6}$$

In the above expression we have neglected all evanescent waves. Equation (3.6) looks like a nasty integral which starts to wildly oscillate for $kr \to \infty$ and we have to be careful in its evaluation.

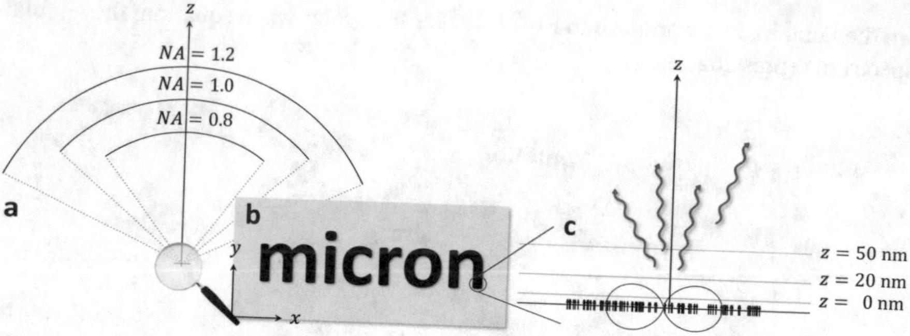

Fig. 3.2 Schematics for angular spectrum representation and imaging of fields emitted from a "micron" object. (**a**) The object is located at the origin, in the center of the magnifying glass. The arcs indicate the range of wavevector components captured by lenses with different numerical apertures *NA*, see also Fig. 3.6. (**b**) The object consists of densely packed dipoles oriented along *z*, see panel (**c**), which form the letters "micron." (**c**) Close-up of the dipole sheet and of the layers where the fields are shown in Fig. 3.6. The red lobe reports the far-field emission pattern of a single dipole

Fortunately, the *stationary phase approximation* [4] provides a recipe for dealing with such integrals. It can be shown that the dominant contributions to the integral come from those regions where the function in the exponential changes least, the so-called stationary points usually associated with extrema of the function in the exponential. Away from these extrema, the strong oscillations of the exponential lead to destructive interference and the contributions to the integral become zero. In the following, we briefly introduce to the method of the stationary phase approximation.

Stationary Phase Approximation As a preliminary step, we consider the one-dimensional integral

$$I \underset{\lambda\to\infty}{\longrightarrow} \int_a^b e^{i\lambda g(u)} f(u)\, du\,.$$

This integral contributes most around the extrema of $g(u)$ [4]. For the sake of simplicity we assume that $g(u)$ has a single extremum which is located at u_0 within the integration range. We then expand $g(u)$ in a Taylor series around u_0 and assume that $f(u)$ changes only moderately there, such that we can approximate

$$I \underset{\lambda\to\infty}{\longrightarrow} f(u_0) \int_{-\infty}^{\infty} e^{i\lambda[g(u_0)+\frac{1}{2}g''(u_0)(u-u_0)^2]}\, du = \left[\frac{2\pi}{\lambda|g''(u_0)|}\right]^{\frac{1}{2}} f(u_0) e^{i\lambda g(u_0) \pm i\frac{\pi}{4}}\,.$$

The upper or lower sign of the last expression in the exponential is taken according to the sign of $g''(u_0)$. To evaluate the integral we have expanded the integration limits to infinity, which is a good approximation because the integrand only contributes significantly around the extremum and otherwise becomes zero because of destructive interference, and have

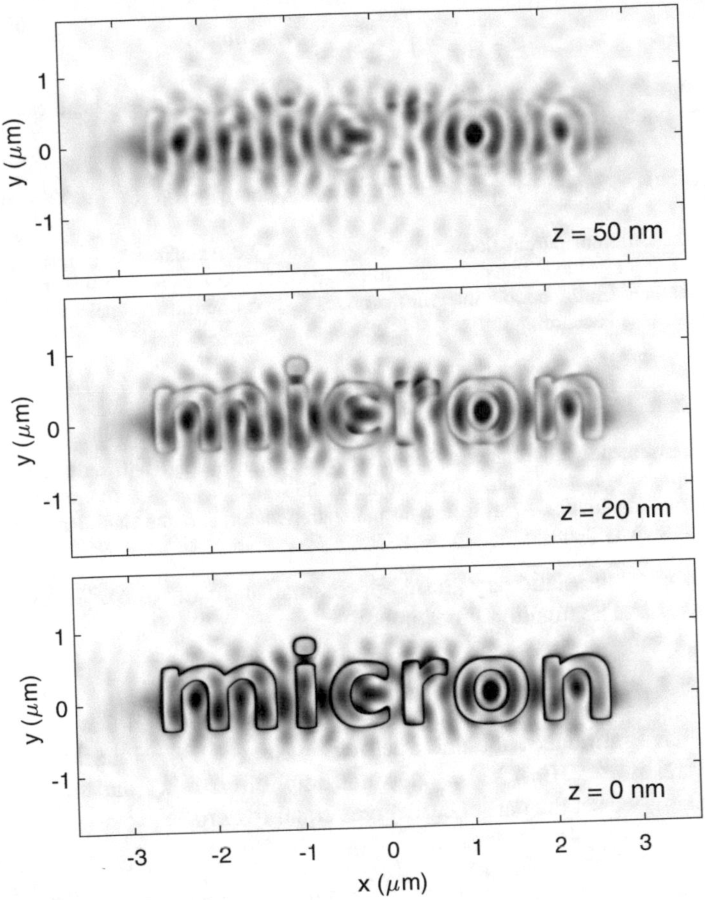

Fig. 3.3 Computation of angular spectrum representation of Eq. (3.4). For the initial field distribution $\boldsymbol{E}_0(x, y)$ we consider the "micron" letters filled with dipoles oscillating at a wavelength of 620 nm and oriented along the z direction, as schematically shown in Fig. 3.2. $\boldsymbol{E}_0(x, y)$ corresponds to the fields 10 nm above the dipole layer. When moving away from the plane (see insets for distances) the fields blur, partially because of the decay of evanescent fields and partially because of defocusing. In Sect. 3.3 we will show that a field image can be obtained from the far-field distribution using a system of lenses

used the properties of Fresnel integrals.[1] A similar procedure can be also used for two-dimensional integrals

[1]More specifically, we use

$$\int_0^\infty \cos\left(t^2\right)\,dt = \int_0^\infty \sin\left(t^2\right)\,dt = \sqrt{\frac{\pi}{8}}\,.$$

$$I \underset{\lambda\to\infty}{\longrightarrow} \int e^{i\lambda g(u,v)} f(u,v)\, du dv\,.$$

We assume again that within the integration regime $g(u, v)$ has a single extremum at position (u_0, v_0) and expand $g(u, v)$ around this point,

$$g(u,v) \approx g(u_0,v_0) + \frac{1}{2} g_{uu}(u-u_0)^2 + \frac{1}{2} g_{vv}(v-v_0)^2 + g_{uv}(u-u_0)(v-v_0)\,,$$

where the different partial derivatives of g have to be evaluated at (u_0, v_0). We next perform a principal axis transformation to rotate the u and v axes such that in the rotated coordinate system the double integral factorizes to two one-dimensional integrals, which can be evaluated according to

$$I \underset{\lambda\to\infty}{\longrightarrow} \pm \frac{2\pi i}{\lambda\sqrt{\Delta}} f(u_0,v_0) e^{i\lambda g(u_0,v_0)}\,, \qquad \Delta = g_{uu} g_{vv} - g_{uv}^2\,. \tag{3.7}$$

We have assumed that the stationary point is an extremum with $\Delta > 0$ and the sign of the above expression has to be chosen according to the sign of $g_{uu} + g_{vv}$ (corresponding to a minimum or maximum). Apart from a normalization factor, Δ is the Gaussian curvature of g at the stationary point.

We next employ the stationary phase approximation to Eq. (3.6). We use r instead of previously λ. The function g becomes

$$g(k_x,k_y) = k_x \hat{r}_x + k_y \hat{r}_y + k_z(k_x,k_y)\hat{r}_z\,,$$

where we have introduced the unit vector $\hat{\boldsymbol{r}} = (\hat{r}_x, \hat{r}_y, \hat{r}_z)$. Note that through the dispersion relation of Eq. (3.3) k_z depends implicitly on k_x and k_y. To find the extremum of g we set the partial derivatives equal to zero,

$$g_x = \frac{\partial g}{\partial k_x} = \hat{r}_x - \frac{k_x}{k_z}\hat{r}_z = 0 \implies \frac{\hat{r}_x}{k_x} = \frac{\hat{r}_z}{k_z}$$

$$g_y = \frac{\partial g}{\partial k_y} = \hat{r}_y - \frac{k_y}{k_z}\hat{r}_z = 0 \implies \frac{\hat{r}_y}{k_y} = \frac{\hat{r}_z}{k_z}\,.$$

Apparently, at the extremum $\boldsymbol{r}$ and $\boldsymbol{k}$ become parallel, as one could have already guessed from $\boldsymbol{r}\cdot\boldsymbol{k} = rk\cos\theta$ which becomes maximal for $\theta = 0$. Since the magnitude of $\boldsymbol{k}$ is fixed by the light frequency, we can set at the extremum $\boldsymbol{k} = k\hat{\boldsymbol{r}}$. We continue to evaluate the second derivatives at the stationary point

$$g_{xx} = -\frac{\hat{r}_x^2 + \hat{r}_z^2}{k\hat{r}_z^2}\,, \qquad g_{yy} = -\frac{\hat{r}_y^2 + \hat{r}_z^2}{k\hat{r}_z^2}\,, \qquad g_{xy} = -\frac{\hat{r}_x \hat{r}_y}{k\hat{r}_z^2}\,.$$

Thus, together with Eq. (3.7) we are led to our final expression for the far-field amplitude of the electromagnetic fields

Far-Field Representation

$$F(\hat{r}) = -2\pi i k \hat{r}_z \, \tilde{E}_0 \left(k\hat{r}_x, k\hat{r}_y\right) \tag{3.8}$$

This expression is quite remarkable. First, despite its rather intricate derivation it is a very simple result stating that the electric field propagating in a given direction is solely determined by a single Fourier component. All other components interfere destructively in the far-field zone. Secondly, it shows that nature can perform Fourier transformations by herself! As we will discuss next, what has to be done in optical imaging is to use lenses or other optical devices to properly put together the Fourier components propagating in different directions in order to get an image of the initial field distribution $E_0(x, y)$. As should be obvious by now, in this process all evanescent waves are lost, which provides a fundamental limit for the resolution of optical imaging.

3.3 Field Imaging and Focusing

The final step in our analysis is imaging where one has to properly put together the far-field components in order to obtain an image of the object fields $E_0(x, y)$. One could try to apply a full wave optical calculation to the complete transformation of the far-fields by all lens elements, however, such an approach would be extremely difficult and in most cases of practical interest not even overly useful. An alternative approach was proposed by Richards and Wolf [5] and is based on the assumption that the objective has perfect aplanatic imaging properties, as shown in Fig. 3.4. We

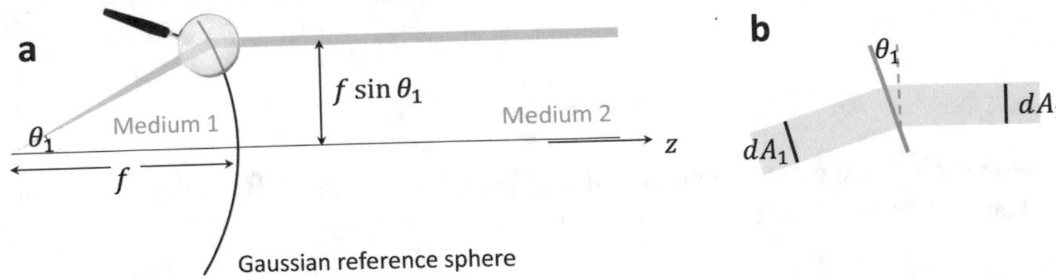

Fig. 3.4 In our theoretical approach we describe lenses through (**a**) Gaussian reference spheres. A ray emerging from the focus and crossing the Gaussian sphere is then transformed to a ray propagating parallel to the optical axis. Conversely, a parallel ray is transformed to a ray propagating towards the focus. (**b**) When crossing the reference sphere the area of the beam along the propagation direction is changed, such that the beam intensity is conserved. For details see text

assume that the object from which the radiation $\boldsymbol{E}_0(x, y)$ is emitted is located in the focus f of the first lens and that the dimensions of the object are much smaller than f. The imaging is performed using the following assumptions:

Geometrical Optics. We start with the optical far-fields $\boldsymbol{F}(\hat{\boldsymbol{r}})$ and trace these fields along rays through the optical system, using the laws of geometrical optics. For the problems we are interested in we will employ the so-called sine condition and model the aplanatic lens through a sphere with radius f, the so-called *Gaussian reference sphere*. A ray emerging from the focus and crossing the Gaussian sphere is then transformed to a ray propagating parallel to the optical axis. Conversely, a parallel ray is transformed to a ray propagating towards the focus.

Intensity Law. The energy flux along each ray must be conserved. The time-averaged energy flux can be computed from the Poynting vector $\boldsymbol{S} = \boldsymbol{E} \times \boldsymbol{H}$, which will be discussed in more detail in Chap. 4, and one obtains for the power transmitted by a ray

$$dP = \frac{1}{2} Z^{-1} |\boldsymbol{E}|^2 \, dA \,. \tag{3.9}$$

Here Z is the impedance of Eq. (2.41) and dA is an infinitesimal cross section perpendicular to the ray propagation. For the ray transition from the object to lens side shown in Fig. 3.4, we then relate the areas in the two media through $\cos\theta_1 = dA_2/dA_1$ and the electric field amplitudes through

$$\frac{1}{2} Z_1^{-1} |\boldsymbol{E}_1|^2 \, dA_1 = \frac{1}{2} Z_2^{-1} |\boldsymbol{E}_2|^2 \cos\theta_1 dA_1 \,.$$

Transmission. At the interface between two media, such as between the embedding medium of the object and the glass of the lens, only part of the light becomes transmitted and the rest is reflected. We will investigate reflection and transmission at planar interfaces in Chap. 8. There we will decompose the incoming electric fields according to Eq. (2.43) into their TE and TM components,

$$\boldsymbol{E} = \boldsymbol{E}^{TE} + \boldsymbol{E}^{TM} \,,$$

where $\boldsymbol{E}^{TE}$ is the component parallel to the interface and $\boldsymbol{E}^{TM}$ is the remainder. The transmitted field then has the form

$$\boldsymbol{E}_{\text{trans}} = T^{TE} \boldsymbol{E}^{TE} + T^{TM} \boldsymbol{E}^{TM} \,,$$

where T^{TE} and T^{TM} are the transmission coefficients. In the following we will not further specify these coefficients, which could be also approximated by one for lenses with an antireflection coating.

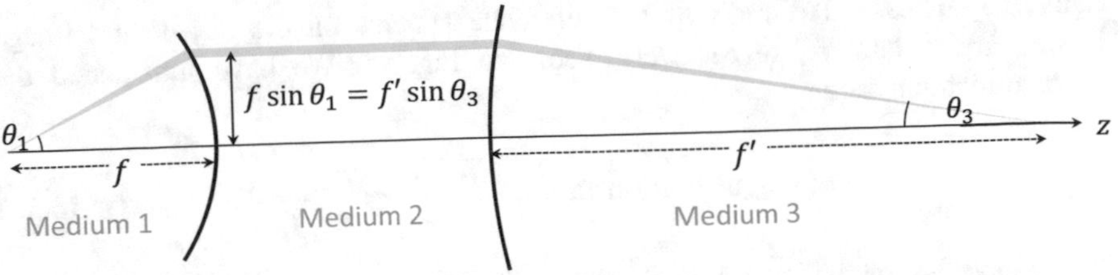

Fig. 3.5 Imaging through aplanatic lens. An object is located in the focus of the first lens (medium 1) described through a Gaussian reference sphere with radius f. After crossing the reference sphere, the beam propagates inside the lens (medium 2) parallel to the optical axis $\hat{z}$. After crossing the second reference sphere with radius $f' \gg f$ the beam is directed towards the focus (medium 3)

Figure 3.5 shows the setup for field imaging, consisting of a collection lens with focal length f and an imaging lens with focal length $f' \gg f$. The derivation of the working equation for such imaging is not overly inspiring and is given for the interested reader in Sect. 3.6. In short, we obtain for the imaging transformation from the incoming far-fields $\boldsymbol{F}_1(\theta, \phi)$ to the image fields $\boldsymbol{E}_3(\rho, \varphi, z)$ the following expression

Imaging of Far-Fields $\boldsymbol{F}_1(\boldsymbol{\theta}, \boldsymbol{\phi})$ by Aplanatic Lens

$$\boldsymbol{E}_3(\rho, \varphi, z) = \sqrt{\frac{n_1}{n_3}} \frac{ik_3 e^{i(k_1 f - k_3 f')}}{2\pi} \frac{f}{f'} \int_0^{\theta_{\max}} \sqrt{\cos\theta}\, \sin\theta\, d\theta \qquad (3.10)$$
$$\times \int_0^{2\pi} d\phi\, \bar{\bar{\mathcal{R}}}^{\text{im}} \cdot \boldsymbol{F}_1(\theta, \phi) e^{-ik_3 \cos\theta_3 z} e^{ik_1 \left(\frac{\rho}{M}\right) \sin\theta \cos(\phi - \varphi)} .$$

The meaning of its different terms is as follows.

Coordinate Systems. On the object side (medium 1) we use spherical coordinates (θ, ϕ) and on the image side (medium 3) cylinder coordinates (ρ, φ, z), with $z = 0$ corresponding to the focal plane.

Media. The refractive indices and wavenumbers are n_1, k_1 on the object side and n_3, k_3 on the image side, the refractive index of the lens turns out to be unimportant here.

Transformation Matrix. The orientation of the electric field vector becomes modified when crossing the Gaussian reference spheres, as described by the transformation matrix $\bar{\bar{\mathcal{R}}}^{\text{im}}$ explicitly given in Eq. (3.34). In principle, this matrix also includes the transmission coefficients, although we shall usually set them equal to one.

Lens Properties. The radii of the reference spheres on the object and image side are f and f', respectively, see also Fig. 3.5. We have introduced the magnification factor

$$\textbf{(Magnification factor)} \qquad M = \frac{n_1}{n_3}\frac{f'}{f}, \tag{3.11}$$

and have assumed $f' \gg f$ to arrive at Eq. (3.10). θ_{max} is the maximal acceptance angle of the lens on the object side. It turns out to be convenient to relate the cutoff angle to the so-called numerical aperture

$$\textbf{(Numerical aperture)} \qquad NA = n \sin\theta_{max}, \tag{3.12}$$

which is a dimensionless number. The higher the NA the better the spatial resolution. In the above expression θ_{max} determines how much of the k-space is captured by the lens, and the refractive index n of the medium outside the lens governs the effective light wavelength in the medium.

Figure 3.6 shows the example of the "micron" letters previously discussed in Fig. 3.3. We observe that the resolution increases with increasing NA values, because a larger portion of the k-space becomes available for the image reconstruction. In all cases the evanescent field components are lost in the imaging process. We will come back to this point in Chap. 6 when discussing point spread functions and the diffraction limit of light.

Field Focusing

An expression similar to Eq. (3.10) can be also obtained for field focusing. The situation we have in mind is depicted in Fig. 3.7, and is given by a beam with electric fields $\boldsymbol{E}_{inc}(\rho, \phi)$ that propagates parallel to the optical axis, impinges on a Gaussian reference sphere with focus f, and becomes focused. The focal fields $\boldsymbol{E}(\rho, \varphi, z)$ can then be related to the incoming fields $\boldsymbol{E}_{inc}$ through

Focal Fields for Incoming Fields $\boldsymbol{E}_{inc}(\boldsymbol{\rho}, \boldsymbol{\phi})$

$$\boldsymbol{E}(\rho, \varphi, z) = \frac{ikfe^{-ikf}}{2\pi}\sqrt{\frac{n_{inc}}{n}}\int_0^{\theta_{max}} \sqrt{\cos\theta}\,\sin\theta\, d\theta \\ \times \int_0^{2\pi} d\phi\, \bar{\bar{\mathcal{R}}}^{foc}\cdot \boldsymbol{E}_{inc}(f\sin\theta, \phi) e^{-ikz\cos\theta} e^{ik\rho\sin\theta\cos(\phi-\varphi)}. \tag{3.13}$$

A detailed derivation of the above expression is given in Sect. 3.6. We use cylinder coordinates (ρ, φ, z) for the focal side, with $z = 0$ corresponding to the focal plane. n_{inc} and n are the refractive indices of the lens and of the medium on the focus side, respectively, and $k = nk_0$ with k_0 being the wavenumber of light in vacuum. $\bar{\bar{\mathcal{R}}}^{foc}$ is

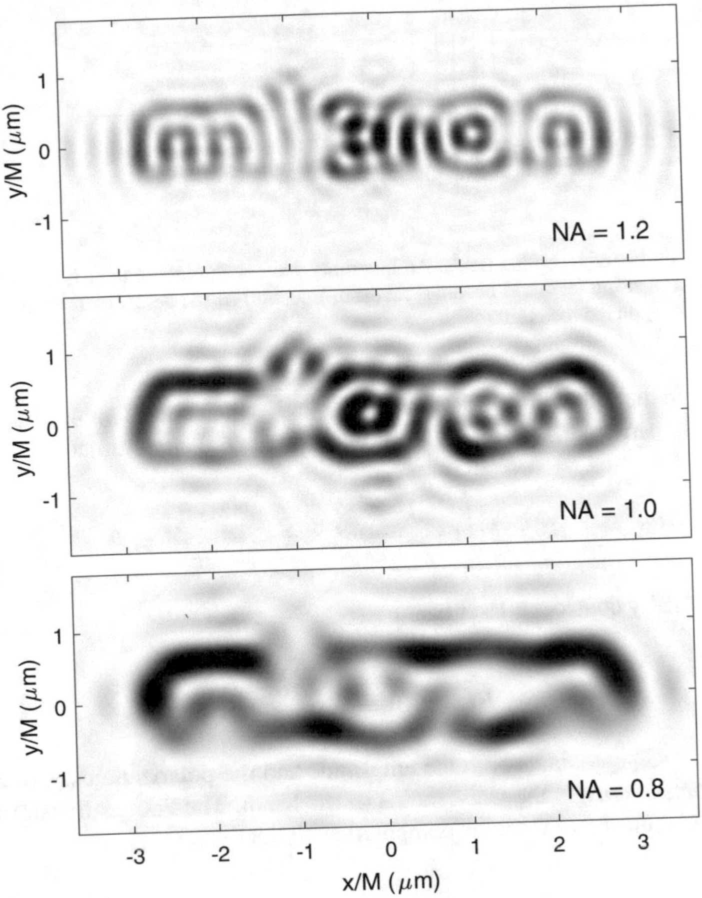

Fig. 3.6 Imaging the far-fields of the "micron" object shown in Fig. 3.3. We use Eq. (3.10) to compute the fields in the focal plane with $n_1 = 1.33$, $n_3 = 1$, and different numerical apertures NA (see insets). For larger NAs a larger portion of the k-space is sampled and the spatial resolution is increased

a transformation matrix for the electric fields, which is explicitly given in Eq. (3.27). The above expression will be used in Sect. 3.5 when investigating the focal fields of laser beams.

3.4 Paraxial Approximation and Gaussian Beams

In many cases of interest light propagates along a certain direction z and spreads out only slowly in the transverse direction. A prominent example is the propagation of laser beams, which will be considered in the next section in the context of tight

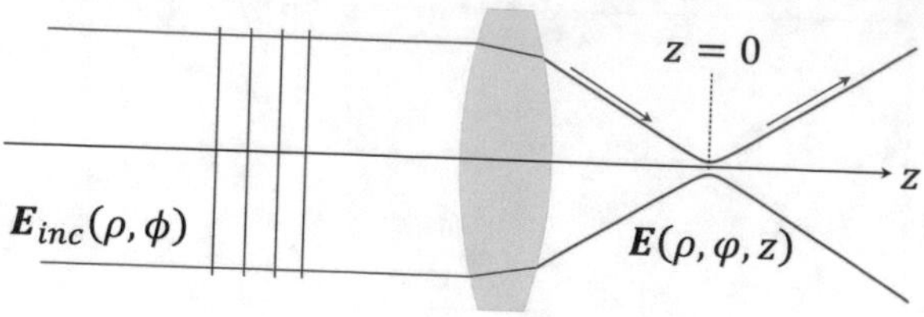

Fig. 3.7 Focusing of incoming fields. An incoming wave with electric field $\boldsymbol{E}_{\rm inc}$ impinges from the left on a focusing lens, and becomes focused. Equation (3.13) describes how to compute the field $\boldsymbol{E}(\rho, \varphi, z)$ on the focus side

focusing. In the following we assume $k_x, k_y \ll k$ which considerably simplifies our analysis. In particular, we employ the so-called paraxial approximation

$$k_z = k\sqrt{1 - \frac{k_x^2 + k_y^2}{k^2}} \approx k - \frac{k_x^2 + k_y^2}{2k} . \tag{3.14}$$

We additionally consider a laser beam

$$\boldsymbol{E}(x, y, 0) = \boldsymbol{E}_0 e^{-\frac{x^2+y^2}{w_0^2}} , \tag{3.15}$$

where $\boldsymbol{E}_0 = E_0\boldsymbol{\epsilon}$ contains both the amplitude and the polarization vector $\boldsymbol{\epsilon}$. In the above expression w_0 is the waist radius of the beam. The Fourier transform of the field distribution at $z = 0$ can be computed analytically,

$$\tilde{\boldsymbol{E}}(k_x, k_y) = (2\pi)^{-2} \int_{-\infty}^{\infty} \boldsymbol{E}_0 e^{-\frac{x^2+y^2}{w_0^2} + i(k_x x + k_y y)} dxdy = \boldsymbol{E}_0 \frac{w_0^2}{4\pi} e^{-\frac{1}{4}w_0^2\left(k_x^2+k_y^2\right)} ,$$

where we have exploited the properties of Gauss integrals. With the choice of a Gaussian envelope, the angular spectrum representation of Eq. (3.4) becomes

$$\boldsymbol{E}(x, y, z) = \boldsymbol{E}_0 \frac{w_0^2}{4\pi} \int_{-\infty}^{\infty} e^{-\frac{1}{4}w_0^2\left(k_x^2+k_y^2\right)} e^{i\left[k_x x + k_y y + \left(k - \frac{k_x^2+k_y^2}{2k}\right)z\right]} dk_x dk_y , \tag{3.16}$$

which can be solved analytically. The detailed derivation is sketched in Exercise 3.5, and it turns out to be convenient to introduce a new parameter $z_0 = \frac{1}{2}kw_0^2$ and to change to polar coordinates with ρ and z. We then obtain for the electric field profile of a Gaussian beam

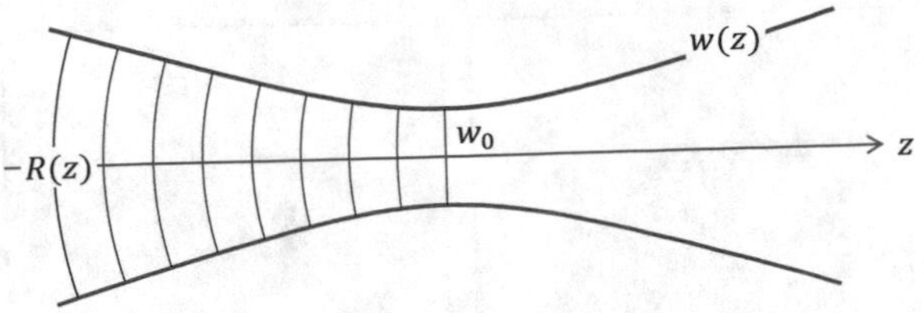

Fig. 3.8 Gaussian beam profile as computed from Eq. (3.17). The red lines show the waist $w(z)$ of the beam and the black lines the radius of curvature $R(z)$ for selected z values

Electric Field Profile of Gaussian Beam

$$\boldsymbol{E}(\rho, z) = \boldsymbol{E}_0 \left[\frac{w_0}{w(z)}\right]^2 \exp\left[-\frac{\rho^2}{w^2(z)}\left(1 - \frac{iz}{z_0}\right) + ikz\right], \tag{3.17}$$

with the beam radius $w(z) = w_0\sqrt{1 + (z/z_0)^2}$. Figure 3.8 gives a schematic sketch of the electric field profile. In principle, the above expression can be brought to an even more transparent form, see, for instance, Sec. 3.3 of Ref. [6] or Exercise 3.6, but we omit the details here. A few points are worth mentioning about Eq. (3.17):

– Gaussian beams are *not* proper solutions of Maxwell's equations. Because of the paraxial approximation they also contain longitudinal wave components.
– Strong focusing of the beam, corresponding to small w_0 values, results in a fast defocusing of the fields.
– With respect to an unfocused beam the Gaussian beam exhibits a phase shift, usually denoted as the Gouy phase shift.

Laser Modes

Gaussian beams often serve as an excellent approximation for laser modes which can propagate over large distances without significant dispersion in the transverse direction. For higher-order laser modes one often distinguishes between different transverse modes.

Hermite–Gaussian Modes. These modes typically emerge from lasers with rectangular cavities and end mirrors. The modes with m and n nodes in the x and y direction can be generated from the fundamental mode of Eq. (3.16) according to

$$E_{mn}^{\mathrm{H}}(x, y, z) = w_0^{m+n} \frac{\partial^m}{\partial x^m} \frac{\partial^n}{\partial y^n} E(x, y, z)\,. \tag{3.18}$$

The evaluation of the derivatives can be simplified by noting that the resulting polynomials in x and y are precisely the Hermite polynomials. Figure 3.9 shows a few selected field profiles.

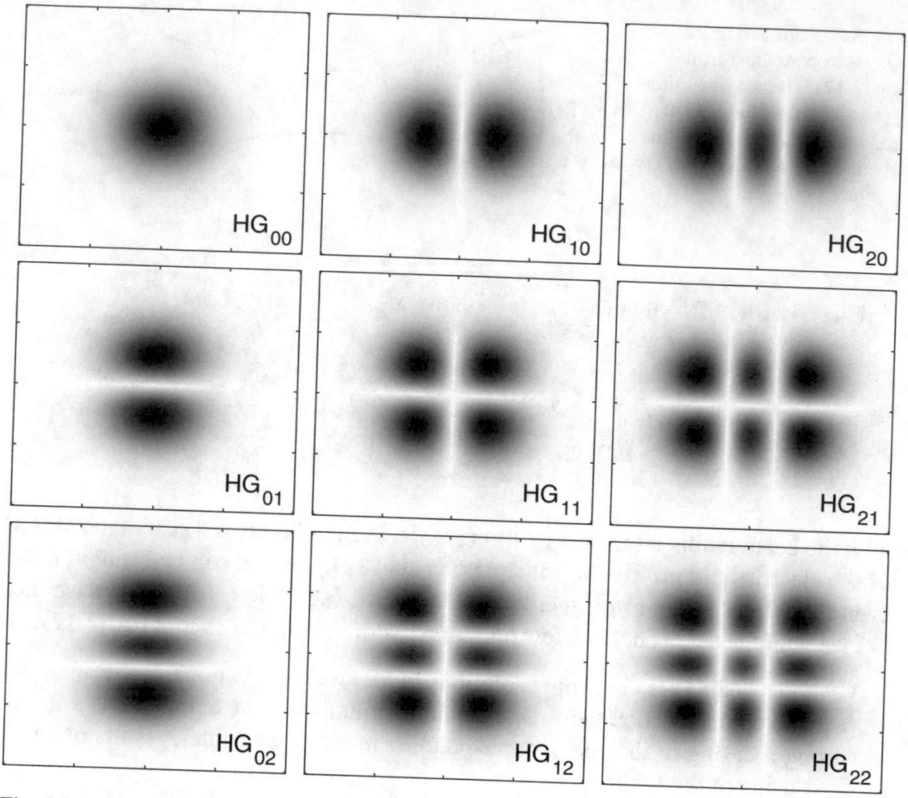

Fig. 3.9 A few selected Hermite–Gauss beams, see Eq. (3.18). We plot the absolute values of the electric fields in the plane $z = 0$

Laguerre–Gaussian Modes. These modes typically emerge from lasers with circular cavities and end mirrors. They can be derived in a similar way through

$$E^{\mathrm{L}}_{mn}(x, y, z) = k^n w_0^{2n+m} e^{ikz} \left(\frac{\partial}{\partial x} + i \frac{\partial}{\partial y} \right)^m \frac{\partial^n}{\partial z^n} e^{-ikz} E(x, y, z) . \qquad (3.19)$$

3.5 Fields of a Tightly Focused Laser Beam

We finally consider tight focusing of a Hermite–Gaussian or Laguerre–Gaussian beam through a lens with a high numerical aperture. We start with the angular spectrum representation of focal fields given by Eq. (3.13) and introduce a few approximations:

- The far-field impinging at the Gaussian reference sphere is approximated by the laser mode $\boldsymbol{E}(x, y, z = 0)$ which is an excellent approximation because the dispersion of the laser beam in the transverse direction is usually very small.
- The transmission coefficients of the lens are approximated by one.
- For the laser beam we consider a polarization vector $\boldsymbol{\epsilon}$.

Consider the transmitted laser fields directly after crossing the Gaussian reference sphere

$$\boldsymbol{E}_{\rm inc}(\theta, \phi) = \sqrt{\frac{n_{\rm inc}}{n}}\cos^{1/2}\theta\, \boldsymbol{E}^{\rm H}_{mn}(f\sin\theta, \phi, 0)\, \bar{\bar{\mathcal{R}}}^{\rm foc}\cdot\boldsymbol{\epsilon},,$$

where $n_{\rm inc}$ and n are the refractive indices on the side of the incoming laser and the focus spot, respectively, and f is the focal length of the lens. $\bar{\bar{\mathcal{R}}}^{\rm foc}$ is the transformation matrix of Eq. (3.27) that propagates the incoming field across the reference sphere, and we have used $\rho = f\sin\theta$ to relate the cylinder coordinates of the incoming laser beam with the spherical coordinates for the reference sphere. Then, the focal fields can be expressed as

Fields of Tightly Focused Laser Beam

$$\boldsymbol{E}(\rho, \varphi, z) = \frac{ikfe^{-ikf}}{2\pi}\sqrt{\frac{n_{\rm inc}}{n}} \times \int_0^{\theta_{\max}} \sqrt{\cos\theta}\,\sin\theta\, d\theta \times \int_0^{2\pi} d\phi\, \boldsymbol{E}_{\rm inc}(\theta, \phi) e^{-ikz\cos\theta} e^{ik\rho\sin\theta\cos(\phi-\varphi)}\,. \tag{3.20}$$

Let us briefly discuss the evaluation of such integrals at the example of the fundamental mode with an incoming polarization vector $\hat{\boldsymbol{x}}$,

$$\boldsymbol{E}^{\rm H}_{00} = E_0\, e^{-f^2\sin^2\theta/w_0^2}\,\hat{\boldsymbol{x}}\,.$$

The evaluation of Eq. (3.20) proceeds as follows:

- The polarization vector is expressed in Cartesian coordinates.
- We use trigonometric identities to bring the terms arising from $\bar{\bar{\mathcal{R}}}^{\rm foc}$ into the form of $\cos n\phi$ and $\sin n\phi$, see Eq. (3.27).
- The integrals over the azimuthal angle can be performed analytically by use of the identities

$$\int_0^{2\pi}\begin{Bmatrix}\sin n\phi\\ \cos n\phi\end{Bmatrix} e^{ix\cos(\phi-\varphi)}\, d\phi = 2\pi i^n J_n(x)\begin{Bmatrix}\sin n\varphi\\ \cos n\varphi\end{Bmatrix}, \tag{3.21}$$

where $J_n(x)$ is the n'th order Bessel function.
– The remaining one-dimensional integrals over θ are solved numerically.

We start by expressing the polarization vector in Cartesian coordinates and use the trigonometric identities given in Exercise 3.4 to obtain for $t^\rho = t^\phi = 1$

$$\bar{\bar{\mathcal{R}}}^{\text{foc}} \cdot \hat{\boldsymbol{x}} = \frac{1}{2}\begin{pmatrix} (1+\cos\theta) - (1-\cos\theta)\cos 2\phi \\ -(1-\cos\theta)\sin 2\phi \\ -2\sin\theta\cos\phi \end{pmatrix}.$$

For our final expression we introduce the functional

$$\mathcal{I}_n\big(g(\theta)\big) = i^n \int_0^{\theta_{\max}} e^{-f^2\sin^2\theta/w_0^2} e^{-ikz\cos\theta} J_n(k\rho\,\sin\theta) g(\theta)\,\sin\theta\cos^{1/2}\theta\,d\theta\,. \tag{3.22}$$

Thus, the field of the tightly focused fundamental laser mode can be written in the form

$$\boldsymbol{E}(\rho,\varphi,z) = ikfe^{-ikf}\sqrt{\frac{n_{\text{inc}}}{n_{\text{foc}}}}\,\frac{1}{2}\begin{pmatrix} \mathcal{I}_0(1+\cos\theta) - \mathcal{I}_2(1-\cos\theta)\cos 2\varphi \\ -\mathcal{I}_2(1-\cos\phi)\sin 2\varphi \\ -\mathcal{I}_1(\sin\theta)\cos\varphi \end{pmatrix}. \tag{3.23}$$

Its evaluation requires the numerical evaluation of three integrals $\mathcal{I}_0$, $\mathcal{I}_1$, and $\mathcal{I}_2$. Figure 3.10 shows the intensity profiles computed from Eq. (3.23) for different NA values. One observes that with increasing numerical aperture the focal fields are confined more tightly, but diverge faster when moving away from $z = 0$. We will come back to tightly focused laser spots in the context of optical trapping in the next chapter.

3.6 Details of Imaging and Focusing Transformations

In this section we provide the details for the derivation of Eqs. (3.10) and (3.13) which describe field imaging and focusing.

3.6.1 *Focusing of Far-Fields*

We start our discussion with the focusing case. The setup we will consider here is depicted in Fig. 3.11 and consists of far-fields impinging on a Gaussian reference sphere, where they become directed towards the focus. For the two media we use the following refractive indices, impedances, and coordinate systems:

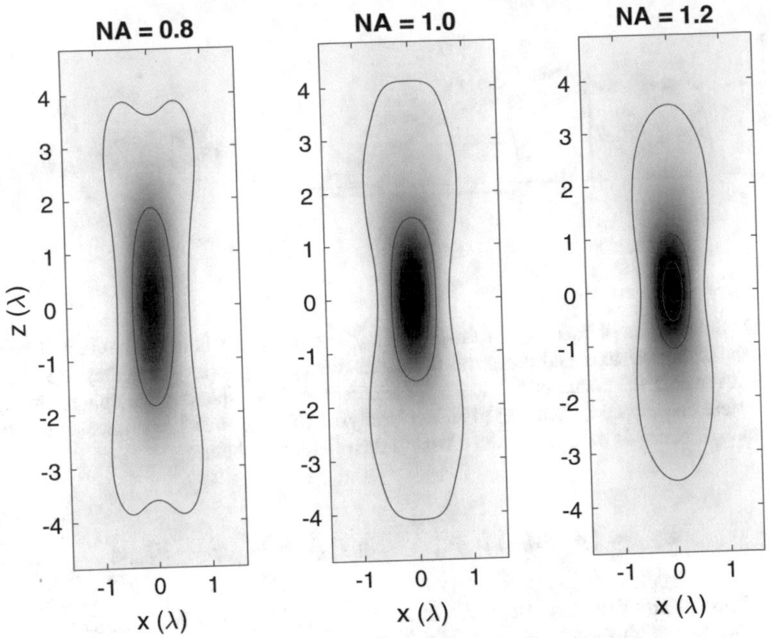

Fig. 3.10 Focusing of the E_{00}^H laser mode for different NA values, as computed from Eq. (3.23). We report the isocontours for 10, 50, and 80% of the maximal intensities together with the intensity profiles in the xz-plane. We use $\lambda_0 = 620\,\text{nm}$ and a refractive index of 1.33 for the embedding medium

Medium 1	...	$n_{\text{inc}}, Z_{\text{inc}}$,	cylinder coordinates with ρ, ϕ, z
Medium 2	...	$n, \quad Z$,	spherical coordinates with r, θ, ϕ.

Importantly, the azimuthal coordinates in both systems coincide. This is because both systems share the z-axis as the optical axis. We start by considering the transformation when going from the lens to the focus side. A detailed view is presented in Fig. 3.11. First, we notice that the cross sections in the two media are related through $\cos\theta = dA_{\text{inc}}/dA$. Thus, from the power law of Eq. (3.9) we get

$$\frac{1}{2} Z_{\text{inc}}^{-1} |\boldsymbol{E}_{\text{inc}}|^2 \, dA = \frac{1}{2} Z^{-1} |\boldsymbol{E}|^2 \, \frac{dA}{\cos\theta} \, .$$

The transmitted ray then has the magnitude

$$|\boldsymbol{E}| = \sqrt{\frac{n_{\text{inc}}}{n}} \cos^{1/2}\theta \, |\boldsymbol{E}_{\text{inc}}| \, , \tag{3.24}$$

where we have set all permeabilities to μ_0 and have expressed the impedances in terms of refractive indices. We next decompose the fields on the lens side (medium 1) into their radial and azimuthal components

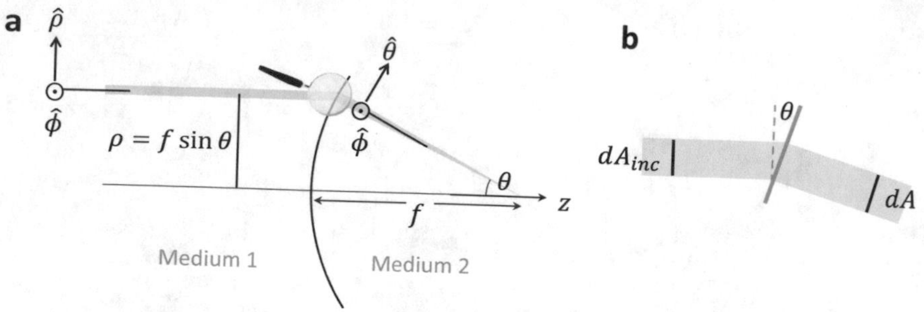

Fig. 3.11 Schematics of focusing of far-fields. (**a**) The incoming fields in medium 1 propagate parallel to the optical axis and cross the Gaussian reference sphere, where they are directed in medium 2 towards the focus of the lens. We indicate on both sides of the sphere the cylinder and spherical coordinate systems used in our analysis. (**b**) The area perpendicular to the beam propagation direction is modified when crossing the reference sphere

$$\boldsymbol{E}_{\text{inc}} = (\hat{\boldsymbol{\rho}} \cdot \boldsymbol{E}_{\text{inc}})\, \hat{\boldsymbol{\rho}} + \left(\hat{\boldsymbol{\phi}} \cdot \boldsymbol{E}_{\text{inc}}\right) \hat{\boldsymbol{\phi}} = E^{\rho}_{\text{inc}}\, \hat{\boldsymbol{\rho}} + E^{\phi}_{\text{inc}}\, \hat{\boldsymbol{\phi}}\,.$$

The z components are zero because electromagnetic waves are transverse. As can be seen in Fig. 3.11, when crossing the Gaussian reference sphere the azimuthal components are transformed into each other, whereas the $\hat{\boldsymbol{\rho}}$ component becomes the $\hat{\boldsymbol{\theta}}$ one. For this transition the $\hat{\boldsymbol{\phi}}$ component has a TE character whereas the $\hat{\boldsymbol{\rho}}$ component has a TM character. Putting together all the results we obtain for the transformed electric fields

$$\boldsymbol{E} = \sqrt{\frac{n_{\text{inc}}}{n}}\, \cos^{1/2}\theta \left(t^{\rho} E^{\rho}_{\text{inc}}\, \hat{\boldsymbol{\theta}} + t^{\phi} E^{\phi}_{\text{inc}}\, \hat{\boldsymbol{\phi}}\right). \tag{3.25}$$

We use t^{ρ}, t^{ϕ} for the TE and TM transmission coefficients. In our calculations we always approximate these coefficients with one.

Transformation Matrix

It turns out to be convenient to do some extra work at this point. More specifically, we seek for a transformation matrix $\bar{\bar{\mathcal{R}}}^{\text{foc}}$ that allows us rewriting Eq. (3.25) in the form

$$\boldsymbol{E} = \sqrt{\frac{n_{\text{inc}}}{n}}\, \cos^{1/2}\theta\, \bar{\bar{\mathcal{R}}}^{\text{foc}} \cdot \boldsymbol{E}_{\text{inc}}\,.$$

Upon acting on a vector, the matrix $\bar{\bar{\mathcal{R}}}^{\text{foc}}$ performs the following tasks:

- it projects the vector onto the basis of the first reference system,
- it adds the TE and TM transmission coefficients, and
- it expands the vector again in the basis of the second reference system.

For rendering the transformation matrix useful, we must additionally relate the two coordinate systems. In our case the azimuthal angles are identical, whereas for the other coordinates we use $\rho = f \sin\theta$. The transformation matrix that fulfills the above tasks can be expressed as follows:

$$\bar{\bar{\mathcal{R}}}^{\text{foc}} = t^p \left(\hat{\boldsymbol{\theta}} \cdot \hat{\boldsymbol{\rho}}^T\right) + t^\phi \left(\hat{\boldsymbol{\phi}} \cdot \hat{\boldsymbol{\phi}}^T\right),$$

where T denotes the transpose of the preceding term. In order to compute the matrix explicitly, we express the different unit vectors in Cartesian coordinates, for details see Exercise 3.3, to arrive at

$$\begin{aligned} \hat{\boldsymbol{\rho}} &= \cos\phi\,\hat{\boldsymbol{x}} + \sin\phi\,\hat{\boldsymbol{y}} \\ \hat{\boldsymbol{\phi}} &= -\sin\phi\,\hat{\boldsymbol{x}} + \cos\phi\,\hat{\boldsymbol{y}} \\ \hat{\boldsymbol{\theta}} &= \cos\phi\cos\theta\,\hat{\boldsymbol{x}} + \sin\phi\cos\theta\,\hat{\boldsymbol{y}} - \sin\theta\,\hat{\boldsymbol{z}}\,. \end{aligned} \tag{3.26}$$

We are then immediately led to the **transformation matrix of the focusing lens**, which relates the fields before and after crossing the Gaussian reference sphere via

$$\bar{\bar{\mathcal{R}}}^{\text{foc}} = \begin{pmatrix} t^p\cos^2\phi\cos\theta + t^\phi\sin^2\phi & \frac{1}{2}(t^p\cos\theta - t^\phi)\sin 2\phi & 0 \\ \frac{1}{2}(t^p\cos\theta - t^\phi)\sin 2\phi & t^p\sin^2\phi\cos\theta + t^\phi\cos^2\phi & 0 \\ -t^p\cos\phi\sin\theta & -t^p\sin\phi\sin\theta & 0 \end{pmatrix}. \tag{3.27}$$

How to Compute the Focal Fields

Suppose that we know the far-fields $\boldsymbol{F}(\hat{\boldsymbol{r}})$ after crossing the Gaussian reference sphere with focal radius f. We next show how to compute the fields at any position on the focus side (medium 2). Our starting point is Eq. (3.4) that relates the electric fields to the Fourier components via

$$\boldsymbol{E}(x, y, z) = \int_{k_x^2+k_y^2<k_{\text{max}}^2} e^{i(k_x x + k_y y - k_z z)}\,\tilde{\boldsymbol{E}}_0(k_x, k_y)\,dk_x dk_y\,,$$

where k_{max} is a cutoff wavenumber determined by the opening angle of the lens. Note that in the exponential we have taken a negative sign of the $k_z z$ term because we have now the reversed propagation in comparison to the angular spectrum representation of Eq. (3.4), namely from the Gaussian reference sphere towards the focus. We next use Eq. (3.8) that relates the far-fields to the Fourier components,

$$\boldsymbol{F}(\hat{\boldsymbol{r}}) = -2\pi i k_z\, \tilde{\boldsymbol{E}}_0\left(k\hat{r}_x, k\hat{r}_y\right)\,.$$

We can combine these two expressions to get

$$\boldsymbol{E}(x, y, z) = \frac{i}{2\pi} \int_{k_x^2+k_y^2<k_{\max}^2} e^{i(k_x x+k_y y-k_z z)}\, \boldsymbol{F}(\hat{\boldsymbol{k}}) \frac{1}{k_z}\, dk_x dk_y\,. \tag{3.28}$$

To proceed further, we perform two coordinate transformations. For the k-space integration we introduce spherical coordinates with

$$k_x = k \sin\theta \cos\phi\,, \quad k_y = k \sin\theta \sin\phi\,, \quad k_z = k\cos\theta\,.$$

The two-dimensional integration $dk_x dk_y$ then has to be transformed using the Jacobian

$$J = \begin{vmatrix} \frac{\partial k_x}{\partial\theta} & \frac{\partial k_x}{\partial\phi} \\ \frac{\partial k_y}{\partial\theta} & \frac{\partial k_y}{\partial\phi} \end{vmatrix} = \begin{vmatrix} k\cos\theta\cos\phi & -k\sin\theta\sin\phi \\ k\cos\theta\sin\phi & k\sin\theta\cos\phi \end{vmatrix} = k^2 \sin\theta\cos\theta\,.$$

Thus, $k_z{}^{-1}\, dk_x dk_y = k\sin\theta\, d\theta d\phi$. For the focus positions, we finally switch to cylinder coordinates with $(x, y, z) = (\rho\cos\varphi, \rho\sin\varphi, z)$. With this, we can re-express the term in the exponential of Eq. (3.28) as

$$k_x x + k_y y = k\sin\theta\cos\phi\, \rho\cos\varphi + k\sin\theta\sin\phi\, \rho\sin\varphi = k\rho\sin\theta\cos(\phi-\varphi)\,.$$

These coordinate transformations allow us rewriting Eq. (3.28) in the form

$$\boldsymbol{E}(\rho, \varphi, z) = \frac{ik}{2\pi} \int_0^{\theta_{\max}} \sin\theta\, d\theta \int_0^{2\pi} d\phi\, \boldsymbol{F}(\theta, \phi) e^{-ikz\cos\theta} e^{ik\rho\sin\theta\cos(\phi-\varphi)}\,, \tag{3.29}$$

where $\theta_{\max}$ is the opening angle of the lens. The final step is to relate the incoming far-fields to those after crossing the Gaussian reference sphere, which can be done through the transformation matrix of Eq. (3.27) via

$$\left(\frac{e^{ikf}}{f}\right) \boldsymbol{F}(\theta, \phi) = \sqrt{\frac{n_{\mathrm{inc}}}{n}} \sqrt{\cos\theta}\, \bar{\bar{\mathcal{R}}}^{\mathrm{foc}} \cdot \boldsymbol{E}_{\mathrm{inc}}(f\sin\theta, \phi)\,.$$

Inserting this far-field into Eq. (3.29) then leads us to the final expression of Eq. (3.13), which relates the incoming fields $\boldsymbol{E}_{\mathrm{inc}}$ to the focal fields.

3.6.2 *Imaging of Far-Fields*

We now return to the more complicated problem of imaging the far-fields by two lenses, see Fig. 3.5. Equipped with the machinery developed for field focusing things turn out to be very similar. For the three media we use the following refractive indices, impedances, and coordinate systems:

Medium 1	...	$n_1, Z_1,$	spherical coordinates with $r_1,\ \theta_1,\ \phi$
Medium 2	...	$n_2, Z_2,$	cylinder coordinates with $\rho_2,\ \phi,\ z_2$
Medium 3	...	$n_3, Z_3,$	spherical coordinates with $r_3,\ \theta_3,\ \phi$,

where again the azimuthal angles coincide in all reference systems. We next trace the ray through the lens system.

Object to Lens. For the ray transition from the object to lens side we relate the cross sections in the two media through $\cos\theta_1 = dA_2/dA_1$. Thus, from the power law of Eq. (3.9) we get

$$\frac{1}{2}Z_1^{-1}|\boldsymbol{E}_1|^2\,dA_1 = \frac{1}{2}Z_2^{-1}|\boldsymbol{E}_2|^2\,\cos\theta_1 dA_1\,.$$

The ray transmitted into medium 2 then has the magnitude

$$|\boldsymbol{E}_2| = \sqrt{\frac{n_1}{n_2}}\,\cos^{-1/2}\theta_1\,|\boldsymbol{E}_1|\,, \tag{3.30}$$

where we have set again all permeabilities to μ_0 and have expressed the impedances in terms of refractive indices. When crossing the Gaussian reference sphere the azimuthal components are transformed into each other, whereas the $\hat{\boldsymbol{\theta}}_1$ component is transformed into the $\hat{\boldsymbol{\rho}}_2$ component. Putting together all results we obtain for the electric field in the lens region

$$\boldsymbol{E}_2 = \sqrt{\frac{n_1}{n_2}}\,\cos^{-1/2}\theta_1\,\tilde{t}\left(E_1^\theta\,\hat{\boldsymbol{\rho}}_2 + E_1^\phi\,\hat{\boldsymbol{\phi}}\right)\,. \tag{3.31}$$

The coefficient $\tilde{t}$ accounts for the transmission of the transversal electric field across the interface.

Lens to Image. For the lens to image transition we proceed in a completely similar fashion as discussed in Sect. 3.3 and obtain

$$\boldsymbol{E}_3 = \sqrt{\frac{n_2}{n_3}}\,\cos^{1/2}\theta_3\,\left(t^\rho E_2^\rho\,\hat{\boldsymbol{\theta}}_3 + t^\phi E_2^\phi\,\hat{\boldsymbol{\phi}}\right)\,. \tag{3.32}$$

Object to Image. The above two equations can be combined to relate the fields on the object and images sides according to

$$\boldsymbol{E}_3 = \sqrt{\frac{n_1}{n_3}}\sqrt{\frac{\cos\theta_3}{\cos\theta_1}}\left(\tilde{t}^\theta E_1^\theta\,\hat{\boldsymbol{\theta}}_3 + \tilde{t}^\phi E_1^\phi\,\hat{\boldsymbol{\phi}}\right). \tag{3.33}$$

Here $\tilde{t}^\theta$, $\tilde{t}^\phi$ are the transition coefficients for the entire structure.

In analogy to the focusing case, we can now introduce a transformation matrix for the imaging system

$$\bar{\bar{\mathcal{R}}}^{\text{im}} = \tilde{t}^\theta\left(\hat{\boldsymbol{\theta}}_3\cdot\hat{\boldsymbol{\theta}}_1^T\right) + t^\phi\left(\hat{\boldsymbol{\phi}}\cdot\hat{\boldsymbol{\phi}}^T\right),$$

which relates the fields at both sides of the imaging system according to

$$\boldsymbol{E}_3 = \sqrt{\frac{n_1}{n_3}}\sqrt{\frac{\cos\theta_3}{\cos\theta_1}}\,\bar{\bar{\mathcal{R}}}^{\text{im}}\cdot\boldsymbol{E}_1\,.$$

Sometimes one can perform integrations over the azimuthal coordinates analytically, provided that the integrand is of the form $\sin n\phi$, $\cos n\phi$ but does not contain squares or products of trigonometric functions. After some simple manipulations, which we sketch in Exercise 3.4, we obtain for the **transformation matrix for the imaging lenses**

$$\begin{aligned}
\mathcal{R}_{xx}^{\text{im}} &= \frac{1}{2}\left\{\left(\tilde{t}^\theta\cos\theta_1\cos\theta_3 + \tilde{t}^\phi\right) + \left(\tilde{t}^\theta\cos\theta_1\cos\theta_3 - \tilde{t}^\phi\right)\cos 2\phi\right\}\\
\mathcal{R}_{xy}^{\text{im}} &= \frac{1}{2}\left(\tilde{t}^\theta\cos\theta_1\cos\theta_3 - \tilde{t}^\phi\right)\sin 2\phi\\
\mathcal{R}_{xz}^{\text{im}} &= -\tilde{t}^\theta\cos\phi\sin\theta_1\cos\theta_3\\[1em]
\mathcal{R}_{yx}^{\text{im}} &= \frac{1}{2}\left(\tilde{t}^\theta\cos\theta_1\cos\theta_3 - \tilde{t}^\phi\right)\sin 2\phi\\
\mathcal{R}_{yy}^{\text{im}} &= \frac{1}{2}\left\{\left(\tilde{t}^\theta\cos\theta_1\cos\theta_3 + \tilde{t}^\phi\right) - \left(\tilde{t}^\theta\cos\theta_1\cos\theta_3 - \tilde{t}^\phi\right)\cos 2\phi\right\}\\
\mathcal{R}_{yz}^{\text{im}} &= -\tilde{t}^\theta\sin\phi\sin\theta_1\cos\theta_3\\[1em]
\mathcal{R}_{zx}^{\text{im}} &= -\tilde{t}^\theta\cos\phi\cos\theta_1\sin\theta_3\\
\mathcal{R}_{zy}^{\text{im}} &= -\tilde{t}^\theta\sin\phi\cos\theta_1\sin\theta_3\\
\mathcal{R}_{zz}^{\text{im}} &= -\tilde{t}^\theta\sin\theta_1\sin\theta_3\,.
\end{aligned} \tag{3.34}$$

We can now put the entire imaging process together. Our starting point is given by Eq. (3.29) which becomes

$$\boldsymbol{E}_3(\rho_3, \varphi_3, z) = \frac{ik_3}{2\pi} \times \int_0^{\theta_{3,\max}} \sin\theta_3\, d\theta_3 \times \int_0^{2\pi} d\phi\, \boldsymbol{F}_3(\theta_3, \phi) e^{-ik_3 \cos\theta_3 z} e^{ik_3\rho_3 \sin\theta_3 \cos(\phi-\varphi_3)} .$$

Here $\boldsymbol{F}_3$ is the far-field on the image side. To render this equation useful, we have to express the far-fields in terms of the object angle θ_1 rather than θ_3. As can be seen in Fig. 3.5, the distance of the conjugate ray from the optical axis is the same on the object and image side, and we obtain

$$f \sin\theta_1 = f' \sin\theta_3 . \tag{3.35}$$

Taking the total derivatives on both sides we get

$$f \cos\theta_1\, d\theta_1 = f' \cos\theta_3\, d\theta_3 \implies d\theta_3 = \frac{f}{f'}\frac{\cos\theta_1}{\cos\theta_3} d\theta_1 \underset{f' \gg f}{\longrightarrow} \frac{f}{f'} \cos\theta_1\, d\theta_1 ,$$

where in the last expression we have approximated $\cos\theta_3 \approx 1$ generally valid for $f' \gg f$. The expression in the exponential can then be rewritten in the form

$$k_3\rho_3 \sin\theta_3 = \left(\frac{n_3}{n_1} k_1 \rho_3\right)\left(\frac{f}{f'} \sin\theta_1\right) = k_1 \frac{\rho}{M} \sin\theta_1 ,$$

where we have introduced the magnification factor

$$M = \frac{n_1}{n_3}\frac{f'}{f} .$$

The far-field amplitudes can be related through

$$\left(\frac{e^{ik_3 f'}}{f'}\right) \boldsymbol{F}_3 \approx \sqrt{\frac{n_1}{n_3}} \cos^{-1/2}\theta_1\, \bar{\bar{\mathcal{R}}}^{\mathrm{im}} \cdot \left(\frac{e^{ik_1 f}}{f}\right) \boldsymbol{F}_1(\theta_1, \phi) ,$$

where we have used again $\cos\theta_3 \approx 1$. We finally put everything together to arrive at Eq. (3.10) for the mapping from the far-fields on the object side to the focal fields on the image side,

Simulating Imaging on a Computer The imaging formula derived in this chapter is a very useful expression for computing the image fields $\boldsymbol{E}^{\mathrm{im}}$ once the object fields $\boldsymbol{E}^{\mathrm{obj}}$ in a given plane $z = 0$ are known. In the following we assume that the electric fields are given at the positions of a rectangular grid, as is usually the case for numerical calculations,

$\boldsymbol{E}_i^{\text{im}}$ is the image field at position (Mx_i, My_i)

$\boldsymbol{E}_i^{\text{obj}}$ is the object field at position (x_i, y_i).

M is the magnification of the lens system. The imaging formula of Eq. (3.10) then provides a linear map between the two fields according to

$$E_{i\alpha}^{\text{im}} = T_{i\alpha,j\beta}\, E_{j\beta}^{\text{obj}}\,, \tag{3.36}$$

where α, β denote the Cartesian coordinates and we have used Einstein's sum convention. One might like to compute the image fields using the above expression, and indeed we will do so in later parts of the book. However, if one applies Eq. (3.36) directly to the imaging of a large number of positions n, say of the order of a million, approximately corresponding to the situation shown in Fig. 3.6, one realizes that the computational mapping becomes extremely slow. The reason is that the above transformation involves a matrix multiplication and thus requires n^2 operations, or, expressed in the terminology of computer science the algorithm has a complexity of $\mathcal{O}(n^2)$.

Fortunately there exists a different approach with a complexity of $\mathcal{O}(n)$. It is based on the fast Fourier transform (FFT), a remarkable algorithm that was included in the top ten algorithms of twentieth century by the IEEE journal Computing in Science and Engineering [7]. This algorithm allows to perform Fourier transforms with a complexity of $\mathcal{O}(n\,\log n)$ which is considerably faster than $O(n^2)$ for matrix multiplications. As the imaging transformation is deeply rooted in Fourier transforms, in the following we revisit imaging and show how to proceed with operations of $\mathcal{O}(n)$ solely. The basic principle is

$$\boldsymbol{E}_i^{\text{obj}} \underset{FFT}{\longrightarrow} \tilde{\boldsymbol{E}}_i \longrightarrow \dots \underset{FFT^{-1}}{\longrightarrow} \boldsymbol{E}_i^{\text{im}}\,.$$

We start with the object fields which are transformed to wavenumber space. Subsequently, we act with a number of operations on the Fourier components, in order to mimic the performance of the lens system, before transforming back to the real-space fields $\boldsymbol{E}_i^{\text{im}}$. The different transformations of the algorithm can be summarized as follows.

Fourier Transform. The algorithm starts with a fast Fourier transform of the object fields

$$\tilde{E}_{i\alpha}^{\text{obj}} = \mathcal{F}\big\{\boldsymbol{E}^{\text{obj}}\big\}\,.$$

There is a subtle point of how to relate the discrete FFT wavenumbers to those of the continuous wavenumber space underlying our analysis, $\boldsymbol{k}_i$, and we refer the interested reader to the specialized literature for a thorough discussion [8].

Far-fields on Object Side. The Fourier transformed fields can be related to the far-field amplitudes using Eq. (3.8), and we get

$$F_{i\alpha} = -2\pi i k_{z,i}\, \theta_{\text{step}}\big(k_{\max} - |\boldsymbol{k}_i|\big)\, \tilde{E}_{i\alpha}^{\text{obj}}\,,$$

with $k_{\max} = n_1 k_0 \sin\theta_{\max} = NA\, k_0$. Here θ_{step} is Heaviside's step function.

Lenses. The fields have to be propagated through the lens system using geometrical optics and the intensity law, see also Eq. (3.33),

$$\left(\frac{e^{ik'f'}}{f'}\right) F_{i\alpha}^{\text{im}} \approx \left(\frac{e^{ikf}}{f}\right)\sqrt{\frac{n}{n'}}\cos^{-1/2}\theta_i\,\left(\bar{t}_i^{\theta}\, F_i^{\theta}\, \hat{\rho}_{i\alpha} + \bar{t}_i^{\phi}\, F_i^{\phi}\, \hat{\phi}_{i\alpha}\right).$$

The different contributions of this expression have the following meaning:

- We use k, n for the object side and k', n' for the image side.
- The unit vectors $\hat{\boldsymbol{\phi}}_i, \hat{\boldsymbol{\theta}}_i, \hat{\boldsymbol{\rho}}_i = \cos\phi_i\,\hat{\boldsymbol{x}} + \sin\phi_i\,\hat{\boldsymbol{y}}$ are defined on the object side (see also Exercise 3.8).
- F_i^θ, F_i^ϕ are the far-field vectors projected on the unit vectors.

Far-fields on Image Side. The Fourier components on the image side can be obtained from

$$\tilde{E}_{i\alpha}^{\text{im}} = \frac{i}{2\pi k'}\,F_{i\alpha}^{\text{im}}\,,$$

where we have approximated $k'_{i,z} \approx k'$ in the spirit of the paraxial approximation.

Inverse Fourier Transform. Finally, we obtain the image fields through an inverse fast Fourier transform

$$E_{i\alpha}^{\text{im}} = \frac{1}{M^2}\,\mathcal{F}^{-1}\big\{\tilde{E}^{\text{im}}\big\}\,,$$

where M is the magnification factor defined in Eq. (3.11). The prefactor is due to the fact that in the expression transforming the far-fields to the image fields, Eq. (3.28), we have to relate $dk'_x dk'_y$ to $dk_x dk_y$ in order to use the inverse Fourier transform, in complete analogy to the derivation of our central imaging expression of Eq. (3.10)

Exercises

NANOPT Toolbox. The book comes along with a collection of MATLAB files, which we will refer to as the NANOPT toolbox (the acronym stands for "Nano Optics") and which can be downloaded from either the publisher's webpage (*for details see page 1*) or the author's webpage at the University of Graz. The toolbox comes along with a description of how to install and use it. The code is kept as simple and elementary as possible, with the main focus of demonstrating how certain methods discussed in this book can be implemented numerically. Several exercises in the book build on the toolbox, and we will usually refer to certain files that have to be modified by the reader.

Exercise 3.1 Use from the NANOPT toolbox the file `demomicron01.m` to obtain Fig. 3.3. Modify the source code to investigate the following points:

(a) What happens for dipoles oriented along x? Compute the fields in the same planes z as in Fig. 3.3.
(b) What happens if the size of the letters is increased by a factor of 10?

Exercise 3.2 Use from the NANOPT toolbox the file `demomicron02.m` to obtain Fig. 3.6. Modify the source code to investigate the following points:

(a) What happens for dipoles oriented along x? Use the same NA values as in Fig. 3.3.
(b) What happens if the size of the letters is increased by a factor of 10?

(c) Set in the initialization of the angular spectrum representation the fields with $k_x \leq 0$ to zero. How is the image modified by this? Interpret the result.

Exercise 3.3 Start from the vector $\boldsymbol{r}$ expressed in spherical coordinates (ρ, ϕ, θ), and compute

$$\boldsymbol{\rho} = \frac{d\boldsymbol{r}}{d\rho}\,, \quad \boldsymbol{\phi} = \frac{d\boldsymbol{r}}{d\phi}\,, \quad \boldsymbol{\theta} = \frac{d\boldsymbol{r}}{d\theta}\,.$$

Divide each vector by its norm in order to get unit vectors, and compare the result with Eq. (3.26).

Exercise 3.4 Use the results of Exercise 3.3 to arrive at the transformation matrix $\bar{\bar{\mathcal{R}}}^{\rm foc}$ of Eq. (3.27). You might like to use the following trigonometric identities:

$$\sin\phi\cos\phi = \frac{1}{2}\sin 2\phi\,, \quad \cos^2\phi = \frac{1}{2}(1+\cos 2\phi)\,, \quad \sin^2\phi = \frac{1}{2}(1-\cos 2\phi)\,.$$

Exercise 3.5 Derive the electric field profile of Eq. (3.16) for the Gaussian beam. Use the Gaussian integral given in Exercise 1.4.

Exercise 3.6 Show that the electric field profile of Eq. (3.17) for a Gaussian beam can be written in the form

$$\boldsymbol{E}(\rho, z) = \boldsymbol{E}_0\left[\frac{w_0}{w(z)}\right]^2 \exp\left[-\frac{\rho^2}{w^2(z)}\left(ikz - \eta(z) + \frac{k\rho^2}{2R(z)}\right)\right],$$

with the wavefront radius $R(z) = z(1 + z_0^2/z^2)$ and the phase correction $\eta(z) = \arctan(z/z_0)$.

Exercise 3.7 Use from the NANOPT toolbox the files `demofocus02.m` and `demofocus04.m` to investigate the focused fields of a laser, see Fig. 3.10.

(a) What happens when changing the NA value?
(b) What happens when changing the order of the laser modes?
(c) Show numerically that the intensity of the focused laser integrated over the xy-plane does not depend on z. Interpret the result.

Exercise 3.8 Show that for the transformation matrix for imaging and for $f' \gg f$ the $\hat{\boldsymbol{\theta}}$ component on the object side is transformed to the $\hat{\boldsymbol{\rho}}$ component on the image side, which follows from Eq. (3.34) by setting $\sin\theta_3 \approx 0$ and $\cos\theta_3 \approx 1$.

Chapter 4
Symmetry and Forces

Where would we be without Emmy Noether and her celebrated theorem? It has given physics an indispensable guiding principle by relating symmetries to conservation laws and vice versa.

In this chapter we discuss the implications of Noether's theorem for Maxwell's equations. In free space, momentum, angular momentum, and energy of electromagnetic waves are conserved quantities. In presence of matter we can benefit from the fact that electrodynamics is a field theory. Suppose that all matter under investigation is located within a given volume Ω. By determining the flux of momentum, angular momentum, or energy through the boundary $\partial\Omega$ enclosing the volume, we immediately get a measure of how much of the conserved quantities is transferred from light to matter, or the other way round. The analysis presented in this chapter builds directly on Maxwell's equations, which has the advantage that it considerably simplifies our approach, but it has the disadvantage that the connection to symmetries is less obvious in comparison to other schemes, in particular those based on Lagrange functions.

4.1 Optical Forces

A photon carries a momentum of $\hbar k$. For a light wavelength of 600 nm we approximately get

$$\hbar k \approx 10^{-34} \frac{2\pi}{600 \times 10^{-9}} \approx 10^{-27}\,\mathrm{m\,kg\,s^{-1}}\,. \tag{4.1}$$

Whenever light becomes scattered or diffracted by some dielectric body, part of the photon momentum is transferred from light to matter. While optical forces play no significant role for macroscopic objects, they can have a noticeable influence

U. Hohenester, *Nano and Quantum Optics*, Graduate Texts in Physics,
https://doi.org/10.1007/978-3-030-30504-8_4

on nano- and microsized objects. Typical forces produced by tightly focused laser beams can be in the range of femto- to piconewtons, which suffices to trap particles entirely by optical means. In the last two decades optical tweezers have emerged as the primary tool to use such optical forces for trapping and manipulating microscopic particles. We will discuss a few selected applications later, but refer the interested reader to the rich literature for further details [9, 10].

The basic principles of optical trapping can be well understood on the basis of Maxwell's equations and its underlying symmetries. In this chapter we will describe optical forces in three different frameworks.

Dipole Approximation. For particles much smaller than the light wavelength one can employ the so-called dipole approximation and describe the particles polarized in presence of strong light fields as point-like dipoles. See Sect. 4.1.1 for details.

Geometrical Optics. For particles much larger than the light wavelength one can employ the framework of geometrical optics, similar to the case of light focusing discussed in the previous chapter. See Sect. 4.1.2 for details.

Wave Optics. For particles of intermediate size one must resort to the full Maxwell's equations. As will be discussed in Sect. 4.5, the forces can be obtained from the so-called Maxwell stress tensor.

4.1.1 Dipole Approximation

Consider first a polarized particle with a size much smaller than the wavelength of light. We will specify later how this particle becomes polarized, for the moment it suffices to assume that it has some dipole moment $\boldsymbol{p}$. As a generic model, we describe the dipole by two particles with charges $\pm q$ that are separated by the distance vector $\boldsymbol{s}$, see Fig. 4.1. The dipole moment is $\boldsymbol{p} = q\boldsymbol{s}$, and we denote the center-of-mass coordinate with $\boldsymbol{r}$. With this, the charges, coordinates, and velocities of the two particles forming the dipole can be expressed as

$$\begin{aligned} q_1 &= -q, \quad \boldsymbol{r}_1 = \boldsymbol{r} - \frac{1}{2}\boldsymbol{s}, \quad \dot{\boldsymbol{r}}_1 = \dot{\boldsymbol{r}} - \frac{1}{2}\dot{\boldsymbol{s}} \\ q_2 &= +q, \quad \boldsymbol{r}_2 = \boldsymbol{r} + \frac{1}{2}\boldsymbol{s}, \quad \dot{\boldsymbol{r}}_2 = \dot{\boldsymbol{r}} + \frac{1}{2}\dot{\boldsymbol{s}}\,. \end{aligned}$$

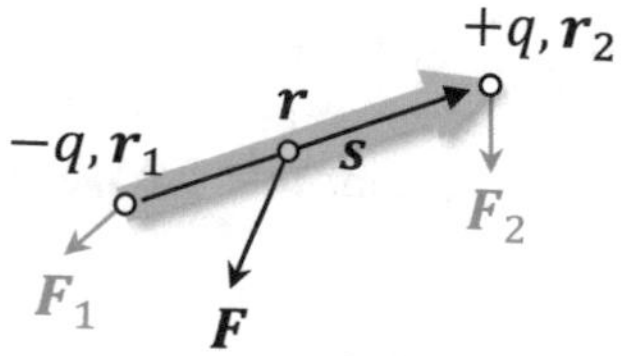

Fig. 4.1 Force acting on dipole. We describe the dipole through two charges $\pm q$ located at positions $\boldsymbol{r}_{1,2}$. The relative distance vector is $\boldsymbol{s}$, and the center-of-mass coordinate $\boldsymbol{r}$

The electromagnetic and interatomic forces $\boldsymbol{f}$ acting on the two particles then read

$$\boldsymbol{F}_1 = -q\big[\boldsymbol{E}(\boldsymbol{r}_1) + \boldsymbol{v}_1 \times \boldsymbol{B}(\boldsymbol{r}_1)\big] + \boldsymbol{f}_{12}$$
$$\boldsymbol{F}_2 = +q\big[\boldsymbol{E}(\boldsymbol{r}_2) + \boldsymbol{v}_2 \times \boldsymbol{B}(\boldsymbol{r}_2)\big] + \boldsymbol{f}_{21}\,.$$

For sufficiently small dipoles we can expand the electric field around the center-of-mass position $\boldsymbol{r}$ and obtain

$$\boldsymbol{E}(\boldsymbol{r} \pm \frac{1}{2}\boldsymbol{s}) \approx \boldsymbol{E}(\boldsymbol{r}) \pm \frac{1}{2}\frac{\partial \boldsymbol{E}(\boldsymbol{r})}{\partial r_k} s_k = \boldsymbol{E}(\boldsymbol{r}) \pm \frac{1}{2}\,(\boldsymbol{s}\cdot\nabla)\,\boldsymbol{E}(\boldsymbol{r})\,,$$

with a similar expression for the magnetic field $\boldsymbol{B}$. To compute the total force acting on the dipole, we notice that the interatomic forces sum up to zero, $\boldsymbol{f}_{12} + \boldsymbol{f}_{21} = 0$, whereas the sum over the electromagnetic forces becomes

$$\begin{aligned}\boldsymbol{F} = &-q\left[\boldsymbol{E} - \frac{1}{2}(\boldsymbol{s}\cdot\nabla)\boldsymbol{E} + \left(\dot{\boldsymbol{r}} - \frac{1}{2}\dot{\boldsymbol{s}}\right) \times \left(\boldsymbol{B} - \frac{1}{2}(\boldsymbol{s}\cdot\nabla)\boldsymbol{B}\right)\right] \\ &+q\left[\boldsymbol{E} + \frac{1}{2}(\boldsymbol{s}\cdot\nabla)\boldsymbol{E} + \left(\dot{\boldsymbol{r}} + \frac{1}{2}\dot{\boldsymbol{s}}\right) \times \left(\boldsymbol{B} + \frac{1}{2}(\boldsymbol{s}\cdot\nabla)\boldsymbol{B}\right)\right].\end{aligned}$$

Here and in the following we suppress the dependence of $\boldsymbol{E}$, $\boldsymbol{B}$ on $\boldsymbol{r}$. Working out the above expression leads us to

$$\boldsymbol{F} = (\boldsymbol{p}\cdot\nabla)\boldsymbol{E} + \dot{\boldsymbol{p}} \times \boldsymbol{B} + \dot{\boldsymbol{r}} \times (\boldsymbol{p}\cdot\nabla)\boldsymbol{B}\,. \tag{4.2}$$

The second term can be rewritten using

$$\frac{d}{dt}\boldsymbol{p} \times \boldsymbol{B} = \dot{\boldsymbol{p}} \times \boldsymbol{B} + \boldsymbol{p} \times \left(\frac{\partial \boldsymbol{B}}{\partial r_k}\dot{r}_k + \frac{\partial \boldsymbol{B}}{\partial t}\right) = \dot{\boldsymbol{p}} \times \boldsymbol{B} - \boldsymbol{p} \times (\nabla \times \boldsymbol{E}) + (\dot{\boldsymbol{r}}\cdot\nabla)\boldsymbol{B}\,.$$

We have used that $\boldsymbol{B}(\boldsymbol{r}, t)$ depends implicitly on the dipole coordinate $\boldsymbol{r}$, and have inserted Faraday's law to arrive at the last expression. In the following we assume that the changes of the fields are small compared to the fields themselves, and the dipole motion is negligible on the time scale of a field oscillation,

$$\left|(\hat{\boldsymbol{p}}\cdot\nabla)\boldsymbol{B}\right| \ll B\,, \quad \left|\dot{\boldsymbol{r}}\cdot\boldsymbol{B}\right| \ll \left|\boldsymbol{r}\cdot\frac{\partial \boldsymbol{B}}{\partial t}\right|.$$

We thus neglect the last term in Eq. (4.2) as well as $(\dot{\boldsymbol{r}}\cdot\nabla)\boldsymbol{B}$. With this, we arrive at

$$\boldsymbol{F} = (\boldsymbol{p}\cdot\nabla)\boldsymbol{E} + \boldsymbol{p} \times (\nabla \times \boldsymbol{E}) + \frac{d}{dt}\boldsymbol{p} \times \boldsymbol{B}\,.$$

When considering time-harmonic fields and averaging over one oscillation period, the time-derivative term becomes zero. The force acting on a dipole $\boldsymbol{p}$ in presence of inhomogeneous electromagnetic fields can then be rewritten as (see also Exercise 4.1)

Optical Force on Dipole

$$\langle \boldsymbol{F} \rangle = \frac{1}{2}\mathrm{Re}\left\{\sum_k p_k \nabla E_k^*(\boldsymbol{r})\right\}. \tag{4.3}$$

For clarity we have not used Einstein's sum convention but have explicitly indicated the summation k over the Cartesian coordinates.

We next consider a polarizable particle that acquires in presence of an electric field the dipole moment

$$\boldsymbol{p} = \alpha \boldsymbol{E}. \tag{4.4}$$

Here α is the polarizability of the particle. In later parts of the book we will show how this quantity can be computed for dielectric or metallic particles, but for the moment it is enough to assume that α is known. Inserting Eq. (4.4) into Eq. (4.3) and working out explicitly the real part of the expression, we are led to

$$\langle \boldsymbol{F} \rangle = \frac{\alpha'}{4}\Big(E_k \nabla E_k^* + E_k^* \nabla E_k\Big) + \frac{i\alpha''}{4}\Big(E_k \nabla E_k^* - E_k^* \nabla E_k\Big).$$

The first term can be rewritten as

$$\nabla\, E_k^* E_k = 2\,\nabla \big\langle \boldsymbol{E} \cdot \boldsymbol{E} \big\rangle,$$

where we have introduced the usual average $\langle \dots \rangle$ for time-harmonic fields, see Eq. (2.35). For the second term we use (see Exercise 4.2)

$$E_k \nabla E_k^* - E_k^* \nabla E_k = \boldsymbol{E} \times (\nabla \times \boldsymbol{E}^*) - \boldsymbol{E}^* \times (\nabla \times \boldsymbol{E}) + \nabla \times (\boldsymbol{E} \times \boldsymbol{E}^*). \tag{4.5}$$

Together with Faraday's law $\nabla \times \boldsymbol{E} = i\omega \boldsymbol{B}$ we are then led to our final expression for the force exerted on a dipole

Optical Force on Polarizable Particle

$$\langle \boldsymbol{F} \rangle = \frac{\alpha'}{2}\nabla \big\langle \boldsymbol{E} \cdot \boldsymbol{E} \big\rangle + \omega\alpha'' \big\langle \boldsymbol{E} \times \boldsymbol{B} \big\rangle + \frac{i\alpha''}{4}\nabla \times \big(\boldsymbol{E} \times \boldsymbol{E}^*\big). \tag{4.6}$$

The first term is known as the dipole force or intensity gradient. For $\alpha' > 0$ the polarizable particle is pushed towards the regions of maximal field intensity. This effect is exploited in optical tweezers for particle trapping. The other two terms are only present for absorbing particles with a finite α''. The second one is known as the scattering force or radiation pressure, and accounts for absorption losses. The third term is a force arising from the presence of spatial polarization gradients [11].

Optical Molasses

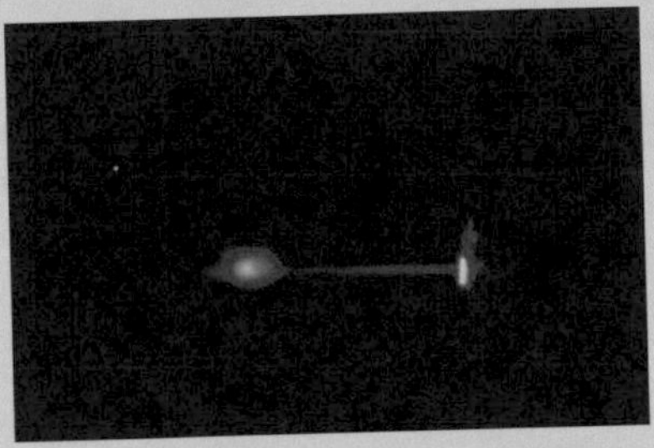

The founder of the field of optical trapping is Arthur Ashkin. Stephen Chu, who was awarded the Nobel Prize in 1997 for the "development of methods to cool and trap atoms with laser light" and who served as the United States Secretary of Energy from 2009 to 2013 under the administration of President Obama, recalls his first discussions with Ashkin on how to trap neutral atoms using light [12].

> During conversations with Art Ashkin, an office neighbor at Holmdel, I began to learn about his dream to trap atoms with light. He found an increasingly attentive listener and began to feed me copies of his reprints.

One of the main problems was to cool the atoms to velocities small enough such that optical forces would be sufficient to keep them in an optical trap, which they achieved with a technique nowadays known as laser cooling. Once accomplished, they could see the trapped atoms with the naked eye inside the vacuum chamber.

> The first signals of atoms confined in optical molasses showed confinement times of a few tens of milliseconds, but shortly afterwards we improved the storage time by over an order of magnitude. Surprisingly, it took us a week after achieving molasses to look inside the vacuum can with our eyes instead of with a photomultiplier tube. When we finally did, we were rewarded with the sight shown in the figure.

Shortly after the announcement of the Nobel Prize in physics 1997, Arthur Ashkin expressed in an interview his joy for the laureate and made the prediction that optical trapping would lead to two more Nobel Prizes, namely "Bose–Einstein condensation (BECs) and the atom laser will get a prize. And I hope that the great things that biologists are doing with optical tweezers will be rewarded with another one." Ashkin was right. The BEC prize followed in 2001, for the second prize it took until 2018 when Ashkin himself got it for "optical tweezers and their application to biological systems." By that time he was 96 years old and the oldest laureate ever.

Figure taken from Ref. [12].

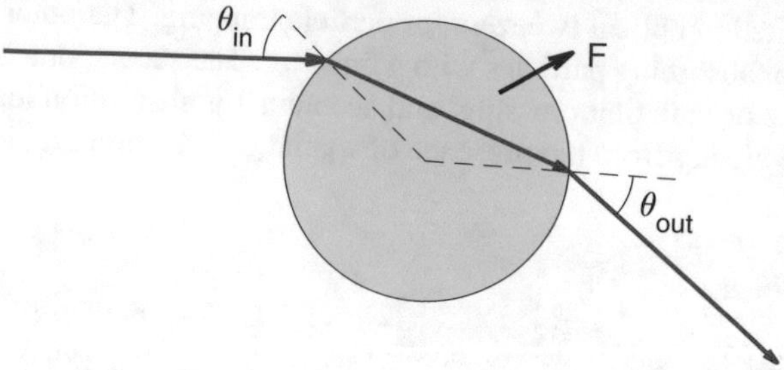

Fig. 4.2 Ray tracing through dielectric sphere. A ray impinges from the left on a dielectric sphere, and becomes diffracted while passing through it. Through the change of the ray direction (here downwards), a force acts on the sphere (here upwards). For a more realistic modelling, one could additionally include the Fresnel reflection and transmission coefficients for the ray entrance and exit at the sphere boundary

4.1.2 Geometrical Optics

In the opposite limit of particles that are substantially larger than the light wavelength, one can employ the framework of geometrical optics. Figure 4.2 shows the principle at the example of a dielectric sphere. Our analysis closely follows the discussion of Sect. 3.3 for ray tracing in optical lens systems. The computation of the forces is done through a simple ray tracing analysis, which can be divided into the following steps.

Rays. We assume that a given laser mode becomes tightly focused through a lens with a high numerical aperture, and start our ray description before the focusing lens. Here the fields propagate in the z-direction, and the total power depends on the field intensity $I(x, y)$

$$P = \int I(x, y)\, dx\, dy \approx \sum_i P_i(x_i, y_i)\,. \tag{4.7}$$

We approximate the incoming light fields by representative rays with power P_i centered at x_i, y_i.

Focus Lens. At the focus lens, the rays are refracted towards the focus. When crossing the Gaussian reference sphere, the incoming fields are transformed according to Eq. (3.32).

Transmission and Reflection. When a ray crosses the boundary of the dielectric sphere, which is located close to the focus of the lens, part of it becomes reflected and part transmitted. In general, we can compute the reflection and transmission probabilities for planar interfaces using the Fresnel coefficients to be discussed in Sect. 8.3.1. The incoming power must be conserved,

$$P_i = P_r + P_t\,, \tag{4.8}$$

where P_r and P_t are the reflected and transmitted power, respectively. In addition, we use Snell's law to compute the angles with respect to the outer surface normals of the sphere,

$$\theta_i = \theta_r\,, \quad n_i \sin\theta_i = n_t \sin\theta_t\,. \tag{4.9}$$

Here n_i and n_t are the refractive indices of the embedding medium and the dielectric sphere, respectively.

Internal Reflections. Any beam that has entered the dielectric sphere becomes either transmitted or reflected at the exit point of the ray. Transmission and reflection can be accounted for in the same manner as discussed above. Quite generally, one has to decide whether multiple internal reflections are considered and, if yes, how many.

Momentum Transfer. For each reflection and transmission of the ray at the sphere boundary, momentum is exchanged between light and particle. The force acting on the particle can be computed from the momentum change (see also discussion of Sect. 4.5)

$$\boldsymbol{F}_{\text{part}} = \frac{n_i P_i}{c}\hat{\boldsymbol{k}}_i - \frac{n_i P_r}{c}\hat{\boldsymbol{k}}_r - \frac{n_t P_t}{c}\hat{\boldsymbol{k}}_t\,, \tag{4.10}$$

where $\hat{\boldsymbol{k}}$ denotes the propagation directions for the different rays. The forces of the individual rays can be summed up to get the total force acting on the particle, or one could use the individual force components to compute the torque exerted on the particle.

Imaging. The outgoing rays can be collected by a second lens to acquire information about the position of the sphere. Imaging is done in complete analogy to the discussion of the previous section.

Figure 4.3 shows the situation of a sphere inside an inhomogeneous light field, as depicted in the bottom of the panels. We assume that initially all light rays propagate upwards, and we are only interested in the force in the transverse direction x. When the sphere is located left with respect to the maximum of the field distribution, see panel (a), light is scattered to the left. According to Newton's action-reaction law, a force in the positive x direction is exerted on the sphere, which pushes it back to the region of maximal field intensity. Similarly, (b) a sphere in the center feels no net force in the transverse direction, and (c) a sphere right to the maximum is again pushed back to the region of maximal field intensity.

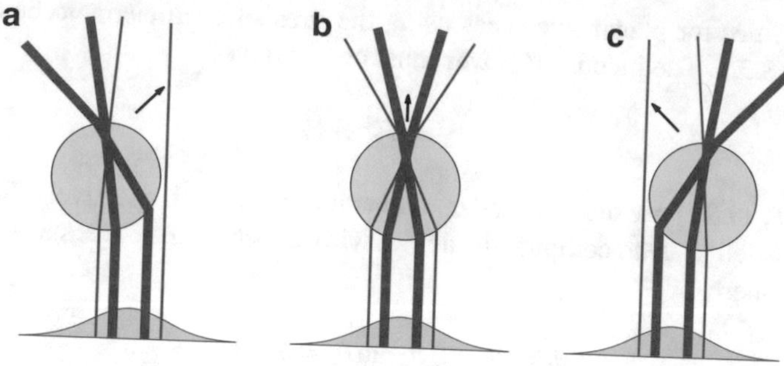

Fig. 4.3 Dielectric sphere in an inhomogeneous light field. (**a**) When the sphere is located left to the maximum of the field distribution, light is diffracted predominantly to the left-hand side. According to Newton's third law, the action-reaction one, a force is exerted on the sphere that pushes it towards the maximum. (**b**) When the sphere is in the center of the inhomogeneous light field, there is no net force in the transverse direction. (**c**) A sphere right to the maximum is again pushed towards the maximum

The First Optical Traps

Shortly after the announcement of the Nobel Prize 1997 to Steven Chu, Claude Cohen-Tannoudji, and William Phillips "for development of methods to cool and trap atoms with laser light," Arthur Ashkin talked in an interview with Bell Labs News about his discovery of optical traps, for which he received more than 20 years later the Nobel Prize in physics 2018.

> Well, I had been interested in radiation pressure for a long time. Even at Columbia, when I used to make megawatt-magnetrons for Millman, I used to think, "What can you do with this power? Maybe you could push small things . . .". So I got myself a microphone and I put pulses on it. This was when I was still a sophomore. I heard some noise and I said, "Aaah, I am seeing the effects of radiation pressure." Turns out, you know, I am not sure whether I was seeing the effects or not. Anyway, I was alerted to this problem.
>
> And I was reminded of it when I went to a conference and there was a guy who did an experiment with lasers and little particles in a laser cavity. He saw the particles staying in the cavity and moving back and forth and doing crazy things. He called them runners and bouncers. People were fascinated by this. I heard this talk and he said at the time, "We think it might be radiation pressure."
>
> When I came home, I did a calculation and realized, given the size of the beam and the particles, it couldn't be radiation pressure. More likely, I thought, is that the particles were heated and that led to the crazy behavior. This made me think of radiation pressure again. I decided to try to see radiation pressure. I made a calculation of how much it would be on a small transparent sphere. That started the whole business for me. What I did was focus a beam down on little spheres in water and watched as they were pushed along and mysteriously collected at the chamber wall. I tried to understand this and figured it out using simple ray diagrams.
>
> Then, I replaced the glass wall with another opposite beam to hold the particles in place with just light. I tried it and it worked. This was the first optical trap. It

(continued)

turned out to be a pretty important discovery. It led to Steve's Nobel Prize and, I believe, it will lead to two more Nobel Prizes.

4.1.3 *Optical Tweezers*

The above principle can be generalized for trapping particles in all three spatial dimensions. The basic principle is depicted in Fig. 4.4 where a dielectric sphere is located in the maximum region of a tightly focused laser beam. Whenever the sphere moves out of the focus, the light exerted by the scattered light pushes it back. Note that for sufficiently large spheres the equilibrium position corresponds to a slightly displaced sphere center, such that the net force of the scattered light sums up to zero.

Optical tweezers have received enormous interest in life sciences within the last decades. In particular with the rapid laser developments and the use of spatial light modulators it is nowadays possible to generate complex optical trapping potentials, with one or several minima that can be displaced, rotated, and deformed at will. For a detailed discussion the interested reader is referred to the literature, see for instance [9–11, 13] and references therein.

Figure 4.5 shows one of the numerous beautiful examples for light trapping [13]. A dielectric sphere is attached to a kinesin motor that moves along a microtubule. By positioning the sphere in an optical trap and following its position, one obtains information about the propagation details of the motor biomolecule. One can also exert a force on the system by pulling away the dielectric sphere through displacement of the trap minimum. The blue and red lines in the figure report the propagation in absence and presence of such a load.

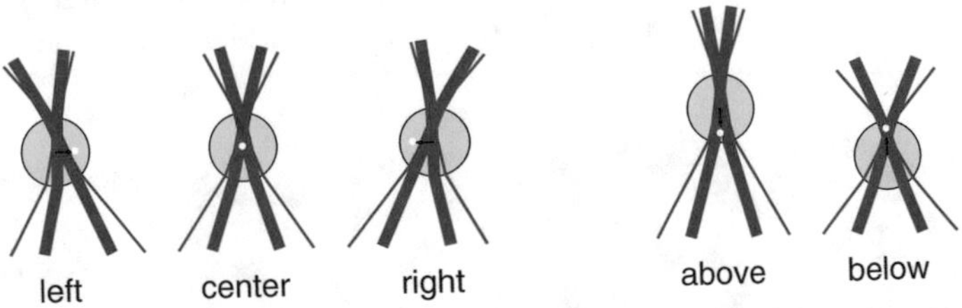

Fig. 4.4 Trapping of a dielectric sphere in all three spatial dimensions. The sphere is located in the tight focus of a sufficiently strong laser beam. Whenever it moves out of the center spot (indicated with a white dot), light becomes asymmetrically scattered and a force is exerted on the sphere which pushes it back to the region of maximal light intensity

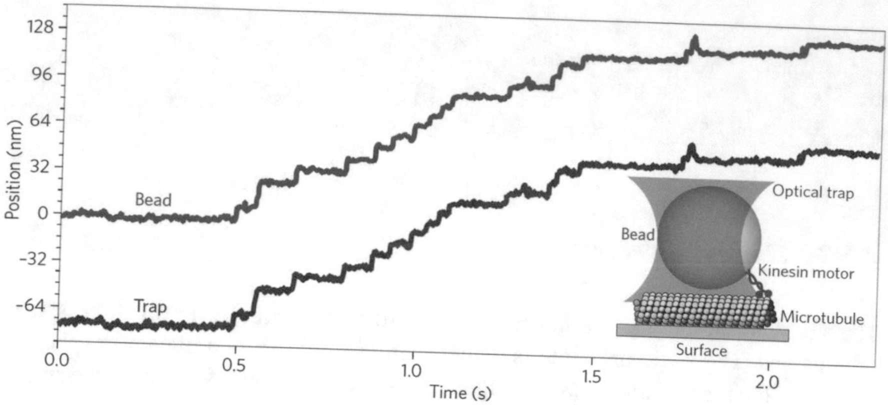

Fig. 4.5 Biological applications of optical tweezers to kinesin motor stepping and RNA folding. Record of motion for a single kinesin motor under force-clamped conditions, displaying discrete, 8 nm steps (blue trace) as it walks along a microtubule (inset, not to scale). The trap position is observed under computer control to maintain a fixed distance behind the bead, thereby imposing a load of a few piconewtons in a direction that hinders movement (red trace). Figure and caption taken from [13]

4.2 Continuity Equation

A prototypical example for conservation laws in physics is given by the continuity equation, which we briefly discuss in the following because it will serve us as a blueprint for other conservation laws. We start from Ampere's law, Eq. (2.15d), and take the divergence on both sides of the equation

$$\nabla \cdot \nabla \times \frac{1}{\mu_0} \boldsymbol{B} = \nabla \cdot \boldsymbol{J} + \varepsilon_0 \nabla \cdot \frac{\partial \boldsymbol{E}}{\partial t} = 0\,.$$

Note that the expression has to be zero since the divergence of a curl field is always zero. Together with Gauss' law $\varepsilon_0 \nabla \cdot \boldsymbol{E} = \rho$ we are thus led to the continuity equation

Continuity Equation

$$\frac{\partial \rho}{\partial t} = -\nabla \cdot \boldsymbol{J}\,. \tag{4.11}$$

To understand this expression better, it is instructive to integrate it over some volume Ω and to employ Gauss' theorem on the right-hand side,

$$\int_\Omega \frac{\partial \rho}{\partial t}\, d^3r = \frac{d}{dt} \int_\Omega \rho\, d^3r = -\int_\Omega \nabla \cdot \boldsymbol{J}\, d^3r = -\oint_{\partial\Omega} \boldsymbol{J} \cdot d\boldsymbol{S}\,.$$

We are thus led to the integral form of Eq. (4.11) which reads

$$\frac{dQ_\Omega}{dt} = -\oint_{\partial\Omega} \boldsymbol{J} \cdot d\boldsymbol{S}\,, \tag{4.12}$$

with Q_Ω being the charge enclosed in the volume Ω. The implication of this equation can be summarized as follows:

Global Charge Conservation. If we extend the volume $\Omega \to \infty$ to infinity the term on the right-hand side becomes zero, and we find that charge is a conserved quantity.

Local Charge Conservation. However, the continuity equation does not only imply global charge conservation, which could still mean that charge is annihilated in some region of space and instantaneously created somewhere else. Such spooky action at a distance is forbidden by Eq. (4.12) which states that whenever the total charge changes in a volume (left-hand side of equation) it must be transported into or out of the volume through a local current density $\boldsymbol{J}$ (right-hand side of equation).

For this reason, electrodynamics is sometimes called a "local field theory." Here "local" means that all quantities, such as fields or charge, change locally and have to be transported by some means to another position. This renders this approach compatible with the theory of relativity, which states that no information can propagate faster than the speed of light. Equations (4.11), (4.12) will serve us as blueprints for other conservation laws.

4.3 Poynting's Theorem

In the following we study energy transport in a linear and lossless material. This does not mean that these restrictions have to apply everywhere in space, but they should hold in the volume where we will explicitly describe the transport. Linear and lossy materials will be investigated in Chap. 7.

We start our analysis with a point-like charge that moves with velocity $\boldsymbol{v}$ in presence of electromagnetic fields. The (infinitesimal) work performed by the fields on the charge while propagating the distance $d\boldsymbol{\ell} = \boldsymbol{v}dt$ in the time interval dt can be expressed as

$$dW = \boldsymbol{F} \cdot d\boldsymbol{\ell} = q(\boldsymbol{E} + \boldsymbol{v} \times \boldsymbol{B}) \cdot \boldsymbol{v}dt = q\boldsymbol{E} \cdot \boldsymbol{v}dt\,.$$

We have used that magnetic fields perform no work because the force $q\boldsymbol{v} \times \boldsymbol{B}$ is perpendicular to $\boldsymbol{v}dt$. We then obtain

$$\frac{dW}{dt} = q\boldsymbol{v} \cdot \boldsymbol{E} \Longrightarrow \frac{dW}{dt} = \int_\Omega \boldsymbol{J} \cdot \boldsymbol{E}\, d^3r\,, \tag{4.13}$$

where we have generalized our result for a single charge to a charge distribution, see also Chap. 7 for details of such averaging. We next express the current distribution through Ampere's law of Eq. (2.30) to the electromagnetic fields

$$\frac{dW}{dt} = \int_\Omega \left(\nabla \times \frac{1}{\mu}\boldsymbol{B} - \varepsilon\frac{\partial \boldsymbol{E}}{\partial t}\right) \cdot \boldsymbol{E}\, d^3r\,.$$

The second term is simplified with

$$\varepsilon\frac{\partial \boldsymbol{E}}{\partial t} \cdot \boldsymbol{E} = \frac{\partial}{\partial t}\left(\frac{\varepsilon}{2}\boldsymbol{E} \cdot \boldsymbol{E}\right),$$

and we use in the first term the vector identity (see Exercise 4.3)

$$\nabla \cdot \frac{1}{\mu}\boldsymbol{E} \times \boldsymbol{B} = \frac{1}{\mu}\boldsymbol{B} \cdot \nabla \times \boldsymbol{E} - \boldsymbol{E} \cdot \nabla \times \frac{1}{\mu}\boldsymbol{B}\,.$$

Putting together the results, we are then led to

$$\frac{dW}{dt} = \int_\Omega \left[\frac{1}{\mu}\boldsymbol{B} \cdot (\nabla \times \boldsymbol{E}) - \nabla \cdot \frac{1}{\mu}(\boldsymbol{E} \times \boldsymbol{B}) - \frac{\partial}{\partial t}\left(\frac{\varepsilon}{2}\boldsymbol{E} \cdot \boldsymbol{E}\right)\right] d^3r\,.$$

For the first term in brackets we use Faraday's law to express $\nabla \times \boldsymbol{E}$ in terms of $\boldsymbol{B}$, and we finally get after a few rearrangements the Poynting's theorem in integral form

Poynting's Theorem in Integral Form

$$\frac{dW}{dt} + \frac{d}{dt}\int_\Omega \left(\frac{\varepsilon}{2}\boldsymbol{E} \cdot \boldsymbol{E} + \frac{1}{2\mu}\boldsymbol{B} \cdot \boldsymbol{B}\right) d^3r = -\oint_{\partial\Omega} \frac{1}{\mu}(\boldsymbol{E} \times \boldsymbol{B}) \cdot d\boldsymbol{S}\,. \tag{4.14}$$

We have used Gauss' theorem to change the integral over the divergence term to a boundary integral. This expression can be interpreted in a similar fashion as the continuity equation of Eq. (4.12). The terms on the left-hand side account for the change of mechanical energy dW/dt stored in the sources and for the electromagnetic energy stored in the fields. The integrand can be interpreted as an electromagnetic energy density $u_{\rm em}$,

$$u_{\rm em} = \frac{\varepsilon}{2}\boldsymbol{E} \cdot \boldsymbol{E} + \frac{1}{2\mu}\boldsymbol{B} \cdot \boldsymbol{B}\,. \tag{4.15}$$

The term on the right-hand side describes the energy flow through the boundary, where the energy flow density is usually called the Poynting vector

Poynting Vector

$$S = \frac{1}{\mu} E \times B = E \times H\,. \tag{4.16}$$

This form of energy conservation is again compatible with a local field theory. Whenever energy changes in a given region of space, it must be transported away through electromagnetic waves or some other mechanical energy transport described by dW/dt.

Plane Wave. For a plane wave we can use the results of Sect. 2.4 for time-harmonic plane waves. With $ZH = \hat{k} \times E$ we obtain for the time-averaged Poynting vector the expression

$$\langle S \rangle = \frac{1}{2}\text{Re}\left(E \times H^*\right) = \frac{1}{2} Z^{-1} |E|^2\, \hat{k}\,. \tag{4.17}$$

We have used $\hat{k} \cdot E = 0$ to arrive at the last expression. This result has been used in the previous chapter in Eq. (3.9) for the power flow of rays in the context of field focusing. We can proceed further and replace in the energy density of Eq. (4.15) the magnetic fields using $Z^{-1}\hat{k} \times E$, and obtain for the time-averaged energy flow the result

$$\langle u_{\text{em}} \rangle = \frac{1}{2}\text{Re}\left(\frac{\varepsilon}{2} E \cdot E^* + \frac{\mu}{2} H \cdot H^*\right) = \frac{\varepsilon}{2} |E|^2\,. \tag{4.18}$$

Thus Eq. (4.17) can be written in the intriguing form

$$\langle S \rangle = \langle u_{\text{em}} \rangle \, \frac{c}{n} \hat{k}\,. \tag{4.19}$$

It shows that for a plane wave the Poynting vector is just the energy density of the wave which is transported with the velocity of the speed of light in the medium in the direction of $\hat{k}$. Here n is the refractive index

$$n = \sqrt{\frac{\mu\varepsilon}{\mu_0 \varepsilon_0}}\,.$$

Evanescent Waves. Consider an electric field of the form ($z > 0$)

$$E(r) = e^{ik\cdot r}\, \hat{y} E_0\,, \quad k = k_x \hat{x} + i\kappa \hat{z}\,,$$

which is polarized along $\hat{\boldsymbol{y}}$, propagates along x, and has an evanescent character along z where the field decays exponentially. For the magnetic field we get

$$Zk\,\boldsymbol{H}\big(k_x\hat{\boldsymbol{x}}+i\kappa\hat{\boldsymbol{z}}\big)\times e^{i\boldsymbol{k}\cdot\boldsymbol{r}}\,\hat{\boldsymbol{y}}E_0=\big(k_x\hat{\boldsymbol{z}}-i\kappa\hat{\boldsymbol{x}}\big)e^{i\boldsymbol{k}\cdot\boldsymbol{r}}\,E_0\,.$$

With this, we get for the time-averaged Poynting vector

$$\langle\boldsymbol{S}\rangle=\frac{1}{2Zk}\mathrm{Re}\left\{\hat{\boldsymbol{y}}\times\big(k_x\hat{\boldsymbol{z}}+i\kappa\hat{\boldsymbol{x}}\big)\right\}\big|E_0\big|^2=\frac{1}{2}Z^{-1}\big|E_0\big|^2\,\hat{k}_x\hat{\boldsymbol{x}}\,. \tag{4.20}$$

The important observation is that no energy is transported in the z direction because of $\mathrm{Re}\{i(\dots)\}=0$. Thus, evanescent waves do not transport energy.

4.4 Optical Cross Sections

Poynting's theorem plays an important role for the calculations of the so-called optical cross sections. The situation we have in mind is sketched in Fig. 4.6 and consists of a plane wave that impinges on a particle. Part of the energy becomes scattered or absorbed by the particle. In the following we consider some boundary $\partial\Omega$ that surrounds the particle, and evaluate the energy flow into the boundary or out of it.

It turns out to be convenient to separate the electromagnetic fields into an incoming part, associated with the plane wave excitation, and a scattered part associated with the response of the particle,

$$\boldsymbol{E}=\boldsymbol{E}_{\rm inc}+\boldsymbol{E}_{\rm sca}\,,\qquad \boldsymbol{H}=\boldsymbol{H}_{\rm inc}+\boldsymbol{H}_{\rm sca}\,. \tag{4.21}$$

It is now easy to see that the absorbed power corresponds to the energy flow $\boldsymbol{E}\times\boldsymbol{H}$ of the total fields into the boundary $\partial\Omega$, whereas the scattered power corresponds

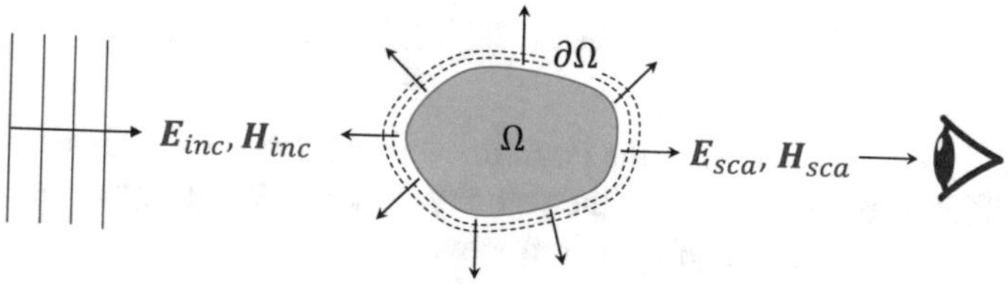

Fig. 4.6 Schematics of optical cross section. A particle is embedded in a background medium and is excited by a plane wave excitation with fields $\boldsymbol{E}_{\rm inc}$, $\boldsymbol{H}_{\rm inc}$. The response of the particle is described by the scattered field $\boldsymbol{E}_{\rm sca}$, $\boldsymbol{H}_{\rm sca}$. By computing the energy flow into and out of the boundary $\partial\Omega$ we obtain the optical cross sections. The attenuation of the fields in the direction of the incoming plane wave, see eye symbol on right-hand side, is related to the extinction cross section which is the sum of absorption and scattering

to the energy flow $\boldsymbol{E}_{\text{sca}} \times \boldsymbol{H}_{\text{sca}}$ of the scattered fields out of the boundary. For time-harmonic fields and by averaging over an oscillation period, we then find for the absorbed and scattered power

Absorption and Scattering Power

$$P_{\text{abs}} = -\frac{1}{2}\oint_{\partial\Omega} \text{Re}\left(\boldsymbol{E} \times \boldsymbol{H}^*\right) \cdot d\boldsymbol{S} \tag{4.22a}$$

$$P_{\text{sca}} = \frac{1}{2}\oint_{\partial\Omega} \text{Re}\left(\boldsymbol{E}_{\text{sca}} \times \boldsymbol{H}_{\text{sca}}^*\right) \cdot d\boldsymbol{S}\,. \tag{4.22b}$$

We have chosen a negative sign in the definition of P_{abs} such that the expression becomes positive. Note that the above definitions also hold for excitations that are not plane waves, for instance, tightly focused laser beams.

For plane waves we can relate the absorbed or scattered power to the intensity of the incoming fields I_{inc},

$$I_{\text{inc}} = \frac{1}{2}\text{Re}\left(\hat{\boldsymbol{k}} \cdot \boldsymbol{E}_{\text{inc}} \times \boldsymbol{H}_{\text{inc}}^*\right) = \frac{1}{2}Z^{-1}\left|\boldsymbol{E}_{\text{inc}}\right|^2\,, \tag{4.23}$$

which has the dimension of power per area, and has previously been computed in Eq. (4.17). We can then define the **optical cross sections** as the ratio between the absorption or scattering power P and I_{inc},

$$\sigma_{\text{abs}} = P_{\text{abs}}/I_{\text{inc}}\,, \qquad \sigma_{\text{sca}} = P_{\text{sca}}/I_{\text{inc}}\,. \tag{4.24}$$

The dimension of the cross section is that of an area. The sum of absorption and scattering is often called the extinction. In experiment it can be measured by placing behind the scatterer a photodetector pointing against the wave propagation direction $\hat{\boldsymbol{k}}_0$, and measuring how much of the incoming power is reduced. Apparently, everything that is absorbed or scattered away from the forward direction contributes to the measured reduction.

It turns out that also theoretically there is a simple way to compute the extinction cross section. We start from the definitions of the absorption and scattering powers in Eq. (4.22),

$$\begin{aligned} P_{\text{ext}} &= \frac{1}{2}\oint_{\partial\Omega} \text{Re}\left(-\boldsymbol{E} \times \boldsymbol{H}^* + \boldsymbol{E}_{\text{sca}} \times \boldsymbol{H}_{\text{sca}}^*\right) \cdot d\boldsymbol{S} \\ &= -\frac{1}{2}\oint_{\partial\Omega} \text{Re}\left(\boldsymbol{E}_{\text{sca}} \times \boldsymbol{H}_{\text{inc}}^* + \boldsymbol{E}_{\text{inc}} \times \boldsymbol{H}_{\text{sca}}^*\right) \cdot d\boldsymbol{S}\,, \end{aligned}$$

where we have used that the energy flux of the incoming fields integrated over the entire boundary is zero (for a plane wave the ingoing flux must equal the outgoing flux). For a plane wave excitation with amplitude E_0, polarization vector $\boldsymbol{\epsilon}_0$, and wavevector $\boldsymbol{k}_0$ the electric and magnetic fields can be written in the form

$$\boldsymbol{E}_{\text{inc}} = E_0 e^{i\boldsymbol{k}_0\cdot\boldsymbol{r}}\,\boldsymbol{\epsilon}_0\,,\quad Z\boldsymbol{H}_{\text{inc}} = E_0 e^{i\boldsymbol{k}_0\cdot\boldsymbol{r}}\,\hat{\boldsymbol{k}}_0\times\boldsymbol{\epsilon}_0\,.$$

The extinction power then becomes

$$P_{\text{ext}} = -\frac{1}{2}\oint_{\partial\Omega}\text{Re}\left[E_0^* e^{-i\boldsymbol{k}_0\cdot\boldsymbol{r}}\left(Z^{-1}\boldsymbol{E}_{\text{sca}}\times\left(\hat{\boldsymbol{k}}_0\times\boldsymbol{\epsilon}_0^*\right)+\boldsymbol{\epsilon}_0^*\times\boldsymbol{H}_{\text{sca}}\right)\right]\cdot d\boldsymbol{S}\,. \tag{4.25}$$

This expression can be related to the scattered far-fields

$$\boldsymbol{E}_{\text{sca}}(\boldsymbol{r}) \underset{kr\to\infty}{\longrightarrow} \frac{e^{ikr}}{r}\boldsymbol{F}(\hat{\boldsymbol{r}})\,. \tag{4.26}$$

As we will demonstrate in the next chapter in Sect. 5.4, the extinction power of Eq. (4.25) can be expressed in terms of the far-field amplitude via

Extinction Power

$$P_{\text{ext}} = \frac{2\pi}{k}Z^{-1}\,\text{Im}\left[E_0^*\boldsymbol{\epsilon}_0^*\cdot\boldsymbol{F}(\hat{\boldsymbol{k}}_0)\right]\,. \tag{4.27}$$

This expression is also known as the *optical theorem*. Similarly to the experimental case, where extinction is obtained from the wave attenuation along the propagation direction, the extinction power can be computed from the far-fields scattered into a single direction, namely the wave propagation direction $\hat{\boldsymbol{k}}_0$.

4.5 Conservation of Momentum

We next analyze how momentum is transported by electromagnetic waves and transferred to mechanical momentum. This has already been discussed at the beginning of this chapter for small and large particles in the context of optical tweezers. Our derivation closely follows Poynting's theorem, but is slightly more complicated. The force exerted on a point-like particle by electromagnetic fields is given by the Lorentz force

$$\boldsymbol{F} = q\left(\boldsymbol{E}+\boldsymbol{v}\times\boldsymbol{B}\right)\,.$$

We can generalize the force for a charge distribution, and express the change of mechanical momentum $\boldsymbol{P}$ through

$$\frac{d\boldsymbol{P}}{dt} = \int_\Omega (\rho\boldsymbol{E} + \boldsymbol{J}\times\boldsymbol{B})\,d^3r = \int_\Omega \boldsymbol{f}\,d^3r\,, \tag{4.28}$$

where we have introduced the force density $\boldsymbol{f}$. As for the derivation of Poynting's theorem, we relate the source terms ρ, $\boldsymbol{J}$ to the electromagnetic fields using the inhomogeneous Maxwell's equations,

$$\boldsymbol{f} = \Big(\varepsilon\nabla\cdot\boldsymbol{E}\Big)\boldsymbol{E} + \Big(\frac{1}{\mu}\nabla\times\boldsymbol{B} - \varepsilon\frac{\partial\boldsymbol{E}}{\partial t}\Big)\times\boldsymbol{B}\,. \tag{4.29}$$

With

$$\frac{\partial}{\partial t}\boldsymbol{E}\times\boldsymbol{B} = \frac{\partial\boldsymbol{E}}{\partial t}\times\boldsymbol{B} + \boldsymbol{E}\times\frac{\partial\boldsymbol{B}}{\partial t} = \frac{\partial\boldsymbol{E}}{\partial t}\times\boldsymbol{B} - \boldsymbol{E}\times(\nabla\times\boldsymbol{E})$$

we can rewrite the last term on the right-hand side of Eq. (4.29), and get

$$\begin{aligned}\frac{d\boldsymbol{P}}{dt} + \frac{d}{dt}\int_\Omega \varepsilon\,\boldsymbol{E}\times\boldsymbol{B}\,d^3r = \int_\Omega \Big[&\varepsilon\boldsymbol{E}(\nabla\cdot\boldsymbol{E}) - \varepsilon\boldsymbol{E}\times(\nabla\times\boldsymbol{E}) \\ &+\frac{1}{\mu}\boldsymbol{B}(\nabla\cdot\boldsymbol{B}) - \frac{1}{\mu}\boldsymbol{B}\times(\nabla\times\boldsymbol{B})\Big]\,d^3r\,.\end{aligned} \tag{4.30}$$

We have added the term $\nabla\cdot\boldsymbol{B}$ which is always zero to make the expression symmetric in $\boldsymbol{E}$ and $\boldsymbol{B}$. The second term on the left-hand side can be assigned to the total electromagnetic momentum $\boldsymbol{P}_{\text{em}}$ in the volume Ω,

$$\boldsymbol{P}_{\text{em}} = \int_\Omega \varepsilon\,\boldsymbol{E}\times\boldsymbol{B}\,d^3r = \int_\Omega \mu\varepsilon\,\boldsymbol{E}\times\boldsymbol{H}\,d^3r\,. \tag{4.31}$$

We next convert the term on the right-hand side of Eq. (4.30) to a boundary integral. Comparison with Poynting's theorem, which connects the electromagnetic energy density u_{em} with the Poynting vector $\boldsymbol{S}$, shows that the momentum density flow has to be a tensor rather than a vector. In the following we adopt a component notation together with Einstein's sum convention for indices that appear twice. The electric field components in the brackets of Eq. (4.30) can be rewritten using the Levi-Civita tensor of Eq. (2.54),

$$\begin{aligned}&E_i\partial_j E_j - \epsilon_{ijk}E_j\epsilon_{klm}\partial_l E_m \\ &= E_i\partial_j E_j - \Big(\delta_{il}\delta_{jm} - \delta_{im}\delta_{jl}\Big)E_j\partial_l E_m \\ &= E_i\partial_j E_j + \Big(E_j\partial_j E_i - E_j\partial_i E_j\Big) = \partial_j E_i E_j - \frac{1}{2}\,\partial_i E_j E_j\,.\end{aligned}$$

We thus find

$$\Big[\boldsymbol{E}(\nabla\cdot\boldsymbol{E})-\boldsymbol{E}\times(\nabla\times\boldsymbol{E})\Big]_i=\partial_j\Big[E_iE_j-\frac{1}{2}\delta_{ij}E_kE_k\Big],$$

and a corresponding expression for $\boldsymbol{B}$. It is now convenient to introduce the symmetric tensor

$$T_{ij}=\varepsilon E_iE_j+\frac{1}{\mu}B_iB_j-\frac{1}{2}\delta_{ij}\left(\varepsilon E_kE_k+\frac{1}{\mu}B_kB_k\right). \tag{4.32}$$

This expression looks even better when it is rewritten in terms of $\boldsymbol{E}$ and $\boldsymbol{H}$, which leads us to the final form of Maxwell's stress tensor

Maxwell's Stress Tensor

$$T_{ij}=\varepsilon E_iE_j+\mu H_iH_j-\frac{1}{2}\delta_{ij}\left(\varepsilon E^2+\mu H^2\right). \tag{4.33}$$

It describes how momentum is transported and transferred by electromagnetic fields. Using the stress tensor[1] we can rewrite Eq. (4.30) in the form

$$\frac{d}{dt}(\boldsymbol{P}+\boldsymbol{P}_{\rm em})=\int_\Omega\nabla\cdot\bar{\bar{T}}\,d^3r=\oint_{\partial\Omega}\bar{\bar{T}}\cdot\hat{\boldsymbol{n}}\,dS\,, \tag{4.34}$$

where $\hat{\boldsymbol{n}}$ is the surface normal of the boundary. Thus, whenever the mechanical or electromagnetic momentum changes in a given volume, it must be transported away in the form of electromagnetic fields. The quantity accounting for the transported momentum is precisely Maxwell's stress tensor.

Abraham–Minkowski Controversy

The Abraham–Minkowski controversy or dilemma, as it is sometimes called, is a rather grotesque story, to say the least. It concerns the question of the momentum carried by a photon propagating in a medium with refractive index n, but, to make this point clear from the beginning, the controversy does not include any quantum aspects and also appears in an analysis solely based on classical electrodynamics. The first contributions to the topic were made by Hermann Minkowski (1908) and Max Abraham (1909), for some details on the history as well as for a comprehensive account of the topic the reader is referred to [14]. While Abraham concluded that the photon momentum should be

[1]Throughout this book we denote tensors with a double bar on top, here $\bar{\bar{T}}$. The components of such tensors are $\big(\bar{\bar{T}}\big)_{ij}=T_{ij}$.

$$\text{(Abraham momentum)} \qquad p_{\text{photon}} = \frac{\hbar k_0}{n}\,,$$

where k_0 is the wavenumber of light in vacuum and n the refractive index of the material, Minkowski concluded that it should be

$$\text{(Minkowski momentum)} \qquad p_{\text{photon}} = n\,\hbar k_0\,.$$

At first sight this looks like a bad joke, and one might think that a simple experiment should decide in favor of one of these definitions. Indeed, there are many conclusive experiments in favor of Abraham's definition. Unfortunately, there exists an equal number of experiments that seems to support Minkowski's definition. So obviously there is something awkward going on here.

Nowadays there is consensus that the debate or dilemma has been resolved by Barnett [15] and is related to something we will discuss in more detail in Chap. 13, namely the distinction between kinetic and canonical momentum. The first one is related to the light momentum alone; the second one is the momentum in presence of electromagnetic fields. The Abraham definition corresponds to the kinetic momentum, whereas the Minkowski form to the canonical one. Yet, the question of which form has to be used in the interpretation of which experiment remains difficult. We here do not enter into any details, but hope that any reader who will step over the problem in the future will be aware of the long history of the topic as well as the potential difficulties in deciding for either the Abraham or Minkowski definition.

Optical Forces and Torques

Maxwell's stress tensor can be used to compute optical forces and torques for particles of arbitrary shape and size. In general, we are interested in situations where a particle is illuminated by optical fields oscillating at a single frequency ω, and the forces are so weak that it takes several field oscillations before the particle has moved somewhere else. Under these conditions it is then permissible to assume time-harmonic fields and to average the net forces over an oscillation period T. The averaged electromagnetic force

$$\left\langle \frac{d\boldsymbol{P}_{\text{em}}}{dt} \right\rangle = \frac{1}{T}\int_0^T \frac{d\boldsymbol{P}_{\text{em}}}{dt}\,dt = 0$$

becomes zero for a time-harmonic field, where $\boldsymbol{P}_{\text{em}}$ is the electromagnetic momentum defined in Eq. (4.31). Maxwell's stress tensor can be averaged over a period, just as discussed for Poynting's theorem and plane waves, and we get

$$\langle T_{ij} \rangle = \frac{1}{2}\text{Re}\left[\varepsilon E_i E_j^* + \mu H_i H_j^* - \frac{1}{2}\delta_{ij}\left(\varepsilon E_k E_k^* + \mu H_k H_k^*\right)\right]. \tag{4.35}$$

We can now compute the forces acting on the particle by defining a boundary $\partial\Omega$ that surrounds the particle. For particles with sharp boundaries we could even take the particle surface itself. The net momentum flow into or out of this boundary then gives the force or torque exerted on the particle,

Optical Force and Torque for Time-Harmonic Fields

$$\langle \boldsymbol{F} \rangle = \oint_{\partial\Omega} \langle \bar{\bar{T}} \rangle \cdot d\boldsymbol{S} \tag{4.36a}$$

$$\langle \boldsymbol{M} \rangle = \oint_{\partial\Omega} \boldsymbol{r} \times \langle \bar{\bar{T}} \rangle \cdot d\boldsymbol{S}\,. \tag{4.36b}$$

Thus, once the electromagnetic fields are known for the problem under study, one can immediately compute optical forces and torques using the above expressions. In later chapters we will show how Maxwell's equations can be solved for particles with sharp boundaries, where we can use the above equations to immediately compute the force or torque exerted by electromagnetic fields on the particle. Equations (4.36) are the general expressions for the fields and torques acting on particles of arbitrary size, for particles much smaller or larger than the light wavelength we can use again the previously discussed more simple dipolar or ray tracing approximations.

4.6 Optical Angular Momentum

In addition to linear momentum, light can also carry angular momentum. The most simple form is light with a circular polarization, as described through the (unnormalized) polarization vector $\boldsymbol{\epsilon}_{\pm} = \hat{\boldsymbol{x}} \pm i\hat{\boldsymbol{y}}$. Figure 4.7 shows an example where a DNA strand is attached on one side to a substrate and on the other side to a quartz cylinder. By placing the system inside an optical trap and using light with circular polarization, one can transfer angular momentum from light to the quartz cylinder and wind up the DNA strand. Through measurement of the cylinder position, one obtains detailed information about the applied torque and the compression of the DNA strand.

Figure 4.8 shows the creation of a light beam that carries an orbital angular momentum (OAM) [16, 17]. A focused light beam passes through a dielectric spiral whose height depends on the azimuthal angle, such that after passage it has acquired an orbital angular momentum. There exist other ways for creating light with OAM, for instance, by using holograms. For the Gauss–Laguerre beams discussed in the previous chapter, the amplitude of the electric field can be expressed as [18]

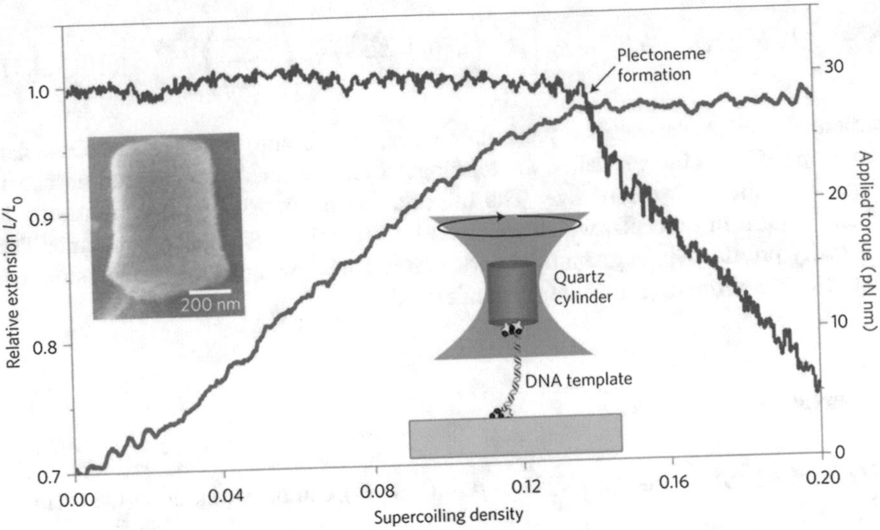

Fig. 4.7 Inset, left: scanning electron micrograph of a nanofabricated quartz cylinder. Inset, right: schematic of the experimental setup (not to scale), in which a DNA molecule is tethered to the cylinder at one end and to a glass surface at the other. Here, the DNA is stretched by a force of 3 pN and twisted at a constant rate of 0.5 turns per second. The relative extension of the DNA (blue trace) and the applied torque (green trace) are plotted as a function of the supercoiling density, which indicates the degree of twist introduced. When the supercoiling density is around 0.14, the coiled DNA undergoes a phase transition from a twisted to a plectonemic form, which is indicated by the plateau in the applied torque and the monotonic decrease in relative extension. Figure and caption taken from [13]

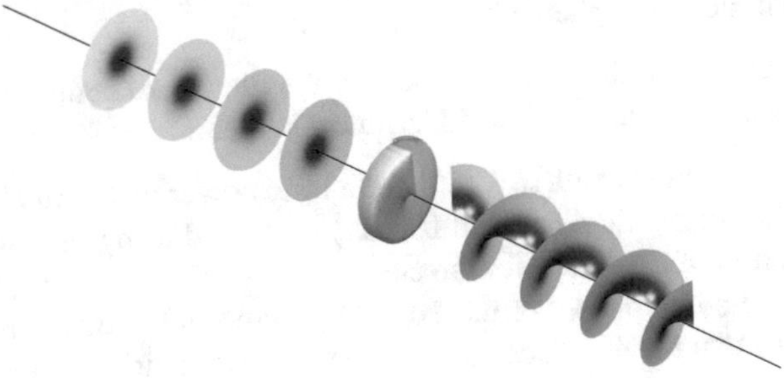

Fig. 4.8 Creation of a light beam with orbital angular momentum. A focused beam with linear or circular polarization impinges from the left-hand side on a phase plate, which is made of a dielectric and has the form of a spiral whose height increases with increasing azimuthal angle. When passing through this device the light acquires a phase dependent delay angle, and after passage the light carries an orbital angular momentum

$$E(\rho, \phi, z = 0) = E_0 \left[e^{i\ell\phi} \right] \tanh\left(\frac{\rho}{w_v}\right) \exp\left(-\frac{\rho^2}{w_0^2}\right), \tag{4.37}$$

where E_0 characterizes the peak amplitude, w_0 is the waist of the Gaussian envelope, ℓ is an integer called the topological charge (or OAM quantum number), and w_v is the vortex core size. The important contribution for our discussion is the $e^{i\ell\phi}$ term in brackets, which accounts for the orbital angular momentum. The intensity profile $|E|^2$ is given by a dark vortex core, owing to the total destructive interference at the origin where ϕ is undefined.

Exercises

Exercise 4.1 Show that $(\boldsymbol{p} \cdot \nabla)\boldsymbol{E} + \boldsymbol{p} \times (\nabla \times \boldsymbol{E})$ can be rewritten in the form of Eq. (4.3).

Exercise 4.2 Derive Eq. (4.2) along the lines discussed in the text.

Exercise 4.3 Show that the relation $\nabla \cdot \boldsymbol{E} \times \boldsymbol{B} = \boldsymbol{B} \cdot \nabla \times \boldsymbol{E} - \boldsymbol{E} \cdot \nabla \times \boldsymbol{B}$ holds.

Exercise 4.4 Consider the situation where a laser beam with a power of 100 mW is focused onto an area of the square wavelength λ^2.

(a) Estimate the field strength in the focus spot. Use a light wavelength of 500 nm and assume a homogeneous field distribution.
(b) Estimate the number of photons passing per second through the area λ^2.
(c) Assume that each photon transfers a momentum $\hbar k$. Estimate the force exerted by the total incoming photon flux.

Exercise 4.5 Use from the NANOPT toolbox the file `demofocus01.m` to compute the electric fields of a tightly focused Hermite–Gaussian laser mode.

(a) Compute the field strength in Vm^{-1} for a laser power of 100 mW.
(b) Compute the magnetic field from Faraday's law and using the finite difference approximation of Eq. (2.10) for the curl.
(c) Compute the z-component of the Poynting vector and integrate it over the xy-plane. Show that the flux does not change when evaluating the fields away from the focus $z = 0$.

Exercise 4.6 The polarizability of a small dielectric sphere is given by

$$\alpha = 4\pi\varepsilon_2 \left(\frac{\varepsilon_1 - \varepsilon_2}{\varepsilon_1 + 2\varepsilon_2}\right) a^3 .$$

Here a is the sphere radius, and ε_1, ε_2 are the permittivities inside and outside the sphere. Consider a glass sphere ($n_1 = 1.5$) embedded in water ($n = 1.33$), and a sphere diameter of 100 nm. Use the results of Exercise 4.5 to compute the dipole force of Eq. (4.6) acting on the glass sphere in the focus spot and away from it. Express the forces in pN.

Exercise 4.7 Same as Exercise 4.6 but for a gold sphere with a permittivity of $\varepsilon_1 = -2.5 + 3.6\,i$ at a wavelength of 500 nm, and a diameter of 50 nm. Compute also the scattering and polarization-gradient forces.

Exercise 4.8 Write the mechanical work in Eq. (4.14) in the form

$$W = \int_\Omega u_{\text{mech}}\, d^3r\,,$$

with the mechanical energy density u_{mech}, and bring Poynting's theorem to a differential form similar to the continuity Eq. (4.11).

Exercise 4.9 Consider a plane electromagnetic wave with a complex polarization vector $\boldsymbol{\epsilon} = \boldsymbol{\epsilon}' + i\boldsymbol{\epsilon}''$. Compute the energy density and the Poynting vector in a form similar to Eq. (4.19).

Exercise 4.10 Consider a plane wave with linear polarization. Show through explicit calculation that the total flux of the Poynting vector through a sphere boundary is zero.

Exercise 4.11 Compute Maxwell's stress tensor of Eq. (4.33) for a plane wave with arbitrary polarization. Interpret the result in terms of momentum and angular momentum carried by the wave.

Exercise 4.12 Use the polarizability given in Exercise 4.6, and a dielectric particle with a real polarizability $\varepsilon_1 = \varepsilon_1'$ that is excited by a time-harmonic field $\boldsymbol{E}_0$.

(a) Compute the induced dipole moment of the sphere.
(b) Compute the fields of a dipole using the results of electrostatics. These results are also valid for an oscillating dipole in the nearfield zone.
(c) Compute in the nearfield zone Maxwell's stress tensor.

Chapter 5
Green's Functions

Green's functions provide a simple and elegant way to solve differential equations, such as the wave equation in electrodynamics, and play an important role in nano optics. In this chapter we start by introducing the basic concepts of Green's functions for the simplified scalar wave equation, and then ponder on the solutions of the full Maxwell's equations.

5.1 What Are Green's Functions?

Consider a linear but otherwise unspecified differential equation for the function $f(\boldsymbol{r})$,

$$L(\boldsymbol{r})f(\boldsymbol{r}) = -Q(\boldsymbol{r})\,, \tag{5.1}$$

where $L(\boldsymbol{r})$ is a differential operator, such as ∇^2 for the Poisson or $\nabla^2 + k^2$ for the Helmholtz equation, and Q is an external source term. In principle L could also be a nonlocal operator containing an integration over space, although in the following we will stick to the more simple form given by Eq. (5.1). To solve the above differential equation, we introduce a Green's function $G(\boldsymbol{r}, \boldsymbol{r}_0)$ defined through

Green's Function Definition

$$L(\boldsymbol{r})G(\boldsymbol{r}, \boldsymbol{r}_0) = -\delta(\boldsymbol{r} - \boldsymbol{r}_0)$$
$$\text{with proper boundary conditions}\,. \tag{5.2}$$

U. Hohenester, *Nano and Quantum Optics*, Graduate Texts in Physics,
https://doi.org/10.1007/978-3-030-30504-8_5

$G(\boldsymbol{r}, \boldsymbol{r}_0)$ describes the response of the system to a point-like source located at position $\boldsymbol{r}_0$, with $\delta(\boldsymbol{r} - \boldsymbol{r}_0)$ being Dirac's delta function (see also Appendix F). Importantly, the Green's function has already built in the proper boundary conditions, such as outgoing waves at infinity for the Helmholtz equation to be discussed further below.

The Green of the Green's Functions

Before pondering on the formalism of Green's functions, it is probably adequate to give credit to the person who invented the formalism, "The Green of the Green's functions" as it was called in an article published a couple of years back in Physics Today [19]. There the author discusses that, as far as it is known, George Green did not receive any mathematical education, rather he spent 5 years in his father's bakery and was then sent to learn to be a miller in the tower mill that his father had built close to Nottingham. The streets of Nottingham were not safe places to walk in after dark, so Green spent many days and nights working in the mill. Probably in this time he also wrote his first work "An Essay on the Application of Mathematical Analysis to the Theories of Electricity and Magnetism," which Green did not dare to send to one of the journals of the learned societies [19]:

> But Green, with no qualifications and no contacts with the scientific establishment, felt that it would be presumptuous to submit his paper to a journal. So he paid for his paper to be published privately in Nottingham. The tentative way in which he approached publication is apparent in his foreword. There, he expressed his hope that "the difficulty of the subject will incline mathematicians to read this work with indulgence, more particularly when they are informed that it was written by a young man, who has been obliged to obtain the little knowledge he possesses, at such intervals and by such means, as other indispensable avocations which offer but few opportunities of mental improvement, afforded. [...]
>
> Green's main purpose in publishing would have been to bring his work to the attention of other mathematicians in the UK and overseas. It appears, however, that with one exception, there was little or no response. That must have been very disheartening.

Had it not been for a few lucky coincidences, Green's work would have probably remained undiscovered. Today Green's functions are abundant in the field of physics and have become one of the primary tools for the solution of differential equations.

The neat thing about $G(\boldsymbol{r}, \boldsymbol{r}_0)$ is that it allows to immediately write down the solution of Eq. (5.1) in the form

$$f(\boldsymbol{r}) = \int G(\boldsymbol{r}, \boldsymbol{r}')Q(\boldsymbol{r}')\, d^3r' \,. \tag{5.3}$$

Applying from the left-hand side the differential operator $L(\boldsymbol{r})$ gives

$$L(\boldsymbol{r})\int G(\boldsymbol{r},\boldsymbol{r}')Q(\boldsymbol{r}')\,d^3r' = -\int \delta(\boldsymbol{r}-\boldsymbol{r}')Q(\boldsymbol{r}')\,d^3r' = -Q(\boldsymbol{r})\,,$$

which shows that Eq. (5.1) is indeed fulfilled. As $G(\boldsymbol{r},\boldsymbol{r}')$ has already built in the proper boundary conditions, also $f(\boldsymbol{r})$ inherits the proper boundary conditions.

What remains to be done is to compute for specific differential operators $L(\boldsymbol{r})$ the corresponding Green's functions. As we will show below, this task is often simpler than it may look at first sight. We conclude this section by emphasizing that the Green's function approach works because of the linearity of the operator $L(\boldsymbol{r})$. $G(\boldsymbol{r},\boldsymbol{r}')$ describes the response of a point-like source, and by superimposing in Eq. (5.3) these responses for all sources $Q(\boldsymbol{r})$ we then obtain the full solution $f(\boldsymbol{r})$.

5.2 Green's Function for the Helmholtz Equation

As a first example, we consider the Helmholtz equation

$$\left(\nabla^2+k^2\right)f(\boldsymbol{r}) = -Q(\boldsymbol{r})\,, \tag{5.4}$$

where k is a wavenumber. Following the prescription of the previous section we are seeking for the Green's function solution

$$\left(\nabla^2+k^2\right)G(\boldsymbol{r},\boldsymbol{r}') = -\delta(\boldsymbol{r}-\boldsymbol{r}')\,. \tag{5.5}$$

For an unbounded homogeneous medium the Green's function can only depend on the distance $|\boldsymbol{r}-\boldsymbol{r}'|$ such that $G(\boldsymbol{r},\boldsymbol{r}') = g(|\boldsymbol{r}-\boldsymbol{r}'|)$. Thus, we get for $r\neq 0$

$$\frac{1}{r^2}\frac{\partial}{\partial r}\left(r^2\frac{\partial g}{\partial r}\right)+k^2g = 0\,,$$

where we have expressed the Laplace operator in spherical coordinates and have neglected all derivatives in the angular directions. As one can easily prove through explicit calculation, the solution of the above equation is provided by out- and ingoing spherical waves

$$g(r) = C\frac{e^{ikr}}{r}+D\frac{e^{-ikr}}{r}\,. \tag{5.6}$$

Here C and D are parameters to be determined from the boundary conditions for $g(r)$. As we are interested in solutions that preserve causality, we only keep the so-called *retarded* solutions [2] consisting of outgoing waves, and set $D = 0$. The parameter C can be determined by integrating Eq. (5.5) over a small sphere with radius a and volume Ω_a which encloses the origin,

$$\lim_{a\to 0}\left(\int_{\Omega_a} \nabla\cdot\nabla\frac{Ce^{ikr}}{r}\,d^3r + k^2\int_{\Omega_a}\frac{Ce^{ikr}}{r}\,d^3r\right) = -1\,.$$

On the right-hand side we have used that the integration over an even infinitesimally small volume gives one if the argument of Dirac's delta function becomes zero within the integration volume. In the second term in parentheses on the left-hand side we introduce spherical coordinates with $d^3r = 4\pi r^2 dr$ and immediately observe that it becomes zero in the limit $a \to 0$. Using Gauss' theorem for the first term, we can transform the volume integration to a surface integration

$$\lim_{a\to 0}\oint \nabla\left(\frac{Ce^{ikr}}{r}\right)\cdot d\boldsymbol{S} = \lim_{a\to 0}\oint \frac{d}{dr}\left(\frac{Ce^{ikr}}{r}\right)dS$$
$$= \lim_{a\to 0} 4\pi a^2\frac{d}{da}\left(\frac{Ce^{ika}}{a}\right) = -4\pi\,.$$

Here we have used that the outer surface normal of $d\boldsymbol{S}$ points for a sphere in the radial direction. Thus, from the above equation we get $C = 1/(4\pi)$. The Green's function for the scalar wave equation and for an unbounded homogeneous medium with outgoing waves at infinity thus becomes

Green's Function for Helmholtz Equation

$$G(\boldsymbol{r},\boldsymbol{r}') = G(\boldsymbol{r}-\boldsymbol{r}') = \frac{e^{ik|\boldsymbol{r}-\boldsymbol{r}'|}}{4\pi|\boldsymbol{r}-\boldsymbol{r}'|}\,. \tag{5.7}$$

5.2.1 *Representation Formula for Helmholtz Equation*

The Green's function can be used to derive an extremely useful expression, the so-called representation formula, which will be used in later parts of this book. The situation we have in mind is depicted in Fig. 5.1 and consists of a volume Ω with a sharp boundary $\partial\Omega$. We shall find it convenient to consider the interior and exterior of volume Ω separately.

- The region inside volume Ω is denoted with the subscript 1 throughout, the wavenumber and Green's function in Ω_1 are k_1 and G_1, respectively.

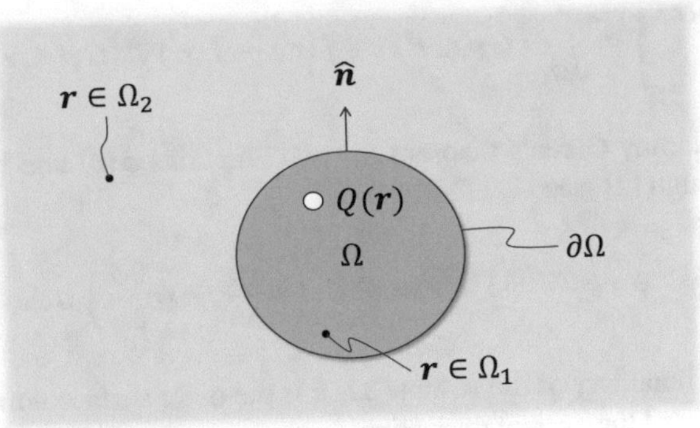

Fig. 5.1 Schematics for representation formula. We consider a volume Ω with a sharp boundary $\partial\Omega$. We denote with $\Omega_1 = \Omega$ and Ω_2 the interior and exterior volume, respectively, and $\hat{\boldsymbol{n}}$ is the outer surface normal of volume Ω. For the representation formula we consider the situations that a point $\boldsymbol{r}$ is located either inside or outside of volume Ω. Q is a source that is located inside volume Ω_1

– The region outside volume Ω is denoted with the subscript 2 throughout, the wavenumber and Green's function in Ω_2 are k_2 and G_2, respectively.

We start with the interior region and multiply Eq. (5.4) with $G_1(\boldsymbol{r},\boldsymbol{r}')$ and Eq. (5.5) with $f(\boldsymbol{r})$. Upon interchanging $\boldsymbol{r} \leftrightarrow \boldsymbol{r}'$ we get

$$G_1(\boldsymbol{r},\boldsymbol{r}')\left(\nabla'^2 + k^2\right) f(\boldsymbol{r}') = -G_1(\boldsymbol{r},\boldsymbol{r}')Q(\boldsymbol{r}') \tag{5.8a}$$

$$f(\boldsymbol{r}')\left(\nabla'^2 + k^2\right) G_1(\boldsymbol{r}',\boldsymbol{r}) = -f(\boldsymbol{r}')\delta(\boldsymbol{r}'-\boldsymbol{r})\,. \tag{5.8b}$$

These equations are integrated over the volume Ω_1. We introduce an *incoming* field that is generated by the source contributions located inside of volume Ω_1,

$$f_1^{\text{inc}}(\boldsymbol{r}) = \int_{\Omega_1} G_1(\boldsymbol{r},\boldsymbol{r}')Q(\boldsymbol{r}')\,d^3r'\,. \tag{5.9}$$

These are the fields that would be the solutions of Helmholtz equation for a homogeneous medium with wavenumber k_1 in absence of additional boundaries. The representation formula, to be derived in a moment, provides the field modifications in presence of such boundaries. When integrating Eq. (5.8) over the volume Ω_1 we have to be careful about the last term in Eq. (5.8b) which vanishes when $\boldsymbol{r}$ and $\boldsymbol{r}'$ are located in different volumes. We next subtract the two expressions in Eq. (5.8) and integrate over volume Ω_1 to arrive at

$$\begin{rcases} \boldsymbol{r} \in \Omega_1: & f(\boldsymbol{r}) \\ \boldsymbol{r} \in \Omega_2: & 0 \end{rcases} = \int_{\Omega_1} \Big(G_1(\boldsymbol{r},\boldsymbol{r}')\nabla'^2 f(\boldsymbol{r}') - f(\boldsymbol{r}')\nabla'^2 G_1(\boldsymbol{r}',\boldsymbol{r}) \Big)\, d^3r' + f_1^{\rm inc}\,.$$

We can now employ Green's theorem for two functions $\phi(\boldsymbol{r})$ and $\psi(\boldsymbol{r})$ which are connected through [2] (see also Exercise 5.1)

$$\int_{\Omega_1} \Big(\phi\nabla^2\psi - \psi\nabla^2\phi\Big)\, d^3r = \oint_{\partial\Omega_1} \left(\phi\frac{\partial\psi}{\partial n} - \psi\frac{\partial\phi}{\partial n}\right) dS\,. \tag{5.10}$$

Here $\partial\Omega$ is the boundary of the volume Ω, $\hat{\boldsymbol{n}}$ is the outer surface normal of Ω_1, and we have introduced the normal derivative

$$\frac{\partial\psi}{\partial n} = \hat{\boldsymbol{n}}\cdot\nabla\psi\,. \tag{5.11}$$

Using Green's theorem for $f(\boldsymbol{r})$ we are finally led to the representation formula

Representation Formula for Integration Over Ω_1

$$\begin{rcases} \boldsymbol{r} \in \Omega_1: & f(\boldsymbol{r}) \\ \boldsymbol{r} \in \Omega_2: & 0 \end{rcases} = f_1^{\rm inc}(\boldsymbol{r}) + \oint_{\partial\Omega} \left[G_1(\boldsymbol{r},\boldsymbol{s}')\frac{\partial f(\boldsymbol{s}')}{\partial n'} - \frac{\partial G_1(\boldsymbol{r},\boldsymbol{s}')}{\partial n'} f(\boldsymbol{s}')\right] dS'. \tag{5.12}$$

It can be interpreted as follows: if we know the values of a function $f(\boldsymbol{s}')$ and of the normal derivative $\partial f(\boldsymbol{s}')/(\partial n')$ at a given boundary, we can obtain the function value $f(\boldsymbol{r})$ everywhere else in the inside region. From here on we denote positions *on* the boundary with the symbol $\boldsymbol{s}$.

An expression similar to Eq. (5.12) can be also obtained for the outside region Ω_2. Proceeding in a similar fashion as for the inside case, we introduce the incoming fields associated with sources in Ω_2, which are now propagated to position $\boldsymbol{r}$ by G_2 rather than G_1,

$$f_2^{\rm inc}(\boldsymbol{r}) = \int_{\Omega_2} G_2(\boldsymbol{r},\boldsymbol{r}')Q(\boldsymbol{r}')\, d^3r'\,. \tag{5.13}$$

Note that these are the fields that would be the solutions of the Helmholtz equation in absence of additional boundaries. In comparison to the inside case two points have to be considered with more care. First, there is an additional boundary $\partial\Omega_\infty$ at the volume outside, see Fig. 5.2, which in most cases of interest can be pushed towards infinity. As the Green's function has already built in the boundary conditions at large distances, in the Helmholtz case the outgoing fields of Eq. (5.7), we do not have to

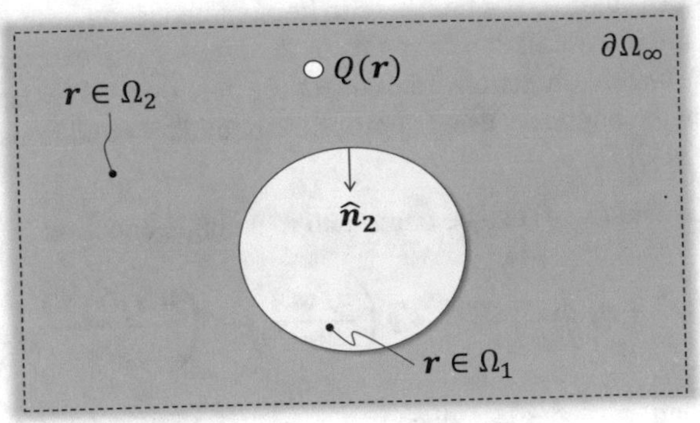

Fig. 5.2 Schematics for representation formula and integration over Ω_2. When integrating over Ω_2, there are two differences with respect to the situation depicted in Fig. 5.1. First, there is also a boundary at infinity, $\partial\Omega_\infty$ (dashed rectangle). In most cases of interest the boundary conditions at infinity are already incorporated properly through the Green's function solution, for instance, by using outgoing rather than ingoing fields. Second, the outer surface normal $\hat{\boldsymbol{n}}_2$ points into volume Ω_1. In most cases of interest, we use in our analysis the surface normal $\hat{\boldsymbol{n}}$ of volume Ω only, with the relation $\hat{\boldsymbol{n}}_2 = -\hat{\boldsymbol{n}}$. Q is a source that is located in Ω_2

account explicitly for this boundary. Second, the outer surface normal $\hat{\boldsymbol{n}}_2$ of volume Ω_2 points into volume Ω_1, see Fig. 5.2, and is anti-parallel to $\hat{\boldsymbol{n}}$. It turns out to be convenient to keep only $\hat{\boldsymbol{n}}$ and to introduce a negative sign in the representation formula instead. With this we are led to

Representation Formula for Integration Over Ω_2

$$\left.\begin{array}{ll} \boldsymbol{r} \in \Omega_1: & 0 \\ \boldsymbol{r} \in \Omega_2: & f(\boldsymbol{r}) \end{array}\right\} = f_2^{\text{inc}}(\boldsymbol{r}) - \oint_{\partial\Omega} \left[G_2(\boldsymbol{r}, \boldsymbol{s}') \frac{\partial f(\boldsymbol{s}')}{\partial n'} - \frac{\partial G_2(\boldsymbol{r}, \boldsymbol{s}')}{\partial n'} f(\boldsymbol{s}') \right] dS' . \tag{5.14}$$

Both representation formulas are reminiscent of the angular spectrum representation introduced in Chap. 3, where we have related the electromagnetic fields in the plane $z = 0$ to the evanescent and propagating fields away from the plane. They are also useful in a number of different contexts such as:

Dirichlet vs. Neumann Problem. In principle, Eqs. (5.12) and (5.14) allow us to express f in terms of its normal derivative and vice versa. When we let $\boldsymbol{r}$ approach the boundary, we obtain a relation that connects the function and its surface derivative. Thus, it either suffices to know the function values (Dirichlet problem) or the surface derivatives (von-Neumann problem) on the boundaries.

Transmission Problem. For bodies with homogeneous material properties k_1, k_2 we can use the representation formulas in the interior and exterior domains with the different Green's functions to connect the function values and their normal derivatives at the boundary using appropriate boundary conditions. We will come back to this in Chap. 9.

Equations (5.12) and (5.14) can be combined to a single equation

$$f(\boldsymbol{r}) = f_j^{\text{inc}}(\boldsymbol{r}) + \tau_j \oint_{\partial\Omega} \left[G_j(\boldsymbol{r}, \boldsymbol{s}') \left(\frac{\partial f(\boldsymbol{s}')}{\partial n'} \right) - \left(\frac{\partial G_j(\boldsymbol{r}, \boldsymbol{s}')}{\partial n'} \right) f(\boldsymbol{s}') \right] dS' \,, \tag{5.15}$$

with $\boldsymbol{r} \in \Omega_j$ and we have introduced $\tau_{1,2} = \pm 1$ to account for the change of the surface normal in case of the exterior solution. The remaining equations with a zero on the left-hand side can be sometimes used for further manipulations of the representation formulas.

5.2.2 *Green's Function vs. Fundamental Solution*

In physics Green's functions are used in different contexts which can sometimes cause confusion. Consider first a source $Q(\boldsymbol{r})$ located in unbounded space. As discussed in the previous section, we can then compute the solution

$$f(\boldsymbol{r}) = \int G(\boldsymbol{r}, \boldsymbol{r}') Q(\boldsymbol{r}') \, d^3 r'$$

directly from the Green's function $G(\boldsymbol{r}, \boldsymbol{r}')$. However, for a photonic environment consisting of sources $Q(\boldsymbol{r})$ together with additional dielectric or metallic nanoparticles, such as sketched in Fig. 5.2, we must resort to Eq. (5.15) or some equivalent form to account for the additional boundary conditions to be fulfilled at the nanoparticle boundaries. In this case, the Green's function of Eq. (5.7) for an unbounded medium no longer suffices to uniquely determine the solution $f(\boldsymbol{r})$.

In principle, we can introduce a new "total" Green's function $G_{\text{tot}}(\boldsymbol{r}, \boldsymbol{r}')$ accounting for these modified boundary conditions, which then relates the solution and sources through

$$f(\boldsymbol{r}) = \int G_{\text{tot}}(\boldsymbol{r}, \boldsymbol{r}') Q(\boldsymbol{r}') \, d^3 r' \,.$$

Suppose that we have a machinery at hand that allows us to solve the differential equation for arbitrary sources $Q(\boldsymbol{r})$. We will learn more about such solution schemes in Chap. 9 on particle plasmons. We could then introduce a point-like unit source $Q_\delta(\boldsymbol{r}) = \delta(\boldsymbol{r} - \boldsymbol{r}_0)$, compute the function values f_δ and derivatives at the boundary using the machinery to be discussed, and compute from Eq. (5.15) the total Green's function according to $G_{\text{tot}}(\boldsymbol{r},\boldsymbol{r}_0) = f_\delta(\boldsymbol{r})$. By construction, the total Green's function then provides the proper response to a point-like source term, and by virtue of the superposition principle we can get the total response through integration over all point-like sources.

The mathematics literature has adopted a more rigorous terminology which calls the solution for the unbounded medium $G(\boldsymbol{r},\boldsymbol{r}')$ the *fundamental solution* and reserves the expression *Green's function* entirely for $G_{\text{tot}}(\boldsymbol{r},\boldsymbol{r}')$. The Green's function thus always fulfills all relevant boundary conditions of the problem under investigation. We will not adopt this terminology here, but will add a word of caution whenever needed.

5.3 Green's Function for the Wave Equation

Consider the wave equation for a time-harmonic electric field

$$-\nabla\times\nabla\times\boldsymbol{E}(\boldsymbol{r})+k^2\boldsymbol{E}(\boldsymbol{r})=-i\mu\omega\boldsymbol{J}(\boldsymbol{r})\,, \tag{5.16}$$

with $k^2 = \varepsilon\mu\,\omega^2$. In principle we could define the Green's function similarly to the Helmholtz equation, but we will here proceed somewhat differently. We first note that in the frequency domain the electric field can be connected to the electromagnetic potentials V, $\boldsymbol{A}$ through $\boldsymbol{E}=i\omega\boldsymbol{A}-\nabla V$. We additionally employ the Lorenz gauge condition $\nabla\cdot\boldsymbol{A}=i\omega\varepsilon\mu\,V$ to obtain for the vector potential the Helmholtz equation

$$\left(\nabla^2+k^2\right)\boldsymbol{A}(\boldsymbol{r})=-\mu\boldsymbol{J}(\boldsymbol{r})\,.$$

Obviously, we can use the Green's function defined in Eq. (5.7) for the Helmholtz equation to solve for $\boldsymbol{A}$ in the form

$$\boldsymbol{A}(\boldsymbol{r})=\mu\int G(\boldsymbol{r},\boldsymbol{r}')\boldsymbol{J}(\boldsymbol{r}')\,d^3r'\,. \tag{5.17}$$

We can now use the Lorenz gauge condition to compute the scalar potential

$$V(\boldsymbol{r})=\frac{\nabla\cdot\boldsymbol{A}(\boldsymbol{r})}{i\varepsilon\mu\omega}=-\frac{i\mu\omega}{k^2}\nabla\cdot\int G(\boldsymbol{r},\boldsymbol{r}')\boldsymbol{A}(\boldsymbol{r}')\,d^3r'\,,$$

with $k^2=\varepsilon\mu\,\omega^2$. The electric field can be expressed as

$$E_i(\boldsymbol{r})=i\omega A_i(\boldsymbol{r})-\partial_i V(\boldsymbol{r})=i\omega\mu\int\left(\delta_{ij}+\frac{\partial_i\partial_j}{k^2}\right)G(\boldsymbol{r},\boldsymbol{r}')J_j(\boldsymbol{r}')\,d^3r'\,, \tag{5.18}$$

where ∂_i denotes the derivative with respect to the i'th Cartesian component and we assume Einstein's sum convention for j. It is now convenient to introduce the so-called dyadic Green's function

Dyadic Green's Function

$$G_{ij}(\boldsymbol{r},\boldsymbol{r}') = \left(\delta_{ij} + \frac{\partial_i\partial_j}{k^2}\right)\frac{e^{ik|\boldsymbol{r}-\boldsymbol{r}'|}}{4\pi|\boldsymbol{r}-\boldsymbol{r}'|}\,, \tag{5.19}$$

and to relate the electric field and the current source via

$$\boldsymbol{E}(\boldsymbol{r}) = i\omega\mu\int\bar{\bar{G}}(\boldsymbol{r},\boldsymbol{r}')\cdot\boldsymbol{J}(\boldsymbol{r}')\,d^3r'\,. \tag{5.20}$$

$G_{ij}(\boldsymbol{r},\boldsymbol{r}')$ has to be interpreted such that it gives for a current source $J_j(\boldsymbol{r}')$ pointing in direction j the electric field $E_i(\boldsymbol{r})$ pointing in direction i. We will occasionally also adopt the compact matrix notation

$$\bar{\bar{G}}(\boldsymbol{r},\boldsymbol{r}') = \left(\mathbb{1} + \frac{\nabla\nabla}{k^2}\right)\frac{e^{ik|\boldsymbol{r}-\boldsymbol{r}'|}}{4\pi|\boldsymbol{r}-\boldsymbol{r}'|}\,,$$

where we implicitly assume $(\nabla\nabla)_{ij} = \partial_i\partial_j$. The matrix $\bar{\bar{G}}$ is often referred to as the dyadic Green's function or Green's dyadics, in short, and plays an important role in the field of nano optics. Similarly to the wave equation for the electric field, Eq. (5.16), the Green's function $\bar{\bar{G}}$ fulfills the equation (see Exercise 5.6)

$$-\nabla\times\nabla\times\bar{\bar{G}}(\boldsymbol{r},\boldsymbol{r}') + k^2\bar{\bar{G}}(\boldsymbol{r},\boldsymbol{r}') = -\delta(\boldsymbol{r}-\boldsymbol{r}')\mathbb{1}\,. \tag{5.21}$$

For the sake of completeness we also show how the magnetic field can be expressed in terms of the Green's dyadics. Faraday's law for time-harmonic fields reads

$$\nabla\times\boldsymbol{E} = i\omega\mu\,\boldsymbol{H}\,.$$

By taking the curl on both sides of Eq. (5.20) we thus find

$$\boldsymbol{H}(\boldsymbol{r}) = \nabla\times\int\bar{\bar{G}}(\boldsymbol{r},\boldsymbol{r}')\cdot\boldsymbol{J}(\boldsymbol{r}')\,d^3r'\,. \tag{5.22}$$

5.3.1 Far-field Limit

In many situations one is interested in the far-fields of a given current source. In principle, one could take the far-field limit of the Green's dyadic. However, we here proceed somewhat differently and start from Eq. (5.17) that relates the current distribution to the vector potential,

$$\boldsymbol{A}(\boldsymbol{r}) = \mu \int \frac{e^{ik|\boldsymbol{r}-\boldsymbol{r}'|}}{4\pi|\boldsymbol{r}-\boldsymbol{r}'|} \boldsymbol{J}(\boldsymbol{r}')\, d^3r' \,.$$

In the following we assume $r \gg r'$ such that we can approximate

$$|\boldsymbol{r}-\boldsymbol{r}'| = r\sqrt{1 - 2\frac{r'}{r}\cos\theta + \left(\frac{r'}{r}\right)^2} \approx r\left(1 - \frac{r'}{r}\cos\theta\right) = r - \hat{\boldsymbol{r}}\cdot\boldsymbol{r}' ,$$

where θ is the angle between $\boldsymbol{r}$ and $\boldsymbol{r}'$, and we have only kept the lowest terms in the Taylor expansion of the square root. Thus, the leading term of the expansion of the vector potential in the limit $r \gg r'$ becomes

$$\boldsymbol{A}(\boldsymbol{r}) \underset{r\gg r'}{\longrightarrow} \left(\frac{e^{ikr}}{4\pi r}\right) \mu \int e^{-ik\hat{\boldsymbol{r}}\cdot\boldsymbol{r}'} \boldsymbol{J}(\boldsymbol{r}')\, d^3r' \,.$$

The magnetic field in the far-field zone can be computed from

$$\boldsymbol{H} = \frac{1}{\mu}\nabla\times\boldsymbol{A} \longrightarrow \frac{1}{\mu} ik\hat{\boldsymbol{r}}\times\boldsymbol{A} ,$$

where we have used that far away from the sources the outgoing electromagnetic fields have the form of plane waves that propagate in the direction of $\hat{\boldsymbol{r}}$. Thus, we find

$$\boldsymbol{H}(\boldsymbol{r}) \underset{r\gg r'}{\longrightarrow} \left(\frac{e^{ikr}}{4\pi r}\right) ik\hat{\boldsymbol{r}} \times \int e^{-ik\hat{\boldsymbol{r}}\cdot\boldsymbol{r}'} \boldsymbol{J}(\boldsymbol{r}')\, d^3r' \,. \tag{5.23}$$

We can finally use Eq. (2.40) to relate in the far-field zone the electric and magnetic fields, and obtain together with $ikZ = i\omega\mu$, where Z is the impedance, the far-field expression for the electric field

$$\boldsymbol{E}(\boldsymbol{r}) \underset{r\gg r'}{\longrightarrow} -\left(\frac{e^{ikr}}{4\pi r}\right) i\omega\mu \int e^{-ik\hat{\boldsymbol{r}}\cdot\boldsymbol{r}'} \hat{\boldsymbol{r}}\times\left[\hat{\boldsymbol{r}}\times\boldsymbol{J}(\boldsymbol{r}')\right] d^3r \tag{5.24}$$

In accordance to the far-field limit for the electric field given in Eq. (3.5) we can pull out the spherical wave amplitude and rewrite the above expression in the form

$$E_i(\boldsymbol{r}) \underset{r\gg r'}{\longrightarrow} \left(\frac{e^{ikr}}{r}\right) i\omega\mu \int \frac{e^{-ik\hat{\boldsymbol{r}}\cdot\boldsymbol{r}'}}{4\pi}\Big(\delta_{ij}-\hat{r}_i\hat{r}_j\Big) J_j(\boldsymbol{r}')\, d^3r' ,$$

where we have explicitly worked out the cross products in the integral. Comparison with the relation between the electric field and the current distribution of Eq. (5.20) allows us to write down the far-field limit of the Green's dyadics in the form

Far-field Limit of Dyadic Green's Function

$$G_{ij}(\boldsymbol{r},\boldsymbol{r}') \underset{r\gg r'}{\longrightarrow} \left(\frac{e^{ikr}}{r}\right) \frac{e^{-ik\hat{\boldsymbol{r}}\cdot\boldsymbol{r}'}}{4\pi}\Big(\delta_{ij}-\hat{r}_i\hat{r}_j\Big) . \tag{5.25}$$

5.3.2 *Representation Formula for Wave Equation*

For the situation depicted in Figs. 5.1 and 5.2 one can obtain, in close analogy to the Helmholtz equation, a representation formula that relates the electric field $\boldsymbol{E}_j(\boldsymbol{r})$ at any given space point to the tangential electric and magnetic fields at the boundary $\partial\Omega$ of a given volume,

Representation Formula for Maxwell's Equations

$$\begin{aligned} \boldsymbol{E}(\boldsymbol{r}) &= \boldsymbol{E}_j^{\rm inc}(\boldsymbol{r}) - \tau_j \oint_{\partial\Omega} \\ &\times \Big\{ i\omega\mu\, \bar{\bar{G}}_j(\boldsymbol{r},\boldsymbol{s}')\cdot\hat{\boldsymbol{n}}'\times\boldsymbol{H}(\boldsymbol{s}') - \Big[\nabla'\times\bar{\bar{G}}_j(\boldsymbol{r},\boldsymbol{s}')\Big]\cdot\hat{\boldsymbol{n}}'\times\boldsymbol{E}(\boldsymbol{s}')\Big\}\, dS' . \end{aligned} \tag{5.26}$$

Here $r \in \Omega_j$ is located either inside or outside the volume Ω, and the sign of $\tau_{1,2} = \pm 1$ must be chosen accordingly. Details about the derivation are given in Sect. 5.5. In the above expression we have introduced the incoming fields

$$\boldsymbol{E}_j^{\rm inc}(\boldsymbol{r}) = i\omega\mu \int_{\Omega_j} \bar{\bar{G}}_j(\boldsymbol{r},\boldsymbol{r}')\cdot\boldsymbol{J}(\boldsymbol{r}')\, d^3r' \tag{5.27}$$

produced by the current sources in Ω_j. Equation (5.26) is a quite remarkable relation which states that from the knowledge of the tangential electric and magnetic fields at the boundaries of a volume, it is possible to compute the electromagnetic fields everywhere else.

5.4 Optical Theorem

In this section we prove the optical theorem of Sect. 4.4. In our proof we proceed in two steps. First, we compute the far-field limit for the representation formula of Eq. (5.26). We then compare this expression with P_{ext} of Eq. (4.25) and show that the two expressions can be related to each other.

Far-field Limit of Representation Formula

From the relation between the electric field and the Green's dyadics, Eq. (5.20), we find that the first term in the curly brackets of the representation formula can be interpreted as an electric field for a source $\hat{\boldsymbol{n}} \times \boldsymbol{H}(\boldsymbol{r})$. Similarly, comparison with Eq. (5.22) shows that the second term corresponds to a magnetic field for a source $\hat{\boldsymbol{n}} \times \boldsymbol{E}(\boldsymbol{r})$. We can thus directly employ the far-field limits for the electromagnetic fields of Eqs. (5.23) and (5.24) and obtain the far-field limit of the representation formula outside of volume Ω

$$\begin{aligned} \boldsymbol{E}(\boldsymbol{r}) &\underset{r\gg r'}{\longrightarrow} \boldsymbol{E}_{\text{inc}}(\boldsymbol{r}) \\ &+\left(\frac{e^{ikr}}{4\pi r}\right)\hat{\boldsymbol{r}}\times\oint_{\partial\Omega} e^{-ik\hat{\boldsymbol{r}}\cdot\boldsymbol{r}'}\Big\{-i\omega\mu\,\hat{\boldsymbol{r}}\times\big[\hat{\boldsymbol{n}}'\times\boldsymbol{H}(\boldsymbol{r}')\big]+ik\hat{\boldsymbol{n}}'\times\boldsymbol{E}(\boldsymbol{r}')\Big\}dS'\,. \end{aligned} \tag{5.28}$$

To facilitate the comparison with the extinction power, we now assume that in the far-field limit the outgoing wave propagates in the direction $\hat{\boldsymbol{k}}_0$ and multiply both sides of the equation with $\boldsymbol{\epsilon}_0^*$. The term with the tangential magnetic field can be simplified using

$$\boldsymbol{\epsilon}_0^*\cdot\hat{\boldsymbol{k}}_0\times\left(\hat{\boldsymbol{k}}_0\times\boldsymbol{u}\right)=\boldsymbol{\epsilon}_0^*\cdot\left[\hat{\boldsymbol{k}}_0\left(\hat{\boldsymbol{k}}_0\cdot\boldsymbol{u}\right)-\boldsymbol{u}\right]=-\boldsymbol{\epsilon}_0^*\cdot\boldsymbol{u}\,,$$

where we have used $\boldsymbol{\epsilon}_0^*\cdot\hat{\boldsymbol{k}}_0=0$ and $\boldsymbol{u}$ is an arbitrary vector. Thus, we get

$$\begin{aligned} \boldsymbol{\epsilon}_0^*\cdot\boldsymbol{E}\big(r\hat{\boldsymbol{k}}_0\big) &\underset{r\gg r'}{\longrightarrow} \boldsymbol{\epsilon}_0^*\cdot\boldsymbol{E}_{\text{inc}}\big(r\hat{\boldsymbol{k}}_0\big) \\ &+\left(\frac{e^{ikr}}{4\pi r}\right)i\omega\mu\oint_{\partial\Omega} e^{-ik_0\cdot\boldsymbol{r}'}\boldsymbol{\epsilon}_0^*\cdot\Big\{\hat{\boldsymbol{n}}'\times\boldsymbol{H}(\boldsymbol{r}')+Z^{-1}\hat{\boldsymbol{k}}_0\times\big[\hat{\boldsymbol{n}}'\times\boldsymbol{E}(\boldsymbol{r}')\big]\Big\}dS'\,, \end{aligned} \tag{5.29}$$

with $ik=i\mu\omega Z^{-1}$. In the spirit of our derivation for the optical cross sections, we decompose the electric field $\boldsymbol{E}=\boldsymbol{E}_{\text{inc}}+\boldsymbol{E}_{\text{sca}}$ into incoming and scattered contributions, with a similar expression for the magnetic field. As discussed in Exercise 5.9 we can replace in the boundary integral of Eq. (5.29) the total fields by the scattered fields. Thus, we get for the far-field amplitude of Eq. (4.26) the final expression

$$\epsilon_0^* \cdot \boldsymbol{F}(\hat{\boldsymbol{k}}_0) = \frac{ikZ}{4\pi} \oint_{\partial\Omega} e^{-i\boldsymbol{k}_0\cdot\boldsymbol{r}'} \epsilon_0^* \cdot \left\{ \hat{\boldsymbol{n}}' \times \boldsymbol{H}_{\rm sca}(\boldsymbol{r}') + Z^{-1}\hat{\boldsymbol{k}}_0 \times \left[\hat{\boldsymbol{n}}' \times \boldsymbol{E}_{\rm sca}(\boldsymbol{r}')\right]\right\} dS' . \quad (5.30)$$

Extinction Power

We next investigate the extinction power of Eq. (4.25)

$$P_{\rm ext} = \frac{1}{2} \oint_{\partial\Omega} {\rm Re}\left[E_0^* e^{-i\boldsymbol{k}_0\cdot\boldsymbol{r}} \left\{ \boldsymbol{H}_{\rm sca}(\boldsymbol{r}) \times \epsilon_0^* + Z^{-1}\left(\hat{\boldsymbol{k}}_0 \times \epsilon_0^*\right) \times \boldsymbol{E}_{\rm sca}(\boldsymbol{r})\right\}\right] \cdot d\boldsymbol{S} ,$$

where for convenience we have reversed the order of the two contributions in brackets as well as the order of the cross products. The last term in brackets can be simplified by use of the cyclic permutation for triple products and through the following manipulation:

$$(\hat{\boldsymbol{k}}_0 \times \epsilon_0^*) \cdot (\boldsymbol{E}_{\rm sca} \times \hat{\boldsymbol{n}}) = \epsilon_0^* \cdot \left[\hat{\boldsymbol{n}}(\hat{\boldsymbol{k}}_0 \cdot \boldsymbol{E}_{\rm sca}) - \boldsymbol{E}_{\rm sca}(\hat{\boldsymbol{k}}_0 \cdot \hat{\boldsymbol{n}})\right] = \epsilon_0^* \cdot \hat{\boldsymbol{k}}_0 \times (\hat{\boldsymbol{n}} \times \boldsymbol{E}_{\rm sca}) .$$

Thus, we are finally led to

$$P_{\rm ext} = \frac{1}{2} \oint_{\partial\Omega} {\rm Re}\left[E_0^* e^{-i\boldsymbol{k}_0\cdot\boldsymbol{r}} \epsilon_0^* \cdot \left\{ \hat{\boldsymbol{n}} \times \boldsymbol{H}_{\rm sca}(\boldsymbol{r}) + Z^{-1}\hat{\boldsymbol{k}}_0 \times \left(\hat{\boldsymbol{n}} \times \boldsymbol{E}_{\rm sca}(\boldsymbol{r})\right)\right\}\right] dS . \quad (5.31)$$

Comparison with Eq. (5.30) shows that $P_{\rm ext}$ can be related to the far-field amplitude according to

$$P_{\rm ext} = {\rm Re}\left[\frac{2\pi}{ikZ} E_0 \epsilon_0^* \cdot \boldsymbol{F}(\hat{\boldsymbol{k}}_0)\right] = \frac{2\pi}{k} Z^{-1} {\rm Im}\left[E_0 \epsilon_0^* \cdot \boldsymbol{F}(\hat{\boldsymbol{k}}_0)\right] ,$$

in accordance to Eq. (4.27). This completes our proof of the optical theorem.

5.5 Details for Representation Formula of Wave Equation

In this section we show how to arrive at the representation formula for Maxwell's equations of Eq. (5.26). Our derivation of the representation formula closely follows the book of Chew [20]. We start with the wave equation for the electric field, which we multiply from the right-hand side with the Green's dyadics,

$$\Big[-\nabla'\times\nabla'\times\boldsymbol{E}(\boldsymbol{r}')\Big]\cdot\bar{\bar{G}}(\boldsymbol{r}',\boldsymbol{r})+k^2\,\boldsymbol{E}(\boldsymbol{r}')\cdot\bar{\bar{G}}(\boldsymbol{r}',\boldsymbol{r})=-i\omega\mu\boldsymbol{J}(\boldsymbol{r}')\cdot\bar{\bar{G}}(\boldsymbol{r}',\boldsymbol{r})\,.$$

For later convenience we have written the wave equation for the primed positions $\boldsymbol{r}'$ rather than the unprimed ones. Similarly, we multiply the wave equation for the Green's dyadics from the left-hand side with $\boldsymbol{E}(\boldsymbol{r}')$,

$$-\boldsymbol{E}(\boldsymbol{r}')\cdot\nabla'\times\nabla'\times\bar{\bar{G}}(\boldsymbol{r}',\boldsymbol{r})+k^2\,\boldsymbol{E}(\boldsymbol{r}')\cdot\bar{\bar{G}}(\boldsymbol{r}',\boldsymbol{r})=-\boldsymbol{E}(\boldsymbol{r}')\delta(\boldsymbol{r}'-\boldsymbol{r})\,.$$

We next integrate the two equations over $\boldsymbol{r}'$ and subtract them, to finally arrive at

$$\int_\Omega\Big\{\Big[-\nabla'\times\nabla'\times\boldsymbol{E}(\boldsymbol{r}')\Big]\cdot\bar{\bar{G}}(\boldsymbol{r}',\boldsymbol{r})+\boldsymbol{E}(\boldsymbol{r}')\cdot\nabla'\times\nabla'\times\bar{\bar{G}}(\boldsymbol{r}',\boldsymbol{r})\Big\}\,d^3r'$$
$$=-\boldsymbol{E}_{\rm inc}(\boldsymbol{r})+\boldsymbol{E}(\boldsymbol{r})\,.$$

We here consider only the case that $\boldsymbol{r}$ and $\boldsymbol{r}'$ are located in the same volume. The term in curly brackets can be rewritten in the form

$$-\nabla'\cdot\Big\{[\nabla'\times\boldsymbol{E}(\boldsymbol{r}')]\times\bar{\bar{G}}(\boldsymbol{r}',\boldsymbol{r})+\boldsymbol{E}(\boldsymbol{r}')\times[\nabla'\times\bar{\bar{G}}(\boldsymbol{r}',\boldsymbol{r})]\Big\}\,. \tag{5.32}$$

To prove this, we use in Eq. (5.32) the vector identity

$$\nabla\cdot\boldsymbol{a}\times\boldsymbol{b}=(\nabla\times\boldsymbol{a})\cdot\boldsymbol{b}-\boldsymbol{a}\cdot(\nabla\times\boldsymbol{b})\,.$$

Suppressing for simplicity the dependence of $\boldsymbol{E}$ and $\bar{\bar{G}}$ on $\boldsymbol{r}$, $\boldsymbol{r}'$ we get

$$-\Big\{\nabla'\times\nabla'\times\boldsymbol{E})\cdot\bar{\bar{G}}-(\nabla'\times\boldsymbol{E})\cdot(\nabla'\times\bar{\bar{G}})+(\nabla'\times\boldsymbol{E})\cdot(\nabla'\times\bar{\bar{G}})-\boldsymbol{E}\cdot(\nabla'\times\nabla'\times\bar{\bar{G}})\Big\}\,,$$

which completes our proof. We next use this modified expression for the term in curly brackets inside the volume integral, and use Gauss' theorem to obtain a boundary integral of the form

$$\boldsymbol{E}(\boldsymbol{r})=\boldsymbol{E}_{\rm inc}(\boldsymbol{r})$$
$$-\oint_{\partial\Omega}\Big\{[\nabla'\times\boldsymbol{E}(\boldsymbol{r}')]\times\bar{\bar{G}}(\boldsymbol{r}',\boldsymbol{r})+\boldsymbol{E}(\boldsymbol{r}')\times\Big[\nabla'\times\bar{\bar{G}}(\boldsymbol{r}',\boldsymbol{r})\Big]\Big\}\cdot\hat{\boldsymbol{n}}'dS'\,, \tag{5.33}$$

where we have explicitly indicated the surface normal $\hat{\boldsymbol{n}}'$ of the boundary. Equation (5.33) has been first derived by Stratton and Chu [21], and correspondingly equations of the above form are often named after these authors. The first term in curly brackets can be simplified by performing a cyclic permutation of the triple product and using Faraday's law for the curl of $\boldsymbol{E}$,

$$\Big[\nabla'\times\boldsymbol{E}(\boldsymbol{r}')\Big]\times\bar{\bar{G}}(\boldsymbol{r}',\boldsymbol{r})\cdot\hat{\boldsymbol{n}}'=i\omega\mu\hat{\boldsymbol{n}}'\times\boldsymbol{H}(\boldsymbol{r}')\cdot\bar{\bar{G}}(\boldsymbol{r}',\boldsymbol{r})\,.$$

In the second term we perform a cyclic permutation of the triple product

$$\boldsymbol{E}(\boldsymbol{r}')\times\left[\nabla'\times\bar{\bar{G}}(\boldsymbol{r}',\boldsymbol{r})\right]\cdot\hat{\boldsymbol{n}}'=\hat{\boldsymbol{n}}'\times\boldsymbol{E}(\boldsymbol{r}')\cdot\nabla'\times\bar{\bar{G}}(\boldsymbol{r}',\boldsymbol{r})\,.$$

Inserting these expressions into Eq. (5.32) then gives

$$\boldsymbol{E}(\boldsymbol{r})=\boldsymbol{E}_{\text{inc}}(\boldsymbol{r})\\-\oint_{\partial\Omega}\Big\{i\omega\mu\hat{\boldsymbol{n}}'\times\boldsymbol{H}(\boldsymbol{r}')\cdot\bar{\bar{G}}(\boldsymbol{r}',\boldsymbol{r})+\hat{\boldsymbol{n}}'\times\boldsymbol{E}(\boldsymbol{r}')\cdot\nabla'\times\bar{\bar{G}}(\boldsymbol{r}',\boldsymbol{r})\Big\}dS'\,.$$

We finally use that $\bar{\bar{G}}$ is a symmetric tensor (see also Sect. 7.4) to exchange the order between the two terms in the scalar products. Together with

$$\left[\nabla'\times\bar{\bar{G}}(\boldsymbol{r}',\boldsymbol{r})\right]_{ij}=\left[\nabla'\times G(\boldsymbol{r}',\boldsymbol{r})\mathbb{1}\right]_{ij}=-\left[\nabla'\times\bar{\bar{G}}(\boldsymbol{r},\boldsymbol{r}')\right]_{ji}\,,$$

where we have used that the curl of a gradient is automatically zero, we finally arrive at the representation formula of Eq. (5.26), which we repeat here for completeness

$$\boldsymbol{E}(\boldsymbol{r})=\boldsymbol{E}_{\text{inc}}(\boldsymbol{r})\\-\tau_j\oint_{\partial\Omega}\Big\{i\omega\mu\,\bar{\bar{G}}(\boldsymbol{r},\boldsymbol{s}')\cdot\hat{\boldsymbol{n}}'\times\boldsymbol{H}(\boldsymbol{s}')-\left[\nabla'\times\bar{\bar{G}}(\boldsymbol{r},\boldsymbol{s}')\right]\cdot\hat{\boldsymbol{n}}'\times\boldsymbol{E}(\boldsymbol{s}')\Big\}\,dS'\,.$$

We have introduced the factor $\tau_{1,2}=\pm1$ in front of the boundary integral to account for the fact that for $\boldsymbol{r}\in\Omega_2$ the normal vector of the boundary points into the volume. This point is discussed in more detail in the context of Eq. (5.14).

Single and Double Layer Potentials

The representation formula can be written in a more compact form. In the following we introduce for the tangential electromagnetic fields the abbreviations $\boldsymbol{u}_E=\hat{\boldsymbol{n}}\times\boldsymbol{E}$ and $\boldsymbol{u}_H=\hat{\boldsymbol{n}}\times\boldsymbol{H}$.

Single Layer. We start with the first expression in brackets, for which we define the integral operator $\mathbb{S}$ defined through (we suppress for $\boldsymbol{u}$ the electric field subscript)

$$[\mathbb{S}\,\boldsymbol{u}](\boldsymbol{r})=\oint_{\partial\Omega}\left[\left(\mathbb{1}+\frac{\nabla\nabla}{k^2}\right)G(\boldsymbol{r},\boldsymbol{s}')\right]\cdot\boldsymbol{u}(\boldsymbol{s}')\,dS'\,.$$

For the second term we use $\nabla G(\boldsymbol{r},\boldsymbol{r}')=-\nabla'G(\boldsymbol{r},\boldsymbol{r}')$ together with

$$(\nabla G)\cdot\boldsymbol{u}=-\left(\nabla'G\right)\cdot\boldsymbol{u}=G\,\nabla'\cdot\boldsymbol{u}-\nabla'\cdot(G\boldsymbol{u})\,.$$

The last term can be transformed with a two-dimensional version of Gauss' law[1], Eq. (2.12), from a boundary to a line integral which becomes zero for a closed boundary. Thus, we arrive at

$$\big[\mathbb{S}\,\boldsymbol{u}\big](\boldsymbol{r}) = \oint_{\partial\Omega}\left[G(\boldsymbol{r},\boldsymbol{s}')\boldsymbol{u}(\boldsymbol{s}') + \frac{1}{k^2}\nabla G(\boldsymbol{r},\boldsymbol{s}')\,\nabla'\cdot\boldsymbol{u}(\boldsymbol{s}')\right] dS'\,. \tag{5.34}$$

In the mathematical literature one usually refers to $\mathbb{S}$ for formal reasons as the *single layer* integral operator.

Double Layer. For the second term in the brackets of Eq. (5.26) we start with

$$\nabla'\times\left(\mathbb{1}+\frac{\nabla\nabla}{k^2}\right)G(\boldsymbol{r},\boldsymbol{r}') = -\nabla\times\left(\mathbb{1}+\frac{\nabla\nabla}{k^2}\right)G(\boldsymbol{r},\boldsymbol{r}') = -\nabla\times\mathbb{1}G(\boldsymbol{r},\boldsymbol{r}')\,.$$

In the last step we have used that the curl of a gradient is automatically zero. We thus end up with the evaluation of the integral

$$\big[\mathbb{D}\,\boldsymbol{u}\big](\boldsymbol{r}) = \oint_{\partial\Omega}\nabla'\times\bar{\bar{G}}(\boldsymbol{r},\boldsymbol{s}')\cdot\boldsymbol{u}(\boldsymbol{s}')\,dS' = -\oint_{\partial\Omega}\nabla\times G(\boldsymbol{r},\boldsymbol{s}')\boldsymbol{u}(\boldsymbol{s}')\,dS'\,, \tag{5.35}$$

where we have introduced $\mathbb{D}$, which in the mathematical literature is usually denoted as the *double layer* integral operator.

Relation Between Operators. By computing the curl of the single layer potential in Eq. (5.35) we arrive at

$$\nabla\times\big[\mathbb{S}\,\boldsymbol{u}\big](\boldsymbol{r}) = \oint_{\partial\Omega}\nabla\times G(\boldsymbol{r},\boldsymbol{s}')\boldsymbol{u}(\boldsymbol{s}')\,dS' = -\big[\mathbb{D}\,\boldsymbol{u}\big](\boldsymbol{r})\,.$$

We have used that in the second term of the single layer operator the curl of a gradient is always zero. Similarly, by computing the curl of the double layer potential in Eq. (5.35) we get

$$\begin{aligned}
\nabla\times\big[\mathbb{D}\,\boldsymbol{u}\big](\boldsymbol{s}) &= -\oint_{\partial\Omega}\nabla\times\nabla\times G(\boldsymbol{r},\boldsymbol{s}')\boldsymbol{u}(\boldsymbol{s}')\,dS' \\
&= -\oint_{\partial\Omega}\Big\{\nabla\big[\nabla\cdot G(\boldsymbol{r},\boldsymbol{s}')\boldsymbol{u}(\boldsymbol{s}')\big] - \nabla^2 G(\boldsymbol{r},\boldsymbol{s}')\boldsymbol{u}(\boldsymbol{s}')\Big\}\,dS' \\
&= -k^2\big[\mathbb{S}\,\boldsymbol{u}\big](\boldsymbol{r})\,,
\end{aligned}$$

[1] In the field of differential geometry there exists the so-called Stokes–Cartan theorem, from which one can derive the laws of Gauss, Stokes, and Green. This framework allows a generalization of these laws to arbitrary dimensions and to curved space. One example is the two-dimensional version of Gauss' law for a curved boundary, which we use in this chapter without explicit proof.

where we have used the defining equation of the scalar Green's function, Eq. (5.5) for $\boldsymbol{r} \neq \boldsymbol{r}'$ to rewrite the second term in brackets. Thus, we find for the relation between the single and double layer integral operators

$$\nabla \times \big[\mathbb{S}\,\boldsymbol{u}\big](\boldsymbol{r}) = -\big[\mathbb{D}\,\boldsymbol{u}\big](\boldsymbol{r})\,, \quad \nabla \times \big[\mathbb{D}\,\boldsymbol{u}\big](\boldsymbol{r}) = -k^2\,\big[\mathbb{S}\,\boldsymbol{u}\big](\boldsymbol{r})\,. \tag{5.36}$$

Using the representation formula of the electric field together with Faraday's law $i\omega\mu\boldsymbol{H} = \nabla \times \boldsymbol{E}$, we can rewrite Eq. (5.26) and the corresponding expression for the magnetic field in the compact form

Representation Formulas for Maxwell's Equations (II)

$$\boldsymbol{E}(\boldsymbol{r}) = \boldsymbol{E}_j^{\rm inc}(\boldsymbol{r}) - \tau_j\Big\{+i\omega\mu\big[\mathbb{S}\,\hat{\boldsymbol{n}} \times \boldsymbol{H}\big](\boldsymbol{r}) - \big[\mathbb{D}\,\hat{\boldsymbol{n}} \times \boldsymbol{E}\big](\boldsymbol{r})\Big\}$$

$$\boldsymbol{H}(\boldsymbol{r}) = \boldsymbol{H}_j^{\rm inc}(\boldsymbol{r}) - \tau_j\Big\{-i\omega\varepsilon\big[\mathbb{S}\,\hat{\boldsymbol{n}} \times \boldsymbol{E}\big](\boldsymbol{r}) - \big[\mathbb{D}\,\hat{\boldsymbol{n}} \times \boldsymbol{H}\big](\boldsymbol{r})\Big\}\,. \tag{5.37}$$

Here $j = 1, 2$ has to be used for $\boldsymbol{r} \in \Omega_{1,2}$, with $\tau_{1,2} = \pm 1$, and the integral operators $\mathbb{S}$, $\mathbb{D}$ are defined in Eqs. (5.34), (5.35). $\boldsymbol{E}_j^{\rm inc}(\boldsymbol{r})$, $\boldsymbol{H}_j^{\rm inc}(\boldsymbol{r})$ are the electromagnetic fields produced by the current sources located in volume Ω_j. We will return to these representation formulas in Chap. 9 when discussing the boundary integral method approach.

Exercises

Exercise 5.1 Prove Green's theorem

$$\int_\Omega \left(\phi\nabla^2\psi - \psi\nabla^2\phi\right)\,d^3r = \oint_{\partial\Omega}\left(\phi\frac{\partial\psi}{\partial n} - \psi\frac{\partial\phi}{\partial n}\right)\,dS\,.$$

Show that the term in parentheses can be written in the form $\nabla \cdot (\phi\nabla\psi - \psi\nabla\phi)$, and use Gauss' theorem of Eq. (2.12) to convert the volume to a boundary integral.

Exercise 5.2 Derive the advanced Green's function for the Helmholtz equations. Start from Eq. (5.6) and keep only the second term with the incoming spherical wave. Repeat the derivation to obtain D.

Exercise 5.3 Decompose within $\boldsymbol{r} \in \Omega_2$ the solution of the Helmholtz equation into incoming and scattered fields

$$f(\boldsymbol{r}) = f_{\rm inc}(\boldsymbol{r}) + f_{\rm sca}(\boldsymbol{r})\,.$$

Show that on the right-hand side of the representation formula of Eq. (5.14), one can use instead of the total fields also the scattered fields. In proving this, you will additionally need Eq. (5.12) for $\boldsymbol{r} \in \Omega_2$.

Exercise 5.4 Transform the Green's function for the Helmholtz equation

$$g(R) = \frac{e^{\pm ikR}}{4\pi R}$$

to the time domain using a Fourier transformation together with the dispersion relation $\omega = kc$. Compare the results for the retarded (positive sign) and advanced (negative sign) Green's functions, and give an interpretation.

Exercise 5.5 Derive the Green's function for the one-dimensional Helmholtz equation

$$\left(\frac{d^2}{dx^2} + k^2\right) G(x) = -\delta(x)\,.$$

Show that by integrating this equation over a small region around the origin one obtains the boundary conditions

$$G(0^+) = G(0^-)\,, \quad \left.\frac{dG}{dx}\right|_{0^+} - \left.\frac{dG}{dx}\right|_{0^-} = -1\,,$$

where $0^\pm$ denote positions infinitesimally larger or smaller than zero. To obtain the retarded Green's function, select the solutions corresponding to outgoing waves at infinity.

Exercise 5.6 Proof the defining equation (5.21) for the Green's function. Rewrite the equation in the form

$$-\Big(\epsilon_{ikl}\,\partial_k\Big)\Big(\epsilon_{lmn}\,\partial_m\Big)G_{ni}(\boldsymbol{r},\boldsymbol{r}') + k^2\,G_{ij}(\boldsymbol{r},\boldsymbol{r}') = -\delta_{ij}\delta(\boldsymbol{r}-\boldsymbol{r}')\,,$$

and express G using Eq. (5.19).

Exercise 5.7 Show that the Green's dyadics $\bar{\bar{G}}(\boldsymbol{r},\boldsymbol{r}')$ of Eq. (5.19) can be written in the form [6, Eq. (8.55)]

$$G_{ij}(\boldsymbol{r},\boldsymbol{r}') = \left[\Big(\delta_{ij} - \hat{R}_i\hat{R}_j\Big) + i\Big(\frac{\delta_{ij} - 3\hat{R}_i\hat{R}_j}{kR}\Big) - \Big(\frac{\delta_{ij} - 3\hat{R}_i\hat{R}_j}{k^2R^2}\Big)\right]\frac{e^{ikR}}{4\pi R}\,,$$

with $\boldsymbol{R} = \boldsymbol{r} - \boldsymbol{r}'$ and the unit vector $\hat{\boldsymbol{R}} = \boldsymbol{R}/R$.

Exercise 5.8 Use the result of Exercise 5.7 to show that the electric field

$$\boldsymbol{E}(\boldsymbol{r}) = \mu_0\omega^2\bar{\bar{G}}(\boldsymbol{r},0)\cdot\boldsymbol{p}$$

of an electric dipole $\boldsymbol{p}$ located at the origin can be rewritten in the form

$$\boldsymbol{E} = \frac{1}{4\pi\varepsilon_0}\left\{k^2(\hat{\boldsymbol{r}}\times\boldsymbol{p})\times\hat{\boldsymbol{r}}\frac{e^{ikr}}{r} + \left[3(\hat{\boldsymbol{r}}\cdot\boldsymbol{p})\hat{\boldsymbol{r}}-\boldsymbol{p}\right]\left(\frac{1}{r^3}-\frac{ik}{r^2}\right)e^{ikr}\right\}.$$

Exercise 5.9 Show that in Eq. (5.29) one can replace the total fields by the scattered fields. Consider the contribution for the incoming fields

$$\oint_{\partial\Omega} e^{-i\boldsymbol{k}_0\cdot\boldsymbol{r}}\boldsymbol{\epsilon}_0^*\cdot\left\{\hat{\boldsymbol{n}}\times\left(\hat{\boldsymbol{k}}_0\times\boldsymbol{E}_{\text{inc}}\right)+\hat{\boldsymbol{k}}_0\times\left(\hat{\boldsymbol{n}}\times\boldsymbol{E}_{\text{inc}}\right)\right\}dS\,,$$

where $\boldsymbol{E}_{\text{inc}} = E_0\boldsymbol{\epsilon}_0 e^{i\boldsymbol{k}_0\cdot\boldsymbol{r}}$. Work out the terms in the brackets using $\hat{\boldsymbol{k}}_0\cdot\boldsymbol{\epsilon}_0 = 0$ and show that the only non-vanishing term is proportional to $\oint_{\partial\Omega} d\boldsymbol{S} = 0$.

Exercise 5.10 Derive the representation formula of Eq. (5.26) for the wave equation using the volume Ω but a position $\boldsymbol{r}\notin\Omega$. You may like to consult the corresponding discussion for the Helmholtz equation. Use the result to show that

$$\oint_{\partial\Omega}\left\{i\omega\mu\,\bar{\bar{G}}(\boldsymbol{r},\boldsymbol{s}')\cdot\hat{\boldsymbol{n}}'\times\boldsymbol{H}_{\text{inc}}(\boldsymbol{s}')-\left[\nabla'\times\bar{\bar{G}}(\boldsymbol{r},\boldsymbol{s}')\right]\cdot\hat{\boldsymbol{n}}'\times\boldsymbol{E}_{\text{inc}}(\boldsymbol{s}')\right\}dS' = 0\,.$$

Chapter 6
Diffraction Limit and Beyond

We have now all ingredients at hand to discuss the diffraction limit of light. In this chapter we start by analyzing imaging of a single oscillating dipole, and then ponder on implications of the diffraction limit for usual optical microscopy. We continue to discuss scanning nearfield optical microscopy (SNOM), a technique that allows imaging of optical nearfields, as well as optical localization microscopy, for which the Nobel Prize in chemistry 2014 has been awarded.

6.1 Imaging a Single Dipole

We start by considering a single dipole located at position $\boldsymbol{r}_0$ and imaged through a lens system, which we describe using the techniques developed in Chap. 3. For the far-fields of the dipole radiation we employ the Green's functions introduced in the last chapter. Altogether, our approach can be broken up into the following steps (see also Fig. 6.1):

– Compute current distribution of dipole,
– compute far-fields of dipole using Eq. (5.24),
– submit the far-fields to the imaging expression of Eq. (3.10), and
– perform for small numerical apertures all integrals analytically.

Our starting expression is the current distribution of an oscillating dipole whose extension is much smaller than any other relevant length scale of the problem under study, which can be written as

U. Hohenester, *Nano and Quantum Optics*, Graduate Texts in Physics,
https://doi.org/10.1007/978-3-030-30504-8_6

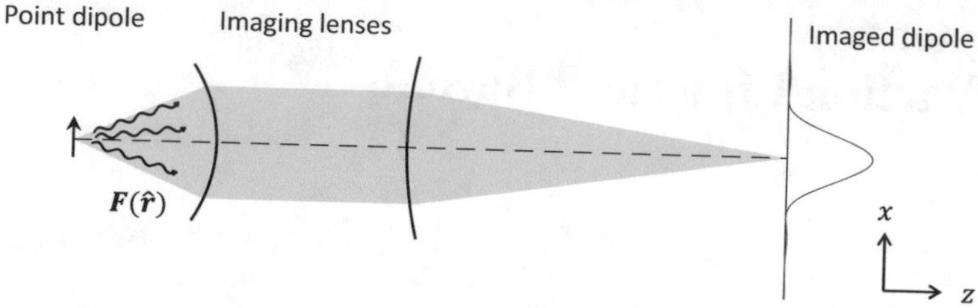

Fig. 6.1 Imaging of point dipole. A point dipole is located in the focus of a lens (sketched through its Gaussian reference sphere, see Fig. 3.4) that collects the far-fields $\boldsymbol{F}(\hat{\boldsymbol{r}})$ emitted by the dipole. A second lens produces an image with a diffraction limited spot due to the removal of evanescent fields. At the beginning of this chapter we show how to describe the imaging process theoretically

Current Distribution of Point Dipole Located at $\boldsymbol{r}_0$

$$\boldsymbol{J}(\boldsymbol{r}) = -i\omega\, \boldsymbol{p}\, \delta(\boldsymbol{r} - \boldsymbol{r}_0)\,. \tag{6.1}$$

Here $\boldsymbol{r}_0$ is the dipole position, $\boldsymbol{p}$ is the dipole moment, and ω the oscillation frequency.

Proof of Eq. (6.1) To compute the current distribution of a dipole, we start from

$$\nabla \cdot x\boldsymbol{J} = \partial_i\, x J_i = \delta_{i1} J_i + x\, \nabla \cdot \boldsymbol{J}\,.$$

The last expression can be related to the charge distribution using the continuity equation for time-harmonic fields

$$i\omega\rho = \nabla \cdot \boldsymbol{J}\,.$$

We next integrate J_1 over a small volume within which the charge distribution of the dipole is located, and get

$$\int_\Omega J_1(\boldsymbol{r})\, d^3r = \oint_{\partial\Omega} x\boldsymbol{J} \cdot d\boldsymbol{S} - i\omega \int_\Omega x\rho(\boldsymbol{r})\, d^3r = -i\omega\, p_1\,.$$

The same result can be obtained with the current distribution given in Eq. (6.1), which completes our proof. In deriving the above expression we have used that the current distribution at the boundary $\partial\Omega$ is zero, because $\boldsymbol{J}$ is confined within Ω, and have introduced the dipole moment

$$\boldsymbol{p} = \int_\Omega \boldsymbol{r}\rho(\boldsymbol{r})\, d^3r\,. \tag{6.2}$$

The electric far-fields of this distribution can be computed using Eq. (5.24), and we obtain

$$\boldsymbol{E}(\boldsymbol{r}) \underset{r \gg r_0}{\longrightarrow} -\left(\frac{e^{ikr}}{4\pi r}\right) \omega^2 \mu e^{-ik\hat{\boldsymbol{r}}\cdot\boldsymbol{r}_0} \left[\hat{\boldsymbol{r}} \times \left(\hat{\boldsymbol{r}} \times \boldsymbol{p}\right)\right] = \left(\frac{e^{ikr}}{r}\right) \boldsymbol{F}(\hat{\boldsymbol{r}}) .$$

Together with

$$-\hat{\boldsymbol{r}} \times \left(\hat{\boldsymbol{r}} \times \boldsymbol{p}\right) = \boldsymbol{p} - \hat{\boldsymbol{r}}\left(\hat{\boldsymbol{r}} \cdot \boldsymbol{p}\right) = \boldsymbol{p}_\perp$$

we then obtain for the far-field amplitude

$$\boldsymbol{F}(\hat{\boldsymbol{r}}) = \frac{k^2}{4\pi\varepsilon} \left(e^{-ik\hat{\boldsymbol{r}}\cdot\boldsymbol{r}_0}\right) \boldsymbol{p}_\perp . \tag{6.3}$$

We next submit Eq. (6.3) to the imaging expression of Eq. (3.10), which we repeat here for clarity

$$\boldsymbol{E}_3(\rho, \varphi, z) = \sqrt{\frac{n_1}{n_3}} \frac{ik_3 e^{i(k_1 f - k_3 f')}}{2\pi} \frac{f}{f'} \int_0^{\theta_{\max}} \sqrt{\cos\theta} \sin\theta \, d\theta$$

$$\times \int_0^{2\pi} d\phi \, \bar{\bar{\mathcal{R}}}^{\text{im}} \cdot \boldsymbol{F}_1(\theta, \phi) e^{-ik_3 \cos\theta_3 z} e^{ik_1 \left(\frac{\rho}{M}\right) \sin\theta \cos(\phi - \varphi)} .$$

$\boldsymbol{F}_1$ is the far-field amplitude of the dipole, $\boldsymbol{E}_3$ is the image field, and for the object and image sides we use the foci f, f', the refractive indices n_1, n_3, and the wavenumbers k_1, k_3. See also Fig. 3.5 and discussion given in Sect. 3.3 for more details. The following calculation can be simplified by noting

$$\bar{\bar{\mathcal{R}}}^{\text{im}} \cdot \boldsymbol{p}_\perp = \bar{\bar{\mathcal{R}}}^{\text{im}} \cdot \boldsymbol{p} ,$$

because the transformation matrix only projects on the directions transverse to $\hat{\boldsymbol{r}}$. We additionally assume $f' \gg f$, such that $\cos\theta_3 \approx 1$, $\sin\theta_3 \approx 0$, and

$$\cos\theta_3 = \sqrt{1 - \left(\frac{f}{f'}\right)^2 \sin^2\theta_1} \approx 1 - \frac{1}{2}\left(\frac{f}{f'}\right)^2 \sin^2\theta_1 .$$

For a dipole at the origin, $\boldsymbol{r}_0 = 0$, and oriented horizontally to the optical axis, we then get from Eq. (3.34)

$$\begin{aligned}
\bar{\bar{\mathcal{R}}}^{\text{im}} \cdot \hat{\boldsymbol{x}} &= \mathcal{R}^{\text{im}}_{xx} \hat{\boldsymbol{x}} + \mathcal{R}^{\text{im}}_{yx} \hat{\boldsymbol{y}} + \mathcal{R}^{\text{im}}_{zx} \hat{\boldsymbol{z}} \\
&\approx \frac{1}{2}\left\{(\cos\theta + 1) + (\cos\theta - 1)\cos 2\phi\right\} \hat{\boldsymbol{x}} + \frac{1}{2}(\cos\theta - 1)\sin 2\phi \, \hat{\boldsymbol{y}} \\
\bar{\bar{\mathcal{R}}}^{\text{im}} \cdot \hat{\boldsymbol{y}} &= \mathcal{R}^{\text{im}}_{xy} \hat{\boldsymbol{x}} + \mathcal{R}^{\text{im}}_{yy} \hat{\boldsymbol{y}} + \mathcal{R}^{\text{im}}_{zy} \hat{\boldsymbol{z}} \\
&\approx \frac{1}{2}(\cos\theta - 1)\sin 2\phi \, \hat{\boldsymbol{x}} + \frac{1}{2}\left\{(\cos\theta + 1) - (\cos\theta - 1)\cos 2\phi\right\} \hat{\boldsymbol{y}} ,
\end{aligned}$$

where we have set $t^\theta = t^\phi = 1$. Similarly, for a dipole oriented vertically to the optical axis we get

$$\bar{\bar{\mathcal{R}}}^{\text{im}} \cdot \hat{\boldsymbol{z}} = \mathcal{R}^{\text{im}}_{xz}\,\hat{\boldsymbol{x}} + \mathcal{R}^{\text{im}}_{yz}\,\hat{\boldsymbol{y}} + \mathcal{R}^{\text{im}}_{zz}\,\hat{\boldsymbol{z}} \approx -\sin\theta\cos\phi\,\hat{\boldsymbol{x}} - \sin\theta\sin\phi\,\hat{\boldsymbol{y}}\,.$$

We next employ Eq. (3.21) for solving the integrals over the azimuthal angles analytically, and introduce for the integration over the polar angle the functional

$$\tilde{\mathcal{I}}_n\big(g(\theta)\big) = i^n \int_0^{\theta_{\max}} e^{-ik_3 z\left[1-\frac{1}{2}\left(\frac{f}{f'}\right)^2\sin^2\theta\right]} J_n(k_1 \frac{\rho}{M}\sin\theta) g(\theta)\,\sin\theta\cos^{1/2}\theta\,d\theta\,. \tag{6.4}$$

Putting together all of the above results, we are led to the electric fields $\boldsymbol{E}_{\hat{\boldsymbol{p}}}$ on the image side produced by an oscillating dipole situated in the focus of the collection lens and oriented along $\hat{\boldsymbol{p}}$,

Imaging of Dipole Fields

$$\begin{aligned}
\boldsymbol{E}_{\hat{\boldsymbol{x}}}(\rho,\varphi,z) &= \frac{1}{2}A\left\{\tilde{\mathcal{I}}_0(1+\cos\theta)\,\hat{\boldsymbol{x}} - \tilde{\mathcal{I}}_2(1-\cos\theta)\left(\cos 2\varphi\,\hat{\boldsymbol{x}} + \sin 2\varphi\,\hat{\boldsymbol{y}}\right)\right\}\\
\boldsymbol{E}_{\hat{\boldsymbol{y}}}(\rho,\varphi,z) &= \frac{1}{2}A\left\{\tilde{\mathcal{I}}_0(1+\cos\theta)\,\hat{\boldsymbol{y}} - \tilde{\mathcal{I}}_2(1-\cos\theta)\left(\sin 2\varphi\,\hat{\boldsymbol{x}} - \cos 2\varphi\,\hat{\boldsymbol{y}}\right)\right\}\\
\boldsymbol{E}_{\hat{\boldsymbol{z}}}(\rho,\varphi,z) &= -A\,\tilde{\mathcal{I}}_1(\sin\theta)\left(\cos\varphi\,\hat{\boldsymbol{x}} + \sin\varphi\,\hat{\boldsymbol{y}}\right)\,.
\end{aligned} \tag{6.5}$$

Here we have introduced the prefactor A, which, apart from an unimportant phase ψ, reads

$$A = e^{i\psi}\,\frac{\sqrt{n_1 n_3}}{4\pi\varepsilon_0}\,k_0^3 p\left(\frac{f}{f'}\right)\,.$$

For a small numerical aperture we can approximately set $\sin\theta \approx \theta$, $\cos\theta \approx 1$. This limit corresponds to the paraxial approximation. The integrals over the Bessel functions can then be solved analytically for $z = 0$ using

$$\int u^{\nu+1} J_\nu(u)\,du = u^{\nu+1} J_{\nu+1}(u)\,.$$

We introduce the dimensionless radius

$$\tilde{\rho} = \frac{n k_0 \theta_{\max}\rho}{M} = \frac{NA}{M}\,k_0\rho\,, \tag{6.6}$$

where NA is the numerical aperture. With this, we obtain

$$\int_0^{\theta_{\max}} \theta^{\nu+1} J_\nu\left(k_1\theta\rho/M\right) d\theta = \left(\frac{M}{k_1\rho}\right)^{\nu+2} \int_0^{\tilde\rho} u^{\nu+1} J_\nu(u)\, du = \theta_{\max}^{\nu+2}\frac{J_{\nu+1}(\tilde\rho)}{\tilde\rho}\,.$$

Thus, we find for the image fields of a dipole in the paraxial approximation

Imaging of Dipole Fields in Paraxial Approximation

$$\begin{aligned}
\boldsymbol{E}_{\hat{\boldsymbol{x}}}(\rho,\varphi,0) &\approx \quad A\theta_{\max}^2\left[\frac{J_1(\tilde\rho)}{\tilde\rho}\right]\hat{\boldsymbol{x}}\\
\boldsymbol{E}_{\hat{\boldsymbol{y}}}(\rho,\varphi,0) &\approx \quad A\theta_{\max}^2\left[\frac{J_1(\tilde\rho)}{\tilde\rho}\right]\hat{\boldsymbol{y}}\\
\boldsymbol{E}_{\hat{\boldsymbol{z}}}(\rho,\varphi,0) &\approx -A\theta_{\max}^3\left[\frac{J_2(\tilde\rho)}{\tilde\rho}\right]\left(\cos\varphi\,\hat{\boldsymbol{x}}+\sin\varphi\,\hat{\boldsymbol{y}}\right)\,.
\end{aligned} \tag{6.7}$$

Figure 6.2 shows the intensity profiles for dipoles oriented (a–c) perpendicular and (d–e) parallel to the optical axis $\hat{\boldsymbol{z}}$. We observe that the point dipole is imaged to a broadened feature with a width comparable to the light wavelength. It turns out to be convenient to introduce the following quantities:

Green's Function. In analogy to the dyadic Green's function introduced in Eq. (5.19), we introduce the point spread Green's function $\bar{\bar{G}}_{\rm PSF}$

$$\boldsymbol{E}(\rho,\varphi,z) = \mu\omega^2\,\bar{\bar{G}}_{\rm PSF}(\rho,\varphi,z)\cdot\boldsymbol{p}\,, \tag{6.8}$$

which gives the image fields at position (ρ,φ,z) for a point dipole located at the origin. Comparison with Eq. (6.5) allows us to express the Green's function in the form

$$\bar{\bar{G}}_{\rm PSF} = \frac{1}{2}\mu\omega^2 A\begin{pmatrix}\tilde{\mathcal{I}}_0-\tilde{\mathcal{I}}_2\cos 2\varphi & -\tilde{\mathcal{I}}_2\sin 2\varphi & -2\tilde{\mathcal{I}}_1\cos\varphi\\ -\tilde{\mathcal{I}}_2\sin 2\varphi & \tilde{\mathcal{I}}_0+\tilde{\mathcal{I}}_2\cos 2\varphi & -2\tilde{\mathcal{I}}_1\sin\varphi\\ 0 & 0 & 0\end{pmatrix}, \tag{6.9}$$

where we have suppressed the arguments of the functions $\tilde{\mathcal{I}}_n$. Alternatively, we could also use the paraxial approximation of Eq. (6.7).

Point Spread Function. The image $I(\rho,\varphi,z)$ of a unit dipole is given by the field intensity

$$I(\rho,\varphi,z) = \left|\bar{\bar{G}}_{\rm PSF}(\rho,\varphi,z)\cdot\hat{\boldsymbol{p}}\right|^2,$$

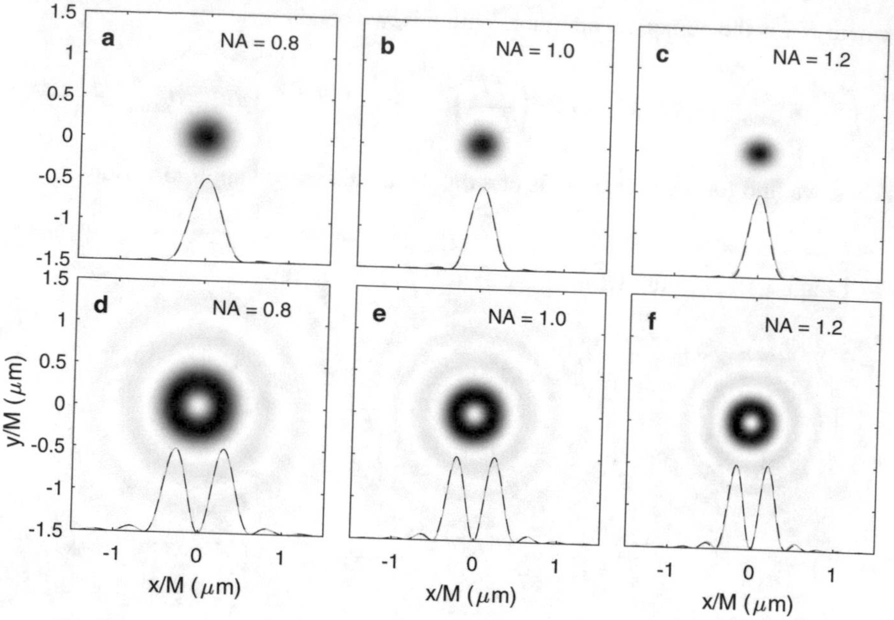

Fig. 6.2 Imaging of point dipole as computed from Eq. (6.5). The upper and lower rows report the field intensities for dipoles oriented along x and z, respectively, and for different NA values. In the bottom of the panels we show cuts through the intensity profiles (solid line) and compare them with the results of the paraxial approximation (dashed lines, almost indistinguishable). The latter ones are computed from Eq. (6.7) and are scaled to the same height. In the calculations we use $n_1 = 1.33$, $n_3 = 1$, and $\lambda = 620\,\text{nm}$

where we have dropped all constant factors. Suppose that the dipole moment is not fixed but rotates randomly, as is often the case for molecules in solution. We can then average over the different dipole orientations using

$$\big\langle I(\rho,\varphi,z)\big\rangle = \frac{1}{4\pi}\oint\Big|\bar{\bar{G}}_{\text{PSF}}(\rho,\varphi,z)\cdot\hat{\boldsymbol{p}}\Big|^2\,d\Omega = \frac{2}{3}\Big(\big|\tilde{\mathcal{I}}_0\big|^2+\big|\tilde{\mathcal{I}}_2\big|^2+2\big|\tilde{\mathcal{I}}_1\big|^2\Big)\,.$$

Here $\hat{\boldsymbol{p}}$ is the unit dipole moment and $\oint\ldots d\Omega$ denotes the integration over all orientations of the dipole. Sometimes one associates $\langle I(\boldsymbol{r},\boldsymbol{r}_0)\rangle$ with the point spread function (for a randomly oriented dipole emitter) that describes how the image of the dipole located at $\boldsymbol{r}_0$ becomes broadened in the image process. In case of multiple dipoles, we should compute from Eq. (6.8) the superposition of the electric fields, and then get the image from the square modulus of the total electric field. If field interferences can be neglected and the dipole orientations are random, one can also convolute the point spread function $\langle I(\boldsymbol{r},\boldsymbol{r}_0)\rangle$ with the positions of the dipoles.

Airy Disk. Consider the Bessel functions in Eq. (6.7) for a dipole oriented along x or y. The first zero $J_1(x_0) = 0$ of the Bessel function is at approximately $x_0 \approx 3.83$. The radius r_{Airy} where the electric field distribution becomes zero is called the Airy radius,

$$\left(\frac{NA}{M}\right)\frac{2\pi r_{\text{Airy}}}{\lambda} \approx 3.83 \implies r_{\text{Airy}} \approx 0.61\,\frac{M\lambda}{NA}\,. \tag{6.10}$$

The disk with radius r_{Airy} is correspondingly called the Airy disk.

6.2 Diffraction Limit of Light

We are now in the position to address the diffraction limit of light. The first rigorous study of the topic is probably due to Ernst Abbe [22], who used a semi-quantitative analysis to get his famous diffraction limit of light stating that two objects have a minimum resolvable distance of

Diffraction Limit of Abbe

$$d = \frac{\lambda}{2\,NA} = \frac{\lambda}{2n\sin\theta_{\text{max}}}\,. \tag{6.11}$$

Ernst Abbe

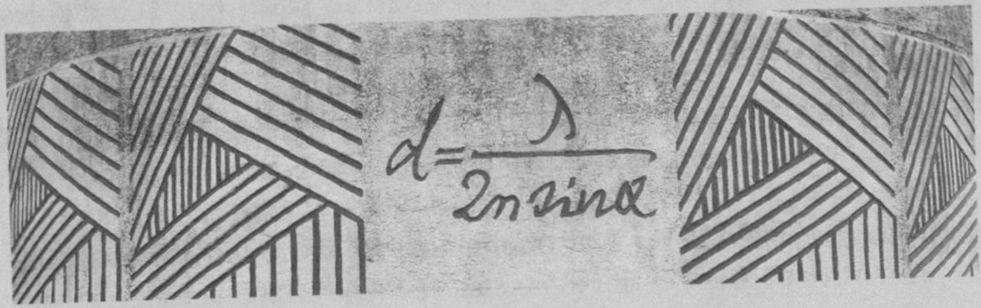

While writing on the last chapters of this book, in spring 2019 I visited Jena. Having worked on the various aspects of nano optics and the diffraction limit of light for quite a while, the sightseeing of Abbe's memorial (shown above) had an almost religious character.

Abbe was professor in Jena when he was hired by Carl Zeiss to improve the manufacturing process of optical instruments, which back then was largely based on trial and error. He devoted a lot of time to the theoretical description of optical instruments and optical imaging, derived the fundamental diffraction limit of light, introduced among many other things the numerical aperture

(continued)

term, and brought optical microscopy to an engineering science. In later years he took over organization duties at the Zeiss company and was seriously concerned with the improvement of the working and living conditions of the employees: he introduced the 8 h workday, created a pension fund and a discharge, and supported the further education of workers in intellectual and political affairs.

I was astonished how much the Zeiss company has shaped the prosperity of the city and continues to contribute to the self-perception of its citizens. Seeing in such a condensed form how a scientific idea can spread into something much bigger and affect the lives of thousands of people, was an immensely positive experience.

In obtaining the last expression we have used the definition of Eq. (3.12) for the numerical aperture. This is also the expression graved into Abbe's memorial in Jena. Sometimes one uses an estimate based on the Airy radius

$$d = 0.61\,\frac{\lambda}{NA}\,. \tag{6.12}$$

It is important to realize that while the definition of the minimal resolvable distance is somewhat arbitrary, the diffraction limit itself is fundamental and is governed by the removal of all evanescent waves from the image.

Figure 6.3 shows an image of two point dipoles that are separated by distances of the order of the wavelength, see values given in panel (a). The dipoles have orientations along x and we compare different NA values of the collection lens. Results for dipoles oriented along z are shown in Fig. 6.4. Without going too much into details, it is obvious that the diffraction limit gives a fundamental lower limit for the resolution, but even when point-like dipole emitters are separated by distances that are larger than the minimal resolvable distance it can be hard to infer from the images the precise number and orientations of these dipoles. An exhaustive and comprehensive account of the topic can be found in the book of Novotny and Hecht [6], where the authors also discuss the following issues which we only briefly address here:

Axial Resolution. So far we have only discussed the resolution that can be achieved for objects located in a given plane z. Unfortunately, the resolution in the z direction, the so-called *axial resolution*, is significantly worse. Consider first the image fields of Eq. (6.5) for $\rho = 0$. We have $J_0(0) = 1$ and $J_{1,2}(0) = 0$, so only the $\tilde{\mathcal{I}}_0$ integral differs from zero. For simplicity, we here evaluate the integral in the paraxial limit with $\sin\theta \approx \theta$, $\cos\theta \approx 1$, and arrive at

$$\tilde{\mathcal{I}}_0(1+\cos\theta) \approx \int_0^{\theta_{\max}} e^{-ik_3z[1-\frac{\eta}{2}\theta^2]}2\theta\,d\theta = \frac{2i}{k_3z\eta}e^{-ik_3z[1-\frac{\eta}{2}\theta^2]}\Big|_0^{\theta_{\max}}\,,$$

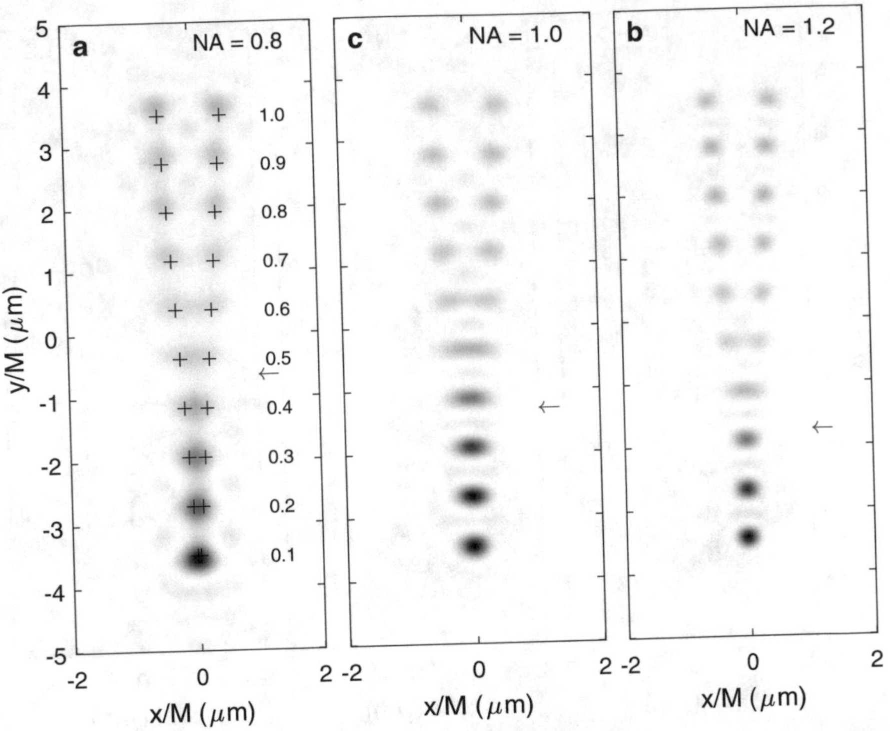

Fig. 6.3 Image of two neighbor dipoles separated by a distance reported in micrometers on the right of panel (a). The two dipoles are oriented along x. We use the same parameters as in Fig. 6.2. The arrows indicate the separation distance of $0.61\,\lambda/\mathrm{NA}$ corresponding to the diffraction limit of light

where we have used the abbreviation $\eta = (f/f')^2$. From this we get for the field intensity of a dipole oriented along x or y the expression

$$\left|\boldsymbol{E}_{\hat{\boldsymbol{x}}}(\rho = 0, z)\right|^2 \approx \frac{A^2}{4}\left(\frac{NA^2}{2n_1^2}\right)^2\left[\frac{\sin\tilde{z}}{\tilde{z}}\right]^2, \qquad \tilde{z} = \frac{k_0 NA^2}{4n_3 M^2}\,z\,. \tag{6.13}$$

We can rewrite the normalized z coordinate in the form $\tilde{z} = z/(\Delta z)$, where Δz is denoted as the depth of field, which can be directly read off from Eq. (6.13). For typical microscope parameters the axial resolution turns out to be a factor of hundred smaller than the radial resolution [6, Eq. (4.17)].

Excitation Point Spread Function. The poor axial resolution is a significant problem for microscopy of three-dimensional objects. A way to improve the resolution is to illuminate not the entire sample but only certain selected spots. By raster-scanning these spots over the sample, one obtains a complete image of the specimen, however, with improved spatial resolution. A way to quantify

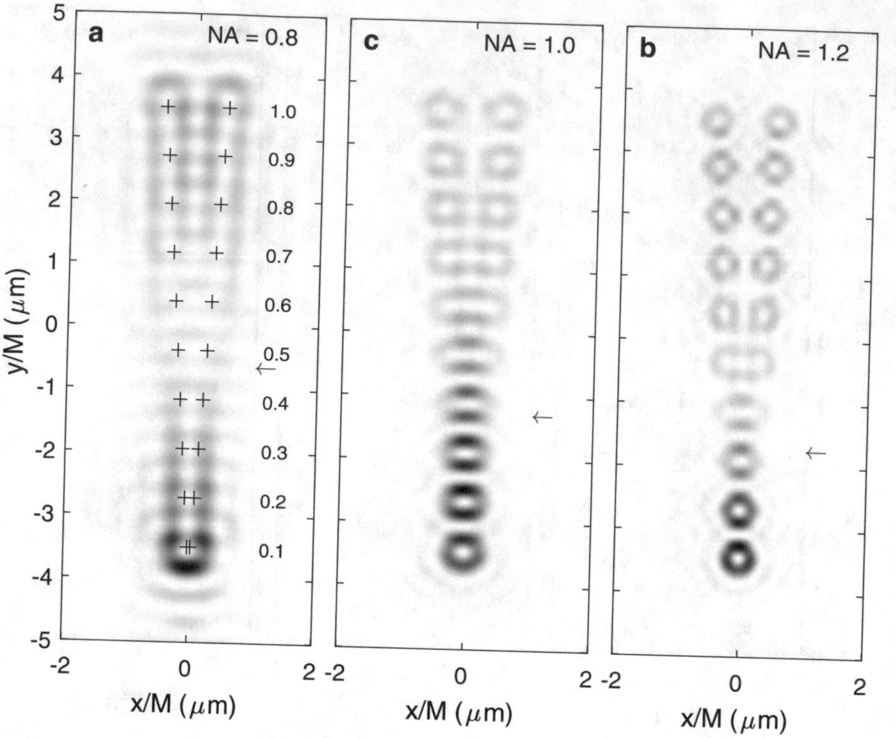

Fig. 6.4 Same as Fig. 6.3 but for dipole orientations along z

the increased resolution is given by the point spread functions (PSF), previously introduced in Eq. (6.8) and discussion thereafter. The detection PSF describes how a point source is spread by the imaging system. Similarly, we can introduce an excitation PSF that describes how the illumination light is focused on a selected spot. The total point spread function is then approximately given by the product of these PSFs,

$$(\text{total PSF}) \approx (\text{excitation PSF}) \times (\text{detection PSF})\,.$$

In general, one should be more specific on how the illuminated and detected light is generated, for instance, through coherent or incoherent scattering processes, and the tools developed in this chapter as well as in later parts of this book allow treating the entire illumination-detection process in full depth. However, in most cases of interest the product of PSFs provides an excellent approximation. To understand the essence of the PSF product, assume that both functions are Gaussians with widths σ_{exc} and σ_{det}. For the product we then have $1/\sigma_{\text{tot}} = 1/\sigma_{\text{exc}} + 1/\sigma_{\text{det}}$, thus, the total width σ_{tot} is *smaller* than the individual ones.

Confocal Microscopy. Confocal microscopy is a popular technique for combining selective excitation with high-resolution detection to achieve an increased

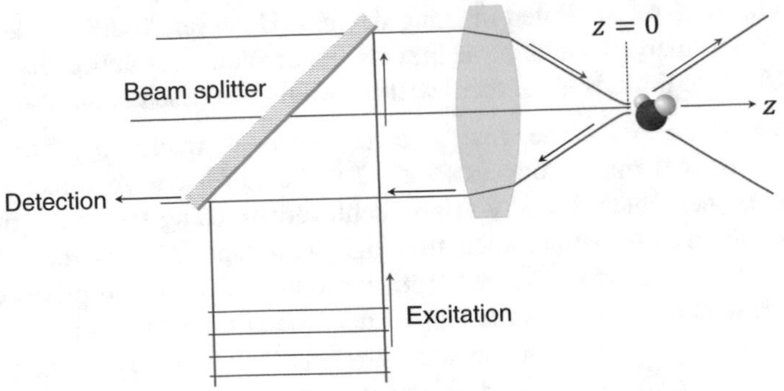

Fig. 6.5 Schematics of confocal microscopy. The sample (here a molecule) is excited through a focusing lens. The incoming light is redirected by a beam splitter that is placed between the focusing lens and the imaging lens (left, not shown). Through the same setup the light scattered by the sample is detected in complete analogy to the imaging system shown in Fig. 3.5

resolution, in particular in the axial direction. The second, probably even more important advantage is that through selective excitation the stray light is strongly suppressed.

The basic principle is identical to the setup of Fig. 3.5 for the detection of an object situated in the focus of a collection lens with a high numerical aperture. However, one additionally introduces a semi-transparent mirror in the region between the collection and imaging lenses, see Fig. 6.5, over which the sample in the focus spot of the first lens is excited. This excitation process can be described in complete analogy to the focusing, see Sect. 3.6.1. Because the additional mirror is semi-transparent, the analysis of the detection process remains practically unaltered to our previous discussion, but the light intensity is reduced by the mirror. Confocal microscopy gives an increased axial resolution, but requires to raster-scan the excitation and detection spots over the specimen.

Multiphoton Microscopy. The resolution in optical microscopy can be also increased by using multiphoton processes, for instance, in the excitation of the molecules to be detected. As previously discussed for Gaussian excitation and detection PSFs, the product of such functions has a smaller width. Similarly, in a multiphoton absorption the effective excitation spot for the multiphoton process can be reduced. Multiphoton processes have the additional advantage that excitation and detection occur at different wavelengths, which allows to dramatically suppress background scatterings.

Before closing the section on the diffraction limit of light, we ask the following questions: Can we improve on the resolution of optical microscopy? And if yes, how? As already mentioned above, the definition of the achievable resolution is somewhat arbitrary and depends on which quantity we are interested in. Yet, the physics underlying the diffraction limit of light is deep and builds on the removal of evanescent fields from the image. As these evanescent fields carry the high spatial resolution, something is lost in the imaging process and there is no way to restore it by means of more sophisticated imaging devices. However, in the last decades two strategies have emerged that allow achieving better spatial resolution than predicted by the diffraction limit. Both of them will be briefly discussed in the following.

Nearfield Microscopy. The first one is nearfield microscopy, or scanning nearfield optical microscopy (SNOM). The basic idea is to directly measure the optical nearfields. Usually this is achieved by using tapered optical fibers with a thin metal coating, such that light can tunnel in and out of the tip. Sometimes one removes the end from the coated fiber tip to produce a small aperture, with a diameter in the tens of nanometer range, but this is not strictly needed since light can also tunnel through thin metal films. By bringing this fiber close to the specimen, the nearfields of the probe approach the fiber tip and are converted to propagating photons, which can be detected at the other end of the fiber. By raster-scanning the tip over the sample, one obtains an optical image of the sample under investigation. In addition to this so-called detection mode, there are other operation modes of SNOM that will be discussed in the following section.

Localization Microscopy. The other approach is quite different in nature and builds on the concept of *preknowledge about the system*. Previously, we have derived the diffraction limit of light under the assumption that we have no preknowledge at all. For this reason, two dipoles must be separated by a critical distance to be observable as separate entities. In confocal microscopy we have already increased our preknowledge about the system in the sense that we know (through specific excitation) from which spot the light is emitted. In the field of localization microscopy, sometimes also referred to as nanoscopy, one decorates the system under study with fluorescence molecules. As we will discuss in Sect. 6.4, one can measure the position of the individual molecules with nanometer spatial accuracy. In this way, one does not measure the object directly but indirectly through the molecules attached to the object. Nanoscopy builds on the preknowledge about the system, namely the fact that fluorescence molecules have been attached to it, and thus allows achieving nanometer spatial resolution.

6.3 Scanning Nearfield Optical Microscopy

Scanning nearfield optical microscopy (SNOM) is a technique based on scanning probe microscopy. An optical fiber is coated with a metal layer, see Fig. 6.6. When the SNOM tip is brought into close vicinity of the probe under study, the evanescent nearfields are converted to propagating photons which can be detected at the fiber end. The throughput of light can be sometimes increased by etching a small aperture into the tip. In addition to this so-called detection mode, there also exists the illumination mode, where light is quenched through the fiber tip illuminating the sample in a sub-wavelength spot, and the scattered light of the probe is detected in the far-field; in the illumination-detection mode both the light

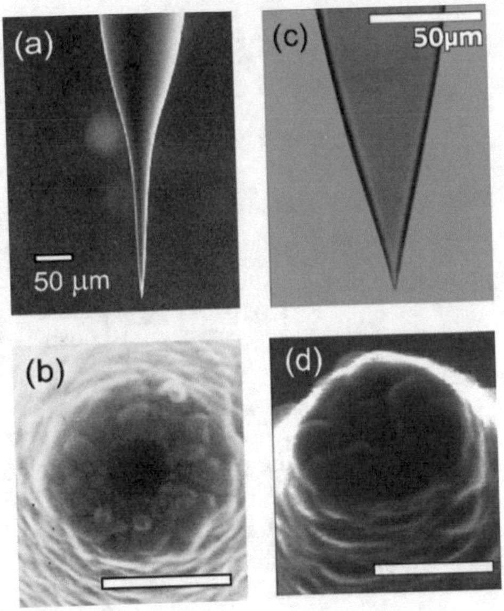

Fig. 6.6 Aluminum-coated aperture probes prepared by pulling (**a**, **b**) and etching (**c**, **d**): (**a**, **c**) macroscopic shape, scanning electron microscopy (SEM), and optical image. (**b**, **d**) SEM close-up of the aperture region, scale bar is 300 nm. Figure and caption taken from [23]

for illumination and detection pass through the fiber. In the following we briefly discuss two simple models for describing SNOM, the Bethe–Bouwkamp model and the even more simple dipole model. For a further discussion of nearfield optical probes the interested reader is referred to Refs. [6, 24].

6.3.1 Bethe–Bouwkamp Model

The Bethe–Bouwkamp model provides an analytic description for the nearfields of a plane wave impinging on a small circular hole in a conducting plane, and has been widely used for SNOM modelling because of its simplicity. The problem was initially tackled by Bethe [25] who, unfortunately, did not treat the boundary conditions at the edges of the hole properly. This error was corrected by Bouwkamp [26], and the model is now named after both of them. The derivation of the working equations relies on a rather exotic coordinate system, namely oblate spheroidal coordinates, and considering that Bethe—one of the most outstanding physicists of the twentieth century and Nobel laureate—did not get things straight from the beginning, we here only state the final results without repeating the somewhat tedious calculations. The interested reader is referred to the very clear and detailed paper of Bouwkamp [26].

The problem considers a plane wave polarized along $\hat{\boldsymbol{x}}$ and propagating along z, with the electric field $\boldsymbol{E}(\boldsymbol{r}) = \hat{\boldsymbol{x}}e^{ikz}$, that impinges on a conducting plane at $z = 0$ that has a small circular hole with radius a. For the solution one uses oblate spheroidal coordinates, which can be expressed either in terms of (μ, ν, ϕ) or (u, v, ϕ) that are related to the Cartesian coordinates through

$$
\begin{aligned}
x &= a\,\cosh\mu\cos\nu\cos\phi = a\sqrt{(1-u^2)(1+v^2)}\,\cos\phi \\
y &= a\,\cosh\mu\cos\nu\sin\phi = a\sqrt{(1-u^2)(1+v^2)}\,\sin\phi \\
z &= a\,\sinh\mu\sin\nu = a\,uv\,,
\end{aligned} \tag{6.14}
$$

with $u = \sin\nu$ and $v = \sinh\mu$. The inverse transformation can be performed most easily with the (μ, ν, ϕ) parameterization

$$
\begin{Bmatrix} \mu \\ \nu \end{Bmatrix} = \begin{Bmatrix} \mathrm{Re} \\ \mathrm{Im} \end{Bmatrix} \left[\cosh^{-1}\left(\frac{\sqrt{x^2+y^2}+iz}{a}\right)\right],
$$

and $\phi = \tan^{-1}(y/x)$. The fields behind the plane, $z > 0$, are expressed by Bouwkamp using the (u, v) parameterization,

$$
\begin{aligned}
E_x &= iku - \frac{2ikau}{\pi}\left\{1 + v\tan^{-1}v + \frac{1}{3}\frac{1}{u^2+v^2} + \frac{x^2-y^2}{3a^2(u^2+v^2)(1+v^2)^2}\right\} \\
E_y &= -\frac{4ik\,xyu}{3\pi a(u^2+v^2)(1+v^2)^2} \\
E_z &= -\frac{4ik\,xv}{3\pi(u^2+v^2)(1+v^2)^2}\,,
\end{aligned} \tag{6.15}
$$

and a similar expression for the magnetic fields. Figure 6.7 shows the nearfield intensity just behind the aperture. One observes that the fields are strongly localized and decay fast when moving away from the hole.

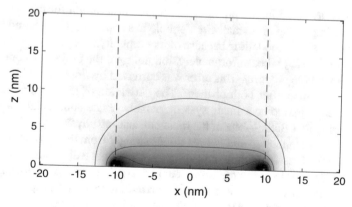

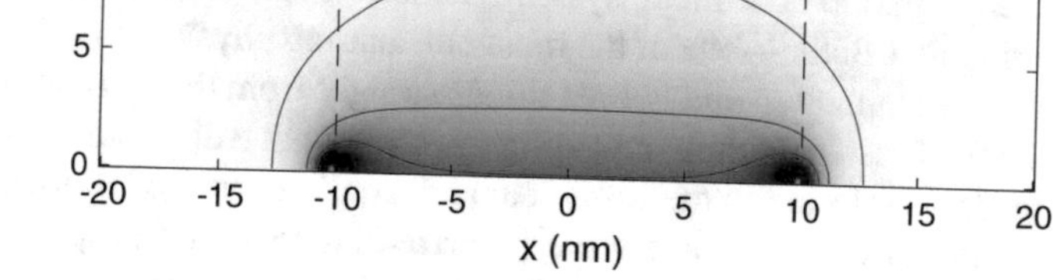

Fig. 6.7 Nearfield intensity as computed from the Bethe–Bouwkamp model, see Eq. (6.15). A plane wave with a unit electric field amplitude and polarized along $\hat{x}$ propagates along z and impinges on a conducting plane at $z = 0$ that has a small hole with radius a. The figure shows the intensity profile after passing the hole. The dashed lines indicate the position of the hole and the solid lines indicate the contours of equal intensity. In the calculations we use $a = 10$ nm and $\lambda = 620$ nm

There exists an even more simple description model for SNOM in terms of effective dipoles which mimic the light emitted (or collected) by the tip. The corresponding effective electric and magnetic dipoles $\boldsymbol{p}_{\text{eff}}$ and $\boldsymbol{m}_{\text{eff}}$ depend on the aperture radius a and on the incident electric field $\boldsymbol{E}_0$ via [25]

$$\boldsymbol{p}_{\text{eff}} = \mp\frac{4}{3}\varepsilon a^3\left(\boldsymbol{E}_0\cdot\hat{z}\right)\hat{z}\,, \quad \boldsymbol{m}_{\text{eff}} = \mp\frac{8}{3}a^3\left[\hat{z}\times\left(\boldsymbol{E}_0\times\hat{z}\right)\right], \tag{6.16}$$

where $\hat{z}$ is the unit vector normal to the plane of the aperture (pointing away from the fiber). The negative and positive sign in front of the expressions refer to the detection and illumination mode, respectively. These dipoles allow for a simple and yet accurate modelling of SNOM. For a more realistic description one usually resorts to computational Maxwell's solvers, which we will discuss in more detail in Chap. 11.

Eric Betzig's Encounter with SNOM

One of the heroes of the early SNOM days, Eric Betzig, recalled the history of one of his celebrated papers on the investigation of localized exciton states in quantum wells [27].

> For my last hurrah with near-field, Harald and I put my near-field probe on his low-temperature tunneling microscope to study quantum well structures, which are the basis of semiconductor lasers, like those in laser pointers. With standard diffraction-limited optics, their spectrum looks like a smooth hill of emission, but we saw a crazy series of super sharp lines. Our probe volume was so small, the light could only be emitted at certain discrete sites. And the wavelength of that light was very sensitive to the local thickness of the quantum well, so they glowed in different colors, which meant we could study them individually.

Betzig was awarded the Nobel Prize of Chemistry in 2014 for a different topic, namely localization microscopy, so his reflection on the SNOM field (see below) might be overly critical. Yet, it is interesting reading the recollections of a Nobel laureate in his bibliography presented at the nobelprize homepage, which finally let him leave science for a couple of years.

> That was a stunning paper, but by this time I was fed up. Although nearfield has proven to be a valuable tool for materials characterization and studying light-matter

(continued)

interactions at the nanoscale, my original goal was to make an optical microscope that could look at living cells with the resolution of an electron microscope. But near-field only worked on samples that were ridiculously flat, where the thing that you wanted to see was ridiculously close to the surface. If you're 20 nanometers away, you lose significant resolution. I knew a cell was a bit rougher than 20 nanometers, so it just wasn't going to happen.

Meanwhile, the field had blown up. There were hundreds of people doing near-field by this time, and much of it was crap. People were fooling themselves with images that had sharp-looking but artifactual structures, and they just didn't want to hear it. I felt like every good result I had provided justification for a hundred lousy papers to follow, and that was a waste of people's time and taxpayers' money. [...] So I quit.

Image taken from Eric Betzig, *"Nobel Lecture: Single molecules, cells, and super-resolution optics"*, Rev. Mod. Phys. 87, 1153 (2015).

6.4 Localization Microscopy

Localization or super-resolution microscopy has been awarded the Nobel Prize in chemistry 2014, and has revolutionized optical microscopy in the life sciences. It comes along and has been commercialized under different names, such as photo-activated localization microscopy (PALM or FPALM) [28], stochastic optical reconstruction microscopy (STORM) [29], or stimulated emission depletion (STED) microscopy [30]. In the following we briefly discuss the basic principles behind these techniques, however, without going into any details. For more information the interested reader is referred to the rich literature on the topic, see for instance, [31].

6.4.1 *Position Accuracy*

Suppose we know that the emission is coming from a *single* quantum emitter. How well can we localize the position of the emitter? For simplicity, we assume that the orientation of the dipole emitter is random, as is usually the case for molecules in solution, such that the emission pattern has a central lobe that can be described by a statistical average over the fields given in Eq. (6.5) or (6.7). In principle, the scheme described below can be also easily adapted to situations where the dipole orientation is fixed, which applies for instance to molecules immobilized in a polymer layer. We also assume that the depth z of the dipole is known, and will comment on how this can be done below.

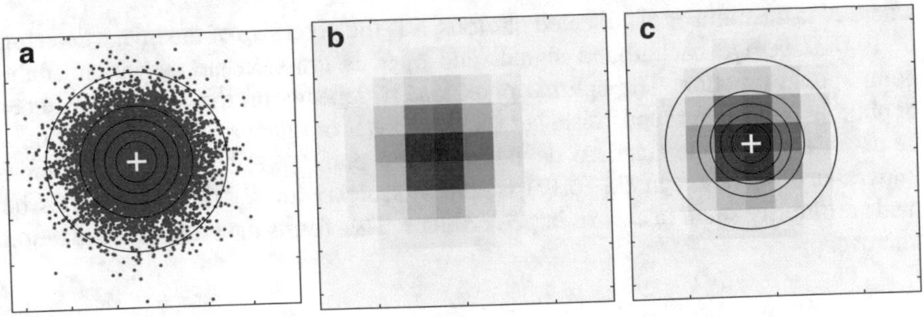

Fig. 6.8 Localization of single dipole emitter. (**a**) A dipole emitter situated at the position marked by the hair cross emits photons. The image of the dipole is broadened due to the point spread function. (**b**) Image of dipole emission for detection through a CCD camera. (**c**) The center spot of the emission is determined from a least square fit to a Gaussian distribution, or some more refined form, which can be done with an accuracy that predominantly depends on the signal-to-noise ratio

Figure 6.8a shows the emission pattern of a single dipole located at the hair cross position. The different points correspond to the detection spots of single photons, and the broadening of the overall distribution is due to the previously discussed point spread function. In typical experiments with CCD cameras one has a resolution governed by the pixels of the detector, which is schematically shown in panel (b). So how well can we determine the position of the dipole emitter from this pixeled image? The somewhat surprising answer is: in principle with arbitrary precision, the only limitation is due to the signal-to-noise ratio which depends on the number of detected photons.

To obtain the position of the dipole, one assumes that the emission pattern on the detector corresponds to the point spread function $f(\boldsymbol{r}, \boldsymbol{r}_0)$ that may additionally contain some modifications, to account for instance for a background signal or the finite pixel size a of the image detector. Suppose that we have measured N total photons, and we are in possession of a histogram accounting for the number of photons n_i measured at pixel position $\boldsymbol{r}_i$. We can then perform a least square fit (or some equivalent procedure) to obtain the *best* dipole position $\boldsymbol{r}_0$ from

$$\sum_i \left| n_i - f(\boldsymbol{r}_i, \boldsymbol{r}_0) \right|^2 \longrightarrow \text{Min}\,.$$

If f contains additional parameters, such as an unknown background signal, they can be additionally submitted to the least square fit. It can be shown that the localization precision σ for such a fit is [32]

$$\sigma = \sqrt{\left(\frac{\sigma_{\text{PSF}}^2 + a^2/12}{N}\right)\left(\frac{16}{9} + \frac{8\pi\sigma_{\text{PSF}}^2 b^2}{a^2 N}\right)}\,, \tag{6.17}$$

where N is the number of collected photons, a is the pixel size of the image detector, b^2 is the average background signal, and σ_{PSF} is the standard deviation of the point spread function. The optimal pixel size a depends on the expected number of photons and background noise, but for most cases of interest the pixel size should be about equal to the standard deviation of the point spread function. The most important observation in Eq. (6.17) is that σ scales with $N^{-1/2}$, and can thus be made arbitrarily small in case of large N values. This forms the basis of localization microscopy.

6.4.2 Photo-Activated Localization Microscopy

In order to render dipole localization useful for microscopy, one needs a few additional ingredients: one should be able (1) to attach fluorescing molecules to the sample one is interested in, (2) turn on a few molecules on demand in order to localize them, (3) and turn them off once they have been localized. Fortunately, the first two tasks were already solved by chemists in the mid-1990s. Martin Chalfie and others had discovered green fluorescent proteins (GFPs) for optical microscopy of cells. He received with two colleagues the Nobel Prize in chemistry 2008 for the

> [...] initial discovery of GFP and a series of important developments which have led to its use as a tagging tool in bioscience. By using DNA technology, researchers can now connect GFP to other interesting, but otherwise invisible, proteins. This glowing marker allows them to watch the movements, positions and interactions of the tagged proteins.

In his Nobel autobiography, Eric Betzig recalls the moment when he became aware around 2002 about these developments.

> I started reading the scientific literature again, and quickly came across Marty Chalfie's paper on green fluorescent protein, which he had published in 1994 as I was leaving Bell. It was like a religious revelation to me. Part of what made imaging with near-field so difficult was that it was hard to label proteins densely enough without also putting the fluorophore on a bunch of nonspecific crap. Here was a way to label with 100% specificity, and you could do it in a live cell. I couldn't believe how amazingly elegant it was. I hadn't wanted to go back to microscopy, but once I learned about GFP, I felt like I had to.

While green fluorescing molecules allow site-specific binding, one still has to address the issue of turning these molecules on and off. Fortunately, these tasks can be solved as follows:

Photoactivation. Control of the fluorescent state of molecules is possible when using photoswitchable molecules. In photoactivation a molecule is switched from an optically dark off-state to an optically bright on-state upon illumination at a specific wavelength. Other molecules exhibit photoswitching, this is the reversible switching between an on-state and an off-state upon illumination at two different wavelengths.

Bleaching. Even if molecules cannot be turned off in a controlled way, they become optically inactive after a short while due to photo bleaching. Whenever a molecule is optically excited, it has a small probability to undergo a conformational change and to become optically inactive. Typically, molecules emit a certain number of photons, say a million, before they bleach and become optically inactive.

With this control over binding and on/off states of fluorescing molecules, the schematics of a localization microscopy can be understood as follows (see Fig. 6.9). Initially, the sample is decorated with fluorescence molecules which are all in the off-state. A weak activation pulse brings a few molecules to the on-state, where they are localized using the previously described localization scheme. After a while they bleach, and all molecules are again in the off-state. The sequence of activation, localization, and bleaching is repeated many times, until one gets an image of the object under investigation. The first photo-activated localization microscope (PALM) was built once it was clear how it could work. Eric Betzig recalls:

> Harald and I built the first PALM microscope in his living room in La Jolla. We were both unemployed, but Harald had some of his equipment from Bell. We pulled that out of storage,

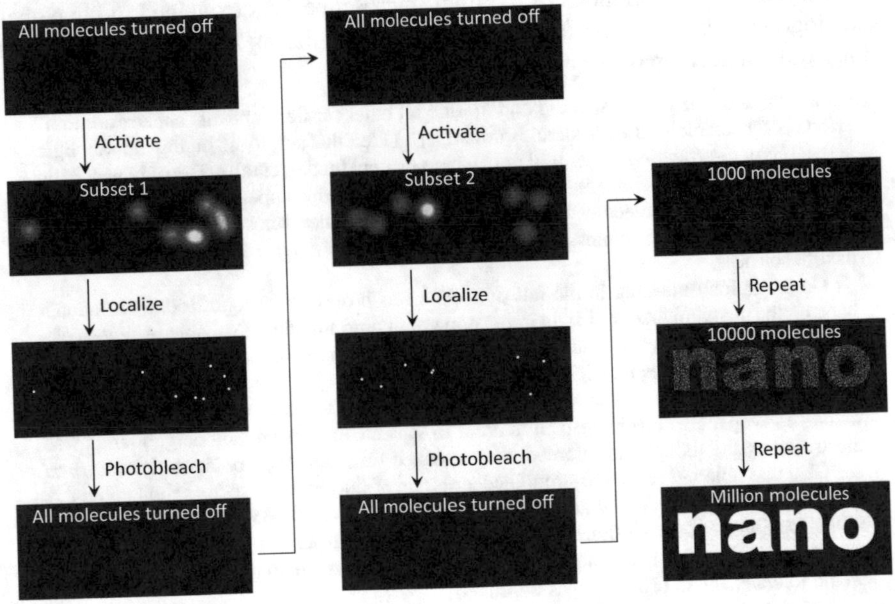

Fig. 6.9 Principle of localization microscopy. One starts with all molecules attached to the sample turned off. A weak activation pulse turns on a few molecules, which are localized before they bleach. This procedure is repeated many times until one gets an image of the sample under investigation, here the letters forming "nano". We assume that the letters are densely packed with dipole emitters, similarly to the setup shown in Figs. 3.2 and 3.3

> and each put in 25 000 dollars to cover everything else we needed. We worked hard, and in September shipped all the parts to rebuild the microscope in the darkroom of Jennifer's lab. The first time we put a cover slip coated with molecules into the microscope and turned on the photoactivating light, the first subset popped up and we knew we had it.

Since then the field has boomed, and localization microscopy has become a key player in the field of life sciences. Three-dimensional images can be obtained by using cylinder lenses to produce light sheets, and to turn on molecules in certain planes of the sample only. Quite generally, there is a trade-off between high resolution and fast scanning, and whereas it is nowadays possible to observe living cells with sub-diffraction resolution in real time, microscopy of fine cell details usually requires longer observation times and immobilized samples.

6.4.3 Stimulated Depletion Microscopy

Stimulated depletion microscopy (STED) is a different localization microscopy technique that was developed in the mid-1990 by Stefan Hell and coworkers. Together with Eric Betzig and William Moerner he received 2014 the Nobel Prize for "surpassing the limitations of the light microscope." As explained in his Nobel autobiography he was interested in the topic already during his doctoral thesis, but things did not develop overly well at the beginning.

> Rumour was that my efforts would end up like all other far-field optical "superresolution" efforts before, namely as an academic curiosity. [...] I felt that simply changing the way light is focused or re-arranging lenses will not change matters fundamentally. The only way to do so would be either via some quantum-optical effects or—what appeared more promising—via the states of the molecules to be imaged. The molecules whose states could be most easily played with were fluorescent ones, which, fortunately, were also those of interest in the life sciences.
>
> On a Saturday morning in the fall of 1993 I was browsing through Rodney Loudon's book on the quantum theory of light in the hope of finding something suitable. A few weeks earlier I had imagined what would happen if the fluorescent molecules would be re-excited from the excited state using slightly offset beams. When my eyes caught a chapter dealing with stimulated emission, it dawned on me: Why excite the molecules, why not de-excite them, i.e., keep them non-fluorescent in order to separate them from their neighbours. I was electrified by the thought and immediately checked Fritz Schäfer's book on dye lasers to see what was reported about the stimulated emission of fluorophores such as rhodamines. A quick assessment showed that an image resolution of at least 30–35 nm could be achieved in the focal plane, i.e. 6–8 times beyond the diffraction barrier. That was amazing. It was also instantly clear that the achievable resolution only depended on the intensity the sample would tolerate, and in principle was unlimited.

The basic principle is shown in Fig. 6.10 and consists of generating an excitation spot that is much smaller than the diffraction limit of light. Panel (a) reports a typical Jablonski diagram of the molecular levels. The molecule is modelled as a system with two electronic states, a ground and excited one, together with a series of vibrational excitations. Assume that the molecule is initially in its groundstate. An excitation pulse then promotes the molecule to an excited electronic and vibronic

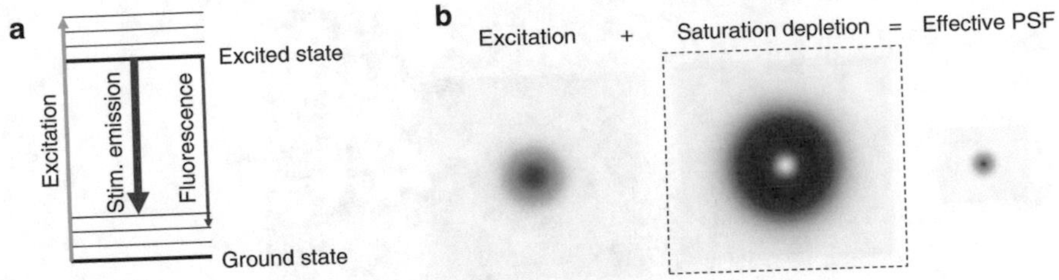

Fig. 6.10 Principle of stimulated emission depletion (STED). (**a**) A fluorescing molecule is excited to an excited electronic and vibrational state, and then undergoes vibrational relaxation until it reaches the vibrational groundstate. It remains there for a few nanoseconds, before emitting a photon and relaxing again to the electronic and vibrational groundstate. If during this waiting period an additional strong depletion pulse is switched on, the molecule undergoes stimulated emission (by adding a photon to the depletion pulse) and returns to the groundstate. (**b**) By combining a Gaussian excitation pulse together with a doughnut-like depletion pulse, the effective excitation spot can be made extremely small

state, where it undergoes relaxation to finally end up in vibronic groundstate of the excited electronic state. Typically, it stays there for a few nanoseconds before emitting a photon, which can be detected as fluorescence. If during this waiting time a second, strong depletion pulse is applied, the molecule is brought in a stimulated emission process to the electronic groundstate, and a photon is added to the depletion pulse. STED employs such molecular excitation and de-excitation as follows:

Excitation Pulse. A first excitation pulse excites the molecules, as schematically shown in panel (a) of Fig. 6.10. The excitation light is focused via far-field optics to the sample, so its extension is of the order of the light wavelength.

Depletion Pulse. A second depletion pulse brings the molecules back to the groundstate, it *depletes* the excited molecular state. The depletion pulse has a doughnut-like shape, which can be achieved with phase plates similar to our previous discussion of the optical angular momentum in Sect. 4.6, and is usually much stronger than the excitation pulse. It thus depletes all molecules with the exception of the center region. This combination of excitation and depletion allows generating excitation spots (excitation PSFs) that are much smaller than the wavelength of light. Finally, the fluorescence of the undepleted molecules is measured as the signal

One of the main differences between PALM or STORM, and STED is that in the former schemes the molecules are excited randomly, whereas in the latter one in a controlled fashion. In STED one has to raster-scan the excitation spot over the specimen, similarly to conventional confocal microscopy. Figure 6.11 shows a comparison between results obtained with confocal and STED microscopy.

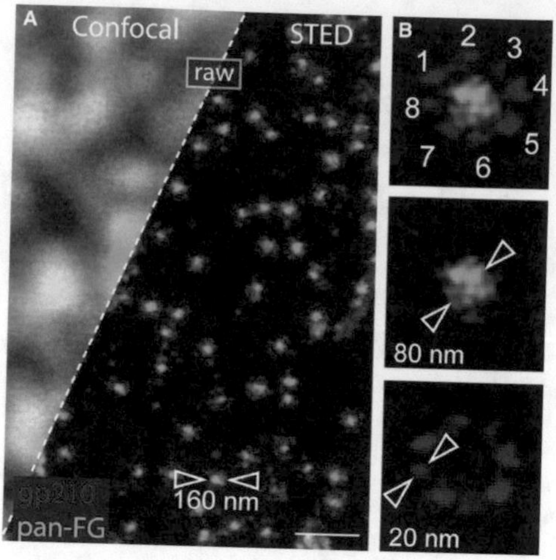

Fig. 6.11 Fluorescence nanoscopy of protein complexes with a compact near-infrared nanosecond-pulsed STED microscope, and comparison with confocal microscopy. (**a**) STED reveals immunolabeled subunits in amphibian NPC. The diameter of the octameric gp210 ring is established as ≈160 nm. Scale bar, 500 nm. (**b**) Individual NPC image showing eight antibody-labeled gp210 homodimers as 20–40 nm sized units and a 80 nm sized localization of the subunits in the central channel. Figure and caption taken from [33]

Can Localization Microscopy Beat the Diffraction Limit?

This chapter has been concerned with the diffraction limit of light, so it might be fair to ask whether the limit is fundamental or can be overcome. Personally I tend to opt for the fundamental nature, but rumor says that there are other, more qualified experts who believe that localization microscopy has beaten the diffraction limit. So I should probably keep the question open for discussion.

Part of the controversy comes from the fact that there seems to exist no unique definition of the diffraction limit, and the answer depends on how one exactly defines it. Arguing about things that depend on definitions are usually not overly rewarding. However, without being too specific it is my understanding that the diffraction limit concerns the loss of evanescent waves from the image, and this lost information can never be restored. Without any preknowledge about the monitored system we thus inevitably have to end up with the diffraction limit or something similar.

To "beat" the diffraction limit, one needs additional knowledge, for instance, that the detected light is coming from fluorescence molecules attached to the sample. It is a great and astonishing achievement that these molecules can be localized so robustly and accurately. However, in my opinion localization microscopy asks different questions to nature than conventional microscopy, and I therefore suggest to not compare conventional and localization microscopy on par with each other.

Exercises

Exercise 6.1 The electric field of an oscillating dipole can be expressed in terms of the dyadic Green's function together with Eq. (6.1) via

$$\boldsymbol{E}(\boldsymbol{r}) = \mu_0 \omega^2 \bar{\bar{G}}(\boldsymbol{r}, \boldsymbol{r}_0) \cdot \boldsymbol{p}\,.$$

With the asymptotic form of the Green's function, Eq. (5.25), compute the far-fields of the dipole.

Exercise 6.2 Use from the NANOPT toolbox the file `demodip05.m` to compute the image (intensity of imaged fields) of two dipoles oriented along x and separated by a distance d. Find for different NA values the distance where the dipoles can be distinguished in the image, and compare with Eq. (6.12).

Exercise 6.3 Use from the NANOPT toolbox the file `demodip06.m` to compute the image I_{12} of two dipoles oriented along x and separated by a distance d. Compute the images I_1 and I_2 for the individual dipoles, and plot the sum $I_1 + I_2$. Do I_{12} and $I_1 + I_2$ differ? And, if yes, why?

Exercise 6.4 Use the same file as in the previous exercise to investigate the situation that the dipoles are not in the same plane $z = 0$, but one dipole is moved away from it. How does this affect the image?

Exercise 6.5 Consider the situation of two dipoles whose orientation is not fixed but changes randomly over time. Discuss how the spectra will look like when the signal is averaged over a sufficiently long time interval.

Chapter 7
Material Properties

Electrodynamics communicates with the material world through the free charge and current distributions ρ, $\boldsymbol{J}$ as well as through the permittivities and permeabilities ε, μ, at least for linear materials. With the advent of modern nanoscience and nanotechnology the field of nano optics has received a strong boost, because the refined control over the sources and material properties offers unprecedented possibilities for novel optical applications.

One could argue that the success of the theory of electrodynamics in matter is due to the fact that, at least for linear materials, all microscopic details can be hidden within the two quantities of ε, μ. In principle, it does not matter how one obtains them, either through phenomenological models, microscopic theories, or through experiment,

$$\left.\begin{array}{l}\text{Phenomenological model}\\[6pt] \text{Microscopic theory}\\[6pt] \text{Experimental measurements}\end{array}\right\} \longrightarrow \quad \varepsilon,\ \mu \quad \longrightarrow \quad \text{Maxwell's equations}\,.$$

All that matters is that we have the quantities at hand, and once they are there we can plug them into Maxwell's equations and solve the problems we are interested in using one of the many techniques that have been developed over centuries and which are partly discussed in this book. In this respect, the separation into the material and electromagnetic worlds through the "linker" of ε, μ is a particularly beautiful solution.

Yet, sometimes we would like to know more about these quantities. In this chapter we will introduce a few phenomenological models for ε, μ, and will discuss general properties of these important quantities. In particular, we will find that there exist different levels of sophistication, which we briefly discuss at the example of the permittivity ε.

U. Hohenester, *Nano and Quantum Optics*, Graduate Texts in Physics,
https://doi.org/10.1007/978-3-030-30504-8_7

Constant Value. Often it suffices to assume a constant value for ε. Examples are glass or water which exhibit only moderate dispersion in the optical frequency range, and can be often (but not always) approximated by constant values. In this book we will usually use $n = 1.5\,(1.33)$ for glass (water) and optical frequencies.

Nonlocality in Time. In many cases of interest the materials have an intrinsic dynamics and one has to consider how, for instance, a material resonance becomes excited by the electric fields acting on the system. This can be described through a relation between the dielectric displacement $\boldsymbol{D}(\boldsymbol{r}, t)$ and the electric field $\boldsymbol{E}(\boldsymbol{r}, t)$ that is nonlocal in time,

$$\boldsymbol{D}(\boldsymbol{r}, t) = \int_0^\infty \varepsilon(\boldsymbol{r}, \tau)\boldsymbol{E}(\boldsymbol{r}, t - \tau)\, d\tau\,. \tag{7.1}$$

Note that this expression preserves causality because only fields in the past influence the matter response at time t. We will discuss in Sect. 7.3.2 that because of this, the real and imaginary parts of ε are strictly related to each other.

Nonlocality in Space. In principle, the optical response could be also nonlocal in space

$$\boldsymbol{D}(\boldsymbol{r}, t) = \int_0^\infty \int \varepsilon(\boldsymbol{r} - \boldsymbol{r}', \tau)\boldsymbol{E}(\boldsymbol{r}', t - \tau)\, d^3r' d\tau\,. \tag{7.2}$$

Examples are metals where a polarization can be created at some position $\boldsymbol{r}'$ and is transported via the conduction electrons to another position $\boldsymbol{r}$, where it influences the response. For typical metals the nonlocality range is less than a nanometer and can thus be often approximated by a local response. Nevertheless, there has been for many years interest in such nonlocal effects, and we will study them more thoroughly in Chap. 14.

Anisotropy. In anisotropic materials an electric field $\boldsymbol{E}$ oriented along a given direction can induce a polarization along a different direction. The material response then has to be described in terms of a tensor $\bar{\bar{\varepsilon}}$,

$$\boldsymbol{D} = \bar{\bar{\varepsilon}} \cdot \boldsymbol{E}\,, \tag{7.3}$$

together with possibly additional convolutions in time or space, as discussed above. In this book we will in general not further investigate such materials, mainly to keep the notation as simple as possible but also because many materials can be well described by isotropic functions $\bar{\bar{\varepsilon}} = \varepsilon\, \mathbb{1}$.

Quite generally, all of the above conclusions also hold for the permeability μ. However, for optical frequencies one can usually set $\mu \approx \mu_0$ for practically all known materials. An exception are metamaterials, these are man-made arrays of nanostructures where the dielectric and magnetic properties can be fully tailored. We will briefly comment on such materials further below.

7.1 Drude–Lorentz and Drude Models

The Drude–Lorentz model is one of the most simple description schemes for a dielectric function. It is based on a harmonic oscillator model, which can be modelled in the form of two oppositely charged particles attached to a spring, and the system is driven by an external electric field $E(t)$. Without specifying any details, we assume that a displacement $x(t)$ leads to a dipole moment $p(t) = ex(t)$. We start by writing down Newton's equations of motion for the driven oscillator

$$m\ddot{x} = -m\omega_0^2 x - m\gamma\dot{x} + eE(t)\,. \tag{7.4}$$

Here m and e are the mass and charge of the oscillator, respectively, ω_0 is the resonance frequency, and γ the damping constant. We assume a time-harmonic electric field of the form

$$E = E_0 e^{-i\omega t}\,.$$

After an initial stage, the system oscillates with the same frequency ω. Then, Eq. (7.4) can be rewritten in the form

$$-\omega^2 x = -\omega_0^2 x + i\gamma\omega x + \frac{eE_0}{m}\,,$$

where we have cancelled the common exponential terms. The displacement can thus be written in the form

$$x = \frac{e}{m}\frac{1}{\omega_0^2 - \omega^2 - i\gamma\omega}E_0\,. \tag{7.5}$$

This is the usual resonance dependence for a harmonic oscillator. It becomes largest when the frequency ω hits the resonance frequency ω_0. On resonance, the amplitude is governed by the damping constant γ and the amplitude of the driving field.

From the above expression we can compute the dipole moment $p = ex$ and, in turn, the polarization $P = nex$ by multiplying with the density of oscillators n (remember that the polarization is a dipole density). We thus get

$$P = nex = \frac{ne^2}{m}\frac{1}{\omega_0^2 - \omega^2 - i\gamma\omega}E_0 = \varepsilon_0\chi_e E_0\,,$$

where we have explicitly written down the relation between P and the electric susceptibility χ_e. Using $\varepsilon = \varepsilon_0(1 + \chi_e)$, we finally obtain the permittivity for a medium of harmonic oscillators with density n in the form

Permittivity of Drude–Lorentz Model

$$\varepsilon(\omega) = \varepsilon_0 \left(1 + \frac{ne^2}{\varepsilon_0 m} \frac{1}{\omega_0^2 - \omega^2 - i\gamma\omega}\right). \tag{7.6}$$

Figure 7.1 shows a typical example for a Drude–Lorentz dielectric function. The imaginary part associated with losses is peaked around the resonance frequency ω_0, where the height and width of the peak are governed by the damping constant γ. The real part exhibits the typical resonance behavior around ω_0, and approaches one for small frequencies where the oscillator follows the driving field and zero for large frequencies where the oscillator is too inert to follow the fast oscillations of the external field.

Despite its simplicity, there is some magic in the Drude–Lorentz model. Quite generally, most material resonances can be described in terms of a harmonic oscillator model, at least close to the resonance. Obviously, one should not take the ingredients of the model too seriously, it is a very generic model that can be described in terms of a few effective parameters (which happen to be those of a mechanical analog). In case of several resonances, one can also generalize the result of Eq. (7.6) and introduce a summation over different oscillator contributions.

For metals we can use the Drude–Lorentz model without any restoring forces, $\omega_0 \to 0$. This leads us to the Drude permittivity for metals

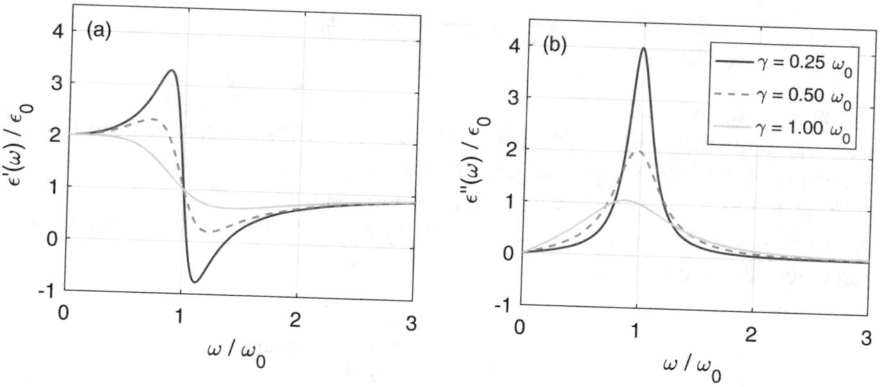

Fig. 7.1 Typical behavior of Drude–Lorentz permittivity for different damping constants γ, see Eq. (7.6). (**a**) Real and (**b**) imaginary part of $\varepsilon(\omega)$. For the oscillator constant we here use $ne^2/(\varepsilon_0 m) = \omega_0^2$

Drude Permittivity for Metals

$$\varepsilon(\omega)/\varepsilon_0 = \kappa_b - \frac{\omega_p^2}{\omega(\omega + i\gamma)}\,. \tag{7.7}$$

Here we have introduced the plasma frequency

$$\omega_p = \sqrt{\frac{ne^2}{\varepsilon_0 m}}\,, \tag{7.8}$$

which can be directly obtained from Eq. (7.6). At the plasma frequency and for $\kappa_b = 1$ we obtain $\varepsilon(\omega_p) \approx 0$ associated with longitudinal plasma oscillations of the conduction electrons. κ_b is an additional contribution accounting for contributions of bound electrons.

Typical values are listed in Table 7.1 and range from one for aluminum to about ten for gold. The solid lines in Fig. 7.2 show the (a) real and (b) imaginary parts of the dielectric function computed for the Drude models of *Au*, *Ag*, and *Al*. The symbols report experimental data. In all cases, we find reasonable to good agreement between the values of the Drude function and the experimental data, with the main exception of ε'' for gold at energies above 2 eV. Here the experimental data show a much larger damping.

To understand the origin of such damping, in the following we consider the microscopic details of metal electrons. In solid state physics, the electron states are described in terms of Bloch functions $u_{n,\boldsymbol{k}}(\boldsymbol{r})$ governed by a wavevector $\boldsymbol{k}$ and

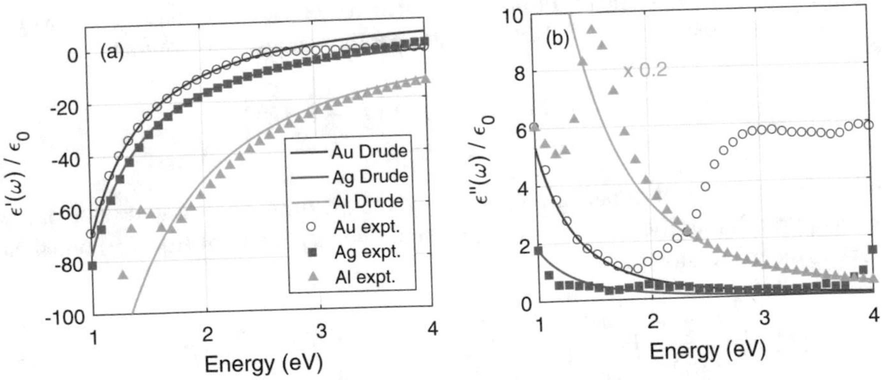

Fig. 7.2 (**a**) Real and (**b**) imaginary part of Drude dielectric function for gold *Au*, silver *Ag*, and aluminum *Al*, and comparison with experimental data. Experimental data for *Au*, *Ag* are taken from [34], and for *Al* from [35]

Table 7.1 List of parameters for Drude model for a few selected metals

Material	κ_b	$\hbar\omega_p$ (eV)	$\hbar/\gamma$ (fs)
Au	10	10	10
Ag	3.3	10	30
Al	1	15	1

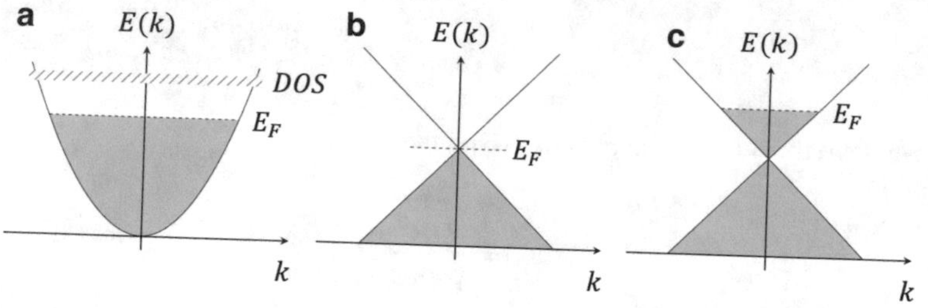

Fig. 7.3 Schematic view of bandstructure for (**a**) parabolic dispersion, (**b**) undoped and (**c**) doped graphene. The dashed lines report the Fermi energies. The dashed area in panel (**a**) reports the density of states (DOS) which corresponds to the number of states available per unit energy and volume

a band index n [36]. The relation between $\boldsymbol{k}$ and the electron energies $E_n(\boldsymbol{k})$ is called the *bandstructure*, which accounts for the electronic and optical properties of electrons in solids. For simple metals the bandstructure can be usually well approximated by a parabolic dispersion

$$E(\boldsymbol{k}) = \frac{\hbar^2 k^2}{2m}, \tag{7.9}$$

where m is the electron mass. Note that this dispersion is identical to that of free electrons. In a metal all states are filled up to a Fermi energy E_F, see Fig. 7.3a, which is determined by the number of electrons in the metal. For reasons to become clear in a moment, we also introduce the density of electron states $g(E)$ which corresponds to the number of electron states per unit energy and volume,

$$g(E) = \lim_{\Omega\to\infty} \Omega^{-1} \sum_{n,\boldsymbol{k}} \delta\Big[E - E_n(\boldsymbol{k})\Big],$$

where it is convenient to perform the thermodynamic limit by letting the size of the solid approach infinity, $\Omega \to \infty$. For the parabolic dispersion of Eq. (7.9) one then obtains the expression [36]

$$g(E) = \frac{2}{(2\pi)^3} \int_{-\infty}^{\infty} \delta\left(E - \frac{\hbar^2 k^2}{2m}\right) d^3k = \left[\frac{(2m)^{3/2}}{2\pi^2\hbar^3}\right] \sqrt{E}\,. \tag{7.10}$$

Thus, the density of states scales with the square root of the energy. In Fig. 7.4 we show a schematic of the bandstructure and density of states for transition metals

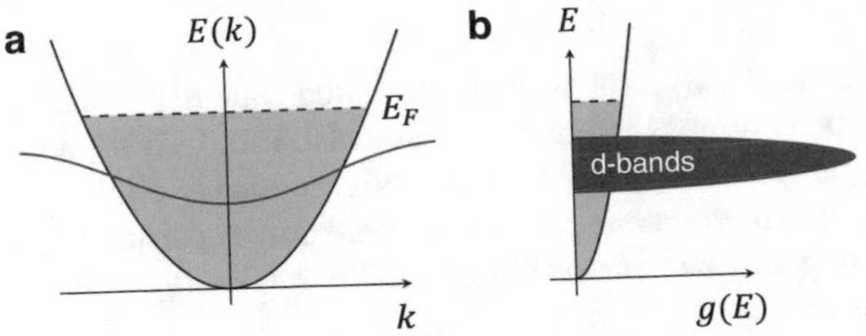

Fig. 7.4 (**a**) Schematics for bandstructure $E(k)$ and (**b**) density of states $g(E)$ for transition metals. In addition to the parabolic band of conduction electrons there is an additional band of localized d-electrons which is situated below the Fermi energy. For energies $\hbar\omega$ where electrons can be promoted from these d-bands to states above E_F, the imaginary part of the permittivity $\varepsilon''(\omega)$ strongly increases

such as gold or silver. In addition to the free-electron like conduction band, there are additional bands below the Fermi energy which are associated with strongly localized d-electrons. The density of states of these d-bands is usually much higher than that of the free electrons, and is responsible for two effects in the dielectric function.

***d*-Band Screening.** When the photon energy $\hbar\omega$ is smaller than the energetic distance between the d-bands and the Fermi energy, no d-band transitions can be induced. However, similar to dielectrics the d-band states can be polarized, leading to the relatively large κ_b values for Au and Ag listed in Table 7.1. In contrast, in aluminum, which is not a transition metal, the atoms only have a single electron in the outer shell. Correspondingly there is no screening contribution of bound electrons, and thus $\kappa_b = 1$.

***d*-Band Transitions.** For energies $\hbar\omega$ where electrons can be promoted from the d-band states to states above E_F, the imaginary part of the permittivity $\varepsilon''(\omega)$ (which is a measure of absorption associated with such transitions) strongly increases. In gold, d-band transitions set in at photon energies above 2 eV, as can be clearly seen from the experimental data in Fig. 7.2, whereas in silver the d-bands are located energetically deeper below the Fermi energy and transitions set in at much higher energies, say around 4 eV. Finally, in aluminum we observe no corresponding transitions. Nevertheless, there is a weak peak around 1.5 eV, which is associated with another type of interband transition.

To summarize, the Drude model provides a viable description scheme for many metals, but should not be used for gold above photon energies of 2 eV where more realistic data from experiment or detailed theoretical calculations including bandstructure effects become indispensable.

Graphene

Graphene and other two-dimensional, so-called van der Waals materials have recently attracted enormous interest [37]. See also Ref. [38] for a discussion of the prospects of graphene for plasmonics and other optics applications. Although we will touch the topic of graphene plasmonics here only superficially, we note that the bandstructure of graphene consists close to the Fermi energy of electron bands with a linear dispersion

$$E_{2D}(k) = \hbar v_F\, k\,, \tag{7.11}$$

as schematically shown in Fig. 7.3. Here v_F is the Fermi velocity. The undoped material is a semi-metal with a vanishing density of states at E_F, and one has to dope it by means of external gates to make the system metallic [38]. We shall denote the Fermi energy of doped graphene with μ.

Because of its bandstructure, doped graphene cannot be described in terms of a simple Drude model. In the most simple approach, the permittivity can be computed using the so-called Lindhard framework [36], and one obtains for the two-dimensional polarization [39, 40]

$$\begin{aligned} P_{2D}(q,\omega) = &-\frac{g\mu}{2\pi\hbar^2 v_F^2} + \frac{F(q,\omega)}{\hbar^2 v_F^2} \\ &\times \Big\{\big[G(x_+) - i\pi\big] - \theta(-x_- - 1)\big[G(-x_-) - i\pi\big] \\ &- \theta(x_- + 1)G(x_-)\Big\}. \end{aligned} \tag{7.12}$$

Here q is a wavenumber, $x_\pm = (\hbar\omega \pm 2\mu)/(\hbar v_F q)$, θ is Heaviside's step function, and $g = g_s g_v = 4$ is the product of the spin and valley degeneracy. We have also introduced the two functions

$$F(q,\omega) = \frac{g}{16\pi}\frac{\hbar v_F^2 q^2}{\sqrt{\omega^2 - v_F^2 q^2}}\,, \qquad G(x) = x\sqrt{x^2-1} - \ln(x + \sqrt{x^2-1})\,.$$

The polarization is related in frequency and wavenumber space to the induced surface charge distribution $\sigma_{\rm ind}$ via

$$\sigma_{\rm ind}(q,\omega) = P_{2D}(q,\omega)\left[\frac{e^2}{2\varepsilon_0 q}\right]\,, \tag{7.13}$$

where the term in brackets is the Fourier transform of the Coulomb potential in two dimensions. In the limit $q \to 0$, Eq. (7.12) can be simplified in the form [39, 40]

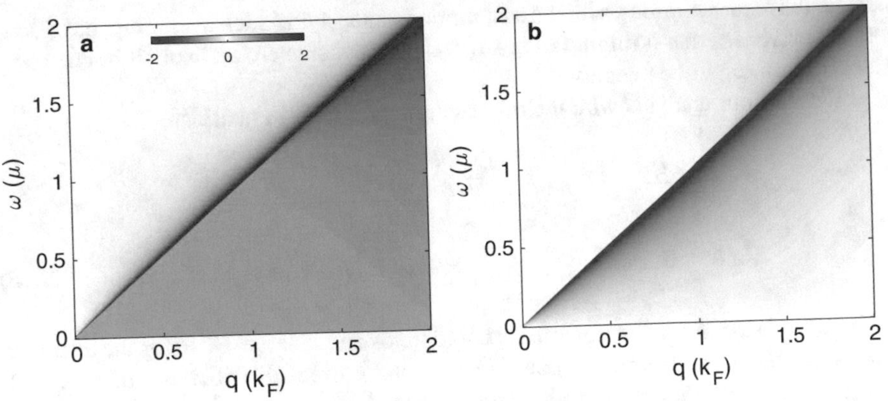

Fig. 7.5 (**a**) Real and (**b**) imaginary part of graphene polarization of Eq. (7.12). The wavenumber is given in units of $k_F = \mu/(\hbar v_F)$, the energy in units of the Fermi energy μ, and the polarization in units of $\mu/(\hbar^2 v_F^2)$

$$P_{2D}(q \to 0, \omega) = \frac{gq^2}{8\pi\hbar\omega}\left[\frac{2\mu}{\hbar\omega} + \frac{1}{2}\ln\left|\frac{2\mu - \hbar\omega}{2\mu - \hbar\omega}\right| - i\frac{\pi}{2}\theta(\hbar\omega - 2\mu)\right]. \tag{7.14}$$

The first term in brackets is associated with intraband transitions, and the other two terms with interband transitions. Figure 7.5 shows the real and imaginary part of the polarization, which exhibits a resonance behavior around $q = \omega$. We will return to the graphene dielectric function in the next chapter when discussing graphene plasmons.

7.2 From Microscopic to Macroscopic Electromagnetism

Electrodynamics in matter relies on the concepts of polarization and magnetization, which can be associated with electric and magnetic dipole densities. This approach works because the characteristic length scale of electromagnetic waves is on the order of micrometers, possibly a few tens to hundreds of nanometers for evanescent waves, whereas the relevant length scale for matter is in the nanometer range. For this reason, the fine details of matter play no important role for the dynamics of electromagnetic waves, which only interact with some kind of averaged matter state.

There exists no clear-cut definition of how an averaging over the microscopic charge and current distributions should be done. One could argue that this is because averaging is so robust that one always gets the same result irrespective of the starting expression. On the other hand, the whole issue of averaging is quite unrewarding as one finally has to end up with the macroscopic Maxwell's equations anyhow. Our discussion of the topic closely follows the book of Jackson [2] and intends to make the reader familiar with the assumptions underlying such averaging, but also to raise awareness that it might fail at small dimensions where an explicit microscopic description might be needed.

We start our analysis with the microscopic Maxwell's equations,

$$\nabla \cdot \boldsymbol{e} = \frac{\varrho}{\varepsilon_0}, \qquad \nabla \times \boldsymbol{e} = -\frac{\partial \boldsymbol{b}}{\partial t}$$

$$\nabla \cdot \boldsymbol{b} = 0, \qquad \nabla \times \boldsymbol{b} = \mu_0 \boldsymbol{j} + \mu_0 \varepsilon_0 \frac{\partial \boldsymbol{e}}{\partial t}, \tag{7.15}$$

where we use $\boldsymbol{e}$, $\boldsymbol{b}$ for the true microscopic fields and reserve $\boldsymbol{E}$, $\boldsymbol{B}$ for the averaged ones. ϱ and $\boldsymbol{j}$ are the microscopic charge and current distributions that we will describe in a semiclassical framework, although there would be no fundamental differences for a quantum mechanical description.

Spatial averaging is done by introducing a sampling function $f(\boldsymbol{r} - \boldsymbol{r}')$ that averages for a given source point $\boldsymbol{r}$ over some small spatial region, as schematically shown in Fig. 7.6. Any physical quantity F can then be averaged according to

$$\big\langle F(\boldsymbol{r}, t) \big\rangle = \int_{-\infty}^{\infty} f(\boldsymbol{r} - \boldsymbol{r}') F(\boldsymbol{r}', t)\, d^3r' = \int_{-\infty}^{\infty} F(\boldsymbol{r} - \tilde{\boldsymbol{r}}, t) f(\tilde{\boldsymbol{r}})\, d^3\tilde{r}, \tag{7.16}$$

where we have used $\tilde{\boldsymbol{r}} = \boldsymbol{r} - \boldsymbol{r}'$ to get the last expression. A few words about the sampling function are at place. First, it should extend over a region L^3 that is so large, say with $L \approx 10$ nm, that it contains a sufficient number of atoms, molecules, or unit cells in case of a solid. On the other hand, L should be small enough that the averaged quantity F does not change noticeable inside the region. Again, for $L \approx 10$ nm this should be a good approximation even for strongly evanescent waves, with the possible exception of sub-nanometer gaps or extremely sharp nanoparticle features.

The averaging prescription of Eq. (7.16) has the advantage that derivatives commute with the averaging,

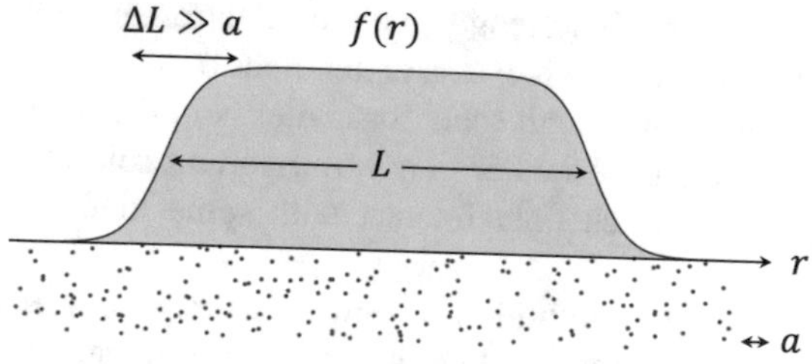

Fig. 7.6 Averaging function $f(\boldsymbol{r})$. The width L should be large enough to average over a sufficiently high number of atoms or molecules (indicated with red dots at bottom of figure), the extent ΔL of the region where f falls to zero should be sufficiently larger than the interatomic distance a, such that atom fluctuations have no significant impact on the averaged values. The function $f(\boldsymbol{r})$ is assumed to be normalized to one, this is $\int_{-\infty}^{\infty} f(\boldsymbol{r})\, d^3r = 1$

$$\frac{\partial}{\partial r_i}\big\langle F(\boldsymbol{r},t)\big\rangle=\int_{-\infty}^{\infty}\frac{F(\boldsymbol{r}-\tilde{\boldsymbol{r}},t)}{\partial r_i}f(\tilde{\boldsymbol{r}})\,d^3\tilde{r}=\left\langle\frac{\partial F(\boldsymbol{r},t)}{\partial r_i}\right\rangle,$$

with a similar expression for time derivatives. For this reason, we can immediately perform the averaging in Eq. (7.15), and by introducing the averaged fields $\boldsymbol{E}=\langle\boldsymbol{e}\rangle$, $\boldsymbol{B}=\langle\boldsymbol{b}\rangle$ we are led to

$$\begin{aligned}\nabla\cdot\boldsymbol{E}&=\frac{\langle\varrho\rangle}{\varepsilon_0}, &\qquad \nabla\times\boldsymbol{E}&=-\frac{\partial\boldsymbol{B}}{\partial t}\\ \nabla\cdot\boldsymbol{B}&=0, &\qquad \nabla\times\boldsymbol{B}&=\mu_0\big\langle\boldsymbol{j}\big\rangle+\mu_0\varepsilon_0\frac{\partial\boldsymbol{E}}{\partial t}\,.\end{aligned}\tag{7.17}$$

For the averaging of the source terms, we concentrate on Gauss' law and consider a charge distribution consisting of point-like charges q_i and dipoles $\boldsymbol{p}_i$,

$$\varrho(\boldsymbol{r})=\sum_i q_i\delta(\boldsymbol{r}-\boldsymbol{r}_i)+\lim_{\eta\to 0}\sum_i\frac{p_i}{\eta}\Big[\delta(\boldsymbol{r}-\boldsymbol{r}_i-\frac{1}{2}\eta\hat{\boldsymbol{p}}_i)-\delta(\boldsymbol{r}-\boldsymbol{r}_i+\frac{1}{2}\eta\hat{\boldsymbol{p}}_i)\Big]\,.$$

For the dipoles we have assumed a generic form consisting of two opposite charges $\pm p_i/\eta$ separated by the distance $\eta\hat{\boldsymbol{p}}_i$, and we will set $\eta\to 0$ at the end of the calculation. The center position of the charges and dipoles is $\boldsymbol{r}_i$. The free charges can be immediately submitted to the averaging procedure and we obtain

$$\left\langle\sum_i q_i\delta(\boldsymbol{r}-\boldsymbol{r}_i)\right\rangle=\sum_i q_i f(\boldsymbol{r}-\boldsymbol{r}_i)=\rho_{\text{ext}}(\boldsymbol{r})\,.$$

For the dipoles we get upon averaging

$$\lim_{\eta\to 0}\sum_i p_i\frac{f(\boldsymbol{r}-\boldsymbol{r}_i-\frac{1}{2}\eta\hat{\boldsymbol{p}}_i)-f(\boldsymbol{r}-\boldsymbol{r}_i+\frac{1}{2}\eta\hat{\boldsymbol{p}}_i)}{\eta}=-\sum_i\boldsymbol{p}_i\cdot\nabla f(\boldsymbol{r}-\boldsymbol{r}_i)\,,$$

The last term can be rewritten in the form

$$-\sum_i\boldsymbol{p}_i\cdot\nabla f(\boldsymbol{r}-\boldsymbol{r}_i)=-\nabla\cdot\left\langle\sum_i\boldsymbol{p}_i\delta(\boldsymbol{r}-\boldsymbol{r}_i)\right\rangle=-\nabla\cdot\boldsymbol{P}(\boldsymbol{r})\,,$$

with $\boldsymbol{P}$ being the macroscopic polarization defined as the dipole density. This leaves us with Gauss' law for Maxwell's equations in matter, Eq. (2.26a). The averaging procedure for the current distribution is similar but more involved, and will not be presented here. Even Jackson seems to be somewhat distressed when it comes to this discussion and notes [2]:

> To complete the discussion we must consider $\langle\boldsymbol{j}\rangle$. Because of its vector nature and the presence of velocities the derivation is considerably more complicated than the earlier treatment of $\langle\varrho\rangle$, even though no new principles are involved. We present only the results, leaving the gory details to a problem for those readers who enjoy such challenges.

7.3 Nonlocality in Time

In the following we consider a dielectric response that is nonlocal in time

$$D(\mathbf{r},t)=\int_0^\infty \varepsilon(\mathbf{r},\tau)\mathbf{E}(\mathbf{r},t-\tau)\,d\tau\,,$$

and a possibly similar relation between $\mathbf{B}$ and $\mathbf{H}$ with a time-dependent permeability μ. In the following we consider time-harmonic fields, see Sect. 2.4. The convolution theorem of Fourier transforms then states that upon Fourier transformation the above relation becomes a product in frequency space,

$$\mathbf{D}(\mathbf{r},\omega)=\varepsilon(\mathbf{r},\omega)\mathbf{E}(\mathbf{r},\omega)\,. \tag{7.18}$$

Note that we use the same symbols for the quantities in the time and frequency domain. In general we will work exclusively in either of these spaces, so there will be little danger of confusion. The neat thing about this transformation behavior is that Maxwell's equations for time-harmonic fields, Eq. (2.34), look almost identical for frequency-dependent permittivities and permeabilities,

Metamaterials

Metals and other related materials have a strong dielectric response in the optical frequency range, but there are no natural materials with a comparable strong magnetic response. In the optical regime one can thus set $\mu = \mu_0$ for practically all materials. The reason for this inequality is the absence of magnetic charges. These observations were summarized by John Pendry quite a while ago [41].

Ideally we should like to proceed in the magnetic case by finding the magnetic analogue of a good electrical conductor: unfortunately there isn't one! Nevertheless we can find some alternatives which we believe do give rise to interesting magnetic effects.

Why should we go to the trouble of microstructuring a material simply to generate a particular μ_{eff}? The answer is that atoms and molecules prove to be a rather restrictive set of elements from which to build a magnetic material. This is particularly true at frequencies in the gigahertz range where the magnetic response of most materials is beginning to tail off. Those materials, such as the ferrites, that remain moderately active are often heavy, and may not have very desirable mechanical properties. In contrast, we shall show, microstructured materials can be

(continued)

designed with considerable magnetic activity, both diamagnetic and paramagnetic, and can if desired be made extremely light.

In the last two decades there have been strong efforts to build such artificial materials, so-called metamaterials. One exploits the fact that a strong response can be achieved in the microwave regime through split-ring resonators. This concept can be translated to the optical range, but one has to be careful that the structures must be significantly smaller than the optical wavelength to get media with effective material parameters, as described in this chapter. The above picture shows an image of a metamaterial with a chiral response [42], one out of a myriad of beautiful examples. Metamaterials have become an important player in the field of nano optics and plasmonics, and with ever increasing sample preparation techniques they will continue to play a key role in the field.

$$\begin{aligned}
\nabla\cdot\varepsilon(\boldsymbol{r},\omega)\boldsymbol{E}(\boldsymbol{r},\omega) &= \rho_{\text{ext}}(\boldsymbol{r},\omega)\\
\nabla\cdot\boldsymbol{B}(\boldsymbol{r},\omega) &= 0\\
\nabla\times\boldsymbol{E}(\boldsymbol{r},\omega) &= i\omega\boldsymbol{B}(\boldsymbol{r},\omega)\\
\nabla\times\mu^{-1}(\boldsymbol{r},\omega)\boldsymbol{B}(\boldsymbol{r},\omega) &= \boldsymbol{J}_{\text{ext}}(\boldsymbol{r},\omega)-i\omega\,\varepsilon\boldsymbol{E}(\boldsymbol{r},\omega)\,.
\end{aligned}\tag{7.19}$$

For this reason, practically everything we have discussed in the previous chapters about time-harmonic fields can be directly carried over to a frequency-dependent system response. Exceptions are the Poynting's theorem, which we will revisit in the next section, and the fact that the permittivities and permeabilities acquire imaginary contributions associated with losses. For propagating and evanescent waves this leads to damping and attenuation.

For conductors and metals it is often convenient to incorporate the response of the conduction carriers into the permittivity. Above we have already seen how this can be done for a Drude dielectric function. In the general case we can generalize Ohm's law for a frequency-dependent conductivity

Optical Conductivity

$$\boldsymbol{J}(\boldsymbol{r},\omega)=\sigma(\boldsymbol{r},\omega)\boldsymbol{E}(\boldsymbol{r},\omega)\,.\tag{7.20}$$

$\sigma(\boldsymbol{r},\omega)$ is often called the optical conductivity which becomes the static conductivity σ_0 for $\omega \to 0$. From the continuity equation in the frequency domain

$$i\omega\rho = \nabla \cdot \boldsymbol{J} = \nabla \cdot \sigma \boldsymbol{E}\,. \tag{7.21}$$

we can establish a relation between ρ and the electric field. Inserting this expression into Gauss' law gives

$$\nabla \cdot \varepsilon_b \boldsymbol{E} = -\frac{i}{\omega}\nabla \cdot \sigma \boldsymbol{E} \implies \nabla \cdot \left(\varepsilon_b + \frac{i\sigma}{\omega}\right)\boldsymbol{E} = 0\,,$$

where ε_b denotes the permittivity of the bound charges. Similarly, Ampere's law can be rewritten in the form

$$\nabla \times \frac{1}{\mu}\boldsymbol{B} = \sigma \boldsymbol{E} - i\omega\varepsilon_b \boldsymbol{E} = -i\omega\left(\varepsilon_b + \frac{i\sigma}{\omega}\right)\boldsymbol{E}\,.$$

Thus, for the optical conductivity of Eq. (7.20) we can absorb the effect of the free carriers into a frequency-dependent permittivity

$$\varepsilon(\boldsymbol{r},\omega) = \varepsilon_b(\boldsymbol{r},\omega) + \frac{i\sigma(\boldsymbol{r},\omega)}{\omega}\,, \tag{7.22}$$

where ε_b is the part associated with the bound charges.

Surface Charges Let us consider an interface, see, for instance, Fig. 2.6, where the upper material is a dielectric with permittivity ε_2, and the lower material is a metal with a permittivity ε_1, whose form is given by Eq. (7.22). We next derive from the boundary conditions of Maxwell's equations the surface charge distributions at the interface. Our starting point is given by $\rho_{\text{ind}} = -\nabla \cdot \boldsymbol{P}$, with ρ_{ind} and $\boldsymbol{P}$ being the induced charge distribution and the polarization, respectively. By integrating this expression over a small volume, in complete analogy to Sect. 2.3.2, we find

$$\sigma_{\text{ind}} = -\hat{\boldsymbol{n}} \cdot \Big(\boldsymbol{P}_2 - \boldsymbol{P}_1\Big) = -\Big(P_2^\perp - P_1^\perp\Big) = -\Big(P_2^\perp - \Big[P_{\text{ind}}^\perp + P_{\text{ext}}^\perp\Big]\Big)\,.$$

In the last experssion we have further decomposed $\boldsymbol{P}_2$ into an induced or bound polarization, and an external one associated with free carriers. To compute the different polarizations we use

$$\boldsymbol{P} = \varepsilon_0 \chi \boldsymbol{E} = (\varepsilon - \varepsilon_0)\,\boldsymbol{E}\,,$$

and finally arrive at

$$\sigma_{\text{ind}} = \begin{cases} -(\varepsilon_2 - \varepsilon_0)\;E_2^\perp & \ldots \text{ surface charge at upper side of interface} \\ (\varepsilon_1 - \varepsilon_0)\;E_1^\perp & \ldots \text{ surface charge at lower side of interface} \\ (\varepsilon_1 - \varepsilon_b)\;E_1^\perp & \ldots \text{ free surface charge at lower side of interface}\,. \end{cases} \tag{7.23}$$

The total polarization charges at an interface can be computed from

$$\sigma_{\text{pol}} = -\left(\varepsilon_2 - \varepsilon_0\right) E_2^{\perp} + \left(\varepsilon_1 - \varepsilon_0\right) E_1^{\perp} = \varepsilon_0 \left(E_2^{\perp} - E_1^{\perp}\right), \tag{7.24}$$

where we have exploited that the normal component of the dielectric displacement is continuous at the interface. We will come back to these expressions in the discussion of surface and particle plasmons.

7.3.1 Poynting's Theorem Revisited

In this section we revisit Poynting's theorem for a linear medium, previously derived in Sect. 4.3, but account for dispersion and absorption effects. Quite generally, we expect two major modifications with respect to our previous result of Eq. (4.14):

- Because of dispersion the velocity of the energy flow becomes modified.
- Because of absorption the energy flow and density become attenuated during propagation.

In complete analogy to our previous derivation of Poynting's theorem, we start with the power performed by the electromagnetic fields on a current distribution but now relate $\boldsymbol{J}_{\text{ext}}$ to the fields $\boldsymbol{D}$, $\boldsymbol{H}$,

$$\frac{dW}{dt} = \int_{\Omega} \boldsymbol{J}_{\text{ext}} \cdot \boldsymbol{E}\, d^3r = \int_{\Omega} \left(\nabla \times \boldsymbol{H} - \frac{\partial \boldsymbol{D}}{\partial t}\right) \cdot \boldsymbol{E}\, d^3r\,.$$

Using in the first term the transformation

$$\nabla \cdot \boldsymbol{E} \times \boldsymbol{H} = \boldsymbol{H} \cdot \nabla \times \boldsymbol{E} - \boldsymbol{E} \cdot \nabla \times \boldsymbol{H} = -\boldsymbol{H} \cdot \frac{\partial \boldsymbol{B}}{\partial t} - \boldsymbol{E} \cdot \nabla \times \boldsymbol{H}$$

we are lead to

$$\frac{dW}{dt} + \int_{\Omega} \left(\boldsymbol{E} \cdot \frac{\partial \boldsymbol{D}}{\partial t} + \boldsymbol{H} \cdot \frac{\partial \boldsymbol{B}}{\partial t}\right) d^3r = -\oint_{\partial\Omega} \boldsymbol{E} \times \boldsymbol{H} \cdot d\boldsymbol{S}\,. \tag{7.25}$$

This is Poynting's theorem for the macroscopic Maxwell's equations. The second term on the left-hand side can be associated with the energy stored in the electromagnetic fields, and the term on the right-hand side describes the energy transport through the Poynting vector $\boldsymbol{S} = \boldsymbol{E} \times \boldsymbol{H}$.

Poynting's Theorem for Linear Medium We now specialize this result for a linear medium with response functions that are nonlocal in time. We make a Fourier decomposition of the electromagnetic fields,

$$\boldsymbol{E}(\boldsymbol{r}, t) = \int_{-\infty}^{\infty} e^{-i\omega t} \boldsymbol{E}(\boldsymbol{r}, \omega)\, d\omega = \int_{0}^{\infty} e^{-i\omega t} \boldsymbol{E}(\boldsymbol{r}, \omega)\, d\omega + \text{c.c.}\,,$$

where c.c. denotes the complex conjugate of the preceding term. In order to arrive at the second expression, we have used $\boldsymbol{E}^*(\boldsymbol{r},\omega)=\boldsymbol{E}(\boldsymbol{r},-\omega)$, which can be directly proven by taking the complex conjugate of the Fourier integral. A product of two functions can then be expanded as follows

$$\boldsymbol{E}(\boldsymbol{r},t)\cdot\boldsymbol{D}(\boldsymbol{r},t)=\int_0^\infty e^{-i(\omega-\omega')t}\,\boldsymbol{E}^*(\boldsymbol{r},\omega')\cdot\boldsymbol{D}(\boldsymbol{r},\omega)\,d\omega d\omega'+\text{c.c.}\,. \tag{7.26}$$

In the following we assume that $\boldsymbol{E}(\boldsymbol{r},\omega)$ is a spectrally narrow distribution (corresponding to a long pulse) centered around frequency ω_0. Quite generally, Eq. (7.26) should also contain terms oscillating with $e^{\pm i(\omega+\omega')t}$, which become zero when averaged over an oscillation period $2\pi/\omega_0$, and will be neglected for simplicity. We start with the first term in parentheses of Eq. (7.25),

$$\boldsymbol{E}\cdot\frac{\partial\boldsymbol{D}}{\partial t}=\int_0^\infty e^{-i(\omega-\omega')t}\,\boldsymbol{E}^*(\boldsymbol{r},\omega')\cdot\big[-i\omega\varepsilon(\omega)\big]\boldsymbol{D}(\boldsymbol{r},\omega)\,d\omega d\omega'\,. \tag{7.27}$$

For a spectrally narrow pulse we can approximate the term in brackets as

$$\omega\varepsilon(\omega)\approx\omega_0\varepsilon(\omega_0)+\frac{d}{d\omega}\Big[\omega\varepsilon(\omega)\Big]_{\omega=\omega_0}(\omega-\omega_0)=g_0+g_1(\omega-\omega_0)\,. \tag{7.28}$$

As we will show next, g_0 gives rise to absorption losses and g_1 to dispersion corrections. Consider the g_0 term first. The integral in Eq. (7.27) becomes

$$\int_0^\infty\Big\{-ig_0\,e^{-i(\omega-\omega')t}\,\boldsymbol{E}^*(\omega')\cdot\boldsymbol{E}(\omega)+ig_0^*\,e^{i(\omega-\omega')t}\,\boldsymbol{E}(\omega')\cdot\boldsymbol{E}^*(\omega)\Big\}\,d\omega d\omega'\,,$$

where we have suppressed the $\boldsymbol{r}$ dependence of the electric fields. Exchanging in the second term $\omega\leftrightarrow\omega'$ we find

$$2g_0''\int_0^\infty e^{-i(\omega-\omega')t}\,\boldsymbol{E}^*(\omega')\cdot\boldsymbol{E}(\omega)\,d\omega d\omega'=g_0''\,\boldsymbol{E}(t)\cdot\boldsymbol{E}(t)\,.$$

For the g_1 term we only consider the real part, which is responsible for dispersion. The imaginary part of g_1 would lead to small corrections of the absorption expression derived above. We then obtain

$$-ig_1'\int_0^\infty e^{-i(\omega-\omega')t}\,\boldsymbol{E}^*(\omega')\cdot\boldsymbol{E}(\omega)\Big[(\omega-\omega_0)-(\omega'-\omega_0)\Big]\,d\omega d\omega'\,,$$

where we have exchanged $\omega\leftrightarrow\omega'$ to arrive at the second term. This can be rewritten in the form

$$g_1'\frac{\partial}{\partial t}\int_0^\infty e^{-i(\omega-\omega')t}\,\boldsymbol{E}^*(\omega')\cdot\boldsymbol{E}(\omega)\,d\omega d\omega'=\frac{1}{2}g_1'\,\frac{\partial}{\partial t}\boldsymbol{E}(t)\cdot\boldsymbol{E}(t)\,.$$

A similar derivation can be done for the magnetic fields in Eq. (7.25).

Putting together all results, we are finally led to Poynting's theorem for dispersive and absorbing media

Poynting's Theorem Including Dispersion and Absorption

$$\frac{dW}{dt}+\frac{d}{dt}\int_\Omega \frac{1}{2}\left(\left[\frac{d\,\omega\varepsilon'(\omega)}{d\omega}\right]_{\omega_0}\boldsymbol{E}\cdot\boldsymbol{E}+\left[\frac{d\,\omega\mu'(\omega)}{d\omega}\right]_{\omega_0}\boldsymbol{H}\cdot\boldsymbol{H}\right)d^3r \quad (7.29)$$
$$=-\oint_{\partial\Omega}\boldsymbol{E}\times\boldsymbol{H}\cdot d\boldsymbol{S}-\int_\Omega \omega_0\Big(\varepsilon''(\omega_0)\boldsymbol{E}\cdot\boldsymbol{E}+\mu''(\omega_0)\boldsymbol{H}\cdot\boldsymbol{H}\Big)d^3r\,.$$

In the above expression $\boldsymbol{E}(\boldsymbol{r},t)$, $\boldsymbol{H}(\boldsymbol{r},t)$ are time-dependent expressions, which are assumed to have a narrow spectrum centered around frequency ω_0. The different terms on the left- and right-hand side can be interpreted as follows:

1st Term on lhs. Power performed by the external sources.
2nd Term on lhs. Effective energy density stored in electromagnetic fields and in the material polarization and magnetization. The above expression accounts for dispersion effects but reduces to Eq. (4.14) for constant ε, μ values.
1st Term on rhs. Energy flow described by Poynting vector.
2nd Term on rhs. Material losses described by the imaginary parts ε'', μ''.

7.3.2 *Kramers–Kronig Relation*

For a response nonlocal in time, the dielectric displacement is related to the electric field through

$$\boldsymbol{D}(\boldsymbol{r},t)=\varepsilon_0\left\{\boldsymbol{E}(\boldsymbol{r},t)+\int_0^\infty \chi_e(\boldsymbol{r},\tau)\boldsymbol{E}(\boldsymbol{r},t-\tau)\,d\tau\right\}.$$

The important point about this expression is that only fields in the past contribute to the system's response, and thus causality is fulfilled. As we will discuss in this section, as a consequence of this there exists a strict relation between the real and imaginary parts of the permittivity, the so-called Kramers–Kronig relation, and from the knowledge of the real part one can obtain the imaginary part and vice versa. The derivation of the relation is quite general, and the only ingredients needed are causality and linearity of the response.

From the Fourier transform of the above equation we find

$$\varepsilon(\omega)/\varepsilon_0=1+\int_0^\infty e^{i\omega\tau}\chi_e(\tau)\,d\tau\,, \tag{7.30}$$

where from now on we suppress the $\boldsymbol{r}$ dependence of χ_e. Taking the complex conjugate of this, we can establish a relation between positive and negative frequencies

$$\varepsilon^*(\omega)/\varepsilon_0 = \varepsilon(-\omega^*)/\varepsilon_0\,, \tag{7.31}$$

where we have considered complex frequencies for reasons to become clear in a moment. In what follows, we will need an important theorem from complex analysis, the so-called Cauchy's theorem, which we briefly discuss in Appendix A. It states that an integration in the complex plane along a closed contour C gives zero if the integrand is analytic within C. In the following we apply Cauchy's theorem to the response function

$$\chi_e(z) = \int_0^\infty e^{iz\tau}\,\chi_e(\tau)\,d\tau\,,$$

where we have considered a complex frequency z. When $\chi_e(\omega)$ exists for real frequencies, nothing severe can happen when extending the frequencies to the upper complex plane with $z'' > 0$. There, the exponential reads

$$e^{i(z'+iz'')\tau} = e^{iz'\tau}e^{-z''\tau}\,.$$

We have always $\tau > 0$ because of the integration limits (causality of the response comes here into play), all what happens is that in the upper complex plane the function values become exponentially damped. For this reason, $\chi_e(z)$ is analytic in the upper complex half-plane. We can thus employ Cauchy's theorem to the following expression

$$\oint_C \frac{\chi_e(z)}{z-\omega}\,dz = 0\,. \tag{7.32}$$

The integration path C in the complex plane is shown in Fig. 7.7. Two points along the path must be considered with special care:

- for $z = \omega$ the denominator in Eq. (7.32) becomes zero,
- for $z \to 0$ the permittivity of metals $\varepsilon(z) \approx \varepsilon_0 + i\sigma_0/z$ can have a pole.

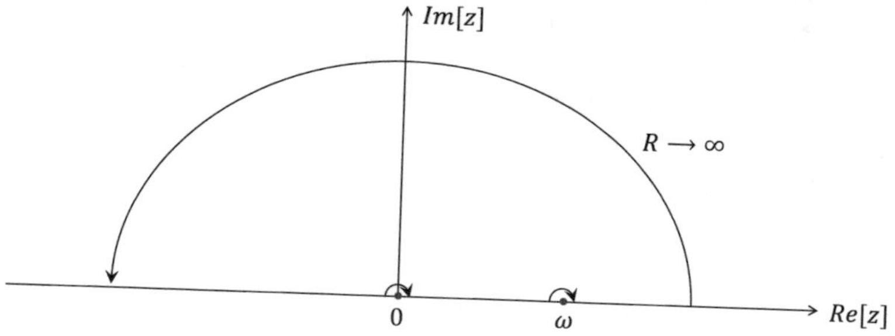

Fig. 7.7 Complex integration path for derivation of Kramers–Kronig relation

As can be seen in the figure, around these special points we move along small semicircles in the upper half-plane. The semicircle around $z = \omega$ can be parameterized in the form

$$z = \omega + re^{i\phi}\,,$$

where ϕ runs from π to zero and we let $r \to 0$ at the end of the calculation. We now assume that $\chi_e(z)$ does not change dramatically around $z = \omega$ and thus pull it out of the integral,

$$\chi_e(\omega) \lim_{r\to 0} \int_\pi^0 \frac{ire^{i\phi}\,d\phi}{\omega + re^{i\phi} - \omega} = i\,\chi_e(\omega) \lim_{r\to 0} \int_\pi^0 d\phi = -i\pi\,\chi_e(\omega)\,.$$

We have used $dz = re^{i\phi}\,d\phi$. A similar procedure can be applied for the pole at $z = 0$. We finally note that the large semicircle in the upper complex half-plane goes to zero because of the $e^{-z''\tau}$ term, and we can rewrite Eq. (7.32) in the form

$$\mathcal{P}\int_{-\infty}^{\infty} \frac{\chi_e(\tilde\omega)}{\tilde\omega - \omega}\,d\tilde\omega - i\pi\Big[\chi_e(\omega) - \frac{i\sigma_0}{\varepsilon_0\omega}\Big] = 0\,.$$

The integral on the right-hand side is a principal value integral where the (infinitesimally small) regions around the critical points $\omega' = 0$ and $\omega' = \omega$ are excluded from the integration. See also Appendix F. Thus, we obtain

$$\chi_e(\omega) = \frac{i\sigma_0}{\varepsilon_0\omega} - \frac{i}{\pi}\mathcal{P}\int_{-\infty}^{\infty} \frac{\chi_e(\tilde\omega)}{\tilde\omega - \omega}\,d\tilde\omega\,. \tag{7.33}$$

By taking the real part on both sides of this equation one obtains a relation between $\chi_e'(\omega)$ on the left-hand side and $\chi_e''(\omega')$ on the right-hand side. Using the dielectric function rather than the electric susceptibility and relating in $\varepsilon(\omega)$ negative frequencies to positive ones by use of Eq. (7.31), we are finally led to the Kramers–Kronig relation

Kramers–Kronig Relation

$$\varepsilon'(\omega) = \varepsilon_0 + \frac{2}{\pi}\mathcal{P}\int_0^{\infty} \frac{\tilde\omega\varepsilon''(\tilde\omega)}{\tilde\omega^2 - \omega^2}\,d\tilde\omega \tag{7.34a}$$

$$\varepsilon''(\omega) = \frac{\sigma_0}{\omega} - \frac{2\omega}{\pi}\mathcal{P}\int_0^{\infty} \frac{\varepsilon'(\tilde\omega) - \varepsilon_0}{\tilde\omega^2 - \omega^2}\,d\tilde\omega\,. \tag{7.34b}$$

The Kramers–Kronig relation shows that the real and imaginary parts of ε are related. Thus, if we know $\varepsilon''(\omega)$ within the entire frequency range, for instance, from absorption measurements, we can immediately compute the corresponding real part. For practical situations the relevant frequency range is often surprisingly large, and one usually has to be careful when exploiting the relation in this way.

Equation (7.34) also puts constraints on possible response functions. Suppose that we have a wish-list for a material's permittivity $\varepsilon(\omega)$ that should have strong variations of $\varepsilon'(\omega)$ within a given frequency range but small losses there. In principle, the Kramers–Kronig relation only states that strong dispersion must be accompanied by losses, which could be possibly located in some other frequency range. However, in many cases dispersion is due to microscopic resonances, such as described by the Drude–Lorentz model in Sect. 7.1, and one often has to pay for dispersion the price of absorption in the same frequency range. This is also a major drawback in the design of metamaterials where large frequency variations of $\varepsilon(\omega)$, $\mu(\omega)$ are accompanied by significant losses at some neighbor frequencies.

7.4 Reciprocity Theorem in Optics

There exists an important theorem in optics, the so-called reciprocity theorem, from which one can derive a symmetry relation for the Green's dyadics. Suppose that

- $\boldsymbol{J}_1$ is a current distribution which produces the fields $\boldsymbol{E}_1$, $\boldsymbol{H}_1$ and
- $\boldsymbol{J}_2$ is a current distribution which produces the fields $\boldsymbol{E}_2$, $\boldsymbol{H}_2$.

Then, one can derive the (Lorentz) reciprocity theorem

Reciprocity Theorem of Optics

$$\int \boldsymbol{J}_1 \cdot \boldsymbol{E}_2 \, d^3r = \int \boldsymbol{J}_2 \cdot \boldsymbol{E}_1 \, d^3r \,. \tag{7.35}$$

Here we assume that the integrals extend over the entire space. This theorem states that the relationship between an oscillating current distribution and the resulting electric field is unchanged if one interchanges the points where the current is placed and where the field is measured.

To prove the theorem, in the following we consider materials with frequency-dependent and anisotropic permittivites and permeabilites $\bar{\bar{\varepsilon}}$, $\bar{\bar{\mu}}$. The curl equations of Maxwell's equations then read

$$\nabla \times \boldsymbol{E} = i\omega\bar{\bar{\mu}} \cdot \boldsymbol{H} \,, \quad \nabla \times \boldsymbol{H} = \boldsymbol{J} - i\omega\bar{\bar{\varepsilon}} \cdot \boldsymbol{E} \,,$$

where for simplicity we do not indicate the spatial and frequency dependence of the fields and material parameters. We next consider the vector identity

$$\nabla \cdot (\boldsymbol{E} \times \boldsymbol{H}) = \boldsymbol{H} \cdot (\nabla \times \boldsymbol{E}) - \boldsymbol{E} \cdot (\nabla \times \boldsymbol{H}) .$$

Applying this identity to $\boldsymbol{E}_1$, $\boldsymbol{H}_2$ and $\boldsymbol{E}_2$, $\boldsymbol{H}_1$, and using Maxwell's equations for the curl terms then gives

$$\begin{aligned} \nabla \cdot (\boldsymbol{E}_1 \times \boldsymbol{H}_2) &= i\omega \boldsymbol{H}_2 \cdot \bar{\bar{\mu}} \cdot \boldsymbol{H}_1 + i\omega \boldsymbol{E}_1 \cdot \bar{\bar{\varepsilon}} \cdot \boldsymbol{E}_2 - \boldsymbol{E}_1 \cdot \boldsymbol{J}_2 \\ \nabla \cdot (\boldsymbol{E}_2 \times \boldsymbol{H}_1) &= i\omega \boldsymbol{H}_1 \cdot \bar{\bar{\mu}} \cdot \boldsymbol{H}_2 + i\omega \boldsymbol{E}_2 \cdot \bar{\bar{\varepsilon}} \cdot \boldsymbol{E}_1 - \boldsymbol{E}_2 \cdot \boldsymbol{J}_1 . \end{aligned} \tag{7.36}$$

In the following we subtract these two equations. For the terms on the right-hand side we then get

$$\boldsymbol{E}_1 \cdot \bar{\bar{\varepsilon}} \cdot \boldsymbol{E}_2 - \boldsymbol{E}_2 \cdot \bar{\bar{\varepsilon}} \cdot \boldsymbol{E}_1 = \left(\boldsymbol{E}_1\right)_i \varepsilon_{ij} \left(\boldsymbol{E}_2\right)_j - \left(\boldsymbol{E}_2\right)_i \varepsilon_{ij} \left(\boldsymbol{E}_1\right)_j ,$$

and a corresponding expression for the magnetic fields. Obviously, for a scalar permittivity $\bar{\bar{\varepsilon}} = \varepsilon \mathbb{1}$ the two terms cancel each other. In the general case it can be shown that $\bar{\bar{\varepsilon}}$, $\bar{\bar{\mu}}$ must be symmetric tensors when the system under investigation exhibits time reversal symmetry [43]. A case where such symmetry would be broken is when a static magnetic field is applied. We shall not be interested in such cases here, and assume in the following that $\bar{\bar{\varepsilon}}$, $\bar{\bar{\mu}}$ are symmetric tensors such that the two terms of the above expression cancel each other. Then, subtraction of the two equations in Eq. (7.36) and integration over the entire space gives

$$\int \Big(\boldsymbol{J}_1 \cdot \boldsymbol{E}_2 - \boldsymbol{J}_2 \cdot \boldsymbol{E}_1\Big)\, d^3r = \oint \Big(\boldsymbol{E}_1 \times \boldsymbol{H}_2 - \boldsymbol{E}_2 \times \boldsymbol{H}_1\Big) \cdot d\boldsymbol{S} . \tag{7.37}$$

To understand why the term on the right-hand side vanishes, we note that far away from the current sources the electromagnetic fields are outgoing waves. According to Eq. (2.40), we can relate the electromagnetic fields in the far-field zone through $Z\boldsymbol{H} = \hat{\boldsymbol{r}} \times \boldsymbol{E}$, and one can readily show that the two terms on the right-hand side cancel each other. This completes our proof for the reciprocity theorem of Eq. (7.35).

Relating the electric fields to the current sources through the Green's dyadics of Eq. (5.20), we get

$$\boldsymbol{J}_1(\boldsymbol{r}) \cdot \bar{\bar{G}}(\boldsymbol{r}, \boldsymbol{r}') \cdot \boldsymbol{J}_2(\boldsymbol{r}') = \boldsymbol{J}_2(\boldsymbol{r}') \cdot \bar{\bar{G}}(\boldsymbol{r}', \boldsymbol{r}) \cdot \boldsymbol{J}_1(\boldsymbol{r}) .$$

Thus, we immediately obtain from the reciprocity theorem the symmetry relation

Symmetry Relation for Dyadic Green's Function

$$G_{ij}(\boldsymbol{r}, \boldsymbol{r}') = G_{ji}(\boldsymbol{r}', \boldsymbol{r}) . \tag{7.38}$$

Exercises

Exercise 7.1 For the Drude–Lorentz permittivity of Eq. (7.6), compute the frequencies where $\varepsilon''(\omega)$ and $|\varepsilon(\omega)|^2$ become maximal. Use the harmonic-oscillator model to argue why the maxima are at different frequencies.

Exercise 7.2 Compute the loss function $\text{Im}[-1/\varepsilon(\omega)]$ for the Drude dielectric function, and plot the function for Ag and Au using the parameters given in Table 7.1. Discuss how the peak positions of the loss function depend on κ_b and the plasma frequency ω_p.

Exercise 7.3 Consider a Drude–Lorentz permittivity of Eq. (7.6) with a sharp resonance, $\gamma \ll \omega_0$. Where does the group velocity v_g become minimal? How does a spectrally narrow wave packet become attenuated (because of material losses) during propagation?

Exercise 7.4 Compute the density of states given in Eq. (7.10). Discuss what happens when the three-dimensional k-space integration is replaced by a two-dimensional integration, corresponding to an electron gas where the electrons can only move in two spatial dimensions. To evaluate the integrals, introduce spherical coordinates in 3D and cylinder coordinates in 2D.

Exercise 7.5 Consider a dielectric function whose imaginary part is given by

$$\varepsilon''(\omega) = \frac{\pi K}{2\omega_0}\delta(\omega - \omega_0)\,,$$

where K is a constant and ω_0 a resonance frequency. Use the Kramers–Kronig relation to compute the corresponding real part $\varepsilon'(\omega)$.

Exercise 7.6 Start from the Drude–Lorentz permittivity of Eq. (7.6) and perform a Fourier transformation to get the time response function. The Fourier integral is evaluated most easily when using the complex integration techniques discussed in Appendix A and in the derivation of the Kramers–Kronig relation.

Exercise 7.7 Show through explicit calculation that the Kramers–Kronig relation is fulfilled for the Drude permittivity of Eq. (7.7).

Exercise 7.8 Show that the right-hand side in Eq. (7.37) becomes zero in the far-field zone, where the magnetic and electric fields are related through $Z\boldsymbol{H} = \hat{\boldsymbol{r}} \times \boldsymbol{E}$.

Exercise 7.9 Show through explicit calculation that the Green's dyadics of Eq. (5.19) fulfills the symmetry relation of Eq. (7.38).

Chapter 8
Stratified Media

The most simple (nano)systems for combining Maxwell's equations with materials are planar layers, which, in case of various layers of materials, are conveniently denoted as *stratified media*. As we will show in this chapter, despite being geometrically simple the physics of planar systems is surprisingly rich. We start by discussing a single interface between a metal and a dielectric, and show that a novel type of excitations, so-called surface plasmons, exists at the interface between the two media. We then continue to develop a general description scheme for stratified media, using a transfer matrix approach, and finally ponder on the calculation of Green's functions for stratified media.

8.1 Surface Plasmons

Consider the interface between two media as depicted in Fig. 8.1. The upper one is a dielectric with permittivity $\varepsilon_1 > 0$, the lower one a metal with a negative permittivity $\varepsilon_2 < 0$. For simplicity, in the following we ignore the imaginary part of ε_2 and discuss implications of lossy materials at the end. The magnetic permeabilities in both materials are set to μ_0.

Suppose that an electromagnetic wave with wavevector $\boldsymbol{k}_1 = (k_x, 0, -k_{1z})$ propagates in the negative z-direction and impinges on the interface. Because of the continuity of the tangential electromagnetic fields, the parallel component k_x of the wavevector must be conserved at the interface. The z-component of the wavevector in the metal is determined from the dispersion relation

$$k_x^2 + k_{2z}^2 = \varepsilon_2 \mu_0 \, \omega^2 \,. \tag{8.1}$$

Because of $\varepsilon_2 < 0$, the right-hand side of the equation is negative and we immediately find $k_{2z}^2 < 0$, which can only be fulfilled for an imaginary wavenumber

U. Hohenester, *Nano and Quantum Optics*, Graduate Texts in Physics,
https://doi.org/10.1007/978-3-030-30504-8_8

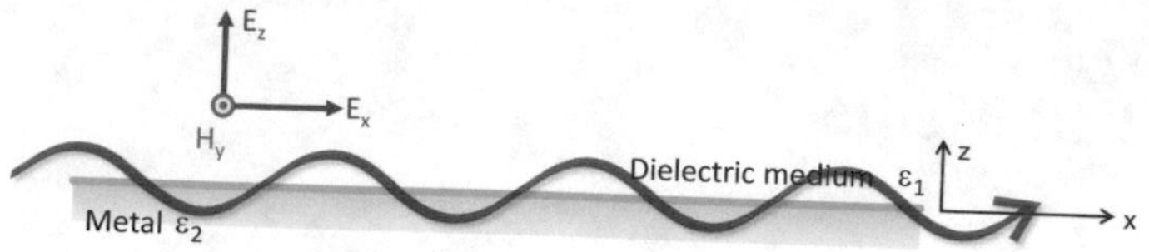

Fig. 8.1 Schematics of a wave that is confined at the interface between a metal and a dielectric and propagates along the x direction. For transverse magnetic (TM) modes the magnetic field is parallel to the interface

k_{2z}. In other words, inside the metal the wave cannot propagate but has an evanescent character with an amplitude that decays exponentially when moving away from the interface. As a result, a wave impinging from the dielectric side on the metal becomes reflected, with only small losses due to ohmic dissipation caused by the exponentially decaying fields inside the metal. In Sect. 8.3.3 we will provide a more thorough discussion of this reflection.

From this analysis it seems that the dielectric–metal interface is a boring object. Fortunately, this hasty judgement is not true. We will show next that a novel type of wave exists at metal–dielectric interfaces, so-called *surface plasmons*, which are bound to the interface and have to be excited optically in a specific manner. In fact, we can distinguish two kinds of guided modes, namely

$$\text{Transverse magnetic (TM):} \quad \boldsymbol{H} = H_y\hat{\boldsymbol{y}} \text{ is parallel to interface}$$
$$\text{Transverse electric (TE):} \quad \boldsymbol{E} = E_y\hat{\boldsymbol{y}} \text{ is parallel to interface.}$$

We first discuss TM modes and will show later that no TE modes exist at metal–dielectric interfaces. For the magnetic fields above and below the interface, we make the following ansatz:

$$\begin{aligned} H_{1y} &= h_1\, e^{ik_x x} e^{-\kappa_{1z} z} = h_1\, e^{i\boldsymbol{k}_1\cdot\boldsymbol{r}}\,, \quad \boldsymbol{k}_1 = (k_x, 0, \;\; i\kappa_{1z}) \\ H_{2y} &= h_2\, e^{ik_x x} e^{\;\kappa_{2z} z} = h_2\, e^{i\boldsymbol{k}_2\cdot\boldsymbol{r}}\,, \quad \boldsymbol{k}_2 = (k_x, 0, -i\kappa_{2z})\,, \end{aligned} \tag{8.2}$$

with the interface being located at position $z = 0$. Here $\kappa_{jz} = (k_x^2 - \varepsilon_j \mu_0 \omega^2)^{\frac{1}{2}}$, and the signs in the exponentials $e^{\pm\kappa_{jz}z}$ have been chosen such that the fields decay exponentially in the upper and lower media. The electric field can be computed from Ampere's law, which leads us to

$$\begin{aligned} \boldsymbol{E}_1 &= \frac{1}{\varepsilon_1\omega} H_{1y}\hat{\boldsymbol{y}} \times \left(k_x\hat{\boldsymbol{x}} + i\kappa_{1z}\hat{\boldsymbol{z}}\right) = \frac{1}{\varepsilon_1\omega} H_{1y}\left(\;\; i\kappa_{1z}\hat{\boldsymbol{x}} - k_x\hat{\boldsymbol{z}}\right) \\ \boldsymbol{E}_2 &= \frac{1}{\varepsilon_2\omega} H_{2y}\hat{\boldsymbol{y}} \times \left(k_x\hat{\boldsymbol{x}} - i\kappa_{2z}\hat{\boldsymbol{z}}\right) = \frac{1}{\varepsilon_2\omega} H_{2y}\left(-i\kappa_{2z}\hat{\boldsymbol{x}} - k_x\hat{\boldsymbol{z}}\right)\,. \end{aligned} \tag{8.3}$$

We next employ the boundary conditions of continuous tangential electric and magnetic fields, and get from Eqs. (8.2) and (8.3)

$$H_{1y}\Big|_{z=0} = H_{2y}\Big|_{z=0} \implies h_1 = h_2$$

$$E_{1x}\Big|_{z=0} = E_{2x}\Big|_{z=0} \implies h_1\frac{\kappa_{1z}}{\varepsilon_1} = -h_2\frac{\kappa_{2z}}{\varepsilon_2}\,.$$

From the expression for the continuity of the tangential electric fields we obtain

$$\frac{\kappa_{1z}}{\kappa_{2z}} = -\frac{\varepsilon_1}{\varepsilon_2}\,. \tag{8.4}$$

Thus, in order to fulfill this expression for real and positive κ-values we find that the signs of the dielectric functions above and below the interface must be different. This leads us to the condition

First Surface Plasmon Condition

$$\varepsilon_1(\omega)\,\varepsilon_2(\omega) < 0\,. \tag{8.5}$$

For the choice of dielectric ($\varepsilon_1 > 0$) and metallic ($\varepsilon_2 < 0$) materials above and below the interface the condition of Eq. (8.5) is automatically fulfilled. We next square both sides of Eq. (8.4) and rewrite the κ_{1z}^2, κ_{2z}^2 terms using the dispersion relation of Eq. (8.1) to arrive at

$$\frac{\varepsilon_1^2}{\varepsilon_2^2} = \frac{k_x^2 - \varepsilon_1\mu_0\,\omega^2}{k_x^2 - \varepsilon_2\mu_0\,\omega^2}\,.$$

Solving for k_x gives

$$k_x = \sqrt{\frac{\varepsilon_1\varepsilon_2}{\varepsilon_0(\varepsilon_1+\varepsilon_2)}}\;k_0\,, \tag{8.6}$$

with $k_0 = \sqrt{\varepsilon_0\mu_0}\,\omega$ being the free-space wavenumber. In order to have a real wavenumber k_x, the following inequality must hold:

Second Surface Plasmon Condition

$$-\,\varepsilon_2(\omega) > \varepsilon_1(\omega)\,. \tag{8.7}$$

In other words, the dielectric function of the metal ε_2 must be stronger negative than ε_1 of the dielectric material. For typical metals with a Drude-like dielectric function this condition can be easily fulfilled.

This new type of excitations are usually called *surface plasmons*, and are propagating waves confined to the interface. The physical nature of these waves is characterized by a number of important features.

Surface Charges. The surface charges induced at the metal–dielectric interface can be computed from Eq. (7.24),

$$\sigma = \varepsilon_0\big(E_{1z} - E_{2z}\big)\Big|_{z=0} = \frac{k_x}{\omega}\left(\frac{\varepsilon_0}{\varepsilon_2} - \frac{\varepsilon_0}{\varepsilon_1}\right) h_1\, e^{ik_x x}\,. \tag{8.8}$$

Apparently, the wave is accompanied by coherent charge density oscillations at the interface, as schematically shown in Fig. 8.2. Similarly to bulk plasmons, the separated charges lead to restoring forces which drive the surface plasmon oscillations.

Evanescent Fields. Because of the confined character of the modes, the electric and magnetic fields have an evanescent character $e^{-\kappa_{jz}|z|}$ and decay exponentially away from the interface, as schematically shown on the right-hand side of Fig. 8.2.

Polarization. From Eq. (8.3) we observe that the waves have an elliptic polarization with the (unnormalized) polarization vector $i\kappa_{jz}\hat{\boldsymbol{x}} - k_x\hat{\boldsymbol{z}}$.

Energy Transport. The averaged Poynting vector for surface plasmons in the dielectric medium can be computed from Eqs. (8.2) and (8.3)

$$\boldsymbol{S}_1 = \frac{1}{2}\,\mathrm{Re}\left(\boldsymbol{E}_1 \times \boldsymbol{H}_1^*\right) = \frac{h_1^2}{2\varepsilon_1\omega} e^{-2\kappa_{1z}z}\mathrm{Re}\left(i\kappa_{1z}\hat{\boldsymbol{z}} + k_x\hat{\boldsymbol{x}}\right) = \frac{h_1^2 k_x}{2\varepsilon_1\omega}\hat{\boldsymbol{x}}\,. \tag{8.9}$$

In agreement to our previous discussions, we find that evanescent waves do not transport energy in the z direction.

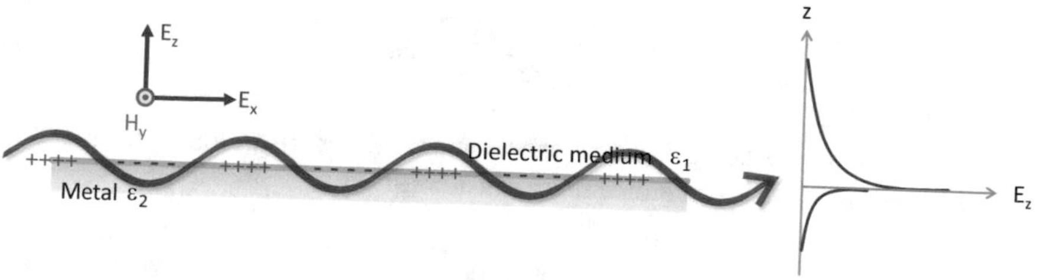

Fig. 8.2 Surface plasmons are electromagnetic waves that propagate at the interface between a metal and a dielectric. They are accompanied with coherent charge density oscillations at the interface. The electric and magnetic fields have an evanescent character and decay exponentially when moving away from the interface

Losses. Losses in the metal can be described through an imaginary part of the permittivity, $\varepsilon_2'' > 0$. When going through our above derivation for surface plasmons, we notice that practically nothing has to be changed for a complex $\varepsilon_2 = \varepsilon_2' + i\varepsilon_2''$. The only main difference is that k_x acquires an imaginary part, which for small losses can be expressed as

$$k_x = \sqrt{\frac{\varepsilon_1\varepsilon_2'}{\varepsilon_0(\varepsilon_1+\varepsilon_2')}}\,k_0\left[1+\frac{i}{2}\left(\frac{\varepsilon_1}{\varepsilon_1+\varepsilon_2'}\right)\frac{\varepsilon_2''}{\varepsilon_2'}+\mathcal{O}\left(\varepsilon''^2\right)\right]. \tag{8.10}$$

This leads to a damping of the propagating plasmon,

$$e^{i(k_x'+ik_x'')x} = e^{-k''x}e^{ik'x} = (\text{damping}) \times (\text{oscillation}).$$

Also κ_{2z} acquires a small imaginary part. As a consequence, the Poynting vector in Eq. (8.9) gets a component in the z-direction, associated with a flow of electromagnetic energy into the metal where it becomes transformed through ohmic losses into heat.

Dispersion. For a Drude dielectric function and $\varepsilon_1 = 1$ the plasmon dispersion of Eq. (8.6) can be inverted, and we get

$$k_x = \sqrt{\frac{1-\frac{\omega_p^2}{\omega^2}}{2-\frac{\omega_p^2}{\omega^2}}}\,\frac{\omega}{c} \implies \omega = \sqrt{\frac{\omega_p^2+2k_x^2c^2-\sqrt{\omega_p^4+4k_x^4c^4}}{2}}. \tag{8.11}$$

The signs of the square roots have been chosen such that ω is positive throughout and $\omega \to 0$ for $k_x \to 0$. Figure 8.3 shows the surface plasmon dispersion $\omega(k_x)$. The black dashed line indicates the light line $\omega = ck$ and the red dotted line the asymptotic value

$$\omega \underset{k_x\to\infty}{\longrightarrow} \frac{\omega_p}{\sqrt{2}}, \tag{8.12}$$

which can be obtained most easily by seeking in the first expression of Eq. (8.11) for the zeros of the denominator $2-\omega_p^2/\omega^2$. Throughout, the light line is above the surface plasmon dispersion and the two curves never cross. This is because the surface plasmon velocity is slowed down in comparison to the speed of light, owing to the hybrid light-matter character of the surface plasmon where the evanescent fields dip into the metal and drive the coherent surface charge oscillations. Figure 8.4 shows the surface plasmon dispersion for a realistic silver dielectric function. The results are similar to the more simple Drude description, but the plasmon frequencies now acquire an imaginary part associated with ohmic losses. For large photon energies these losses become so large that the dispersion bends back.

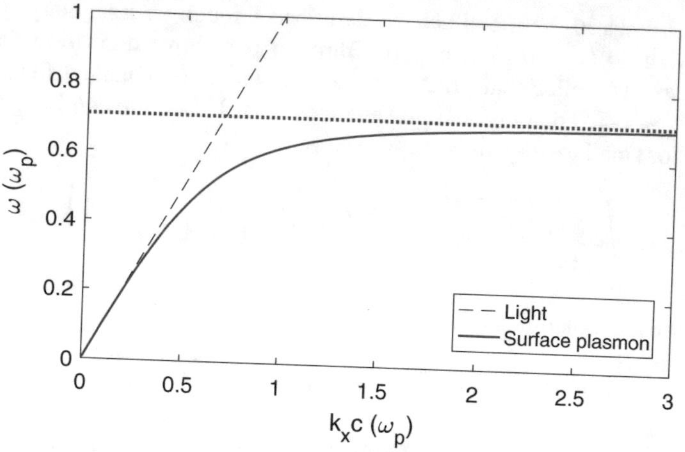

Fig. 8.3 Dispersion for surface plasmon, as computed from Eq. (8.6) using a Drude dielectric function $\varepsilon_2(\omega) = 1 - \omega_p^2/\omega^2$ and $\varepsilon_1 = 1$. The asymptotic value $\omega_p/\sqrt{2}$ is indicated by the red dotted line, the light dispersion $\omega = ck$ by the black dashed line

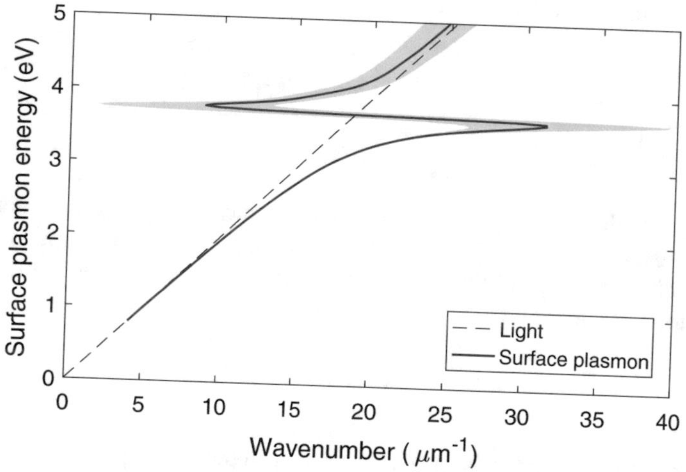

Fig. 8.4 Same as Fig. 8.4 but for a silver dielectric function extracted from experiment [34]. The gray shaded area indicates the broadening $k'_x \pm \frac{1}{2}k''_x$ associated with plasmon losses. For energies above say 3.7 eV plasmons are strongly damped and the dispersion bends back

Light Confinement. Owing to the hybrid light-matter character, the wavelength of surface plasmons can be significantly shorter than the free-space wavelength of light. This allows for light confinement in sub-wavelength volumes.

Propagation Length. Figure 8.5 shows the propagation length $\text{Im}(k_x)^{-1}$ caused by surface plasmon damping. In general, one is only interested in surface plasmons with a sufficiently small attenuation, say surface plasmons for silver

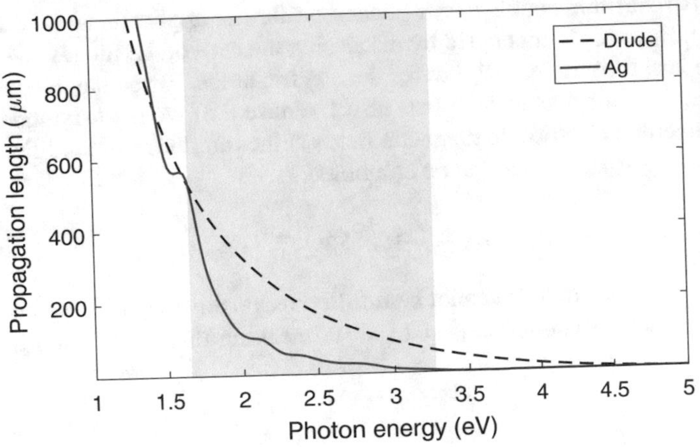

Fig. 8.5 Propagation length $\mathrm{Im}(k_x)^{-1}$ for surface plasmons in silver using a Drude dielectric function (dashed line) and a dielectric function extracted from experiment [34] (solid line). The gray shaded area indicates the visible regime. In general, one is only interested in surface plasmons with sufficiently long propagation lengths

with energies below 3.5 eV, which can propagate over large distances without being strongly damped.

In the paper "The promise of plasmonics" [44], which appeared in 2007 in Scientific American and became influential for the entire research field nowadays called *plasmonics*, Harry Atwater beautifully describes the physical nature of surface plasmons and envisions a number of possible applications.

> In the 1980s researchers experimentally confirmed that directing light waves at the interface between a metal and a dielectric (a nonconductive material such as air or glass) can, under the right circumstances, induce a resonant interaction between the waves and the mobile electrons at the surface of the metal. (In a conductive metal, the electrons are not strongly attached to individual atoms or molecules.) In other words, the oscillations of electrons at the surface match those of the electromagnetic field outside the metal. The result is the generation of surface plasmons—density waves of electrons that propagate along the interface like the ripples that spread across the surface of a pond after you throw a stone into the water.
>
> Over the past decade investigators have found that by creatively designing the metal-dielectric interface they can generate surface plasmons with the same frequency as the outside electromagnetic waves but with a much shorter wavelength. This phenomenon could allow the plasmons to travel along nanoscale wires called interconnects, carrying information from one part of a microprocessor to another. Plasmonic interconnects would be a great boon for chip designers, who have been able to develop ever smaller and faster transistors but have had a harder time building minute electronic circuits that can move data quickly across the chip.

Or, as expressed by Harry Atwater in a single sentence [45]:

> Plasmonics has given photonics the ability to go to the nanoscale.

In later parts of this book we will ponder on further applications. Our analysis of surface plasmonics is concluded by discussing the case of TE modes. We start with the same ansatz as in Eq. (8.2) but for the electric fields. Repeating the derivation in a completely analogous fashion (see also Exercise 8.3) we get from the continuity of the tangential electric and magnetic fields at the interface the conditions $e_1 = e_2$ and $e_1\kappa_{1z} = -e_2\kappa_{2z}$, which can be combined to

$$e_1\left(\kappa_{1z} + \kappa_{2z}\right) = 0\,. \tag{8.13}$$

Apparently, this expression cannot be fulfilled for positive κ_{1z} and κ_{2z} values. Thus, the only possible solution is $e_1 = e_2 = 0$, meaning that no TE modes exist at a metal-dielectric interface.

8.1.1 Kretschmann and Otto Geometry

In the previous section we have discussed that at the interface between a metal and a dielectric a novel type of excitations exists, so-called surface plasmons, associated with coherent surface charge excitations bound to light fields. However, we have also seen that these surface plasmons cannot be excited directly by optical means, or, as Harry Atwater has expressed it carefully, can only be excited "under the right circumstances." To understand what these right circumstances are, we first analyze energy and momentum conservation in an optical excitation process.

Light carries energy and momentum. In the following we adopt a photon language where the photon carries energy $\hbar\omega$ and momentum $\hbar\boldsymbol{k}$, but our reasoning works equally well for a purely classical electromagnetic description. Consider light oscillating with angular frequency ω and propagating in a medium with dielectric constant ε_1 that impinges on a metal surface located at $z = 0$. The momentum carried by the photon is

$$\hbar\boldsymbol{k} = \Big[\sin\theta\,\hat{\boldsymbol{x}} + \cos\theta\,\hat{\boldsymbol{z}}\Big]\frac{\hbar k_0}{n_1}\,, \tag{8.14}$$

where θ is the angle of the incoming light with respect to the z-axis, $k_0 = \frac{\omega}{c}$, and n_1 is the refractive index of the dielectric medium. In order to excite a surface plasmon the following quantities must be conserved:

- energy $\hbar\omega$,
- parallel momentum $\hbar k_x = (\hbar k_0/n_1)\sin\theta$.

Note that for the slab structure the translational symmetry along z is broken and according to Noether's theorem the momentum along z is not conserved.

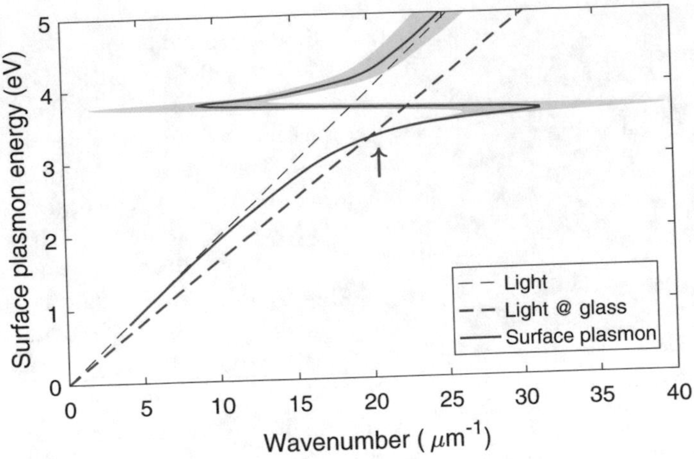

Fig. 8.6 Principle of light excitation of surface plasmons. The surface plasmon dispersion is the same as in Fig. 8.6, but the light line is tilted for light propagating through a dielectric material such as glass. As a result, the surface plasmon dispersion and the light dispersion in glass cross at the point indicated by the arrow where surface plasmon excitation through light becomes possible

Here comes the problem of light excitation of surface plasmons: the light line, indicated by the black dashed line in Fig. 8.6, *never* crosses the surface plasmon dispersion (we do not consider the back-bending at higher photon energies). This is because surface plasmons are slowed down with respect to freely propagating light, as an effect of the hybrid light-matter character. For this reason, even for grazing incidence with $\sin\theta = 1$ energy and momentum cannot be simultaneously conserved. Light impinging on a dielectric–metal interface becomes reflected but cannot excite surface plasmons.

So we must play a trick to optically excite surface plasmons. The basic principle is sketched in Fig. 8.6 and relies on slowing down the incoming light, using a geometry to be discussed in a moment. As a result of this slow down, the tilted light line now crosses the surface plasmon dispersion, see arrow in Fig. 8.6, and a surface plasmon is launched. By varying the frequency ω and the angle θ of the incoming light one can map out the entire surface plasmon dispersion. The problem we have to address next is how to slow down the incoming light without affecting significantly the surface plasmon dispersion, which, according to Eq. (8.6), depends on both ε_1 and ε_2. For this reason, one cannot simply increase ε_1 as this would slow down both the incoming light as well as the propagating surface plasmons in the same manner. Two geometries were suggested by Otto [46] and Kretschmann [47] that allow solving this problem.

Extraordinary Transmission

In 1998 Thomas Ebbesen and coworkers published a paper called "Extraordinary optical transmission through sub-wavelength hole arrays" [48], which received enormous interest and is often considered as the seminal paper in the field of plasmonics. The authors took a 200 nm thin silver film and drilled a periodic array of sub-wavelength holes with diameters of 150 nm into it. When illuminating the array with light, they observed for specific values of wavelength and propagation angle a dramatic increase of the transmitted light intensity. In the paper the authors write [48]:

> In particular, sharp peaks in transmission are observed at wavelengths as large as ten times the diameter of the cylinders. At these maxima the transmission efficiency can exceed unity (when normalized to the area of the holes), which is orders of magnitude greater than predicted by standard aperture theory. Our experiments provide evidence that these unusual optical properties are due to the coupling of light with plasmons—electronic excitations—on the surface of the periodically patterned metal film.

In short, the reason for this strong increase in transmission is that the periodic array of holes breaks the translational symmetry of the silver layer and provides a momentum $G = 2\pi/a$ determined by the lattice constant a of the array. Thus, when light impinges on the air–silver interface, it can pick up this additional momentum G and excite a surface plasmon at the illuminated metal surface. For sufficiently thin films the plasmon tunnels from one side of the silver film to the other one, where it becomes transformed again to light.

The importance of this experiment is the demonstration that optical properties of metals can be completely tailored in nanostructured devices. This has initiated the research field of *plasmonics*, where surface plasmons are the workhorse which are tailored and controlled by means of nanofabrication.

Otto Geometry. In the Otto geometry light propagates through a prism and impinges with an angle θ on a glass–air interface where it suffers total internal reflection, see left panel of Fig. 8.7. When a metal slab is placed close to this interface, the evanescent fields of the incoming and reflected light can excite surface plasmons which propagate at the interface between the metal and air.

Kretschmann Geometry. In the Kretschmann geometry a metal layer with a thickness of a few tens of nanometers is placed on top of the prism. Light impinging from the prism side on the metal becomes reflected, but through the evanescent fields in the metal surface plasmons can be excited at the metal–air interface. This geometry can be realized more easily in experiment, but one has

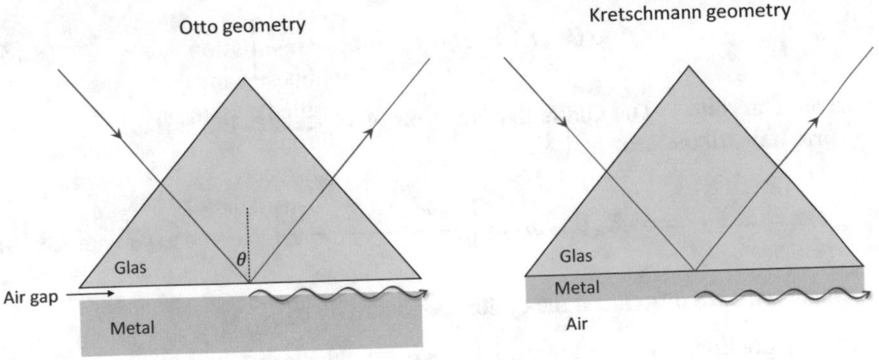

Fig. 8.7 Otto and Kretschmann configuration for light excitation of surface plasmons. In the Otto geometry light propagates through a prism and impinges with an angle θ on a glass–air interface where it suffers total internal reflection. When a metal slab is placed close to this interface, the evanescent fields of the incoming and reflected light can excite surface plasmons which propagate at the interface between the metal and air. In the Kretschmann geometry a metal layer with a thickness of a few tens of nanometers is placed on top of the prism, and the evanescent fields of the reflected light penetrate through the layer and excite surface plasmons at the metal–air interface

to ensure that the metal film is thin enough such that the evanescent fields are still large enough at the adjacent side of the metal film.

To describe these geometries in the framework of Maxwell's equations, further below we will introduce a generic solution scheme for stratified media, the so-called transfer matrix approach. Using this approach, we will then analyze the Kretschmann geometry and will discuss the working principle of plasmon sensors that exploit such geometries.

8.2 Graphene Plasmons

In the previous chapter we have briefly addressed the optical properties of doped graphene. We next adapt our analysis of surface plasmons to the case of such a two-dimensional electron gas with a linear band structure dispersion, Eq. (7.11). Suppose that the graphene sheet is deposited on a substrate with permittivity ε_2, and the material above the sheet has a permittivity ε_1. In analogy to the previous section, we consider excitations with a TM character and make for the magnetic fields the same ansatz as in Eq. (8.2). However, we have to modify the boundary conditions at $z = 0$ to account for the singular surface charge and current distributions of the doping electrons in graphene.

Surface Charge. From Eq. (7.13), which relates the induced surface charge distribution σ to an external potential, we can express σ in terms of the potential $V = -ik_x E_x$ associated with the propagating TM mode through

$$\sigma(k_x, \omega) = P_{2D}(k_x, \omega) \left[\frac{ie^2 E_x}{k_x} \right] . \tag{8.15}$$

Surface Current. The continuity equation (4.11) relates the surface charge and current distributions σ and $\boldsymbol{K}$,

$$\omega\sigma = k_x K_x \implies K_x(k_x, \omega) = \left[\frac{i\omega e^2 P_{2D}(k_x, \omega)}{k_x^2} \right] E_x = \sigma_{2D} E_x \,, \tag{8.16}$$

where we have introduced the optical conductivity σ_{2D}.

From the continuity of the tangential electric field we get in accordance to our previous discussion

$$E_{1x}\Big|_{z=0} = E_{2x}\Big|_{z=0} \implies h_1 \frac{\kappa_{1z}}{\varepsilon_1} = -h_2 \frac{\kappa_{2z}}{\varepsilon_2} .$$

For matching the tangential magnetic fields we have to additionally consider the surface current distribution K_x, and we arrive at

$$\Big[H_{1y} - H_{2y} \Big]_{z=0} = -K_x\Big|_{z=0} \implies h_1 - h_2 = -\sigma_{2D} \frac{1}{2} \left(h_1 \frac{i\kappa_{1z}}{\varepsilon_1 \omega} - h_2 \frac{i\kappa_{2z}}{\varepsilon_2 \omega} \right) .$$

The last expression can be rewritten in the form

$$h_1 \frac{\kappa_{1z}}{\varepsilon_1} \left(\frac{\varepsilon_1}{\kappa_{1z}} + \frac{i\sigma_{2D}}{2\omega} \right) = h_2 \frac{\kappa_{2z}}{\varepsilon_2} \left(\frac{\varepsilon_2}{\kappa_{2z}} + \frac{i\sigma_{2D}}{2\omega} \right) .$$

Together with the boundary condition for the tangential electric field we arrive at the surface plasmon condition for graphene plasmons [49]

Surface Plasmon Condition for Graphene Plasmons

$$\frac{\sigma_{2D}(k_x, \omega)}{i\omega} = \frac{\varepsilon_1}{\kappa_{1z}} + \frac{\varepsilon_2}{\kappa_{12}} . \tag{8.17}$$

In its most simple form, we take for the optical conductivity the limit $k_x \to 0$ and consider in Eq. (7.14) only intraband transitions

$$\sigma_{2D}(k_x \to 0, \omega) = \frac{i\omega e^2}{k_x^2} P_{2D}(k_x \to 0, \omega) \approx \frac{i\omega e^2}{k_x^2} \left[\frac{g k_x^2 \mu}{4\pi(\hbar\omega)^2} \right] = i \frac{g e^2 \mu}{4\pi\hbar^2 \omega} .$$

In the non-retarded limit $c \to \infty$ we can replace in Eq. (8.17)

$$\kappa_{jz} = \sqrt{k_x^2 - \frac{\varepsilon_j}{\varepsilon_0}\frac{\omega^2}{c^2}} \approx k_x \,.$$

With this, we get a plasmon dispersion

$$\hbar\omega \approx \sqrt{\frac{ge^2\mu}{4\pi}\frac{k_x}{\varepsilon_1+\varepsilon_2}}\,, \tag{8.18}$$

which is proportional to the square root of the wavenumber, contrary to metal surface plasmons that exhibit a linear dependence for small wavenumbers.

In 2012 two paper appeared back to back in Nature [50, 51] that reported the first observation of graphene plasmons. Figure 8.8 shows optical spectra for graphene nanoribbons measured with a scanning nearfield optical microscope in the infrared regime. The different panels report spectra taken at different wavelengths, showing intensity maxima at the wavelengths for confined plasmon modes (see also scans in outer panels). Starting with these seminal papers, graphene plasmonics has turned into a vivid and hot research fields.

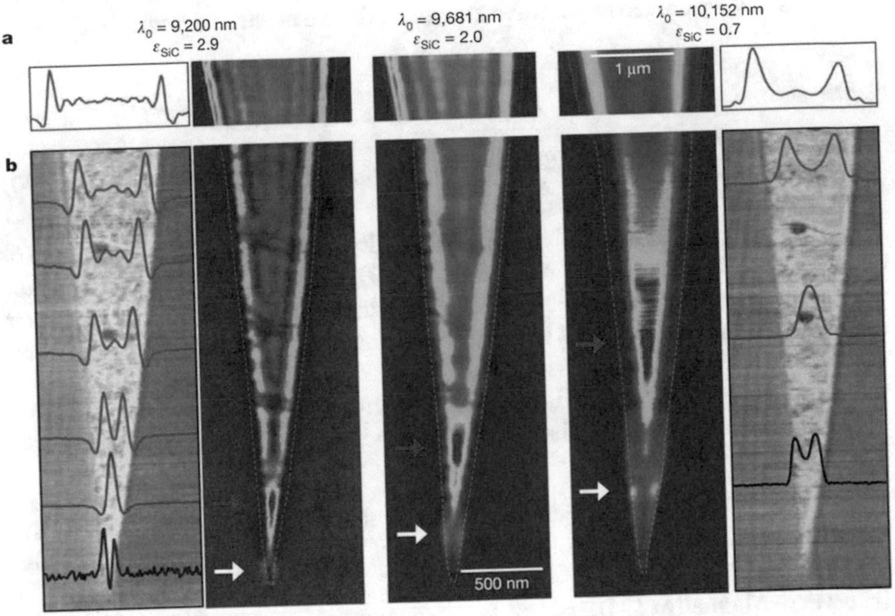

Fig. 8.8 Nearfield optical images taken with imaging wavelengths (λ_0) of 9200 nm (left), 9681 nm (middle), and 10,152 nm (right), corresponding, respectively, to SiC dielectric constants of 2.9, 2.0, and 0.7. (**a**) Images of a graphene ribbon ≈1 μm wide, revealing a strong dependence of the fringe spacing, and thus plasmon wavelength, on the excitation wavelength; (**b**) images of a tapered graphene ribbon; both ribbons are on the same 6H-SiC substrate. The topography (obtained by atomic force microscopy) is shown in greyscale in the leftmost and rightmost panels, and outlined by dashed lines in the central, colored panels. Image and caption taken from [50]

8.3 Transfer Matrix Approach

The transfer matrix approach is a generic solution scheme of Maxwell's equations for stratified media. We first discuss reflection and transmission of electromagnetic fields at a single planar interface, and then ponder on the solution of Maxwell's equations for a layered system.

8.3.1 *Fresnel Coefficients*

We start with Faraday's law for time-harmonic fields $\nabla \times \boldsymbol{E} = i\mu\omega\boldsymbol{H}$. Taking the curl on both sides of the equation and using Ampere's law with Maxwell's displacement current, we are led to

$$\nabla \times \frac{1}{\mu}\nabla \times \boldsymbol{E} = i\omega(\nabla \times \boldsymbol{H}) = i\omega(-i\omega\varepsilon\boldsymbol{E})\,. \tag{8.19}$$

Here and in the following we consider media with no charge and current distributions, $\rho = \boldsymbol{J} = 0$. From Eq. (8.19) and by taking in a similar fashion the curl of Ampere's law we are led to the wave equations for the electromagnetic fields

$$\begin{aligned}\mu\nabla \times \frac{1}{\mu}\nabla \times \boldsymbol{E} - \omega^2\mu\varepsilon\boldsymbol{E} &= 0\\ \varepsilon\nabla \times \frac{1}{\varepsilon}\nabla \times \boldsymbol{H} - \omega^2\mu\varepsilon\boldsymbol{H} &= 0\,.\end{aligned} \tag{8.20}$$

These two wave equations exhibit a kind of duality in the sense that one equation is obtained from the other one by replacing $\boldsymbol{E} \leftrightarrow \boldsymbol{H}$, $\varepsilon \leftrightarrow \mu$ [20]. This duality is not deep in a physical sense and does not reflect an underlying symmetry, in fact other replacements can be found that transform one equation to the other one. However, we will find such replacement useful below.

Consider the setup depicted in Fig. 8.9 where a plane wave impinges on an interface between two materials with ε_1, μ_1 and ε_2, μ_2. We can distinguish two situations for the polarization.

Transverse Electric (TE). Here the electric field $\boldsymbol{E}$ vector is parallel to the interface, or, in other words, transverse to the symmetry axis (here the z-axis) of the layer.

Transverse Magnetic (TM). Here the magnetic field vector $\boldsymbol{H}$ is parallel to the interface. It has been shown in Eq. (2.43) that a general wave can always be decomposed into its TE and TM components.

We first ponder on the TE polarization case, the results for TM polarization will then directly follow from the duality principle. Without loss of generality, we assume that

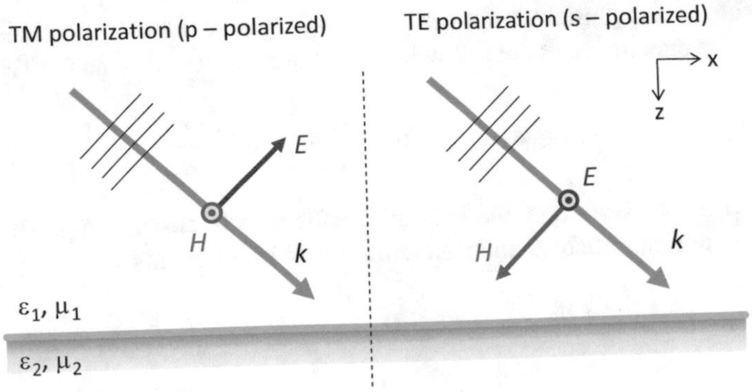

Fig. 8.9 TM versus TE excitation. A plane wave with wavevector $\boldsymbol{k}$ impinges on an interface between two materials with permittivities $\varepsilon_{1,2}$ and permeabilities $\mu_{1,2}$. When the magnetic field vector $\boldsymbol{H}$ is parallel to the interface, the wave is transverse magnetic (TM or p polarization), when the electric field vector $\boldsymbol{E}$ is parallel to the interface the wave is transverse electric (TE or s polarization). Note that the alternative notations of s and p polarizations, which are widely used in literature but will not be adopted in this book, refer to the German terms *senkrecht* (perpendicular) and *parallel*, corresponding to the orientations of the electric fields with respect to the plane formed by the interface and the incoming wave depicted in the above figure

the propagation direction of the wave is in the xz-plane and the electric field vector points along y, such that $\boldsymbol{E} = E_y\hat{\boldsymbol{y}}$. From Gauss' law we then find

$$\nabla\cdot\varepsilon(z)\boldsymbol{E} = \frac{\partial}{\partial y}\varepsilon(z)E_y = 0 \implies \varepsilon(z)\frac{\partial E_y}{\partial y} = 0\,,$$

where we have used that for a stratified medium $\varepsilon(z)$ only depends on the z-coordinate. We see that E_y does not depend on y. Using this and working out the spatial derivatives in Eq. (8.20), we are led to

$$\left(\frac{\partial^2}{\partial x^2} + \mu(z)\frac{\partial}{\partial z}\mu^{-1}(z)\frac{\partial}{\partial z} + \omega^2\mu\varepsilon\right)E_y = 0\,. \tag{8.21}$$

The above equation factorizes in x and z. For an incoming plane wave we can thus make the ansatz $E_y = e_y(z)e^{\pm ik_xx}$. This leads us to

$$\left(\mu(z)\frac{\partial}{\partial z}\mu^{-1}(z)\frac{\partial}{\partial z} + \omega^2\mu\varepsilon - k_x^2\right)e_y(z) = 0\,. \tag{8.22}$$

At the interface between two materials the parallel component of the electric field is conserved, $e_{1y}(z) = e_{2y}(z)$. For the second boundary condition we have to invoke the magnetic field, which we compute from Faraday's law

$$i\omega\mu\,\boldsymbol{H} = \nabla\times\left(e^{\pm ik_xx}e_y(z)\,\hat{\boldsymbol{y}}\right) = \left(-\frac{de_y(z)}{dz}\,\hat{\boldsymbol{x}} \pm ik_xe_y(z)\,\hat{\boldsymbol{z}}\right)e^{\pm ik_xx}\,.$$

Using that the tangential component of the magnetic field is continuous at the interface, we then arrive at the following two boundary conditions for TE waves

$$e_{1y} = e_{2y}\,, \qquad \mu_1^{-1}\frac{de_{1y}}{dz} = \mu_2^{-1}\frac{de_{2y}}{dz}\,. \tag{8.23}$$

The rest of the derivation of the Fresnel coefficients is rather straightforward. We consider incoming, reflected and transmitted field components

$$\begin{aligned} e_{1y}(z) &= e_0\left(e^{ik_{1z}z} + R^{TE}e^{-ik_{1z}z}\right), \quad & k_{1z} &= \sqrt{\omega^2\,\mu_1\varepsilon_1 - k_x^2} \\ e_{2y}(z) &= e_0 \qquad\qquad T^{TE}e^{\;ik_{2z}z}\quad , & k_{2z} &= \sqrt{\omega^2\,\mu_2\varepsilon_2 - k_x^2}\,. \end{aligned} \tag{8.24}$$

Here R^{TE} and T^{TE} are the Fresnel reflection and transmission coefficients that describe how much of the wave amplitude is reflected and transmitted, and k_{1z}, k_{2z} are the z-components of the wavevectors in the two media. Note that because of Maxwell's boundary conditions the parallel component of the wavevector k_x is conserved at the interface. Inserting Eq. (8.24) into the boundary conditions of Eq. (8.23) gives (we set $z = 0$)

$$1 + R^{TE} = T^{TE}\,, \quad \frac{k_{1z}}{\mu_1}\left(1 - R^{TE}\right) = \frac{k_{2z}}{\mu_2}T^{TE}\,, \tag{8.25}$$

which finally leads us to the Fresnel reflection and transmission coefficients

Fresnel Coefficients

$$\begin{aligned} R^{TE} &= \frac{\mu_2 k_{1z} - \mu_1 k_{2z}}{\mu_2 k_{1z} + \mu_1 k_{2z}}\,, \qquad & T^{TE} &= \frac{2\mu_2 k_{1z}}{\mu_2 k_{1z} + \mu_1 k_{2z}} \\ R^{TM} &= \frac{\varepsilon_2 k_{1z} - \varepsilon_1 k_{2z}}{\varepsilon_2 k_{1z} + \varepsilon_1 k_{2z}}\,, & T^{TM} &= \frac{2\varepsilon_2 k_{1z}}{\varepsilon_2 k_{1z} + \varepsilon_1 k_{2z}}\,\frac{Z_2}{Z_1}\,. \end{aligned} \tag{8.26}$$

The expression for TM polarization can be obtained through explicit calculation or by employing the duality principle. In the latter case one obtains coefficients for the magnetic fields. To get coefficients that match the electric fields at the interface rather than the magnetic ones, one has to multiply T^{TM} with the ratio of impedances of Eq. (2.41), as we have done in Eq. (8.26). In the above form, the Fresnel coefficients have the following properties.

Boundary Conditions. By construction, the boundary conditions of Maxwell's equations are fulfilled at the interface.

Flux Conservation. The flux of the incoming, reflected, and transmitted waves is conserved. Using Eq. (4.17) for the time-averaged Poynting vector we find for the z-component

$$\frac{1}{2}Z_1^{-1}\frac{k_{1z}}{k_1}-\frac{1}{2}Z_1^{-1}\left|R\right|^2\frac{k_{1z}}{k_1}=\frac{1}{2}Z_2^{-1}\left|T\right|^2\frac{k_{2z}}{k_2}\,,$$

where the terms on the left-hand side correspond to the energy flux of the incoming and reflected waves, respectively, and the term on the right-hand side to that of the transmitted wave. The above expression holds for both TE and TM polarizations. From $k=\sqrt{\mu\varepsilon}\,\omega$ we then get the flux conservation

$$\left|R\right|^2+\left|T\right|^2\left(\frac{\mu_1 k_{2z}}{\mu_2 k_{1z}}\right)=1\,. \tag{8.27}$$

Poles. The denominators of the TM Fresnel coefficients become zero at

$$\frac{k_{1z}}{k_{2z}}=-\frac{\varepsilon_1}{\varepsilon_2}\,,$$

with a corresponding expression for TE polarizations. The above equation is identical with Eq. (8.4) previously derived for surface plasmons. Thus, the poles of the reflection coefficients provide information about the eigenmodes confined to the interface. A hand waving argument for this is that the reflected and transmitted waves are given by e_0R and e_0T, respectively, with e_0 being the amplitude of the incoming field. When $R,\,T\to\infty$ reflected and transmitted waves can exist even for $e_0\to 0$, which can be interpreted in terms of eigenmodes oscillating in absence of any external excitation.

8.3.2 *Transfer Matrices*

We now turn to the stratified medium depicted in Fig. 8.10. Let us introduce a two-component vector where the components e_m^+ and e_m^- describe the fields that propagate in medium m into the positive and negative z-direction, respectively. For simplicity, in the following we drop the subscript y for the electric field component and the superscript TE for the Fresnel coefficients. Let us consider the first interface. The field in medium 1 is $e_1^++e_1^-$ and the field in medium 2 is $e_2^++e_2^-$. Submitting these fields to the boundary condition of Eq. (8.23) gives

$$e_1^++e_1^-=e_2^++e_2^-\,,\qquad \frac{k_{1z}}{\mu_1}\left(e_1^+-e_1^-\right)=\frac{k_{2z}}{\mu_2}\left(e_2^+-e_2^-\right)\,.$$

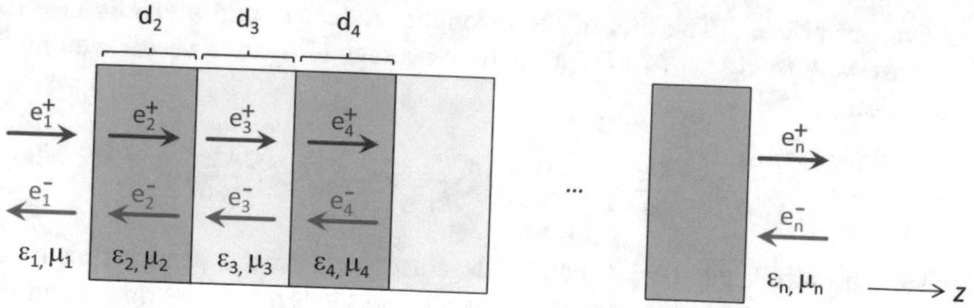

Fig. 8.10 Schematics of transfer matrix method. A plane wave impinges on a system consisting of different layers of thickness d_m, each having a different permittivity ε_m and permeability μ_m. Within the transfer matrix approach, the wave is propagated through the layer system. At each interface we match the fields using the boundary conditions of Maxwell's equations, and employ free wave propagation inside the layers

Multiplying the first equation with k_{1z}/μ_1 and adding or subtracting the second one leads us to

$$T_{12}e_1^+ = e_2^+ + R_{12}e_2^-\,, \quad T_{12}e_1^- = R_{12}e_2^+ + e_2^-\,, \tag{8.28}$$

where we have explicitly indicated the dependence of the Fresnel coefficients on the media indices. We can thus relate the field components on both sides of the interface through a matrix equation

$$\begin{pmatrix} e_1^+ \\ e_1^- \end{pmatrix} = \frac{1}{T_{12}}\begin{pmatrix} 1 & R_{12} \\ R_{12} & 1 \end{pmatrix}\begin{pmatrix} e_2^+ \\ e_2^- \end{pmatrix}. \tag{8.29}$$

We next introduce a matrix

$$\bar{\bar{M}}_{m-1,m} = \frac{1}{T_{m-1,m}}\begin{pmatrix} 1 & R_{m-1,m} \\ R_{m-1,m} & 1 \end{pmatrix} \tag{8.30}$$

that describes how the fields are connected over an interface through the boundary conditions of Maxwell's equations. In a multilayer system we additionally have to consider the phases $e^{\pm ik_{mz}d_m}$ acquired by the fields when propagating through the m'th layer with thickness d_m. This can be done by introducing the propagation matrices $\bar{\bar{P}}_m$

$$\bar{\bar{P}}_m = \begin{pmatrix} e^{-ik_{mz}d_m} & 0 \\ 0 & e^{ik_{mz}d_m} \end{pmatrix}. \tag{8.31}$$

The sign of the exponentials in the propagation matrix is due to the fact that the fields on the right-hand side of the matrix correspond to larger z-values. We can finally combine the interface and propagation matrices in the form,

$$\begin{pmatrix} e_1^+ \\ e_1^- \end{pmatrix} = \bar{\bar{M}}_{1,n}^{\text{tot}} \begin{pmatrix} e_n^+ \\ e_n^- \end{pmatrix} .$$

Here we have introduced the *transfer matrix* $\bar{\bar{M}}^{\text{tot}}$ which propagates a wave through the stratified layer system,

Transfer Matrix

$$\bar{\bar{M}}_{1,n}^{\text{tot}} = \bar{\bar{M}}_{1,2} \cdot \prod_{m=2}^{n-1} \bar{\bar{P}}_m \cdot \bar{\bar{M}}_{m,m+1} . \tag{8.32}$$

The concept of transfer matrices is surprisingly powerful, yet it is rather simple because we only have to deal with the multiplication of 2×2 matrices.

Illumination of Waves Propagating Along the +*z*-Direction Consider a wave $e_1^+ = e_0 e^{ik_{1z}}$ that propagates in the positive z-direction. When impinging on the lowest interface, part of the wave becomes reflected and another one transmitted,

$$e_0 \begin{pmatrix} 1 \\ \tilde{R}_{1,n} \end{pmatrix} = e_0 \begin{pmatrix} [\![M_{1,n}^{\text{tot}}]\!]_{11} & [\![M_{1,n}^{\text{tot}}]\!]_{12} \\ [\![M_{1,n}^{\text{tot}}]\!]_{21} & [\![M_{1,n}^{\text{tot}}]\!]_{22} \end{pmatrix} \begin{pmatrix} \tilde{T}_{1,n} \\ 0 \end{pmatrix} . \tag{8.33}$$

Here $[\![M_{1,n}^{\text{tot}}]\!]_{ij}$ denotes the matrix elements of the total transfer matrix. $\tilde{R}$ and $\tilde{T}$ are the generalized reflection and transmission coefficients which account for the reflection and transmission properties of the entire stratified medium, contrary to the Fresnel coefficients (without the tilde) which only match fields at the individual interfaces. In the above expression we have exploited that for the setup of an incoming wave propagating along $+z$ the component e_n^- in region n propagating along $-z$ has to be zero. We can solve for the generalized coefficients and get

$$\begin{aligned} \tilde{T}_{1,n} &= [\![M_{1,n}^{\text{tot}}]\!]_{11}^{-1} \\ \tilde{R}_{1,n} &= [\![M_{1,n}^{\text{tot}}]\!]_{21} [\![M_{1,n}^{\text{tot}}]\!]_{11}^{-1} . \end{aligned} \tag{8.34}$$

When computing the reflected and transmitted waves explicitly, we have to be careful about an additional phase factor. Let z_1 and z_n denote the positions of the first and last interface, respectively, and consider an incoming field of the form

$$e_1^+(z) = e_0\, e^{ik_{1z}z} \implies e_1^+(z_1) = e_0\, e^{ik_{1z}z_1} .$$

Using the generalized reflection and transmission coefficients, the reflected and transmitted field components then become

$$\begin{aligned} z < z_1 \quad &\ldots \quad e_1^-(z) = e_0\, \tilde{R}_{1,n} \exp\Big[-ik_{1z}(z - z_1)\Big] \\ z > z_n \quad &\ldots \quad e_n^+(z) = e_0\, \tilde{T}_{1,n} \exp\Big[\; i(k_{nz}z + k_{1z}z_1)\Big] . \end{aligned} \tag{8.35}$$

Note that by a similar token we can also compute the fields everywhere else inside the stratified medium.

Illumination of Waves Propagating Along the $-z$-Direction We can use the same transfer matrix $\bar{\bar{M}}$ for a wave $e_0 e^{-ik_{nz}}$ propagating along the $-z$-direction. The generalized reflection and transmission coefficients can be computed from

$$e_0 \begin{pmatrix} 0 \\ \tilde{T}_{n,1} \end{pmatrix} = e_0 \begin{pmatrix} [\![M_{1,n}^{\rm tot}]\!]_{11} & [\![M_{1,n}^{\rm tot}]\!]_{12} \\ [\![M_{1,n}^{\rm tot}]\!]_{21} & [\![M_{1,n}^{\rm tot}]\!]_{22} \end{pmatrix} \begin{pmatrix} \tilde{R}_{n,1} \\ 1 \end{pmatrix}.$$

Note that in comparison with Eq. (8.33) we have exchanged the layer indices of $\tilde{R}_{n,1}$, $\tilde{T}_{n,1}$ to emphasize that the incoming wave propagates in medium n rather than medium 1. Solving for the generalized reflection and transmission coefficients we get

$$\begin{aligned} \tilde{R}_{n,1} &= -[\![M_{1,n}^{\rm tot}]\!]_{11}^{-1}[\![M_{1,n}^{\rm tot}]\!]_{12} \\ \tilde{T}_{n,1} &= [\![M_{1,n}^{\rm tot}]\!]_{22} - [\![M_{1,n}^{\rm tot}]\!]_{21}[\![M_{1,n}^{\rm tot}]\!]_{11}^{-1}[\![M_{1,n}^{\rm tot}]\!]_{12}\,. \end{aligned} \tag{8.36}$$

For an incoming field of the form

$$e_n^-(z) = e_0\, e^{-ik_{nz}z} \implies e_n^-(z_n) = e_0\, e^{-ik_{nz}z_n}\,,$$

the reflected and transmitted field components then become

$$\begin{aligned} z > z_n \quad &\ldots \quad e_n^+(z) = e_0\, \tilde{R}_{n,1} \exp\Big[\;\; ik_{nz}(z - z_n)\Big] \\ z < z_1 \quad &\ldots \quad e_1^-(z) = e_0\, \tilde{T}_{n,1} \exp\Big[-i(k_{1z}z + k_{nz}z_n)\Big]\,. \end{aligned} \tag{8.37}$$

Reflection and Transmission Coefficients of Slab

As an illustrative example, we consider a system consisting of three media, where the outer media 1 and 3 are semi-infinite, and the inner region has a finite thickness d_2. Combining the transfer and propagation matrices leads us to

$$\begin{pmatrix} e_1^+ \\ e_1^- \end{pmatrix} = \frac{1}{T_{12}} \begin{pmatrix} 1 & R_{12} \\ R_{12} & 1 \end{pmatrix} \begin{pmatrix} e^{-i\phi_2} & 0 \\ 0 & e^{i\phi_2} \end{pmatrix} \frac{1}{T_{23}} \begin{pmatrix} 1 & R_{23} \\ R_{23} & 1 \end{pmatrix} \begin{pmatrix} e_3^+ \\ e_3^- \end{pmatrix},$$

where $\phi_2 = k_{2z}d_2$ is the phase associated with the propagation through the slab. The elements of the transfer matrix can be evaluated to

$$\bar{\bar{M}}_{1,3}^{\rm tot} = \frac{1}{T_{12}T_{23}} \begin{pmatrix} e^{-i\phi_2} + R_{12}R_{23}e^{i\phi_2} & e^{-i\phi_2}R_{23} + R_{12}e^{i\phi_2} \\ e^{-i\phi_2}R_{12} + R_{23}e^{i\phi_2} & e^{-i\phi_2}R_{12}R_{23} + e^{i\phi_2} \end{pmatrix}.$$

From this expression and Eq. (8.34) we then obtain for the generalized reflection and transmission coefficients of a slab

Generalized Reflection and Transmission Coefficients for Slab

$$\tilde{R}_{1,3} = R_{12} + \frac{T_{12}R_{23}T_{21}e^{2i\phi_2}}{1 - R_{21}R_{23}e^{2i\phi_2}}, \quad \tilde{T}_{1,3} = \frac{T_{12}T_{23}e^{i\phi_2}}{1 - R_{21}R_{23}e^{2i\phi_2}}. \tag{8.38}$$

In the denominator we have replaced $R_{12} = -R_{21}$ which follows from Eq. (8.26). To get physical insight into this expression, we expand the denominator in $\tilde{R}_{13}$ into a geometric series

$$\tilde{R}_{1,3} = R_{12} + T_{12}R_{23}T_{21}e^{2i\phi_2} + T_{12}R_{23}\left(R_{21}R_{23}\right)T_{21}e^{4i\phi_2} + T_{12}R_{23}\left(R_{21}R_{23}\right)^2 T_{21}e^{6i\phi_2} + \cdots .$$

This allows for a particularly transparent interpretation schematically shown in Fig. 8.11. The first term R_{12} accounts for the case that the incoming wave is directly reflected at the upper interface. The remaining contributions then account for multiple internal reflections, for instance, the second term with $T_{12}R_{23}T_{21}$ accounts for one reflection at the lower interface, as is also apparent from the propagation phase $e^{2i\phi_2}$.

8.3.3 *Surface Plasmons Revisited*

Using the transfer matrix formalism, we can now easily compute the generalized reflection and transmission coefficients for stratified media. The black dashed line indicated with (i) in Fig. 8.12 shows the reflection coefficient for a single glass–gold interface. As previously discussed in Sect. 8.1.1, the dispersion relations for photons and surface plasmons do not cross, and consequently no surface plasmons can be directly launched by light: the reflection coefficient R^{TM} is close to unity for all incoming angles, with only small losses attributed to minor field penetration and ohmic losses in the metal.

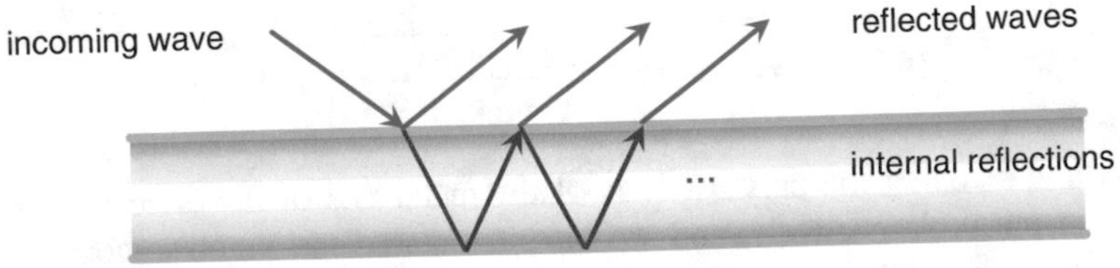

Fig. 8.11 Schematic representation for the generalized reflection coefficients of a slab. The incoming wave becomes reflected or transmitted, and the transmitted wave can undergo multiple internal reflections where on the upper interface it has some probability to be transmitted. In the figure we have not indicated the transmitted waves at the lower interface

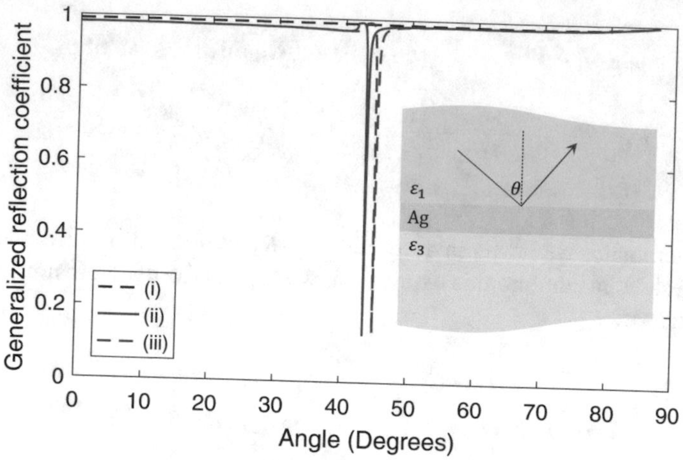

Fig. 8.12 Generalized reflection coefficient $|\tilde{R}^{TM}|$ for (**i**) semi-infinite slab and (**ii**, **iii**) layer structure. See inset for a schematic representation. In the simulations we use a free-space wavelength of $\lambda = 620\,\text{nm}$ for the incoming light, corresponding to a photon energy of 2 eV, a silver dielectric function extracted from optical experiments [34], a refractive index of $n_1 = 1.5$ (glass) for the medium above the layer through which the structure is excited, and a refractive index of $n_1 = 1$ for the medium below the layer. (**i**) For a single interface no surface plasmons can be excited. (**ii**) For the slab structure (thickness $d = 50$ nm), surface plasmons are launched at a specific angle, where energy and momentum of the incoming photon and the surface plasmon at the lower interface coincide. (**iii**) Simulation where the permittivity ε_3 of the lowest medium is increased by 5%. Because of the strongly localized, evanescent fields of the surface plasmon, the plasmon energy is very sensitive to changes of the local dielectric environment

Things change considerably for the Kretschmann geometry, see red solid line indicated with (ii) in Fig. 8.12. For an angle of approximately 43° the energy and momentum of the incoming light (through the upper glass medium) coincide with those of the surface plasmon at the lower silver interface, and a surface plasmon is launched. This can be clearly seen as a pronounced dip of the generalized reflection coefficient where practically all energy of the incoming light is transferred to the surface plasmons. Because the fields of surface plasmons are strongly confined, their dispersion depends very sensitively on changes of the local dielectric environment. For a semi-infinite slab the plasmon dispersion is given by Eq. (8.6),

$$k_x = k_1 \sin\theta = \sqrt{\frac{\varepsilon_2\varepsilon_3}{\varepsilon_0(\varepsilon_2+\varepsilon_3)}}\,k_0\,,$$

with ε_2 and ε_3 being the permittivities of the metal and of the dielectric material below the metal, respectively, and we have expressed the parallel momentum of the incoming light at the reflection dip in the form $k_1 \sin\theta$. When the dielectric permittivity is changed by a small amount $\varepsilon_3 + \delta\varepsilon_3$, also the angle of the reflection dip changes by a small amount $\theta + \delta\theta$. Linearization of the above expression yields

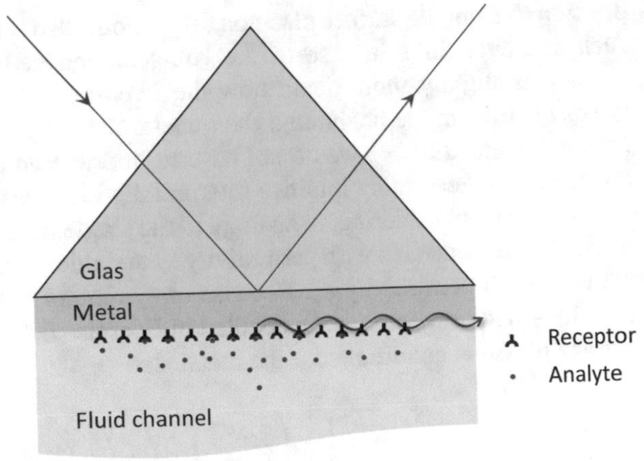

Fig. 8.13 Schematics of surface plasmon sensor based on a Kretschmann configuration. The metal surface is functionalized with receptors. Upon binding of analytes the permittivity of the lower medium is slightly changed , causing a modification of the surface plasmon dispersion, which can be optically detected

$$k_1\Big[\sin\theta + \cos\theta\,\delta\theta\Big] \approx \sqrt{\frac{\varepsilon_2\varepsilon_3}{\varepsilon_0(\varepsilon_2+\varepsilon_3)}}\left[1+\frac{1}{2}\frac{\varepsilon_2}{\varepsilon_2+\varepsilon_3}\frac{\delta\varepsilon_3}{\varepsilon_3}\right]k_0\,.$$

Thus, to the lowest order of approximation the change $\delta\theta$ is given by

$$\delta\theta \approx \frac{1}{2}\frac{\varepsilon_2}{\varepsilon_3+\varepsilon_2}\frac{\delta\varepsilon_3}{\varepsilon_3}\tan\theta\,. \tag{8.39}$$

The red dashed line indicated with (iii) in Fig. 8.12 shows the reflection coefficient for a setup where the permittivity of the lowest medium has been increased by 5%. This change causes a significant shift of the angle θ for which a surface plasmon is launched, demonstrating the strong sensitivity of surface plasmons to changes in the local dielectric environment. This sensitivity is due to the strongly localized, evanescent character of the surface plasmon fields, which become significantly affected by dielectric modifications in close proximity to the metal surface.

Surface Plasmon Sensors This surface sensitivity can be exploited for surface plasmon (bio)sensors, which have found widespread application in various fields of research, for instance for clinical pregnancy tests [44, 52]. The basic principle is shown in Fig. 8.13: the lower metal surface is functionalized with receptors that can bind specific analytes to be detected. The lower medium is connected to a fluid channel which transports some liquid, such as water. If no analytes are present in the fluid, nothing binds to the receptors and the reflectivity dip of the Kretschmann geometry remains unchanged. On the other hand, if analytes are present they can bind to the receptors, causing a small change $\delta\varepsilon$ of the dielectric environment (the refractive index of the analytes typically somewhat differs from that of the liquid), causing a shift $\delta\theta$ of the reflectivity dip which is observed through an optical detection scheme. Typical sensitivities of sensors based on surface plasmons are of the order of milli-degrees for $\delta\theta$.

Coupled Surface Plasmons

At the beginning of this chapter we have discussed surface plasmons for a single metal–dielectric interface. With the transfer matrix approach we are now in the position to compute surface plasmon dispersions also for more complicated systems, such as metal films in case of the Kretschmann geometry. In the following we discuss in slightly more detail how the two approaches are connected. To make things clear from the beginning: the transfer matrix provides a rigorous solution of Maxwell's equation, so we do not have to improve on it. However, we shall find it convenient to get a more intuitive interpretation for these rigorous results.

We start by considering, in analogy to the single metal–dielectric interface shown in Fig. 8.1, a metal slab with permittivity ε_2 and thickness d embedded in a dielectric medium with permittivity ε_1. Because of symmetry, the solutions are even or odd functions with respect to z. As detailed in Exercise 8.6, one can derive the modified surface plasmon conditions for the metal slab

$$\tanh\left(\frac{\kappa_{2z}d}{2}\right) = \begin{cases} -\left(\dfrac{\kappa_{1z}\varepsilon_2}{\kappa_{2z}\varepsilon_1}\right) & \text{for } H_y \text{ even } E_z \text{ odd} \\ -\left(\dfrac{\kappa_{2z}\varepsilon_1}{\kappa_{1z}\varepsilon_2}\right) & \text{for } H_y \text{ odd, } \; E_z \text{ even}\,, \end{cases} \tag{8.40}$$

which must be solved numerically. Figure 8.14 shows the dispersions for a 30 nm thick silver film. As can be seen, the frequencies of the even and odd modes are split, which can be understood in terms of coupling between the modes confined to the upper and lower interface of the metal slab. When the film thickness is increased, see Fig. 8.15 for a 80 nm film, the coupling is reduced and the mode splitting is considerably smaller.

For a qualitative understanding of this behavior, we introduce a most simple model based on two coupled oscillators, which can be associated with the surface

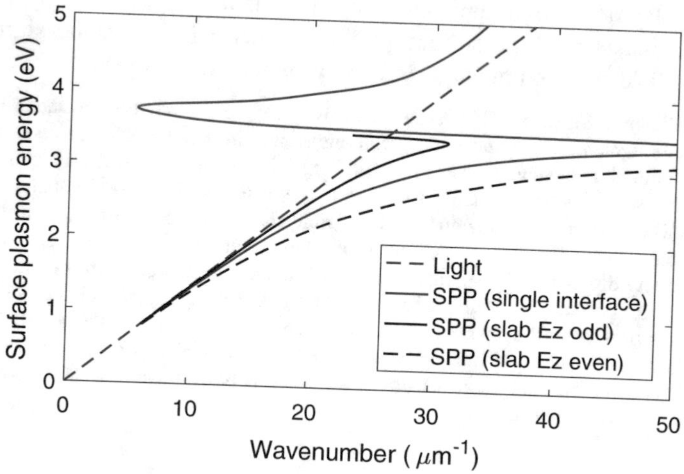

Fig. 8.14 Surface plasmon dispersion for 30 nm thick silver film embedded in a material with $\varepsilon_1 = 2.25$. The dashed line reports the light line, the solid red line the surface plasmon dispersion for a single interface. The black lines report the surface plasmon dispersions for the metal slab, as computed from Eq. (8.40)

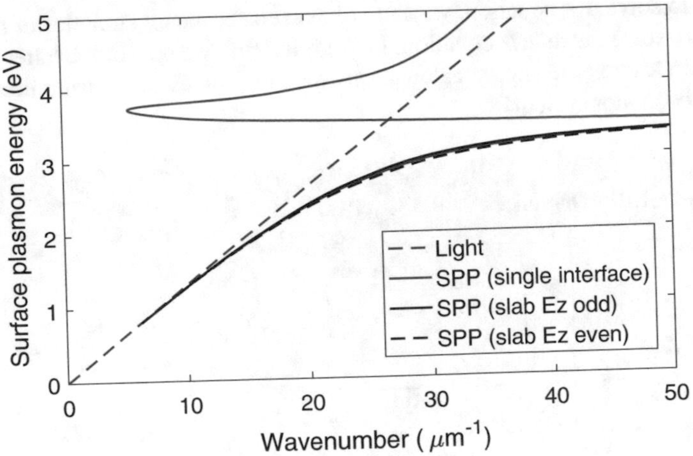

Fig. 8.15 Same as Fig. 8.15 but for 80 nm thick silver film

plasmons on the two sides of the metal slab propagating with the same parallel momentum $\hbar k_x$ (which is a conserved quantity for the slab geometry). We start from Newton's equations of motion

$$\frac{d^2x_a}{dt^2} + \gamma\frac{dx_a}{dt} + \omega_a^2\,x_a = g\,x_b + f_0e^{-i\omega t}$$
$$\frac{d^2x_b}{dt^2} + \gamma\frac{dx_b}{dt} + \omega_b^2\,x_b = g\,x_a\,.$$

The amplitudes of the oscillators are denoted with x_a, x_b, the resonance frequencies are ω_a, ω_b, and γ is a generic damping constant. We have also introduced a coupling constant g between the two oscillators, as well as a driving term for oscillator a, with frequency ω and amplitude f_0. Upon external driving, the two oscillators start to oscillate after an initial stage with the same frequency ω as the external excitation. We introduce a complex amplitude vector

$$\begin{pmatrix} x_a(t) \\ x_b(t) \end{pmatrix} = e^{-i\omega t}\begin{pmatrix} \bar{x}_a \\ \bar{x}_b \end{pmatrix},$$

where as usual we implicitly assume that one has to take the real part of x_a, x_b in order to get the physical displacement. Inserting this vector in Newton's equations of motion and cancelling the common exponential factors then leads us to the matrix equation

$$\begin{pmatrix} \omega_a^2 - \omega^2 - i\gamma\omega & -g \\ -g & \omega_b^2 - \omega^2 - i\gamma\omega \end{pmatrix}\begin{pmatrix} \bar{x}_a \\ \bar{x}_b \end{pmatrix} = \begin{pmatrix} f_0 \\ 0 \end{pmatrix}. \tag{8.41}$$

In order to solve this equation, we can either seek for the eigenmodes of the system or directly solve the matrix equation. In the following we opt for the latter approach, and obtain after some simple calculations the amplitudes of a harmonically driven coupled oscillator system

Harmonically Driven Coupled Oscillators

$$\bar{x}_a = \left[\frac{\omega_b^2 - \omega^2 - i\gamma\omega}{\left(\omega_a^2 - \omega^2 - i\gamma\omega\right)\left(\omega_b^2 - \omega^2 - i\gamma\omega\right) - g^2}\right] f_0$$

$$\bar{x}_b = \left[\frac{g}{\left(\omega_a^2 - \omega^2 - i\gamma\omega\right)\left(\omega_b^2 - \omega^2 - i\gamma\omega\right) - g^2}\right] f_0. \qquad (8.42)$$

We next analyze the coupled plasmon system in terms of this simplified oscillator model. Figure 8.16 shows a schematic view of the system. Consider first the situation that both surface plasmons have the same frequency $\omega_a = \omega_b$, corresponding to the case that the permittivities of the materials above and below the metal film are identical, $\varepsilon_1 = \varepsilon_3$. In this case the eigenmodes of the matrix in Eq. (8.41) are symmetric and anti-symmetric superpositions of the individual surface plasmon modes. The eigenfrequencies for $\omega_a = \omega_b = \omega_0$ can be computed for $\omega_0 \gg \gamma$, $\omega_0 \gg g$ as

$$\omega_\pm \approx \omega_0 - i\frac{\gamma\omega_0}{2} \pm \frac{g}{2}.$$

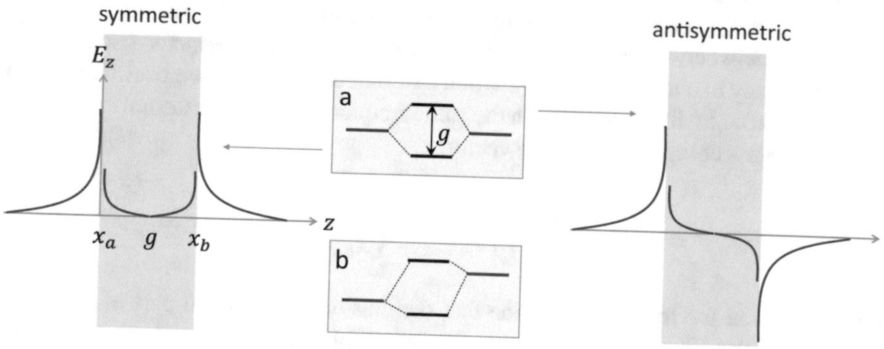

Fig. 8.16 Schematics of coupled surface plasmons. Two surface plasmons are confined to the interfaces of a metal slab, which is modelled by the harmonic oscillators x_a and x_b that become coupled through the evanescent surface plasmon fields, as described by the coupling constant g. (**a**) For equal plasmon energies the eigenmodes of the coupled system consist of symmetric and anti-symmetric modes, schematically shown on the left- and right-hand side, with an energy splitting of g. (**b**) For unequal plasmon energies the mode splitting is reduced, and the eigenmode character is governed by the properties of the uncoupled oscillators x_a, x_b

They are split by the coupling constant g and have the same damping as the individual oscillators. The symmetric mode, whose frequency is lowered, is sometimes called the *bonding* mode, and the anti-symmetric mode the *anti-bonding* mode. For sufficiently thick metal films the coupling g is usually much smaller than γ.

In principle, a similar analysis also holds for oscillators with different frequencies, corresponding, for instance, to the usual Kretschmann geometry with different surface plasmon dispersions at the two slab interfaces, as schematically shown in Fig. 8.16b. The incoming light then drives oscillator a, and the excitation is transferred via coupling to oscillator b. This provides us with a simple but viable interpretation scheme for surface plasmon couplings in metallic layer systems.

8.4 Negative Refraction

We have seen in this chapter that even such a simple system as a metal–dielectric interface exhibits interesting physics. One may wonder whether similar effects may exist for other material combinations, in particular systems with a large magnetic response $\mu(\omega)$ in the optical range. As we have discussed in the previous chapter such materials do not exist in nature but can be manufactured artificially. Suppose that we could manufacture metamaterials with arbitrary $\varepsilon(\omega)$, $\mu(\omega)$ values, as schematically shown in Fig. 8.17. What could we expect?

$\varepsilon > 0$, $\mu > 0$. This situation corresponds to normal dielectrics, although there one usually has $\mu \approx \mu_0$. From the dispersion relation we get

$$k^2 = \varepsilon(\omega)\mu(\omega)\,\omega^2 \implies \omega = \frac{kc}{n(\omega)}, \qquad n(\omega) = \sqrt{\frac{\varepsilon(\omega)\mu(\omega)}{\varepsilon_0\mu_0}}\,. \tag{8.43}$$

Thus the speed of light is altered by the refractive index $n(\omega)$ in the medium with respect to free space, which gives rise to refraction of plane waves (or rays in case of geometric optics) at the interface between materials of different $n(\omega)$ values.

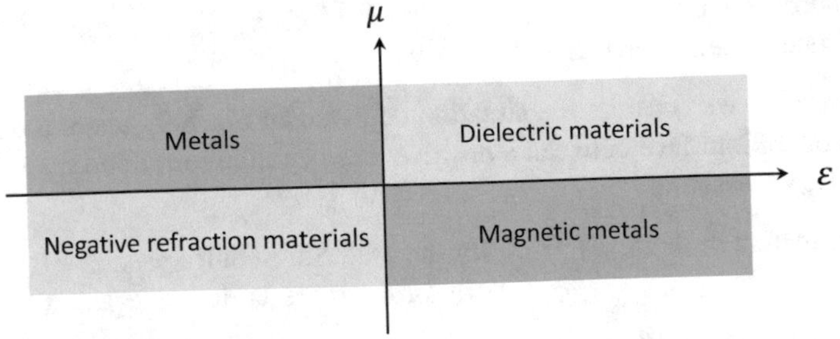

Fig. 8.17 Classification of materials with respect to their permittivities ε and permeabilities μ

$\varepsilon < 0, \mu > 0$. This situation corresponds to metals. From the dispersion relation $\mu\varepsilon\,\omega^2 = k_x^2 + k_z^2$ we observe that only evanescent waves with an imaginary wavenumber exist inside the metal. For this reason, light becomes reflected at the interface between a dielectric and a metal. Additionally, there exists a novel type of excitation at such interfaces, the previously discussed surface plasmons.

$\varepsilon > 0, \mu < 0$. This situation corresponds to magnetic metals. Owing to the duality principle, the physical properties are similar to normal metals; however, the role of electric and magnetic fields is exchanged. Again light cannot propagate inside the magnetic metal, and there exist surface plasmons (now with TE character) at the interface between a dielectric and a magnetic metal.

$\varepsilon < 0, \mu < 0$. At first this looks like a not overly interesting case. Submitting these material parameters to the dispersion relation, we obtain the same result as in Eq. (8.43) with the refractive index

$$n = \pm\sqrt{\frac{(-|\varepsilon|)(-|\mu|)}{\varepsilon_0\mu_0}}\,,$$

which appears to be identical to the $\varepsilon > 0$, $\mu > 0$ case. However, as already noted in the equation we have to be careful about the sign of the square root. Usually one does not bother about the sign, it is chosen positive and indeed this is correct for normal dielectrics. However, for $\varepsilon < 0$, $\mu < 0$ one has to choose the negative sign, as we will demonstrate in a moment, and this has given these materials the name *negative refraction* materials.

So how can we decide which sign has to be chosen for the refractive index? As always in electrodynamics, the answer is provided by Maxwell's equations. The curl equations without external source terms and for plane waves are given by Eq. (2.38) and read

$$\boldsymbol{k}\times\boldsymbol{E} = \mu\omega\,\boldsymbol{H}\,,\quad \boldsymbol{k}\times\boldsymbol{H} = -\varepsilon\omega\,\boldsymbol{E}\,.$$

From this it can be immediately seen that

- for materials with $\varepsilon > 0$, $\mu > 0$ the vectors $\boldsymbol{k}$, $\boldsymbol{E}$, $\boldsymbol{H}$ form a right-handed coordinate system, this is $\hat{\boldsymbol{k}}\times\hat{\boldsymbol{E}} = \hat{\boldsymbol{H}}$, whereas
- for materials with $\varepsilon < 0$, $\mu < 0$ the vectors $\boldsymbol{k}$, $\boldsymbol{E}$, $\boldsymbol{H}$ form a left-handed coordinate system, this is $\hat{\boldsymbol{k}}\times\hat{\boldsymbol{E}} = -\hat{\boldsymbol{H}}$.

In the following we consider the situation depicted in Fig. 8.9, where a plane wave impinges on an interface between a positive and negative refraction material. More specifically, we assume

- the parameters $\varepsilon_1 = \varepsilon_0$, $\mu_1 = \mu_0$ for the material in half-space $z < 0$, and
- the parameters $\varepsilon_2 = -\varepsilon_0$, $\mu_2 = -\mu_0$ for the material in half-space $z > 0$.

We discuss the case of an incoming plane wave with TM character. For the wavevector and fields in the positive index material we use

$$\boldsymbol{k}_1 = \begin{pmatrix} k_x \\ 0 \\ k_z \end{pmatrix}, \quad \boldsymbol{H}_1 = \begin{pmatrix} 0 \\ H_y \\ 0 \end{pmatrix}, \quad \boldsymbol{E}_1 = -Z_0\,\hat{\boldsymbol{k}}_1 \times \boldsymbol{H}_1 = \frac{Z_0 H_y}{k}\begin{pmatrix} k_z \\ 0 \\ -k_x \end{pmatrix}.$$

Here Z_0 is the impedance in free space. In order to get the fields in the negative index material, we employ the boundary conditions of Maxwell's equations with continuity of the tangential electromagnetic fields and of the normal component of the dielectric displacement. This gives

$$E_{1x} = E_{2x}\,, \quad \varepsilon_0 E_{1z} = -\varepsilon_0 E_{2z}\,, \quad H_{1y} = H_{2y}\,.$$

Thus, when transferring the fields from the positive to the negative refraction material we get

$$\boldsymbol{H}_2 = \boldsymbol{H}_1\,, \quad \boldsymbol{E}_2 = \begin{pmatrix} E_{1x} \\ 0 \\ -E_{1z} \end{pmatrix} = \frac{Z_0 H_y}{k}\begin{pmatrix} k_z \\ 0 \\ k_x \end{pmatrix}. \tag{8.44}$$

We can now use the fact that in the negative index material the vectors $\boldsymbol{k}$, $\boldsymbol{E}$, $\boldsymbol{H}$ form a left-handed coordinate system,

$$\boldsymbol{k}_2 = -k\hat{\boldsymbol{E}}_2 \times \hat{\boldsymbol{H}}_2 = \begin{pmatrix} k_z \\ 0 \\ k_x \end{pmatrix} \times \begin{pmatrix} 0 \\ 1 \\ 0 \end{pmatrix} = \begin{pmatrix} k_x \\ 0 \\ -k_z \end{pmatrix}. \tag{8.45}$$

Thus, the parallel component of the wavevector is conserved but the z-component points in the opposite direction. In other words, the phase velocities in *both* materials point towards the interface! This is a highly unusual situation corresponding to what we have previously denoted as "negative refraction." See Fig. 8.18 for a graphical representation. We can additionally compute the Poynting vectors in both materials

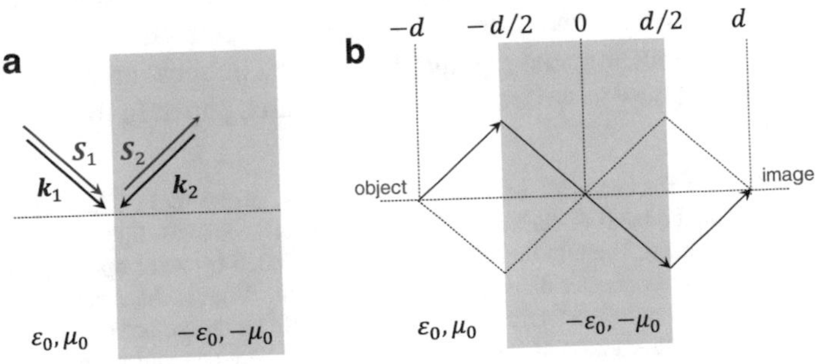

Fig. 8.18 Negative refraction and Veselago lens. (**a**) Plane wave impinging on interface between positive and negative index material. The black (red) arrows report the wavevectors (Poynting vectors) in the two materials. (**b**) Veselago lens. Light is emitted from an object, becomes redirected at a slab filled with negative index material, and finally refocuses at the mirror position of the object to form an image

$$S_1 = \frac{1}{2} E_1 \times H_1 = \frac{Z_0 H_y^2}{2k} \begin{pmatrix} k_x \\ 0 \\ k_z \end{pmatrix}, \quad S_2 = \frac{1}{2} E_2 \times H_2 = \frac{Z_0 H_y^2}{2k} \begin{pmatrix} -k_x \\ 0 \\ k_z \end{pmatrix},$$

which shows that the energy flows continuously in the positive z-direction, and it is only the phase velocity that is opposite to the energy flow direction in the negative index material.

8.4.1 Veselago Lens

Negative index materials were first considered by Veselago in 1964 [53], and remained something like a theoretical curiosity. Veselago also realized that a slab of negative refraction material could act as a lens. Consider a point source at position $z = -d$ in front of a slab with thickness d, as depicted in Fig. 8.18. As can be easily seen by tracing the rays emitted from the object, they refocus on the other side of slab at the mirror position $z = d$. Quite generally, the lens is not an overly useful device and comes along with a number of rather unspectacular features.

- The Veselago lens has no curvature,
- has no optical axis,
- has no magnification,
- cannot focus parallel rays of light.

8.4.2 The Perfect Lens

The Veselago lens passed into silence, if it ever received noticeable interest, but came to spotlight in 2000 when John Pendry analyzed transmission of evanescent fields in a Veselago lens and found that “negative refraction makes a perfect lens” [54]. I strongly recommend reading through this publication which serves as a beautiful example of how to write an excellent research paper. In the first paragraph Pendry states:

> Optical lenses have for centuries been one of scientists’ prime tools. Their operation is well understood on the basis of classical optics: curved surfaces focus light by virtue of the refractive index contrast. Equally their limitations are dictated by wave optics: no lens can focus light onto an area smaller than a square wavelength. What is there new to say other than to polish the lens more perfectly and to invent slightly better dielectrics? In this Letter I want to challenge the traditional limitation on lens performance and propose a class of “superlenses,” and to suggest a practical scheme for implementing such a lens.

To understand why a Veselago lens is a “superlens” we will now analyze wave focusing using the transfer matrix approach previously developed. We consider a slab with thickness d and filled with a negative refraction material that is located in vacuum. Suppose that an evanescent wave with wavenumber

$$k_{1z} = \sqrt{k_0^2 - k_x^2} = i\sqrt{k_x^2 - k_0^2} \equiv i\kappa_0$$

impinges on the first interface. We use $k_0 = \frac{\omega}{c}$ and assume $k_x > k_0$ for the evanescent wave. Inside the negative refraction material we have

$$k_{2z} = \sqrt{\varepsilon\mu\omega^2 - k_x^2} = i\sqrt{k_x^2 - \varepsilon\mu\omega^2} \equiv i\kappa\,,$$

and will set $\varepsilon \to -\varepsilon_0$ and $\mu \to -\mu_0$ at the end of the calculation. Note that we have dropped all subscripts for the material indices. To compute the transmission coefficient of the slab structure, we use Eq. (8.38) and obtain

$$\tilde{T}_{13} = \lim_{\varepsilon\to-\varepsilon_0}\lim_{\mu\to-\mu_0} \frac{\left(\dfrac{2\mu\kappa_0}{\mu\kappa_0+\mu_0\kappa}\right)\left(\dfrac{2\mu_0\kappa}{\mu_0\kappa+\mu\kappa_0}\right)e^{-\kappa d}}{1+\left(\dfrac{\mu\kappa_0-\mu_0\kappa}{\mu\kappa_0+\mu_0\kappa}\right)\left(\dfrac{\mu_0\kappa-\mu\kappa_0}{\mu_0\kappa+\mu\kappa_0}\right)e^{-2\kappa d}}\,. \tag{8.46}$$

Note that the arguments in the exponentials have been chosen such that the evanescent waves decay successively for multiple reflections, as can be seen by expanding the denominator into a geometric series in complete analogy to the previously discussed case of propagating waves.

We next note that in the limit $\varepsilon \to -\varepsilon_0$ and $\mu \to -\mu_0$ we get $\kappa \to \kappa_0$, and the Fresnel transmission and reflection coefficients diverge because the denominators $\mu\kappa + \mu_0\kappa'$ become zero. For this reason, we can neglect in the denominator of Eq. (8.46) the constant factor of 1. Thus, we find for the generalized transmission coefficient of a slab filled with negative refraction material

Generalized Transmission Coefficient for Perfect Lens

$$\tilde{T}_{13} = \lim_{\varepsilon\to-\varepsilon_0}\lim_{\mu\to-\mu_0} \frac{4\mu\mu_0\,\kappa_0\kappa}{(\mu\kappa_0-\mu_0\kappa)(\mu_0\kappa-\mu\kappa_0)}e^{\kappa d} = e^{\kappa_0 d}\,. \tag{8.47}$$

In other words, the Veselago lens amplifies evanescent waves! It even amplifies them such that at the image position the amplitude of the evanescent wave precisely equals that at the object position. This is indeed an unexpected and striking result. In his original paper Pendry concludes:

> Thus, even though we have meticulously carried through a strictly causal calculation, our final result is that the medium does amplify evanescent waves. Hence we conclude that with this new lens both propagating and evanescent waves contribute to the resolution of the image. Therefore there is no physical obstacle to perfect reconstruction of the image beyond practical limitations of apertures and perfection of the lens surface.

There is some magic in the perfect imaging of the perfect lens, and if one carefully goes through the theoretical calculation leading to Eq. (8.47) it does not become obvious where and why the magic happens. The Veselago lens does something one would not have expected, namely the amplification of evanescent waves. There exists an intuitive interpretation model for the Veselago lens that was put forward in Ref. [55], and which helps understanding part of the magic. In this paper the authors start by discussing the confined modes of an interface between a positive and negative index material, similar to the analysis of surface plasmons at the beginning of this chapter. It turns out that the interface supports both TE and TM modes, which in case of ideal material parameters are undamped. We can thus employ again the coupled oscillator model of Eq. (8.42), but we now have to consider the situation that coupling dominates over damping. The first oscillator x_a corresponds to the confined modes at the $-d/2$ interface, which becomes driven by the evanescent fields of the object. On resonance, practically all excitation is transferred to the coupled oscillator x_b, see Exercise 8.10, which starts to strongly oscillate. The magic suggested, but not fully explained by the model is that the confined mode at the $d/2$ interface (corresponding to x_b) is excited in such a manner that its evanescent field decays to a value at the image position that corresponds exactly to the value at the object position.

Recalling the Kramers–Kronig relations of Eq. (7.34) it is obvious that material parameters of $-\varepsilon_0$, $-\mu_0$ cannot be achieved for the entire frequency range, but must be accompanied by losses at other frequencies. Currently also no materials or metamaterials exist that would represent a viable realization of negative index materials in the optical regime. For these reasons, the perfect lens has attracted a lot of interest but applications have remained scarce so far. In Sect. 8.5.2 we will discuss approximate strategies to implement the perfect lens in experiment.

8.5 Green's Function for Stratified Media

We conclude this chapter by discussing how to compute the Green's function for a stratified medium. In doing so, we first note that the dyadic Green's function $\bar{\bar{G}}(\boldsymbol{r},\boldsymbol{r}')$ gives the electric field at position $\boldsymbol{r}$ produced by an oscillating dipole at position $\boldsymbol{r}'$ via (see Fig. 8.19)

$$\boldsymbol{E}(\boldsymbol{r}) = \mu\omega^2\,\bar{\bar{G}}(\boldsymbol{r},\boldsymbol{r}')\cdot\boldsymbol{p}\,.$$

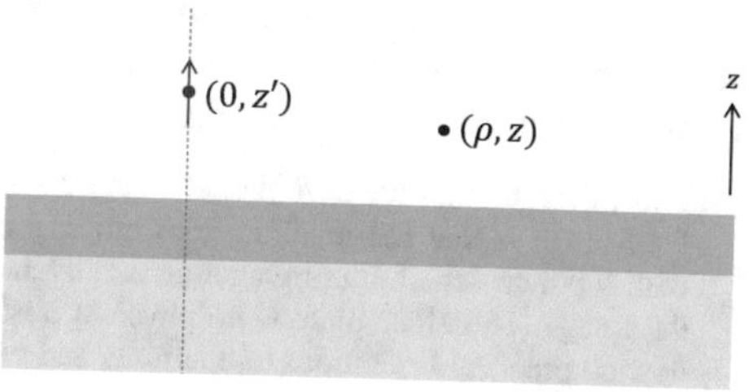

Fig. 8.19 Dipole above stratified medium. A dipole is located at position $(0, z')$ above a stratified medium, and the reflected field is computed at position (ρ, z)

The above expression is correct for an unbounded medium when using the Green's function of Eq. (5.19). To compute the electric field for a stratified medium, we additionally have to introduce a reflected part of the Green's function

$$E(r) = \mu\omega^2 \left[\bar{\bar{G}}(r, r') + \bar{\bar{G}}_{\text{refl}}(r, r')\right] \cdot p\,, \tag{8.48}$$

which has to be chosen such that the boundary conditions of Maxwell's equations are fulfilled at the interfaces of the stratified medium. Conceptually, the computation of the reflected Green's function is simple and consists of the following steps.

Optical Cloaking

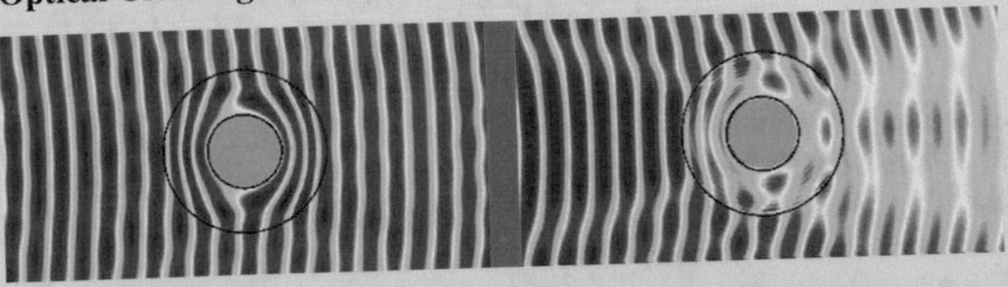

In 2006 two papers appeared back to back in Science which addressed the question of what could be done with metamaterials where the permittivities and permeabilities $\bar{\bar{\varepsilon}}(r, \omega)$, $\bar{\bar{\mu}}(r, \omega)$ can be manufactured at will [56, 57]. In fact, they approached the problem from a slightly different perspective and asked: How do Maxwell's equations look in curved space? It turns out that they look identical to uncurved space provided that $\bar{\bar{\varepsilon}}$, $\bar{\bar{\mu}}$ become renormalized such that they include factors due to the transformation of the coordinate system. Thus, by designing $\bar{\bar{\varepsilon}}$, $\bar{\bar{\mu}}$ one can mimic the effects of curved space. In the abstract of Ref. [56] the authors write:

> Using the freedom of design that metamaterials provide, we show how electromagnetic fields can be redirected at will and propose a design strategy. The conserved fields—electric displacement field D, magnetic induction field B, and Poynting vector S—are all displaced in a consistent manner. A simple illustration is given of the cloaking of a proscribed volume of space to exclude completely all electromagnetic fields. Our work has relevance to exotic lens design and to the cloaking of objects from electromagnetic fields.

Cloaking of an object was demonstrated shortly after the theoretical proposal in an experiment using microwave radiation because manufacturing of metamaterials is significantly easier for such wavelengths [58]. The figure above shows how a plane wave becomes directed around an object in simulations with ideal material parameters (left panel) and in the microwave experiment (right panel). This demonstrates that it is indeed possible to control electromagnetic fields and to steer them around objects. The topic has received tremendous interest in recent years, but producing high-quality metamaterials has remained one of the major challenges in the field.

Image courtesy of David R. Smith.

Plane Wave Decomposition. We start by decomposing the Green's dyadics $\bar{\bar{G}}$ for an unbounded medium into plane waves.

Fresnel Coefficients. Each incoming plane wave of this decomposition becomes reflected and transmitted at the interfaces of the stratified medium, using the generalized reflection and transmission coefficients previously discussed.

Integration. We finally integrate over all reflected and transmitted waves to obtain the reflected Green's function.

Technically, various steps are more difficult. The starting point is the plane wave decomposition of the Green's function derived in Appendix B,

$$G_{ij}(\boldsymbol{r},\boldsymbol{r}') = \frac{1}{8\pi^3 k_m^2}\int_{-\infty}^{\infty} e^{i\boldsymbol{k}\cdot(\boldsymbol{r}-\boldsymbol{r}')}\left(\frac{k_m^2\delta_{ij}-k_ik_j}{k^2-k_m^2}\right)d^3k\,,$$

where k_m is the wavenumber of the medium within which the source (dipole) is located. It turns out to be convenient to add a small imaginary part $i\eta$ to the wavenumber k_m and to let approach $\eta\to 0$ at the end of the calculation. The k_z integration can then be performed analytically using complex analysis, and in the remaining integration over the parallel wavevector components one introduces polar coordinates and integrates analytically over the azimuthal angle. The final expression of Eq. (B.9) becomes an integral over the radial wavevector component k_ρ only, which reads

$$\bar{\bar{G}}_{ij}(\boldsymbol{r},\boldsymbol{r}') = -\frac{\hat{z}_i\hat{z}_j}{k_m^2}\delta(\boldsymbol{r}-\boldsymbol{r}') + \frac{i}{4\pi}$$
$$\times\int_0^\infty \frac{e^{ik_{mz}|z-z'|}}{k_{mz}}\left\{\left\langle\epsilon_i^{TE}(\boldsymbol{k}_m^\pm)\epsilon_j^{TE}(\boldsymbol{k}_m^\pm)\right\rangle+\left\langle\epsilon_i^{TM}(\boldsymbol{k}_m^\pm)\epsilon_j^{TM}(\boldsymbol{k}_m^\pm)\right\rangle\right\}k_\rho dk_\rho\,. \tag{8.49}$$

Here $\boldsymbol{\epsilon}^{TE}$ and $\boldsymbol{\epsilon}^{TM}$ are polarization vectors with TE and TM character, respectively, and the brackets denote an average over the azimuthal angle. See Eq. (B.8) for the explicit form of these terms. The sign of $k_{mz}^\pm=\pm k_{mz}$ has to be chosen positive for $z>z'$ and negative else. In the above form

- $\bar{\bar{G}}$ fulfills inside the medium with wavenumber k_m the defining equation for the dyadic Green's function, see Eq. (5.21),
- but does not fulfill Maxwell's boundary conditions at the interfaces of a stratified medium.

To account for these boundary conditions, we additionally introduce a reflected Green's function $\bar{\bar{G}}_{\text{refl}}$ consisting of solutions of the homogeneous wave equation.

8.5.1 Source Located Above Topmost Layer

We first discuss the situation depicted in Fig. 8.19 where both the source and observation points z' and z are located above the topmost interface of the stratified medium. We denote the topmost medium with n and the position of the interface with z_n. The exponential term in Eq. (8.49)

$$\exp ik_{nz}|z - z'| = \begin{cases} \exp ik_{nz}(z - z') & \text{for } z > z' \\ \exp ik_{nz}(z' - z) & \text{for } z < z' \end{cases}$$

corresponds to an upgoing wave for $z > z'$ and a downgoing wave for $z < z'$. To account properly for the boundary conditions at z_n, we introduce the total Green's function

$$\begin{aligned} G_{ij}^{\text{tot}}(\boldsymbol{r}, \boldsymbol{r}') &= -\frac{\hat{z}_i \hat{z}_j}{k_n^2} \delta(\boldsymbol{r} - \boldsymbol{r}') \\ &+ \frac{i}{4\pi} \int_0^\infty \frac{1}{k_{nz}} \left\{ \left\langle \left[e^{ik_{nz}|z-z'|} \epsilon_i^{TE}(\boldsymbol{k}_n^\pm) + A_n^{TE} e^{ik_{nz}z} \epsilon_i^{TE}(\boldsymbol{k}_n^+) \right] \epsilon_j^{TE}(\boldsymbol{k}_n^\pm) \right\rangle \right. \\ &\left. + \left\langle \left[e^{ik_{nz}|z-z'|} \epsilon_i^{TM}(\boldsymbol{k}_n^\pm) + A_n^{TM} e^{ik_{nz}z} \epsilon_i^{TM}(\boldsymbol{k}_n^+) \right] \epsilon_j^{TE}(\boldsymbol{k}_n^\pm) \right\rangle \right\} k_\rho dk_\rho \,. \end{aligned}$$

The reasons for adding the two upgoing waves with TE and TM character are as follows.

Electric Field. The Green's dyadics gives the electric field at position $\boldsymbol{r}$, and it is the electric field which must fulfill Maxwell's boundary condition at $z = z_n$. For this reason we have added a reflected field for the unprimed coordinate z and not the primed one.

Homogeneous Solution. The additional terms are plane waves which are solutions of the homogeneous wave equation. For this reason, the total Green's function still fulfills the defining equation with a delta-like source at $\boldsymbol{r} = \boldsymbol{r}'$.

Boundary Conditions. The coefficients A_n^{TE}, A_n^{TM} must be chosen such that the boundary conditions of Maxwell's equations are fulfilled at the topmost interface.

For matching the fields at $z = z_n$ we generally have $z < z'$ and the exponential term $e^{ik_{nz}(z'-z)}$ corresponds to a downgoing wave. Apart from a normalization constant, the TE electric field in medium n slightly above the interface (but below the source position) can be expressed as

$$e_n(z) = e^{ik_{nz}(z'-z)} + A_n^{TE} e^{ik_{nz}z} \,.$$

To compute the unknown coefficient A_n^{TE} we notice that the upgoing reflected field can be related to the downgoing incoming field via the generalized reflection coefficient $\tilde{R}$ of the transfer matrix approach, see Eq. (8.36),

$$e_n^+ = \tilde{R}_{n,1}^{TE}\, e_n^- \implies \left[A_n^{TE} e^{ik_{nz}z_n}\right] = \tilde{R}_{n,1}^{TE}\left[e^{ik_{nz}(z'-z_n)}\right].$$

This allows us to express the reflected wave in the form

$$A_n^{TE} e^{ik_{nz}z} = \tilde{R}_{n,1}^{TE}\, e^{ik_{nz}(z+z'-2z_n)}. \tag{8.50}$$

The same procedure can be applied to the TM fields. Summing up the two contributions, we arrive at the reflected part of the Green's function for a stratified medium

Reflected Green's Function for z, z' Above Stratified Medium

$$G_{ij}^{\text{refl}}(\boldsymbol{r},\boldsymbol{r}') = \frac{i}{4\pi}\int_0^\infty \frac{e^{ik_{nz}(z+z'-2z_n)}}{k_{nz}}$$
$$\times \left\{\tilde{R}_{n,1}^{TE}\left\langle\epsilon_i^{TE}(\boldsymbol{k}_n^+)\epsilon_j^{TE}(\boldsymbol{k}_n^\pm)\right\rangle + \tilde{R}_{n,1}^{TM}\left\langle\epsilon_i^{TM}(\boldsymbol{k}_n^+)\epsilon_j^{TM}(\boldsymbol{k}_n^\pm)\right\rangle\right\} k_\rho dk_\rho\,, \tag{8.51}$$

where the positive sign of $k_{nz}^\pm$ has to be chosen for $z > z'$ and the negative one for $z < z'$. The above expression combines the generalized reflection coefficients of the transfer matrix approach with the (angle-averaged) polarization vector products with TE and TM character, respectively, which are explicitly given in Eq. (B.8). As discussed in Appendix B, one can safely deform the integration path for k_ρ away from the real axis into the complex plane, which is generally needed for an efficient computation of the reflected Green's function.

The above formalism has to be slightly adapted to compute the Green's dyadics for the source located in the uppermost medium, and the observation point located inside or below the stratified medium. We here only discuss the latter case, and will sketch the computation scheme for general source and observation points at the end of this chapter. The downgoing wave in the lowest medium can be related to the "incoming" field of the uppermost medium through

$$e_1^- = \tilde{T}_{n,1}\, e_n^- \implies A_1 e^{-ik_{1z}z_1} = \tilde{T}_{n,1}\left[e^{ik_{nz}(z'-z_n)}\right].$$

Thus, the Green's dyadics for the source and observation points z', z above and below the stratified medium can be written in the form

$$G_{ij}^{\text{refl}}(\boldsymbol{r},\boldsymbol{r}') = \frac{i}{4\pi}\int_0^\infty \frac{e^{ik_{1z}(z_1-z)+ik_{nz}(z'-z_n)}}{k_{nz}}$$
$$\times \left\{\tilde{T}_{n,1}^{TE}\left\langle\epsilon_i^{TE}(\boldsymbol{k}_1^-)\epsilon_j^{TE}(\boldsymbol{k}_n^-)\right\rangle + \tilde{T}_{n,1}^{TM}\left\langle\epsilon_i^{TM}(\boldsymbol{k}_1^-)\epsilon_j^{TM}(\boldsymbol{k}_n^-)\right\rangle\right\} k_\rho dk_\rho\,. \tag{8.52}$$

Note that in the layers away from the source point the total Green's function consists of the reflected part only. To show that this form is indeed correct, it is sufficient to convince oneself that

- the expression fulfills the homogeneous wave equation for z being located in the lowest medium (because plane waves are solutions of the homogeneous wave equation), and that
- the boundary conditions of Maxwell's equations are fulfilled at the lowest interface (by construction).

In principle, by propagating the fields through the stratified medium, using the transfer matrix approach previously discussed, we can also compute the fields in all other layers.

8.5.2 *Imaging with an Imperfect Veselago Lens*

With the reflected Green's functions we are now in the position to simulate imaging through an imperfect Veselago lens. Indeed, it was already proposed in Pendry's original work [54] to replace for a proof-of-principle experiment the perfect lens with an ordinary silver slab with $\mu = \mu_0$, and operating at a frequency where the permittivity becomes $-\varepsilon_0$. A first experimental realization was reported by Fang and coworkers [59]. Figure 8.20 shows the imaging with such a non-perfect superlens, using (a) a negative refractive index material with small losses, (b) a metal slab with stronger losses and a permeability of μ_0, and (c) no slab at all. One clearly observes that even the non-perfect metal slab leads to a relatively good imaging of the point dipoles.

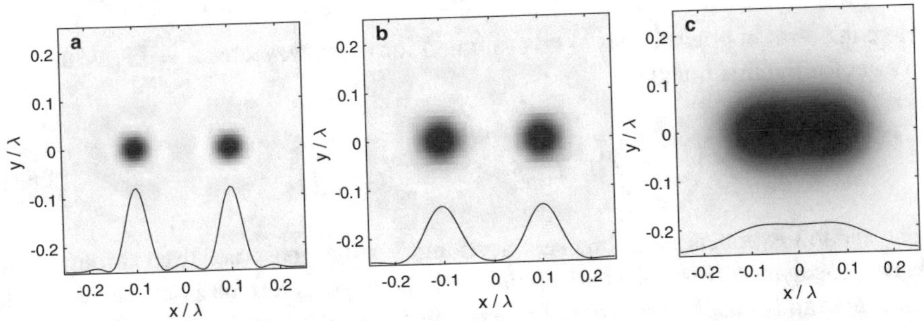

Fig. 8.20 Imaging of two point dipoles with orientation along z and separated by a distance of 80 nm through a slab of material filled with (**a**) negative index material with small losses, $\varepsilon : \varepsilon_0 = -1 + 0.01\,i$, $\mu : \mu_0 = -1$, (**b**) metal with $\varepsilon : \varepsilon_0 = -1 + 0.1\,i$, $\mu : \mu_0 = 1$, and (**c**) vacuum. The thickness of the slab is 40 nm, the object and image are located 20 nm away from the slab, see sketch of Veselago lens in Fig. 8.18, and the light wavelength is $\lambda = 400$ nm

8.5.3 Far-field Limit

Suppose that a dipole is located above a stratified medium, see Fig. 8.19, and we are interested in the radiation emitted to the far-field. From Eqs. (8.51) and (8.52) we observe that the reflected Green's functions contain integrals of the form

$$\mathcal{I}_1 = \frac{i}{4\pi}\int_0^\infty e^{ik_{nz}(z+z'-2z_n)}\big\langle\dots\big\rangle k_\rho dk_\rho$$

$$\mathcal{I}_2 = \frac{i}{4\pi}\int_0^\infty e^{ik_{1z}(z_1-z)+ik_{nz}(z'-z_n)}\big\langle\dots\big\rangle k_\rho dk_\rho\,.$$

$\mathcal{I}_{1,2}$ are for observation points z located above or below the stratified medium. We first undo the average over the azimuthal angle, see Eq. (B.6) and discussion thereafter, and rewrite the integrals in the form

$$\mathcal{I}_1 = \frac{i}{8\pi^2}\int_{-\infty}^\infty e^{ik_x(x-x')+ik_y(y-y')}e^{ik_{nz}(z+z'-2z_n)}\big[\dots\big] dk_x dk_y$$

$$\mathcal{I}_2 = \frac{i}{8\pi^2}\int_{-\infty}^\infty e^{ik_x(x-x')+ik_y(y-y')}e^{ik_{1z}(z_1-z)+ik_{nz}(z'-z_n)}\big[\dots\big] dk_x dk_y\,.$$

We are now seeking for the far-field limit of these expressions where $\boldsymbol{r} = (x, y, z)$ in the exponential becomes very large. The evaluation of this integral is identical to our previous discussion in Chap. 3 about the far-field representation of electromagnetic fields, where the stationary phase approximation has been used to obtain

$$\frac{i}{8\pi^2}\int_{-\infty}^\infty e^{i\boldsymbol{k}\cdot\boldsymbol{r}}\Big(\dots\Big) dk_x dk_y \underset{kr\gg 1}{\longrightarrow} \frac{i}{8\pi^2}\left(\frac{e^{ikr}}{r}\right)\Big[-2\pi i k_z\Big(\dots\Big)\Big]_{\boldsymbol{k}=k\hat{\boldsymbol{r}}}\,.$$

Here the term in brackets has to be evaluated for the wavevector $\boldsymbol{k} = k\hat{\boldsymbol{r}}$. With this, we get for the first integral

$$\mathcal{I}_1 \longrightarrow \left(\frac{e^{ik_n r}}{4\pi r}\right)\Big[e^{-i\boldsymbol{k}\cdot\boldsymbol{r}'}e^{2ik_z(z'-z_n)}k_z\Big(\dots\Big)\Big]_{\boldsymbol{k}=k_n\hat{\boldsymbol{r}}}\,. \tag{8.53}$$

The second exponential in brackets accounts for the phase acquired through the wave propagation from the source point z' to the upper interface z_n of the stratified medium and back. For the integral where the observation point z is located below the stratified medium we correspondingly get

$$\mathcal{I}_2 \longrightarrow \left(\frac{e^{ik_1 r}}{4\pi r}\right)\Big[e^{-i\boldsymbol{k}\cdot\boldsymbol{r}'}e^{i\psi}k_z\Big(\dots\Big)\Big]_{\boldsymbol{k}=k_1\hat{\boldsymbol{r}}}\,. \tag{8.54}$$

Here $\psi = k_1 z(z' - z_1) + k_{nz}(z' - z_n)$ is a phase that is usually irrelevant. To get the final expressions for the far-fields of the reflected Green's functions, we continue to simplify the vectors $\boldsymbol{\epsilon}^{TE}$, $\boldsymbol{\epsilon}^{TM}$.

z Above Stratified Medium. When z is located above the stratified medium, we can express the wavevector $\boldsymbol{k}_n$ in spherical coordinates, where the angles are given by the propagation directions of the electromagnetic far-fields. We then get from Eq. (B.5)

$$\boldsymbol{\epsilon}^{TE}(\boldsymbol{k}_n^+) = \begin{pmatrix} \sin\phi \\ -\cos\phi \\ 0 \end{pmatrix} = \hat{\boldsymbol{\phi}}\,, \quad \boldsymbol{\epsilon}^{TM}(\boldsymbol{k}_n^+) = \begin{pmatrix} \cos\phi\cos\theta \\ \sin\phi\cos\theta \\ -\sin\theta \end{pmatrix} = \hat{\boldsymbol{\theta}}\,,$$

where $\hat{\boldsymbol{\phi}}$, $\hat{\boldsymbol{\theta}}$ are unit vectors in the directions of the azimuthal and polar angles, respectively. Note that in the upper half-space $\boldsymbol{k}_n = \boldsymbol{k}_n^+$ always points upwards.

z below Stratified Medium. When z is located below the stratified medium, we can expand $\boldsymbol{k}_1$ in spherical coordinates and find that $\boldsymbol{\epsilon}^{TE}(\boldsymbol{k}_1^-)$, $\boldsymbol{\epsilon}^{TM}(\boldsymbol{k}_1^-)$ can again be expressed in terms of $\hat{\boldsymbol{\phi}}$, $\hat{\boldsymbol{\theta}}$. Note that in the lower half-space $\boldsymbol{k}_1 = \boldsymbol{k}_1^-$ always points downwards. We additionally need the vector

$$\boldsymbol{\epsilon}^{TM}(\boldsymbol{k}_n^-) = k_n^{-1}\begin{pmatrix} k_1\cos\phi\sin\theta \\ k_1\sin\phi\sin\theta \\ -k_{nz} \end{pmatrix} \times \begin{pmatrix} \sin\phi \\ -\cos\phi \\ 0 \end{pmatrix} = k_n^{-1}\begin{pmatrix} k_{nz}\cos\phi \\ k_{nz}\sin\phi \\ -k_1\sin\theta \end{pmatrix},$$

which relates the spherical coordinates in the lower medium to the wavevector in the upper medium, where the source is located. k_{nz} is the z-component of the wavevector, which has to be computed from the dispersion relation.

We can finally express the far-field limit of the reflected ($z > 0$) and transmitted ($z < 0$) Green's functions in the form

$$G_{ij}^{\text{refl}}(\boldsymbol{r}, \boldsymbol{r}') \to \left(\frac{e^{ik_n r}}{4\pi r}\right)\left[e^{-i\boldsymbol{k}\cdot\boldsymbol{r}'}e^{2ik_z(z'-z_n)}\left(\tilde{R}_{n,1}^{TE}\,\hat{\phi}_i\hat{\phi}_j + \tilde{R}_{n,1}^{TM}\,\hat{\theta}_i\hat{\theta}_j\right)\right]_{\boldsymbol{k}=k_n\hat{\boldsymbol{r}}}$$
$$G_{ij}^{\text{trans}}(\boldsymbol{r}, \boldsymbol{r}') \to \left(\frac{e^{ik_1 r}}{4\pi r}\right)\left[e^{-i\boldsymbol{k}\cdot\boldsymbol{r}'}e^{i\psi}\frac{k_{1z}}{k_{nz}}\left(\tilde{T}_{n,1}^{TE}\,\hat{\phi}_i\hat{\phi}_j + \tilde{T}_{n,1}^{TM}\,\hat{\theta}_i\epsilon_j^{TM}(\boldsymbol{k}_n^-)\right)\right]_{\boldsymbol{k}=k_1\hat{\boldsymbol{r}}}. \tag{8.55}$$

The angles ϕ, θ are given by the propagation directions of the electromagnetic far-fields, and we have termed the Green's dyadics in the lower medium as transmitted.

Consider next a dipole with moment $\boldsymbol{p}$ that is located at position $\boldsymbol{r}_0$ above the stratified medium. According to Eq. (6.3), the far-fields of the oscillating dipole can be expressed as

$$\boldsymbol{F}^{\rm dip}(\hat{\boldsymbol{r}}) = \frac{k_n^2}{4\pi\varepsilon_n}\left(e^{-ik_n\hat{\boldsymbol{r}}\cdot\boldsymbol{r}_0}\right)\Big[p_\phi\,\hat{\boldsymbol{\phi}} + p_\theta\,\hat{\boldsymbol{\theta}}\Big],$$

with $p_\phi = \hat{\boldsymbol{\phi}}\cdot\boldsymbol{p}$ and $p_\theta = \hat{\boldsymbol{\theta}}\cdot\boldsymbol{p}$. Similarly, from Eq. (8.55) we get the far-field amplitude for the radiation into the upper half-space,

$$\boldsymbol{F}^{\rm refl}(\hat{\boldsymbol{r}}) = \frac{k_n^2}{4\pi\varepsilon_n}\left(e^{-ik_n\hat{\boldsymbol{r}}\cdot\boldsymbol{r}_0}\right)\Big[e^{2ik_{nz}(z_0-z_n)}\Big(\tilde{R}_{n,1}^{TE}\,p_\phi\,\hat{\boldsymbol{\phi}} + \tilde{R}_{n,1}^{TM}\,p_\theta\,\hat{\boldsymbol{\theta}}\Big)\Big].$$

Let us express the sum of the two contributions as

$$\boldsymbol{F}(\hat{\boldsymbol{r}}) = \frac{k_n^2}{4\pi\varepsilon_n}\left(e^{-ik_n\hat{\boldsymbol{r}}\cdot\boldsymbol{r}_0}\right)\Big[\mathscr{P}_\phi\,\hat{\boldsymbol{\phi}} + \mathscr{P}_\theta\,\hat{\boldsymbol{\theta}}\Big],$$

where $\mathscr{P}_\phi$, $\mathscr{P}_\theta$ can be directly read off from the above equations. In the far-field the power radiated in a given direction $\hat{\boldsymbol{r}}$ is given by Eq. (4.17),

$$r^2\langle\boldsymbol{S}\rangle = \frac{1}{2}Z^{-1}|\boldsymbol{F}|^2\,\hat{\boldsymbol{r}} = \frac{1}{2}\left(\frac{k_n^2}{4\pi\varepsilon_n}\right)^2\Big[|\mathscr{P}_\phi|^2 + |\mathscr{P}_\theta|^2\Big]\hat{\boldsymbol{r}}.$$

Thus, for a dipole located above a stratified medium the power radiated into a given direction $\hat{\boldsymbol{r}}$ within the upper half-space can be expressed as

Power Radiated by Dipole Above Stratified Medium ($z > 0$)

$$r^2\langle\boldsymbol{S}\rangle = \frac{1}{2}\left(\frac{k_n^2}{4\pi\varepsilon_n}\right)^2 \times\left(\Big|1+e^{2ik_{nz}(z_0-z_n)}\,\tilde{R}_{n,1}^{TE}\Big|^2\,p_\phi^2 + \Big|1+e^{2ik_{nz}(z_0-z_n)}\,\tilde{R}_{n,1}^{TM}\Big|^2\,p_\theta^2\right). \tag{8.56}$$

We can proceed similarly for the radiation to the lower half-space. The far-field amplitude of the transmitted wave reads

$$\boldsymbol{F}^{\rm trans}(\hat{\boldsymbol{r}}) = \frac{k_1^2}{4\pi\varepsilon_1}\left(e^{-ik_1\hat{\boldsymbol{r}}\cdot\boldsymbol{r}_0}\right)\Big[e^{i\psi}\frac{k_{1z}}{k_{nz}}\Big(\tilde{T}_{n,1}^{TE}\,p_\phi\,\hat{\boldsymbol{\phi}} + \tilde{T}_{n,1}^{TM}\,p^{TM}\,\hat{\boldsymbol{\theta}}\Big)\Big],$$

where we have introduced $p^{TM} = \boldsymbol{\epsilon}^{TM}(\boldsymbol{k}_n^-)\cdot\boldsymbol{p}$. Note that when the dipole position and observation point are located in different layers, we only have to consider the transmitted wave and there is no direct contribution from the dipole itself. Thus, in the lower half-space we get for the power radiated by the dipole into direction $\hat{\boldsymbol{r}}$ the expression

Power Radiated by Dipole Above Stratified Medium ($z < 0$)

$$r^2\langle S\rangle = \frac{1}{2}\left(\frac{k_1^2}{4\pi\varepsilon_1}\right)^2 \left|\frac{k_{1z}}{k_{nz}}\right|^2 \left(\left|\tilde{T}_{n,1}^{TE}\,p_\phi\right|^2 + \left|\tilde{T}_{n,1}^{TM}\,p^{TM}\right|^2\right). \tag{8.57}$$

Figure 8.21 shows the angle resolved pattern of the power radiated by an oscillating dipole. We chose dipole-surface distances of (a) 10 nm, (b) $\lambda/4$, and (c) $\lambda/2$, with $\lambda = 600$ nm, and for dipoles oriented parallel and perpendicular to the interface. As can be seen from the figure, the emission pattern depends strongly on z_0, which can be attributed to the interference between the direct emission and the reflected wave in Eq. (8.56). Figure 8.22 shows the emitted power for a glass substrate, using the same simulation parameters as in Fig. 8.21. One observes that most of

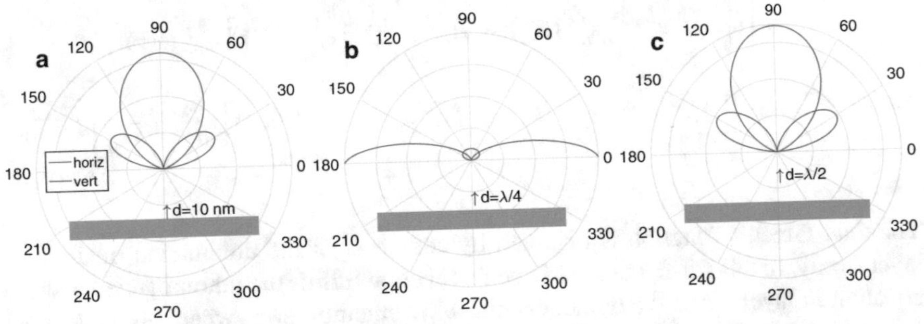

Fig. 8.21 Emission pattern of dipole located above a gold surface. The panels report dipole-surface distances z_0 of (**a**) 10 nm, (**b**) $\lambda/4$, and (**c**) $\lambda/2$, with a wavelength of 600 nm. We set the permittivity of the upper medium to ε_0 and use tabulated values for the gold permittivity [34]. The emission patterns are computed using Eq. (8.56)

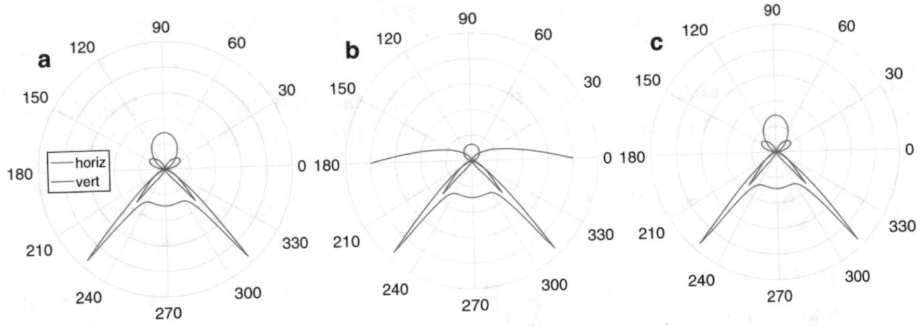

Fig. 8.22 Same as Fig. 8.21 but for glass substrate. The panels report dipole-surface distances z_0 of (**a**) 10 nm, (**b**) $\lambda/4$, and (**c**) $\lambda/2$, with a wavelength of 600 nm. We set the permittivity of the upper medium to ε_0 and use a refractive index of 1.5 for glass

the radiation goes into the glass substrate. The angle at which the emission into the lower medium is peaked corresponds to the angle of total internal reflection for the glass-air interface. A more thorough discussion of the emission properties of dipoles located above a substrate or other types of stratified media, and of effects such as "forbidden light" is given in Ref. [6].

8.5.4 *Source Located Inside Layer*

We finally consider the situation that the source point z' is located inside layer ℓ, and the observation point in layer m. Using the same reasoning as for the source above the stratified medium, we can express the reflected Green's function as

$$G_{ij}^{\text{refl}}(\boldsymbol{r},\boldsymbol{r}') = \frac{i}{4\pi}\int_0^\infty \frac{1}{k_{mz}}$$
$$\times \Bigg[\Big\langle\Big[A_m^{TE}\,e^{ik_{mz}z}\,\epsilon_i^{TE}(\boldsymbol{k}_m^+) + B_m^{TE}\,e^{-ik_{mz}z}\,\epsilon_i^{TE}(\boldsymbol{k}_m^-)\Big]\,\epsilon_j^{TE}(\boldsymbol{k}_\ell^\pm)\Big\rangle$$
$$+\Big\langle\Big[A_m^{TM}\,e^{ik_{mz}z}\,\epsilon_i^{TM}(\boldsymbol{k}_m^+) + B_m^{TM}\,e^{-ik_{mz}z}\,\epsilon_i^{TM}(\boldsymbol{k}_m^-)\Big]\,\epsilon_j^{TM}(\boldsymbol{k}_\ell^\mp)\Big\rangle\Bigg]\,k_\rho dk_\rho. \tag{8.58}$$

The total Green's function is obtained by adding $\bar{\bar{G}}$ of the unbounded medium in layer ℓ only. In this form, the total Green's function fulfills the inhomogeneous wave equation in layer ℓ and the homogeneous wave equation everywhere else. We must compute the coefficients A_m, B_m for upgoing and downgoing waves such that the boundary conditions of Maxwell's equations are fulfilled at all interfaces. Let z_ℓ^+, z_ℓ^- denote the positions of the upper and lower interfaces of layer ℓ, respectively. Inside layer ℓ we have for the TE waves

$$e_\ell(z) = e^{ik_{\ell z}|z-z'|} + A_\ell^{TE}\,e^{ik_{\ell z}z} + B_\ell^{TE}\,e^{-ik_{\ell z}z}\,.$$

For $z<z'$ the first term $e^{ik_{\ell z}(z'-z)}$ corresponds to a downgoing wave. We thus relate the fields at the lower interface of medium ℓ at position z_ℓ^- through

$$e_\ell^+ = \tilde{R}_{\ell,1}^{TE}\,e_\ell^- \implies \Big[A_\ell^{TE}\,e^{ik_{\ell z}z_\ell^-}\Big] = \tilde{R}_{\ell,1}^{TE}\Big[e^{ik_{\ell z}(z'-z_\ell^-)} + B_\ell^{TE}\,e^{-ik_{\ell z}z_\ell^-}\Big]\,.$$

For $z>z'$ the first term $e^{ik_{\ell z}(z-z')}$ corresponds to an upgoing wave. We thus relate the fields at the upper interface of medium ℓ at position z_ℓ^+ through

$$e_\ell^- = \tilde{R}_{\ell,n}^{TE}\,e_\ell^+ \implies \Big[B_\ell^{TE}\,e^{-ik_{\ell z}z_\ell^+}\Big] = \tilde{R}_{\ell,n}^{TE}\Big[e^{ik_{\ell z}(z_\ell^+-z')} + A_\ell^{TE}\,e^{ik_{\ell z}z_\ell^+}\Big]\,.$$

From these two expressions we can compute A_ℓ^{TE}, B_ℓ^{TE} and, in turn, propagate the fields through the different layers of the stratified medium. A corresponding procedure can be also used for the TM fields.

Slab Structure As an example, we here consider the situation where the source dipole is located inside a slab. For this geometry, we can replace in the above expressions the generalized reflection coefficients by the Fresnel coefficients. We assume that the interfaces are located at positions 0 and d, and thus get

$$R_{21}^{TE}\Big[e^{ik_z(z'-z)} + B_2^{TE}\,e^{-ik_z z}\Big]_{z=0} = \Big[A_2^{TE}\,e^{ik_z z}\Big]_{z=0}$$
$$R_{23}^{TE}\Big[e^{ik_z(z-z')} + A_2^{TE}\,e^{ik_z z}\Big]_{z=d} = \Big[B_2^{TE}\,e^{-ik_z z}\Big]_{z=d},$$

with the Fresnel reflection coefficients R_{21}^{TE}, R_{23}^{TE} for the interfaces at $z = 0$ and d. A corresponding expression can be obtained for the TM modes. We can now solve for the unknown wave amplitudes

$$A_2^{TE} = \frac{e^{ik_z d}\,R_{21}^{TE}}{1 - e^{2ik_z d}\,R_{21}^{TE}\,R_{23}^{TE}}\Big[e^{ik_z(z'-d)} + e^{ik_z(d-z')}\,R_{23}^{TE}\Big] \tag{8.59}$$
$$B_2^{TE} = \frac{e^{ik_z d}\,R_{23}^{TE}}{1 - e^{2ik_z d}\,R_{21}^{TE}\,R_{23}^{TE}}\Big[e^{ik_z(d-z')} + e^{ik_z(d+z')}\,R_{21}^{TE}\Big].$$

By expanding the denominator into a geometric series, we obtain a representation of the waves in terms of multiple reflections inside the slab, similarly to the situation shown in Fig. 8.11. These coefficients have to be finally inserted into Eq. (8.58) in order to compute the reflected Green's function for the inner region of the slab.

Exercises

Exercise 8.1 Consider for the surface plasmons complex κ_{1z} and κ_{2z} values. Compute the averaged Poynting vectors $\boldsymbol{S}_1$, $\boldsymbol{S}_2$ and interpret the result. In which direction does the energy flow?

Exercise 8.2 Compute the asymptotic values $k_x \to \infty$ of the surface plasmon dispersion from Eq. (8.6) by finding the values of ω where the real part of the denominator under the square root becomes zero.

Exercise 8.3 Consider the interface between two materials with permittivities ε_1, ε_2 and permeabilities μ_0. Repeat the derivation of surface plasmons, but now for TE modes with $\boldsymbol{E} = E_y\hat{\boldsymbol{y}}$. Show that no such waves exist.

Exercise 8.4 A Drude dielectric function representative for silver is given in the form

$$\varepsilon(\omega) = \varepsilon_0\left(\kappa_b - \frac{\omega_p^2}{\omega(\omega + i\gamma)}\right),$$

with the parameters $\kappa_b = 3.3$, $\hbar\omega_p = 9\,\text{eV}$, $\hbar\gamma = 0.022\,\text{eV}$. Use the plasmon dispersion $k_x(\omega)$ to compute the wavelength $\lambda = 2\pi/k_x'$ and the propagation length $\delta = 2\pi/k_x''$ for a few selected photon energies in the range between 0.5 and 3 eV.

Exercise 8.5 Use from the NANOPT toolbox the file `demostrat1.m` for the simulation of surface plasmon excitation in a Kretschmann geometry.

(a) Change the permittivity of the lowest medium and compute the angle for the plasmon dip in the reflection spectra. Plot the change of permittivity versus the change of angle.
(b) Same as a. but for a 20 nm thick additional layer placed on top of the metal surface. Change the permittivity of the layer only, and set the permittivity of the upper half-space to ε_0.

Exercise 8.6 Consider a metal slab with thickness d and permittivity ε_1 embedded in a background material with ε_2. For TM modes we can make the following ansatz for the magnetic fields:

$$\boldsymbol{H}(z) = \hat{\boldsymbol{y}} e^{ik_x x} \begin{cases} h_1 e^{-\kappa_{1z} z} & \text{for } z > +\frac{d}{2} \\ h_2 \left(e^{\kappa_{2z} z} \pm e^{-\kappa_{2z} z}\right) & \text{for } -\frac{d}{2} < z < \frac{L}{2} \\ \pm h_1 e^{+\kappa_{1z} z} & \text{for } z < -\frac{d}{2}\,, \end{cases}$$

where we have used that because of symmetry the solutions must be odd or even functions of z.

(a) Repeat the analysis of surface plasmons for the above ansatz, and show that such modes exist.
(b) Show that for large L values the dispersion of these modes coincides with that of a single interface. The modified surface plasmon dispersions are now given by Eq. (8.40).
(c) Discuss the even and odd character for the magnetic and electric fields.
(d) Which modes have a higher energy, the even or odd ones? Try to motivate the results using simple arguments.

Exercise 8.7 Consider two media with permittivities ε_1, ε_2 and permeabilities μ_0. Determine for TM polarized waves the *Brewster angle* for which the reflected wave amplitude vanishes. How large is this angle for $n_2 : n_1 = 1.5$? Suppose that a light with arbitrary polarization impinges on the interface. How is the reflected wave polarized?

Exercise 8.8 Consider reflection/transmission at an interface with $n_1 > n_2$. Find the angle at which *total internal reflection* occurs, this is when the wavenumber k_z in the second medium becomes purely imaginary.

Exercise 8.9 Compute for the perfect lens the generalized reflection coefficient and show that $\tilde{R}_{13} = 0$.

Exercise 8.10 As a simple model for the perfect lens, start from Eq. (8.42) for the amplitudes of two harmonic oscillators, where one of them is driven by an external field, and compute the displacements $\bar{x}_a$, $\bar{x}_b$ for $\omega = \omega_a = \omega_b$ and $\gamma_a = \gamma_b \approx 0$. Interpret the result. You might also like to refer to Ref. [55].

Exercise 8.11 An object is located at distance δ away from a Veselago lens with thickness d.

(a) At which distance from the lens is the image formed at the other side?
(b) Trace the evanescent fields through the lens and show that they are equal at the object and image positions.

Exercise 8.12 Use from the NANOPT toolbox the file `demostrat05.m` for the emission pattern of a dipole above a metal substrate. Investigate the pattern for various dipole-layer distances, and interpret the result.

Exercise 8.13 Use from the NANOPT toolbox the file `demostrat05.m` for the emission pattern of a dipole above a metal substrate. Replace the metal substrate by a thin metal film (use thicknesses of 20, 50, and 100 nm) located on top of a glass substrate. Vary the thickness of the metal film, and compare the results with those of a semi-infinite metal layer.

Chapter 9
Particle Plasmons

In the last chapter we have discussed surface plasmons. These are coherent charge oscillations at the interface between a metal and a dielectric, which can be excited optically and propagate along the interface. They come along with evanescent electromagnetic fields, which decay exponentially when moving away from the interface. In this chapter we analyze metallic nanoparticles embedded in dielectrics and show that corresponding excitations exist there, which we will denote as particle plasmons. We start our discussion with nanoparticles much smaller than the wavelength of light, where the quasistatic approximation can be employed, and generalize the results later for larger particles.

9.1 Quasistatic Limit

Consider the wave equation for the scalar potential, see Eq. (2.22),

$$\left(\nabla^2 + k^2\right) V(\boldsymbol{r}) = -\frac{\rho(\boldsymbol{r})}{\varepsilon}\,,$$

which we have written down for time-harmonic fields using $k^2 = \mu\varepsilon\,\omega^2$. We have previously derived this expression by employing the Lorenz gauge condition

$$\nabla \cdot \boldsymbol{A}(\boldsymbol{r}) = i\mu\varepsilon\omega\, V(\boldsymbol{r}) = i\sqrt{\mu\epsilon}\, kV(\boldsymbol{r})\,.$$

To understand the essence of the quasistatic approximation, we assume the following points:

- The nanoparticles are much smaller than the light wavelength. Let L be a characteristic length scale of the nanoparticles, typically a few tens of nanometers, and λ the light wavelength, typically on the order of micrometers.

U. Hohenester, *Nano and Quantum Optics*, Graduate Texts in Physics,
https://doi.org/10.1007/978-3-030-30504-8_9

– The Laplace operator ∇^2 gives the curvature of the function it is acting on. We now assume that the electromagnetic fields and potentials vary on the length scale of L, such that the following estimate holds:

$$\left|\nabla^2 V\right| \sim \frac{1}{L^2}\left|V\right| \gg \left|k^2 V\right| \sim \frac{1}{\lambda^2}\left|V\right|.$$

For these reasons, we can ignore in the wave equation for $V(\boldsymbol{r})$ the k^2 term and end up with the Poisson equation of electrostatics. The "quasi" of the quasistatic approximation refers to the fact that in the matching of the electric fields at the particle boundary we keep the frequency-dependent permittivities $\varepsilon(\boldsymbol{r},\omega)$.

Quasistatic Approximation

$$\nabla^2 V(\boldsymbol{r}) = -\frac{\rho(\boldsymbol{r})}{\varepsilon(\boldsymbol{r},\omega)}, \tag{9.1}$$

Additionally we consider the frequency-dependent permittivities $\varepsilon(\omega)$ in evaluating the boundary conditions.

From the Lorenz gauge condition we find

$$L\left|\nabla\cdot\boldsymbol{A}\right| \sim \left|\boldsymbol{A}\right| \sim L\left|i\sqrt{\mu\varepsilon}\,kV(\boldsymbol{r})\right| \sim \frac{L}{\lambda c}\left|V\right| \ll \frac{1}{c}\left|V\right|.$$

Thus, the vector potential is much smaller than the scalar potential and can be approximately neglected. Altogether, it turns out that the quasistatic approximation works extremely well for small nanoparticles, say smaller than about 50–100 nanometers. The solution of the Poisson equation is considerably simpler than that of the wave equation, and one can immediately employ the full machinery developed for electrostatics. Throughout we relate the scalar potential and the electric field through

$$\boldsymbol{E}(\boldsymbol{r}) = -\nabla V(\boldsymbol{r}). \tag{9.2}$$

When it comes to light scattering of small nanoparticles treated in the quasistatic approximation one has to be careful:

– As the light wavelength is much larger than the size of the nanoparticle, the particle is driven by an electric field constant in space but varying in time. For the time-harmonic fields the corresponding potential is

$$V(\boldsymbol{r}) = -E_0\boldsymbol{\epsilon}_0\cdot\boldsymbol{r}, \tag{9.3}$$

where E_0 is the electric field amplitude and $\boldsymbol{\epsilon}_0$ the light polarization.

- For scattering, absorption, and extinction of light we must resort to the full Maxwell's equations. We will do so by computing the induced dipole moment of the nanoparticle, and evaluating the far-fields for an oscillating dipole.

We finish this section by stating the boundary conditions (b.c.) of the electrostatic potential at the interface between two materials with permittivities $\varepsilon_1(\omega)$, $\varepsilon_2(\omega)$. Continuity of the tangential electric field is obtained through continuity of the potential at the boundary

$$\text{1st b.c.}\quad V\left(\boldsymbol{r}-0^+\hat{\boldsymbol{n}}\right)\Big|_{\partial\Omega}=V\left(\boldsymbol{r}+0^+\hat{\boldsymbol{n}}\right)\Big|_{\partial\Omega}, \tag{9.4a}$$

where $\hat{\boldsymbol{n}}$ is a vector perpendicular to the boundary pointing from the inside to the outside, and 0^+ is an infinitesimally small quantity. Thus, if we take the derivative in a direction parallel to the boundary, in order to get $\boldsymbol{E}_\parallel$, we obtain the same values on both sides. For the second boundary condition we assume that no free surface charge contributions are present at the boundary, and thus get from the continuity of the dielectric displacement in the direction perpendicular to the boundary the condition

$$\text{2nd b.c.}\quad \varepsilon_1(\omega)\frac{\partial V\left(\boldsymbol{r}-0^+\hat{\boldsymbol{n}}\right)}{\partial n}\Bigg|_{\partial\Omega}=\varepsilon_2(\omega)\frac{\partial V\left(\boldsymbol{r}+0^+\hat{\boldsymbol{n}}\right)}{\partial n}\Bigg|_{\partial\Omega}, \tag{9.4b}$$

where $\partial/\partial n$ is the derivative along the direction of the outer surface normal.

9.2 Spheres and Ellipsoids in the Quasistatic Limit

In this section we consider light scattering of small spheres and ellipsoids. We call the sphere solutions the "Mie solutions in the quasistatic approximation" and the ellipsoid solutions the "Mie–Gans solutions." Both of them exploit special coordinates, spherical ones for the Mie case and ellipsoidal ones for the Mie–Gans case, in order to express the boundary conditions in such a simple form that they can be exploited analytically. Unfortunately, such an approach cannot be generalized to arbitrary particle geometries, and we will resort to different computation schemes in later parts of this chapter.

9.2.1 *Quasistatic Mie Theory*

In the following we consider a sphere illuminated by light with an electric field of the form of Eq. (9.3) and a polarization along $\hat{z}$ (see Fig. 9.1),

$$V(\boldsymbol{r})=-E_0\hat{z}\cdot\boldsymbol{r}=-E_0\,r\cos\theta\,.$$

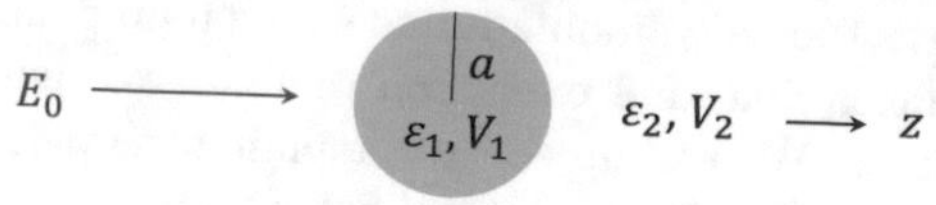

Fig. 9.1 Schematics of quasistatic Mie theory. A dielectric or metallic sphere with radius a and permittivity ε_1 is embedded in a background medium with permittivity ε_2. An external electric field with strength E_0 is applied along the z direction, and induces a dipole moment in the sphere. V_1, V_2 are solutions of Laplace's equation with coefficients that are determined by matching the potentials and their surface derivatives at the sphere boundary

In the following we introduce spherical coordinates and use that the solutions do not depend on the azimuthal angle ϕ. Laplace's equation in spherical coordinates is then of the form

$$\frac{1}{r}\frac{\partial^2}{\partial r^2}\Big[rV(r,\theta)\Big]+\frac{1}{r^2\sin\theta}\frac{\partial}{\partial\theta}\left[\sin\theta\frac{\partial V(r,\theta)}{\partial\theta}\right]=0\,. \tag{9.5}$$

It can be shown that the solution factorizes into two contributions which only depend on r and θ, respectively, and is of the form [2]

$$V(r,\theta)=\sum_{\ell=0}^{\infty}\left(A_\ell r^\ell+\frac{B_\ell}{r^{\ell+1}}\right)P_\ell(\cos\theta)\,. \tag{9.6}$$

Here A_ℓ, B_ℓ are coefficients to be determined from the boundary conditions. P_ℓ are the Legendre polynomials which are discussed in more detail in Appendix C. Using Eq. (9.6) we now tackle the problem of light excitation of a nanosphere with radius a. We first write down the solutions inside and outside the sphere in the form

$$V(r,\theta)=\begin{cases} V_1(r,\theta)=\sum_\ell A_\ell r^\ell\,P_\ell(\cos\theta) & \text{for } r\le a\\ V_2(r,\theta)=\sum_\ell \dfrac{B_\ell}{r^{\ell+1}}\,P_\ell(\cos\theta)-E_0\,r\cos\theta & \text{for } r>a\,,\end{cases}$$

where the subscripts 1, 2 refer to the solutions inside and outside the sphere. We have used that inside the sphere V remains finite at the origin (and we thus set all B_ℓ coefficients to zero), whereas outside the sphere we require $V\to -E_0\,r\cos\theta$ for large radii (and we thus set all A_ℓ coefficients to zero). We now observe from the definition of the Legendre polynomials that

$$-E_0\,r\cos\theta=-E_0\,r\,P_1(\cos\theta)\,,$$

and evaluate the boundary conditions. From the continuity of the potentials at the sphere boundary we find

$$\left(V_1 - V_2\right)_{r=a} = 0 \Longrightarrow \begin{cases} A_\ell a^\ell = \dfrac{B_\ell}{a^{\ell+1}} & \text{for } \ell \neq 1 \\ A_1 a = \dfrac{B_1}{a^2} - E_0 a & \text{for } \ell = 1\,. \end{cases}$$

Similarly, from the continuity of the dielectric displacement in the radial direction we obtain

$$\left(\varepsilon_1 \frac{\partial V_1}{\partial r} - \varepsilon_2 \frac{\partial V_2}{\partial r}\right)_{r=a} = 0 \Longrightarrow \begin{cases} \varepsilon_1 \ell A_\ell a^{\ell-1} = -\varepsilon_2 \dfrac{(\ell+1)B_\ell}{a^{\ell+2}} & \text{for } \ell \neq 1 \\ \varepsilon_1 A_1 = -\varepsilon_2 (2\dfrac{B_1}{a^3} + E_0) & \text{for } \ell = 1\,. \end{cases}$$

The equations with $\ell \neq 1$ can be satisfied simultaneously only with $A_\ell = B_\ell = 0$ for all ℓ.[1] For $\ell = 1$ we find

$$V(r,\theta) = \begin{cases} -\left(\dfrac{3\varepsilon_2}{\varepsilon_1 + 2\varepsilon_2}\right) E_0 r \cos\theta & \text{for } r \leq a \\ \left(\dfrac{\varepsilon_1 - \varepsilon_2}{\varepsilon_1 + 2\varepsilon_2}\right) \dfrac{a^3}{r^2} E_0 \cos\theta - E_0 r \cos\theta & \text{for } r > a\,. \end{cases} \tag{9.7}$$

The electric fields inside the sphere can be computed by taking the derivative of the scalar potential. Together with $z = r\cos\theta$ we get

$$\boldsymbol{E}_1 = \left(\frac{3\varepsilon_2}{\varepsilon_1 + 2\varepsilon_2}\right) E_0 \hat{\boldsymbol{z}}\,. \tag{9.8}$$

As for the field outside the sphere, we find that the induced part is of the form of a dipole potential

$$V_2(\boldsymbol{r}) = \frac{1}{4\pi\varepsilon_2} \frac{\boldsymbol{p} \cdot \hat{\boldsymbol{r}}}{r^2}\,.$$

The dipole moment of the optically excited sphere then becomes

Dipole Moment of Optically Excited Sphere

$$\boldsymbol{p} = 4\pi\varepsilon_2 \left(\frac{\varepsilon_1 - \varepsilon_2}{\varepsilon_1 + 2\varepsilon_2}\right) a^3 E_0 \hat{\boldsymbol{z}}\,. \tag{9.9}$$

[1] The resulting condition $\varepsilon_1 \ell = -\varepsilon_2(\ell + 1)$ can neither be fulfilled for dielectric particles with $\varepsilon_1 > 0$, $\varepsilon_2 > 0$ nor for metallic particles with a permittivity that has a non-zero imaginary part.

So far nothing has been overly exciting. Equation (9.7) is usually derived in a beginner's course of electrostatics, and although its derivation forces us to make contact with the Laplace equation in spherical coordinates most readers will be familiar with the final result. Similarly, the finding of a constant field inside the sphere and an induced dipolar field at its outside are well-known. Yet, there is something special about Eq. (9.7). To see this, we next discuss the expression separately for dielectric and metallic nanospheres.

Dielectric Sphere

Consider a dielectric sphere with $\varepsilon_1 > \varepsilon_2$. From Eq. (9.8) we observe that the electric field inside the sphere is smaller than outside because of the induced polarization field inside the sphere, which screens the external field. To obtain the polarization charges, we first compute the normal components of the electric field

$$E^\perp = -\frac{\partial V}{\partial r} = \begin{cases} E_1^\perp = \left(\dfrac{3\varepsilon_2}{2\varepsilon_2+\varepsilon_1}\right) E_0 \cos\theta & \text{for } r = a^- \\ E_2^\perp = \left(\dfrac{3\varepsilon_1}{2\varepsilon_2+\varepsilon_1}\right) E_0 \cos\theta & \text{for } r = a^+ \,, \end{cases} \tag{9.10}$$

where $a^\mp$ denotes positions infinitesimally inside or outside the sphere. We then obtain the polarization charge $\sigma_{\rm pol}$ using Eq. (7.24),

$$\sigma_{\rm pol}(\theta) = -\varepsilon_0 \left(E_2^\perp - E_1^\perp\right) = 3\varepsilon_0 \left(\frac{\varepsilon_1 - \varepsilon_2}{\varepsilon_1 + 2\varepsilon_2}\right) E_0 \cos\theta \,. \tag{9.11}$$

$\sigma_{\rm pol}$ produces an electric field that is oriented oppositely to the applied field. When we consider the above scenario in the time domain, see Fig. 9.2, we observe that the polarization charge density oscillates in phase with the driving electric field, thus screening the external field at any instant of time.

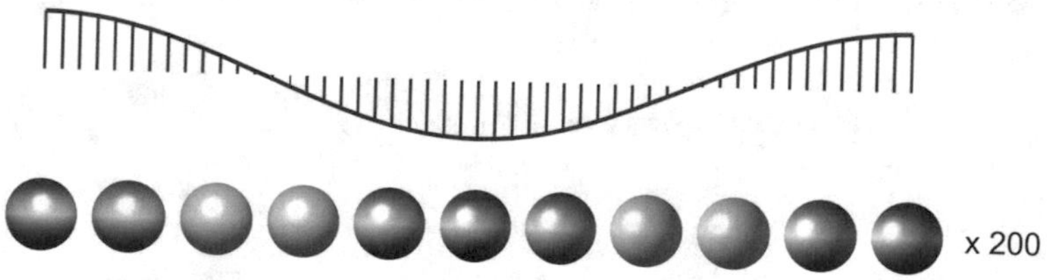

Fig. 9.2 Polarization charge oscillations at the surface of a dielectric sphere with $\varepsilon_1 = 3\varepsilon_0$ and $\varepsilon_2 = \varepsilon_0$. The charge oscillates in phase with the exciting field, which is shown at the top of the figure. The scaling factor of 200 is with respect to the charge oscillations of a metallic nanosphere shown in Fig. 9.3

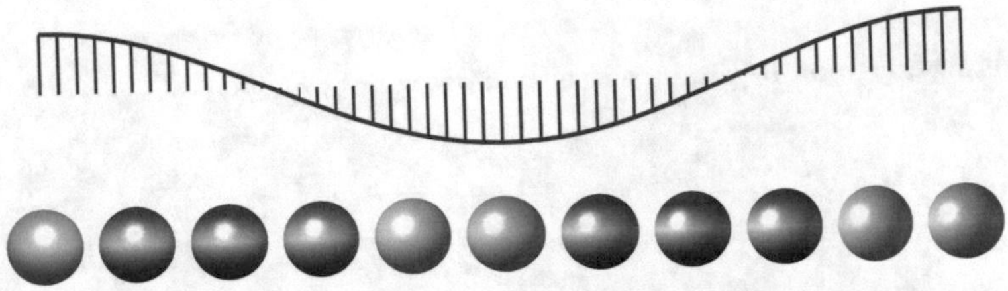

Fig. 9.3 Same as Fig. 9.2 but for a metallic nanosphere with material parameters representative for silver. The surface charges are larger by a factor of 200 in comparison to the dielectric sphere shown in Fig. 9.2, and have a phase difference of 90° at resonance

Metallic Sphere

For metallic spheres the dielectric function $\varepsilon_1(\omega)$ becomes negative for frequencies smaller than the plasma frequency ω_p, see discussion in Sect. 7.1. As a consequence, the denominator in Eq. (9.7) can become extremely small at certain frequencies ω_{SP}, and we find that

$$\left(\frac{\varepsilon_1 - \varepsilon_2}{\varepsilon_1 + 2\varepsilon_2}\right) \gg 1 \quad \text{for} \quad \varepsilon_1(\omega_{\text{SP}}) + 2\varepsilon_2 \approx 0\,. \tag{9.12}$$

As a consequence, the induced dipole moment of Eq. (9.9) becomes extremely large there. The only reason why it remains finite at ω_{SP} are metal losses described through $\varepsilon_1''(\omega_{\text{SP}})$. It is important to realize that for a Drude-type dielectric function $\varepsilon_1'(\omega)$ scans below the plasma frequency the entire negative range

$$-\infty < \varepsilon_1'(\omega) \leq 0 \quad \text{for} \quad 0 < \omega \leq \omega_p\,.$$

Thus the resonance condition of Eq. (9.12) will be certainly fulfilled at some frequency. To better understand the nature of these so-called surface plasmon (SP) resonances, in the following we consider inside the sphere a Drude dielectric function

$$\varepsilon_1(\omega)/\varepsilon_0 = \kappa_b - \frac{\omega_p^2}{\omega(\omega + i\gamma)}\,,$$

with κ_b being the dielectric response of bound valence electrons in the metal. The resonance condition now becomes

$$\varepsilon_0\left(\kappa_b - \frac{\omega_p^2}{\omega_{\text{SP}}^2}\right) + 2\varepsilon_2 \approx 0\,. \tag{9.13}$$

We thus get for the plasmon frequency of a metallic sphere described through a Drude permittivity

Surface Plasmon Resonance Frequency for Sphere

$$\omega_{\rm SP} \approx \frac{\omega_p}{\sqrt{\kappa_b + 2\dfrac{\varepsilon_2}{\varepsilon_0}}} = \beta\,\omega_p\,. \tag{9.14}$$

Here β is a dimensionless number that gives the surface plasmon frequency in units of the plasma frequency ω_p. For an ideal free-electron gas with $\kappa_b = 1$ and a background permittivity of $\varepsilon_2 = \varepsilon_0$ we get

$$\beta = \frac{\omega_{\rm SP}}{\omega_p} \longrightarrow \frac{1}{\sqrt{3}}\,.$$

Oscillator Model. If we assume that the losses in the metal are sufficiently small, $\gamma \ll \omega_{\rm SP}$, we can approximate close to $\omega_{\rm SP}$ the Drude dielectric function as

$$\frac{\varepsilon_1(\omega)}{\varepsilon_0} \approx \kappa_b - \frac{\omega_p^2}{\omega_{\rm SP}^2}\left[1 + \frac{\Delta\omega^2}{\omega_{\rm SP}^2} + i\frac{\gamma}{\omega_{\rm SP}}\right]^{-1} \approx \frac{\varepsilon_1(\omega_{\rm SP})}{\varepsilon_0} + \frac{\omega_p^2\Delta\omega^2}{\omega_{\rm SP}^4} + i\frac{\omega_p^2\gamma}{\omega_{\rm SP}^3}\,,$$

where $\Delta\omega^2 = \omega^2 - \omega_{\rm SP}^2$ and we have expanded the terms in brackets in a Taylor series. We next compute the induced dipole moment of Eq. (9.9). The numerator can be approximated in the form

$$\varepsilon_1 - \varepsilon_2 = (\varepsilon_1 + 2\varepsilon_2) - 3\varepsilon_2 \approx -3\varepsilon_2\,,$$

where the terms in parentheses becomes zero at the resonance frequency. The dipole moment can then be written as

$$\boldsymbol{p} \approx \frac{4\pi\varepsilon_2^2}{\varepsilon_0}\left(\frac{3\beta^2\,\omega_{\rm SP}^2}{\omega_{\rm SP}^2 - \omega^2 - i\gamma\omega_{\rm SP}}\right) a^3\,E_0\hat{\boldsymbol{z}}\,. \tag{9.15}$$

The term in parentheses is directly related to the resonance term of a driven harmonic oscillator, previously derived in Eq. (7.5) for the Drude–Lorentz permittivity.

Surface Charges. In contrast to the dielectric sphere, the surface charges are due to free and bound carriers. In the following we consider $\kappa_b = 1$, $\varepsilon_2 = \varepsilon_0$ and leave the general case to Exercise 9.2. The surface charge is now entirely due to free carriers, and we obtain from Eq. (9.11) at the resonance frequency $\omega = \omega_{\rm SP}$

$$\sigma(\theta) = 3\varepsilon_0\left(\frac{\varepsilon_1 - \varepsilon_2}{\varepsilon_1 + 2\varepsilon_2}\right)E_0\cos\theta \approx -3i\varepsilon_0\frac{\omega_{\rm SP}}{\gamma}\,E_0\cos\theta\,. \tag{9.16}$$

From this expression it is obvious that only the damping term γ keeps the surface charges finite, but the expression can become otherwise very large. Localized surface plasmons can be associated with resonances of metallic nanoparticles where a surface charge distribution oscillates coherently at the boundary between a metal and a dielectric. The situation is similar to the surface plasmons considered in the previous chapter; however, because of the confinement to the sphere surface we get a discrete resonance frequency rather than a continuum of modes.

The behavior of localized surface plasmons is depicted in Fig. 9.3. An external light field separates carriers at the particle boundary. In addition to the driving light field, there are now also restoring forces due to the separated carriers. Thus, if the sphere is driven at the resonance frequency ω_{SP}, an oscillation with a large amplitude builds up, where in the stationary state the energy absorbed by the light field equals the energy dissipated by the metal losses (see also Exercise 9.1). Note that in Eq. (9.16) the surface charge σ is completely imaginary and the phase difference between the driving field and the sphere polarization is 90°, in accordance to the resonance behavior of a driven harmonic oscillator.

Particle Plasmons. Localized surface plasmons are known under different names, sometimes combined with the supplement "polariton" in order to emphasize the hybrid nature of the material and electrodynamic excitations. In the following we adopt the term "particle plasmons" for any kind of surface plasmon excitation of a particle with finite size. We will discuss further below that the resonance frequencies and field enhancements can be tailored through nanoparticle shape and size. This offers a flexible and versatile platform for confining electromagnetic fields at the nanoscale, which can be exploited for numerous applications.

Maxwell's Equations. There is some discrepancy between the innocently looking expressions derived from Maxwell's equations and the considerably richer resonance physics governing particle plasmons. At first sight it may seem somewhat unexpected that one has to put additional work into the interpretation of the results from Maxwell's theory to unravel its deeper physics. However, this is because Maxwell's equations do contain all the necessary ingredients to describe particle plasmons, and one only has to take a fresh view on the results to discover what is already hidden within them.

9.2.2 *Mie–Gans Theory*

The second type of geometries for which the electrostatic problem can be solved analytically are ellipsoids. We denote the semiaxes with a_k and assume

$$a_1 \geq a_2 \geq a_3 \,.$$

In the excellent textbook of Bohren and Huffman [60] the authors start by discussing:

The natural coordinates, albeit unfamiliar and not without their disagreeable features, for formulating the dipole moment of an ellipsoidal particle induced by an uniform electrostatic field are the *ellipsoidal coordinates*.

Readers interested in the details of such exotic coordinate systems are referred to the above-cited book, in the following I simply state the final results. It turns out that the dipole moment can be expressed in a form similar to Eq. (9.9) for spherical nanoparticles, and one finds for a light polarization along one of the principal axes $\hat{\boldsymbol{e}}_k$ the result

Dipole Moment of Optically Excited Ellipsoid

$$\boldsymbol{p} = 4\pi\varepsilon_2\left(\frac{\varepsilon_1-\varepsilon_2}{3\varepsilon_2+3L_k\left(\varepsilon_1-\varepsilon_2\right)}\right)a_1a_2a_3\,E_0\hat{\boldsymbol{e}}_k\,. \tag{9.17}$$

Here we have introduced the depolarization factors

$$L_k = \frac{1}{2}a_1a_2a_3\int_0^\infty \frac{dq}{\left(a_k^2+q^2\right)\sqrt{\left(a_1^2+q^2\right)\left(a_2^2+q^2\right)\left(a_3^2+q^2\right)}}\,,$$

which fulfill the sum rule $L_1 + L_2 + L_3 = 1$. For spheres one finds $L_k = \frac{1}{3}$ and Eq. (9.17) reduces to the previously derived result. For a prolate ellipsoid with $a_1 \geq a_2 = a_3 = a_\perp$ one can solve the above integral analytically, and obtains for the depolarization factor along the long axis

$$L_1 = \frac{1-e^2}{e^2}\left(-1+\frac{1}{2e}\ln\frac{1+e}{1-e}\right)\,,\quad e^2 = 1-\frac{a_\perp^2}{a_1^2}\,.$$

Similarly to the sphere, we can compute the surface plasmon resonance frequencies $\omega_{\rm SP}$ by seeking for the zeros of the denominator given in Eq. (9.17),

$$3\varepsilon_2 + 3L_k\left[\varepsilon_1'(\omega_{\rm SP})-\varepsilon_2\right] = 0\,. \tag{9.18}$$

The entire discussion regarding the resonance behavior of particle plasmon excitations in spheres also applies to ellipsoidal particles, however, due to the depolarization factors L_k we have an additional means to control the resonance frequency by choosing different nanoparticle geometries. Before demonstrating this point in more detail, we derive expressions for the optical cross sections in the quasistatic regime.

9.2.3 Optical Cross Sections

We consider a dipole moment induced by an external light field $\boldsymbol{E}_{\text{inc}}$

$$\boldsymbol{p} = \bar{\bar{\alpha}}(\omega) \cdot \boldsymbol{E}_{\text{inc}} = 4\pi\varepsilon\,\bar{\bar{\alpha}}_{\text{vol}}(\omega) \cdot \boldsymbol{E}_{\text{inc}}\,, \tag{9.19}$$

where $\bar{\bar{\alpha}}$ is the *polarizability* tensor of the particle. In SI units the polarizability has the units $\text{C}\,\text{m}^2\text{V}^{-1}$, and it is often more convenient to pull out a factor of $4\pi\varepsilon$, where ε is the permittivity of the background medium, to get a polarizability α_{vol} that has the dimension of a volume. Obviously, the polarizability of Eq. (9.19) applies for spheres (with $\bar{\bar{\alpha}} = \alpha\mathbb{1}$), ellipsoids (with $\bar{\bar{\alpha}}$ being a diagonal matrix), as well as other nanoparticle geometries. We start from the far-fields of an oscillating dipole given in Eq. (6.3),

$$\boldsymbol{E}_{\text{sca}} \underset{kr\gg 1}{\longrightarrow} \left(\frac{e^{ikr}}{r}\right)\frac{k^2}{4\pi\varepsilon}\,\boldsymbol{p}_\perp$$

$$\boldsymbol{H}_{\text{sca}} \underset{kr\gg 1}{\longrightarrow} Z^{-1}\left(\frac{e^{ikr}}{r}\right)\frac{k^2}{4\pi\varepsilon}\,\hat{\boldsymbol{r}}\times\boldsymbol{p}_\perp\,,$$

where the fields propagate along direction $\hat{\boldsymbol{r}}$ and $\boldsymbol{p}_\perp = \boldsymbol{p} - \hat{\boldsymbol{r}}(\hat{\boldsymbol{r}}\cdot\boldsymbol{p})$. We now submit these far-fields to the scattering power of Eq. (4.22),

$$P_{\text{sca}} = \frac{1}{2}\oint_{\partial\Omega} \text{Re}\left(\boldsymbol{E}_{\text{sca}} \times \boldsymbol{H}^*_{\text{sca}}\right)\cdot\hat{\boldsymbol{r}}\,dS\,.$$

In what follows we consider for the boundary a sphere located in the far-field zone. Inserting the far-fields into this expression then leads us to

$$P_{\text{sca}} = \frac{1}{2}Z^{-1}\left(\frac{k^2}{4\pi\varepsilon}\right)^2\oint \text{Re}\big[\boldsymbol{p}_\perp \times (\hat{\boldsymbol{r}}\times\boldsymbol{p}^*_\perp)\big]\cdot\hat{\boldsymbol{r}}\,d\Omega\,,$$

where $dS = r^2 d\Omega$ and $d\Omega$ denotes an infinitesimally small surface element over the azimuthal and polar angles. Performing a cyclic permutation of the triple product we are led to

$$\text{Re}\big[\boldsymbol{p}_\perp \times (\hat{\boldsymbol{r}}\times\boldsymbol{p}^*_\perp)\big]\cdot\hat{\boldsymbol{r}} = \big|\hat{\boldsymbol{r}}\times\boldsymbol{p}_\perp\big|^2 = \big|\hat{\boldsymbol{r}}\times\boldsymbol{p}\big|^2\,.$$

Putting together the results we find

$$P_{\text{sca}} = Z^{-1}\frac{k^4}{32\pi^2\varepsilon^2}\int\big|\hat{\boldsymbol{r}}\times\boldsymbol{p}\big|^2\,d\Omega\,. \tag{9.20}$$

The integration can be performed by choosing a coordinate system where $\boldsymbol{p}$ points in the z-direction,

$$\int \left|\hat{\boldsymbol{r}} \times \boldsymbol{p}\right|^2 d\Omega = 2\pi \int_0^\pi p^2 \sin^2\theta \, \sin\theta d\theta = \frac{8\pi}{3} p^2 .$$

To arrive at our final expression, we consider an incoming light field of the form $E_0\boldsymbol{\epsilon}_0$ with an intensity $I_{\text{inc}} = \frac{1}{2}Z^{-1}\left|E_0\right|^2$. When dividing the scattered power by I_{inc} we arrive at our final expression for the scattering cross section of a polarizable particle

Scattering Cross Section of Polarizable Particle

$$C_{\text{sca}} = \frac{k^4}{6\pi\varepsilon^2}\left|\bar{\bar{\alpha}} \cdot \boldsymbol{\epsilon}_0\right|^2 = \frac{8\pi}{3}k^4\left|\bar{\bar{\alpha}}_{\text{vol}} \cdot \boldsymbol{\epsilon}_0\right|^2 . \tag{9.21}$$

For the extinction cross section we start from the optical theorem of Eq. (4.27),

$$P_{\text{ext}} = \frac{2\pi}{k}Z^{-1}\,\text{Im}\left[E_0^*\boldsymbol{\epsilon}_0^* \cdot \boldsymbol{F}(\hat{\boldsymbol{k}}_0)\right] = \frac{2\pi}{k}Z^{-1}\frac{k^2}{4\pi\varepsilon}\text{Im}\left[\left|E_0\right|^2 \boldsymbol{\epsilon}_0^* \cdot \bar{\bar{\alpha}} \cdot \boldsymbol{\epsilon}_0\right] .$$

Division by the incoming intensity I_{inc} then leads us to the extinction cross section

Extinction Cross Section of Polarizable Particle

$$C_{\text{ext}} = \frac{k}{\varepsilon}\text{Im}\left[\boldsymbol{\epsilon}_0^* \cdot \bar{\bar{\alpha}} \cdot \boldsymbol{\epsilon}_0\right] = 4\pi k\,\text{Im}\left[\boldsymbol{\epsilon}_0^* \cdot \bar{\bar{\alpha}}_{\text{vol}} \cdot \boldsymbol{\epsilon}_0\right] . \tag{9.22}$$

We next use the above results to compute optical spectra for metallic nanospheres and ellipsoids. Figures 9.4 and 9.5 show extinction cross sections for silver and gold nanospheres and nanoellipsoids, respectively. We observe that with increasing axis ratio the plasmon modes shift to lower energies and the resonances become sharper. In all figures the light polarization is along the long axis z of the ellipsoids. There exist additional plasmon resonances associated with charge oscillations in the x and y directions, whose resonance frequencies, however, do not depend significantly on the axis ratio, as can be seen from the depolarization factors L_k for the ellipsoid.

When comparing silver and gold nanoparticles, we observe that the plasmon frequencies are much stronger damped for gold because of the larger imaginary parts of the gold dielectric function. In particular for resonances above 2 eV the spectra are governed by losses associated with $\varepsilon_1''(\omega)$. Figure 9.6 reports the resonance peak positions of ellipsoids with different axis ratios, together with the maximal value of C_{ext} (corresponding to the size of the dots). One observes that for both silver and gold the plasmon peaks shift with increasing axis ratios to smaller energies.

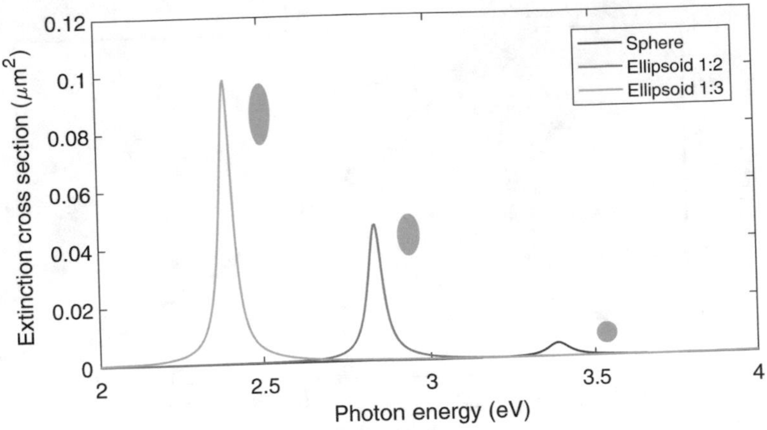

Fig. 9.4 Extinction cross sections for silver sphere and ellipsoids with different axis ratios. We use $a_1 = a_2 = 20$ nm and change a_3. The extinction cross section is computed for plane wave excitation and a polarization along the long axis, the silver dielectric function is taken from Ref. [34]. For the refractive index of the embedding medium we use $n_2 = 1.33$

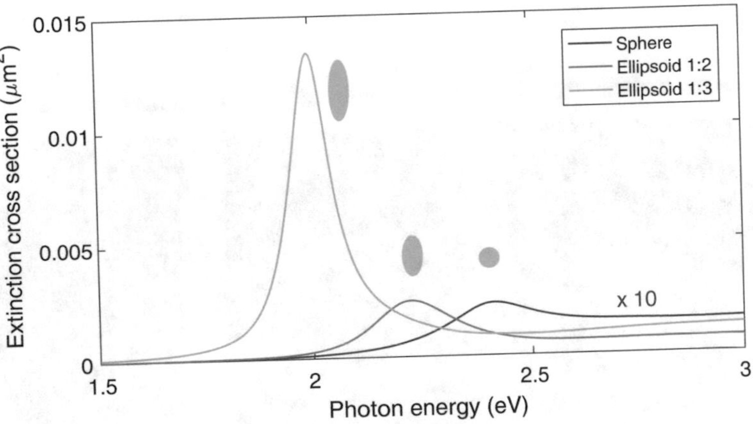

Fig. 9.5 Same as Fig. 9.4 but for gold. The values of the sphere are scaled by a factor of 10 for better visibility. Note that in comparison to silver the cross sections are much smaller because of the larger damping of gold caused by its larger imaginary part of the permittivity

Finally, in Fig. 9.7 we show the field enhancements for ellipsoids with different axis ratios and for light polarization along z. With increasing axis ratio the field enhancement becomes dramatically increased. This is due to both the tighter field confinement at sharp feature of the nanostructures, as well as due to the smaller damping for resonances at lower energies. Because of the latter effect, the plasmons can be excited more strongly, as has been previously discussed in Eq. (9.15) for the simple oscillator model. Indeed, it is the combination of pronounced resonances and strong field enhancements that makes plasmonic nanoparticles so interesting for various applications.

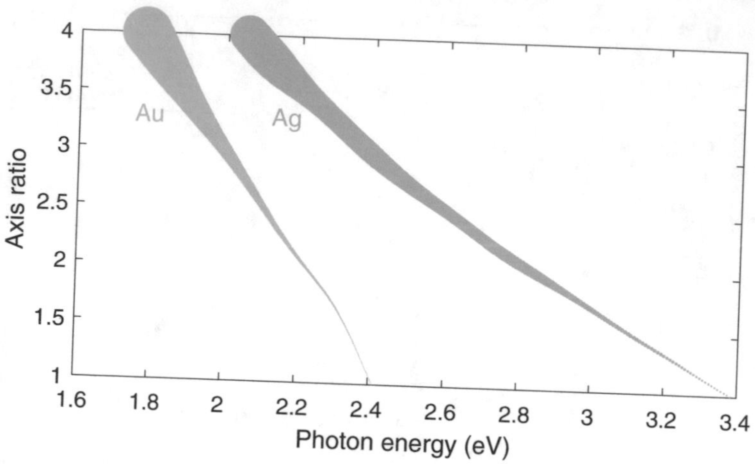

Fig. 9.6 Resonance positions of gold and silver nanoellipsoids for different axis ratios. The positions of the filled circles correspond for each axis ratio to the photon energy where C_{ext} is largest, the sizes correspond to the values of C_{ext}. Values for gold scaled by a factor of 5 for better visibility

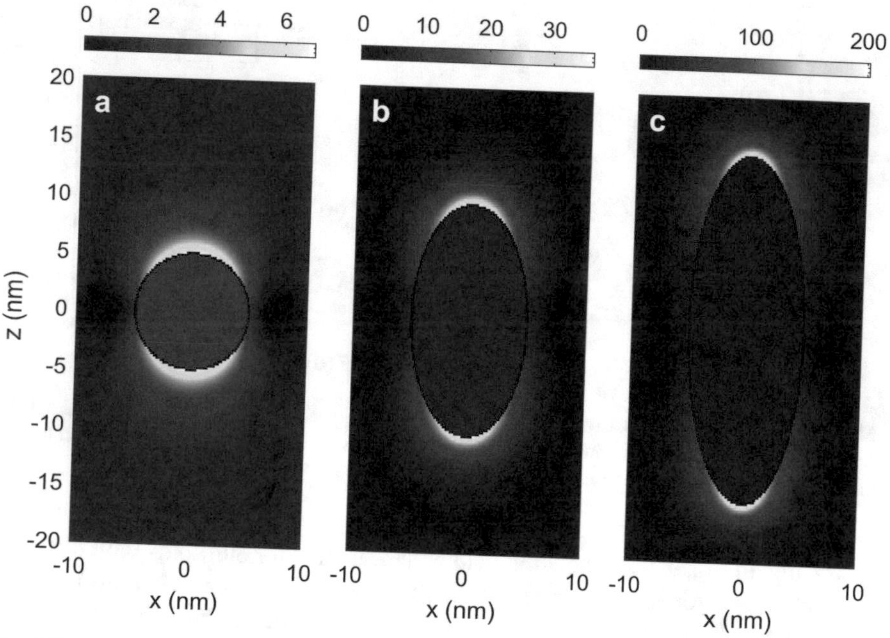

Fig. 9.7 Intensity enhancement for silver sphere and ellipsoids. The nearfield enhancement $|\boldsymbol{E}(\boldsymbol{r})|^2$ is plotted at the resonance energies extracted of Fig. 9.6 for (**a**) sphere, (**b**) a_1:a_3 =1:2, and (**c**) a_1:a_3 =1:3, and for a light polarization in the $\hat{z}$ direction. The color bars with the respective ranges of the enhancement factors are reported on the top of the panels

9.3 Boundary Integral Method for Quasistatic Limit

When the nanoparticle shape is neither a sphere nor an ellipsoid, one generally has to resort to some numerical scheme. In the following we introduce to the boundary integral method approach, from which a number of important conclusions can be drawn. Its numerical implementation will be discussed in Chap. 11.

For the general solution of Poisson's equation we employ a Green's function scheme. The quasistatic Green's function follows from that of the Helmholtz equation, see Eq. (5.7), by setting $k \to 0$ and we arrive at

$$G(\boldsymbol{r}, \boldsymbol{r}') = \frac{1}{4\pi|\boldsymbol{r} - \boldsymbol{r}'|}\,. \tag{9.23}$$

To compute for a given external potential V_{inc} the induced potential we could start from the representation formula of Eq. (5.15) and let $\boldsymbol{r}$ approach the boundary. This approach will be pursued further below for the solution of the full Maxwell's equations (without the quasistatic approximation). However, for the quasistatic case we can introduce a more simple approach based on the surface charge distribution σ, which captures the physical essence of particle plasmons. We start by writing down the total potential inside and outside the particle in the form

$$V(\boldsymbol{r}) = V_{\text{inc}}(\boldsymbol{r}) + \oint_{\partial\Omega} G(\boldsymbol{r}, \boldsymbol{s}')\sigma(\boldsymbol{s}')\, dS'\,, \tag{9.24}$$

where the integral extends over the particle boundary. V_{inc} is again the potential associated with an external source or an incoming wave. From here on we denote positions located off and on the boundary with $\boldsymbol{r}$ and $\boldsymbol{s}$, respectively. By applying the Laplace operator on both sides of the above equation we find

$$\nabla^2 V(\boldsymbol{r}) = \nabla^2 V_{\text{inc}}(\boldsymbol{r}) - \oint_{\partial\Omega} \delta(\boldsymbol{r} - \boldsymbol{s}')\sigma(\boldsymbol{s}')\, dS' = -\frac{\rho(\boldsymbol{r})}{\varepsilon} \quad \text{for } \boldsymbol{r} \notin \partial\Omega\,.$$

In deriving this result, we have used the defining equation for the quasistatic Green's function

$$\nabla^2 G(\boldsymbol{r}, \boldsymbol{r}') = -\delta(\boldsymbol{r} - \boldsymbol{r}')\,, \tag{9.25}$$

and have assumed that $\nabla^2 V_{\text{inc}}$ gives the charge distribution inside the volume under consideration. For an incoming light field this term would be absent and we would end up with Laplace's equation. By construction, Eq. (9.24) fulfills the Poisson equation and, owing to the definition of the Green's function, it has built in the proper boundary condition of decaying fields at infinity. The unknown surface charge distribution $\sigma(\boldsymbol{s})$ is next computed such that the boundary conditions of Eq. (9.4) at the particle boundary are fulfilled.

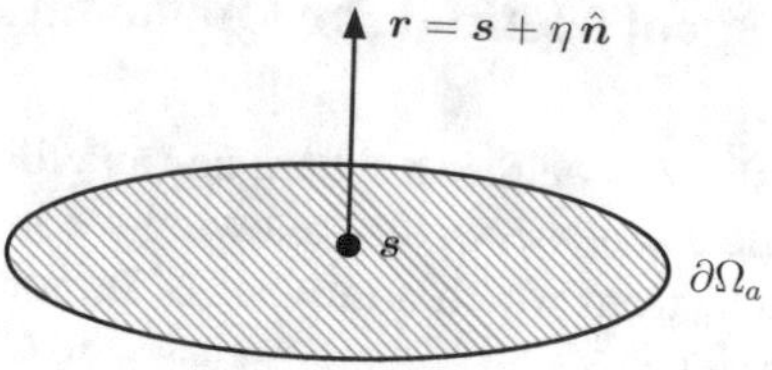

Fig. 9.8 Schematics of Dirichlet and Neumann trace. The observation point is located at $s + \eta\hat{n}$, and we approach the boundary from either the outside or inside by performing the limit $\eta \to 0$. The quantities of interest are integrated over a small disk-like boundary $\partial\Omega_a$

Dirichlet Trace. We start by computing the potential at the boundary in such a way that we let approach $\boldsymbol{r} \to \boldsymbol{s}$ the boundary from the outside or inside,

$$V(\boldsymbol{s}^{\pm}) = V_{\text{inc}}(\boldsymbol{s}^{\pm}) + \lim_{\eta\to 0} \oint_{\partial\Omega} \frac{\sigma(\boldsymbol{s}')\,dS'}{4\pi\left|\boldsymbol{s} \pm \eta\hat{\boldsymbol{n}} - \boldsymbol{s}'\right|}\,.$$

In the mathematical literature this limiting procedure is known as the Dirichlet trace. The limit can be performed without difficulties for V_{inc}. For the second term we have to be more careful because of the $1/r$ dependence of the Green's function. As we will demonstrate in the following, the integration over a small boundary region is essential to obtain well-behaved results. We chose a coordinate system where $\boldsymbol{s}$ is located at the origin and where $\boldsymbol{r}$ approaches this point along the z axis, see Fig. 9.8. For the small boundary region we consider a disk with radius a that is located in the xy-plane. Then, the term involving the Green's function can be expressed as

$$\lim_{\eta\to 0} \int_{\partial\Omega_a} \frac{\sigma(\boldsymbol{s}')\,dS'}{4\pi\left|\boldsymbol{s} \pm \eta\hat{\boldsymbol{n}} - \boldsymbol{s}'\right|} \approx \frac{1}{2}\sigma(\boldsymbol{s}) \lim_{\eta\to 0} \int_0^a \frac{\rho\,d\rho}{\sqrt{\rho^2+\eta^2}} = \frac{1}{2}\sigma(\boldsymbol{s})a \underset{a\to 0}{\longrightarrow} 0\,.$$

We have assumed that $\sigma(\boldsymbol{s}')$ changes slowly inside the disk, such that it can be approximated by its value at the central position $\boldsymbol{s}$ and can be taken out of the integral. Additionally, the integral has been evaluated in polar coordinates. The above expression shows that the Dirichlet trace can be performed safely, and one obtains the same value independent on whether one approaches the boundary from the outside or inside.

Neumann Trace. We next apply the same analysis to

$$\frac{\partial V(\boldsymbol{s}^{\pm})}{\partial n} = \frac{\partial V_{\text{inc}}(\boldsymbol{s}^{\pm})}{\partial n} - \lim_{\eta\to 0} \hat{\boldsymbol{n}} \cdot \oint_{\partial\Omega} \frac{\boldsymbol{s} \pm \eta\hat{\boldsymbol{n}} - \boldsymbol{s}'}{4\pi\left|\boldsymbol{s} \pm \eta\hat{\boldsymbol{n}} - \boldsymbol{s}'\right|^3}\,\sigma(\boldsymbol{s}')\,dS'\,.$$

This limiting procedure is known as the Neumann trace. In the second term on the right-hand side we have already evaluated the normal derivative with respect to $\boldsymbol{r}$. The limit of the incoming contribution can be safely performed. Introducing again polar coordinates for the second term we are led to

$$\mp\frac{1}{2}\sigma(\boldsymbol{s})\lim_{\eta\to 0}\int_0^a\frac{\eta\,\rho d\rho}{(\rho^2+\eta^2)^{\frac{3}{2}}}=\mp\frac{1}{2}\sigma(\boldsymbol{s})\lim_{\eta\to 0}\left[-\frac{\eta}{\sqrt{\rho^2+\eta^2}}\right]_0^a\xrightarrow[a\to 0]{}\mp\frac{1}{2}\sigma(\boldsymbol{s}),$$

where the negative or positive sign has to be used for the approach of the boundary from the outside or inside. We have also used $(\boldsymbol{s}-\boldsymbol{s}')\cdot\hat{\boldsymbol{n}}=0$ which is valid for the small, planar disk shown in Fig. 9.8. From this discussion we find that the Neumann trace gives different results depending on whether we approach the boundary from the outside or inside.

We have now all ingredients at hand to evaluate the boundary conditions at $\partial\Omega$. Because of the properties of the Dirichlet trace, the continuity of the potential is automatically fulfilled

$$V(\boldsymbol{s}^+)=V(\boldsymbol{s}^-).$$

For the second type of boundary conditions, Eq. (9.4b), we have to evaluate the potential derivatives

$$\frac{\partial V(\boldsymbol{s}^\pm)}{\partial n}=\frac{\partial V_{\rm inc}(\boldsymbol{s})}{\partial n}\mp\frac{1}{2}\sigma(\boldsymbol{s})+\oint_{\partial\Omega}\frac{\partial G(\boldsymbol{s},\boldsymbol{s}')}{\partial n}\sigma(\boldsymbol{s}')\,dS'. \tag{9.26}$$

From the continuity of the dielectric displacement we then obtain after some simple manipulations the final result

Quasistatic Boundary Integral Equation

$$\Lambda(\omega)\sigma(\boldsymbol{s})+\oint_{\partial\Omega}\frac{\partial G(\boldsymbol{s},\boldsymbol{s}')}{\partial n}\sigma(\boldsymbol{s}')\,dS'=-\frac{\partial V_{\rm inc}(\boldsymbol{s})}{\partial n}. \tag{9.27}$$

Here we have introduced the quantity

$$\Lambda(\omega)=\frac{1}{2}\frac{\varepsilon_1(\omega)+\varepsilon_2(\omega)}{\varepsilon_1(\omega)-\varepsilon_2(\omega)}.$$

Equation (9.27) is an integral equation which allows computing the surface charge distribution σ for a given external excitation. Once we have computed σ, we obtain the potential and in turn the electric fields everywhere else using Eq. (9.24). Equation (9.27) is a very appealing expression for a number of reasons.

Simple Derivation. First, its derivation is rather simple and has the central object of plasmonics at its core, namely the surface charge distribution. For this reason results can be interpreted quite intuitively.

Simple Computation. Equation (9.27) can be easily implemented for numerical use, as we will show explicitly in Sect. 11.2.

Eigenmodes. The probably most attractive point is that the boundary integral equation can be solved by introducing geometric eigenmodes, which allow us to separate geometry from material properties. We start by introducing the abbreviation

$$F(s,s') = \frac{\partial G(s,s')}{\partial n} = \frac{\partial}{\partial n}\left(\frac{1}{4\pi|s-s'|}\right)$$

for the surface derivative of the Green's function. Unfortunately, F is symmetric only for selected geometries such as spheres, and in general one has to be careful in the eigenmode analysis. We can seek for the right and left eigenmodes $u_k(s)$ and $\tilde{u}_k(s)$ of F, respectively, which are defined as[2]

Plasmonic Eigenmodes (Quasistatic Limit)

$$\oint_{\partial\Omega} F(s,s')u_k(s')\,dS' = \lambda_k\, u_k(s)$$

$$\oint_{\partial\Omega} \tilde{u}_k(s')F(s',s)\,dS' = \lambda_k\, \tilde{u}_k(s)\,. \tag{9.28}$$

They possess the same eigenvalues λ_k and form a complete and biorthogonal set with the orthogonality relation

$$\oint_{\partial\Omega} \tilde{u}_k(s)u_{k'}(s)\,dS = \delta_{kk'}\,, \tag{9.29}$$

as will be discussed in more detail in Sect. 9.7. There we will also show that one can use the symmetry relation

$$\oint_{\partial\Omega} G(s,s_1)F(s_1,s')\,dS_1 = \oint_{\partial\Omega} G(s',s_1)F(s_1,s'')\,dS_1 = 0 \tag{9.30}$$

[2] With this choice the eigenvalues are bound to values in the range $-1/2 \le \lambda_k \le 0$. In the literature one often rescales the eigenvalues such that they lie in the range of $-1 \le \lambda_k \le 0$ [61, 62], but we here prefer to stay with the choice of Eq. (9.28).

to prove that the eigenvalues are always real-valued, and that u_k, $\tilde{u}_k$ are related to each other via

$$\tilde{u}_k(s) = \oint_{\partial\Omega} G(s, s')u_k(s')\,dS' . \tag{9.31}$$

The neat thing about u_k, $\tilde{u}_k$ is that their properties only depend on the nanoparticle geometry, whereas all material properties are captured in $\Lambda(\omega)$. Once the eigenmodes u_k, $\tilde{u}_k$ have been computed, for instance, through numerical computation, we can expand the surface charges for some given external perturbation V_{inc} in terms of these eigenmodes

$$\sigma(s) = \sum_k C_k\, u_k(s) .$$

Inserting this expression into the boundary integral Eq. (9.27) and using the fact that u_k is an eigenmode of F, we are led to

$$\oint_{\partial\Omega} \tilde{u}_k(s) \sum_{k'} \big[\Lambda(\omega) + \lambda_{k'}\big] C_{k'} u_{k'}(s)\,dS = -\oint_{\partial\Omega} \tilde{u}_k(s) \frac{\partial V_{\text{inc}}(s)}{\partial n}\,dS ,$$

where we have multiplied both sides of the equation with $\tilde{u}_k$ and have integrated them over the particle boundary. Using the biorthogonality of the eigenmodes, we can immediately compute the expansion coefficients and are led to the eigenmode solution of the quasistatic boundary integral equation in the form

$$\sigma(s) = -\sum_k \big(\Lambda(\omega) + \lambda_k\big)^{-1} \left[\oint_{\partial\Omega} \tilde{u}_k(s') \frac{\partial V_{\text{inc}}(s')}{\partial n'}\,dS'\right] u_k(s) . \tag{9.32}$$

This expression is particularly useful for many applications and allows solving the boundary integral equation in terms of the evaluation of simple integrals.

Green's Function. As shown in more detail in Sect. 9.7, the total Green's function for a plasmonic nanoparticle in the quasistatic limit can be expressed in the form

Eigenmode Expansion of Total Green's Function

$$G_{\text{tot}}(\boldsymbol{r}, \boldsymbol{r}') = G(\boldsymbol{r}, \boldsymbol{r}') - \sum_k V_k(\boldsymbol{r}) \left[\frac{\lambda_k \pm \frac{1}{2}}{\Lambda(\omega) + \lambda_k}\right] V_k(\boldsymbol{r}') , \tag{9.33}$$

The positive or negative sign has to be chosen depending on whether $\boldsymbol{r}$, $\boldsymbol{r}'$ are located either outside or inside the particle, and we have introduced the eigenpotentials

$$V_k(\boldsymbol{r}) = \oint_{\partial\Omega} G(\boldsymbol{r}, \boldsymbol{s}')u_k(\boldsymbol{s}')\, dS' . \tag{9.34}$$

The response to an external charge distribution $\rho(\boldsymbol{r})$ can then be written in the form

$$V(\boldsymbol{r}) = \varepsilon_2^{-1} \int G(\boldsymbol{r}, \boldsymbol{r}')\rho(\boldsymbol{r}')\, d^3r' ,$$

where $\boldsymbol{r}$ is assumed to be located outside the particle, and we correspondingly have to take the positive sign in Eq. (9.33).

Perturbation Theory. As we will show for coupled particles below, the eigenmode analysis can be easily submitted to a perturbation theory, in close analogy to quantum mechanics.

9.3.1 Plasmonic Eigenmodes

Figure 9.9 shows the plasmonic eigenmodes for the example of a nanodisk. The modes of lowest eigenvalues have (a) dipolar, (b) quadrupolar, and (c) hexapolar character. Here the surface charges are concentrated at the particle edges. In addition, there exist (d, e) modes confined to the flat surfaces, where the surface charge oscillates in the radial direction. Note that in optical experiments only the dipole mode can be excited, whereas all other modes possess no net dipole moment and thus remain optically dark [63].

In the following we discuss how to determine the resonance frequency ω_k for a given eigenvalue λ_k. From Eq. (9.32) we observe that the surface charge distribution σ becomes largest when the resonance denominator approaches zero,

$$\Lambda'(\omega_k) + \lambda_k = 0 .$$

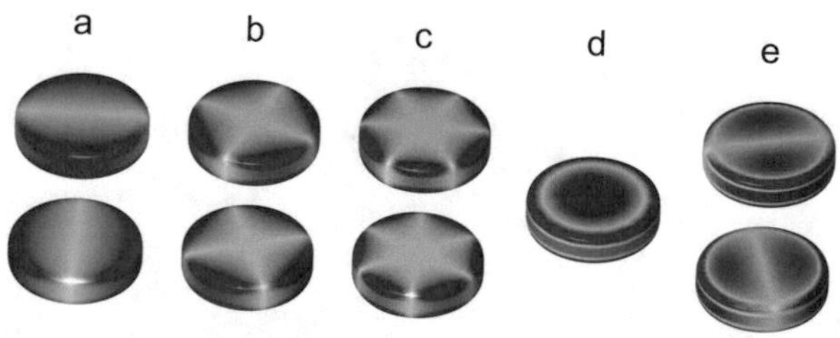

Fig. 9.9 A few selected eigenmodes for a nanodisk. (**a**) Dipole, (**b**) quadrupole, and (**c**) hexapole mode of lowest energy, which are all twofold degenerate and have nodes in the azimuthal directions. Higher excited modes additionally have (**d, e**) nodes in the radial direction

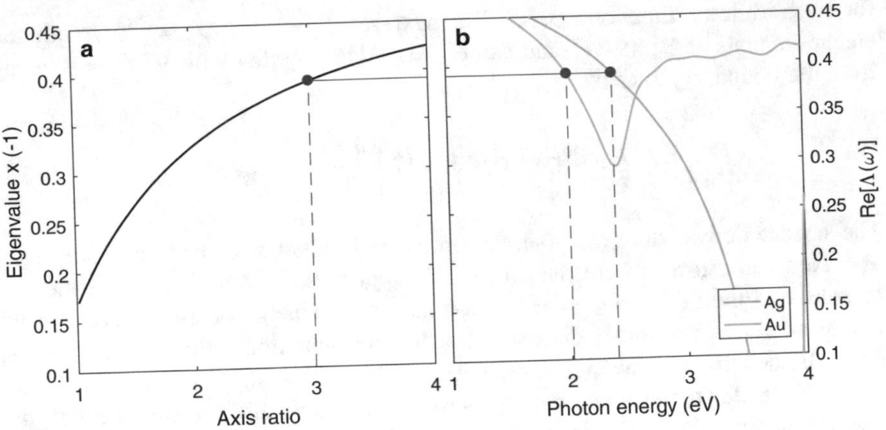

Fig. 9.10 Construction of how to obtain the resonance frequencies for a given eigenvalue λ_k. Panel (**a**) reports the eigenvalues λ_1 of the energetically lowest modes for ellipsoids with different axis ratios. For a specific axis ratio, here three, the resonance condition reads $\lambda_1 + \Lambda'(\omega) = 0$. Panel (**b**) reports the real part $\Lambda'(\omega)$ for silver and gold, together with $n_2 = 1.33$. The crossing points of $-\lambda_1$ and $\Lambda'(\omega)$ give the resonance positions. For gold, clear resonances only exist for larger axis ratios because of the strong interband damping above 2 eV

As previously discussed for spheres, it is the imaginary part $\Lambda''(\omega_k)$ that keeps σ finite at resonance. We emphasize again that in the above expression $\Lambda(\omega)$ only depends on the material properties, whereas λ_k only depends on the nanoparticle geometry. Figure 9.10 shows how to determine the resonance frequency ω_k for an ellipsoid made of silver or gold. Panel (a) reports for different axis ratios the (negative) eigenvalues λ_1 of the dipolar mode. In addition, there exist further modes, such as quadrupolar and hexapolar ones, which we do not consider here. Panel (b) shows $\Lambda(\omega)$ for silver and gold, and for different frequencies ω. At resonance the values for $-\lambda_1$ and $\Lambda'(\omega)$ must equal each other, as sketched in the figure by the red lines for an axis ratio of 1:3. Here the resonances are at about 2 eV for gold and 2.4 eV for silver, respectively (see also Figs. 9.4 and 9.5). From the construction we additionally observe that for gold pronounced resonances only exist for sufficiently large aspect ratios where the photon energy stays below 2 eV. For larger photon energies, the strong losses of gold lead to a significant damping of the plasmon resonances, which, in turn, become strongly broadened in the optical spectra. See, for instance, the faint peak in Fig. 9.5 for the sphere.

Plasmonic eigenmodes have been studied in detail in Refs. [62, 65], and in the following we repeat some of the results presented there in order to get an intuitive understanding of the eigenvalues. Our starting point is Green's first identity [2]

$$\int_\Omega \Big(\phi\nabla^2\psi + \nabla\phi\cdot\nabla\psi\Big)\, d^3r = \int_{\partial\Omega} \phi\frac{\partial\psi}{\partial n}\, dS$$

for two arbitrary functions ϕ, ψ. We next set $\phi = V_k$, $\psi = V_{k'}$, with the eigenpotentials of Eq. (9.34), and use $\boldsymbol{E}_k = -\nabla V_k$ together with $\nabla^2 V_k = 0$ (away from the boundary) to arrive at

$$\int_{\Omega_{1,2}} \boldsymbol{E}_k \cdot \boldsymbol{E}_{k'}\, d^3r = \tau_{1,2} \oint_{\partial\Omega} V_k \left[\frac{\partial V_{k'}}{\partial n}\right]_{1,2} dS\,. \tag{9.35}$$

The surface derivative of the potential on the right-hand side has to be interpreted as a Neumann trace, where the position is approached from either the inside or outside, and the factor $\tau_{1,2} = \pm 1$ accounts for the change of the surface normal in case of Ω_2, as previously discussed for the representation formula in Sect. 5.2.1. From the definition of the eigenmodes, Eq. (9.28), and using an expression similar to Eq. (9.26) for the evaluation of the surface derivative of the eigenpotential, the integral on the right-hand side can be evaluated to

Plasmonic Colors

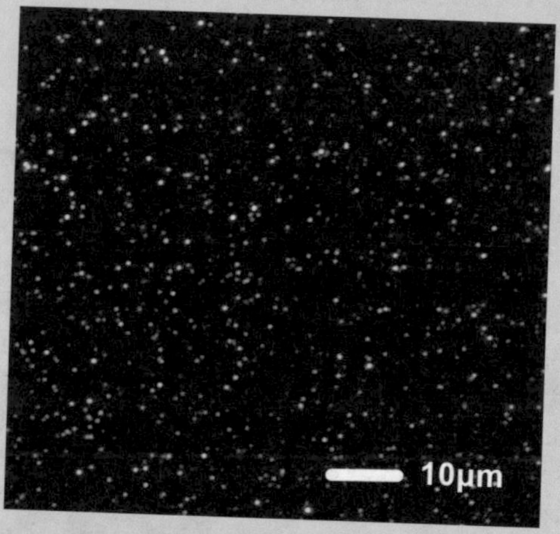

Depending on the shape and size of metallic nanoparticles, the plasmon resonances can be in the visible or infrared regime. The picture on the left shows a darkfield microscopy image where metallic nanoparticles are situated on a glass substrate, which is illuminated from below under an angle of total internal reflection. Only at the nanoparticle positions light is scattered, and the colors depend on the resonance frequencies for the different geometries. Similar color effects are also known from stained glass windows in medieval churches, where metallic nanoparticles are embedded in a glass matrix and give rise to bright and shiny colors. Because metallic particles are not degraded by light scattering, contrary to dyestuff, the colors do not change over time.

Plasmonics exploits that these resonances, which are associated with the coherent electron charge oscillations at the interface between a metallic nanoparticle and an embedding dielectric, come together with strong, evanescent fields. By resonantly driving the particle plasmons, one can deliver high field intensities at volumes with nanometer dimensions. This simple principle of plasmonics for light confinement at the nanoscale holds promise for many applications, such as sensorics, light harvesting, or cancer therapy [44, 64].

Image courtesy of Carsten Sönnichsen.

$$\pm\oint_{\partial\Omega} V_k\left[\pm\frac{1}{2}u_{k'}(s)+\oint_{\partial\Omega} F(s,s')u_{k'}(s')\,dS'\right]dS=\left(\frac{1}{2}\pm\lambda_k\right)\delta_{kk'}\,.$$

To arrive at the last expression, we have used the eigenmode definition of Eq. (9.28) together with the orthogonality relation

$$\oint_{\partial\Omega} u_k(s)G(s,s')u_{k'}(s')\,dSdS'=\delta_{kk'}\,,$$

which follows from Eqs. (9.29) and (9.31). Thus, we find

$$\int_{\Omega_{1,2}} \boldsymbol{E}_k\cdot\boldsymbol{E}_{k'}\,d^3r=\left(\frac{1}{2}\pm\lambda_k\right)\delta_{kk'}\,.$$

In other words, the fields of the eigenmodes are orthogonal to each other, in both the inside and outside volumes $\Omega_{1,2}$. By adding and subtracting the expressions for Ω_1, Ω_2 we are then led to

Relation Between Eigenvalues and Electrostatic Energies

$$\lambda_k=\frac{1}{2}\frac{W_k^1-W_k^2}{W_k^1+W_k^2}\,,\quad W_k^{1,2}=\frac{\varepsilon_0}{2}\int_{\Omega_{1,2}}\boldsymbol{E}_k\cdot\boldsymbol{E}_k\,d^3r\,. \tag{9.36}$$

Here $W_{1,2}$ are the electrostatic energies of the surface charge distribution $u_k(s)$ at the particle inside and outside, respectively. This allows for a simple interpretation of the eigenvalues in terms of the electrostatic energies for the eigenmode surface charge distributions.

Electrostatic Limit. In the electrostatic limit all fields are expelled from the metal, corresponding to $W^1=0$. This corresponds to an eigenvalue $\lambda_0=-1/2$ translating to a resonance frequency $\omega\to 0$. The surface charge distribution has a net charge, similarly to a charged metallic boundary, and can usually not be excited optically.

Dipolar Modes. For the dipolar excitations in spheres and ellipsoids there is a constant electric field inside the particle, see also Fig. 9.7. With increasing aspect ratio the positive and negative parts of the charge distribution are pushed to the opposite ends of the ellipsoids. As a consequence, the field distribution W_2 outside the particle increases in comparison to W_1, and the eigenvalues shift to lower values resulting in a redshift of the plasmon resonances (see also Fig. 9.6).

High Excitation Numbers. Finally, for high excitation numbers the plasmon eigenmodes have a large number of nodes, similarly to surface plasmons with a short wavelength. As a consequence, the electric fields are tightly confined to the particle boundary and extend equally to the particle inside and outside. For $W^1\approx W^2$ the plasmon eigenvalue then approaches zero.

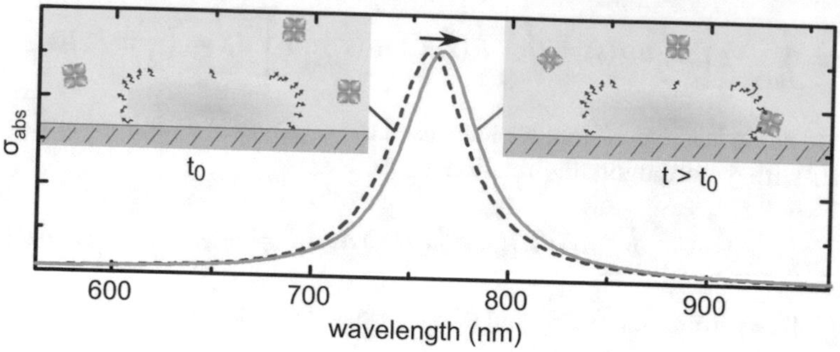

Fig. 9.11 Working principle of plasmon based (bio)sensor. A metallic nanoparticle becomes functionalized with receptors. Upon binding of analytes in a fluid channel, the dielectric environment changes and the plasmon resonance slightly shifts (compare dashed and solid lines for spectra). This shift can be detected optically, sometimes with a sensitivity at the single molecule level. Image taken from Ref. [66]

Plasmon Sensors

Similarly to the case of surface plasmons, the shift of the plasmon resonance upon changes of the dielectric environment can be used for plasmon (bio)sensors with an extremely high sensitivity. The working principle is sketched in Fig. 9.11 and consists of a plasmonic nanoparticle functionalized with receptors that can only bind specific analytes. When the analyte to be detected binds to the receptor the dielectric environment undergoes a tiny change, which can be observed as a (small) shift of the plasmon resonance position. In comparison to surface plasmons at flat metal surfaces, particle plasmons have much tighter field confinements which typically makes detectors based on nanoparticles extremely sensitive. In some cases it even becomes possible to detect the binding of *single* molecules.

In general, one defines the sensitivity S of the sensor as the change of resonance frequency $d\omega_{\rm res}$ upon a small change dn_2 of the refractive index of the embedding medium [67],

$$\text{(Sensitivity)} \quad \dots \quad S_\omega = \left(\frac{d\omega_{\rm res}}{dn_2}\right) .$$

To detect small changes of $\omega_{\rm res}$ efficiently, the plasmon peak should be sufficiently narrow. Let Γ denote the width of the plasmon peak. We can then define a figure of merit FOM as the ratio between S_ω and Γ,

$$\text{(Figure of merit)} \quad \dots \quad \text{FOM} = S_\omega : \Gamma = \frac{1}{\Gamma}\left(\frac{d\omega_{\rm res}}{dn_2}\right) .$$

We can estimate this quantity from the resonance denominator $\Lambda(\omega) + \lambda_{\rm dip}$ of the dipole resonance. Linearizing this expression close to resonance gives

$$\Lambda(\omega) \approx \Lambda_{\text{res}} + \left[\frac{\delta\Lambda}{\delta\varepsilon_1}\right]\left(\frac{\delta\varepsilon_1}{\delta\omega}\right)\delta\omega + \left[\frac{\delta\Lambda}{\delta\varepsilon_2}\right]\delta\varepsilon_2 .$$

We now exploit that the terms in brackets only differ in sign, as can be immediately seen from the definition of Λ, to obtain from $\text{Re}[\Lambda(\omega) + \lambda_{\text{dip}}] = 0$ the change of resonance frequency,

$$\delta\omega \approx \left(\frac{\delta\varepsilon_1'}{\delta\omega}\right)^{-1}\delta\varepsilon_2 .$$

If we assume that the width of the plasmon peak approximately depends on the imaginary part of the metal dielectric function, $\Gamma \propto \varepsilon_1''$, we get

$$\text{FOM} \propto \left(\varepsilon_1''\frac{\delta\varepsilon_1'}{\delta\omega}\right)^{-1} .$$

Importantly, this expression only depends on the material properties ε_1', ε_1'' of the metal evaluated at the particle plasmon resonance. It is largest for small damping and at a working point where ε_1' has a strong frequency dependence. By tailoring the geometry of the particle, one can tune the resonance frequency and, in turn, the figure of merit for plasmon sensors.

9.3.2 *Coupled Particles*

A successful approach for increasing the field enhancement of plasmonic nanoparticles is interparticle coupling. Figure 9.12 shows the situation where two nanospheres approach each other and become optically excited by a plane wave with a polarization vector oriented along the symmetry axis. As can be seen from the figure, the field intensity at the plasmon resonance frequency associated with the coupled structure increases when the particles approach each other. In the following we analyze such interparticle coupling in the framework of eigenmodes.

The eigenvalue equation for the coupled particles reads

$$\oint_{\partial\Omega} F(\mathbf{s},\mathbf{s}')U_k(\mathbf{s}')\,dS' = \Lambda_k\,U_k(\mathbf{s}) , \tag{9.37}$$

where the integral extends over the boundaries of both particles, which we denote in the following as left (L) and right (R). Here $U_k(\mathbf{s})$ are the eigenmodes with eigenenergies Λ_k. In what follows, we denote the eigenmodes of the individual particles with u_m^L, u_n^R and the corresponding eigenenergies with λ_m^L, λ_n^R, and exploit that the eigenmodes u_m^L, u_n^R form a complete set. For this reason, the eigenvalue

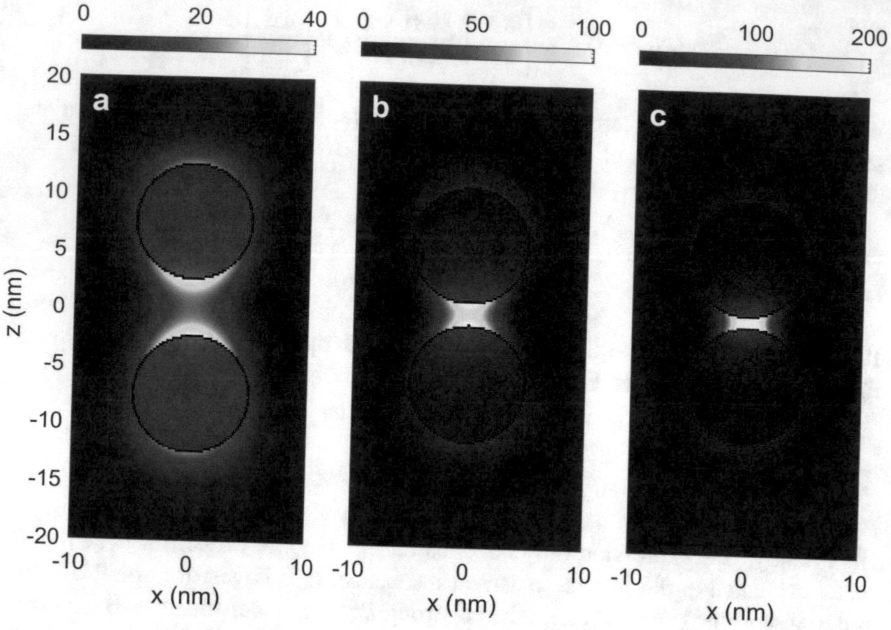

Fig. 9.12 Intensity enhancement for coupled silver spheres with diameters of 10 nm. The nearfield enhancement $|\boldsymbol{E}(\boldsymbol{r})|^2$ is plotted at the resonance energies for gap distances of (**a**) 5 nm, (**b**) 2 nm, and (**c**) 1 nm, and for a light polarization in the $\hat{z}$ direction. The color bars with the respective ranges of the enhancement factors are reported on the top of the panels

equation for the coupled particles can be expanded in terms of the eigenmodes of the uncoupled particles.

Eigenvalue Equation for Coupled Particles. We start by expanding the eigenmodes of the coupled particles through the eigenmodes of the uncoupled particles

$$U_k(\boldsymbol{s}) = \begin{cases} \sum_m C_{m,k}^L \, u_m^L(\boldsymbol{s}) & \text{for } \boldsymbol{s} \in \partial\Omega^L \\ \sum_n C_{n,k}^R \, u_n^R(\boldsymbol{s}) & \text{for } \boldsymbol{s} \in \partial\Omega^R \,. \end{cases}$$

Here $C_{m,k}^L$, $C_{n,k}^R$ are the expansion coefficients to be determined. We next insert this expression into Eq. (9.37), multiply the equation from the left with the adjoint eigenmodes $\tilde{u}_m^L$, $\tilde{u}_n^R$, and integrate over the boundaries to arrive at

$$\begin{aligned} \lambda_m^L \, C_{m,k}^L + \sum_{n'} C_{n',k}^R \oint \tilde{u}_m^L(\boldsymbol{s}_L) F(\boldsymbol{s}_L, \boldsymbol{s}_R) u_{n'}^R(\boldsymbol{s}_R) \, dS_L dS_R &= \Lambda_k \, C_{m,k}^L \\ \lambda_n^R \, C_{n,k}^R + \sum_{m'} C_{m',k}^L \oint \tilde{u}_n^R(\boldsymbol{s}_R) F(\boldsymbol{s}_R, \boldsymbol{s}_L) u_{m'}^L(\boldsymbol{s}_L) \, dS_L dS_R &= \Lambda_k \, C_{n,k}^R \,. \end{aligned} \tag{9.38}$$

We have used the eigenvalue expressions of Eq. (9.28) to relate the boundary integrals with $F(s_L, s_L)$, $F(s_R, s_R)$ to the eigenvalues λ^L, λ^R. For the remaining boundary integrals we introduce the abbreviation

$$\mathscr{F}_{mn} = \oint \tilde{u}_m^L(s_L) F(s_L, s_R) u_n^R(s_R)\, dS_L dS_R\,. \tag{9.39}$$

As discussed in Exercise 9.10, the integral is real-valued and has the symmetry property $\mathscr{F}_{mn} = \mathscr{F}_{nm}$. Using Green's first identity, we can relate this expression along the same lines of our previous derivation of Eq. (9.35) to the electric fields $\boldsymbol{E}_m^L = -\nabla V_m^L$, $\boldsymbol{E}_n^R = -\nabla V_n^R$. This gives

$$\int_{\Omega_2} \boldsymbol{E}_m^L \cdot \boldsymbol{E}_n^R \, d^3r = -\sum_{j=L,R} \oint_{\partial\Omega j} V_m^L(s_j) \left[\frac{\partial V_n^R(s_j)}{\partial n} \right] dS_j = -2\mathscr{F}_{mn}\,,$$

showing that the matrix elements $\mathscr{F}_{mn}$ can be expressed in terms of the electrostatic interaction energy between the surface charge distributions associated with the eigenmodes of the individual particles. We continue to introduce a matrix $\bar{\bar{\mathscr{F}}}$ formed by these matrix elements, as well as the matrices $\bar{\bar{\lambda}}^L$, $\bar{\bar{\lambda}}^R$ with the single-particle eigenvalues on the diagonal. Together with the matrices $\bar{\bar{C}}^L$, $\bar{\bar{C}}^R$ for the expansion coefficients we rewrite Eq. (9.38) in the compact form

$$\begin{pmatrix} \bar{\bar{\lambda}}^L & \bar{\bar{\mathscr{F}}} \\ \bar{\bar{\mathscr{F}}} & \bar{\bar{\lambda}}^R \end{pmatrix} \cdot \begin{pmatrix} \bar{\bar{C}}^L \\ \bar{\bar{C}}^R \end{pmatrix} = \begin{pmatrix} \bar{\bar{C}}^L & \bar{\bar{C}}^R \end{pmatrix} \cdot \bar{\bar{\Lambda}}\,. \tag{9.40}$$

Thus, by diagonalizing the first matrix on the left-hand side we can compute the eigenenergies Λ and eigenfunctions of the coupled particle, using the modes of the individual particles as basis functions.

As a particularly simple example we choose two identical particles, such as the spheres shown in Fig. 9.12, and consider for each particle a single mode $u_{L,R}$ with eigenenergy λ. The matrix equation of Eq. (9.40) then reduces to

$$\begin{pmatrix} \lambda & \mathscr{F} \\ \mathscr{F} & \lambda \end{pmatrix} \cdot \begin{pmatrix} C_L \\ C_R \end{pmatrix} = \begin{pmatrix} C_L & C_R \end{pmatrix} \cdot \begin{pmatrix} \Lambda_1 & 0 \\ 0 & \Lambda_2 \end{pmatrix},$$

where $\mathscr{F}$ accounts for the interaction between the two particles. Thus, the eigenmodes of the coupled particle can be associated with bonding and anti-bonding modes, $C_L = \pm C_R$, and the corresponding eigenvalues are

Eigenvalues of Coupled Particles

$$\Lambda_{1,2} = \lambda \pm \mathscr{F}\,. \tag{9.41}$$

Fig. 9.13 Schematics for coupling between two plasmonic nanoparticles. The dipole resonances of the individual particles (eigenvalue λ) become coupled through $\mathscr{F}$, and split into a bonding (parallel dipole moments) and anti-bonding (anti-parallel dipole moments) configurations

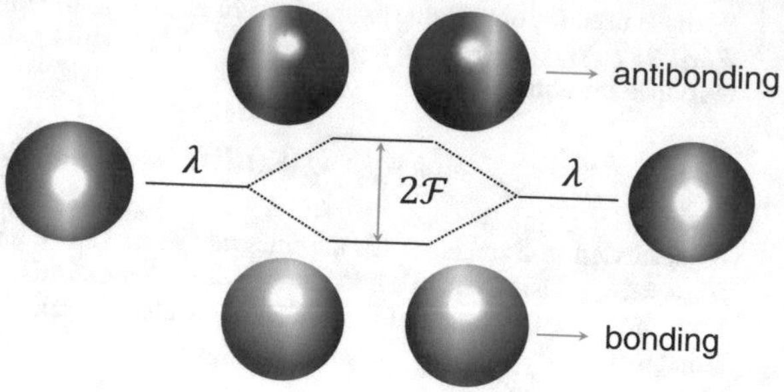

A schematic sketch of this is shown in Fig. 9.13. The two dipole modes of the isolated spheres couple and form bonding (parallel orientation $\rightarrow\rightarrow$ of dipole moments) and anti-bonding (anti-parallel orientation $\rightarrow\leftarrow$ of dipole moments) modes. Only the bonding configuration has a net dipole moment and can be optically excited. For this configuration, the charges in the gap region have opposite signs on the two spheres, thus giving rise to the strong gap field previously discussed in Fig. 9.12

Figure 9.14 shows the lowest eigenvalues of the coupled nanospheres as a function of gap distance. As can be seen from the exact results (red lines), the lowest eigenvalue for the bonding mode $\rightarrow\rightarrow$ becomes smaller when the spheres approach each other, and thus the plasmon resonance peak redshifts. One additionally observes other modes, which can be associated with dipole moments oriented perpendicularly to the symmetry axis, $\uparrow\downarrow$, and the anti-bonding configurations $\uparrow\uparrow$, $\leftarrow\rightarrow$ of these modes. The approximate eigenvalues computed from Eq. (9.41) agree well with the exact results, at least for sufficiently large gap distances. For small interparticle distances it no longer suffices to only consider the dipole modes. When including in addition to the dipole modes with eigenvalue λ_1 also quadrupole modes with λ_2 (all aligned parallel to the symmetry axis), the interaction matrix of Eq. (9.40) reads

$$\begin{pmatrix} \lambda_1 & 0 & \mathscr{F}_{11} & \mathscr{F}_{12} \\ 0 & \lambda_2 & \mathscr{F}_{12} & \mathscr{F}_{22} \\ \mathscr{F}_{11} & \mathscr{F}_{12} & \lambda_1 & 0 \\ \mathscr{F}_{12} & \mathscr{F}_{22} & 0 & \lambda_2 \end{pmatrix}.$$

Here $\mathscr{F}_{11}$, $\mathscr{F}_{12}$ are the interactions between the genuine dipole and quadrupole modes, and $\mathscr{F}_{12}$ is the mutual dipole-quadrupole coupling. The above matrix can be diagonalized, and we obtain

$$\begin{aligned} \Lambda_{1,2} &= \frac{1}{2}\left[(\lambda_1^- + \lambda_2^-) \pm \sqrt{(\lambda_1^- + \lambda_2^-)^2 + 4\mathscr{F}_{12}^2}\right] \\ \Lambda_{3,4} &= \frac{1}{2}\left[(\lambda_1^+ + \lambda_2^+) \pm \sqrt{(\lambda_1^+ + \lambda_2^+)^2 + 4\mathscr{F}_{12}^2}\right], \end{aligned} \tag{9.42}$$

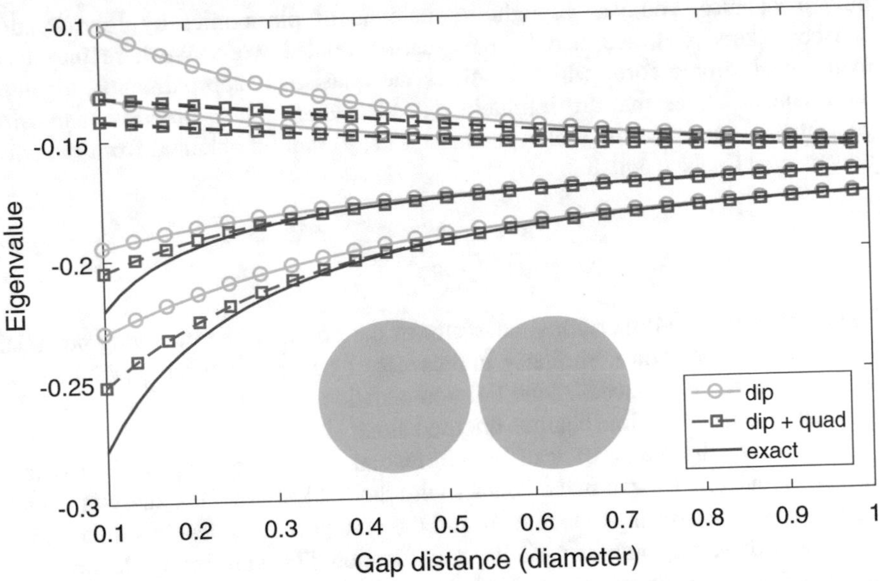

Fig. 9.14 Exact (red line) and approximate (symbols) eigenvalues for coupled spheres and different gap distances. The circles report results based on a perturbation theory that only includes the dipole modes of the individual spheres, the squares report results including dipole and quadrupole modes

with $\lambda_1^\pm = \lambda_1 \pm \mathscr{F}_{11}$, $\lambda_2^\pm = \lambda_2 \pm \mathscr{F}_{22}$. Expressed in words, the bonding and anti-bonding modes are shifted further apart as an effect of $\mathscr{F}_{12}$. The squares in Fig. 9.14 show the corresponding results. Indeed, the agreement with the exact results is much better than for the dipole case of Eq. (9.41). The mixing of dipole and quadrupole modes can be also seen in the charge distributions shown in Fig. 9.13 for the bonding and anti-bonding modes, where the distribution is modified with respect to the uncoupled spheres: in the bonding configuration more charge is accumulated in the gap region, leading to an increase of the gap fields and a redshift of the corresponding plasmon resonance, whereas in the anti-bonding configuration charge is pushed out of the gap region. When the distance between the spheres is further reduced in Fig. 9.14, the exact and approximate eigenvalues start to deviate again, indicating a significant contribution of further multipolar states. Hybridization models of more complex plasmonic nanoparticles have been studied in some detail in the literature, see for instance Ref. [68].

9.4 Conformal Mapping

Transformation optics and conformal mapping are fun topics, which allow obtaining analytic solutions for nanoparticle geometries where one would not expect that such solutions exist. Conformal mapping has some history in electrostatics and fluid mechanics, and was brought to the field of plasmonics by John Pendry and coworkers [69], see also [70] for earlier related work. We here introduce to its most simple form which assumes the quasistatic approximation together with nanostructures that are infinitely extended along the z direction (nanowire geometries), such that the scalar potential $V(x, y)$ can be obtained from the two-dimensional Laplace equation

$$\frac{\partial^2 V}{\partial x^2} + \frac{\partial^2 V}{\partial y^2} = 0\,.$$

To preserve the two-dimensional character of the problem, the incoming potential V_{inc} must not depend on z. Similarly, in presence of a charge distribution the Laplace equation has to be replaced by the Poisson equation, and thus we assume that the distribution consists of line charges oriented along z.

Here comes the magic trick of transformation optics and conformal mapping: in two dimensions we can make contact with the field of complex analysis, which we briefly summarize in Appendix A, and the powerful tools developed there. Let $z = x + iy$ denote a complex variable. Any function $f(z)$ can then be decomposed into its real and imaginary parts according to

$$f(x, y) = u(x, y) + iv(x, y)\,.$$

In the following we are interested in *analytic* functions, these are functions that can be expanded in a Taylor series around any point z_0 and the limit $\lim_{z\to z_0} f(z)$ gives $f(z_0)$ irrespectively of how we approach z_0. Analytic functions are smooth and infinitely differentiable. We will also consider functions with a pole, such as $1/z$, where $f(z)$ is analytic everywhere except at the pole $z = 0$. As shown in appendix A, for analytic functions one can derive the Cauchy–Riemann equations, Eq. (A.2), that connect the derivatives of the real and imaginary parts via

$$\frac{\partial u}{\partial x} = \frac{\partial v}{\partial y}\,, \quad \frac{\partial v}{\partial x} = -\frac{\partial u}{\partial y}\,. \tag{9.43}$$

As a direct consequence of this analyticity u, v are so-called harmonic functions, these are functions that fulfill Laplace's equation in two dimensions. Indeed, we find

$$\frac{\partial^2 u}{\partial x^2} = \frac{\partial}{\partial x}\left(\frac{\partial u}{\partial x}\right) = \frac{\partial}{\partial x}\left(\frac{\partial v}{\partial y}\right) = \frac{\partial}{\partial y}\left(\frac{\partial v}{\partial x}\right) = \frac{\partial}{\partial y}\left(-\frac{\partial u}{\partial y}\right) = -\frac{\partial^2 u}{\partial y^2}\,.$$

By a similar token we can also show that v is a harmonic function. We next state a few remarkable properties of analytic functions, which we will prove later in this section. If we perform a coordinate transformation $z' = g(z)$, then also

$$u'(x', y') = u\Big(x'(x, y), y'(x, y)\Big) \tag{9.44}$$

is a harmonic function. An important geometrical property of coordinate transformations with analytic functions $g(z)$ is that they preserves angles, and therefore define a *conformal mapping*. This has the following important consequence. Suppose that

- $V(x, y)$ is a solution of the Laplace or Poisson equation, with
- a given value V at the boundary ∂D of a two-dimensional domain D,
- and/or a given normal derivative $\frac{\partial V}{\partial n}$ at ∂D.

Then also the transformed potential

$$V'(x', y') = V\Big(x'(x, y), y'(x, y)\Big)$$

is a solution of the Laplace or Poisson equation. Owing to the preservation of angles (more specifically, because the outer surface normals $\hat{\boldsymbol{n}}$ remain perpendicular to the boundary), the new solution fulfills the *same* boundary conditions, however, for the transformed boundary $\partial D'$. Within the field of plasmonics this close connection between the solutions can be exploited as follows.

Reference Structure. Let D be a two-dimensional domain of a wire-like structure with an infinite extension along z, which is filled with a material that has some permittivity ε. We denote with $V(x, y)$ the solution of the Laplace or Poisson equation which has the proper boundary conditions of Eq. (9.4) at the boundary ∂D.

Transformed Structure. Let $x' = x'(x, y)$, $y' = y'(x, y)$ be a coordinate transformation that preserves angles. Then also $V'(x', y')$ is a solution of the Laplace or Poisson equation, which, by construction, fulfills the proper boundary conditions of Eq. (9.4) at the transformed boundary $\partial D'$.

Indeed, this is a quite remarkable finding. It suffices to solve the Laplace or Poisson equation once for a single structure with boundary ∂D, a myriad of solutions for transformed boundaries $\partial D'$ then immediately follows through simple coordinate transformations.

9.4.1 A Few Selected Examples

Before we return to the mathematical details of conformal mapping and its implications for the solution of Laplace's equation, it is probably instructive to present a few selected examples to give an idea of what the approach can do. In the following we consider the mapping

$$z' = g(z) = \frac{1}{z^*}, \tag{9.45}$$

which defines the inversion of the complex plane. Expressed in its real and imaginary parts, we get (note that without the complex conjugate in the denominator the y coordinate would simply get a negative sign)

$$x' = \frac{x}{x^2 + y^2}\,, \quad y' = \frac{y}{x^2 + y^2}\,.$$

Consider in the complex plane a straight line parallel to the real axis

$$z(t) = y_0\Big(\tan t + i\Big)\,, \tag{9.46a}$$

where $t \in (-\pi/2, \pi/2)$ is a real-valued parameter such that $x(t)$ scans all values. With this we immediately get

$$\begin{aligned} x'(t) &= y_0^{-1}\frac{\tan t}{\tan^2 t + 1} = \frac{\sin t \cos t}{y_0} = \frac{\sin 2t}{2y_0} \\ y'(t) &= y_0^{-1}\frac{1}{\tan^2 t + 1} = \frac{\cos^2 t}{y_0} = \frac{1 + \cos 2t}{2y_0}\,. \end{aligned} \tag{9.46b}$$

In words, a straight line with impact parameter iy_0 transforms to a circle with radius $1/(2y_0)$ centered at $i/(2y_0)$. Figure 9.15 shows a few examples of such a mapping. Panel (a) depicts how a semi-infinite slab filled with a dielectric material is mapped on a cylinder. Correspondingly, two slabs are mapped on (b) two kissing cylinders. The structure is quite remarkable because it contains a singular point, the touching point of the two cylinders, which would be very hard to study with non-analytic models. Finally, panel (c) reports the mapping of a slab to a crescent-shaped cylinder, which was suggested in Ref. [69] as a plasmonic light-harvesting device.

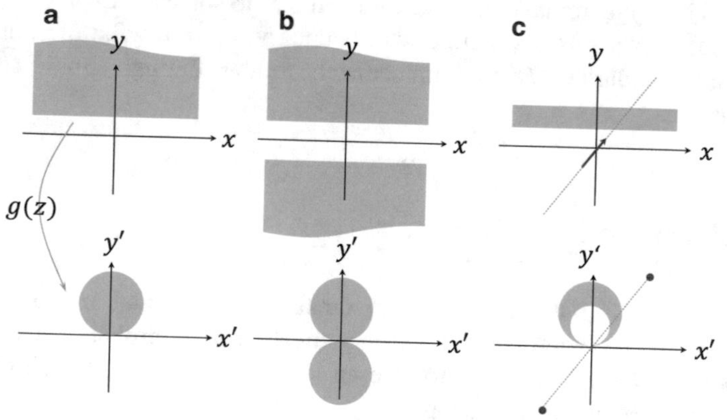

Fig. 9.15 Conformal mapping through inversion of complex plane. (**a**) The semi-infinite half-space is mapped on a cylinder. (**b**) The two half-spaces separated by a gap are mapped on two "kissing" cylinders. (**c**) A slab is mapped on a crescent-shape cylinder. In the panel we also report the transformation of the external source, here a dipole, which is mapped to two charges located at infinity. The corresponding electric field is a constant field oriented along the dipole direction

We also show the mapping of the excitation, where a dipole at the origin (original frame) is mapped to two charges located at infinity (transformed frame). The corresponding field is constant and mimics a plane wave excitation in the quasistatic limit, with a constant electric field along the dipole direction. In a reversed manner, we could use two charges at infinity in the original frame to mimic the excitation of an oscillating dipole in the transformed frame.

9.4.2 Details About Conformal Mapping

Before presenting results for the structures shown in Fig. 9.15, we sketch the proofs that the Laplace equation remains valid under conformal mapping and that such mapping preserves the angles. We assume that the mapping $g(z)$ is one-to-one, which means that there exists an inverse transformation. This is certainly true for our example of the inversion of the complex plane given in Eq. (9.45). From Eq. (9.44) relating the functions u, v with primed and unprimed coordinates we find

$$\frac{\partial u'}{\partial x'} = \frac{\partial u}{\partial x}\frac{\partial x}{\partial x'} + \frac{\partial u}{\partial y}\frac{\partial y}{\partial x'}$$

$$\frac{\partial u'}{\partial y'} = \frac{\partial u}{\partial x}\frac{\partial x}{\partial y'} + \frac{\partial u}{\partial y}\frac{\partial y}{\partial y'}\,.$$

Continuing along the same lines we get

$$\frac{\partial^2 u'}{\partial x'^2} = \frac{\partial^2 u}{\partial x^2}\left(\frac{\partial x}{\partial x'}\right)^2 + 2\frac{\partial^2 u}{\partial x \partial y}\left(\frac{\partial x}{\partial x'}\right)\left(\frac{\partial y}{\partial x'}\right) + \frac{\partial^2 u}{\partial y^2}\left(\frac{\partial y}{\partial x'}\right)^2$$

$$\frac{\partial^2 u'}{\partial y'^2} = \frac{\partial^2 u}{\partial x^2}\left(\frac{\partial x}{\partial y'}\right)^2 + 2\frac{\partial^2 u}{\partial x \partial y}\left(\frac{\partial x}{\partial y'}\right)\left(\frac{\partial y}{\partial x'}\right) + \frac{\partial^2 u}{\partial y^2}\left(\frac{\partial y}{\partial y'}\right)^2\,.$$

Using the Cauchy–Riemann equations

$$\frac{\partial x}{\partial x'} = \frac{\partial y}{\partial y'}\,, \qquad \frac{\partial y}{\partial x'} = -\frac{\partial x}{\partial y'}$$

for the complex function $z = x + iy$, then gives

$$\frac{\partial^2 u'}{\partial x'^2} + \frac{\partial^2 u'}{\partial y'^2} = \left[\left(\frac{\partial x}{\partial x'}\right)^2 + \left(\frac{\partial y}{\partial y'}\right)^2\right]\left[\frac{\partial^2 u}{\partial x^2} + \frac{\partial^2 u}{\partial y^2}\right]. \tag{9.47}$$

Thus, whenever u solves the Laplace equation, so does u'. To show that conformal maps preserve angles, we proceed as follows.

Phase and Gradient Angle. We start by introducing the phase or argument of a complex number

$$\ln\left(re^{i\theta}\right) = \ln r + i\theta \implies \theta = \arg z = \operatorname{Im}\bigl(\ln z\bigr)\,.$$

Consider a curve $z(t)$ that depends on a parameter t. The derivative with respect to this parameter t is denoted with $\dot{z}(t)$ and gives the tangent to the curve. The gradient angle of this tangent is then given through

$$\theta = \arg\left[\dot{z}(t)\right]\,.$$

Crossing of Curves. We next introduce two curves $z_1(t)$, $z_2(t)$, which are assumed to intersect at point $z = z_1(t_1) = z_2(t_2)$. For each t value we can define a tangent to the curve, and the angle between the two curves at the intersection point z can be expressed as

$$\theta = \theta_2 - \theta_1 = \arg\left[\dot{z}_2(t_2)\right] - \arg\left[\dot{z}_1(t_1)\right]\,.$$

Conformal Mapping. Upon conformal mapping, the curve $z(t)$ is transformed to $z' = g[z(t)]$. Employing the chain rule, we then find

$$\dot{z}'(t) = \frac{dz'}{dt} = \frac{dg}{dz}\frac{dz}{dt} = g'\bigl(z(t)\bigr)\dot{z}(t)\,.$$

For the angle between the transformed curves we get

$$\theta' = \operatorname{Im}\Bigl\{\ln\bigl[g'(z)\dot{z}_2(t_2)\bigr]\Bigr\} - \operatorname{Im}\Bigl\{\ln\bigl[g'(z)\dot{z}_2(t_2)\bigr]\Bigr\} = \arg\dot{z}_2 - \arg\dot{z}_1\,.$$

Thus, although the angles of the tangent vectors $\dot{z}_{1,2}(t)$ become changed upon mapping, the *relative* angle between the curves remains identical. This is what we have previously called the preservation of angles.

Most importantly for our context of plasmonics, the direction $\hat{\boldsymbol{n}}$ perpendicular to a given particle boundary ∂D is mapped to a new direction $\boldsymbol{n}'$ which is again perpendicular to the transformed boundary $\partial D'$. This has the important consequence that a potential $V(x, y)$ fulfilling the Dirichlet and Neumann boundary conditions of Eq. (9.4) at the boundary ∂D is mapped on a potential $V'(x', y')$ that also fulfills the boundary conditions at $\partial D'$.

9.4.3 *Kissing Cylinders*

Here the fun ends and the real work starts. In the following we consider the case of kissing cylinders, see Fig. 9.15b, with a constant electric field applied along the y-direction. We shall be particularly interested in the potential at the touching point.

Dipole Excitation. Our starting expression is the potential produced by a unit dipole at the origin and oriented along $\hat{\boldsymbol{y}}$

$$V_{\rm dip}(\boldsymbol{r}) = \frac{1}{4\pi\,\varepsilon_b}\int_{-\infty}^{\infty}\frac{\hat{\boldsymbol{y}}\cdot\hat{\boldsymbol{r}}}{r^2}\,dz = \frac{1}{2\pi\,\varepsilon_b}\frac{y}{r^2}\,,$$

with the integration extending over the entire z-axis. Here ε_b is the permittivity of the embedding medium. As this dipole is located in the gap region between two metal slabs, see Fig. 9.15b, we can solve Poisson's equation through a Fourier transformation along x. The Fourier transform of the potential with respect to x reads

$$\tilde{V}_{\rm dip}(k,y) = \frac{1}{2\pi\,\varepsilon_b}\int_{-\infty}^{\infty} e^{-ikx}\frac{y}{x^2+y^2}\,dx = \frac{1}{2\varepsilon_b}\,\mathrm{sgn}(y)e^{-|ky|}\,. \tag{9.48}$$

Poisson's Equation for Infinite Half-Spaces. The induced potential is a solution of Laplace's equation which is of the form

$$\left(\frac{\partial^2}{\partial x^2}+\frac{\partial^2}{\partial y^2}\right)e^{ikx}\,e^{\pm ky} = 0\,.$$

Because $V_{\rm dip}$ in Eq. (9.48) is an odd function of y, also the induced potential must be an odd function. With this, we can make the following ansatz for the induced potential ($y > 0$):

$$\tilde{V}_{\rm ind}(k,y) = \begin{cases} a(k)\,e^{ikx}\,\sinh(|k|y) & \text{for } y < \delta/2 \\ b(k)\,e^{ikx}\,e^{-|k|(y-\delta/2)} & \text{for } y > \delta/2\,, \end{cases}$$

where $-\delta/2 < y < \delta/2$ corresponds to the gap region between the semi-infinite half-spaces filled with a material of permittivity ε, see Fig. 9.15b. For $y > \delta/2$ we have chosen a potential that decays exponentially when moving away from the interface. The coefficients $a(k)$, $b(k)$ must be obtained from the boundary conditions at the layer interface $y = \delta/2$. Continuity of the potential and of the normal derivative of the dielectric displacement, Eq. (9.4), leads us to

$$\frac{1}{2}\varepsilon_b^{-1}\,e^{-\frac{1}{2}|k|\delta} + a(k)\sinh(\frac{1}{2}|k|\delta) - b(k)$$

$$-\frac{1}{2}e^{-\frac{1}{2}|k|\delta} + \varepsilon_b\,a(k)\cosh(\frac{1}{2}|k|\delta) = -\varepsilon\,b(k)\,.$$

Solving for $a(k)$ then gives

$$a(k) = -\frac{1}{\varepsilon_b}\frac{e^{\alpha}}{e^{|k|\delta}-e^{\alpha}}\,,\qquad e^{\alpha} = \frac{\varepsilon-\varepsilon_b}{\varepsilon+\varepsilon_b}\,.$$

The induced potential can be obtained from the inverse Fourier transform, and we arrive at [71, Eq. (5)]

$$V_{\text{ind}}(x, y) = -\frac{1}{\varepsilon_b}\int_{-\infty}^{\infty}\left[\frac{e^{\alpha}}{e^{|k|\delta} - e^{\alpha}}\right] e^{ikx}\sinh(|k|y)\,\frac{dk}{2\pi}\,. \tag{9.49}$$

Approximate Integral Solution. In Refs. [71, 72] some information is given of how to approximately solve this integral using complex integration. For $\varepsilon' < -\varepsilon_b$ the integral is dominated by the pole at $|k|\delta = \alpha$ close to the real axis, although in principle there exist also other poles and branch cuts which lead to corrections, such as creeping waves [71]. For $x > 0$ we can add for the integration path in Eq. (9.49) a semicircle with radius R in the upper complex plane, see Fig. B.1 for a related path, where R goes to infinity at the end of the calculation. Then, either the exponential e^{ikx} or the asymptotic $e^{-|k|\delta}$ form of the term in brackets leads to an exponential damping of the integrand (note that y is bound to values smaller than $\delta/2$, so the sinh contribution is always smaller). Thus, we find from Eq. (9.49) by use of the residue theorem

$$V_{\text{ind}}(x, y) = \int_C f(x, y; k)\, dk \approx 2\pi i \lim_{k\to k_0}(k - k_0)\, f(x, y; k)\,,$$

with the pole at $k_0 = \alpha/\delta$. This gives

$$V_{\text{ind}}(x, y) \approx -\frac{i}{\delta\varepsilon_b} e^{ik_0 x}\sinh(k_0 y)\,. \tag{9.50}$$

Conformal Mapping. We now submit this expression to the conformal mapping. From Eq. (9.46) we find that $\delta = a^{-1}$ is given by the inverse of the cylinder radii a, and the positions $(x', y') = a(\sin\theta, 1 + \cos\theta)$ at the circumference of the transformed cylinder originate from $(x, y) = \frac{1}{2}a^{-1}(\tan\frac{\theta}{2}, 1)$.

Putting together the results, we arrive at the induced potential of the kissing cylinders for a constant electric field applied along the symmetry axis,

Potential of Kissing Cylinders (Plane Wave Excitation)

$$V_{\text{ind}}(a, \theta) \approx -\frac{ia}{\varepsilon_b}\exp\left[i\alpha\tan\frac{\theta}{2}\right]\sinh\alpha\,. \tag{9.51}$$

Figure 9.16 shows the potential for (a) the dipole excitation of the slab structure, Eq. (9.50), and (b) the mapped kissing cylinder structure, Eq. (9.51), where the

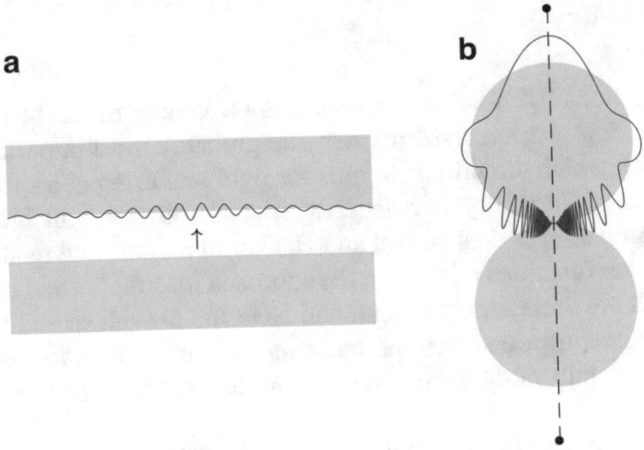

Fig. 9.16 Sketch of potential (real part) for kissing cylinders. (**a**) An oscillating dipole launches surface plasmons which propagate along the interface of the slab structure and become damped because of ohmic losses of the metal, as described through the imaginary part of the permittivity ε''. (**b**) Upon conformal mapping through inversion of the complex plane, the slab structure is mapped to two kissing cylinders. The waves at infinity are mapped to the origin, where the exponential damping is due to ε''. For clarity, we only show the plasmon oscillations for the upper particles

oscillating dipole is mapped on a plane wave excitation with a constant electric field along y. In (a) the oscillating dipole launches surface plasmons, which propagate along the metal interface and dissipate energy because of metal losses. In the mapped structure, the constant electric field launches particle plasmons which propagate towards the touching point, where they become adiabatically slowed down. The exponential decay of the amplitude close to the touching point is caused by the small imaginary part of α. In Ref. [72] the authors also discuss the optical properties of kissing cylinders, and obtained for the extinction cross section the surprisingly simple result $C_{\text{ext}} \propto \text{Re}(\ln \alpha)$. The wavelength of the plasmon oscillation becomes increasingly short when approaching the touching point of the cylinders, because of the adiabatic slowing down, thus the cylinders cannot act as a resonator. As a consequence, C_{ext} does not exhibit any resonance features.

Transformation optics and conformal mapping are powerful techniques, in particular for analyzing singular points where the prediction power of numerical schemes becomes questionable. On the other hand, in the vicinity of such singular points the plasmon wavelength becomes increasingly short and the field strengths extremely high, thus the validity of classical electrodynamics and local permittivities becomes questionable. For this reason, it might be important to consider in the vicinity of singular points corrections, such as nonlocal dielectric functions or quantum tunneling [73], as will be discussed in later parts of the book.

9.5 Mie Theory

Maxwell's equations (without the quasistatic approximation) can be solved analytically only for a few restricted geometries. The probably most important example is Mie theory [74], which provides solutions for light scattering at a spherical particle of arbitrary size. Mie theory is a feast of special functions, including spherical harmonics as well as spherical Bessel and Hankel functions, and readers not overly familiar with such functions might find the approach inscrutable and confusing. Yet, the final results are surprisingly simple and have found widespread use in various fields. In the following we sketch the basic ingredients of Mie theory and provide its main results, details of the derivation are summarized for the interested reader in Appendix E.

The situation we have in mind is depicted in Fig. 9.17 and consists of a spherical nanoparticle with radius R, permittivity ε_1, and permeability μ_1, which is embedded in a medium with material properties ε_2, μ_2. The sphere is excited by an incoming plane wave with electric field $\boldsymbol{E}_2^{\rm inc}$, and the response of the nanoparticle is described by the scattered fields $\boldsymbol{E}_1^{\rm sca}$, $\boldsymbol{E}_2^{\rm sca}$ inside and outside the sphere. The Mie solution is then obtained by expanding the electromagnetic fields inside and outside the sphere in terms of spherical waves, which are solutions of the wave equation, and by matching the fields at the sphere boundary using the boundary conditions of Maxwell's equations. The transverse electric fields are expanded using Eq. (E.4),

$$\boldsymbol{E}_1^{\rm sca}(\boldsymbol{r}) = Z_1 \sum_{\ell,m} \left[d_{\ell m}\, j_\ell(k_1 r) \quad \boldsymbol{X}_{\ell m}(\theta,\phi) + \frac{i}{k_1} c_{\ell m} \nabla \times j_\ell(k_1 r) \quad \boldsymbol{X}_{\ell m}(\theta,\phi) \right]$$

$$\boldsymbol{E}_2^{\rm sca}(\boldsymbol{r}) = Z_2 \sum_{\ell,m} \left[b_{\ell m} h_\ell^{(1)}(k_2 r) \boldsymbol{X}_{\ell m}(\theta,\phi) + \frac{i}{k_2} a_{\ell m} \nabla \times h_\ell^{(1)}(k_2 r) \boldsymbol{X}_{\ell m}(\theta,\phi) \right],$$

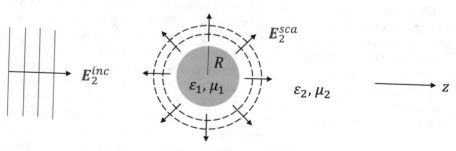

Fig. 9.17 Schematics of Mie problem. A spherical particle with radius R and material properties ε_1, μ_1 is embedded in a medium with ε_2, μ_2. The particle is excited by a plane wave with electric field $\boldsymbol{E}_2^{\rm inc}$, and the response of the sphere is described by the scattered fields $\boldsymbol{E}_2^{\rm sca}$ and $\boldsymbol{E}_1^{\rm sca}$ outside and inside the spherical particle, respectively

with k, Z being the wavenumbers and impedances, respectively. The different terms entering this equation can be described as follows.

$c_{\ell m}, d_{\ell m}$ are coefficients for the electromagnetic fields inside the sphere.

$a_{\ell m}, b_{\ell m}$ are coefficients for the electromagnetic fields outside the sphere. These coefficients must be determined from the boundary conditions of Maxwell's equations.

$j_\ell(k_1 r)$ are the spherical Bessel functions of order ℓ which are solutions of the spherical wave equation, see Eq. (C.25). These functions remain finite at the origin, Eq. (C.28), and are thus chosen inside the sphere.

$h_\ell^{(1)}(k_2 r)$ are the spherical Hankel functions of order ℓ. They have the form of outgoing waves for large arguments, Eq. (C.29), and are thus chosen outside the sphere.

$\boldsymbol{X}_{\ell m}(\theta, \phi)$ are the vector spherical harmonics of Eq. (D.8), which provide together with $\nabla \times \boldsymbol{X}_{\ell m}(\theta, \phi)$ a complete basis for the angular part of transverse vector functions.

In the following we omit the details of the actual, somewhat technical calculation, for details see Appendix E, and simply state the final results. It turns out to be convenient to introduce the Riccati–Bessel functions of Eq. (E.11),

$$\psi_\ell(x) = x j_\ell(x)\,, \qquad \xi_\ell(x) = x h_\ell(x)\,,$$

and their derivatives $\psi_\ell'(x)$, $\xi_\ell'(x)$. The Mie coefficients characterizing the electromagnetic fields outside the sphere can then be expressed in the form of Eq. (E.22),

Mie Coefficients

$$a_\ell = \frac{Z_2\psi_\ell(x_1)\psi_\ell'(x_2) - Z_1\psi_\ell'(x_1)\psi_\ell(x_2)}{Z_2\psi_\ell(x_1)\xi_\ell'(x_2) - Z_1\psi_\ell'(x_1)\xi_\ell(x_2)}$$

$$b_\ell = \frac{Z_2\psi_\ell'(x_1)\psi_\ell(x_2) - Z_1\psi_\ell(x_1)\psi_\ell'(x_2)}{Z_2\psi_\ell'(x_1)\xi_\ell(x_2) - Z_1\psi_\ell(x_1)\xi_\ell'(x_2)}\,, \tag{9.52}$$

with similar expressions for the coefficients c_ℓ, d_ℓ characterizing the fields inside the sphere. We have introduced $x_{1,2} = k_{1,2}R$. With these coefficients, we can immediately express the scattering and extinction cross sections for a plane wave excitation in the form of Eqs. (E.30) and (E.26),

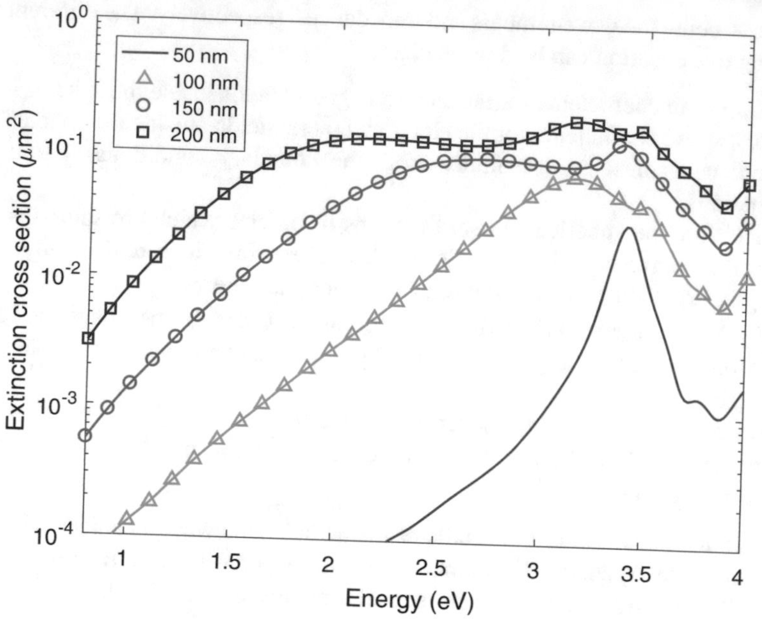

Fig. 9.18 Extinction cross sections for silver nanospheres and for different sphere diameters, as computed within Mie theory from Eq. (9.53). The dielectric function of silver is taken from Ref. [34] and the permittivity of the embedding medium is ε_0

Scattering and Extinction Cross Sections for Sphere

$$C_{\text{sca}} = \frac{2\pi}{k_2^2} \sum_\ell (2\ell+1)\left(|a_\ell|^2 + |b_\ell|^2\right)$$

$$C_{\text{ext}} = \frac{2\pi}{k_2^2} \sum_\ell (2\ell+1)\text{Re}\Big[a_\ell + b_\ell\Big]. \tag{9.53}$$

The neat thing about Eq. (9.53) is that it is a relatively simple expression that allows computing the optical cross sections in a very efficient manner.

Figure 9.18 shows the optical extinction spectra for silver nanospheres with different diameters. For the smallest sphere with a diameter of 50 nm (red line) the spectra are very similar to the quasistatic case, see Fig. 9.4. A number of features can be observed in the figure.

Red Shift. With increasing size of the nanoparticle the dipole peak shifts to the red. In a qualitative picture, the electric force acting on the surface charge distribution of the particle plasmons becomes retarded, as an effect of the finite speed of light. If one naively pictures the particle plasmon as a mechanical oscillator, retardation can be associated with a softening of the spring constant, which thus leads to a redshift of the resonance frequency.

Radiation Damping. With increasing size the peaks additionally broaden as a result of radiation damping. This is because a large dipole emits light much more efficiently. For the largest sphere, the dipole resonance becomes so broad that it is even hard to identify it at all.

Breaking of Selection Rules. Finally, with increasing size the usual optical selection rules (derived in the quasistatic limit) become softened, and new peaks appear in the spectra. They can be associated with quadrupole and other multipolar modes.

The above features are of general nature, and can be also observed for other nanoparticle geometries. Quite generally, for bulky systems, such as spheres or cubes, the quasistatic approximation works well for nanoparticle dimensions smaller than say 50 nm. For flat or strongly elongated structures the quasistatic approximation remains valid up to dimensions of about 100 nm. For larger structures with shapes different from spheres no analytic solutions are available, and one typically has to resort to numerical simulation approaches. As a first step towards this, in the following we introduce to the boundary integral method approach for the full Maxwell's equations. A detailed account of numerical techniques in nano optics will be given in Chap. 11.

9.6 Boundary Integral Method for Wave Equation

For the full Maxwell's equations one can also derive a boundary integral method in order to compute the electromagnetic fields for particles of arbitrary size and shape. However, things turn out to be more complicated than for the quasistatic case.

Our starting point is the representation formula of Eq. (5.26) which relates the electromagnetic fields $\boldsymbol{E}(\boldsymbol{r})$, $\boldsymbol{H}(\boldsymbol{r})$ in the entire volume to the (tangential) fields $\hat{\boldsymbol{n}}\times\boldsymbol{E}$, $\hat{\boldsymbol{n}}\times\boldsymbol{H}$ at the boundary. In the following we will perform a limiting procedure $\boldsymbol{r}\to\boldsymbol{s}$ and use the boundary conditions of Maxwell's equations to get an expression from which the electromagnetic fields at the boundary can be computed for given incoming fields $\boldsymbol{E}_{\rm inc}$, $\boldsymbol{H}_{\rm inc}$. As we are only interested in the tangential fields, we take on both sides of the representation formula of Eq. (5.26) the cross product with the outer surface normal $\hat{\boldsymbol{n}}$, and arrive at

$$\begin{aligned}
\boldsymbol{u}_E(\boldsymbol{s}) &= \hat{\boldsymbol{n}}\times\boldsymbol{E}_j^{\rm inc}(\boldsymbol{s}) - \tau_j \lim_{\boldsymbol{r}\to\boldsymbol{s}} \hat{\boldsymbol{n}}\times\Big\{+i\omega\mu\big[\mathbb{S}\,\boldsymbol{u}_H\big](\boldsymbol{r}) - \big[\mathbb{D}\,\boldsymbol{u}_E\big](\boldsymbol{r})\Big\} \\
\boldsymbol{u}_H(\boldsymbol{s}) &= \hat{\boldsymbol{n}}\times\boldsymbol{H}_j^{\rm inc}(\boldsymbol{s}) - \tau_j \lim_{\boldsymbol{r}\to\boldsymbol{s}} \hat{\boldsymbol{n}}\times\Big\{-i\omega\varepsilon\big[\mathbb{S}\,\boldsymbol{u}_E\big](\boldsymbol{r}) - \big[\mathbb{D}\,\boldsymbol{u}_H\big](\boldsymbol{r})\Big\}, \qquad (9.54)
\end{aligned}$$

with the tangential electromagnetic fields $\boldsymbol{u}_E=\hat{\boldsymbol{n}}\times\boldsymbol{E}$, $\boldsymbol{u}_H=\hat{\boldsymbol{n}}\times\boldsymbol{H}$.

Single Layer. We start with the single layer potential of Eq. (5.34)

$$\lim_{r\to s}\hat{\boldsymbol n}\times\big[\mathbb{S}\,\boldsymbol u\big](\boldsymbol r)=\lim_{r\to s}\hat{\boldsymbol n}\times\oint_{\partial\Omega}\left[G(\boldsymbol r,\boldsymbol s')\boldsymbol u(\boldsymbol s')+\frac{1}{k^2}\nabla G(\boldsymbol r,\boldsymbol s')\,\nabla'\cdot\boldsymbol u(\boldsymbol s')\right]dS',$$

where the first term can be analyzed similarly to the Dirichlet trace of the quasistatic approach, and the second term only involves tangential derivatives which give the same value irrespective whether approaching the boundary from the outside or inside. Thus, the value of the single layer potential does not depend on the direction of the limiting procedure.

Double Layer. For the double layer potential we get

$$\begin{aligned}\lim_{r\to s}\hat{\boldsymbol n}\times\big[\mathbb{D}\,\boldsymbol u\big](\boldsymbol r)&=-\lim_{r\to s}\hat{\boldsymbol n}\times\oint_{\partial\Omega}\nabla\times G(\boldsymbol r,\boldsymbol r')\boldsymbol u(\boldsymbol r')\,dS'\\&=-\lim_{r\to s}\oint_{\partial\Omega}\left\{\nabla\big[\hat{\boldsymbol n}\cdot G(\boldsymbol r,\boldsymbol s')\boldsymbol u(\boldsymbol s')\big]-\frac{\partial G(\boldsymbol r,\boldsymbol s')}{\partial n}\boldsymbol u(\boldsymbol s')\right\}dS',\end{aligned}$$

where we have simplified the outer products. The first term in brackets becomes zero for $\boldsymbol r\to\boldsymbol s'$ because for a tangential vector field $\boldsymbol u$ we have $\hat{\boldsymbol n}\cdot\boldsymbol u=0$. The evaluation of the second term is then identical to our previous discussion of the quasistatic case, and we find

$$\lim_{r\to s}\hat{\boldsymbol n}\times\big[\mathbb{D}\,\boldsymbol u\big](\boldsymbol r)=\frac{1}{2}\tau_j\hat{\boldsymbol n}\times\boldsymbol u(\boldsymbol s)+\hat{\boldsymbol n}\times\big[\mathbb{D}\,\boldsymbol u\big](\boldsymbol s),$$

where the positive or negative sign in $\tau_{1,2}=\pm1$ has to be chosen when approaching the boundary from the inside or outside.

Putting together all results, we get from Eq. (9.54) the set of equations

$$\begin{aligned}\frac{1}{2}\boldsymbol u_E(\boldsymbol s)&=\hat{\boldsymbol n}\times\boldsymbol E_j^{\rm inc}(\boldsymbol s)-\tau_j\hat{\boldsymbol n}\times\Big\{+i\omega\mu\big[\mathbb{S}\,\boldsymbol u_H\big](\boldsymbol s)-\big[\mathbb{D}\,\boldsymbol u_E\big](\boldsymbol s)\Big\}\\\frac{1}{2}\boldsymbol u_H(\boldsymbol s)&=\hat{\boldsymbol n}\times\boldsymbol H_j^{\rm inc}(\boldsymbol s)-\tau_j\hat{\boldsymbol n}\times\Big\{-i\omega\varepsilon\big[\mathbb{S}\,\boldsymbol u_E\big](\boldsymbol s)-\big[\mathbb{D}\,\boldsymbol u_H\big](\boldsymbol s)\Big\}.\end{aligned}\tag{9.55}$$

We have now all ingredients at hand to formulate a boundary integral method for the solution of Maxwell's equation. The situation we have in mind is depicted in Fig. 9.19 and consists of one or several nanoparticles subject to some incoming excitation, for instance, a plane wave. The basic quantities of the boundary integral method approach are the tangential electromagnetic fields

$$\boldsymbol u_E=\hat{\boldsymbol n}\times\boldsymbol E\,,\quad \boldsymbol u_H=\hat{\boldsymbol n}\times\boldsymbol H\,,$$

which are continuous at the particle boundary. We also assume that the particle excitation is only through the exterior region, such that $\boldsymbol E_1^{\rm inc}$, $\boldsymbol H_1^{\rm inc}$ are both zero.

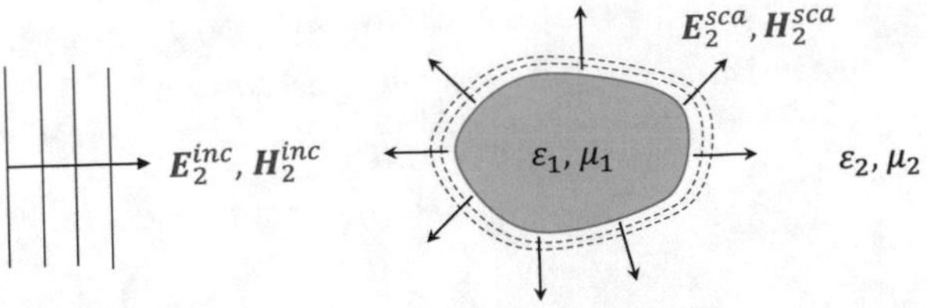

Fig. 9.19 Schematics for transmission problem for boundary integral method. One or several nanoparticles are excited by some incoming fields, for instance, those of a plane wave excitation. The response of the system is described by the scattered electromagnetic fields $\boldsymbol{E}_{1,2}^{\text{sca}}$, $\boldsymbol{H}_{1,2}^{\text{sca}}$ inside and outside the nanoparticle. In the boundary integral method approach one matches the tangential electromagnetic fields $\hat{\boldsymbol{n}} \times \boldsymbol{E}$, $\hat{\boldsymbol{n}} \times \boldsymbol{H}$ at the particle boundary $\partial\Omega$, where $\hat{\boldsymbol{n}}$ is the outer surface normal. Once the tangential fields are known at $\partial\Omega$, the fields everywhere else can be computed using the representation formula. In the figure ε_1, μ_1 denote the material parameters of the nanoparticle, and ε_2, μ_2 the material parameters of the embedding medium

With this, we get from Eq. (9.55) at the particle inside the equations

$$\begin{aligned}\frac{1}{2}\boldsymbol{u}_E(\boldsymbol{s}) &= -i\omega\mu_1\,\hat{\boldsymbol{n}} \times \big[\mathbb{S}_1\,\boldsymbol{u}_H\big](\boldsymbol{s}) + \hat{\boldsymbol{n}} \times \big[\mathbb{D}_1\,\boldsymbol{u}_E\big](\boldsymbol{s})\\ \frac{1}{2}\boldsymbol{u}_H(\boldsymbol{s}) &= i\omega\varepsilon_1\,\hat{\boldsymbol{n}} \times \big[\mathbb{S}_1\,\boldsymbol{u}_E\big](\boldsymbol{s}) + \hat{\boldsymbol{n}} \times \big[\mathbb{D}_1\,\boldsymbol{u}_H\big](\boldsymbol{s}),\end{aligned} \tag{9.56}$$

where $\mathbb{S}_{1,2}$, $\mathbb{D}_{1,2}$ are the integral operators with the Green's functions evaluated for wavenumbers $k_{1,2}$. Similarly, at the particle outside we get the equations

$$\begin{aligned}\frac{1}{2}\boldsymbol{u}_E(\boldsymbol{s}) &= \hat{\boldsymbol{n}} \times \boldsymbol{E}^{\text{inc}}(\boldsymbol{s}) + i\omega\mu_2\,\hat{\boldsymbol{n}} \times \big[\mathbb{S}_2\,\boldsymbol{u}_H\big](\boldsymbol{s}) - \hat{\boldsymbol{n}} \times \big[\mathbb{D}_2\,\boldsymbol{u}_E\big](\boldsymbol{s})\\ \frac{1}{2}\boldsymbol{u}_H(\boldsymbol{s}) &= \hat{\boldsymbol{n}} \times \boldsymbol{H}^{\text{inc}}(\boldsymbol{s}) - i\omega\varepsilon_2\,\hat{\boldsymbol{n}} \times \big[\mathbb{S}_2\,\boldsymbol{u}_E\big](\boldsymbol{s}) - \hat{\boldsymbol{n}} \times \big[\mathbb{D}_2\,\boldsymbol{u}_H\big](\boldsymbol{s}).\end{aligned} \tag{9.57}$$

From the two sets of Eqs. (9.56) and (9.57) we must extract two equations in order to compute the unknowns $\boldsymbol{u}_{E,H}$ for given incoming fields $\boldsymbol{E}^{\text{inc}}$, $\boldsymbol{H}^{\text{inc}}$. In the literature there exist a myriad of different combinations of these equations, or of slightly modified representation formulas derived from Maxwell's wave equations. The approaches are often named after the chosen equations, such as the electric field integral equations (EFIE), the magnetic field integral equations (MFIE), or the combined field integral equations (CFIE), and the interested reader is referred to the specialized literature. We here introduce one variant which results from the subtraction of the two sets of equations. This yields the boundary integral equations for the full Maxwell's equations

Boundary Integral Equation for Maxwell's Equations

$$\hat{n} \times \Big(\mathbb{D}_1[u_E] + \mathbb{D}_2[u_E]\Big) - \hat{n} \times \Big(i\omega\mu_1\mathbb{S}_1[u_H] + i\omega\mu_2\mathbb{S}_2[u_H]\Big) = \hat{n} \times E^{\text{inc}}$$
$$\hat{n} \times \Big(\mathbb{D}_1[u_H] + \mathbb{D}_2[u_H]\Big) + \hat{n} \times \Big(i\omega\varepsilon_1\mathbb{S}_1[u_E] + i\omega\varepsilon_2\mathbb{S}_2[u_E]\Big) = \hat{n} \times H^{\text{inc}}, \tag{9.58}$$

which is known as the Poggio–Miller–Chang–Harrington–Wu–Tsai formulation [75–77]. Equation (9.58) defines two integral equations for the unknown tangential electric and magnetic fields. Once these fields have been computed, we can immediately calculate the fields everywhere else using the representation formulas. In Chap. 11 we will show how to solve the above operator equation using a numerical Galerkin scheme.

9.6.1 Eigenmodes for Full Maxwell's Equations

In principle, we can also compute eigenmodes for the solutions of the full Maxwell's equations, but things turn out to be considerably more complicated. Let us rewrite Eq. (9.58) in the compact form

$$\mathbb{A}(\omega)\big[u\big] = f^{\text{inc}},$$

where $\mathbb{A}(\omega)$ is the integral operator for the transmission problem, u is the solution vector consisting of the tangential electromagnetic fields, and f^{inc} is the inhomogeneity associated with the incoming fields. In contrast to Eq. (9.27) for the quasistatic boundary integral method, it is no longer possible to separate $\mathbb{A}(\omega)$ into geometry and material parts. Furthermore, $\mathbb{A}(\omega)$ is neither symmetric nor Hermitian. In the eigenmode analysis of the full Maxwell's equations one thus either seeks for the eigenmodes for a given resonance frequency ω,

$$\mathbb{A}(\omega)\big[u_k\big] = \lambda_k u_k,$$

and of the adjoint modes $\tilde{u}_k$. However, this approach has the disadvantage that the eigenmode decomposition has to be performed for each frequency ω separately. Alternatively, one can compute the so-called quasinormal modes [78, 79]

$$\mathbb{A}(\omega_k)\big[u_k\big] = 0,$$

associated with complex eigenfrequencies ω_k. The real part gives the resonance frequency and the imaginary part is related to the lifetime of the mode. These modes have several of the appealing features of the eigenmodes of the quasistatic approach, but are often of rather limited use. We do not enter here into the details and refer the interested reader to the literature for further details [78, 79].

9.7 Details of Quasistatic Eigenmode Decomposition

In this section we provide further details for the eigenmode approach in the quasistatic limit, and demonstrate the eigenmode decomposition of Eq. (9.33) for the quasistatic Green's function.

9.7.1 *Eigenmode Analysis of Ouyang and Isaacson*

We start from the seminal work of Ouyang and Isaacson [80] and prove the basic properties of the eigenmode decomposition of Eq. (9.28). First, we introduce the integration operator $\mathbb{G}$ with the properties

$$[\mathbb{G}f](\boldsymbol{s}) = \oint_{\partial\Omega} G(\boldsymbol{s},\boldsymbol{s}')f(\boldsymbol{s}')\,dS'$$

$$[f\mathbb{G}]\,(\boldsymbol{s}) = \oint_{\partial\Omega} f(\boldsymbol{s}')G(\boldsymbol{s}',\boldsymbol{s})\,dS'\,,$$

and a corresponding operator $\mathbb{F}$ for the surface derivative of the Green's function. Importantly, the combined operator

$$[\mathbb{G}\mathbb{F}](\boldsymbol{s},\boldsymbol{s}') = [\mathbb{G}\mathbb{F}](\boldsymbol{s}',\boldsymbol{s}) \tag{9.59}$$

is symmetric with respect to $\boldsymbol{s}, \boldsymbol{s}'$. This can be shown by evaluating Green's identity

$$\int_\Omega (\phi\nabla^2\psi - \psi\nabla^2\phi)d^3r = \oint_{\partial\Omega}\left(\phi\frac{\partial\psi}{\partial n} - \psi\frac{\partial\phi}{\partial n}\right)dS$$

for $\phi(\boldsymbol{r}) = G(\boldsymbol{r}',\boldsymbol{r})$ and $\psi(\boldsymbol{r}) = G(\boldsymbol{r},\boldsymbol{r}'')$. Using the definition for the Green's function, see Eq. (9.25), and assuming that both $\boldsymbol{r}', \boldsymbol{r}''$ are located outside (or inside) the particle, we can equate the left hand side to zero. We then obtain

$$\oint_{\partial\Omega}\left[G(\boldsymbol{r}',\boldsymbol{s})\frac{\partial G(\boldsymbol{s},\boldsymbol{r}'')}{\partial n} - G(\boldsymbol{r}'',\boldsymbol{s})\frac{\partial G(\boldsymbol{s},\boldsymbol{r}')}{\partial n}\right]dS = 0\,. \tag{9.60}$$

We next let approach $\boldsymbol{r}', \boldsymbol{r}''$ the surface from the outside or inside, in complete analogy to our previous discussion of the von-Neumann trace on page 222. Both limiting procedures give $\pm 1/2$ contributions which cancel each other when taking the difference of the two terms. Thus, we arrive at

$$\oint_{\partial\Omega} \Big[G(s', s)F(s, s'') - G(s'', s)F(s, s')\Big] dS = 0\,. \tag{9.61}$$

This proves that $\mathbb{G}\mathbb{F}$ is a symmetric operator. Also the operator $\mathbb{G}$ is symmetric and can be diagonalized in the form

$$\mathbb{G} = \mathbb{Q}^T \Lambda \mathbb{Q} = (\Lambda^{1/2}\mathbb{Q})^T (\Lambda^{1/2}\mathbb{Q}) = \mathbb{U}^T\mathbb{U}\,, \tag{9.62}$$

where $\mathbb{Q}$ is a real operator and Λ is diagonal. In the last equalities we have exploited that $\mathbb{G}$ is positive definite and thus all eigenvalues are larger than zero. We can now construct another symmetric operator

$$\mathbb{D} = \left(\mathbb{U}^{-1}\right)^T \mathbb{G}\mathbb{F}\,\mathbb{U}^{-1} = \left(\mathbb{U}^{-1}\right)^T \mathbb{U}^T\mathbb{U}\mathbb{F}\,\mathbb{U}^{-1} = \mathbb{U}\mathbb{F}\mathbb{U}^{-1}\,,$$

which has real eigenvalues and the eigenfunctions form a complete orthogonal set. Then,

$$\begin{aligned}
\big[\mathbb{D}x_k\big](s) = \lambda_k\, x_k(s) &\implies \Big[\mathbb{F}\,\mathbb{U}^{-1}x_k\Big](s) = \lambda_k \Big[\mathbb{U}^{-1}x_k\Big](s) \\
\big[x_k\mathbb{D}\big](s) = \lambda_k\, x_k(s) &\implies \big[x_k\mathbb{U}\,\mathbb{F}\big](s) = \lambda_k \big[x_k\mathbb{U}\big](s)\,,
\end{aligned}$$

and the eigenmodes u_k, $\tilde{u}_k$ of Eq. (9.28) are related to x_k through

$$\begin{aligned}
u_k(s) &= \Big[\mathbb{U}^{-1}x_k\Big](s) \\
\tilde{u}_k(s) &= \big[x_k\mathbb{U}\big](s)\,.
\end{aligned} \tag{9.63}$$

From these expressions one can easily demonstrate the biorthogonality of u_k, $\tilde{u}_k$, which completes our proof for the eigenmode properties stated previously. Using Eq. (9.63) we can also prove the following orthogonality relation (see Exercise 9.9)

$$\oint_{\partial\Omega} \frac{u_k(s)u_{k'}(s')}{4\pi|s - s'|}\, dS dS' = \delta_{kk'}\,. \tag{9.64}$$

Comparison with the biorthogonality relation of Eq. (9.29) then allows us to relate $\tilde{u}_k(s)$ to $u_k(s)$ via

$$\tilde{u}_k(s) = \oint_{\partial\Omega} G(s, s')u_k(s')\, dS'\,. \tag{9.65}$$

9.7.2 Eigenmode Decomposition of Green's Function

Consider the representation formula of Eq. (9.24),

$$V(\boldsymbol{r}) = V_{\text{inc}}(\boldsymbol{r}) + \oint_{\partial\Omega} G(\boldsymbol{r}, \boldsymbol{s}')$$

$$\left\{ -\sum_k u_k(\boldsymbol{s}')\big(\Lambda(\omega) + \lambda_k\big)^{-1} \left[\oint_{\partial\Omega} \tilde{u}_k(\boldsymbol{s}'') \frac{\partial V_{\text{inc}}(\boldsymbol{s}'')}{\partial n''}\, dS'' \right] \right\} dS' , \qquad (9.66)$$

where we have inserted the eigenmode solution of Eq. (9.32). We assume that the external potential is produced by some charge distribution $\rho(\boldsymbol{r})$,

$$V_{\text{inc}}(\boldsymbol{r}) = \varepsilon_2^{-1} \int G(\boldsymbol{r}, \boldsymbol{r}')\rho(\boldsymbol{r}')\, d^3r' .$$

As discussed in Chap. 5 for Green's functions, in the above case G only accounts for the boundary conditions at infinity, and should thus be rather called a fundamental solution than a Green's function (although we will here stick to the usual Green's function terminology used in physics). However, Eq. (9.66) suggests introducing a total Green's function for Maxwell's equations in the quasistatic limit

Eigenmode Representation of Total Green's Function (Biorth.)

$$G_{\text{tot}}(\boldsymbol{r}, \boldsymbol{r}_0) = G(\boldsymbol{r}, \boldsymbol{r}_0) - \sum_k \left[\oint_{\partial\Omega} G(\boldsymbol{r}, \boldsymbol{s}') u_k(\boldsymbol{s}')\, dS' \right] \big(\Lambda(\omega) + \lambda_k\big)^{-1}$$

$$\times \left[\oint_{\partial\Omega} \tilde{u}_k(\boldsymbol{s}'') \frac{\partial G(\boldsymbol{s}'', \boldsymbol{r}_0)}{\partial n''}\, dS'' \right] . \qquad (9.67)$$

This allows us to write the solution for an external charge distribution in the usual form

$$V(\boldsymbol{r}) = \varepsilon_2^{-1} \int G_{\text{tot}}(\boldsymbol{r}, \boldsymbol{r}')\rho(\boldsymbol{r}')\, d^3r' .$$

We can rewrite this expression in a more symmetric form. The last term in brackets of Eq. (9.67) can be expressed with the relation of Eq. (9.65) between $\tilde{u}_k(\boldsymbol{s})$ and $u_k(\boldsymbol{s})$ as

$$\oint_{\partial\Omega} \left[u_k(\boldsymbol{s}_1) G(\boldsymbol{s}_1, \boldsymbol{s}'')\right] \frac{\partial G(\boldsymbol{s}'', \boldsymbol{r}_0)}{\partial n''}\, dS'' dS_1 \,.$$

From the symmetry relation of the Ouyang–Isaacson approach of Eq. (9.60) we find

$$\oint_{\partial\Omega} G(\boldsymbol{r}_1, \boldsymbol{s}'') \frac{\partial G(\boldsymbol{s}'', \boldsymbol{r}_0)}{\partial n}\, dS'' = \oint_{\partial\Omega} \frac{\partial G(\boldsymbol{r}_1, \boldsymbol{s}'')}{\partial n''} G(\boldsymbol{s}'', \boldsymbol{r}_0)\, dS'' \,,$$

where we additionally have to perform the limit $\boldsymbol{r}_1 \to \boldsymbol{s}_1$ on both sides of the equation. While this can be safely done on the left-hand side, the limiting on the right-hand side has to be performed similarly to the von-Neumann trace procedure, and we arrive at

$$\lim_{\boldsymbol{r}_1 \to \boldsymbol{s}_1} \oint_{\partial\Omega} \frac{\partial G(\boldsymbol{r}_1, \boldsymbol{s}'')}{\partial n''} G(\boldsymbol{s}'', \boldsymbol{r}_0)\, dS'' = \pm\frac{1}{2} G(\boldsymbol{s}_1, \boldsymbol{r}_0) + \oint_{\partial\Omega} \frac{\partial G(\boldsymbol{s}_1, \boldsymbol{s}'')}{\partial n''} G(\boldsymbol{s}'', \boldsymbol{r}_0)\, dS'' .$$

The sign depends on whether $\boldsymbol{r}_1$ approaches the boundary from the outside or inside. Using the fact that $u_k(\boldsymbol{s})$ is an eigenfunction of the surface derivative of the Green's function, we get

$$\oint_{\partial\Omega} u_k(\boldsymbol{s}_1) \frac{\partial G(\boldsymbol{s}_1, \boldsymbol{s}'')}{\partial n''} G(\boldsymbol{s}'', \boldsymbol{r}_0)\, dS'' dS_1 = \lambda_k \oint_{\partial\Omega} u_k(\boldsymbol{s}'') G(\boldsymbol{s}'', \boldsymbol{r}_0)\, dS'' \,.$$

Putting together all results we arrive at the final expression for the quasistatic Green's function expanded in terms of eigenfunctions given in Eq. (9.33). For a more detailed discussion of the eigenmodes and the expansion of the Green's function in these modes see Ref. [62].

Exercises

Exercise 9.1 Model a particle plasmon as a driven harmonic oscillator with resonance frequency ω_0 and damping γ.

(a) Compute the solution for a harmonically driven oscillator.
(b) Compute the solution for the free oscillator.
(c) Suppose that at time zero an external field $E_0 e^{-i\omega t}$ is turned on. Write down the solution as the sum of the homogeneous (free oscillator) and particular (driven oscillator) solutions. Discuss how the driven oscillator approaches the steady state.

(d) Show that in the steady-state the work performed by the external field equals the dissipated power.
(e) Find the driving frequencies at which the (1) amplitude and (2) absorbed power are largest.

Exercise 9.2 Compute in analogy to Eq. (9.16) the surface charge distribution $\sigma(\theta)$ for a metallic nanosphere with a Drude dielectric function, Eq. (7.7), and separate $\sigma(\theta)$ into free and bound contributions using Eq. (7.23).

Exercise 9.3 Start from Eq. (9.6) for the quasistatic potential inside and outside the sphere. Along the lines discussed for the quasistatic Mie theory, consider inside (outside) the sphere the coefficients A_ℓ (B_ℓ) only.

(a) Seek for the resonance modes corresponding to solutions in absence of external excitations by working out the boundary conditions. Use the Drude dielectric function of Eq. (7.7) and show that these modes exist for certain complex frequencies ω_ℓ only. Discuss the physical meaning of the various modes.
(b) Expand ω_ℓ in a Taylor series for small values of the Drude damping term γ, and compare the resulting mode frequency with the surface plasmon resonance frequency of Eq. (9.14).
(c) Compute the resonance modes also for the Drude–Lorentz permittivity of Eq. (7.6).

Exercise 9.4 Solve Maxwell's equations in the quasistatic limit for a coated sphere with a core (radius a, permittivity ε_1) and a coating layer (thickness δ, permittivity ε_2), which is embedded in a background medium with permittivity ε_3. Suppose that an electric field is applied along the z-direction. Along the same lines as for the sphere discussed in Sect. 9.2.1, write down an ansatz for the solution in the three regions and determine the unknown coefficients from the boundary conditions of Maxwell's equations.

Exercise 9.5 Compute the scattering and extinction cross sections for a small sphere in the quasistatic limit. Determine for a Drude dielectric function the frequencies at which the cross sections become maximal. Interpret the result using the result of Exercise 9.1e.

Exercise 9.6 Relate the depolarization factor L_k from Mie–Gans theory to the eigenvalue λ_k of the boundary integral method approach.

Exercise 9.7 Use from the NANOPT toolbox the files `demobem03.m` and `demobem04.m` which compute and plot the eigenvalues λ_k and eigenmodes $u_k(s)$ of an ellipsoid, using the boundary element method (BEM) to be described in Chap. 11.

(a) Compare the numerical eigenvalues with the analytic expressions of Mie–Gans theory.
(b) Plot the eigenmodes and give an interpretation in terms of dipolar and multipolar modes.

(c) Investigate prolate and oblate ellipsoids. What are the differences? What are the similarities?
(d) Compare the results for the oblate ellipsoids with the disk eigenmodes shown in Fig. 9.9. Can you identify all modes?

Exercise 9.8 The electric field in the direction normal to the particle boundary can be obtained from $E_\perp = -\partial_n V$, with $\partial_n = \hat{\boldsymbol{n}} \cdot \nabla$. Starting from the boundary integral equation of Eq. (9.27), show how to express the induced field $E_\perp$ in terms of the surface charge distribution σ.

Exercise 9.9 Use the eigenmodes of Ouyang and Isaacson, Eq. (9.62), to prove the orthogonality relation of Eq. (9.64).
Hint: start from the orthogonality relation for x_k, and then relate x_k to u_k.

Exercise 9.10 Show that for $\mathscr{F}_{mn}$ defined in Eq. (9.39) the symmetry relation $\mathscr{F}_{mn} = \mathscr{F}_{nm}$ holds. Start from the definition of Eq. (9.65) for the adjoint eigenfunctions, and use the symmetry properties of Eq. (9.61).

Exercise 9.11 Use from the NANOPT toolbox the files `demobem05.m` and `demobem06.m` which compute an plot the eigenvalues λ_k and eigenmodes $u_k(\boldsymbol{s})$ of coupled spheres or ellipsoids.

(a) Start from the case of coupled spheres, and plot the eigenmodes. Interpret the modes with lowest eigenvalue λ_k in terms of bonding and anti-bonding modes using as basis functions the single-sphere modes. Refer to the discussion given for Fig. 9.14.
(b) Is the effect of coupling more pronounced for the spheres or ellipsoids? Interpret the distance dependence and give an interpretation.

Exercise 9.12 Use Eq. (7.24) to compute the surface charge distribution for the kissing cylinders, Eq. (9.51).

Exercise 9.13 Start from the scattering and extinction cross sections of Mie's theory, Eq. (9.53), and perform the limit $kR \ll 1$ using Eq. (C.28).

(a) Show that in the lowest order of approximation one recovers the results of the quasistatic Mie theory.
(b) Derive the leading-order corrections for the scattering cross sections.

Exercise 9.14 Use from the NANOPT toolbox the file `demospecmie01.m`.

(a) Investigate the scattering and extinction spectra for sphere diameters in the range between 50 nm and 1 μm, using gold and silver spheres.
(b) Investigate for the smallest and largest sphere the minimal cutoff value $\ell_{\max}$ of angular orders which is needed to get converged results.
(c) Investigate for selected frequencies and sphere radii the emission pattern using `demospecmie03.m`. Interpret the results.
(d) Investigate the scattering spectra and emission patterns for a dielectric sphere with $n_1 = 1.5$.

Exercise 9.15 Use from the NANOPT toolbox the file `demoimmie01.m` for the light excitation of a metallic nanosphere. The propagation direction of the exciting light is along z, and the scattered light is observed in the x direction using an imaging lens, see Chap. 3. Investigate the influence of linear and circular polarization of the exciting light on the image, and interpret the result. You may like to consult Ref. [81].

Chapter 10
Photonic Local Density of States

In this chapter we investigate the coupling of quantum emitters and other local probes to plasmonic nanoparticles. We shall find it convenient to introduce the concept of a *photonic local density of states* (LDOS), which is a measure of how efficiently an oscillator transfers energy to its environment. In later parts of the chapter we will also discuss surface-enhanced Raman spectroscopy (SERS) and electron energy loss spectroscopy (EELS), which play important roles in the fields of nano optics and plasmonics. This chapter combines many of the concepts introduced in previous chapters, including Green's functions, stratified media, and particle plasmons. Our analysis will be based on the framework of classical electrodynamics, but we will show in later parts of this book that the results only need to be slightly adapted to account for quantum effects.

10.1 Decay Rate of Quantum Emitter

We start by considering the power dissipated by an oscillating dipole, possibly located close to a plasmonic nanoparticle or in some non-trivial photonic environment. As we will show in the following, our derivation is quite general and the only assumption we make about the photonic environment is that its response is linear. The power performed by a current distribution $\boldsymbol{J}$ against the electromagnetic fields has been previously derived in Eq. (4.13) and reads

$$\frac{dW}{dt} = -\int \boldsymbol{J}(\boldsymbol{r},t)\cdot\boldsymbol{E}(\boldsymbol{r},t)\,d^3r\,,$$

where $\boldsymbol{E}$ is the electric field. The negative sign with respect to Eq. (4.13) is because we now seek for the power performed by the current distribution against the electric field (and not by the electric field on the current distribution). We assume that the

U. Hohenester, *Nano and Quantum Optics*, Graduate Texts in Physics,
https://doi.org/10.1007/978-3-030-30504-8_10

dipole oscillates with an angular frequency ω and obtain for the dissipated power averaged over one oscillation period

$$\left\langle \frac{dW}{dt} \right\rangle = -\frac{1}{2} \int \mathrm{Re}\Big\{ \boldsymbol{J}^*(\boldsymbol{r}) \cdot \boldsymbol{E}(\boldsymbol{r}) \Big\} \, d^3r = -\frac{1}{2} \mathrm{Re}\Big\{ \Big(i\omega \, \boldsymbol{p}^* \Big) \cdot \boldsymbol{E}(\boldsymbol{r}_0) \Big\} .$$

In the last expression we have used Eq. (6.1) to relate the current distribution to the dipole moment $\boldsymbol{p}$ of the quantum emitter. What comes next is a trick that is almost too beautiful to be true. We can use the dyadic Green's function to relate the electric field to the source of the field, namely the quantum emitter itself. From Eq. (5.20) we get

$$\boldsymbol{E}(\boldsymbol{r}) = i\omega\mu \int \bar{\bar{G}}_{\mathrm{tot}}(\boldsymbol{r}, \boldsymbol{r}') \cdot \boldsymbol{J}(\boldsymbol{r}') \, d^3r' = \mu\omega^2 \, \bar{\bar{G}}_{\mathrm{tot}}(\boldsymbol{r}, \boldsymbol{r}_0) \cdot \boldsymbol{p} \, .$$

Here $\bar{\bar{G}}_{\mathrm{tot}}$ is the total Green's function that includes the free-space part as well as a reflected part in case of a non-trivial photonic environment

$$\bar{\bar{G}}_{\mathrm{tot}}(\boldsymbol{r}, \boldsymbol{r}') = \bar{\bar{G}}(\boldsymbol{r}, \boldsymbol{r}') + \bar{\bar{G}}_{\mathrm{refl}}(\boldsymbol{r}, \boldsymbol{r}') \, . \tag{10.1}$$

Putting together all results, we arrive at the time-averaged power dissipated by a dipole

Averaged Power Dissipated by Dipole Emitter

$$P = \left\langle \frac{dW}{dt} \right\rangle = \frac{\omega^3}{2c^2\varepsilon} \, \mathrm{Im}\Big\{ \boldsymbol{p}^* \cdot \bar{\bar{G}}_{\mathrm{tot}}(\boldsymbol{r}_0, \boldsymbol{r}_0, \omega) \cdot \boldsymbol{p} \Big\} . \tag{10.2}$$

This is a quite remarkable result showing that the dissipated power can be computed from the dyadic Green's function and the dipole moment alone. The imaginary part of $\boldsymbol{p}^* \cdot \boldsymbol{E}$ accounts for the out-of-phase component between the oscillating dipole and the field generated by the dipole. As we will discuss in the following, in general only part of the dissipated power is radiated by the dipole, the other part is usually transferred to the photonic environment where it becomes transformed to heat, for instance, through ohmic losses. We can derive from Eq. (10.2) a number of useful expressions.

Free-Space. For computing the power radiated by an oscillating dipole in free space we assume that the dipole is oriented along $\hat{z}$. We then get from the explicit expression for the dyadic Green's function given in Exercise 5.7

$$\begin{aligned}\operatorname{Im}\Big\{G_{zz}(\boldsymbol{r},\boldsymbol{r}')\Big\} &= \operatorname{Im}\Bigg\{\Bigg[1-\hat{R}_z^2+i\left(\frac{1-3\hat{R}_z^2}{kR}\right)-\left(\frac{1-3\hat{R}_z^2}{k^2R^2}\right)\Bigg] \\ &\quad\times\frac{1}{4\pi R}\left(1+ikR-\frac{1}{2}k^2R^2-\frac{i}{6}k^3R^3+\dots\right)\Bigg\} \\ &= \frac{k}{6\pi}+\mathcal{O}(kR)\,.\end{aligned} \tag{10.3}$$

We have introduced $\boldsymbol{R}=\boldsymbol{r}-\boldsymbol{r}'$ and have expanded the exponential e^{ikR} in a Taylor series. We can now safely perform the limit $R\to 0$ and get

$$P_0=\frac{\omega^3p^2}{2c^2\varepsilon}\left(\frac{k}{6\pi}\right)=\frac{\mu\omega^4p^2}{12\pi c}\,. \tag{10.4}$$

Comparison with Eq. (9.20) for the scattered power of an oscillating dipole

$$P_{\text{sca}}=Z^{-1}\frac{k^4}{32\pi^2\varepsilon^2}\left(\frac{8\pi}{3}p^2\right)=\frac{\mu\omega^4p^2}{12\pi c}$$

shows that the results are identical.

Photonic Enhancement. For the Green's function decomposition of Eq. (10.1) we can express the enhancement of the dissipated power in the form

$$\frac{P}{P_0}=1+\frac{6\pi}{k}\operatorname{Im}\Big\{\hat{\boldsymbol{p}}^*\cdot\bar{\bar{G}}_{\text{refl}}(\boldsymbol{r}_0,\boldsymbol{r}_0,\omega)\cdot\hat{\boldsymbol{p}}\Big\}\,. \tag{10.5}$$

Here $\hat{\boldsymbol{p}}$ is the unit vector of the dipole moment. As we will show below, in some cases this enhancement factor can be also smaller than one, corresponding to an inhibition of dissipated power.

Radiated Power. In general, only part of the dissipated power is converted to radiation, whereas the rest becomes absorbed in the photonic environment and is eventually transformed to heat, such as for the ohmic losses of plasmonic nanoparticles. To compute the scattered power, we can use Eq. (4.22) which computes P_{sca} derived from Poynting's theorem using the electromagnetic fields at a boundary $\partial\Omega$ enclosing the dipole as well as all relevant particles of the photonic environment. If necessary, one can also shift $\partial\Omega$ towards infinity.

Decay Rate. Consider a quantum emitter which emits photons with energy $\hbar\omega$. Although our above derivation has been entirely based on classical arguments, we can try to compute the decay rate Γ via

$$\Gamma=\frac{P}{\hbar\omega}\,. \tag{10.6}$$

The rationale behind this expression is as follows: P gives the energy dissipated per unit time. According to quantum mechanics, the system can only emit photons with an energy quantum of $\hbar\omega$. For this reason, the above expression is expected to give the decay probability per unit time, the so-called decay rate, and indeed we will show in later parts of this book that a full quantum approach gives the same result.

As a specific example we here use Eq. (10.4) to compute the scattering rate of a quantum emitter embedded in free space

$$\Gamma_0 = \left(\frac{\mu\omega^4 p^2}{12\pi c}\right)\frac{1}{\hbar\omega} = \frac{\mu\omega^3}{3\pi\hbar c}\left(\frac{p}{2}\right)^2 . \tag{10.7}$$

The same result can be obtained from a full quantum mechanical treatment and is usually referred to as the Wigner–Weisskopf decay rate. There is a rather subtle point concerning the dipole moment, which, however, is of only technical nature. In classical electrodynamics the time-dependent dipole moment is usually obtained from

$$\boldsymbol{p}(t) = \mathrm{Re}\left\{e^{-i\omega t}\boldsymbol{p}\right\} = \frac{1}{2}\left(e^{-i\omega t}\boldsymbol{p} + \text{c.c.}\right),$$

whereas in quantum mechanics $\boldsymbol{p}$ typically refers to the dipole transition moment from the ground to the excited state. See Sect. 15.4 for a full quantum description. The expectation value of the dipole operator then differs by a factor of 1/2 from the classical result, as we have indicated in the term on the right-hand side of Eq. (10.7). In any case, we can safely use

$$\frac{\Gamma}{\Gamma_0} = \frac{P}{P_0}$$

to compute for a quantum emitter the enhancement of the decay rate in a given photonic environment with respect to its free-space decay rate. Below we will also use the lifetimes $\tau_0 = 1/\Gamma_0$ for the oscillating dipole in free space and $\tau = 1/\Gamma$ for the dipole situated in a different photonic environment.

Purcell Factor. In the field of quantum electrodynamics it is convenient to express the enhancement of the radiative decay rate in terms of the so-called Purcell factor [82]. The enhancement typically refers to the decay of a quantum emitter in a cavity, where light is confined to a volume Ω_{cav}, for instance, by means of two mirrors. It bounces back and forth between the mirrors until it escapes from the cavity through losses. One usually introduces the quality factor $Q = \delta\omega/\omega$ that accounts for the time spent by the light in the cavity. With the cavity parameters Ω_{cav} and Q the Purcell factor reads [82]

$$F = \frac{3}{4\pi^2}\left(\frac{\lambda}{n}\right)^3\left(\frac{Q}{\Omega_{\text{cav}}}\right), \tag{10.8}$$

where λ/n is the effective wavelength of the light in the cavity. From this we find that the light–matter interaction can be enhanced either by making the volume smaller or by increasing the time spent by the light in the cavity. Although Ω_{cav} is difficult to define for plasmonic nanoparticles, it turns out that the quality factor is typically low (because of the large ohmic losses of metals), but on the other hand the effective mode volume can be extremely small. For these reasons, quite large Purcell enhancement factors can be achieved for plasmonic nanoparticles.

Effective Polarizability

This might be the right place to address a subtle point concerning the proper choice of the effective polarizability. In the previous chapter we derived in Eq. (9.19) an expression for the induced dipole moment of a small polarizable particle,

$$\boldsymbol{p} = \bar{\bar{\alpha}}(\omega) \cdot \boldsymbol{E}_{\text{inc}}(\boldsymbol{r}_0)\,.$$

Here $\bar{\bar{\alpha}}$ is the polarizability tensor and $\boldsymbol{E}_{\text{inc}}$ is the incoming electric field which polarizes the particle. As noted in Ref. [83], the induced dipole suffers radiation damping which should be additionally taken into account. We can easily do so using the leading-order contribution $i\,\text{Im}\{G_{zz}(\boldsymbol{r}_0,\boldsymbol{r}_0)\}$ previously derived for the free-space decay rate, and we are led to

$$\boldsymbol{p} = \bar{\bar{\alpha}}(\omega) \cdot \left[\boldsymbol{E}_{\text{inc}}(\boldsymbol{r}_0) + \omega^2\mu\left(\frac{ik}{6\pi}\right)\bar{\bar{\alpha}}\cdot\boldsymbol{p}\right].$$

The first term in brackets is the external field, the second one the field produced by the polarized particle on itself. The above equation can be solved for $\boldsymbol{p}$ and we obtain

$$\boldsymbol{p} = \left(\mathbb{1} - \frac{ik^3}{6\pi\varepsilon}\bar{\bar{\alpha}}\right)^{-1} \cdot \bar{\bar{\alpha}} \cdot \boldsymbol{E}_{\text{inc}}(\boldsymbol{r}_0) = \bar{\bar{\alpha}}_{\text{eff}} \cdot \boldsymbol{E}_{\text{inc}}(\boldsymbol{r}_0)\,. \tag{10.9}$$

When the polarizable particle is placed in a non-trivial photonic environment, we can also consider additional losses by introducing the enhancement factor $P : P_0$ in the effective polarizability

$$\bar{\bar{\alpha}}_{\text{eff}} = \left(\mathbb{1} - \frac{ik^3}{6\pi\varepsilon}\left[\frac{P}{P_0}\right]\bar{\bar{\alpha}}\right)^{-1} \cdot \bar{\bar{\alpha}}\,.$$

In principle, the effective polarizability tensor $\bar{\bar{\alpha}}_{\text{eff}}$ should be used for all optical cross sections derived in the previous chapter. For small particles the correction term is typically small and can be safely neglected. However, for larger particles the additional damping term can lead to a noticeable improvement in comparison to solutions of the full Maxwell's equations.

10.1.1 Photonic Local Density of States

The title of this chapter is the "photonic local density of states," so we should clarify its meaning. The density of states (DOS) is a useful concept usually introduced in solid state physics, where it typically refers to the number of electronic states per energy and per unit volume. Let E_i denote the electronic energy levels of some solid state system, such as free electrons in a metal. Then, the density of states $D(E)$ is defined as [36]

$$D(E) = \frac{1}{\Omega}\sum_i \delta(E - E_i)\,,$$

where Ω is the volume of the system and i labels the different states. For a finite system $D(E)$ typically consists of a series of sharp delta-like peaks, whereas in the thermodynamic limit $\Omega \to \infty$ the density of states becomes a continuous function.

We can try to adopt a similar concept to photonic systems. A useful definition of the photonic density of states is

$$\rho(\omega) = \frac{1}{\Omega}\sum_{\boldsymbol{k},s} \delta(\omega - \omega_{\boldsymbol{k}s})\,, \tag{10.10}$$

where $\boldsymbol{k}$ labels the wavenumbers of the photon states and s the different polarizations. In the above expression we assume that the electromagnetic fields are confined within some volume Ω, for instance, by enclosing the volume with mirrors. In the limit $\Omega \to \infty$ we can replace the summation over $\boldsymbol{k}$ by an integration

$$\rho(\omega) \underset{\Omega\to\infty}{\longrightarrow} \frac{2}{\Omega}\frac{\Omega}{(2\pi)^3}\int \delta(\omega - ck)\,d^3k = \frac{1}{4\pi^3}\int_0^\infty \delta(\omega - ck)\,4\pi k^2 dk\,, \tag{10.11}$$

where we have added a factor of two for the polarization degrees of freedom and the additional factor $\Omega/(2\pi)^3$ is due to the replacement of the sum by an integral [36]. Working out the integration, we finally obtain the free-space photonic density of states

Free-Space Photonic Density of States

$$\rho^0(\omega) = \frac{\omega^2}{\pi^2 c^3}\,. \tag{10.12}$$

We have added a superscript zero to indicate that the result refers to a free-space photonic environment. For different photonic environments, for instance, in presence of nanoparticles or stratified media, it turns out to be convenient to introduce a local density of states which can be related to the Green's dyadics. Unfortunately, the definition of this so-called local density of states is not unique and there exist different definitions in literature. In accordance to Ref. [84] we introduce the local density of states projected along some direction $\hat{\boldsymbol{p}}$

Photonic Local Density of States Projected Along $\hat{\boldsymbol{p}}$

$$\rho_{\hat{\boldsymbol{p}}}(\boldsymbol{r},\omega)=\frac{6\omega}{\pi c^2}\,\mathrm{Im}\Big\{\hat{\boldsymbol{p}}^*\cdot\bar{\bar{G}}_{\mathrm{tot}}(\boldsymbol{r},\boldsymbol{r},\omega)\cdot\hat{\boldsymbol{p}}\Big\}\,. \tag{10.13}$$

Together with Eq. (10.3) we then obtain

$$\rho^0_{\hat{z}}(\boldsymbol{r},\omega)=\frac{6\omega}{\pi c^2}\,\mathrm{Im}\Big\{G_{zz}(\boldsymbol{r},\boldsymbol{r},\omega)\Big\}=\frac{6\omega}{\pi c^2}\frac{\omega}{6\pi c}=\frac{\omega^2}{\pi^2c^3}\,,$$

which coincides with the free-space density of states $\rho^0(\omega)$ of Eq. (10.12). If we average over all angles (see Exercise 10.1 for details), we get the averaged local density of states

$$\rho(\boldsymbol{r},\omega)=\Big\langle\rho_{\hat{\boldsymbol{p}}}(\boldsymbol{r},\omega)\Big\rangle_{\hat{\boldsymbol{p}}}=\frac{2\omega}{\pi c^2}\,\mathrm{Im}\Big\{\mathrm{tr}\Big[\bar{\bar{G}}_{\mathrm{tot}}(\boldsymbol{r},\boldsymbol{r},\omega)\Big]\Big\}\,, \tag{10.14}$$

where tr denotes the trace over the Green's dyadics, this is the sum of the diagonal elements. The decay rate of a quantum emitter can then be related to the photonic density of states via

$$\Gamma_{\mathrm{tot}}=\frac{P}{\hbar\omega}=\frac{\omega}{3\hbar\varepsilon}\left(\frac{p}{2}\right)^2\rho_{\hat{\boldsymbol{p}}}(\boldsymbol{r},\omega)\,. \tag{10.15}$$

In accordance to the discussion of the Wigner–Weisskopf decay rate of Eq. (10.7) the term in parentheses corresponds to the quantum mechanical expectation value of the dipole operator. Similarly, if one averages over the orientations of the oscillating dipole one obtains the angle-averaged decay rate

$$\Big\langle\Gamma_{\mathrm{tot}}\Big\rangle_{\hat{\boldsymbol{p}}}=\frac{\omega}{\hbar\varepsilon}\left(\frac{p}{2}\right)^2\rho(\boldsymbol{r},\omega)\,. \tag{10.16}$$

Figure 10.1 shows the photonic LDOS of a dimer of two silver nanoparticles, as extracted from electron energy loss spectroscopy experiments. Here we shall not be interested in how the data have been obtained, but use the results only to discuss why the concept of a photonic LDOS is often extremely useful. In the figure the

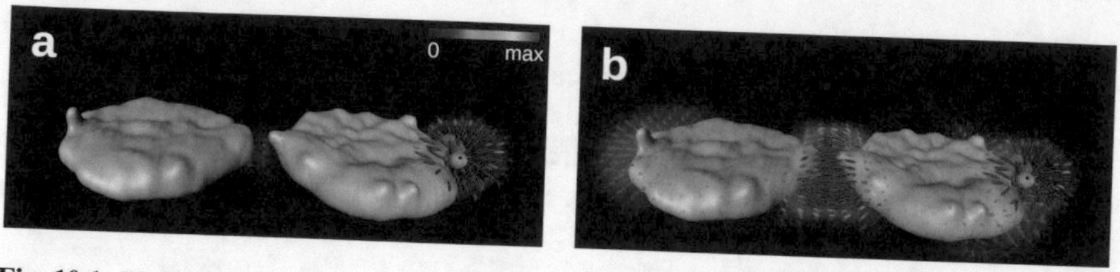

Fig. 10.1 Photonic LDOS as obtained from electron microscopy, for two coupled plasmonic nanoparticles and two different plasmon modes [85]. Panel (**a**) shows a mode that is confined at a protrusion on the right particle, panel (**b**) shows the bonding dipole mode where the dipole moments on the individual particles are oriented parallel to each other. The color of the stencils corresponds to the enhancement of the photonic LDOS and the direction to the orientation of oscillating dipoles with the largest $\rho_{\hat{\boldsymbol{p}}}(\boldsymbol{r},\omega)$ values. Figure adapted from Ref. [85]

color of the stencils corresponds to the enhancement of the photonic LDOS and the direction to the orientation of oscillating dipoles with the largest $\rho_{\hat{\boldsymbol{p}}}(\boldsymbol{r},\omega)$ values. Panel (a) reports a plasmon mode that is confined at a protrusion on the right particle. The strong nearfield enhancement of this sharp feature leads to a strong LDOS enhancement. In other words, an oscillating dipole situated close to this feature would decay much faster than in free space. The orientations of the stencils close to the nanoparticle approximately correspond to the directions of the electric field. Similarly, in panel (b) we show a bonding dipole mode where the dipole moments of the individual nanoparticles are oriented parallel to each other. With this configuration, we observe a strong LDOS enhancement in the gap region between the two particles, in close analogy to our discussion of coupled particles in Chap. 9. As a final remark, we note that in general there is also a magnetic contribution to the LDOS, as discussed in more detail in Ref. [86].

10.2 Quantum Emitter in Photonic Environment

With the tools developed in the last chapters, we can compute the photonic LDOS for a number of different setups. In the following we start with stratified media, and then investigate plasmonic nanoparticles for a few selected geometries.

10.2.1 Quantum Emitter Above Metal Slab

We first consider the situation where a dipole is located on top of a metal layer, as schematically depicted in the inset of Fig. 10.2. The expression of Eq. (10.5) for the enhancement of the power dissipated by the dipole can be immediately combined with the reflected Green's function of Eq. (8.51) for a stratified medium. Figure 10.2 shows the resulting lifetime enhancement or reduction $\tau:\tau_0$ for a dipole situated above a silver layer. Close to the interface τ drops dramatically. Here

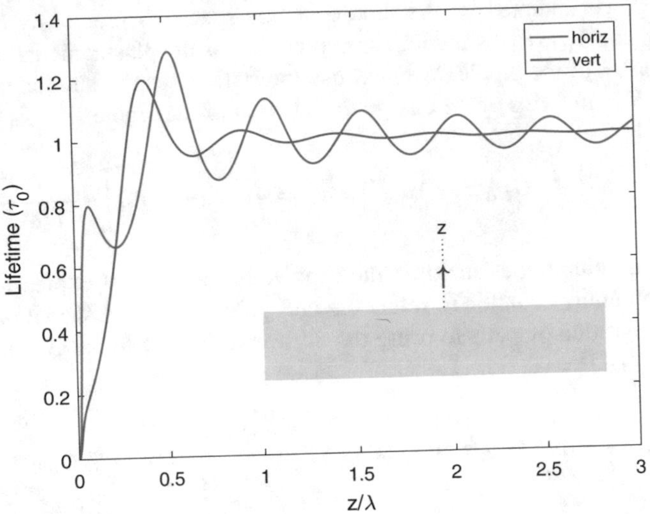

Fig. 10.2 Lifetime reduction and enhancement for dipole placed above a silver slab, with the dipole-slab distance z. When approaching the silver surface, one observes lifetime oscillations attributed to the constructive and destructive back-action of the reflected fields on the oscillating dipole. Close to the metal surface the lifetime dramatically drops because the dipole launches surface plasmons and loses energy. In the calculation we use $\lambda = 400$ nm and the silver dielectric function given in Ref. [34]

the dipole couples to the surface plasmons of the silver slab, where the transferred energy is converted via ohmic losses to heat. For larger dipole-metal separations the lifetime increases, owing to the reduced nearfield coupling to plasmons, and the dissipated power is mainly due to emitted radiation. For distances comparable to the wavelength of the emitted radiation we observe oscillations of τ, associated with constructive and destructive back-action of the reflected fields on the oscillating dipole. These effects were first demonstrated by Drexhage [87] and are discussed in detail in Ref. [88].

10.2.2 *Quasistatic Approximation*

As a second example, we consider a quantum emitter interacting with plasmonic nanoparticles, which are described within the quasistatic approximation. We first derive the expression for the dissipated power of an oscillating dipole within the quasistatic approximation. In complete analogy to our previous discussion, we start from the work done by the dipole against an electric field

$$P = P_0 - \frac{1}{2}\mathrm{Re}\Big\{\Big(i\omega\, \boldsymbol{p}^*\Big)\cdot\Big[-\nabla V_{\mathrm{refl}}(\boldsymbol{r}_0)\Big]\Big\}\,.$$

P_0 is the power radiated by the dipole in free space, as previously computed in Eq. (10.4), and $V_{\rm refl}(\boldsymbol{r}_0)$ is the reflected potential of the plasmonic nanoparticles at the position $\boldsymbol{r}_0$ of the dipole. We next use the static Green's function of Eq. (9.23) to express $V_{\rm refl}$ in terms of the charge distribution of the dipole

$$V_{\rm refl}(\boldsymbol{r}) = \varepsilon^{-1}\int_\Omega G_{\rm refl}(\boldsymbol{r},\boldsymbol{r}')\rho(\boldsymbol{r}')\,d^3r' = \int_\Omega G_{\rm refl}(\boldsymbol{r},\boldsymbol{r}')\left[\frac{\nabla'\cdot\boldsymbol{J}(\boldsymbol{r}')}{i\varepsilon\omega}\right]d^3r'\,.$$

Here Ω is a volume that encloses the dipole, and in the last expression we have used the continuity equation to relate the charge and current distributions. We next perform integration by parts to bring the derivative from $\boldsymbol{J}$ to the Green's function. Using $\nabla'\cdot(G\boldsymbol{J}) = (\nabla' G)\cdot\boldsymbol{J} + G(\nabla\cdot\boldsymbol{J})$ we get

$$V_{\rm refl}(\boldsymbol{r}) = -\frac{i}{\varepsilon\omega}\oint_{\partial\Omega} G_{\rm refl}(\boldsymbol{r},\boldsymbol{s}')\boldsymbol{J}(\boldsymbol{s}')\cdot d\boldsymbol{S}' + \frac{i}{\varepsilon\omega}\int_\Omega \boldsymbol{J}(\boldsymbol{r}')\cdot\nabla' G_{\rm refl}(\boldsymbol{r},\boldsymbol{r}')\,d^3r'\,.$$

The first term has been converted to a boundary integral using Gauss' theorem of Eq. (2.12). $\boldsymbol{J}$ becomes zero for a boundary located outside the dipole charge distribution, and the first term correspondingly vanishes. Relating $\boldsymbol{J}$ to the dipole moment $\boldsymbol{p}$ by use of Eq. (6.1), we then arrive at

$$P = P_0 - \frac{\omega}{2\varepsilon}\mathrm{Im}\Big\{(\boldsymbol{p}^*\cdot\nabla)(\boldsymbol{p}\cdot\nabla')G_{\rm refl}(\boldsymbol{r},\boldsymbol{r}')\Big\}_{\boldsymbol{r}=\boldsymbol{r}'=\boldsymbol{r}_0}\,.$$

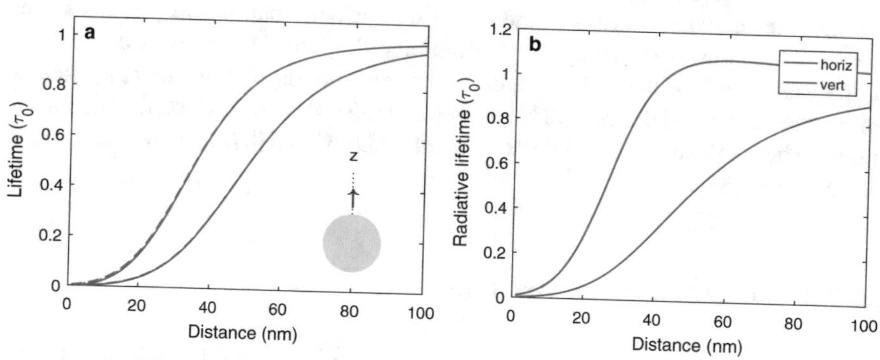

Fig. 10.3 Lifetime reduction for an oscillating dipole located close to a silver nanosphere, as computed within the quasistatic approximation at the dipole resonance wavelength of 355 nm. (**a**) Total lifetime for horizontal dipole oriented along x and vertical dipole oriented along z. The solid lines report simulation results where all eigenmodes are considered, the dashed (dashed-dotted) lines report simulation results where only dipole (dipole and quadrupole) modes are considered. (**b**) same as panel (**a**) but for radiate lifetime. In both panels the distance is measured from the nanosphere. We use the silver dielectric function of Ref. [34], a dielectric constant of one for the embedding medium, and a sphere radius of 50 nm

We continue by introducing the eigenmode decomposition of Eq. (9.33) for the reflected Green's function

$$G_{\rm refl}(\boldsymbol{r},\boldsymbol{r}') = -\sum_k V_k(\boldsymbol{r})\frac{\lambda_k+\frac{1}{2}}{\Lambda(\omega)+\lambda_k}V_k(\boldsymbol{r}')\,,$$

where $V_k(\boldsymbol{r})$ is the potential associated with the k'th eigenmode that has the eigenvalue λ_k, and $\Lambda(\omega)$ is the function containing the frequency-dependent permittivities of the nanoparticle and the embedding medium. With this decomposition we then arrive at the averaged power dissipated by a dipole emitter in the quasistatic approximation

Averaged Power Dissipated by Dipole Emitter (Quasistatic)

$$P = P_0 + \frac{\omega}{2\varepsilon}\sum_k \mathrm{Im}\left\{\frac{\lambda_k+\frac{1}{2}}{\Lambda(\omega)+\lambda_k}\right\}\left|\boldsymbol{p}\cdot\nabla V_k(\boldsymbol{r})\right|^2_{\boldsymbol{r}=\boldsymbol{r}_0}. \qquad (10.17)$$

Figure 10.3a shows the reduction of the lifetime τ for a dipole situated above a silver nanosphere, as computed from Eq. (10.17). Far away from the sphere τ approaches the free-space lifetime τ_0. With decreasing distance the dipole couples to more modes, resulting in a drop of τ. The dashed (dashed-dotted) lines report results for simulations where only the dipole (dipole and quadrupole) modes are considered. The agreement with the full simulation results (solid lines) including all modes is extremely good at large dipole-sphere separations, but becomes worse for smaller distances where the dipole starts coupling to higher multipolar modes. Panel (b) reports the radiative lifetime enhancement $\tau{:}\tau_0$, which is computed from the total dipole moment (dipole plus plasmonic nanoparticle) together with Eq. (9.20) for the scattered power of an oscillating dipole. Here the lifetime reduction (or enhancement) depends on the dipole orientation. On resonance the dipole with orientation along z couples strongest to the plasmon dipole mode oriented along z, whose dipole moment is out of phase with the driving dipole. The contributions from the two dipole moments thus add up in the emission process, and the coupled system emits light more efficiently. In contrast, the horizontal dipole oriented along x induces a mirror dipole oriented in the opposite direction with respect to the driving dipole. As a consequence, the emitted radiation becomes smaller because of the lowering of the total dipole moment, and τ even slightly exceeds the free space value τ_0.

A similar behavior is observed in Fig. 10.4 for a silver ellipsoid with an axis ratio of 1:2, where most of the results can be interpreted along the same lines as for the nanosphere. The most obvious difference is the strong increase of the radiative

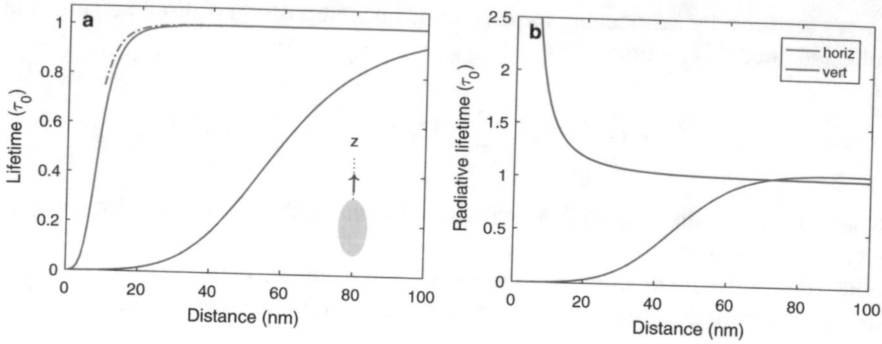

Fig. 10.4 Same as Fig. 10.3 but for an ellipsoid with an axis ratio of 1:2, a long axis of 50 nm, and the dipole resonance wavelength of 400 nm

lifetime for small distances, which is due to the different resonance frequencies of the plasmonic dipole modes oriented along z and x, y. Because of symmetry the horizontal dipole can only excite the x, y plasmon modes, which are at a higher energy in comparison to the z mode, and the dipole is thus off-resonant with respect to the plasmon modes it is coupled to. As a consequence, the induced surface charge distribution can be described effectively in terms of a mirror dipole which lowers the total dipole moment, leading to reduced scattering and a larger radiative lifetime.

10.2.3 *Mie Theory*

The radiated and total dissipated powers P_{rad} and P_{tot} can be computed analytically within Mie theory. We omit here the derivation of the results, but give a detailed account in Appendix E. Equations (E.37) and (E.38) give the power radiated by an oscillating dipole located in the vicinity of a metallic (or dielectric) sphere, and Eqs. (E.41) and (E.42) give the total dissipated power. In general, the two expressions differ because the oscillating dipole couples to plasmonic modes that dissipate power either through radiation or through ohmic losses.

Figure 10.5 shows the ratio of radiated and total dissipated power $P_{\text{rad}}:P_{\text{tot}}$, the so-called *quantum yield*, for a dipole situated above a silver nanosphere, as shown in the figure insets. We use the same parameters as for the extinction spectra previously shown in Fig. 9.18, and consider different sphere diameters and dipole-sphere separations. It can be seen that far away from the sphere practically all power goes into radiation, and only close to the sphere non-radiative losses start to dominate.

There exists a beautiful experiment by the Novotny group [84] which investigates the coupled molecule–nanosphere system in a textbook-like manner. In the experiment, a gold nanosphere is attached to a fiber, and the fiber is raster-scanned over a thin film in which dye molecules are embedded. In this way, it becomes possible to experimentally investigate the system we have previously analyzed within Mie

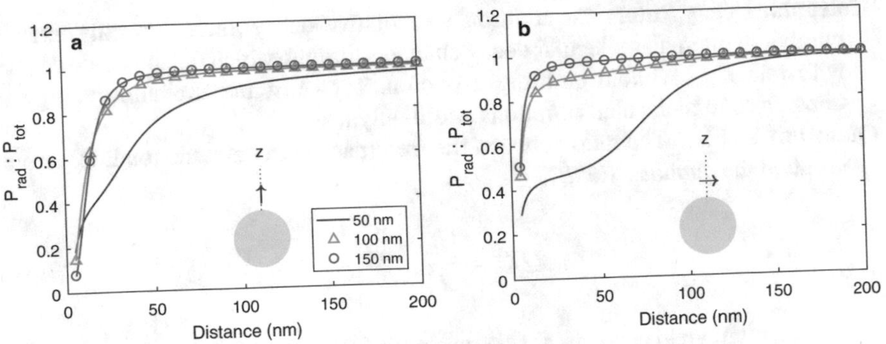

Fig. 10.5 Quantum yield for dipole located above silver sphere, as computed from Mie theory. We use the same parameters as in Fig. 9.18 and compute the radiative and total dissipated powers P_{rad} and P_{tot} at the dipole resonances of 3.4 eV (50 nm), 3.15 eV (100 nm), and 2.6 eV (150 nm). Panels (**a**) and (**b**) report results for vertical and horizontal dipole orientation, respectively

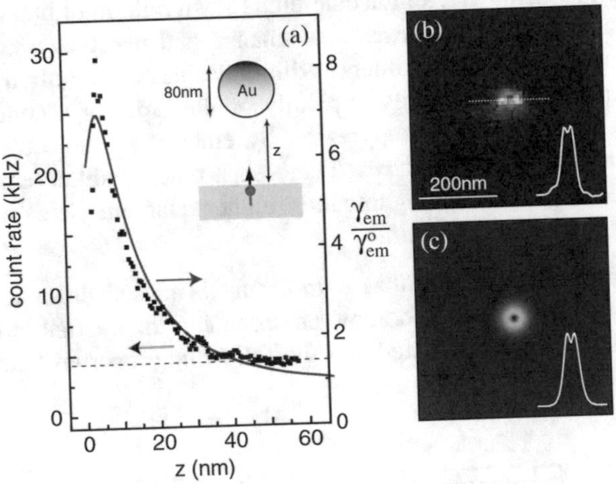

Fig. 10.6 Measured fluorescence of a coherently driven system consisting of a dipole (dye molecule) and gold nanoparticle [84]. (**a**) Upon decreasing the dipole-sphere distance the emission first increases, because of the larger molecule–nanoparticle coupling, and quenches at small distances, because of the increased non-radiative decay and ohmic losses. (**b**) Experimental and (**c**) simulated emission pattern for scan of gold nanoparticle over molecule. Figure taken from Anger et al. [84]

theory. Before pondering on the results shown in Fig. 10.6, we first discuss the situation of an oscillating dipole in free space. We shall find it convenient to adopt the notions of radiative and total *decay rates*, rather than radiated and total dissipated powers, where Eq. (10.6) provides the link between these quantities.

Radiative Decay Rate. An excited molecule in free space decays by emitting a photon. We denote the corresponding decay rate, this is the number of decays per unit time, as Γ_{rad}^{0}. The superscript zero is a reminder for the free space decay.

Molecular Decay Rate. In addition to radiative decay, there typically exist a number of internal, molecular decay channels through which the molecule decays with rate $\Gamma_{\rm mol}$ without emitting a photon. Typically, the excitation energy is transferred to molecular vibrations and finally heat.

Quantum Yield. The ratio between the free space decay and the total decay rate is called the *quantum yield*,

$$QY^0 = \frac{\Gamma^0_{\rm rad}}{\Gamma_{\rm mol} + \Gamma^0_{\rm rad}}\,. \tag{10.18}$$

The quantum yield accounts for the probability that the excited molecule decays by emitting a photon, rather than through internal decay. For typical molecules this quantum yield can be relatively small, say of the order of a few 10% or sometimes even less.

Bleaching. Typical molecules *bleach* with a certain probability. It is generally reasonable to assume that a molecule emits say a million of photons on average before it undergoes a conformational change and becomes optically inactive. For this reason, optical experiments with molecules can be often performed for a certain amount of time only, typically of the order of seconds to minutes. Bleaching can be strongly suppressed by embedding molecules in a matrix and working at low temperatures. There exist other quantum emitters, such as quantum dots, which have significantly higher quantum yields and stay stable over much longer periods of time.

Consider next the situation that a molecule is placed close to a plasmonic nanoparticle. By changing the photonic environment, the internal molecular decay channel remains usually unaffected, but the following properties become modified (see also Fig. 10.7):

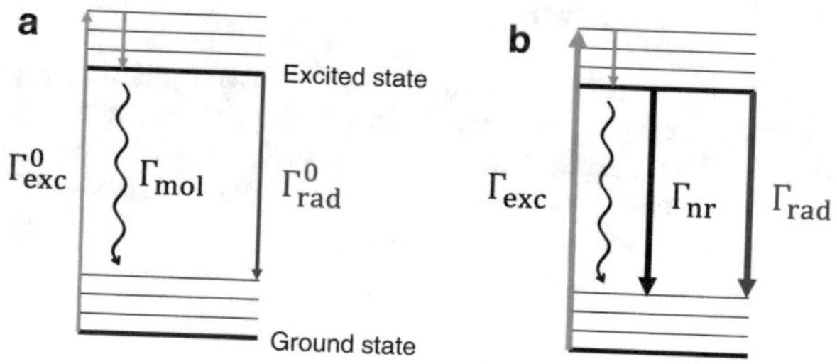

Fig. 10.7 Jablonski diagram of optically excited molecule. (**a**) Molecule in free space. $\Gamma^0_{\rm exc}$ is the excitation rate of the molecule, typically into an excited vibronic level of the excited state. After fast vibronic relaxation, it remains in the vibronic groundstate of the excited molecular state, from where it decays either through internal relaxation ($\Gamma_{\rm mol}$) or by emitting a photon ($\Gamma^0_{\rm rad}$). (**b**) For a molecule coupled to a plasmonic nanoparticle, the excitation and radiative decay rates become modified. In addition, a non-radiative decay channel is opened where the molecule decays by transferring its energy to the plasmonic nanoparticle, where it is transformed through ohmic losses to heat

Radiative Decay. The coupled molecule–nanoparticle typically has an enhanced radiative decay rate $\Gamma_{\rm rad}$, see Figs. 10.3 and 10.4, although in some cases one also observes a decrease of $\Gamma_{\rm rad}$.

Non-radiative Decay. In addition to the radiative decay, the plasmonic nanoparticle can also decay through ohmic losses. The resulting non-radiative decay is described through the rate $\Gamma_{nr} = \Gamma_{\rm tot} - \Gamma_{\rm rad}$, where $\Gamma_{\rm tot}$ is the total decay rate to be computed from the total dissipated power of Eq. (10.2).

Quantum Yield. With these rate modifications, the quantum yield of the coupled dipole-nanoparticle system becomes

$$QY = \frac{\Gamma_{\rm rad}}{\left(\Gamma_{\rm mol} + \Gamma_{\rm nr}\right) + \Gamma_{\rm rad}} . \tag{10.19}$$

The modification with respect to the free space value QY^0 is not obvious and depends on the details of the rate changes. In absence of internal relaxation the quantum yield always drops because through the coupling to the plasmonic nanoparticle a non-radiative decay channel ($\Gamma_{\rm nr}$) is opened. For molecules with a small QY^0 value, the quantum yield usually increases in presence of a plasmonic nanoparticle, at least for molecules sufficiently far away from the particle where radiative decay dominates over the non-radiative one.

Fluorescence Rate. In addition, also the excitation rate $\Gamma_{\rm exc}$ changes in presence of a plasmonic nanoparticle, see Fig. 10.7. The fluorescence rate, or equivalently the number of photons emitted per unit time, then changes according to $\Gamma_{\rm exc} \times QY$ or expressed in words

$$(\text{Fluorescence rate}) = (\text{Excitation rate}) \times (\text{Quantum yield}) .$$

The net change of fluorescence intensity depends on the details of the various rates. In Fig. 10.6 one observes that with decreasing molecule–nanosphere separation the fluorescence intensity first increases, because radiative decays are more frequent and outplay internal molecular decay, and only for the smallest distances the fluorescence quenches because of the increase of Γ_{nr}. Here the molecule predominantly excites multipolar plasmon modes, which decay via ohmic losses as previously shown in Fig. 10.5.

10.3 Surface-Enhanced Raman Scattering

Raman scattering and surface-enhanced Raman scattering (SERS) are genuine quantum processes. Yet, one can learn already a lot about these processes using a classical approach, as we will do in this section. Quantum descriptions are typically more difficult, see, for instance, [89], but lead to similar results with the exception of thermal population factors missing in a classical approach. Consider an oscillator that is driven by an external electric field $\boldsymbol{E}_L$ via

$$p(t) = \left[\bar{\bar{\alpha}}_0 + \bar{\bar{\alpha}}_1 \cos(\Omega_{\text{vib}} t)\right] \cdot \left(E_L \cos(\omega_L t)\right), \tag{10.20}$$

where $p(t)$ is the time-dependent dipole moment of the oscillator. We have decomposed the polarizability into a time-independent term $\bar{\bar{\alpha}}_0$ and a term $\bar{\bar{\alpha}}_1 \cos \Omega_{\text{vib}} t$ oscillating with some vibronic frequency $\Omega_{\text{vib}} \ll \omega_L$ that is much smaller than the frequency ω_L of the external fields. This model corresponds, for instance, to a molecule whose polarizability is modulated by a molecular vibration. If we now consider only the term proportional to $\bar{\bar{\alpha}}_1$, we arrive at

$$\begin{aligned} p_1(t) &= \bar{\bar{\alpha}}_1 \cdot E_L \cos(\omega_L t) \cos(\Omega_{\text{vib}} t) \\ &= \bar{\bar{\alpha}}_1 \cdot E_L \frac{1}{2} \Big\{ \cos[(\omega_L + \Omega_{\text{vib}})t] + \cos[(\omega_L - \Omega_{\text{vib}})t] \Big\}. \end{aligned}$$

Thus, the oscillator not only oscillates at the driving frequency ω_L (described by the term proportional to $\bar{\bar{\alpha}}_0$ that we have not considered here) but also at the modulated frequencies, the so-called Raman frequencies

Raman Frequencies

$$\omega_R^{\pm} = \omega_L \pm \Omega_{\text{vib}}. \tag{10.21}$$

In the last chapter we have derived the scattering cross section for a polarizable particle, Eq. (9.21). We can thus directly use this result to express the (classical) Raman cross section in the form

$$C_R = \frac{k^4}{6\pi \varepsilon^2} \left| \frac{1}{2} \bar{\bar{\alpha}}_1 \cdot \boldsymbol{\epsilon}_L \right|^2, \tag{10.22}$$

where k is the wavenumber of the emitted light (we assume $\omega_R^{\pm} \approx \omega_L$, such that k is the wavenumber of the driving field), ε the permittivity of the embedding medium, and $\boldsymbol{\epsilon}_L$ the polarization of the driving field. In fact, a full quantum mechanical description gives a very similar result; however, the scattered light intensities for the Stokes frequency ω_R^- and anti-Stokes frequency ω_R^+ differ in general.

The Raman effect was predicted theoretically by Adolf Smekal in 1923 and was observed experimentally 5 years later by Indian physicist Chandrasekhara Venkata Raman. Since then it has found widespread use because it allows for the detection of vibrational frequencies, which serve as fingerprints for specific molecules, however, not in the infrared but optical frequency regime. In 1974 the group of Fleischmann reported the observation of strong Raman signals from a single layer of molecules that was deposited on electrochemically roughened metal electrodes [91]. This paper is considered as the first observation of SERS. Another important step forward is the work of Nie and Emory [90] who observed SERS

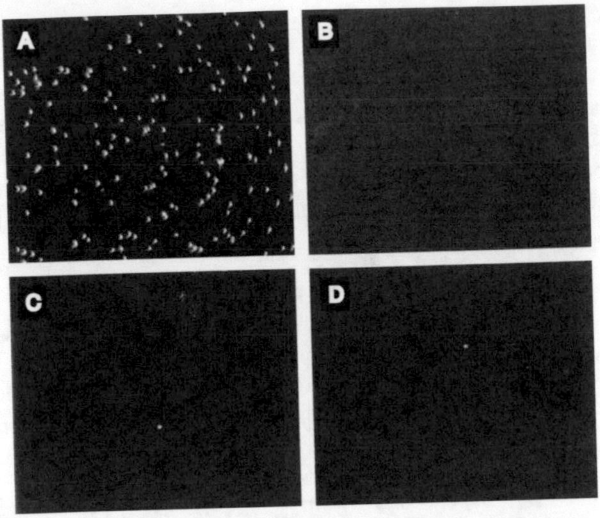

Fig. 10.8 Single *Ag* nanoparticles imaged with evanescent-wave excitation. Total internal reflection of the laser beam at the glass–liquid interface was used to reduce the laser scattering background. (**a**) Unfiltered photograph showing scattered laser light from all particles immobilized on a polylysine-coated surface. (**b**) Filtered photographs taken from a blank Ag colloid sample (incubated with 1 mM NaCl and no R6G analyte molecules). (**c**) and (**d**) Filtered photographs taken from a Ag colloid sample incubated with a few molecules, showing at least one Raman scattering particle. Figure taken from Ref. [90]

signals from *single* molecules, as shown in Fig. 10.8. In the abstract of their paper the authors write:

> Optical detection and spectroscopy of single molecules and single nanoparticles have been achieved at room temperature with the use of surface-enhanced Raman scattering. Individual silver colloidal nanoparticles were screened from a large heterogeneous population for special size-dependent properties and were then used to amplify the spectroscopic signatures of adsorbed molecules. For single rhodamine 6G molecules adsorbed on the selected nanoparticles, the intrinsic Raman enhancement factors were on the order of 10^{14} to 10^{15}, much larger than the ensemble-averaged values derived from conventional measurements. This enormous enhancement leads to vibrational Raman signals that are more intense and more stable than single-molecule fluorescence.

Although the SERS enhancement by up to 15 orders of magnitude remains somewhat dubious, numerous follow-up work has confirmed very large enhancement factors. In what follows, we analyze Raman scattering in presence of a modified photonic environment. As we will show, the modification influences both (1) the excitation process of the oscillator, as well as (2) the emission properties of the oscillator. The combination of the two effects gives rise to huge enhancements of the SERS signals in accordance to experimental observations.

Emission Enhancement. We start with the modification of the emission process. From Eq. (10.1) we observe that the electric field generated by an oscillating dipole in presence of a modified photonic environment can be described as

$$E(\mathbf{r}) = \mu\omega^2 \left[\bar{\bar{G}}(\mathbf{r}, \mathbf{r}_{\rm dip}) + \bar{\bar{G}}_{\rm refl}(\mathbf{r}, \mathbf{r}_{\rm dip})\right] \cdot \mathbf{p}\,,$$

where $\bar{\bar{G}}$ accounts for the emission of the oscillator in free space and $\bar{\bar{G}}_{\rm refl}$ for the modifications associated with, say, a nearby plasmonic nanoparticle. $\mathbf{r}_{\rm dip}$ is the location of the driven dipole. Far away from the dipole and the plasmonic nanoparticle, the electromagnetic fields are outgoing waves of the form

$$E(\mathbf{r}) \underset{r\gg r_{\rm dip}}{\longrightarrow} \mu\omega^2 \left(\frac{e^{ikr}}{r}\right)\left[\bar{\bar{f}}(\hat{\mathbf{r}}, \mathbf{r}_{\rm dip}) + \bar{\bar{f}}_{\rm refl}(\hat{\mathbf{r}}, \mathbf{r}_{\rm dip})\right] \cdot \mathbf{p}\,.$$

Here $\bar{\bar{f}}$ and $\bar{\bar{f}}_{\rm refl}$ are the asymptotic forms of the free space and reflected Green's function, respectively. The free space part can be read off from Eq. (5.25),

$$f_{ij}(\hat{\mathbf{r}}, \mathbf{r}_{\rm dip}) = \frac{e^{-ik\hat{\mathbf{r}}\cdot\mathbf{r}_{\rm dip}}}{4\pi}\Big(\delta_{ij} - \hat{r}_i\hat{r}_j\Big)\,, \tag{10.23}$$

and is of the order of unity. In contrast, the enhancement of $\bar{\bar{f}}_{\rm refl}$, which can be obtained from the reflected Green's function in the far-field limit, can become large in presence of plasmonic nanoparticles, with enhancement factors up to orders of hundreds. The far-field amplitude of the light emitted from the oscillating dipole at the Raman frequencies can thus be written in the form

$$F(\hat{\mathbf{r}}) = \mu\omega^2 \left[\bar{\bar{f}}(\hat{\mathbf{r}}, \mathbf{r}_{\rm dip};, \omega_R^\pm) + \bar{\bar{f}}_{\rm refl}(\hat{\mathbf{r}}, \mathbf{r}_{\rm dip};, \omega_R^\pm)\right] \cdot \mathbf{p}_1\,, \tag{10.24}$$

where for clarity we have indicated the Raman frequencies $\omega_R^\pm$ at which the light is emitted.

Excitation Enhancement. In addition to the enhancement of the emission process, also the driving field of the oscillator becomes modified. Instead of Eq. (10.20) we get

$$\mathbf{p}(t) = \left[\bar{\bar{\alpha}}_0 + \bar{\bar{\alpha}}_1\cos(\Omega_{\rm vib}t)\right] \cdot \Big(\mathbf{E}_L(\mathbf{r}_{\rm dip}, t) + \mathbf{E}_{\rm refl}(\mathbf{r}_{\rm dip}, t)\Big)\,,$$

where $\mathbf{E}_{\rm refl}(\mathbf{r}_{\rm dip}, t)$ is the reflected part of the driving field at the dipole position. We can next use the far-field amplitude $\bar{\bar{f}}_{\rm refl}$ of the reflected Green's function to express this enhancement of the excitation process. As a preliminary step, we assume that the incoming field $\mathbf{E}_L$ is produced by another oscillating dipole with moment $\mathbf{p}_L$ which is located at position $\mathbf{r}$ far away from the dipole and the plasmonic nanoparticle. The electric field at $\mathbf{r}_{\rm dip}$ can then be expressed in the usual form

$$\mathbf{E}(\mathbf{r}_{\rm dip}) = \mu\omega^2 \left[\bar{\bar{G}}(\mathbf{r}_{\rm dip}, \mathbf{r}) + \bar{\bar{G}}_{\rm refl}(\mathbf{r}_{\rm dip}, \mathbf{r})\right] \cdot \mathbf{p}_L\,.$$

From the symmetry relation of Eq. (7.38) for the Green's dyadic, which we have derived using the reciprocity theorem of optics, we can rewrite the Green's functions as

$$\bar{\bar{G}}(\boldsymbol{r}_{\rm dip}, \boldsymbol{r}) = \bar{\bar{G}}^T(\boldsymbol{r}, \boldsymbol{r}_{\rm dip}) \underset{r \gg r_{\rm dip}}{\longrightarrow} \mu\omega^2 \left(\frac{e^{ikr}}{r}\right) \bar{\bar{f}}^T(\hat{\boldsymbol{r}}, \boldsymbol{r}_{\rm dip}),$$

where the superscript T denotes the transpose of the matrix. A corresponding expression can be obtained for $\bar{\bar{G}}_{\rm refl}$. We next consider an incoming plane wave with amplitude $\boldsymbol{E}_L$ that propagates along the direction $\hat{\boldsymbol{k}}_L$. Using the asymptotic form of Eq. (10.23) for the free space Green's function, we can easily show that the incoming field can be expressed as

$$\boldsymbol{E}_L\, e^{i\boldsymbol{k}_L\cdot \boldsymbol{r}_{\rm dip}} = \bar{\bar{f}}^T(-\hat{\boldsymbol{k}}_L, \boldsymbol{r}_{\rm dip}) \cdot 4\pi \boldsymbol{E}_L\,.$$

Thus, the total driving field at the position of the dipole can be written in the form

$$\boldsymbol{E}_{\rm tot}(\boldsymbol{r}_{\rm dip}) = \left[\bar{\bar{f}}(-\hat{\boldsymbol{k}}_L, \boldsymbol{r}_{\rm dip}; \omega_L) + \bar{\bar{f}}_{\rm refl}(-\hat{\boldsymbol{k}}_L, \boldsymbol{r}_{\rm dip}; \omega_L)\right]^T \cdot 4\pi \boldsymbol{E}_L\,. \tag{10.25}$$

Putting together Eqs. (10.24) and (10.25) for the enhancements of the scattering and excitation processes, we get for the far-field amplitude of the scattered Raman process

$$\boldsymbol{F}(\hat{\boldsymbol{r}}) = \mu\omega^2 \left[\bar{\bar{f}}(\hat{\boldsymbol{r}}, \boldsymbol{r}_{\rm dip}; , \omega_R^{\pm}) + \bar{\bar{f}}_{\rm refl}(\hat{\boldsymbol{r}}, \boldsymbol{r}_{\rm dip}; , \omega_R^{\pm})\right] \cdot \bar{\bar{\alpha}}_1$$
$$\cdot \left[\bar{\bar{f}}(-\hat{\boldsymbol{k}}_L, \boldsymbol{r}_{\rm dip}; \omega_L) + \bar{\bar{f}}_{\rm refl}(-\hat{\boldsymbol{k}}_L, \boldsymbol{r}_{\rm dip}; \omega_L)\right]^T \cdot 4\pi \boldsymbol{E}_L\,.$$

We finally use Eq. (4.17), which relates $\boldsymbol{F}(\hat{\boldsymbol{r}})$ to the Poynting vector in the far-field zone, divide by the intensity $I_{\rm inc} = \frac{1}{2}Z^{-1}\,|\boldsymbol{E}_L|^2$ of the driving field, and integrate over all angles to get the cross section for surface-enhanced Raman scattering

Cross Section for Surface-Enhanced Raman Scattering (SERS)

$$C_{\rm SERS} = \oint \left| 2\pi\mu\omega^2 \left[\bar{\bar{f}}(\hat{\boldsymbol{r}}, \boldsymbol{r}_{\rm dip}) + \bar{\bar{f}}_{\rm refl}(\hat{\boldsymbol{r}}, \boldsymbol{r}_{\rm dip})\right] \cdot \bar{\bar{\alpha}}_1 \right.$$
$$\left. \cdot \left[\bar{\bar{f}}(-\hat{\boldsymbol{k}}_L, \boldsymbol{r}_{\rm dip}) + \bar{\bar{f}}_{\rm refl}(-\hat{\boldsymbol{k}}_L, \boldsymbol{r}_{\rm dip})\right]^T \cdot \boldsymbol{\epsilon}_L \right|^2 d\Omega\,. \tag{10.26}$$

We have suppressed the frequency dependence of the far-field amplitudes, which for $\Omega_{\rm vib} \ll \omega_L$ can all be evaluated at the driving frequency ω_L. In the first line the term in brackets accounts for the enhancement of the scattering process, whereas in the second line the term in brackets accounts for the enhancement of the excitation process. Altogether the Raman cross section scales with the fourth power of the enhancement factor

$$C_{\rm SERS} \propto \left|f_{\rm refl}\right|^4 . \tag{10.27}$$

Thus, for field enhancement factors $|f_{\rm refl}|$ of the order of hundreds the enhancement of SERS can be of the order of billions. One may wonder whether similar enhancement factors can be also obtained for a resonantly driven oscillator which emits at the driving frequency ω_L. Repeating the same analysis as just presented, we indeed recover a similar enhancement. However, in this case we additionally have to consider the light directly emitted by the plasmonic nanoparticle, which scatters at the same frequency and has a much higher intensity, thus completely masking the resonant dipole emission. Thus, the magic of SERS is that the driven oscillator acts as a frequency modulator, and the scattered Raman light is spectroscopically detuned from the driving fields.

SERS combines plasmonics with chemical sensitivity that allows to identify molecules through their specific Raman spectra. This has opened a rich and interdisciplinary research field with many exciting developments, such as tip-enhanced Raman scattering (TERS). We will here not enter into details but refer the interested reader to the pertinent literature, such as, for instance, [89, 92].

10.4 Förster Resonance Energy Transfer

Förster resonance energy transfer (FRET) is a process where a donor molecule located at position $\boldsymbol{r}_D$ transfers energy to a nearby acceptor molecule located at position $\boldsymbol{r}_A$ via a non-radiative dipole coupling. Such processes play an important role in photosyntheses, and were first studied by Förster [93]. We here discuss FRET in a classical framework. The work performed by the donor molecule on the acceptor can be expressed through (see also discussion at the beginning of this chapter)

$$P_{D\to A} = \frac{1}{2}\int \mathrm{Re}\Big\{\boldsymbol{J}_A^*(\boldsymbol{r})\cdot\boldsymbol{E}_D(\boldsymbol{r})\Big\}\,d^3r = -\frac{\omega}{2}\mathrm{Im}\Big\{\boldsymbol{p}_A^*\cdot\boldsymbol{E}_D(\boldsymbol{r}_A)\Big\}.$$

Here $\boldsymbol{p}_A$ is the dipole moment of the acceptor molecule. Quite generally, once energy is transferred from the donor to the acceptor the reversed process would start because of the reciprocity theorem of optics. However, in FRET complexes the excitation in the acceptor molecule usually suffers fast intra-molecular relaxation, hereby effectively removing it from the inter-molecular FRET channel. We here do not consider such relaxation explicitly but simply discard any back-action from the acceptor on the donor instead. Suppose that the dipole moment of the acceptor $\boldsymbol{p}_A$ is induced by the acceptor by means of a polarizability $\bar{\bar{\alpha}}_A$ via

$$\boldsymbol{p}_A = \bar{\bar{\alpha}}_A \cdot \boldsymbol{E}_D(\boldsymbol{r}_A) = \left[\sum_a \boldsymbol{n}_a \alpha_a \boldsymbol{n}_a^T\right] \cdot \boldsymbol{E}_D(\boldsymbol{r}_A) .$$

In the last expression we have performed a principal axis transformation for the polarizability, and assume that the eigenvectors $\boldsymbol{n}_a$ can be chosen real. Thus, the power transferred from the donor to the acceptor can be written in the form

$$P_{D\to A} = \frac{\omega}{2} \sum_a \alpha_a'' \left|\boldsymbol{n}_a \cdot \boldsymbol{E}_D(\boldsymbol{r}_A)\right|^2 .$$

We finally use the dyadic Green's function to relate the electric field of the donor to the source of the field, namely the donor dipole. With this, we get for the power performed by the donor on the acceptor the expression

Power Performed by Donor on Acceptor

$$P_{D\to A} = \frac{\mu^2 p_D^2 \omega^5}{2} \sum_a \alpha_a'' \left|\boldsymbol{n}_a \cdot \bar{\bar{G}}(\boldsymbol{r}_A, \boldsymbol{r}_D) \cdot \boldsymbol{n}_D\right|^2 . \tag{10.28}$$

It is convenient to further simplify this expression. First, we assume that the orientation of the acceptor molecule is random and average over the orientation of the molecule

$$\frac{1}{4\pi} \oint \left|\hat{\boldsymbol{r}} \cdot E\hat{\boldsymbol{z}}\right|^2 d\Omega = \frac{1}{2}\left|E\right|^2 \int_0^\pi \cos\theta^2 \sin\theta d\theta = \frac{1}{3}\left|E\right|^2 .$$

Here we have assumed without loss of generality that the electric field of the donor points in the z-direction. By a similar token we can average over the dipole moment of the donor to finally get

$$P_{D\to A} = \frac{\mu^2 p_D^2 \bar{\alpha}_A'' \omega^5}{6} \mathrm{tr}\left\{\left|\bar{\bar{G}}(\boldsymbol{r}_A, \boldsymbol{r}_D)\right|^2\right\} , \tag{10.29}$$

where $\bar{\alpha}'' = \frac{1}{3}\mathrm{tr}(\bar{\bar{\alpha}}'')$ is the imaginary part of the averaged polarizability tensor. We next relate this expression to the emission properties of the donor, more specifically its scattered power, and the absorption properties of the acceptor, more specifically its extinction power. From Eq. (10.4) we find for the power emitted by the donor

$$P_D = \frac{\mu\omega^4 p_D^2}{12\pi c}\,.$$

From Eq. (9.22) we obtain for the absorbed power of the randomly oriented acceptor

$$C_{\rm abs} = \mu\omega c\,\bar\alpha_A''\,.$$

We have used that the extinction cross section equals the absorption cross section in the quasistatic limit, at least when not introducing the effective polarizability of Eq. (10.9). Putting together the results we arrive at

$$\frac{P_{D\to A}}{P_D} = 2\pi C_{\rm abs}(\omega)\,\mathrm{tr}\left\{\left|\bar{\bar G}(\boldsymbol{r}_A,\boldsymbol{r}_D)\right|^2\right\}. \tag{10.30}$$

The expression can be further rewritten if we assume that the donor not only emits at a single frequency but has a non-trivial lineshape $f_D(\omega)$ possibly due to some vibronic sidebands or a more complicated molecular level structure. Similarly, the acceptor absorption can be also broadband, thus forcing us to introduce an integral over frequencies. The FRET rate then depends also on the overlap between the donor emission and acceptor absorption. For a more detailed discussion see, for instance, [6].

For dipoles located in free space and close to each other we can use the free space Green's function worked out in Exercise 5.7, where the leading term for small separations $\boldsymbol{R} = \boldsymbol{r}_A - \boldsymbol{r}_D$ becomes

$$G_{ij}(\boldsymbol{r}_D,\boldsymbol{r}_A) \approx \frac{3\hat R_i\hat R_j - \delta_{ij}}{4\pi k^2 R^3}\,.$$

With this we obtain the result

$$\frac{P_{D\to A}}{P_D} \approx \left[\frac{R_0}{R}\right]^6, \tag{10.31}$$

where R is the donor–acceptor distance and R_0 is the Förster radius, which is typically of the order of a few nanometers. Because of this strong distance dependence, the FRET probability is typically very high for separations smaller than R_0 and then drops drastically.

When placing the donor–acceptor complex in a different environment, it is difficult to compete with the extremely efficient FRET transfer at distances smaller than the Förster radius R_0. However, in presence of plasmonic nanoparticles the transfer can occur over large distances when donor and acceptor both couple to the plasmonic nearfields [95]. Figure 10.9 shows an experiment of the Barnes group demonstrating energy transfer over a 120 nm thick silver film [94]. When only donor molecules are placed on top of the silver film the fluorescence exhibits an almost monoexponential decay. Conversely, the sample with the acceptor molecules

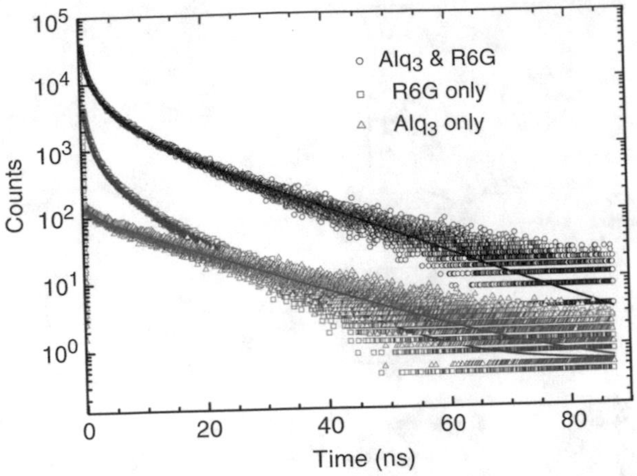

Fig. 10.9 Förster resonance energy transfer across a 120 nm thick silver film [94]. The donor dye molecules Alq3 are placed on one side of the film, the acceptor dye molecules rhodamine 6G (R6G) on the opposite side. The sample consisting of donor molecules placed on top of the silver film shows an almost monoexponential fluorescence decay. The sample consisting of acceptor molecules on top of the silver film shows a different decay characteristic with a multi-exponential decay curve. When both molecules are placed on opposite sides of the film, the fluorescence decay is reminiscent of the acceptor decay only, thus demonstrating an efficient excitation transfer from the donor to the acceptor. Figure taken from [94]

shows a different decay characteristic with a multi-exponential decay curve. When both donor and acceptor molecules are placed on opposite sides of the film, the fluorescence decay is reminiscent of the acceptor decay only, thus demonstrating an efficient excitation transfer from the donor to the acceptor.

10.5 Electron Energy Loss Spectroscopy

Electron energy loss spectroscopy (EELS) is a technique based on electron microscopy that allows measuring the fields of plasmonic nanoparticles with nanometer spatial and 5–600 meV energy resolution. The basic principle is shown in Fig. 10.10. A swift electron with a kinetic energy in the 50–300 keV range, corresponding to electron velocities between 0.4 and 0.7 times the speed of light, passes by or penetrates through a metallic nanoparticle. With a certain probability it excites a particle plasmon and thereby loses a tiny fraction of its kinetic energy. By spectrally analyzing the energy loss one can map out the plasmonic spectrum, where the usual optical selection rules do not apply for the electron probe and thus the entire spectrum can be accessed. Additionally, by raster-scanning the beam over the specimen one obtains detailed spatial information about the plasmonic fields.

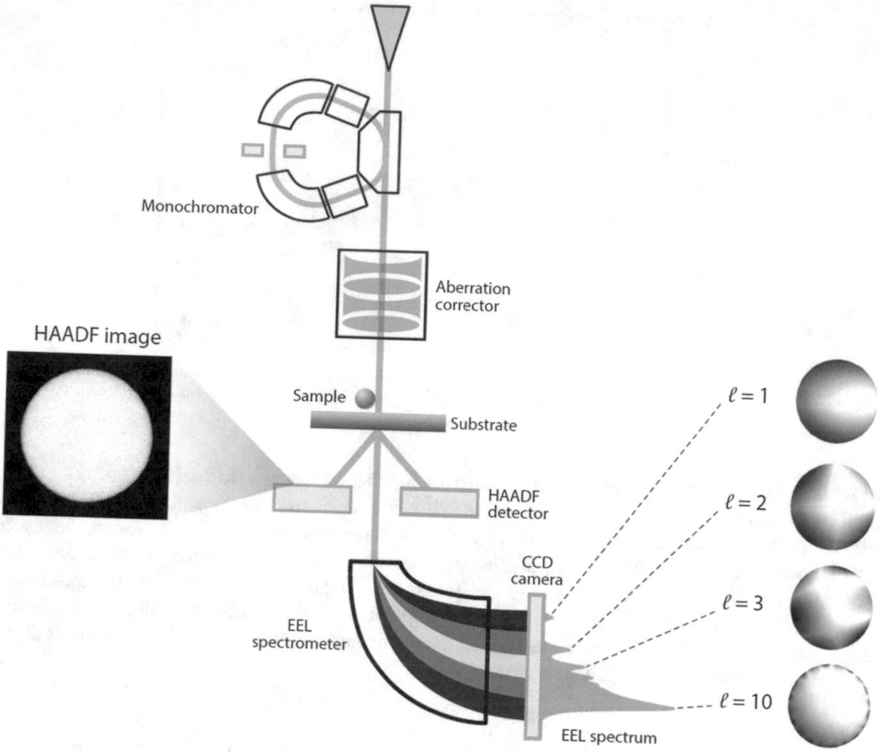

Fig. 10.10 Schematics of electron energy loss spectroscopy (EELS), a technique based on electron microscopy [96]. A swift electron with a velocity of about half the speed of light passes by or penetrates through a metallic nanoparticle, and excites with a certain probability a particle plasmon. By spectrally analyzing the energy loss of the electron and raster-scanning the electron beam over the specimen, one obtains information about the spectral and spatial properties of the plasmon modes. EELS combines the sub-nanometer spatial resolution and 10–100 meV spectral resolution of electron microscopy. Figure courtesy of David J. Masiello

EELS has a long tradition in the field of plasmonics. In fact, the first observation of surface plasmons was made using EELS [97]. Although the field has remained more or less active since then, the next breakthrough has been achieved in 2007 when two groups showed independently that with further improvements it is possible to map the fields of plasmonic nanoparticles with nanometer resolution [98, 99]. These papers mark the beginning of new era where EELS has become a highly accurate and versatile measurement technique for plasmonic nearfields.[1] A detailed account for recent developments can be found in several review articles [100, 101].

[1]Initially I thought that the breakthrough was only possible with the new generation of electron microscopes, but as Mathieu Kociak, one of the authors of these papers, explains this is not correct: "Actually, both [electron microscopes] were already extremely old (VG machines, 35 years old at that time!). The key in both work was data processing (PCA for Michel, deconvolution for

The proper theoretical description of EELS requires a full quantum mechanical framework, which we will present in Chap. 14. However, by combining classical electromagnetic theory with one aspect of quantum physics, namely its particle nature which dictates that the electron can lose energy only in fractions of $\hbar\omega$, allows us to derive a semiclassical formula that captures the essentials of a viable EELS theory. In the following we assume that the swift electron propagates along direction $\hat{z}$ with velocity v. Its trajectory can be parameterized through

$$\boldsymbol{r}_e(t) = \boldsymbol{R}_0 + \hat{z}\, vt\,, \tag{10.32}$$

with the impact parameter $\boldsymbol{R}_0 = (x_0, y_0)$. Here and in the following we assume that the tiny electron energy loss does not change the electron trajectory significantly. Below we will compute the electromagnetic fields produced by the current distribution $\boldsymbol{J}(\boldsymbol{r}, t)$ of the swift electron, but for the time being we assume that the electromagnetic response $\boldsymbol{E}(\boldsymbol{r}, t)$, $\boldsymbol{H}(\boldsymbol{r}, t)$ of the plasmonic nanoparticle is at hand (for instance, by employing the boundary integral method approach of Sect. 9.6). The work performed by the electron against the electric field is then given by

$$dW = -q\Big(\boldsymbol{E}[\boldsymbol{r}_e(t), t] + \boldsymbol{v}\times\boldsymbol{B}[\boldsymbol{r}_e(t), t]\Big)\cdot\boldsymbol{v}\,dt = -q\boldsymbol{E}[\boldsymbol{r}_e(t), t]\cdot\boldsymbol{v}\,dt\,,$$

where $q = -e$ is the charge and $\boldsymbol{v} = v\hat{z}$ the velocity of the electron. We have exploited the fact that magnetic fields cannot directly perform work because of $(\boldsymbol{v}\times\boldsymbol{B})\cdot\boldsymbol{v} = 0$. The total work performed by the electron against the electromagnetic fields corresponds to the energy loss ΔE of the swift electron, and is obtained by integrating over the entire electron trajectory,

$$\Delta E = e\int_{-\infty}^{\infty}\boldsymbol{v}\cdot\boldsymbol{E}[\boldsymbol{r}_e(t), t]\,dt\,.$$

The energy loss can be spectrally decomposed into different frequency components using the Fourier transform of the electric field. This gives

$$\Delta E = \frac{e}{\pi}\int_{-\infty}^{\infty}\mathrm{Re}\left\{\int_0^{\infty} e^{-i\omega t}\,\boldsymbol{v}\cdot\boldsymbol{E}[\boldsymbol{r}_e(t), \omega]\,d\omega\right\}dt\,, \tag{10.33}$$

where we have used $\boldsymbol{E}^*(\boldsymbol{r}, \omega) = \boldsymbol{E}(\boldsymbol{r}, -\omega)$ that holds for any real-valued function $\boldsymbol{E}(\boldsymbol{r}, t)$. At this point we have to deviate from a purely classical approach and introduce some quantum aspects.

Classical approach. In a purely classical approach the energy loss ΔE is a mere number which is given by summing up the individual energy losses to the different plasmon modes. However, ΔE contains no spectroscopic information.

---

us). Then only the works were reproduced with new generations of microscopes, which, indeed, unleashed the EELS-plasmonic beast."

Quantum approach. In a quantum approach, an electron can only lose energy in fractions of $\hbar\omega$. Thus, by spectrally analyzing the energy loss of a single electron we obtain the *probabilities* that the electron has lost a certain amount of energy, or has lost no energy.

Semiclassical approach. Similarly to our previous discussion about the relation between the dissipated power of an oscillating dipole and the corresponding decay rate, Eq. (10.6), we can employ a semiclassical description and relate the energy loss ΔE to a loss probability $\mathcal{P}_{\rm EELS}(\boldsymbol{R}_0,\omega)$ via

$$\Delta E = \int_0^\infty \hbar\omega\,\mathcal{P}_{\rm EELS}(\boldsymbol{R}_0,\omega)\,d\omega\,. \tag{10.34}$$

The reasoning behind this expression is the aforementioned particle nature of quantum mechanics that dictates that an electron can lose energy only in fractions of $\hbar\omega$.

By combining Eqs. (10.33) and (10.34) we then get for the electron energy loss probability

Electron Energy Loss Probability

$$\mathcal{P}_{\rm EELS}(\boldsymbol{R}_0,\omega) = \frac{e}{\pi\hbar\omega}\int_{-\infty}^{\infty} \mathrm{Re}\Big\{e^{-i\omega t}\boldsymbol{v}\cdot\boldsymbol{E}[\boldsymbol{r}_e(t),\omega]\Big\}\,dt\,. \tag{10.35}$$

This expression forms the starting point for deriving the working equation for the electron energy loss probability. In the following we first compute the electromagnetic fields of a swift electron, and then use the reflected Green's function to relate the electric field to the current source of the swift electron.

10.5.1 Fields Produced by Swift Electrons

The charge distribution of a swift electron propagating along the trajectory of Eq. (10.32) is given by

$$\rho(\boldsymbol{r},t) = -e\delta(\boldsymbol{R}-\boldsymbol{R}_0)\delta(z-vt)\,, \tag{10.36}$$

where $\boldsymbol{R}=(x,y)$ are the in-plane coordinates of the electron. The corresponding current distribution is $\boldsymbol{J}(\boldsymbol{r},t)=\boldsymbol{v}\rho(r,t)$. To compute the electric fields associated with this distribution in an unbounded medium (we will account later for the presence of plasmonic nanoparticles), we transform Maxwell's equation to frequency and wavenumber space. We first compute the Fourier transform in time

$$\rho(\boldsymbol{r},\omega) = \int e^{i\omega t}\rho(\boldsymbol{r},\omega)\,dt = -\frac{e}{v}\delta(\boldsymbol{R}-\boldsymbol{R}_0)e^{iqz}\,, \tag{10.37}$$

where we have introduced the wavenumber $q = \omega/v$. We continue to additionally compute the spatial Fourier transform

$$\rho(\boldsymbol{k},\omega) = -\frac{e}{v}\int_{-\infty}^{\infty} e^{-i\boldsymbol{k}\cdot\boldsymbol{r}}\delta(\boldsymbol{R}-\boldsymbol{R}_0)e^{iqz}\,d^3r = -\frac{2\pi e}{v}e^{-i\boldsymbol{k}\cdot\boldsymbol{R}_0}\delta\big(k_z-q\big)\,.$$

To get the electric field, we combine this result with the dyadic Green's function in reciprocal space (see term in parentheses below), as derived in Sect. B.2, to get

$$E_i(\boldsymbol{k},\omega) = i\omega\mu\left(-\frac{1}{k_1^2}\frac{k_ik_j-k_1^2\delta_{ij}}{k^2-k_1^2}\right)\Big[-\frac{2\pi e}{v}e^{-i\boldsymbol{k}\cdot\boldsymbol{R}_0}\delta\big(k_z-q\big)\Big]v\delta_{jz}\,,$$

where k_1 is the wavenumber inside the medium where the electron is propagating. We next compute E_z, E_x parallel and perpendicular to the electron propagation direction in real space.

Parallel Component. Through the inverse Fourier transform we obtain the electric field in real space

$$E_z(\boldsymbol{r},\omega) = \left(\frac{i\omega\mu e}{4\pi^2k_1^2}\right)\int_{-\infty}^{\infty} e^{i\boldsymbol{k}\cdot(\boldsymbol{r}-\boldsymbol{R}_0)}\left(\frac{k_z^2-k_1^2}{k^2-k_1^2}\right)\delta\big(k_z-q\big)\,d^3k\,. \tag{10.38}$$

To simplify this expression, we first introduce

$$q^2-k_1^2 = \frac{\omega^2}{v^2}\Big(1-v^2\,\varepsilon\mu\Big) = q^2\gamma_\varepsilon^{-2}\,,\quad \gamma_\varepsilon = \frac{1}{\sqrt{1-v^2\,\varepsilon\mu_0}}\,. \tag{10.39}$$

Here and in the following we set the permeability to μ_0. Introducing polar coordinates the above integral becomes

$$\begin{aligned}&\int_0^\infty\int_0^{2\pi} e^{i(qz+k_\rho\rho\cos\varphi)}\frac{\gamma_\varepsilon^{-2}q^2}{k_\rho^2+\gamma_\varepsilon^{-2}q^2}\,d\varphi k_\rho dk_\rho\\ &= 2\pi\gamma_\varepsilon^{-2}q^2\,e^{iqz}\int_0^\infty \frac{J_0(k_\rho\rho)}{k_\rho^2+\gamma_\varepsilon^{-2}q^2}\,k_\rho dk_\rho = 2\pi\gamma_\varepsilon^{-2}q^2\left[e^{iqz}K_0\left(\frac{q\rho}{\gamma_\varepsilon}\right)\right],\end{aligned}$$

where K_0 is the modified Bessel functions of order zero.

Perpendicular Components. Similarly, for the perpendicular component E_x we get

$$E_x(\boldsymbol{r},\omega) = \left(\frac{i\omega\mu e}{4\pi^2k_1^2}\right)\int_{-\infty}^{\infty} e^{i\boldsymbol{k}\cdot(\boldsymbol{r}-\boldsymbol{R}_0)}\left(\frac{k_xk_z}{k^2-k_1^2}\right)\delta\big(k_z-q\big)\,d^3k\,.$$

The integral on the right-hand side can be evaluated as

$$\int_0^\infty \int_0^{2\pi} e^{i(qz+k_\rho \rho \cos\varphi)} \frac{qk_\rho \cos\varphi}{k_\rho^2+\gamma_\varepsilon^2 q^2} d\varphi k_\rho dk_\rho$$
$$= 2\pi i q\, e^{iqz} \int_0^\infty \frac{J_1(k_\rho \rho)}{k_\rho^2+\gamma_\varepsilon^{-2} q^2} k_\rho^2 dk_\rho = 2\pi i \gamma_\varepsilon^{-1} q^2 \left[e^{iqz} K_1\left(\frac{q\rho}{\gamma_\varepsilon}\right)\right],$$

where K_1 is the modified Bessel functions of order one.

Combining the above results, we obtain for the electric field associated with the swift electron the expression

Electric Field of Swift Electron

$$\boldsymbol{E}_{\rm inc}(\boldsymbol{r},\omega) = \left(\frac{e\omega}{2\pi v^2 \gamma_\varepsilon}\right) e^{iqz} \left[\frac{i}{\gamma_\varepsilon} K_0\left(\frac{q\rho}{\gamma_\varepsilon}\right)\hat{z} - K_1\left(\frac{q\rho}{\gamma_\varepsilon}\right)\hat{\boldsymbol{\rho}}\right], \tag{10.40}$$

where we have introduced $\boldsymbol{\rho} = \boldsymbol{R} - \boldsymbol{R}_0$. Figure 10.11 shows the z and radial components of the electric field given by Eq. (10.40). As can be seen, the radial component is much larger than the z component. For sufficiently small radial distances ρ, the radial component exhibits a $1/\rho$ dependence reminiscent of the electric field produced by a charged wire. At large distances the electric field decays exponentially.

10.5.2 Decomposition Into Bulk and Surface Losses

We now start analyzing the loss contributions of the swift electron. Quite generally, one has to be careful that an electron propagating in a lossy medium with $\varepsilon'' > 0$ can suffer losses even without exciting particle plasmons. These losses can be evaluated by computing Eq. (10.35) for the electron energy loss probability using the electric field of Eq. (10.38) for the swift electron together with $\boldsymbol{R} = \boldsymbol{R}_0$, to get the electric field at the electron position,

$$\mathcal{P}_{\rm bulk}(\omega) = \frac{e}{\pi\hbar\omega}\int_0^L {\rm Re}\left\{\left(\frac{i\omega\mu e}{2\pi k_1^2}\right)\int_0^{q_{\rm max}} \left(\frac{q^2\gamma_\varepsilon^{-2}}{k_\rho^2+q^2\gamma_\varepsilon^{-2}}\right) k_\rho dk_\rho\right\} dz\,.$$

In the above expression we have replaced the integration over t by an integration over $z = vt$, and consider only a finite propagation distance L. This situation

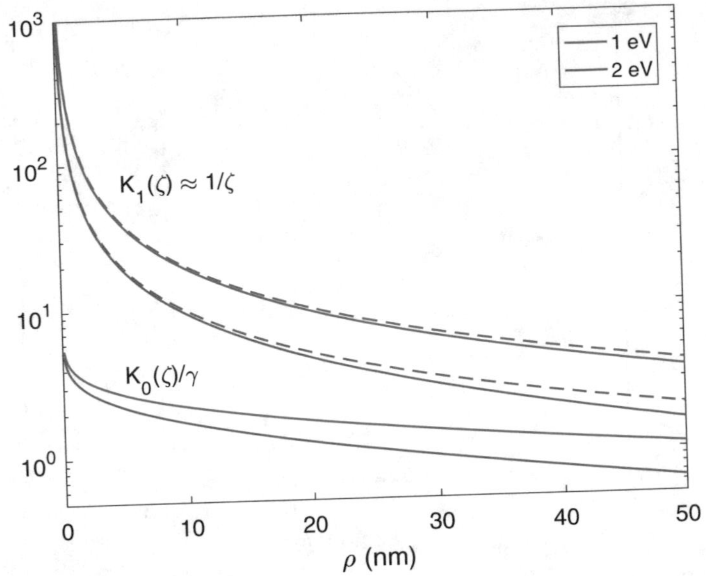

Fig. 10.11 Electric field components of Eq. (10.40) in z (K_0) and radial (K_1) directions, and for two selected loss energies. We use the abbreviation $\zeta = q\rho/\gamma_\varepsilon$. The dashed lines report the $1/\rho$ dependence for the radial electric field of a charged wire, which agrees well with K_1 at sufficiently small distances

corresponds to typical EELS experiments where the electron passes through a metallic nanoparticle of finite size. In addition, we have replaced the upper integration limit of k_ρ by a cutoff parameter $q_{\max}$ that is related to the half-aperture collection angle φ_{out} of the microscope spectrometer, similar to the numerical aperture of Eq. (3.12) in optics. Typical values of φ_{out} are of the order of a few tens of mrad [100]. Performing the above integration finally leads us to the expression for bulk losses

Bulk Losses

$$\mathcal{P}_{\text{bulk}}(\omega) = \frac{e^2 L}{4\pi^2\hbar v^2} \text{Im}\left\{\left(\frac{v^2}{c^2} - \frac{\varepsilon_0}{\varepsilon}\right) \ln\left[\frac{\gamma_\varepsilon^2 q_{\max}^2}{q^2} + 1\right]\right\}. \tag{10.41}$$

In the nonretarded limit $c \to \infty$ this expression reduces to

$$\mathcal{P}^{\text{NR}}_{\text{bulk}}(\omega) = \frac{e^2 L}{4\pi^2\hbar v^2} \text{Im}\left\{-\frac{\varepsilon_0}{\varepsilon(\omega)}\right\} \ln\left[\frac{q_{\max}^2}{q^2} + 1\right]. \tag{10.42}$$

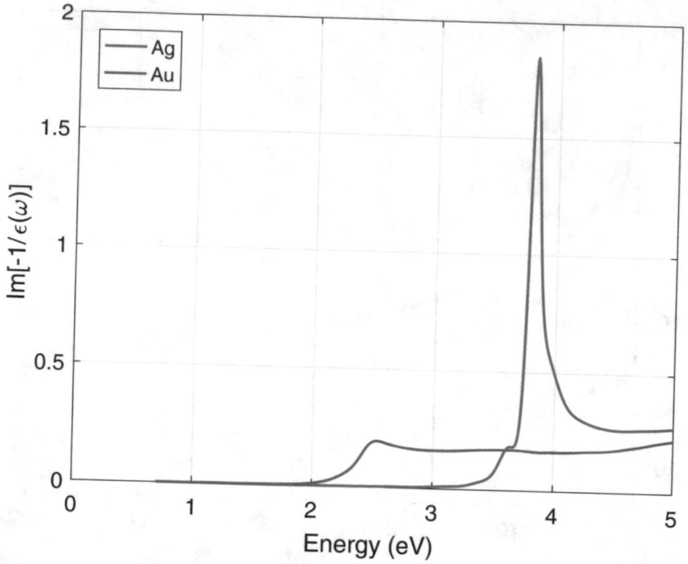

Fig. 10.12 Loss function $\mathrm{Im}[-1/\varepsilon(\omega)]$ for silver (Ag) and gold (Au), respectively, for the dielectric function tabulated in Ref. [34]. For silver the loss function is dominated by a single peak associated with bulk plasmon excitations, for gold one observes a broad distribution above 2 eV associated with d-band absorption

For the case of a Drude type dielectric function $\varepsilon(\omega)/\varepsilon_0 = 1 - \omega_p^2/[\omega(\omega + i\gamma)]$ with ω_p and γ being the plasma and collision frequencies of the free electron gas, the loss function can be evaluated analytically and we obtain

$$\mathrm{Im}\left\{-\frac{\varepsilon_0}{\varepsilon(\omega)}\right\} = \frac{2\omega\gamma\omega_p^2}{(\omega_p^2 - \omega^2)^2 + 4\omega^2\gamma^2}\,.$$

This expression corresponds to a Lorentzian peak at the plasma frequency $\omega \approx \omega_p$ which is broadened by the collision frequency γ. For more realistic dielectric functions, such as those extracted from optical experiments [34], we find similar shapes with practically no bulk losses in the low-energy regime relevant for particle plasmons, and a pronounced peak at the bulk plasmon energy, see Fig. 10.12. In case of gold one observes a broad distribution above 2 eV associated with d-band absorption, as previously discussed in Chap. 7.

For an electron propagating in a more complicated photonic environment, including one or several plasmonic nanoparticles, the energy loss can be split into the following contributions:

Bulk Losses. For each lossy material through which the electron propagates one has a bulk contribution $\mathcal{P}_{\text{bulk}}$ that depends on the propagation length L and the material properties encoded in $\varepsilon(\omega)$.

Begrenzung Effect. The bulk losses have been computed for an unbounded medium. Quite generally, when the electron propagates through a finite volume we expect that corrections appear close to the volume boundaries, associated, for instance, with a less effective screening close to the interface. These modifications lead to the so-called *begrenzung* (German expression for restriction) effect.

Surface Losses. In addition, for plasmonic or dielectric nanoparticles there exist other loss channels where the swift electron excites a surface plasmon or some other type of surface excitations. In many cases one is most interested in these loss contributions as they specifically depend on the nanoparticle geometry.

10.5.3 Expressing EELS Through the Green's Dyadics

In what follows, we present a formalism that accounts for the begrenzung and surface contributions on the same footing. We start from Eq. (10.35) and relate the electric field to the current distribution of the swift electron using the dyadic Green's function,

$$\mathcal{P}_{\text{EELS}}(\boldsymbol{R}_0,\omega)=\frac{e}{\pi\hbar\omega}\times\int_{-\infty}^{\infty}\text{Re}\left\{e^{-iq(z-z')}\hat{\boldsymbol{z}}\cdot\left[i\omega\mu_0\,\bar{\bar{G}}(\boldsymbol{R}_0,z,\boldsymbol{R}_0,z')\cdot\left(-e\hat{\boldsymbol{z}}\right)\right]\right\}\,dzdz'.$$

In the Green's function we have explicitly indicated the dependence on the parallel and perpendicular components (with respect to the electron propagation direction) z and $\boldsymbol{R}_0$, respectively. The prefactor can be simplified according to

$$\frac{e^2\mu_0\varepsilon_0}{\pi\hbar\varepsilon_0}=\left(\frac{1}{4\pi\varepsilon_0}\frac{e^2}{\hbar c}\right)\frac{4}{c}=\frac{4\alpha}{c},$$

where the term in parentheses is the fine structure constant $\alpha\approx 1/137$. We finally decompose the Green's function into a free space part $\bar{\bar{G}}$ and a reflected part $\bar{\bar{G}}_{\text{refl}}$ that accounts for the begrenzung and surface effects. Then, putting together all results we are led to the total energy loss probability expressed in terms of the reflected Green's function

Electron Energy Loss Probability (Green's Function Expression)

$$\mathcal{P}_{\text{EELS}}(\boldsymbol{R}_0,\omega)=\frac{4\alpha}{c}\int_{-\infty}^{\infty}\text{Im}\left\{e^{-iq(z-z')}\hat{\boldsymbol{z}}\cdot\bar{\bar{G}}_{\text{refl}}(\boldsymbol{R}_0,z,\boldsymbol{R}_0,z')\cdot\hat{\boldsymbol{z}}\right\}\,dzdz'+\mathcal{P}_{\text{bulk}}(\omega)\,. \tag{10.43}$$

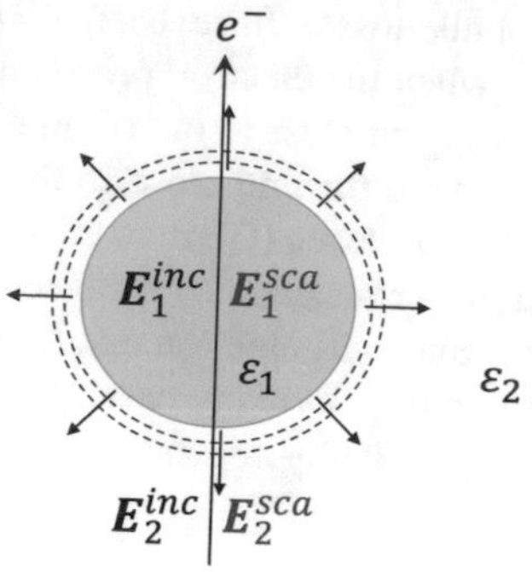

Fig. 10.13 Schematics of calculation of EELS probability. A swift electron passes along z through a plasmonic nanosphere. The "incoming" fields of the electron are $\boldsymbol{E}_{1,2}^{\text{inc}}$ inside and outside the sphere, see Eq. (10.40), the "scattered" fields are $\boldsymbol{E}_{1,2}^{\text{sca}}$ and account for begrenzung and surface effects. The electron energy loss is obtained by evaluating the work performed by the electron against the electric fields along the entire electron trajectory, see Eq. (10.35). The scattered electromagnetic far-fields can be measured as a cathodoluminescence

Let us stop here for a moment and recall what has been done so far. The problem we are considering consists of a swift electron that passes by a metallic or dielectric nanoparticle (aloof geometry) or penetrates through it, as schematically shown in Fig. 10.13, and loses a tiny fraction of its kinetic energy. To compute the energy loss probability we proceed as follows:

Semiclassical Loss Formula. Our starting expression is the semiclassical expression of Eq. (10.35), which gives the frequency decomposition of the power performed by the electron against the electric fields acting on the electron. These fields are induced by the electron itself. The quantum aspect of the semiclassical formula is that the electron can lose its energy only in fractions of $\hbar\omega$. All other ingredients of the theory are purely classical.

Incoming Fields. The "incoming" fields of the electron are computed for an unbounded medium filled with permittivities ε_1, ε_2, depending on whether the electron propagates inside or outside the nanoparticle under consideration. These fields fulfill Maxwell's equations inside the two media, including the charge and current sources of the swift electron, but not the boundary conditions at the particle boundary $\partial\Omega$.

Scattered Fields. To account for the boundary conditions at $\partial\Omega$, we additionally introduce the "scattered" fields $\boldsymbol{E}_{1,2}^{\text{sca}}$, $\boldsymbol{H}_{1,2}^{\text{sca}}$, using, for instance, the boundary integral method approach introduced in Sect. 9.6. These fields are solutions of the Maxwell equations without source terms. Thus, with the combination of incoming and scattered fields Maxwell's equations inside the media are fulfilled and also the boundary conditions at $\partial\Omega$ are properly accounted for. For this reason, $\boldsymbol{E}_{\text{inc}} + \boldsymbol{E}_{\text{sca}}$ and $\boldsymbol{H}_{\text{inc}} + \boldsymbol{H}_{\text{sca}}$ are the proper solutions of the problem under study.

Bulk and Boundary Losses. Although the separation into incoming and scattered fields is somewhat artificial, it turns out to be convenient to split the losses in Eq. (10.35) into bulk losses due to $\boldsymbol{E}_{\text{inc}}$ and boundary losses due to $\boldsymbol{E}_{\text{sca}}$. The latter ones can be further separated into begrenzung and surface losses, but this separation is somewhat inexplicit in the approach we have taken.

Nonlocality. As can be seen most clearly in Eq. (10.43), the electron energy loss is a nonlocal process where the electron first induces at position z' a particle plasmon, or some other type of surface excitation, and the induced fields act back on the electron at position z. This genuine nonlocality complicates the interpretation of EELS in terms of a photonic LDOS or some related quantities.

Cathodoluminescence. Finally, by evaluating the scattered fields far away from the particle, or equivalently the electromagnetic energy carried away by the Poynting vector through a boundary surrounding all particles, we can compute the so-called cathodoluminescence probability [100, 101]. Such luminescence provides complementary information about the optical properties of plasmonic or dielectric nanoparticles.

Figure 10.14 shows simulated EELS spectra for a silver nanotriangle with a base length of approximately 80 nm and a height of 10 nm. We observe at least four pronounced peaks, labeled with (a–d), in addition to the bulk plasmon peak around 3.8 eV (see also Fig. 10.12). In Fig. 10.15 we show EELS maps at the resonance energies. They are obtained by raster-scanning the electron beam over the sample, and agree well with the experimental maps reported in the seminal paper of Ref. [99]. The observed modes can be ascribed to (a) dipolar and (b, c) multipolar plasmon modes confined to the nanoparticle edges, as well as (d) face modes

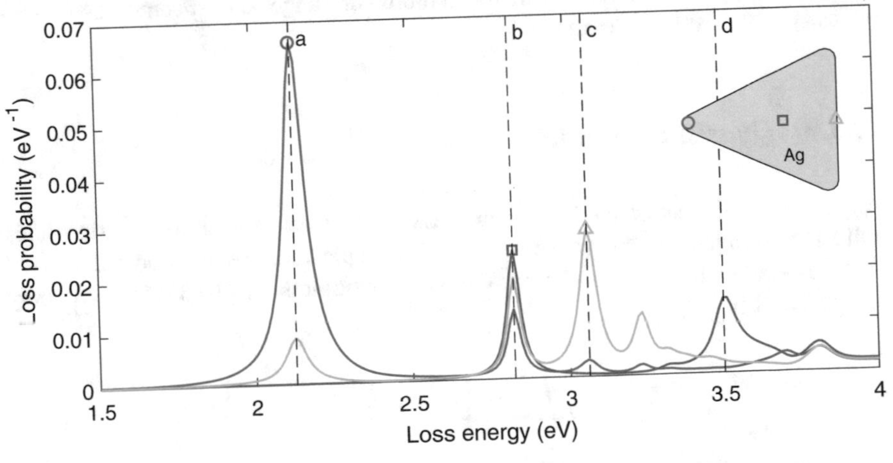

Fig. 10.14 EELS spectra for a silver nanotriangle with a base length of approximately 80 nm and a height of 10 nm, and for three different positions of the electron beam, indicated in the inset. In our simulations we use dielectric functions extracted from optical experiments [34]. The dashed lines report the energetic positions of the plasmon resonances where the spatial EELS maps of Fig. 10.15 are computed

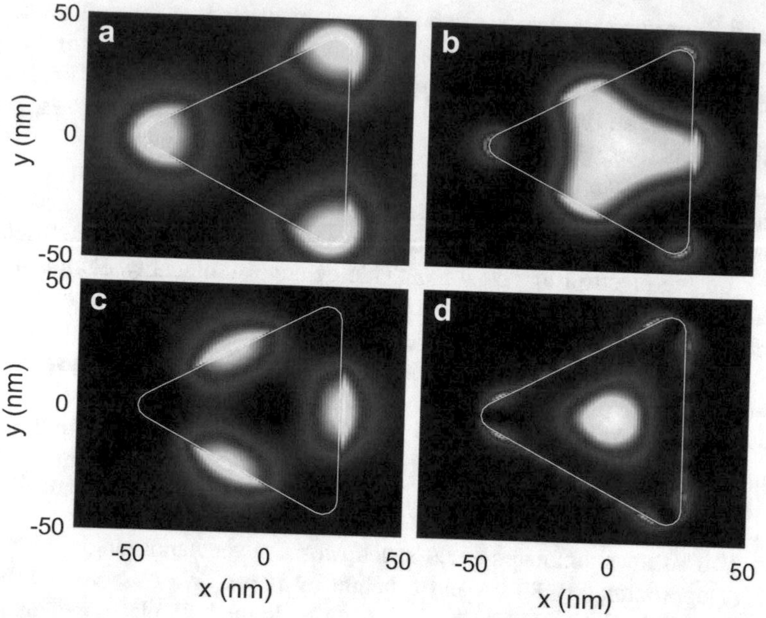

Fig. 10.15 Spatial EELS maps for same nanotriangle as investigated in Fig. 10.14 at the plasmon resonances indicated by dashed lines. The modes can be assigned to (**a**) dipole, (**b, c**) hexapole, and (**d**) breathing modes, as discussed in text

confined to the upper and lower triangle surfaces. Without going into any details, the above example nicely demonstrates the capability of EELS to map particle plasmon modes with spatial resolutions in the nanometer range and spectral resolutions typically in the 100 meV range.

10.5.4 Quasistatic Limit

We conclude this section by discussing EELS in the quasistatic approximation. This will allow us in particular to make contact with the plasmonic eigenmodes discussed in some detail in the previous chapter. Our starting point is Eq. (10.35) for the energy loss probability

$$\mathcal{P}_{\rm EELS}(\boldsymbol{R}_0,\omega)=-\frac{1}{\pi\hbar\omega}\int_{-\infty}^{\infty}\mathrm{Re}\Big\{\boldsymbol{J}^*(\boldsymbol{r},\omega)\cdot\boldsymbol{E}_{\rm ind}(r,\omega)\Big\}\,d^3r+\mathcal{P}_{\rm bulk}\,,$$

where we have used $\boldsymbol{J}(\boldsymbol{r})=\boldsymbol{v}\,\rho(\boldsymbol{r},\omega)$ with the charge distribution of Eq. (10.37). We have also separated the electric field into an induced or "scattered" part, and an "incoming" part that has been absorbed into the bulk contribution. In the quasistatic approximation $\mathcal{P}_{\rm bulk}$ has to be replaced by Eq. (10.42), and we additionally relate the induced electric field to a potential via $\boldsymbol{E}_{\rm ind} = -\nabla V_{\rm ind}$. We next employ the vector identity

$$\nabla\cdot(\boldsymbol{J}^*V_{\rm ind}) = (\nabla\cdot\boldsymbol{J}^*)V_{\rm ind} + \boldsymbol{J}^*\cdot\nabla V_{\rm ind}\,,$$

where, upon integration over the entire space and use of the divergence theorem, the term on the left-hand side vanishes because $V_{\rm ind}$ approaches zero at infinity. Together with the continuity equation $i\omega\rho = \nabla\cdot\boldsymbol{J}$ we then find

$$\mathcal{P}_{\rm EELS}(\boldsymbol{R}_0,\omega) = -\frac{1}{\pi\hbar}\int \mathrm{Im}\Big\{\rho^*(\boldsymbol{r},\omega)V_{\rm ind}(\boldsymbol{r},\omega)\Big\}\,d^3r + \mathcal{P}_{\rm bulk}^{\rm NR}\,. \tag{10.44}$$

In the following we assume that the electron trajectory does not penetrate the plasmonic nanoparticle (aloof geometry). The induced potential can then be related to the electron charge distribution $\rho(\boldsymbol{r})$ via the quasistatic Green's function. Together with the eigenmode decomposition of Eq. (9.33) for the induced Green's function we get

$$\mathcal{P}_{\rm EELS} = \frac{1}{\pi\hbar\varepsilon_0}\int \mathrm{Im}\Big\{\rho^*(\boldsymbol{r},\omega)\sum_k V_k(\boldsymbol{r})\left[\frac{\lambda_k+\frac{1}{2}}{\Lambda(\omega)+\lambda_k}\right]V_k(\boldsymbol{r}')\rho(\boldsymbol{r}',\omega)\Big\}\,d^3r d^3r'\,,$$

where λ_k is the eigenvalue of the eigenpotential $V_k(\boldsymbol{r})$. This leads us to the electron energy loss probability in the quasistatic limit

Electron Energy Loss Probability (Quasistatic Limit)

$$\mathcal{P}_{\rm EELS}(\boldsymbol{R}_0,\omega) = \frac{e^2}{\pi\hbar v^2\varepsilon_0}\sum_k \mathrm{Im}\left\{\frac{\lambda_k+\frac{1}{2}}{\Lambda(\omega)+\lambda_k}\right\}\left|\int_{-\infty}^{\infty} e^{iqz}V_k(\boldsymbol{R}_0,z)\,dz\right|^2. \tag{10.45}$$

This expression provides a quite transparent decomposition of the loss function into lineshape functions and oscillator strengths. The term in curly brackets gives a Lorentzian lineshape at the resonance frequencies where $\mathrm{Re}[\Lambda(\omega)+\lambda_k]\approx 0$, and the integral gives a coefficient describing how well a given eigenmode can be excited by the electron beam.

Exercises

Exercise 10.1 Start from Eq. (10.13) for a real-valued dipole moment $\boldsymbol{p}$. Show that the average over all dipole orientations corresponds to the replacement of the dyadic Green's function with $\frac{1}{3}\mathrm{tr}\bar{\bar{G}}$.

Exercise 10.2 Use from the NANOPT toolbox the file `demostrat03.m` that computes the lifetime of a dipole situated above a gold substrate.

(a) What happens if the metal substrate is replaced by a thin metal film (study thicknesses between 10 and 100 nm) on top of a glass substrate.
(b) Modify the program such that the lifetime is computed for a fixed dipole distance but for different transition frequencies. What happens when changing the thickness of the metal film? Can you identify the coupled surface plasmon modes for a slab shown in Figs. 8.14 and 8.15?

Exercise 10.3 Use from the NANOPT toolbox the file `demodipmie02.m` that computes the decay rate of a dipole situated above a gold sphere within Mie theory. Investigate what happens if you change the sphere diameter in the range from 50 to 500 nm, and determine the minimal cutoff value $\ell_{\max}$ for angular orders needed to get converged results.

Exercise 10.4 Show that Eq. (10.17) for the power dissipated by a dipole emitter in the quasistatic limit can be written in the form

$$\frac{P}{P_0}=1+\frac{6\pi}{k^3}\sum_k \operatorname{Im}\left\{\frac{\lambda_k+\frac{1}{2}}{\Lambda(\omega)+\lambda_k}\right\}\left|\hat{\boldsymbol{p}}\cdot\nabla V_k(\boldsymbol{r})\right|^2_{\boldsymbol{r}=\boldsymbol{r}_0}\,.$$

Use Eq. (9.64) to obtain the physical dimension of the eigenmodes $u_k(\boldsymbol{s})$ and to prove that the second term on the right-hand side is dimensionless.

Exercise 10.5 Use from the NANOPT toolbox the file `demobem07.m` that computes the decay rates for an oscillating dipole above a nanosphere for a fixed frequency and varying dipole-sphere distances, see Fig. 10.3, and `demobem08.m` which computes the decay rates for a fixed dipole-sphere distances and varying oscillation frequencies.

(a) Modify the program in order to compute the decay rates shown in Fig. 10.3 for an ellipsoid.
(b) Plot for a dot-sphere separation of 10 nm the surface charge distribution for a horizontal and vertical dipole, and compare the results.
(c) Plot for a horizontal dipole the total scattering rate as a function of transition frequency, and find the resonance frequency where the decay rate is largest. Plot the corresponding surface charge distribution.

Exercise 10.6 Start from the same files as in Exercise 10.5 and rewrite them for coupled spheres. You might like to use the file `demobem02.m` as a template. For sphere diameters of 50 nm and gap distances of 5, 10, 25, 50 nm, find the resonance frequencies for dipoles oriented vertically and horizontally to the symmetry axis, and compute the enhancements for the total and radiative decay rates on resonance.

Exercise 10.7 Consider $\mathcal{P}_{\rm EELS}(\boldsymbol{R}_0,\hbar\omega)$ defined in Eq. (10.34), which is the probability that a swift electron with impact parameter $\boldsymbol{R}_0$ loses an energy of $\hbar\omega$. Use the induced or "scattered" fields $\boldsymbol{E}_{\rm sca}$, $\boldsymbol{H}_{\rm sca}$ to define the cathodoluminescence probability $\mathcal{P}_{\rm CL}(\boldsymbol{R}_0, \hbar)$, this is the probability that a photon with energy $\hbar\omega$ is produced by a swift electron with impact parameter $\boldsymbol{R}_0$. Express the final result in terms of the far-field amplitude $\boldsymbol{F}(\hat{\boldsymbol{r}})$ of Eq. (3.5).

Hint: start from the scattered power of Eq. (4.22).

Exercise 10.8 Consider the EELS bulk loss probability of Eq. (10.41). Under which conditions do losses occur for an entirely real $\varepsilon(\omega)$? You might like to consult Refs. [2, 100] for a more thorough discussion of Cherenkov radiation responsible for these losses.

Exercise 10.9 Use Eq. (9.34) to rewrite the last term on the right-hand side of Eq. (10.45) in the form

$$\int_{-\infty}^{\infty} e^{iqz} V_k(\boldsymbol{r})\, dz = \oint_{\partial\Omega} \left[\int_{-\infty}^{\infty} e^{iqz} G(\boldsymbol{r}, \boldsymbol{s}')\, dz\right] u_k(\boldsymbol{s}')\, dS'.$$

Use

$$\int_{-\infty}^{\infty} \frac{e^{iqz}}{\sqrt{\rho^2 + z^2}}\, dz = 2\, K_0(q\rho)$$

to perform the integral in brackets analytically. With this one obtains an expression where the integration extends over the particle boundary only, rather than over the entire z-axis.

Chapter 11
Computational Methods in Nano Optics

In this book we are mainly concerned with the concepts of nano optics and plasmonics using analytic solution schemes for selected problems. However, in the last decades numerical simulation approaches have become the method of choice for various nano optics problems. The advantages of numerical techniques are that they are more versatile and allow for a much closer comparison with experiment. In this chapter we briefly introduce to the most popular simulation techniques, however, without going too much into technical details.

11.1 Finite Difference Time Domain Simulations

We start with the approach that can be probably considered as the mother of all computational Maxwell's solvers, the so-called finite difference time domain (FDTD) simulation. Readers interested in a more detailed account are referred to other textbooks such as [102, 103]. The FDTD approach is a general simulation scheme suited for the solution of Maxwell's equations with no particular restriction on the system. It was first introduced by Yee [104] in 1966 and implemented successfully by Taflove and Browdin almost 10 years later [105, 106].

We start our discussion of the FDTD scheme by expressing the curl equations in terms of time-dependent differential equations

$$\varepsilon\frac{\partial \boldsymbol{E}}{\partial t} = \nabla\times\boldsymbol{H} - \boldsymbol{J}\,, \qquad \mu\frac{\partial \boldsymbol{H}}{\partial t} = -\nabla\times\boldsymbol{E}\,, \tag{11.1}$$

which have to be solved together with the constraints

$$\nabla\cdot\varepsilon\boldsymbol{E} = \rho\,, \quad \nabla\cdot\mu\boldsymbol{H} = 0\,.$$

U. Hohenester, *Nano and Quantum Optics*, Graduate Texts in Physics,
https://doi.org/10.1007/978-3-030-30504-8_11

The Early Years of FDTD

In an interview in Nature Photonics in 2015, Allen Taflove, one of the inventors of the FDTD method, recalled that in the 1970s when he wrote his PhD thesis numerical simulations techniques did not receive too much attention.

> *[My supervisor] Brodwin agreed to let me pursue this topic for my PhD thesis, even though it was not in the mainstream of his research. By 1975, my algorithm development, coding and validations had progressed to the point where a fully three-dimensional model of the microwave irradiated human eye could be implemented. That year I published my results in two papers in IEEE Transactions on Microwave Theory and Techniques and earned my PhD. But, like Yee's paper, my work remained uncited.*

However, even ten years later FDTD simulations still had to be performed on supercomputers, such as the at that time well-known Cray, but often found little recognition by the community. In the interview Taflove remembers such an episode:

> FDTD *was ignored, and at times even ridiculed. One such episode remains burned into my memory. In 1986, an internationally known professor actually laughed at me in an open meeting that had been called to consider the future of computational electromagnetics, because of my use of supercomputers. Pointing directly at me, he joked: "Look at Taflove over there. When he wakes up in the morning, he gets down on his knees, bows his head, and says, 'And now, let us Cray'." I was speechless and terrified.*

It can be shown that if the initial conditions satisfy these constraints and the fields are evolved according to Eqs. (11.1), the solutions satisfy the divergence equations at all times. To prove this, we take the divergence on both sides of Eq. (11.1) to get

$$\frac{\partial}{\partial t}\Big(\nabla\cdot\varepsilon\boldsymbol{E}\Big)=-\nabla\cdot\boldsymbol{J}=\frac{\partial\rho}{\partial t},\quad \frac{\partial}{\partial t}\Big(\nabla\cdot\mu\boldsymbol{H}\Big)=0,$$

where we have used that time and space derivatives can be exchanged, and that the divergence of a curl is automatically zero. We have also employed the continuity equation to relate $\nabla\cdot\boldsymbol{J}$ to the time derivative of ρ. Rewriting the first equation in the form

$$\frac{\partial}{\partial t}\Big(\nabla\cdot\varepsilon\boldsymbol{E}-\rho\Big)=0,$$

we observe that if $\nabla \cdot \varepsilon \boldsymbol{E} = \rho$ holds at the initial time, it also holds at later times because the curl equations properly propagate the divergence constraints.[1] A similar conclusion can be drawn for the magnetic field. Thus, it suffices to consider in the time evolution of $\boldsymbol{E}, \boldsymbol{H}$ only the curl equations. In what follows, we will investigate Maxwell's equations in three, two, and one spatial dimensions.

Three Spatial Dimensions. In three spatial dimensions Maxwell's equations can be expressed in component form as follows:

$$\begin{aligned}
\varepsilon\frac{\partial E_x}{\partial t} &= \frac{\partial H_z}{\partial y} - \frac{\partial H_y}{\partial z} - J_x, & \mu\frac{\partial H_x}{\partial t} &= \frac{\partial E_y}{\partial z} - \frac{\partial E_z}{\partial y} \\
\varepsilon\frac{\partial E_y}{\partial t} &= \frac{\partial H_x}{\partial z} - \frac{\partial H_z}{\partial x} - J_y, & \mu\frac{\partial H_y}{\partial t} &= \frac{\partial E_z}{\partial x} - \frac{\partial E_x}{\partial z} \\
\varepsilon\frac{\partial E_z}{\partial t} &= \frac{\partial H_y}{\partial x} - \frac{\partial H_x}{\partial y} - J_z, & \mu\frac{\partial H_z}{\partial t} &= \frac{\partial E_x}{\partial y} - \frac{\partial E_y}{\partial x}\,.
\end{aligned} \tag{11.2}$$

Two Spatial Dimensions. In two spatial dimensions we assume that ε, μ only depend on x, y. The set of Eqs. (11.2) then separates into two sets of equations, which we shall denote as transverse electric (TE) and transverse magnetic (TM),

$$\begin{aligned}
&(\text{TE}) & \varepsilon\frac{\partial E_x}{\partial t} &= \frac{\partial H_z}{\partial y} - J_x & &(\text{TM}) & \mu\frac{\partial H_x}{\partial t} &= -\frac{\partial E_z}{\partial y} \\
&(\text{TE}) & \varepsilon\frac{\partial E_y}{\partial t} &= -\frac{\partial H_z}{\partial x} - J_y & &(\text{TM}) & \mu\frac{\partial H_y}{\partial t} &= \frac{\partial E_z}{\partial x} \\
&(\text{TM}) & \varepsilon\frac{\partial E_z}{\partial t} &= \frac{\partial H_y}{\partial x} - \frac{\partial H_x}{\partial y} - J_z & &(\text{TE}) & \mu\frac{\partial H_z}{\partial t} &= \frac{\partial E_x}{\partial y} - \frac{\partial E_y}{\partial x}\,.
\end{aligned} \tag{11.3}$$

The TE equations only involve the components E_x, E_y, and H_z, and the TM ones the components H_x, H_y, and E_z.

One Spatial Dimension. In one spatial dimension we assume that ε, μ only depend on x. The sets of Eqs. (11.3) then become further simplified

$$\begin{aligned}
& & \varepsilon\frac{\partial E_x}{\partial t} &= -J_x & & & \mu\frac{\partial H_x}{\partial t} &= 0 \\
&(\text{TE}) & \varepsilon\frac{\partial E_y}{\partial t} &= -\frac{\partial H_z}{\partial x} - J_y & &(\text{TM}) & \mu\frac{\partial H_y}{\partial t} &= \frac{\partial E_z}{\partial x} \\
&(\text{TM}) & \varepsilon\frac{\partial E_z}{\partial t} &= \frac{\partial H_y}{\partial x} - J_z & &(\text{TE}) & \mu\frac{\partial H_z}{\partial t} &= -\frac{\partial E_y}{\partial x}\,.
\end{aligned} \tag{11.4}$$

The first two equations decouple from the rest, the TE equations only involve the components E_y, H_z, and the TM ones the components H_y, E_z.

[1]The divergence constraints hold unconditionally for the exact solutions. In case of an approximate solution, as is the case in computational electrodynamics, the constraints are not fulfilled automatically, and one has to check for their applicability for each algorithm separately.

11.1.1 The Magic of the FDTD Approach

In the finite difference time domain approach we closely stay with Maxwell's equations, but approximate the time and spatial derivatives through finite differences. The magic of the FDTD approach is due to the insight of Yee [104] who realized that the coupled set of equations for the electromagnetic fields can be solved efficiently by two staggered grids. We start by discussing the situation for the TE equations in the one-dimensional case of Eq. (11.4). First, we introduce two staggered spatial grids (see Fig. 11.1)

Staggered Spatial Grids (One Spatial Dimension)

$$x_i = i\,\Delta x$$
$$x_{i+\frac{1}{2}} = (i+\tfrac{1}{2})\,\Delta x, \qquad i = 1, 2, \dots N_x, \tag{11.5}$$

which are shifted with respect to each other by half a lattice constant Δx. Similarly, we introduce two staggered time grids

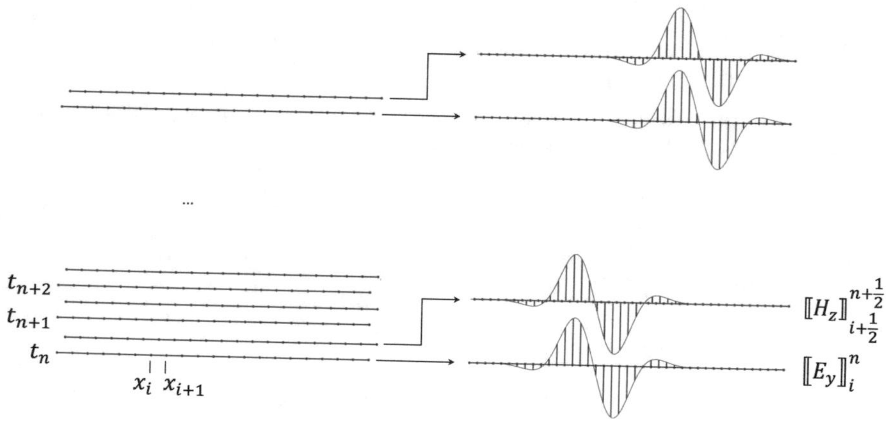

Fig. 11.1 Schematics of FDTD in one dimension. The electric and magnetic fields are given at the discretized positions of two staggered grids. By repeatedly updating the electromagnetic fields according to the FDTD algorithm of Eq. (11.9), the fields are propagated forward in time

Staggered Time Grids

$$t_n = n\,\Delta t$$
$$t_{n+\frac{1}{2}} = (n + \tfrac{1}{2})\,\Delta t, \qquad n = 1, 2, \ldots, \tag{11.6}$$

which are shifted with respect to each other by half a time step Δt. In what follows, we approximate the electric field at the integer grid positions

$$\text{(Electric field)} \qquad [\![E_y]\!]_i^n = E_y\big(i\,\Delta x, n\,\Delta t\big)\,. \tag{11.7}$$

We use the notation $[\![\ldots]\!]$ for the discretized electromagnetic fields within the FDTD approach. Similarly, we approximate the magnetic field at the half-integer grid positions

$$\text{(Magnetic field)} \qquad [\![H_z]\!]_{i+1/2}^{n+1/2} = H_z\big([i+1/2]\,\Delta x, [n+1/2]\,\Delta t\big)\,. \tag{11.8}$$

With these staggered grids, the spatial and time derivatives in Eq. (11.4) can be approximated in the form

FDTD Equations in One Spatial Dimension (TE)

$$\mu_{i+1/2}\frac{[\![H_z]\!]_{i+1/2}^{n+1/2} - [\![H_z]\!]_{i+1/2}^{n-1/2}}{\Delta t} = -\frac{[\![E_y]\!]_{i+1}^{n} - [\![E_y]\!]_{i}^{n}}{\Delta x} \tag{11.9a}$$

$$\varepsilon_i\frac{[\![E_y]\!]_{i}^{n+1} - [\![E_y]\!]_{i}^{n}}{\Delta t} = -\frac{[\![H_z]\!]_{i+1/2}^{n+1/2} - [\![H_z]\!]_{i-1/2}^{n+1/2}}{\Delta x} - [\![J_y]\!]_i^{n+1/2}\,. \tag{11.9b}$$

If the initial electric and magnetic fields are known, these expressions can be used to successively compute the fields at later times. The algorithm for solving Maxwell's equations in the time domain can be formulated as follows:

- specify magnetic field $[\![H_z]\!]^{1/2}$ at initial time $t_{1/2}$,
- specify electric field $[\![E_y]\!]^{1}$ at initial time t_1,
- use Eq. (11.9a) to compute the magnetic field at time $t_{n+1/2}$,
- use Eq. (11.9b) to compute the electric field at time t_{n+1}.

By successively repeating the last two steps, one can propagate the fields in time. The beautiful thing about FDTD is that it is particularly transparent, easy to implement, and computationally very efficient. Let me add a few comments about the scheme.

Accuracy. The approximation of derivatives through finite differences is not exact but comes with an error that is of order Δx^3 and Δt^3. For this reason, the algorithm is second-order accurate.

Stability. The update scheme for the electromagnetic fields, which is often called a *leapfrog* scheme, has some hidden magic. As we will discuss in Sect. 11.1.2, the time propagation is unconditionally stable for sufficiently small values of Δx, Δt, contrary to other finite difference schemes where such stability is not guaranteed [102].

Divergence Behavior. Equation (11.9) preserves the divergence constraints of the solutions. Thus, if the initial fields have the correct divergence properties, the FDTD solutions at later times also have them [102].

Higher Dimensions. The FDTD scheme can be easily generalized to two- and three-dimensional problems, as will be discussed further below.

Staircase Effects. Within FDTD, a metallic or dielectric nanoparticle is modelled in terms of space dependent material properties ε_i, $\mu_{i+1/2}$. Because of the rectangular grid underlying the FDTD approach, the nanoparticles become approximated by volumes with staircase-like boundaries. This approximation introduces additional errors, which are often referred to as "staircase errors."[2]

Figure 11.2 shows a simple example for field propagation in FDTD. An electromagnetic field pulse impinges on a slab consisting of a material that has a larger permittivity. Similar to our previous discussion in Chap. 8 on stratified media, part of the field becomes reflected and transmitted at the interface. One also clearly observes the different velocities outside and inside the layer.

Two-Dimensional Case

The FDTD formalism can be easily extended to two and three spatial dimensions. We here only discuss the two-dimensional case with TE polarization. We define the electric and magnetic fields at the grid positions

$$[\![E_x]\!]^n_{i+\frac{1}{2},j}, \quad [\![E_y]\!]^n_{i,j+\frac{1}{2}}, \quad [\![H_z]\!]^{n+1/2}_{i+1/2,j+1/2},$$

where j labels the y coordinates of the two-dimensional grid. See Fig. 11.3 for a graphical representation. From Eq. (11.3) we find for the TE case

[2]In the context of staircase effects, one additionally has to be careful about the staggered grids. Typically one is interested in dielectric or metallic bodies with a spatially varying permittivity, rather than permeability. For this reason, one defines the boundary of the body through ε_i and sets, if needed, $\mu_{i+1/2}$ to the particle's permeability only inside the so-defined boundary.

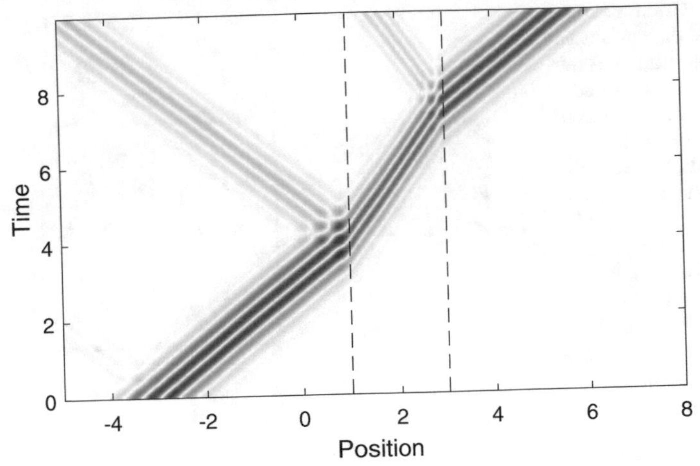

Fig. 11.2 Example for one-dimensional FDTD simulation. An electromagnetic pulse impinges on a slab with a material that has the dielectric constant $\kappa = 3$. Part of the field becomes reflected and transmitted at the interfaces of the layer. Positions and times are given in dimensionless units

$$\mu_{i+1/2,j+1/2}\frac{[\![H_z]\!]^{n+1/2}_{i+1/2,j+1/2}-[\![H_z]\!]^{n-1/2}_{i+1/2,j+1/2}}{\Delta t} = \frac{[\![E_y]\!]^{n}_{i,j+1/2}-[\![E_y]\!]_{i,j-1/2}}{\Delta y}-\frac{[\![E_x]\!]^{n}_{i+1/2,j}-[\![E_x]\!]_{i-1/2,j}}{\Delta x} \tag{11.10}$$

$$\varepsilon_{i+1/2,j}\frac{[\![E_x]\!]^{n+1}_{i+1/2,j}-[\![E_x]\!]^{n}_{i+1/2,j}}{\Delta t} = \frac{[\![H_z]\!]^{n+1/2}_{i+1/2,j+1/2}-[\![H_z]\!]^{n+1/2}_{i+1/2,j-1/2}}{\Delta y}-[\![J_x]\!]^{n+1/2}_{i+1/2,j}$$

$$\varepsilon_{i,j+1/2}\frac{[\![E_y]\!]^{n+1}_{i,j+1/2}-[\![E_y]\!]^{n}_{i,j+1/2}}{\Delta t} = -\frac{[\![H_z]\!]^{n+1/2}_{i+1/2,j+1/2}-[\![H_z]\!]^{n+1/2}_{i-1/2,j+1/2}}{\Delta x}-[\![J_y]\!]^{n+1/2}_{i,j+1/2},$$

with a similar expression for the TM fields. A closer inspection of the equations reveals that all derivatives can be approximated by centered finite differences, where all fields required at the neighbor grid positions appear naturally within the Yee cell scheme of the discretized electromagnetic fields. We will show results for two-dimensional problems further below. Yee cells can be also defined for the three-dimensional case, as discussed in full length for instance in Refs. [102, 103].

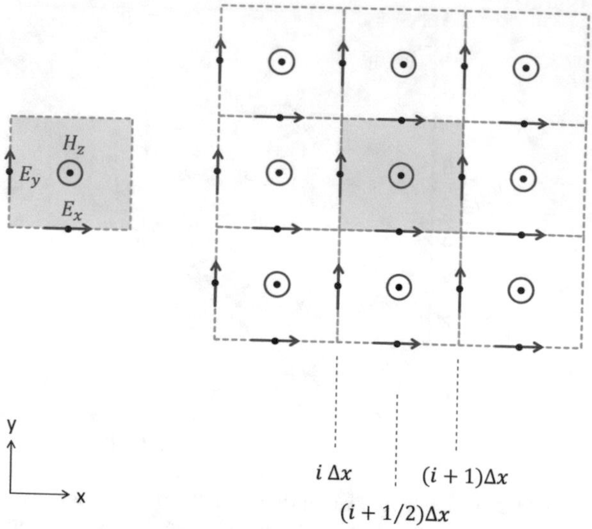

Fig. 11.3 Grid for FDTD in two spatial dimensions and for TE polarization. Again we introduce two staggered spatial grids for the electric and magnetic field components, respectively. On the left-hand side of the figure we show a "unit cell' of the two-dimensional Yee grid, the so-called Yee cell. By working out the finite differences given in Eq. (11.10), we find that all expressions result in second-order centered differences

11.1.2 Stability and Dispersion

In the following we consider one spatial dimension and the situation of a wave propagation in a homogeneous medium with constant ε, μ values. For further simplification we assume periodic boundary conditions. We can thus expand the solutions for the electric and magnetic fields in a Fourier series

$$[\![H_z]\!]_{i+1/2}^{n-1/2} = \sum_k e^{ikx_{i+1/2}} [\![\mathscr{H}_z]\!]_k^{n-1/2}, \qquad [\![E_y]\!]_i^n = \sum_k e^{ikx_i} [\![\mathscr{E}_y]\!]_k^n,$$

with the Fourier coefficients $[\![\mathscr{H}_z]\!]_k^{n-1/2}$ and $[\![\mathscr{E}_y]\!]_k^n$. The update expressions of Eq. (11.9) for the electromagnetic fields then translate to

$$[\![Z\mathscr{H}_z]\!]_k^{n+1/2} = [\![Z\mathscr{H}_z]\!]_k^{n-1/2} - S\left(e^{i\varphi/2} - e^{-i\varphi/2}\right) [\![\mathscr{E}_y]\!]_k^n$$

$$[\![\mathscr{E}_y]\!]_k^{n+1} = [\![\mathscr{E}_y]\!]_k^n - S\left(e^{i\varphi/2} - e^{-i\varphi/2}\right) [\![Z\mathscr{H}_z]\!]_k^{n+1/2},$$

where we have introduced the impedance Z, the phase $\varphi = k\Delta x$, and the stability factor $S = c\Delta t/\Delta x$, with c being the speed of light in the medium. The above set of equations can be rewritten in matrix form

$$\begin{pmatrix} 1 & 0 \\ 2iS\sin\frac{\varphi}{2} & 1 \end{pmatrix} \cdot \begin{pmatrix} [\![Z\mathscr{H}_z]\!]_k^{n+1/2} \\ [\![\mathscr{E}_y]\!]_k^{n+1} \end{pmatrix} = \begin{pmatrix} 1 & -2iS\sin\frac{\varphi}{2} \\ 0 & 1 \end{pmatrix} \cdot \begin{pmatrix} [\![Z\mathscr{H}_z]\!]_k^{n-1/2} \\ [\![\mathscr{E}_y]\!]_k^n \end{pmatrix},$$

which can be cast to the form

$$\begin{pmatrix} [\![Z\mathscr{H}_z]\!]_k^{n+1/2} \\ [\![\mathscr{E}_y]\!]_k^{n+1} \end{pmatrix} = \begin{pmatrix} 1 & -2iS\sin\frac{\varphi}{2} \\ -2iS\sin\frac{\varphi}{2} & 1-4S^2\sin^2\frac{\varphi}{2} \end{pmatrix} \cdot \begin{pmatrix} [\![Z\mathscr{H}_z]\!]_k^{n-1/2} \\ [\![\mathscr{E}_y]\!]_k^{n} \end{pmatrix}. \tag{11.11}$$

The matrix on the right-hand side is the so-called *amplification matrix*, which at each time step is applied on the electromagnetic field vector to propagate the fields in time. According to the von-Neumann condition, the eigenvalues of this matrix must be smaller or equal to one in order to get a numerically stable algorithm. Simple algebra yields for the eigenvalues of the amplification matrix

$$\lambda_\pm = 1 - 2\left(S\sin\frac{\varphi}{2}\right)^2 \pm 2\sqrt{\left(S\sin\frac{\varphi}{2}\right)^4 - \left(S\sin\frac{\varphi}{2}\right)^2}, \tag{11.12}$$

whose absolute value equals one for $S \leq 1$. Thus, the leapfrog update scheme of the FDTD approach is unconditionally stable for $S \leq 1$. A similar conclusion can be drawn for higher-dimensional update schemes [102, 103].

Yet, there is a price we have to pay for the simple FDTD propagation scheme. For a plane wave the electromagnetic fields should acquire a phase

$$e^{-i\omega\Delta t} = e^{-ikc\,\Delta t} = 1 - ikc\,\Delta t - \tfrac{1}{2}(kc\,\Delta t)^2 + \tfrac{i}{6}(kc\,\Delta t)^3 + \mathcal{O}\left(\Delta t^4\right)$$

for the propagation in the time interval Δt. In the last equality we have expanded the exponential for small arguments. If we expand the eigenvalues $\lambda_\pm$ for small arguments $k\Delta x$ through

$$\lambda_\pm = 1 \mp ikc\,\Delta t - \tfrac{1}{2}(kc\,\Delta t)^2 \pm \frac{i}{24}\left(\frac{1}{S^2} + 3\right)(kc\,\Delta t)^3 + \mathcal{O}\left(\Delta t^4\right),$$

we observe that only the lowest terms of the series coincide with those of the exponential. Thus, the dispersion relation $\omega(k)$ of the discretized FDTD update scheme differs from the $\omega = kc$ dispersion of the exact wave solution. As a consequence, when simulating wavepacket propagation in a homogeneous medium, the FDTD solutions give rise to wavepacket broadening because of the artificial dispersion caused by the finite difference discretization of Maxwell's equations. The situation is even more adverse in higher dimensions where the degree of dispersion depends on the propagation direction of the wave [102, 103].

We finally note that for $S = 1$ there exists a magic combination of Δx, Δt values where dispersion becomes completely suppressed. However, since in realistic FDTD simulations one typically has regions of different material parameters, associated with different c and S values, the existence of such a magic time step is usually of little practical use.

11.1.3 Perfectly Matched Layers

The computational domain of FDTD simulations has to be chosen of finite size. For this reason, one has to be careful about the behavior of the electromagnetic fields at the domain boundaries. In the unlikely case of a sufficiently large domain, the electromagnetic fields, which are either imprinted on the system through the initial conditions or created by current sources, will stay inside the domain. However, as time goes on the fields will always hit the boundary. Since the grid points at the domain boundaries are not connected to exterior grid points, a naive solution of the FDTD equations would lead to spurious reflections at the domain boundaries, thus spoiling the simulation results. In the context of FDTD several techniques exist for suppressing spurious reflections, the most important ones being the absorbing boundary conditions (ABCs) and the perfectly matched layers (PMLs). In particular the latter ones, initially introduced by Berenger [107], constitute a truly groundbreaking improvement for FDTD.

Consider a plane wave that impinges on a dielectric slab, as discussed in the context of the transfer matrix approach. For TM polarization and normal incidence the Fresnel reflection coefficient of Eq. (8.26) becomes

$$R^{\rm TM}=\frac{\varepsilon_2k_1-\varepsilon_1k_2}{\varepsilon_2k_1+\varepsilon_1k_2}=\frac{\varepsilon_2\sqrt{\varepsilon_1\mu_1}\,\omega-\varepsilon_1\sqrt{\varepsilon_2\mu_2}\,\omega}{\varepsilon_2\sqrt{\varepsilon_1\mu_1}\,\omega+\varepsilon_1\sqrt{\varepsilon_2\mu_2}\,\omega}=\frac{Z_1-Z_2}{Z_1+Z_2},$$

where we have used $k=\sqrt{\varepsilon\mu}\,\omega$ and have pulled out $\varepsilon_1\varepsilon_2\,\omega$ in the numerator and denominator to express the reflection coefficient in terms of the impedances

$$Z_1=\sqrt{\frac{\mu_1}{\varepsilon_1}},\qquad Z_2=\sqrt{\frac{\mu_2}{\varepsilon_2}}\,. \tag{11.13}$$

The grand idea underlying PMLs is that we enclose the simulation domain of interest by an artificial medium, see Fig. 11.4, whose permittivity and permeability is chosen such that (1) the impedance Z_1 of the inner medium matches Z_2 of the artificial outer material, in order to suppress back-reflections, and to (2) make the outer material dissipative in order to damp the outgoing waves. There exists a huge bunch of literature devoted to PMLs and possible generalizations and improvements. In the following we only sketch the main ideas without going into any details. The interested reader is referred to the literature [102, 103].

Consider for the PML material the following permittivity and permeability:

$$\varepsilon_2=\varepsilon_1\left(1+i\frac{\sigma}{\omega}\right),\quad \mu_2=\mu_1\left(1+i\frac{\sigma}{\omega}\right)\,. \tag{11.14}$$

One immediately observes that with this choice we get $Z_1=Z_2$ and the PML material is dissipative. There still remains one problem. In the above discussion we have assumed normal incidence of the wave at the interface, for the general case of oblique incidence the situation is more complicated. In what follows we discuss

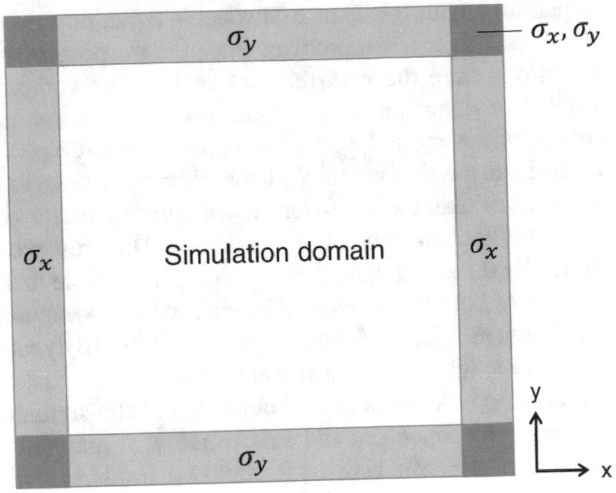

Fig. 11.4 Schematics for FDTD simulations in two spatial dimensions using perfectly matched layers. The simulation domain of interest is surrounded by perfectly matched layers (PML) with conductivities σ_x, σ_y which are chosen such that fields entering from the simulation domain (1) are not reflected (impedance matching), and are (2) absorbed in the PML material

the case of a TM wave in two spatial dimensions, with E_z, H_x, and H_y as dynamic variables, and refer to the original literature [107] for three spatial dimensions. In close analogy to Eq. (11.3), we rewrite the curl equations of Maxwell's equations in the form

$$\begin{aligned}
\varepsilon\frac{\partial E_{zx}}{\partial t} &= \frac{\partial H_y}{\partial x} - \tfrac{1}{2}J_x, & \mu\frac{\partial H_x}{\partial t} &= -\left(\frac{\partial E_{zx}}{\partial y} + \frac{\partial E_{zy}}{\partial y}\right) \\
\varepsilon\frac{\partial E_{zy}}{\partial t} &= -\frac{\partial H_y}{\partial x} - \tfrac{1}{2}J_x, & \mu\frac{\partial H_y}{\partial t} &= \left(\frac{\partial E_{zx}}{\partial x} + \frac{\partial E_{zy}}{\partial x}\right),
\end{aligned} \tag{11.15}$$

where we have introduced a somewhat artificial splitting of E_z into E_{zx} and E_{zy}. As the magnetic field components H_x and H_y are coupled to the sum of $E_z = E_{zx} + E_{zy}$, this splitting is at this point purely formal. However, we now continue to add to Eq. (11.15) dissipative terms of the form

$$\begin{aligned}
\varepsilon\frac{\partial E_{zx}}{\partial t} + \sigma_x E_{zx} &= \frac{\partial H_y}{\partial x} - \tfrac{1}{2}J_x, & \mu\frac{\partial H_x}{\partial t} + \sigma_y H_x &= -\left(\frac{\partial E_{zx}}{\partial y} + \frac{\partial E_{zy}}{\partial y}\right) \\
\varepsilon\frac{\partial E_{zy}}{\partial t} + \sigma_y E_{zy} &= -\frac{\partial H_y}{\partial x} - \tfrac{1}{2}J_x, & \mu\frac{\partial H_y}{\partial t} + \sigma_x H_y &= \left(\frac{\partial E_{zx}}{\partial x} + \frac{\partial E_{zy}}{\partial x}\right),
\end{aligned} \tag{11.16}$$

where σ_x and σ_y are conductivities. We next investigate the situation depicted in Fig. 11.4 where the two-dimensional simulation domain of interest is surrounded by a thin layer filled with PML materials, which are chosen as follows:

$$\text{Left and right PML with } \sigma_x > 0,\ \sigma_y = 0 \implies \text{damping of } E_{xz}$$
$$\text{Top and bottom PML with } \sigma_x = 0,\ \sigma_y > 0 \implies \text{damping of } E_{yz}\,.$$

Apparently, with this choice Maxwell's equations are not fulfilled inside the PML. However, this is not a problem since the purpose of PMLs is only (1) to suppress reflections from the PML material back to the computational simulation domain, and (2) to damp the fields in the PML material, which is achieved through the conductivities σ_x and σ_y. In the corners we set $\sigma_x = \sigma_y$ to remove all outgoing waves from the simulation. With the addition of these PML layers we can perform the simulations extremely efficiently, without any major complications in comparison to FDTD simulations without PML layers. The only exception is the slight overhead caused by the splitting of E_z into E_{zx} and E_{zy}. Note that the total simulation domain consists of both the domain of interest (whose dynamics is governed by Maxwell's equations) and the artificial PML material (whose dynamics is governed by unphysical conductivities). We also mention that there exists an alternative approach for suppressing reflections at the boundaries of the simulation domain, which is based on frequency-dependent and anisotropic $\bar{\bar{\varepsilon}}$, $\bar{\bar{\mu}}$ material parameters [102, 103]. One also has to be careful because in most PML implementations the layers do not absorb evanescent waves, so one has to make the simulation domain sufficiently large such that evanescent waves have decayed before they reach the boundary.

When using PML layers in FDTD simulations an additional complication arises because of the discretized version of Eq. (11.16). Here a PML layer of finite width can still lead to noticeable reflections at the simulation boundaries. To suppress these reflections, it is often advantageous to use graded $\sigma_x(x)$ and $\sigma_y(y)$ profiles, which start for the outgoing waves with a small σ value that is then gradually increased. A typical choice is $\sigma(d) = \sigma_0 d^m$, where d is the distance from the PML boundary to the simulation domain, σ_0 is some properly chosen constant, and m is a value between three and four. With this approach reflections at the FDTD simulation boundaries can be strongly suppressed. Figure 11.5 shows the example of an outgoing wave for simulations (a–c) without and (d–e) with PML layers.

11.1.4 Material Properties

Being a solution scheme in the time domain, the inclusion of frequency-dependent material parameters is not straightforward in FDTD. For Drude or Drude–Lorentz type permittivities, which usually suffice for a proper description of metals, we can rely on the microscopic description models introduced in Sect. 7.1. For instance, within the Drude–Lorentz model we use Eq. (7.4) for a driven harmonic oscillator,

$$m\ddot{x} = -m\omega_0^2 x - m\gamma\dot{x} + eE(t)\,.$$

where m and e are the mass and charge of the oscillator, respectively, ω_0 is the resonance frequency, and γ a damping constant. The oscillator displacement is related to the material polarization via $P = nex$, where n is the density of oscillators. The above model can be easily incorporated into the FDTD formalism by introducing a suitable discretization of the oscillator's equation of motion. For

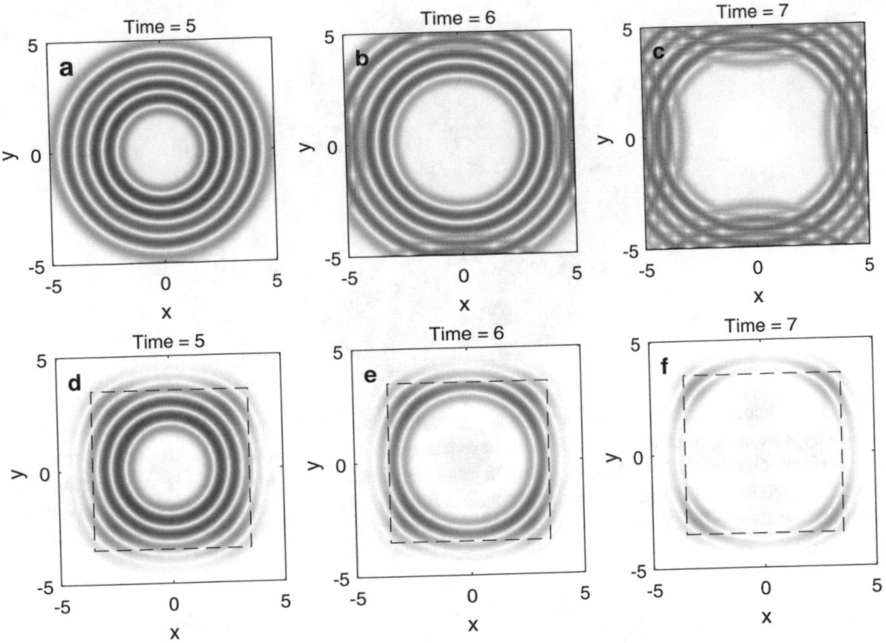

Fig. 11.5 Snapshots of electric field $E_z(x, y)$ for two-dimensional FDTD simulations (TM polarization) of an outgoing wave **(a–c)** without and **(d–e)** with perfectly matched layers. Without PML layers the wave becomes reflected at the simulation boundary causing spurious effects in the simulations. With PML layers the outgoing wave becomes absorbed in the PML material and back-reflections are strongly suppressed because of impedance matching. x, y, and t are given in dimensionless units

a more detailed discussion of such implementations, as well as of more advanced techniques for the implementation of other frequency-dependent permittivities and permeabilities, we refer the reader to the literature [102, 103].

11.2 Boundary Element Method

In previous chapters we have introduced the boundary integral method approach in some detail. When discussing its discretized version, the so-called *boundary element method* (BEM), we will build on many of the previously derived results. The systems we have in mind are depicted in Fig. 11.6 and consist of a metallic or dielectric particle with permittivity ε_1 embedded in a background medium with permittivity ε_2. The definition of denoting the particle inside and outside with subscripts 1 and 2 will be adopted throughout. Within the quasistatic approximation we can use Eq. (9.27) to compute the surface charges $\sigma(s)$ at the particle boundary $\partial\Omega$ from

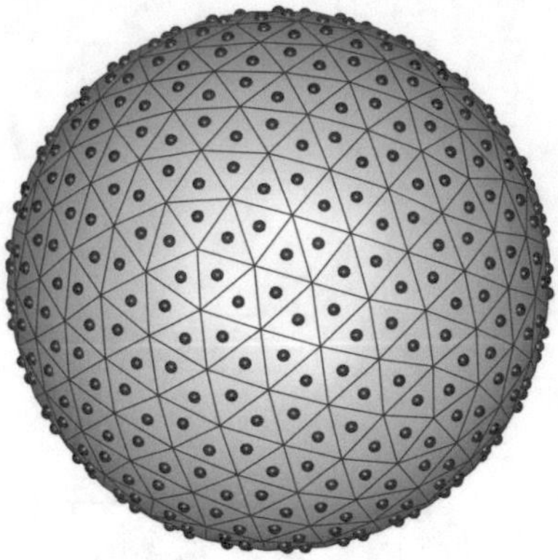

Fig. 11.6 Discretization of boundary for quasistatic BEM. The boundary of a sphere is approximated by a number of boundary elements, here triangles. In its most simple form, we approximate the surface charge distribution by surface charges located at the centroids of the boundary elements (see red dots), and match the potential and its surface derivative at the centroid positions (collocation method)

$$\Lambda(\omega)\sigma(\boldsymbol{s}) + \oint_{\partial\Omega} \frac{\partial G(\boldsymbol{s},\boldsymbol{s}')}{\partial n}\sigma(\boldsymbol{s}')\,dS' = -\frac{\partial V_{\text{inc}}(\boldsymbol{s})}{\partial n}\,.$$

Here, $\Lambda(\omega) = \frac{1}{2}(\varepsilon_1 + \varepsilon_2)/(\varepsilon_1 - \varepsilon_2)$, $G(\boldsymbol{s},\boldsymbol{s}')$ is the Green's function connecting positions $\boldsymbol{s}$ and $\boldsymbol{s}'$ on the particle boundary, and V_{inc} is the scalar potential of the incoming excitation, for instance a plane wave or an oscillating dipole. Once $\sigma(\boldsymbol{s})$ is known, we can compute the potential everywhere else using Eq. (9.24).

To render this expression suitable for numerical evaluation, we discretize the particle boundary in terms of sufficiently small boundary elements $\mathcal{T}_i$, as schematically shown in Fig. 11.6. These elements should be small enough to provide a meaningful approximation of the true particle boundary, with smaller elements at sharp edges or corners and larger elements at the flatter regions of the particle. The accuracy of the chosen grid is typically checked by running simulations with different numbers of boundary elements n, and comparing the corresponding simulation results. If the simulations give similar results we say that they are *converged* and keep the results, otherwise we further increase the number of boundary elements and run the simulations again. Each boundary element $\mathcal{T}_i$ has

- a centroid position $\boldsymbol{s}_i$,
- a normal vector $\boldsymbol{n}_i$,
- an area $\mathcal{A}_i$, and
- a surface charge distribution σ_i.

We next assume that the surface charge $q_i = \sigma_i \mathcal{A}_i$ within a given boundary element is located at the centroid position $\boldsymbol{s}_i$. More accurate approximation schemes will be discussed below. For the surface derivative of the Green's function we then get

$$\left[\!\left[F_{\text{stat}}\right]\!\right]_{ii'} = \left[\!\left[\frac{\partial G_{\text{stat}}}{\partial n}\right]\!\right]_{ii'} = -\boldsymbol{n}_i \cdot \frac{\boldsymbol{s}_i - \boldsymbol{s}_{i'}}{4\pi|\boldsymbol{s}_i - \boldsymbol{s}_{i'}|^3}, \tag{11.17}$$

where we have indicated the quasistatic approximation of F_{stat}. Here and in the following we use the notation $[\![\dots]\!]$ for vectors or matrices resulting from the discretization of quantities within the BEM approach. The boundary integral equation can then be expressed in terms of a matrix equation

$$\sum_{i'}\left(\Lambda(\omega)\delta_{ii'} + F_{ii'}^{\text{stat}}\mathcal{A}_{i'}\right)\sigma_{i'} = -\left(\frac{\partial V_{\text{inc}}}{\partial n}\right)_i,$$

whose solution provides us with the unknown surface charges. Within the quasistatic approximation, the surface charge distribution is computed from the solution of the matrix equation

Boundary Element Method (Quasistatic)

$$[\![\sigma]\!] = -\left[\!\left[\Lambda(\omega)\mathbb{1} + \mathscr{F}_{\text{stat}}\right]\!\right]^{-1} \cdot \left[\!\left[\frac{\partial V_{\text{inc}}}{\partial n}\right]\!\right], \tag{11.18}$$

where we have introduced the matrix $\mathscr{F}_{ii'}^{\text{stat}} = F_{ii'}^{\text{stat}}\mathcal{A}_{i'}$. To summarize, the computation of the surface charge distribution within the boundary element method approach consists of the following steps:

- discretize particle boundary using boundary elements of finite size,
- compute the surface derivative of the Green's function using Eq. (11.17),
- and solve the matrix equation of Eq. (11.18).

This approach can be easily implemented. However, despite its simplicity it gives surprisingly accurate results, as we will demonstrate in a moment. There exists one additional point that needs to be considered more carefully, which was first noted by Fuchs and Liu [108] and which concerns the diagonal elements F_{ii}^{stat}. Suppose that $\boldsymbol{s}'$ is a point on the particle boundary $\partial\Omega$, and $\boldsymbol{r}$ a point on a boundary S that encloses $\partial\Omega$. Owing to Gauss' theorem, which states that the flux of the electric field through some closed boundary S is proportional to the total charge Q enclosed in the volume limited by the boundary, we get

$$\oint_S \boldsymbol{E}\cdot d\boldsymbol{S} = \frac{Q}{\varepsilon_0} \implies -\oint_S \frac{\partial G_{\text{stat}}(\boldsymbol{r},\boldsymbol{s}')}{\partial n}dS = 1\,.$$

If we approach the boundaries $S \to \partial\Omega$ from the outside, we have to be careful about the limit $\boldsymbol{r} \to \boldsymbol{s}$ that we have previously discussed for the Neumann trace on page 222, and which gives an additional factor $-\frac{1}{2}\delta(\boldsymbol{s}-\boldsymbol{s}')$. We then obtain

$$\oint_{\partial\Omega} \frac{\partial G_{\text{stat}}(\boldsymbol{s}, \boldsymbol{s}')}{\partial n} dS = -\tfrac{1}{2} \,.$$

Thus, upon discretization of the boundary integral in terms of boundary elements we arrive at the sum rule

$$\sum_i F_{ii'}^{\text{stat}} \mathcal{A}_i = -\tfrac{1}{2}, \tag{11.19}$$

which allows us to compute the diagonal elements F_{ii}^{stat} once the off-diagonal elements have been computed from Eq. (11.17). This sum rule is of paramount importance for obtaining accurate results even for a relatively small number of boundary elements. In the following we discuss a few representative examples, and leave further examples as exercises to the interested reader.

Dipole Moment. Once the surface charge distribution is at hand, we can compute the dipole moment of the nanoparticle from

$$\boldsymbol{p} = \sum_i \boldsymbol{s}_i \Big(\sigma_i \mathcal{A}_i \Big), \tag{11.20}$$

which, in turn, allows us to compute the optical scattering and extinction spectra from Eqs. (9.21) and (9.22).

Convergence. As a representative example, we compute optical spectra for spherical and ellipsoidal nanoparticles for which analytic results can be obtained from Mie–Gans theory. Figure 11.7 shows a comparison between the numerical results (symbols) and the analytic expressions of Mie–Gans theory (solid lines). Despite the rather small number of a few hundred boundary elements, the agreement is excellent throughout.

Eigenmodes. The eigenvalues and eigenmodes of the quasistatic approach based on Eq. (9.28) can be easily computed within the BEM approach through

$$[\![\mathscr{F}_{\text{stat}}]\!] \cdot [\![u_k]\!] = \lambda_k [\![u_k]\!] \,. \tag{11.21}$$

The solution of the above equation can be easily obtained by using standard numerical methods for eigenvalue problems, as provided by most linear algebra software packages.

Full Maxwell's Equations

There exists a variant of the BEM approach discussed above, which allows solving the full Maxwell's equations. The approach has been developed by Garcia de Abajo and coworkers [109] and has found widespread use in the plasmonics community. It differs from the usual field-based approach based on Eq. (9.58), which

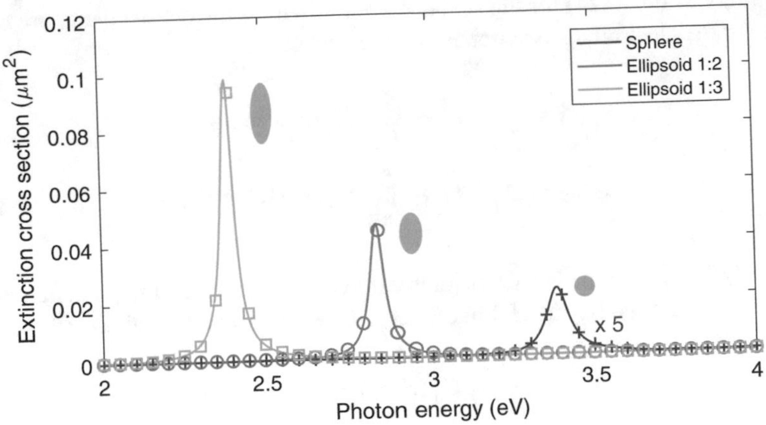

Fig. 11.7 Extinction cross sections for silver nanospheres and nanoellipsoids, as computed from Eq. (9.22), and comparison of BEM results (symbols) with analytic results of quasistatic Mie and Mie–Gans theory. We use the same material parameters and geometries as in Fig. 9.4. We observe excellent agreement between the analytic and numerical results, despite the rather small number of about 500 boundary elements

we will discuss further below, and uses as basic quantities the scalar and vector potentials $V, \boldsymbol{A}$ rather than the electromagnetic fields $\boldsymbol{E}, \boldsymbol{B}$. An open source toolbox implementing this scheme has been co-developed by the author of this book [110].

Our starting point are the Helmholtz equations of Eq. (2.22), which, for time-harmonic fields, read

$$\left(\nabla^2 + k_j^2\right) V_j(\boldsymbol{r}) = -\frac{\rho(\boldsymbol{r})}{\varepsilon_j}$$

$$\left(\nabla^2 + k_j^2\right) \boldsymbol{A}_j(\boldsymbol{r}) = -\mu_0 \boldsymbol{J}(\boldsymbol{r}) .$$

Here $k_j^2 = \mu_0 \varepsilon_j \omega^2$, and $j = 1, 2$ denotes the regions inside and outside the nanoparticle, respectively, which are separated by the boundary $\partial\Omega$. Throughout we set all permeabilities equal to μ_0. The above equations have been obtained using the Lorenz gauge condition of Eq. (2.21),

$$\nabla \cdot \tilde{\boldsymbol{A}}_j = i k_0 \kappa_j V_j , \qquad (11.22)$$

with $\varepsilon_j = \varepsilon_0 \kappa_j$, and k_0 is the wavenumber of light in vacuum. We shall find it convenient to work with the vector potential $\tilde{\boldsymbol{A}}_j = c\boldsymbol{A}_j$ which has the same dimensions as the scalar potential. Here c is the speed of light in vacuum. The electromagnetic fields can be obtained from the potentials via

$$\boldsymbol{E}_j = i k_0 \tilde{\boldsymbol{A}}_j - \nabla V_j , \quad c\boldsymbol{B}_j = \nabla \times \tilde{\boldsymbol{A}}_j .$$

In analogy to Eq. (9.24) for the scalar potential within the quasistatic approach, we make for the electromagnetic potentials the ansatz

$$V_j(\boldsymbol{r}) = V_j^{\text{inc}}(\boldsymbol{r}) + \oint_{\partial\Omega} G_j(\boldsymbol{r}, \boldsymbol{s}')\sigma_j(\boldsymbol{s}')\, dS'$$
$$\tilde{\boldsymbol{A}}_j(\boldsymbol{r}) = \tilde{\boldsymbol{A}}_j^{\text{inc}}(\boldsymbol{r}) + \oint_{\partial\Omega} G_j(\boldsymbol{r}, \boldsymbol{s}')\boldsymbol{h}_j(\boldsymbol{s}')\, dS', \tag{11.23}$$

where V_j^{inc}, $\tilde{\boldsymbol{A}}_j^{\text{inc}}$ are the electromagnetic potentials of the incoming excitation, and the Green's function for the Helmholtz equation is defined through Eq. (5.5),

$$\left(\nabla^2 + k_j^2\right) G_j(\boldsymbol{r}, \boldsymbol{r}') = -\delta(\boldsymbol{r} - \boldsymbol{r}')\,.$$

The ansatz of Eq. (11.23) is such that the Helmholtz equations inside and outside the particle are automatically fulfilled. We have introduced the surface charge σ_1, σ_2 and current $\boldsymbol{h}_1, \boldsymbol{h}_2$ distributions. In comparison to the quasistatic approach, the latter surface current distributions have to be introduced and we can no longer set equal σ_1, σ_2 at the particles' inside and outside. The unknown surface charge and current distributions must be determined such that the boundary conditions of Maxwell's equations are fulfilled at the particle boundary $\partial\Omega$.

Within the BEM approach we again approximate the boundary by elements $\mathcal{T}_i$ of finite size, and consider the surface charges $[\![\sigma_j]\!]$ and currents $[\![\boldsymbol{h}_j]\!]$ at the centroids of the boundary elements. As discussed in more detail in Sect. 11.6, these surface charge and current distributions can be obtained from the working equations of Eq. (11.66),

$$[\![\sigma_2]\!] = [\![\mathscr{G}_2]\!]^{-1} \cdot [\![\Sigma]\!]^{-1} \cdot \left([\![\tilde{D}^e]\!] + ik_0[\![\hat{\boldsymbol{n}}]\!] \cdot \left(\kappa_1 - \kappa_2\right)[\![\Delta]\!]^{-1} \cdot [\![\tilde{\boldsymbol{\alpha}}]\!]\right)$$
$$[\![\boldsymbol{h}_2]\!] = [\![\mathscr{G}_2]\!]^{-1} \cdot [\![\Delta]\!]^{-1} \cdot \left([\![\tilde{\boldsymbol{\alpha}}]\!] + ik_0[\![\hat{\boldsymbol{n}}]\!] \cdot \left(\kappa_1 - \kappa_2\right)[\![\mathscr{G}_2]\!] \cdot [\![\sigma_2]\!]\right),$$

together with Eq. (11.58) to compute the surface charge and current distributions $[\![\sigma_1]\!]$, $[\![\boldsymbol{h}_1]\!]$ at the particles' insides. For a definition of the various matrices see Sect. 11.6. Altogether, the solution scheme is considerably more complicated than in the quasistatic case, but is still numerically tractable and efficient for boundary discretizations with say a few thousand boundary elements.

11.3 Galerkin Scheme

In the previous section we have introduced a BEM approach where we have approximated the surface charge distribution $\sigma(\boldsymbol{s})$ by a finite number of values $[\![\sigma]\!]_i$ located at the centroids $\boldsymbol{s}_i$ of the boundary elements. There is nothing wrong with this approach, and in fact it has been used (and is still used) by many researchers in the field of plasmonics. Yet, there are situations where one would like to do better, where "better" should be understood in two different ways.

Interpolation. First, we may expect that an improved approximation scheme based on interpolation, in its most simple form linear interpolation of $\sigma(s)$ within a boundary element, will give more accurate simulation results. A systematic approach for this is the Galerkin scheme, to be discussed in a moment.

Convergence. The Galerkin scheme has the additional advantage that it can be investigated more rigorously in the field of numerical mathematics. Although we will not discuss these issues here, one can generally prove that certain types of boundary element approaches are guaranteed to converge when refining the boundary discretization, in contrast to centroid based approaches where often no rigorous conclusions can be drawn. For this reason, the field-based boundary element method approach, to be discussed in this section, is the "standard" approach usually employed in computational electrodynamics.

In the following, we first introduce the Galerkin scheme in a general manner, and then ponder on its implementation within the boundary element method framework. The same Galerkin scheme can be also used for other Maxwell's solver schemes based on the finite element method (FEM), as will be discussed later in this chapter.

11.3.1 Grand Idea of Galerkin Scheme

Consider first the equation

$$\int_{\mathscr{D}} K(\boldsymbol{r},\boldsymbol{r}')u(\boldsymbol{r}')\,dV' = b(\boldsymbol{r}), \tag{11.24}$$

where K is a kernel, u is the solution we are seeking for, and b is some inhomogeneity. $\boldsymbol{r}$ is restricted to a finite domain $\mathscr{D}$, which is two-dimensional for the BEM approach and three-dimensional for the FEM approach. The integration over the domain $\mathscr{D}$ is denoted with dV. The Galerkin scheme provides a means of converting an operator equation, such as Eq. (11.24), to a matrix representation that can be solved through matrix inversion. We first expand the solution $u(\boldsymbol{r})$ using a truncated basis $\varphi_\nu(\boldsymbol{r})$ of dimension n,

Approximate Solution

$$u(\boldsymbol{r}) \approx u^e(\boldsymbol{r}) = \sum_{\nu=1}^{n} \varphi_\nu(\boldsymbol{r})u_\nu^e, \tag{11.25}$$

where $u^e(\boldsymbol{r})$ is the approximate solution with expansion coefficients u^e_ν. The superscript e is a reminder of the finite *elements* to be introduced below, which support these basis functions. Because of the truncated basis, the defining Eq. (11.24) is no longer fulfilled,

$$\int_{\mathscr{D}} K(\boldsymbol{r},\boldsymbol{r}')\left[\sum_{\nu'=1}^{n}\varphi_{\nu'}(\boldsymbol{r}')u^e_{\nu'}\right]dV' - b(\boldsymbol{r}) = \text{res}, \tag{11.26}$$

but gives a finite residuum. We now seek for the best possible solution vector u^e_ν within the truncated function space. To this end, we multiply Eq. (11.26) with a test function

$$w^e(\boldsymbol{r}) = \sum_{\nu=1}^{n}\tilde{\varphi}_\nu(\boldsymbol{r})w^e_\nu\,. \tag{11.27}$$

In the Galerkin scheme we use for $w^e(\boldsymbol{r})$ the *same* functions $\tilde{\varphi}_\nu(\boldsymbol{r}) = \varphi_\nu(\boldsymbol{r})$ as for the expansion of the approximate solution vector, but for generality we here state the more general case where expansion and test functions might be different. To provide a viable computational approach, the number of expansion and test functions should be the same. We next integrate Eq. (11.26) multiplied with the test function over the entire domain $\mathscr{D}$ and get

$$\sum_{\nu=1}^{n} w^e_i\left[\sum_{\nu'=1}^{n}K_{\nu\nu'}u^e_{\nu'} - b_\nu\right] = \sum_{\nu=1}^{n} w^e_\nu\int_{\mathscr{D}}\tilde{\varphi}_i(\boldsymbol{r})\,\text{res}\,dV,$$

where we have introduced the abbreviations

$$\begin{aligned} K_{\nu\nu'} &= \iint_{\mathscr{D}}\tilde{\varphi}_\nu(\boldsymbol{r})K(\boldsymbol{r},\boldsymbol{r}')\varphi_{\nu'}(\boldsymbol{r}')\,dVdV' \\ b_\nu &= \int_{\mathscr{D}}\tilde{\varphi}_\nu(\boldsymbol{r})b(\boldsymbol{r})\,dV\,. \end{aligned} \tag{11.28}$$

At the optimal solution the value of the above expression should not change for small variations δw^e_ν of the test function. Apparently, the residuum term is minimized when the expression in brackets becomes zero, which can be written in explicit vector and matrix notation as

$$\bar{\bar{K}}\cdot\boldsymbol{u}^e = \boldsymbol{b}\,. \tag{11.29}$$

This equation can be solved through simple matrix inversion. The method of solving an operator equation by inverting a truncated version of a matrix representation is known as the Petrov–Galerkin method, the method of weighted residuals, or the method of moments. In the specialized literature one often calls the defining

Eq. (11.24) the *strong formulation* of the problem, whereas the expansion in terms of testing functions is called the Galerkin *weak formulation*.

Variational Formulation

The Galerkin scheme can be also formulated as a variational problem. We again start from the defining equation, Eq. (11.24), together with the expansion of the solution in terms of some basis functions $\varphi_\nu(s)$, Eq. (11.25). Introducing the function

$$\mathcal{S} = \int_{\mathscr{D}} w^e(\boldsymbol{r}) \left[\int K(\boldsymbol{r}, \boldsymbol{r}') u^e(\boldsymbol{r})\, dV' - b(\boldsymbol{r}) \right] dV, \tag{11.30}$$

the approximate solution can be obtained by minimizing this function with respect to the coefficients w_i^e of the test function. This leads us to the equations

$$\frac{\delta \mathcal{S}}{\delta w_\nu^e} = \sum_{\nu'=1}^{n} K_{\nu\nu'} u_{\nu'}^e - b_\nu = 0, \tag{11.31}$$

where ν runs over all degrees of freedom of the problem. In exercise 11.8 we show that the collocation approach of the BEM approach can be formulated using such a variational approach.

11.3.2 *Unstructured Grid*

The Galerkin scheme usually comes together with a decomposition of the computational domain $\mathscr{D}$ into elements $\mathcal{T}_i$ of finite size,

$$\mathscr{D} = \bigcup_{i=1}^{\mathcal{N}} \mathcal{T}_i\,. \tag{11.32}$$

In case of the boundary element method approach $\mathcal{T}_i$ are usually triangles, in case of the finite element method approach tetrahedrons. We assume that neighbor elements share a common vertex, edge, or face, but make otherwise no additional assumptions about the grid. The grid will be denoted as an *unstructured grid*, in contrast to the structured grid of the FDTD approach where each grid point has a well-defined neighborhood, with neighbors along all Cartesian directions. For this unstructured grid we introduce the following quantities:

Local Degrees of Freedom. Within each element $\mathcal{T}_i$ we introduce the local degrees of freedom u_{ia}^e which characterize the solution. For instance, in a triangle

u_{ia}^e might be associated with the function values at the triangle corners, with $a = 1, 2, 3$.

Local Shape Functions. We introduce the local shape functions $N_{ia}(\boldsymbol{r})$ that approximate the solution within $\mathcal{T}_i$ through

$$u^e(\boldsymbol{r}) = \sum_a N_{ia}^e(\boldsymbol{r})\, u_{ia}^e \quad \text{for } \boldsymbol{r} \in \mathcal{T}_i\,.$$

Outside the element $\mathcal{T}_i$ the shape functions $N_{ia}(\boldsymbol{r})$ are assumed to be zero. This property is sometimes denoted as a "local support." In a triangle, the shape functions might perform a linear interpolation from the triangle corners to the interior of the triangle.

Global Degrees of Freedom. In general, we impose certain constraints on the solution, such as that $u^e(\boldsymbol{r})$ is continuous when going from one element $\mathcal{T}_i$ to a neighbor one. This can be achieved by sharing information between neighbor elements of the grid, for instance by assigning the same values of u^e to the vertices of neighbor triangles. Thus, the total number of degrees of freedom is not governed by the number $\mathcal{N}$ of grid elements (triangles or tetrahedrons), but, in this example, by the number of grid vertices. To account for this, we introduce the global degrees of freedom $[\![u^e]\!]_\nu$, where ν runs from one to the total number n of global degrees of freedom.

Translation Matrix. We finally need a prescription of how to translate the global degrees of freedom $[\![u^e]\!]_\nu$ to the local ones. To this end, we introduce the translation matrix $T_{ia,\nu}$ with the translation property

$$u_{ia}^e = \sum_r T_{ia,\nu}\, [\![u^e]\!]_\nu\,.$$

Explained verbally, T takes the elements of the global degrees of freedom $[\![u^e]\!]_\nu$ and assigns them to the local degrees u_{ia}^e within a given element $\mathcal{T}_i$. See also Fig. 11.8. Typically, the matrix T has only a few matrix elements that differ from zero, and it is thus sparse.

We have now all ingredients at hand to formulate the Galerkin approach for an unstructured grid and for shape functions with a local support. The solution is expanded according to Eq. (11.25) in the form

$$u^e(\boldsymbol{r}) = \sum_{ia}\sum_\nu N_{ia}^e(\boldsymbol{r})\Big(T_{ia,\nu}[\![u^e]\!]_\nu\Big), \tag{11.33}$$

which can be interpreted as follows: the term in parentheses translates the global degrees of freedom $[\![u^e]\!]$ to the local ones, u_{ia}^e, and the shape functions $N_{ia}(\boldsymbol{r})$ perform interpolation in element $\mathcal{T}_i$. A completely similar approach can be pursued for the test functions, with possibly different shape functions $\tilde{N}_{ia}^e(\boldsymbol{r})$. Inserting these functions into the matrix elements of Eq. (11.28) for the Galerkin scheme, we get

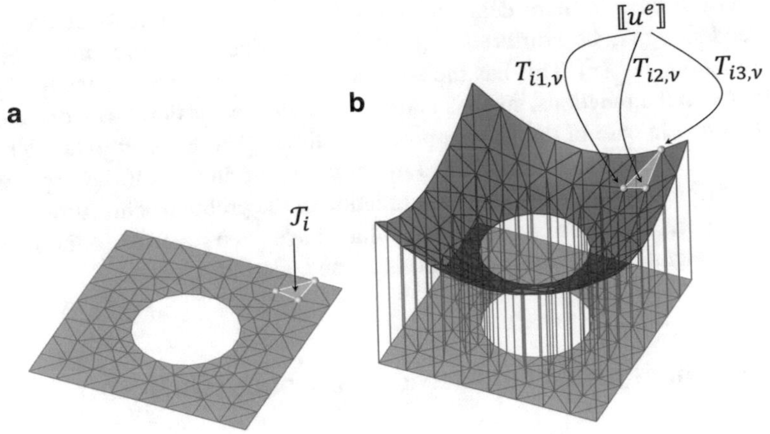

Fig. 11.8 Schematics of Galerkin scheme. (**a**) The computational domain $\mathscr{D}$ is approximated by an unstructured grid consisting of elements $\mathcal{T}_i$ of finite size. (**b**) The solution $u(\boldsymbol{r})$ is approximated by the function values u_{ia}^e at the vertices of the elements, and by performing (linear) interpolation within the elements. To get a smooth function, we assign the *same* values u^e to vertices shared by neighbor elements. This is achieved through the translation matrix $T_{ia,\nu}$ that assigns the elements of the global solution vector $[\![u^e]\!]_\nu$ to the vertices of the local elements

Matrix Elements for Galerkin Scheme

$$[\![K]\!]_{\nu\nu'} = \sum_{ia}\sum_{i'a'} T_{ia,\nu}\left[\int_{\mathcal{T}_i}\int_{\mathcal{T}_{i'}} \tilde{N}_{ia}^e(\boldsymbol{r})K(\boldsymbol{r},\boldsymbol{r}')N_{i'a'}^e(\boldsymbol{r}')\,dV\,dV'\right] T_{i'a',\nu'}$$

$$[\![b]\!]_\nu = \sum_{ia} T_{ia,\nu}\left[\int_{\mathcal{T}_i} \tilde{N}_{ia}^e(\boldsymbol{r})b(\boldsymbol{r})\,dV\right]. \qquad (11.34)$$

Note that these matrix elements can be easily computed because the integrals extend over the elements $\mathcal{T}_i$, $\mathcal{T}_{i'}$ only, as a result of the local support of the shape functions. In complete analogy to our previous discussion, we get a matrix equation $[\![K]\!] \cdot [\![u^e]\!] = [\![b]\!]$ that again can be solved through matrix inversion.

Vector Solution

In computational electrodynamics we typically have to solve equations for vector functions, such as the tangential electric and magnetic fields in case of the BEM approach. Quite generally, this can be done along the lines discussed in this section, although we have to account for the different components of the vector solution.

As we will discuss in more detail in the following, it turns out to be convenient to proceed along slightly different lines and to introduce from the outset vectorial shape functions $\boldsymbol{N}_{ia}^{e}(\boldsymbol{r})$. This has the advantage that we can impose easily physical constraints on the functions, such as continuity of the vector flow from one element to another one in case of the BEM approach. Although these constraints are always fulfilled for the exact solutions, things might go wrong in case of the approximate solutions originating from the weak formulation of the problem. This turns out to be vital for suppressing spurious solutions, which have been reported in the literature for more straightforward implementations of the Galerkin scheme.

11.4 Boundary Element Method Approach (Galerkin)

In this section we apply the Galerkin scheme to the boundary integral method for the solution of the full Maxwell's equations. Quite generally, the approach is quite involved and contains a lot of technicalities. However, with the machinery developed in Chaps. 5, 9, and 11 we are in the position of deriving the working equations without too much additional work. In the following we first ponder on the vector shape elements, the so-called Raviart–Thomas or Rao–Wilton–Glisson basis functions, and then sketch the implementation of the boundary integral method equations introduced in Sect. 9.6.

11.4.1 Raviart–Thomas Basis Functions

Suppose that $\boldsymbol{u}(s)$ is a vector field where all vectors are tangential to the particle boundary. This case corresponds to the field-based BEM approach of Eq. (9.58). As discussed in Sect. 9.6, the tangential magnetic and electric fields can be also interpreted as surface currents and magnetizations, respectively.

For a discretized particle boundary, we then have to make sure that the approximate solution $\boldsymbol{u}^e(s)$ fulfills certain boundary conditions at the edges between two neighbor triangles. Let $A = lh$ be a two-dimensional domain which encloses a given edge with length l, see Fig. 11.9, and h is the height of the area which we let approach zero at the end. From the two-dimensional version of Gauss' theorem we get

$$\int_A \nabla\cdot\boldsymbol{u}\,dS = \oint_{\partial A}\boldsymbol{u}\cdot d\boldsymbol{\eta} = 0 \implies \Big[\boldsymbol{u}(s^+) - \boldsymbol{u}(s^-)\Big]\cdot\boldsymbol{\eta} = 0, \tag{11.35}$$

where $\boldsymbol{\eta}$ is a vector perpendicular to the edge. $\boldsymbol{u}(s^{\pm})$ are the values of the vector field at the two adjacent sides of the edge. Equation (11.35) states that the flux (in the direction of $\boldsymbol{\eta}$) of the tangential field $\boldsymbol{u}$ leaving one triangle must equal the flux

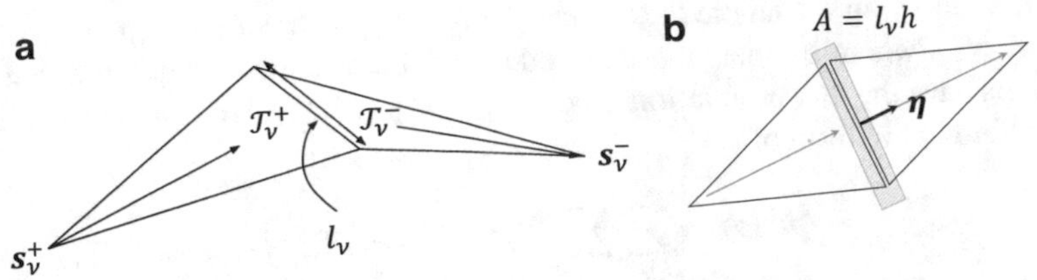

Fig. 11.9 Raviart–Thomas basis elements. (**a**) Each edge has two neighbor triangles $\mathcal{T}_\nu^+$ and $\mathcal{T}_\nu^-$, situated left and right of the (oriented) edge. In $\mathcal{T}_\nu^+$ the shape functions originate from the vertex s_ν^+, located oppositely to the edge and flow towards the edge. In $\mathcal{T}_\nu^-$ they flow in the opposite direction. The vector functions are defined such that the flow leaving triangle $\mathcal{T}_\nu^+$ equals the flow entering triangle $\mathcal{T}_\nu^-$. (**b**) Schematics for application of the two-dimensional Gauss' theorem. A is the domain enclosing the edge, with a height $h \to 0$, and $\boldsymbol{\eta}$ is a vector perpendicular to the edge. For details see text

entering the neighbor triangle. For the surface current distribution of Eq. (9.58) this means that no singular line-charge distribution builds up at the edges.

We next introduce a table $\mathcal{E}_\nu$ of unique triangle edges (without consideration of orientation) with $\nu = 1, \dots n$, where n is the total number of edges. In addition, we denote with $\mathcal{T}_\nu^\pm$ the triangles located on the two sides of the edges, and with $s_\nu^\pm$ the vertices located oppositely to the edges, as schematically shown in Fig. 11.9. With this we can define the following vector shape functions:

Basis Functions BEM Approach (Galerkin)

$$\boldsymbol{\varphi}_\nu(s) = \begin{cases} N_{\nu+}^e(s) = \dfrac{l_\nu}{2\mathcal{A}_\nu^+}\left(s - s_\nu^+\right) & \text{for } s \in \mathcal{T}_\nu^+ \\ N_{\nu-}^e(s) = -\dfrac{l_\nu}{2\mathcal{A}_\nu^-}\left(s - s_\nu^-\right) & \text{for } s \in \mathcal{T}_\nu^- \\ 0 & \text{else}\,. \end{cases} \tag{11.36}$$

Interpolation from edges ν of grid to interiors of triangles $\mathcal{T}_\nu^\pm$ located at both edge sides using the Raviart–Thomas shape functions $N_{\nu\pm}^e(s)$.

Here l_ν is the length of the edge, and $\mathcal{A}_\nu^\pm$ are the triangle areas. Note that the choice of which shape function has a positive or negative sign is arbitrary, the only thing that matters is that the two functions have different signs. The prefactors of the shape functions have been chosen such that the flux through the edge is unity, as discussed in more detail in Exercise 11.9, which guarantees the conservation of the flux at the edges. One often refers to $N_{\nu\pm}^e$ as the Raviart–Thomas or Rao–Wilton–Glisson basis elements.

In the following we denote the Raviart–Thomas basis functions in a given triangle $\mathcal{T}_i$ with $\boldsymbol{N}^e_{ia}$, where a labels the three edges of the triangle. We assume that these functions have the flux orientation according to Eq. (11.36). We can now expand the vector field in the form of

$$\boldsymbol{u}^e(\boldsymbol{s}) = \sum_\nu \sum_{ia} \boldsymbol{N}^e_{ia}(\boldsymbol{s}) \Big(T_{ia,\nu} [\![u^e]\!]_\nu \Big), \tag{11.37}$$

where the coefficients $[\![u^e]\!]_\nu$ characterize the solution $\boldsymbol{u}^e(\boldsymbol{s})$, and $T_{ia,\nu}$ is the previously defined translation matrix from the unique edges to the edges of the local element. Expressed verbally, the term in parentheses translates from the global edges to the local triangle edges. The shape functions $\boldsymbol{N}^e_{ia}$ then interpolate the tangential fields from the edges to the triangle interior. The matrix elements and the inhomogeneity of Eq. (11.28) for the discretized kernel can be written in the form

$$\begin{aligned} [\![K]\!]_{\nu\nu'} &= \sum_{ia} \sum_{i'a'} T_{ia,\nu} T_{i'a',\nu'} \int_{\mathcal{T}_i} \int_{\mathcal{T}_{i'}} \boldsymbol{N}^e_{ia}(\boldsymbol{s}) \cdot K(\boldsymbol{s},\boldsymbol{s}') \boldsymbol{N}^e_{i'a'}(\boldsymbol{s}')\, dS dS' \\ [\![b]\!]_\nu &= \sum_{ia} T_{ia,\nu} \int_{\mathcal{T}_i} \boldsymbol{N}^e_{ia}(\boldsymbol{s}) \cdot \boldsymbol{b}(\boldsymbol{s})\, dS\,. \end{aligned} \tag{11.38}$$

Note that in the spirit of the Galerkin scheme we have used the same test and expansion functions $\boldsymbol{N}^e_{ia}$.

11.4.2 Galerkin Scheme for Full Maxwell's Equations

In Sect. 9.6 we have derived a field-based boundary integral method approach for the solution of the full Maxwell's equations. Its central expression of Eq. (9.58) reads

$$\begin{aligned} \hat{\boldsymbol{n}} \times \Big(\mathbb{D}_1[\boldsymbol{u}_E] + \mathbb{D}_2[\boldsymbol{u}_E] \Big) - \hat{\boldsymbol{n}} \times \Big(i\omega\mu_1 \mathbb{S}_1[\boldsymbol{u}_H] + i\omega\mu_2 \mathbb{S}_2[\boldsymbol{u}_H] \Big) &= \hat{\boldsymbol{n}} \times \boldsymbol{E}^{\rm inc} \\ \hat{\boldsymbol{n}} \times \Big(\mathbb{D}_1[\boldsymbol{u}_H] + \mathbb{D}_2[\boldsymbol{u}_H] \Big) + \hat{\boldsymbol{n}} \times \Big(i\omega\varepsilon_1 \mathbb{S}_1[\boldsymbol{u}_E] + i\omega\varepsilon_2 \mathbb{S}_2[\boldsymbol{u}_E] \Big) &= \hat{\boldsymbol{n}} \times \boldsymbol{H}^{\rm inc}, \end{aligned}$$

where $\boldsymbol{u}_E = \hat{\boldsymbol{n}} \times \boldsymbol{E}$, $\boldsymbol{u}_H = \hat{\boldsymbol{n}} \times \boldsymbol{H}$ denote the tangential electromagnetic fields, and the single and double layer integral operators $\mathbb{S}_j$, $\mathbb{D}_j$ are defined through

$$\begin{aligned} \big[\mathbb{S}_j\, \boldsymbol{u}\big](\boldsymbol{s}) &= \oint_{\partial\Omega} \Big[G_j(\boldsymbol{s},\boldsymbol{s}')\boldsymbol{u}(\boldsymbol{s}') + \frac{1}{k_j^2} \nabla G_j(\boldsymbol{s},\boldsymbol{s}')\, \nabla' \cdot \boldsymbol{u}(\boldsymbol{s}') \Big]\, dS' \\ \big[\mathbb{D}_j\, \boldsymbol{u}\big](\boldsymbol{s}) &= \oint_{\partial\Omega} \Big[\nabla' \times G_j(\boldsymbol{s},\boldsymbol{s}') \Big] \cdot \boldsymbol{u}(\boldsymbol{s}')\, dS'\,. \end{aligned} \tag{11.39}$$

See Sect. 5.5 for the derivation of these expressions. To obtain within the Galerkin scheme a weak formulation of the problem, we multiply the boundary integral equations with test functions $\boldsymbol{w}^e_{E,H}(\boldsymbol{s})$ and integrate over the boundary. We additionally use

$$\oint_{\partial\Omega} \boldsymbol{w}^e_{E,H}(\boldsymbol{s})\cdot\hat{\boldsymbol{n}}\times\Big[\dots\Big]\,dS = -\oint_{\partial\Omega}\hat{\boldsymbol{n}}\times\boldsymbol{w}^e_{E,H}(\boldsymbol{s})\cdot\Big[\dots\Big]\,dS$$

to shuffle the cross product with the outer surface normal to the test function. Next, we expand both the tangential electromagnetic fields $\boldsymbol{u}_{E,H}$ and the test functions $\hat{\boldsymbol{n}}\times\boldsymbol{w}^e_{E,H}$ in the basis of Thomas–Raviart elements. This gives

$$\boldsymbol{u}^e_{E,H}(\boldsymbol{s}) = \sum_\nu\sum_{ia}\boldsymbol{N}^e_{ia}(\boldsymbol{s})\left(T_{ia,\nu}[\![u^e_{E,H}]\!]_\nu\right),$$

and a corresponding expression for $\boldsymbol{w}^e_{E,H}$. Upon discretization within the Galerkin scheme we get the single and double layer matrices

$$\begin{aligned}
[\![\mathbb{S}_j]\!]_{\nu\nu'} &= \sum_{ia}\sum_{i'a'} T_{ia,\nu}T_{i'a',\nu'}\oint_{T_i}\oint_{T_{i'}} \\
&\quad\times\left\{\boldsymbol{N}^e_{ia}(\boldsymbol{s})\cdot\boldsymbol{N}^e_{i'a'}(\boldsymbol{s}') - \frac{\left[\nabla\cdot\boldsymbol{N}^e_{ia}(\boldsymbol{s})\right]\left[\nabla'\cdot\boldsymbol{N}^e_{i'a'}(\boldsymbol{s}')\right]}{k_j^2}\right\}G_j(\boldsymbol{s},\boldsymbol{s}')\,dSdS' \\
[\![\mathbb{D}_j]\!]_{\nu\nu'} &= \sum_{ia}\sum_{i'a'} T_{ia,\nu}T_{i'a',\nu'}\oint_{T_i}\oint_{T_{i'}}\boldsymbol{N}^e_{ia}(\boldsymbol{s})\cdot\left[\nabla' G_j(\boldsymbol{s},\boldsymbol{s}')\right]\times\boldsymbol{N}^e_{i'a'}(\boldsymbol{s}')\,dSdS',
\end{aligned}\tag{11.40}$$

where we have performed for $\mathbb{S}_j$ an integration by parts in order to bring the derivative from the Green's function to the Raviart–Thomas basis element. With the inhomogeneities

$$[\![q^{\text{inc}}_{E,H}]\!]_\nu = \sum_{ia} T_{ia,\nu}\oint_{T_i}\boldsymbol{N}^e_{ia}(\boldsymbol{s})\cdot\begin{Bmatrix}\boldsymbol{E}^{\text{inc}}(\boldsymbol{s})\\ \boldsymbol{H}^{\text{inc}}(\boldsymbol{s})\end{Bmatrix}dS,$$

we then arrive at the working equations for the field-based BEM approach [111]

Working Equation for Field-Based BEM (Galerkin)

$$\begin{pmatrix}[\![\mathbb{D}_1+\mathbb{D}_2]\!] & -i\omega[\![\mu_1\mathbb{S}_1+\mu_2\mathbb{S}_2]\!]\\ i\omega[\![\varepsilon_1\mathbb{S}_1+\varepsilon_2\mathbb{S}_2]\!] & [\![\mathbb{D}_1+\mathbb{D}_2]\!]\end{pmatrix}\cdot\begin{pmatrix}[\![u^e_E]\!]\\ [\![u^e_H]\!]\end{pmatrix} = \begin{pmatrix}[\![q^{\text{inc}}_E]\!]\\ [\![q^{\text{inc}}_H]\!]\end{pmatrix}. \tag{11.41}$$

Upon inversion of the matrix equation, we get the vectors $[\![u^e_{E,H}]\!]$ characterizing the solution. In the evaluation of the integrals we have to be careful about the divergent behavior of the Green's function and its surface derivative for $s \to s'$. Although all integrals are well behaved, one has to proceed carefully for integrations over a single triangle, as well as for integrations over triangles that share a common edge or vertex. These integrals are conveniently solved by either analytic integration [112] or by using a full numerical quadrature, where suitable coordinate transformations allow to render the singular contributions feasible for numerical integration [113, 114]. For details the interested reader is referred to the literature.

The symbols in Fig. 11.10 show simulation results of the extinction cross section for silver nanospheres of different size. The tangential fields are computed according to Eq. (11.41). For the extinction cross sections we compute the Poynting vector directly at the particle boundary. We also compare with the analytic results of Mie theory (solid lines), finding perfect agreement throughout.

11.5 Finite Element Method

The finite element method (FEM) is a general solution scheme for Maxwell's equations, and can deal with arbitrary sources and inhomogeneous dielectric and magnetic environments. Similarly to the FDTD approach, it starts with the curl equations, Eq. (11.1), and solves them either in the time or frequency domain. Many features of the FEM approach are along the same lines as for FDTD, for instance the consideration of perfectly matched layers. However, the FEM scheme starts from the outset with an unstructured grid, typically coming from the discretization of the three-dimensional computational domain $\Omega = \bigcup_i \mathcal{T}_i$ into tetrahedrons $\mathcal{T}_i$, and thus allows avoiding the staircase approximation errors inherent to FDTD. Additionally, within FEM one can use basis functions with an arbitrary polynomial degree, which opens the possibility for highly accurate simulations. One might argue that FEM is the most versatile and accurate simulation approach considered here, although its implementation is somewhat demanding and there exist many technicalities which will not be addressed in our brief discussion.

In the following we start by discussing FEM in the frequency domain, and then ponder on the proper choice of the basis elements, the so-called Nedelec elements. We finally address the solution of Maxwell's equations in the time domain within the discontinuous Galerkin scheme.

11.5.1 Finite Element Method in Frequency Domain

Explained briefly, the FEM approach is a direct implementation of the Galerkin scheme for the full Maxwell's equations. We start with the curl equations

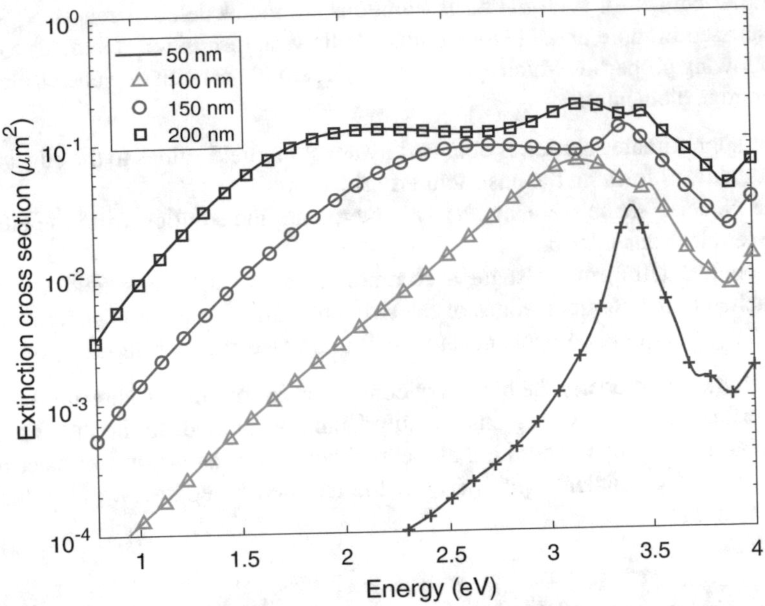

Fig. 11.10 BEM Galerkin simulations for silver nanospheres of different size. We compare the numerical results (symbols) with the analytic results of Mie theory (solid lines), see also Fig. 9.18, finding perfect agreement throughout. The boundary discretization of the sphere used in the simulation is shown in Fig. 11.6

$$\nabla \times \boldsymbol{E} = -\bar{\bar{\mu}} \cdot \frac{\partial \boldsymbol{H}}{\partial t}, \qquad \nabla \times \boldsymbol{H} = \boldsymbol{J} + \bar{\bar{\varepsilon}} \cdot \frac{\partial \boldsymbol{E}}{\partial t},$$

and consider for generality anisotropic material parameters $\bar{\bar{\varepsilon}}$, $\bar{\bar{\mu}}$. To submit these equations to the Galerkin scheme we have to specify the basis elements. Naively, we could assign the electric and magnetic fields to the vertices of the tetrahedrons and perform a linear interpolation within the elements. Unfortunately it turns out that this implementation sometimes leads to convergence to wrong solutions or spurious modes [115, 116]. This failure is attributed to the lack of enforcing the constraint of divergence free fields. In the comprehensive review article of Hesthaven it is discussed that it turns out to be advantageous to use vectorial basis functions instead [116].

> The main motivation for seeking vector basis functions is the observation that the boundary conditions for Maxwell's equations are vectorial, i.e., it is natural when seeking a conforming discretization to utilize vector basis functions. Such basis functions, often known as curl conforming elements, should satisfy fundamental properties of the solutions to Maxwell's equations, e.g., support tangential continuity of the solutions. This allows for imposing tangential continuity between elements with different materials as well as impose boundary conditions in a natural way. Furthermore, the use of such elements guarantees the absence of spurious modes in frequency-domain finite element schemes.

A popular choice for vectorial basis functions are the Nedelec elements $\boldsymbol{N}_{ia}^{e}(\boldsymbol{r})$, to be discussed in more detail below. For our following discussion it suffices to know the following properties which apply to the most simple version of Nedelec elements (linear edge elements):

- The approximate solution is obtained by attaching field values to the edges of the tetrahedrons forming the unstructured grid.
- The Nedelec shape elements $\boldsymbol{N}_{ia}^{e}(\boldsymbol{r})$ interpolate the solution from the edges to the tetrahedrons interior.
- When two tetrahedrons share a common face and the *same* field values are attached to the common edges of the two tetrahedrons, then it is guaranteed that the tangential field components at the common face are continuous.

Through the last property the boundary conditions of continuous tangential fields at the interface between two regions are automatically fulfilled. In the FEM approach the global degrees of freedom are associated with the time-dependent electric and magnetic field coefficients $[\![u_{E,H}^{e}(t)]\!]$ at the tetrahedron edges. We introduce the basis functions

Basis Functions FEM Approach (Galerkin)

$$\boldsymbol{\varphi}_\nu(\boldsymbol{r}) = \sum_{ia} \boldsymbol{N}_{ia}^{e}(\boldsymbol{r})\, T_{ia,\nu}\,. \tag{11.42}$$

Translate from unique edges ν of grid to local edges a of tetrahedron $\mathcal{T}_i$ which contain edge ν, and perform interpolation using the Nedelec shape functions $\boldsymbol{N}_{ia}^{e}(\boldsymbol{r})$.

The approximate solution $\boldsymbol{E}^e$ then reads

$$\boldsymbol{E}^e(\boldsymbol{r},t) = \sum_{ia}\sum_{\nu} \boldsymbol{N}_{ia}^{e}(\boldsymbol{r}) \left(T_{ia,\nu}[\![u_E^e(t)]\!]_\nu\right), \tag{11.43}$$

with a corresponding expression for the magnetic field. Here $\boldsymbol{N}_{ia}^{e}$ are the Nedelec elements that are assumed to have a local support, which means that they are zero for $\boldsymbol{r}$ being located outside of tetrahedron $\mathcal{T}_i$. Explained verbally, the term in parentheses brings the global degrees of freedom $[\![u_{E,H}^{e}]\!]$ (defined at the unique edges of the grid) to the edges of the local element $\mathcal{T}_i$, where $T_{ia,\nu}$ is the translation matrix. The basis elements $\boldsymbol{N}_{ia}^{e}$ perform an interpolation within each tetrahedron. Inserting this ansatz into Faraday's law then gives

$$\bar{\bar{\mu}} \cdot \sum_{a'\nu} \boldsymbol{N}^e_{ia'}(\boldsymbol{r}) \left(T_{ia',\nu} [\![\partial_t u^e_H(t)]\!]_\nu \right) = -\sum_{a'\nu} \nabla \times \boldsymbol{N}^e_{ia'}(\boldsymbol{r}) \left(T_{ia',\nu} [\![u^e_E(t)]\!]_\nu \right),$$

where we have assumed that $\boldsymbol{r}$ is located inside of $\mathcal{T}_i$. A similar expression can be obtained for Ampere's law. We next multiply these expressions with the test functions

$$\boldsymbol{w}^e_{E,H}(\boldsymbol{r},t) = \sum_{ia} \sum_{\nu} \boldsymbol{N}^e_{ia}(\boldsymbol{r}) \left(T_{ia,\nu} [\![w^e_{E,H}(t)]\!]_\nu \right),$$

and integrate over the tetrahedron volumes. Because of the local support of the basis functions we only have to consider integrals where both shape functions are located inside the same tetrahedron. To simplify our final expression, we introduce the mass matrices

$$\begin{Bmatrix} M^\varepsilon_{ia,ia'} \\ M^\mu_{ia,ia'} \end{Bmatrix} = \int_{\mathcal{T}_i} \boldsymbol{N}^e_{ia}(\boldsymbol{r}) \cdot \begin{Bmatrix} \bar{\bar{\varepsilon}} \\ \bar{\bar{\mu}} \end{Bmatrix} \cdot \boldsymbol{N}^e_{ia'}(\boldsymbol{r})\, d^3r \tag{11.44}$$

together with the stiffness matrix S and the inhomogeneity J, defined through

$$\begin{aligned} S_{ia,ia'} &= \int_{\mathcal{T}_i} \boldsymbol{N}^e_{ia}(\boldsymbol{r}) \cdot \nabla \times \boldsymbol{N}^e_{ia'}(\boldsymbol{r})\, d^3r \\ J_{ia}(t) &= \int_{\mathcal{T}_i} \boldsymbol{N}^e_{ia}(\boldsymbol{r}) \cdot \boldsymbol{J}(\boldsymbol{r},t)\, d^3r\,. \end{aligned} \tag{11.45}$$

With this, the curl equations of Maxwell's equations read

$$\sum_{i,aa'} \sum_{\nu'} T_{ia,\nu}\, T_{ia',\nu'} \left(M^\mu_{ia,ia'} [\![\partial_t u^e_H(t)]\!]_{\nu'} + S_{ia,ia'} [\![u^e_E(t)]\!]_{\nu'} \right) = 0$$

$$\sum_{i,aa'} \sum_{\nu'} T_{ia,\nu}\, T_{ia',\nu'} \left(M^\varepsilon_{ia,ia'} [\![\partial_t u^e_E(t)]\!]_{\nu'} - S_{ia,ia'} [\![u^e_H(t)]\!]_{\nu'} \right) = -\sum_{ia} T_{ia,\nu} J_{ia}(t)\,.$$

The working equations of the FEM approach directly follow from these expressions. We get

Working Equations FEM Approach

$$\begin{aligned} [\![M^\mu]\!] \cdot [\![\partial_t u^e_H(t)]\!] + [\![S]\!] \cdot [\![u^e_E(t)]\!] &= 0 \\ [\![M^\varepsilon]\!] \cdot [\![\partial_t u^e_E(t)]\!] - [\![S]\!] \cdot [\![u^e_H(t)]\!] &= -[\![J(t)]\!], \end{aligned} \tag{11.46}$$

where we have introduced

$$[\![A]\!]_{\nu\nu'} = \sum_{i,aa'} T_{ia,\nu} T_{ia',\nu'} \, A_{ia,ia'}$$

for the mass and stiffness matrices, together with $[\![J]\!]_\nu = \sum_{ia} T_{ia,\nu} \, J_{ia}$. In the frequency domain the time-derivative operator has to be replaced by $-i\omega$, and the electric and magnetic field coefficients can be obtained through matrix inversion. Because of the local support of the basis functions the matrices within the FEM approach are sparse, which can be exploited for a fast and efficient solution of the working equations.

11.5.2 Nedelec Elements

The basis elements of the FEM scheme are usually known as edge elements, Nedelec elements, Whitney forms, or curl conforming vector elements [115–118]. In the following we briefly discuss its most simple form, which allows representing a vector field $\boldsymbol{u}(\boldsymbol{r}) = \boldsymbol{a} + \boldsymbol{b} \times \boldsymbol{r}$ within a given tetrahedron, and refer for a thorough discussion as well as for the derivation of basis elements with a higher polynomial order to the specialized literature. Within a given tetrahedron $\mathcal{T}$ the vector function is given by

$$\boldsymbol{u}^e(\boldsymbol{r}) = \sum_{a=1}^{6} \boldsymbol{N}_a^e(\boldsymbol{r}) \, u_a^e,$$

where $\boldsymbol{N}_a^e$ are the Nedelec elements and u_a^e the coefficients for the approximated vector field given at the six edges a of the tetrahedron. The recipe for obtaining the shape elements then proceeds as follows:

- Let $\boldsymbol{r}_k$ denote the four vertices of the tetrahedron.
- We introduce four linear basis functions $\lambda_k(\boldsymbol{r}) = a_k + b_k x + c_k y + d_k z$, where the coefficients are chosen such that they are one at a given vertex and zero otherwise, corresponding to $\lambda_k(\boldsymbol{r}_{k'}) = \delta_{kk'}$.
- Each edge $\mathcal{E}_a$ is directed from vertex $\boldsymbol{r}_{a1}$ to vertex $\boldsymbol{r}_{a2}$ in a direction that has to be specified once at the beginning.

With this, the Nedelec elements of lowest order are defined through

Curl Conforming Nedelec Elements

$$\boldsymbol{N}_a^e(\boldsymbol{r}) = \big[\nabla\lambda_{a1}(\boldsymbol{r})\big]\lambda_{a2}(\boldsymbol{r}) - \lambda_{a1}(\boldsymbol{r})\big[\nabla\lambda_{a2}(\boldsymbol{r})\big]. \tag{11.47}$$

For a given vector field $\boldsymbol{u}(\boldsymbol{r})$ the field coefficients u_a^e at the edges can then be obtained from

$$u_a^e = \int_{\mathcal{E}_a} \boldsymbol{u}\cdot\boldsymbol{\tau}_a\, ds,$$

where $\boldsymbol{\tau}_a$ is the unit vector of edge $\mathcal{E}_a$, and ds denotes the integration along the edge.

Unit Tetrahedra. As a representative example, we consider the unit tetrahedron with vertices $\boldsymbol{r}_1 = (0,0,0)$, $\boldsymbol{r}_2 = (1,0,0)$, $\boldsymbol{r}_3 = (0,1,0)$, and $\boldsymbol{r}_4 = (0,0,\pm 1)$. The coefficients of the linear basis functions can be obtained from

$$\begin{pmatrix} a_1 & b_1 & c_1 & d_1 \\ a_2 & b_2 & c_2 & d_2 \\ a_3 & b_3 & c_3 & d_3 \\ a_4 & b_4 & c_4 & d_4 \end{pmatrix} \cdot \begin{pmatrix} 1 & 1 & 1 & 1 \\ 0 & 1 & 0 & 0 \\ 0 & 0 & 1 & 0 \\ 0 & 0 & 0 & \pm 1 \end{pmatrix} = \mathbb{1}\,.$$

This leads us to the linear basis functions

$$\lambda_1(\boldsymbol{r}) = 1 - x - y - z, \quad \lambda_2(\boldsymbol{r}) = x, \quad \lambda_3(\boldsymbol{r}) = y, \quad \lambda_4(\boldsymbol{r}) = \pm z,$$

and their derivatives

$$\nabla\lambda_1(\boldsymbol{r}) = -\hat{\boldsymbol{x}} - \hat{\boldsymbol{y}} - \hat{\boldsymbol{z}}, \quad \nabla\lambda_2(\boldsymbol{r}) = \hat{\boldsymbol{x}}, \quad \nabla\lambda_3(\boldsymbol{r}) = \hat{\boldsymbol{y}}, \quad \nabla\lambda_4(\boldsymbol{r}) = \pm\hat{\boldsymbol{z}}\,.$$

Thus, the Nedelec shape functions become

$$\begin{aligned}
&\text{Edge } \mathcal{E}_1 \text{ from } \boldsymbol{r}_1 \text{ to } \boldsymbol{r}_2 \quad \ldots \quad && \boldsymbol{N}_1^e(\boldsymbol{r}) = -(1-y-z)\hat{\boldsymbol{x}} - x\hat{\boldsymbol{y}} - x\hat{\boldsymbol{z}} \\
&\text{Edge } \mathcal{E}_2 \text{ from } \boldsymbol{r}_1 \text{ to } \boldsymbol{r}_3 \quad \ldots \quad && \boldsymbol{N}_2^e(\boldsymbol{r}) = -y\hat{\boldsymbol{x}} - (1-x-z)\hat{\boldsymbol{y}} - y\hat{\boldsymbol{z}} \\
&\text{Edge } \mathcal{E}_4 \text{ from } \boldsymbol{r}_1 \text{ to } \boldsymbol{r}_4 \quad \ldots \quad && \boldsymbol{N}_3^e(\boldsymbol{r}) = \mp z\hat{\boldsymbol{x}} \mp z\hat{\boldsymbol{y}} \mp (1-x-y)\hat{\boldsymbol{z}} \\
&\text{Edge } \mathcal{E}_5 \text{ from } \boldsymbol{r}_2 \text{ to } \boldsymbol{r}_3 \quad \ldots \quad && \boldsymbol{N}_4^e(\boldsymbol{r}) = \;\; y\hat{\boldsymbol{x}} - x\hat{\boldsymbol{y}} \\
&\text{Edge } \mathcal{E}_5 \text{ from } \boldsymbol{r}_2 \text{ to } \boldsymbol{r}_4 \quad \ldots \quad && \boldsymbol{N}_5^e(\boldsymbol{r}) = \pm z\hat{\boldsymbol{x}} \mp x\hat{\boldsymbol{z}} \\
&\text{Edge } \mathcal{E}_6 \text{ from } \boldsymbol{r}_3 \text{ to } \boldsymbol{r}_4 \quad \ldots \quad && \boldsymbol{N}_6^e(\boldsymbol{r}) = \pm z\hat{\boldsymbol{y}} \mp y\hat{\boldsymbol{z}}\,.
\end{aligned} \tag{11.48}$$

See Fig. 11.11 for a graphical representation. Consider next the two tetrahedrons where $\boldsymbol{r}_4$ is located either above or below the xy-plane. They share a common face formed by the vertices $\boldsymbol{r}_1$, $\boldsymbol{r}_2$, $\boldsymbol{r}_3$. As can be easily inferred from the Nedelec elements given in Eq. (11.48), the tangential vector components of $\boldsymbol{N}_a^e$ in the plane $z=0$ are identical for the two tetrahedrons, in agreement with our assumptions about these curl conforming basis elements. We here omit the proof that this is also true for arbitrary tetrahedrons.

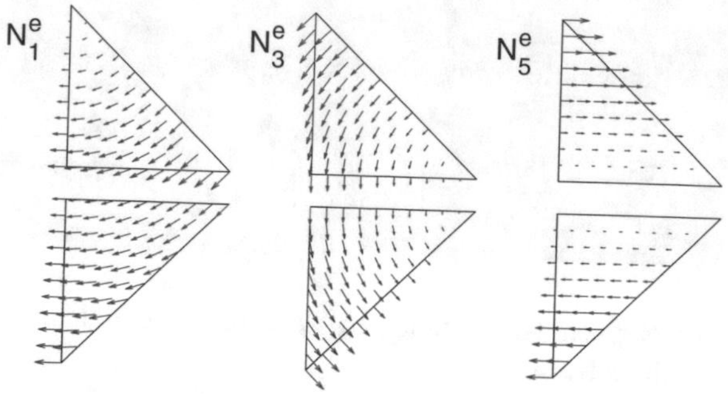

Fig. 11.11 Nedelec basis elements in xz-plane for unit tetrahedrons discussed in text. The basis elements ensure tangential continuity of the vector fields. We only show those elements that are non-zero in the given plane

We continue to compute $\hat{\boldsymbol{n}} \times \boldsymbol{N}_a^e$ for the tetrahedron below the xy-plane and for the face lying in the xy-plane. Obviously, the normal face vector pointing outwards of the tetrahedron is $\hat{\boldsymbol{n}} = \hat{\boldsymbol{z}}$. Simple algebra yields

$$\hat{\boldsymbol{z}} \times \boldsymbol{N}_1^e = x\hat{\boldsymbol{x}} - (1-y)\hat{\boldsymbol{y}}, \quad \hat{\boldsymbol{z}} \times \boldsymbol{N}_2^e = (1-x)\hat{\boldsymbol{x}} - y\hat{\boldsymbol{y}}, \quad \hat{\boldsymbol{z}} \times \boldsymbol{N}_4^e = x\hat{\boldsymbol{x}} + y\hat{\boldsymbol{y}},$$

where all other combinations give zero at $z = 0$.

It is interesting to compare this result with the Raviart–Thomas basis elements of Eq. (11.36) for the BEM approach using the unit triangle in the xy-plane. Apart from some normalization constants, which we will not be interested in here, it turns out that $\hat{\boldsymbol{n}} \times \boldsymbol{N}_a^e$ precisely coincides with the Raviart–Thomas elements given on the face with outer surface normal $\hat{\boldsymbol{n}}$.

This is not just a lucky coincidence, and indeed one can devise a construction rule for the Raviart–Thomas elements that closely resembles that for Nedelec elements. As a consequence of this close relation between Nedelec and Raviart–Thomas basis elements, it becomes possible to efficiently couple BEM and FEM simulations. This can be done most easily within the variational formulation of the Galerkin approach, which allows treating both approaches on the same footing. We will not further inquire into this topic here.

11.5.3 Discontinuous Galerkin Scheme

In principle, the working Eqs. (11.46) of the FEM approach can be also solved in the time domain. Contrary to the FDTD approach we are not bound to finite differences for time stepping, but can resort to more refined schemes such as the Runge–Kutta one. However, when solving Eqs. (11.46) in the time domain we suffer from the large, albeit sparse mass matrices $[\![M]\!]$ which have to be inverted at each time step. This can make simulations extremely slow.

The discontinuous Galerkin scheme is an approach that avoids the inversion of such mass matrices, and which allows making FEM simulations in the time domain fast and feasible. As a first step, we enlarge the number of global degrees of freedom and introduce *separate* basis functions $\boldsymbol{\varphi}_{ia}(\boldsymbol{r})$ for each tetrahedron $\mathcal{T}_i$,

Basis Functions FEM Approach (Discontinuous Galerkin)

$$\boldsymbol{\varphi}_{ia}(\boldsymbol{r}) = \begin{cases} \boldsymbol{N}_{ia}^e(\boldsymbol{r}) & \text{for } \boldsymbol{r} \in \mathcal{T}_i \\ 0 & \text{else}\,. \end{cases} \tag{11.49}$$

Nedelec shape functions $\boldsymbol{N}_{ia}^e(\boldsymbol{r})$ for all edges a of tetrahedron $\mathcal{T}_i$.

The approximate solution $\boldsymbol{E}^e$ then reads

$$\boldsymbol{E}^e(\boldsymbol{r},t) = \sum_{ia} \boldsymbol{N}_{ia}^e(\boldsymbol{r})\, [\![u_E^e(t)]\!]_{ia}\,, \tag{11.50}$$

with a corresponding expression for the magnetic field. At first sight this does not look like an overly good idea. First, we have significantly enlarged the space of global degrees of freedom. Secondly, the basis functions are defined only locally, and functions in neighbor elements do in general not communicate with each other. Thus, it is no longer guaranteed that the tangential electromagnetic fields are continuous when passing from one tetrahedron to a neighbor one. This possible discontinuity has given the approach its name, the *discontinuous* Galerkin scheme. However, with some extra work we can again enforce continuity. To demonstrate this, we insert the ansatz of Eq. (11.50) into Faraday's law, multiply with the test functions $\boldsymbol{w}_{E,H}^e$, and integrate over the volume of $\mathcal{T}_i$, to finally arrive at

$$\int_{\mathcal{T}_i} \boldsymbol{N}_{ia}^e(\boldsymbol{r}) \cdot \sum_{a'} \left[\bar{\bar{\mu}} \cdot \boldsymbol{N}_{ia'}^e(\boldsymbol{r})\, [\![\partial_t u_H^e(t)]\!]_{ia'} + \nabla \times \boldsymbol{N}_{ia'}^e(\boldsymbol{r})\, [\![u_E^e(t)]\!]_{ia'} \right] d^3r = 0\,.$$

This expression is similar to the working equations of the FEM approach, Eq. (11.46). Obviously, it is the second term in brackets, the curl expression, which couples through the spatial derivatives to the neighbor elements. In the FEM scheme described in Sect. 11.5.1 we have accounted for this coupling and the resulting continuity of the tangential electromagnetic fields (which follow from Maxwell's equations) through the basis functions of Eq. (11.42) that have built in such continuity from scratch. For the local basis functions of Eq. (11.49) we have to proceed differently. We first perform integration by parts of the second term in brackets,

$$\int_{\mathcal{T}_i} \boldsymbol{N}_{ia}^e \cdot \nabla \times \boldsymbol{N}_{ia'}^e \, d^3r = \oint_{\partial\mathcal{T}_i} \boldsymbol{N}_{ia}^e \times \boldsymbol{N}_{ia'}^e \cdot \hat{\boldsymbol{n}} dS - \int_{\mathcal{T}_i} \nabla \times \boldsymbol{N}_{ia}^e \cdot \boldsymbol{N}_{ia'}^e \, d^3r \,.$$

Here we have converted the term with $\nabla \cdot \boldsymbol{N}_{ia}^e \times \boldsymbol{N}_{ia'}^e$ into a boundary integral by means of Gauss' theorem, with $\hat{\boldsymbol{n}}$ being the outer surface normal of tetrahedron $\mathcal{T}_i$. By performing cyclic permutation in the triple product we then get for the boundary term

$$-\oint_{\partial\mathcal{T}_i} \boldsymbol{N}_{ia}^e(\boldsymbol{s}) \cdot \sum_{a'} \left(\hat{\boldsymbol{n}} \times \boldsymbol{N}_{ia'}^e(\boldsymbol{s}) \, [\![u_E^e(t)]\!]_{ia'}\right) dS, \tag{11.51}$$

where for clarity we have introduced again the field coefficients $[\![u_E^e(t)]\!]$ and have summed over a'. The term in parentheses can thus be interpreted as $\hat{\boldsymbol{n}} \times \boldsymbol{E}^e(\boldsymbol{s}, t)$. Here comes the magic trick of the discontinuous Galerkin scheme: because of the tangential continuity of the electromagnetic fields we can mix in the boundary term the tangential field contributions of neighbor elements, thereby coupling these elements and re-enforcing continuity of the tangential field components. See also Fig. 11.12.

Let us postpone for a moment the question of how to actually choose the coupling term, and introduce for the expression in parentheses of Eq. (11.51) a flux term $\boldsymbol{F}_E(\boldsymbol{s}, t)$ that depends on the tangential field components only. For the exact solution this term must equal the flux of the neighbor element, so there is some degree of freedom in its choice. For Faraday's law we then get

$$\int_{\mathcal{T}_i} \boldsymbol{N}_{ia}^e(\boldsymbol{r}) \cdot \left[\partial_t \boldsymbol{H}^e(\boldsymbol{r}, t) + \nabla \times \boldsymbol{E}^e(\boldsymbol{r}, t)\right] d^3r = \oint_{\partial\mathcal{T}_i} \boldsymbol{N}_{ia}^e(\boldsymbol{s}) \cdot \boldsymbol{F}_E(\boldsymbol{s}, t) \, dS \,.$$

In analogy to the previously discussed Galerkin scheme we introduce the mass and stiffness matrices of Eqs. (11.44) and (11.45), and obtain for the curl equations

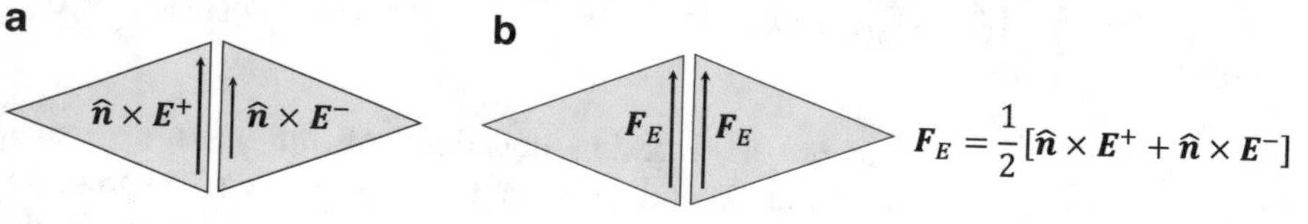

Fig. 11.12 Schematics of discontinuous Galerkin scheme. (**a**) We start with local degrees of freedom defined for each volume element separately. For this reason, the tangential flux at the interfaces between neighbor elements can be discontinuous. (**b**) To ensure tangential continuity, we mix the local fluxes, using centered or upwind flux schemes, thereby coupling elements in the stiffness matrix $[\![S]\!]$ but leaving the mass matrices $[\![M^\varepsilon]\!]$, $[\![M^\mu]\!]$ (which must be diagonalized) block-diagonal

$$\sum_{a'} \left(M^{\mu}_{ia,ia'} [\![\partial_t u^e_H(t)]\!]_{ia'} + S_{ia,ia'} [\![u^e_E(t)]\!]_{ia'} \right) = \oint_{\partial \mathcal{T}_i} \boldsymbol{N}^e_{ia} \cdot \boldsymbol{F}_E \, dS$$

$$\sum_{a'} \left(M^{\varepsilon}_{ia,ia'} [\![\partial_t u^e_E(t)]\!]_{ia'} - S_{ia,ia'} [\![u^e_H(t)]\!]_{ia'} \right) = \oint_{\partial \mathcal{T}_i} \boldsymbol{N}^e_{ia} \cdot \boldsymbol{F}_H \, dS - J_{ia}(t) \,.$$

The electric and magnetic flux terms on the right-hand side of the equations can be expanded in the same basis as used for the electromagnetic fields. This gives us

$$\oint_{\partial \mathcal{T}_i} \boldsymbol{N}^e_{ia} \cdot \boldsymbol{F}_{E,H} \, dS = \oint_{\partial \mathcal{T}_i} \boldsymbol{N}^e_{ia} \cdot \sum_{a'} \boldsymbol{N}^e_{ia} [\![F^e_{E,H}]\!]_{ia'} \, dS = \sum_{a'} \mathscr{F}_{ia,ia'} [\![F^e_{E,H}]\!]_{ia'},$$

where $\mathscr{F}$ is a short-hand notation for the boundary integral over the Nedelec basis functions. In the following we use $[\![A]\!]_{ia,ia'} = A_{ia,ia'}$ for the matrices in the discontinuous Galerkin scheme. The working equations for the time-dependent FEM approach then read

Working Equations FEM Approach (Discontinuous Galerkin)

$$[\![M^{\mu}]\!] \cdot [\![\partial_t u^e_H(t)]\!] + [\![S]\!] \cdot [\![u^e_E(t)]\!] = [\![\mathscr{F}]\!] \cdot [\![F^e_E(t)]\!] \tag{11.52}$$

$$[\![M^{\varepsilon}]\!] \cdot [\![\partial_t u^e_E(t)]\!] - [\![S]\!] \cdot [\![u^e_H(t)]\!] = [\![\mathscr{F}]\!] \cdot [\![F^e_H(t)]\!] - [\![J(t)]\!] \,.$$

We still need to invert the mass matrices, which, however, are now of block-diagonal form and fall apart into sub-matrices of the order of the number of Nedelec basis elements within each tetrahedron. The diagonalization of these sub-matrices is fast and no longer constitutes a major bottleneck for the solution of the time-dependent FEM equations.

We finally briefly comment on the proper choice of the flux terms $\boldsymbol{F}_{E,H}$. Let $\boldsymbol{E}^-$, $\boldsymbol{H}^-$ denote the electromagnetic fields in tetrahedron $\mathcal{T}_i$, and $\boldsymbol{E}^+$, $\boldsymbol{H}^+$ the fields in the neighbor elements. In the centered flux approach we then set

$$\boldsymbol{F}_E(\boldsymbol{s}) = \tfrac{1}{2} \hat{\boldsymbol{n}} \times \left[\boldsymbol{E}^+(\boldsymbol{s}) + \boldsymbol{E}^-(\boldsymbol{s}) \right], \quad \boldsymbol{F}_H(\boldsymbol{s}) = \tfrac{1}{2} \hat{\boldsymbol{n}} \times \left[\boldsymbol{H}^+(\boldsymbol{s}) + \boldsymbol{H}^-(\boldsymbol{s}) \right].$$

There exist other choices, such as the upwind flux scheme, for which we refer the interested reader to the more specialized literature [115, 116, 118]. In general, the stability and convergence of the approach significantly depends on the chosen flux scheme.

11.6 Details of Potential Boundary Element Method

In this section we provide the details for obtaining the working equations of the potential-based boundary element method approach. We start from the expressions of Eq. (11.23) for the scalar and vector potentials

$$
\begin{aligned}
V_j(\boldsymbol{r}) &= V_j^{\text{inc}}(\boldsymbol{r}) + \oint_{\partial\Omega} G_j(\boldsymbol{r},\boldsymbol{s}')\sigma_j(\boldsymbol{s}')\,dS' \\
\tilde{\boldsymbol{A}}_j(\boldsymbol{r}) &= \tilde{\boldsymbol{A}}_j^{\text{inc}}(\boldsymbol{r}) + \oint_{\partial\Omega} G_j(\boldsymbol{r},\boldsymbol{s}')\boldsymbol{h}_j(\boldsymbol{s}')\,dS'\,.
\end{aligned}
\tag{11.53}
$$

If we take on both sides of the equations the derivatives $\partial_n = \boldsymbol{n}\cdot\nabla$, we get

$$
\begin{aligned}
\partial_n V_j(\boldsymbol{r}) &= \partial_n V_j^{\text{inc}}(\boldsymbol{r}) + \oint_{\partial\Omega} \partial_n G_j(\boldsymbol{r},\boldsymbol{s}')\sigma_j(\boldsymbol{s}')\,dS' \\
\partial_n\tilde{\boldsymbol{A}}_j(\boldsymbol{r}) &= \partial_n\tilde{\boldsymbol{A}}_j^{\text{inc}}(\boldsymbol{r}) + \oint_{\partial\Omega} \partial_n G_j(\boldsymbol{r},\boldsymbol{s}')\boldsymbol{h}_j(\boldsymbol{s}')\,dS'\,.
\end{aligned}
\tag{11.54}
$$

We next consider the limit $\boldsymbol{r}\to\boldsymbol{s}$. This limit can be safely performed in Eq. (11.53) where we simply have to replace $\boldsymbol{r}$ by $\boldsymbol{s}$. In Eq. (11.54) the limiting procedure has to be done with more care, as previously discussed for the Neumann trace on page 222, and we arrive at

$$
\begin{aligned}
\partial_n V_j(\boldsymbol{s}) &= \partial_n V_j^{\text{inc}}(\boldsymbol{s}) + \oint_{\partial\Omega} \partial_n F_j(\boldsymbol{s},\boldsymbol{s}')\sigma_j(\boldsymbol{s}')\,dS' \pm \tfrac{1}{2}\sigma_j(\boldsymbol{s}) \\
\partial_n\tilde{\boldsymbol{A}}_j(\boldsymbol{s}) &= \partial_n\tilde{\boldsymbol{A}}_j^{\text{inc}}(\boldsymbol{s}) + \oint_{\partial\Omega} \partial_n F_j(\boldsymbol{s},\boldsymbol{s}')\boldsymbol{h}_j(\boldsymbol{s}')\,dS' \pm \tfrac{1}{2}\boldsymbol{h}_j(\boldsymbol{s})\,.
\end{aligned}
\tag{11.55}
$$

Here the upper or lower sign has to be chosen for the potentials at the particle inside or outside, respectively, and $F_j = \partial_n G_j$ is the surface derivative of the Green's function. Within the boundary element method approach we approximate the boundary by elements of finite size, and consider the potentials at the centroids of these boundary elements. With this we get for the potentials of Eq. (11.53)

$$
\begin{aligned}
[\![V_j]\!] &= [\![V_j^{\text{inc}}]\!] + [\![\mathscr{G}_j]\!]\cdot[\![\sigma_j]\!] \\
[\![\tilde{\boldsymbol{A}}_j]\!] &= [\![\tilde{\boldsymbol{A}}_j^{\text{inc}}]\!] + [\![\mathscr{G}_j]\!]\cdot[\![\boldsymbol{h}_j]\!],
\end{aligned}
\tag{11.56}
$$

where $[\![\mathscr{G}_j]\!]_{ii'} = [\![G_j]\!]_{ii'}\mathcal{A}_{i'}$ are the matrix elements of the discretized Green's function (see exercise 11.5 for details). Similarly, for the surface derivatives of Eq. (11.55) we obtain

$$
\begin{aligned}
[\![\partial_n V_j]\!] &= [\![\partial_n V_j^{\text{inc}}]\!] + [\![\mathscr{H}_j]\!]\cdot[\![\sigma_j]\!] \\
[\![\partial_n\tilde{\boldsymbol{A}}_j]\!] &= [\![\partial_n\tilde{\boldsymbol{A}}_j^{\text{inc}}]\!] + [\![\mathscr{H}_j]\!]\cdot[\![\boldsymbol{h}_j]\!]\,.
\end{aligned}
\tag{11.57}
$$

Here $[\![\mathscr{H}_{1,2}]\!] = [\![\mathscr{F}_{1,2} \pm \frac{1}{2}\mathbb{1}]\!]$, and $[\![\mathscr{F}_j]\!]_{ii'} = [\![F_j]\!]_{ii'}\mathcal{A}_{i'}$ are the matrix elements for the surface derivatives of the Green's functions. We next employ the boundary conditions of Maxwell's equations to compute the unknown surface charge and current distributions. In deriving the working equations, we suppress the explicit notation $[\![\ldots]\!]$ for matrices discretized within our BEM approach.

Continuity of Potentials. We first assume that the scalar and vector potentials are continuous at the particle boundary. With this, it is guaranteed that also the tangential components of $\boldsymbol{E}$ and the normal component of $\boldsymbol{B}$ are continuous. From Eq. (11.56) we get

$$\begin{aligned} \mathscr{G}_1 \cdot \sigma_1 &= \mathscr{G}_2 \cdot \sigma_2 + \delta V^{\text{inc}}, \quad \delta V^{\text{inc}} = V_2^{\text{inc}} - V_1^{\text{inc}} \\ \mathscr{G}_1 \cdot \boldsymbol{h}_1 &= \mathscr{G}_2 \cdot \boldsymbol{h}_2 + \delta \tilde{\boldsymbol{A}}^{\text{inc}}, \quad \delta \tilde{\boldsymbol{A}}^{\text{inc}} = \tilde{\boldsymbol{A}}_2^{\text{inc}} - \tilde{\boldsymbol{A}}_1^{\text{inc}}. \end{aligned} \tag{11.58}$$

Continuity of Dielectric Displacement. In addition, the normal component of the dielectric displacement

$$\hat{\boldsymbol{n}} \cdot \boldsymbol{D}_j = \varepsilon_j \Big(ik_0\, \hat{\boldsymbol{n}} \cdot \tilde{\boldsymbol{A}}_j - \partial_n V_j \Big)$$

is continuous at the particle boundary. In its discretized version, this continuity translates to (we cancel a factor of ε_0 on both sides of the equation)

$$\begin{aligned} &\kappa_1 \Big\{ ik_0 \hat{\boldsymbol{n}} \cdot \Big(\tilde{\boldsymbol{A}}_1^{\text{inc}} + \mathscr{G}_1 \cdot \boldsymbol{h}_1 \Big) - \Big(\partial_n V_1^{\text{inc}} + \mathscr{H}_1 \cdot \sigma_1 \Big) \Big\} = \\ &\kappa_2 \Big\{ ik_0 \hat{\boldsymbol{n}} \cdot \Big(\tilde{\boldsymbol{A}}_2^{\text{inc}} + \mathscr{G}_2 \cdot \boldsymbol{h}_2 \Big) - \Big(\partial_n V_2^{\text{inc}} + \mathscr{H}_2 \cdot \sigma_2 \Big) \Big\}. \end{aligned} \tag{11.59}$$

Continuity of Lorenz Gauge Condition. We finally exploit the continuity of

$$\partial_n \tilde{\boldsymbol{A}}_j - ik_0 \kappa_j V_j, \tag{11.60}$$

which can be derived from the Lorenz gauge condition of Eq. (11.22), as shown in Exercise 11.6. In discretized form we get

$$\begin{aligned} &\partial_n \tilde{\boldsymbol{A}}_1^{\text{inc}} + \mathscr{H}_1 \cdot \boldsymbol{h}_1 - ik_0 \kappa_1 \Big(V_1^{\text{inc}} + \mathscr{G}_1 \cdot \sigma_1 \Big) \\ &= \partial_n \tilde{\boldsymbol{A}}_2^{\text{inc}} + \mathscr{H}_2 \cdot \boldsymbol{h}_2 - ik_0 \kappa_2 \Big(V_2^{\text{inc}} + \mathscr{G}_2 \cdot \sigma_2 \Big). \end{aligned} \tag{11.61}$$

Equations (11.58)–(11.61) are eight equations for the unknowns σ_j, $\boldsymbol{h}_j$, which can be rewritten in the form

$$\mathscr{G}_1 \cdot \sigma_1 - \mathscr{G}_2 \cdot \sigma_2 = \delta V^{\text{inc}} \tag{11.62a}$$

$$\mathscr{G}_1 \cdot \boldsymbol{h}_1 - \mathscr{G}_2 \cdot \boldsymbol{h}_2 = \delta \tilde{\boldsymbol{A}}^{\rm inc} \tag{11.62b}$$

$$\kappa_1 \mathscr{H}_1 \cdot \sigma_1 - \kappa_2 \mathscr{H}_2 \cdot \sigma_2 - ik_0 \hat{\boldsymbol{n}} \cdot \Big\{ \kappa_1 \mathscr{G}_1 \cdot \boldsymbol{h}_1 - \kappa_2 \mathscr{G}_2 \cdot \boldsymbol{h}_2 \Big\} = D^e \tag{11.62c}$$

$$\mathscr{H}_1 \cdot \boldsymbol{h}_1 - \mathscr{H}_2 \cdot \boldsymbol{h}_2 - ik_0 \hat{\boldsymbol{n}} \cdot \Big\{ \kappa_1 \mathscr{G}_1 \cdot \sigma_1 - \kappa_2 \mathscr{G}_2 \cdot \sigma_2 \Big\} = \boldsymbol{\alpha}, \tag{11.62d}$$

with the abbreviations

$$D^e = \left(\kappa_2 \partial_n V_2^{\rm inc} - \kappa_1 \partial_n V_1^{\rm inc} \right) - ik_0 \hat{\boldsymbol{n}} \cdot \left(\kappa_2 \tilde{\boldsymbol{A}}_2^{\rm inc} - \kappa_1 \tilde{\boldsymbol{A}}_1^{\rm inc} \right)$$

$$\boldsymbol{\alpha} = ik_0 \hat{\boldsymbol{n}} \cdot \left(\kappa_2 V_2^{\rm inc} - \kappa_1 V_1^{\rm inc} \right) + \left(\partial_n \tilde{\boldsymbol{A}}_2^{\rm inc} - \partial_n \tilde{\boldsymbol{A}}_1^{\rm inc} \right).$$

In the following we briefly discuss how this set of equations can be solved in an efficient manner. We first eliminate σ_1, $\boldsymbol{h}_1$, using the first two equations, and employ

$$\mathscr{H}_j \cdot \ldots = \left(\mathscr{H}_j \cdot \mathscr{G}_j^{-1} \right) \cdot \mathscr{G}_j \cdot \ldots = \Sigma_j \cdot \mathscr{G}_j \cdot \ldots,$$

with the matrix $\Sigma_j = \mathscr{H}_j \cdot \mathscr{G}_j^{-1}$. From the last two equations in Eq. (11.62) we then get

$$\begin{aligned} &\kappa_1 \Sigma_1 \cdot \left(\mathscr{G}_2 \cdot \sigma_2 + \delta V^{\rm inc} \right) - \kappa_2 \Sigma_2 \cdot \mathscr{G}_2 \cdot \sigma_2 \\ &\qquad - ik_0 \hat{\boldsymbol{n}} \cdot \Big\{ \kappa_1 \left(\mathscr{G}_2 \cdot \boldsymbol{h}_2 + \delta \tilde{\boldsymbol{A}}^{\rm inc} \right) - \kappa_2 \mathscr{G}_2 \cdot \boldsymbol{h}_2 \Big\} = D^e \end{aligned} \tag{11.63a}$$

$$\begin{aligned} &\Sigma_1 \cdot \left(\mathscr{G}_2 \cdot \boldsymbol{h}_2 + \delta \tilde{\boldsymbol{A}}^{\rm inc} \right) - \Sigma_2 \cdot \mathscr{G}_2 \cdot \boldsymbol{h}_2 \\ &\qquad - ik_0 \hat{\boldsymbol{n}} \cdot \Big\{ \kappa_1 \left(\mathscr{G}_2 \cdot \sigma_2 + \delta V^{\rm inc} \right) - \kappa_2 \mathscr{G}_2 \cdot \sigma_2 \Big\} = \boldsymbol{\alpha}\,. \end{aligned} \tag{11.63b}$$

We next introduce the auxiliary matrix

$$\Delta = \Sigma_1 - \Sigma_2, \tag{11.64}$$

to rewrite Eq. (11.63b) as

$$\begin{aligned} &\Delta \cdot \mathscr{G}_2 \cdot \boldsymbol{h}_2 - ik_0 \hat{\boldsymbol{n}} \cdot \left(\kappa_1 - \kappa_2 \right) \mathscr{G}_2 \cdot \sigma_2 \\ &\qquad = \boldsymbol{\alpha} - \Sigma_1 \cdot \delta \tilde{\boldsymbol{A}}^{\rm inc} + ik_0 \hat{\boldsymbol{n}} \cdot \delta V^{\rm inc} = \tilde{\boldsymbol{\alpha}}\,. \end{aligned}$$

This expression can be solved for $\mathscr{G}_2 \cdot \boldsymbol{h}_2$. Upon insertion into Eq. (11.63a) we then get

$$\Sigma \cdot \mathscr{G}_2 \cdot \sigma_2 = \tilde{D}^e + ik_0 \hat{\boldsymbol{n}} \cdot \left(\kappa_1 - \kappa_2 \right) \Delta^{-1} \cdot \tilde{\boldsymbol{\alpha}},$$

where we have introduced the auxiliary quantities

$$\Sigma = \kappa_1 \Sigma_1 - \kappa_2 \Sigma_2 + k_0^2(\kappa_1 - \kappa_2)^2 \hat{\boldsymbol{n}} \cdot \Delta^{-1} \cdot \hat{\boldsymbol{n}}$$
$$\tilde{D}^e = D^e - \kappa_1 \Sigma_1 \cdot \delta V^{\text{inc}} + i k_0 \kappa_1 \hat{\boldsymbol{n}} \cdot \delta \tilde{\boldsymbol{A}}^{\text{inc}} . \tag{11.65}$$

The working equations of the potential-based boundary element method approach can be finally summarized as

Working Equations for Potential-Based BEM Approach

$$[\![\sigma_2]\!] = [\![\mathscr{G}_2]\!]^{-1} \cdot [\![\Sigma]\!]^{-1} \cdot \Big([\![\tilde{D}^e]\!] + i k_0 [\![\hat{\boldsymbol{n}}]\!] \cdot (\kappa_1 - \kappa_2)[\![\Delta]\!]^{-1} \cdot [\![\tilde{\boldsymbol{\alpha}}]\!]\Big)$$
$$[\![\boldsymbol{h}_2]\!] = [\![\mathscr{G}_2]\!]^{-1} \cdot [\![\Delta]\!]^{-1} \cdot \Big([\![\tilde{\boldsymbol{\alpha}}]\!] + i k_0 [\![\hat{\boldsymbol{n}}]\!] \cdot (\kappa_1 - \kappa_2)[\![\mathscr{G}_2]\!] \cdot [\![\sigma_2]\!]\Big), \tag{11.66}$$

together with Eq. (11.58) to compute the surface charge and current distributions $[\![\sigma_1]\!]$, $[\![\boldsymbol{h}_1]\!]$ at the boundary insides. The solution of the working equations requires four matrix inversions and two matrix multiplications, together with the computationally less expensive addition of matrices and multiplication of matrices by vectors.

Exercises

Exercise 11.1 Compute the eigenvalues of Eq. (11.12) for the stability matrix of the FDTD approach using the "magic" time step $\Delta t = \Delta x / c$. Show that for this choice the numerical dispersion within FDTD is completely suppressed.

Exercise 11.2 Consider a TM wave, described by the z-component of the electric field E_z, that is moving to the left-hand side such that

$$\left(\frac{\partial}{\partial x} - \frac{1}{v}\frac{\partial}{\partial t}\right) E_z = 0,$$

where v is the speed of light in the corresponding medium. In the FDTD framework the solution of this equation is approximated by centered differences for the spatial and time coordinates, resulting in

$$\frac{[\![E_z]\!]_{3/2}^{n+1} - [\![E_z]\!]_{3/2}^{n}}{\Delta t} = v\,\frac{[\![E_z]\!]_2^{n+1/2} - [\![E_z]\!]_1^{n+1/2}}{\Delta x}\,. \tag{11.67}$$

We have evaluated the wave equation at a mesh point $i = 3/2$ just inside the boundary, where $i = 1$ is assumed to be the left domain boundary. Since the electric fields E_z are only defined for integer time and space indices, we approximate the half-integer terms of E_z through $[\![E_z]\!]_{3/2} = \frac{1}{2}([\![E_z]\!]_1 + [\![E_z]\!]_2)$.

(a) Compute the resulting discretized wave equation of Eq. (11.67).
(b) Solve for the unknown $[\![E_z]\!]_1^{n+1}$.

If the electric field is updated according to the equation derived in (b), a wave impinging on the left boundary moves out of the simulation domain without any reflections. This is the essence of the so-called *absorbing boundary conditions* [102].

Exercise 11.3 Within FDTD the perfectly matched layers can be also implemented by using anisotropic permittivities $\bar{\bar{\varepsilon}}$ and permeabilities $\bar{\bar{\mu}}$.

(a) Follow the derivation of the Fresnel coefficients presented in Chap. 8 for an interface between a material with local material parameters ε_1, μ_1 and a nonlocal one with

$$\bar{\bar{\varepsilon}}_2 = \begin{pmatrix} \varepsilon_2^{\perp} & & \\ & \varepsilon_2^{\perp} & \\ & & \varepsilon_2^{z} \end{pmatrix}, \qquad \bar{\bar{\mu}}_2 = \begin{pmatrix} \mu_2^{\perp} & & \\ & \mu_2^{\perp} & \\ & & \mu_2^{z} \end{pmatrix}.$$

(b) Along the same lines as discussed in Sect. 11.1.3, determine the values of $\varepsilon_2^{\perp}$, ε_2^{z} and $\mu_2^{\perp}$, μ_2^{z} such that the Fresnel reflection coefficient for an incoming wave is zero (impedance matching) and the wave is damped in medium 2.

Exercise 11.4 Start from the curl equations of Eq. (11.1) and split the fields into incoming and scattered fields $\boldsymbol{E} = \boldsymbol{E}_{\text{inc}} + \boldsymbol{E}_{\text{sca}}$, $\boldsymbol{H} = \boldsymbol{H}_{\text{inc}} + \boldsymbol{H}_{\text{sca}}$.

(a) Suppose that the incoming fields $\boldsymbol{E}_{\text{inc}}$, $\boldsymbol{H}_{\text{inc}}$ are solutions of the homogeneous wave equation for a background medium with ε_b, μ_b.
(b) Derive the equations of motion for the scattered fields and for inhomogeneous material parameters $\varepsilon(\boldsymbol{r})$, $\mu(\boldsymbol{r})$, and show that the incoming fields can be formally considered as electric and magnetic current distributions.

Exercise 11.5 For the diagonal elements of the potential based BEM approach one has to be careful about the singular behavior of the Green's function of Eq. (5.7) for $\boldsymbol{r} \to \boldsymbol{r}'$. Let $\boldsymbol{s}_i$ be the centroid of a triangular boundary element $\mathcal{T}_i$, and consider

$$[\![\mathcal{G}_{\text{stat}}]\!]_{ii} = \int_{\mathcal{T}_i} \frac{1}{4\pi|\boldsymbol{s}_i - \boldsymbol{s}'|}\,dS'\,.$$

Show that the integral can be performed safely by introducing polar coordinates with the origin at s_i. Perform the radial integral analytically, and derive the one-dimensional integral over the polar angle which must be solved numerically.

Exercise 11.6 Use the continuity of the tangential component of $\boldsymbol{B}$ at a boundary to show that

$$\hat{\boldsymbol{n}} \times (\boldsymbol{B}_2 - \boldsymbol{B}_1) = \nabla(\hat{\boldsymbol{n}} \cdot \delta \boldsymbol{A}) - (\hat{\boldsymbol{n}} \cdot \nabla)\delta \boldsymbol{A} = 0,$$

with $\delta \boldsymbol{A} = \boldsymbol{A}_2 - \boldsymbol{A}_1$. Because the tangential derivative of $\delta \boldsymbol{A}$ is zero, the first term on the right-hand side can be rewritten as $\nabla(\hat{\boldsymbol{n}} \cdot \delta \boldsymbol{A}) = \hat{\boldsymbol{n}}(\nabla \cdot \delta \boldsymbol{A})$. Use the Lorenz gauge condition of Eq. (11.22) to derive the continuity of the expression given in Eq. (11.60).

Exercise 11.7 Derive the matrix elements and the inhomogeneity of Eq. (11.34) for the Galerkin scheme. Use for the translation matrix $T_{[ia],\nu}$ a combined index $[ia]$ to rewrite the expressions in terms of matrix multiplications.

Exercise 11.8 The variational formulation of the Galerkin approach can be applied to the collocation formulation of the boundary element method approach. The functional for the quasistatic approach based on Eq. (9.27) reads

$$\mathcal{S} = \oint_{\partial\Omega} w^e(\boldsymbol{s}) \left\{ \oint_{\partial\Omega} \Big[\Lambda(\omega)\delta(\boldsymbol{s} - \boldsymbol{s}') + F^{\text{stat}}(\boldsymbol{s}, \boldsymbol{s}') \Big] u^e(\boldsymbol{s}')\, dS' + \frac{\partial V^{\text{inc}}(\boldsymbol{s})}{\partial n} \right\} dS\,.$$

Chose for the solution $u^e(\boldsymbol{s})$ functions that are constant within a given boundary element and zero otherwise, and for the test functions Dirac delta functions located at the centroids,

$$\tilde{\varphi}_i(\boldsymbol{s}) = \delta(\boldsymbol{s} - \boldsymbol{s}_i), \quad \varphi_i(\boldsymbol{s}) = \begin{cases} 1 & \text{for } \boldsymbol{s} \in \mathcal{T}_i \\ 0 & \text{else}\,. \end{cases}$$

Show that $\delta\mathcal{S}/\delta w_i^e$ leads to Eq. (11.18), however, with matrix elements

$$[\![\mathscr{F}^{\text{stat}}]\!]_{ii'} = \int_{\mathcal{T}_{i'}} F^{\text{stat}}(\boldsymbol{s}_i, \boldsymbol{s}')\, dS'$$

that are averaged over the triangle. Such an integration scheme for the Green's functions is also used in Refs. [109, 110].

Exercise 11.9 Consider the Raviart–Thomas basis elements of Eq. (11.36) for the BEM approach. Show through explicit calculation that the flux $\boldsymbol{\varphi}_\nu \cdot \boldsymbol{\eta}$ is conserved at the edge between two adjacent triangles.

Exercise 11.10 Derive the Nedelec elements of Eq. (11.48) for an arbitrary tetrahedron rather than a unit one. Devise the transformation rules that allow converting the basis functions for a unit tetrahedron to those of an arbitrary tetrahedron.

Chapter 12
Quantum Effects in Nano Optics

There are many good reasons to bring nano optics to the quantum domain. For instance, a single quantum emitter interacting with plasmonic nanoparticles emits *single* photons only. One could also try to estimate the number of photons impinging on a plasmonic nanoparticle for a typical laser excitation, to find that these are just *a few* photons on the time scale of the plasmon lifetime ($\approx$10 fs). So plasmonics is deeply rooted in the quantum domain. Yet, for plasmonics modelling it often suffices to employ classical electrodynamics, as we have done almost exclusively in the first part of this book. The second part of this book deals with the questions: why is classical electrodynamics so successful for describing plasmonics? Under which conditions does a classical description break down and a quantum treatment is needed? And what are the tools needed for describing quantum effects in nano optics and plasmonics?

In quantum physics particles exhibit a wave behavior that is unfamiliar to us from everyday life. Genuine quantum phenomena, such as interference, quantization, or tunneling, have no classical analogues, and correspondingly they often appear strange to us. On the other hand, the particle nature, namely that particles such as electrons or atoms are either observed as one entity or not at all, appears more natural—probably because we are familiar with this from our macroscopic world. In the same way as particles exhibit wave behavior, wave phenomena exhibit particle behavior. The probably best known manifestation of this is Einstein's celebrated equation

$$E = h\nu, \tag{12.1}$$

which states that the smallest portion of energy one can extract from a light field (or add to it) is given by Planck's constant h multiplied by the light frequency. These energy portions are conveniently called **photons** and play a central role in the quantum approach to electrodynamics, the so-called field of quantum electrodynamics. So while the particle nature of light is unfamiliar to us from our

U. Hohenester, *Nano and Quantum Optics*, Graduate Texts in Physics,
https://doi.org/10.1007/978-3-030-30504-8_12

everyday life (our eye operates on the few- to multi-photon level), the wave nature of light remains essentially unchanged when going from the classical to the quantum domain. For this reason, the novel quantum aspects of light concern the particle nature only, and give rise to less spectacular phenomena in comparison to the wave aspects of matter—at least as long as not too exotic photon states come into play. So this is one of the central messages of the quantum part of this book: most of the results derived in the first "classical" part of this book remain practically unchanged. They only have to be complemented by thermal Bose–Einstein occupation factors, or some other type of distribution function, to properly account for the photons' noise properties.

As simple as this message may be, the theoretical tools needed to arrive at this conclusion are manifold and quite demanding; my educated guess is that readers unfamiliar with the topic will feel lost at some point, or will wonder whether the whole topic is really interesting as such (yes, it really is!) or is simply presented in an intractable manner here (which still might be the case). However, let me assure that the tools developed in this quantum part provide a versatile and powerful armory, for use in both nano optics on the quantum level and quantum optics, with applications that go far beyond the topics of this book. In short, from here on we introduce the following concepts.

Quantum Electrodynamics. To quantize Maxwell's equations as well as the matter part needed for Maxwell's theory, we use the concept of canonical quantization, which provides a prescription of how to submit a classical model to quantization. For the light fields this leads to an operator form of Maxwell's equations, which must be supplemented with a photon wavefunction, whereas for the matter part it leads to Schrödinger's equation with the light–matter interaction given in terms of the minimal coupling or multipolar Hamiltonian.

Correlation Functions. In linear response, a perturbed system can be described in terms of the fluctuations of the equilibrium system solely. This is the essence of the fluctuation–dissipation theorem. Prominent examples are the dielectric function and the optical conductivity, which can be related to density–density and current–current correlations, as well as the correlations for the electromagnetic fields, which can be related to the dyadic Green's function of classical electrodynamics.

Fluctuational Electrodynamics. In thermal equilibrium, the linear system response can be expressed through the above-mentioned correlation functions. From the cross-spectral densities of the electromagnetic fields we can compute a variety of interesting quantities, such as the decay rate of a quantum emitter placed within a non-trivial photonic environment, its energy renormalization (Lamb shift), and the Casimir–Polder force acting on the quantum emitter. We will also investigate Casimir forces between macroscopic bodies, and thermal radiation and heat flux at the nanoscale.

Quantum Optics Toolbox. For a linear photonic environment and the nonlinear response of a quantum emitter placed within this environment, we can employ the quantum optics toolbox, including the master equation of Lindblad form, or generalizations of it, as well as the quantum regression theorem that allows computing optical spectra or photon correlations.

In the remainder of this introductory quantum chapter we discuss these various techniques in slightly more detail, before presenting them in greater depth in later chapters. Whenever possible, we add some comments of how they can be used in the field of nano optics and plasmonics.

12.1 Going Quantum in Three Steps

Step 1. Canonical Quantization
Canonical quantization gives a prescription of how to submit a classical model to quantization. In the context of Maxwell's equations, its starting point is the description of light, matter, and the mutual interaction in terms of a classical Lagrange function, as will be discussed in some length in Chap. 13. Once the Lagrange function is at hand, we can submit it to canonical quantization, which replaces the dynamic variables and their canonical momenta by non-commuting operators ready for use within quantum mechanics. The approach of obtaining a quantum model out of a classical model appears somewhat awkward at first sight, as the classical model has no prior knowledge of how its quantum version may look like. However, the approach becomes more transparent if one changes the perspective and considers the classical model as an approximation of the (more fundamental) quantum version. From this reversed viewpoint it is now much clearer that the classical approximation already contains many useful hints of how the quantum version might look like. For most systems of interest one can then apply the standard procedure of canonical quantization to arrive at the quantum model, which, as has been emphasized by Paul Dirac, must stand on its own feet and has to be ultimately tested against experiment without making any contact to the initial classical model.

For Maxwell's equations this procedure gives operators for the electromagnetic fields

$$\boldsymbol{E}(\boldsymbol{r}),\ \boldsymbol{H}(\boldsymbol{r}) \longrightarrow \hat{\boldsymbol{E}}(\boldsymbol{r}),\ \hat{\boldsymbol{H}}(\boldsymbol{r}), \tag{12.2}$$

with similar expressions for the material operators, as schematically depicted in Fig. 12.1. For the field quantization we consider a quantization box within which the electromagnetic fields are confined, and denote the eigenfrequencies and mode functions of the corresponding wave equation with ω_λ and $\boldsymbol{u}_\lambda(\boldsymbol{r})$, respectively. In this eigenbasis the electric field operator can be expressed as [4, 119]

$$\hat{\boldsymbol{E}}(\boldsymbol{r}) = \sum_\lambda \left(\frac{\hbar\omega_\lambda}{2\varepsilon_0}\right)^{\frac{1}{2}} i\left(\boldsymbol{u}_\lambda(\boldsymbol{r})\hat{a}_\lambda - \boldsymbol{u}_\lambda^*(\boldsymbol{r})\hat{a}_\lambda^\dagger\right), \tag{12.3}$$

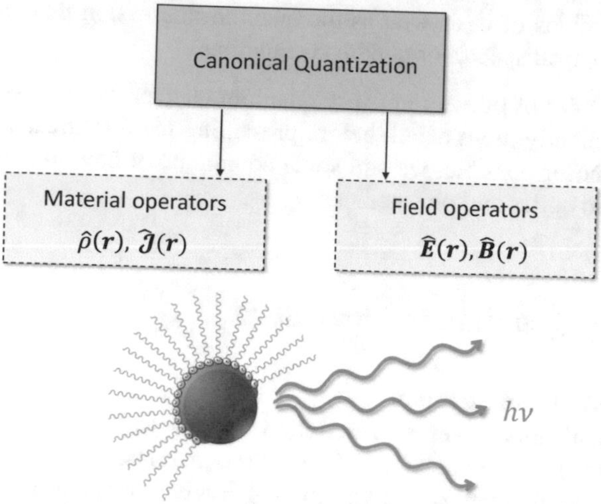

Fig. 12.1 The Lagrange formalism provides a unified platform for the description of light, matter, and its mutual interaction. Through canonical quantization we bring the light–matter system to the quantum realm, with operators for the matter excitations and the electromagnetic fields, which must be additionally applied to a wavefunction. Within this approach we can describe problems such as the decay of a quantum emitter through emission of a photon, as schematically indicated on the bottom of the figure for a colloidal quantum dot

where $\hat{a}_\lambda$, $\hat{a}_\lambda^\dagger$ are the photon operators with bosonic commutation relations, and the prefactor has been chosen for reasons to become clearer in the next chapter. The important observation at this point is that the propagation properties of the quantized electric fields, which are encapsulated in the frequencies ω_λ and mode functions $\boldsymbol{u}_\lambda(\boldsymbol{r})$, are obtained from the solutions of the *classical* wave equation, and it is only the occupation of these modes, described by the photon operators $\hat{a}_\lambda$, $\hat{a}_\lambda^\dagger$, that accounts for the novel quantum properties of light.

If we were only interested in electromagnetic fields in free space, our quantization journey would end here. The same applies to Maxwell's equations in presence of non-absorbing dielectrics, where we simply have to replace the eigenfrequencies and modes by those of the more complicated photonic environment. However, the quantization procedure fails in presence of absorbing materials, such as metals or doped semiconductors, since the eigenfrequencies ω_λ of the wave equation acquire an imaginary part due to losses described by the imaginary part of the permittivity, and with this the norm of the photon wavefunction can no longer be preserved. As a consequence, the field quantization in presence of absorbing media has to be done differently, and we are forced to add one or two additional steps before going fully quantum.

Step 2. Correlation Functions

The second step is often omitted when quantizing Maxwell's equations in presence of absorbing media, but will turn out to be useful when making contact with ab-initio descriptions for the material response. In linear response we can establish within Kubo's formalism a rigid relation between the fluctuations of the unperturbed system in thermal equilibrium and the expectation value of observables in the weakly perturbed system. This forms the essence of the so-called fluctuation–dissipation theorem, which will be discussed in more detail in Chap. 14.

Within Kubo's formalism one considers a quantum-mechanical system, described by the Hamiltonian $\hat{H}_0$, that is coupled via

$$\hat{V}(t) = \hat{v}\,X(t) \tag{12.4}$$

to an external classical field $X(t)$, with $\hat{v}$ being an operator that accounts for this coupling. In linear response, the change of some observable $\hat{u}$ can be computed from

$$\delta u(t) = -\frac{i}{\hbar}\int_0^t \Big\langle \Big[\hat{u}(t), \hat{v}(t')\Big]\Big\rangle_{\text{eq}} X(t')\,dt', \tag{12.5}$$

as will be proven in Chap. 14. Here the operators are given in an interaction picture with respect to the system Hamiltonian $\hat{H}_0$, and the expectation value for the commutator has to be evaluated in thermal equilibrium. The neat thing about Eq. (12.5) is that it relates the properties of the *perturbed* system (left-hand side) to the fluctuations of the *unperturbed* system (right-hand side), where the latter are often much easier to compute. It is convenient to denote the left-hand side of the equation as the dissipation part, and the right-hand side as the fluctuation part. Together they form the fluctuation–dissipation theorem, see also Fig. 12.2.

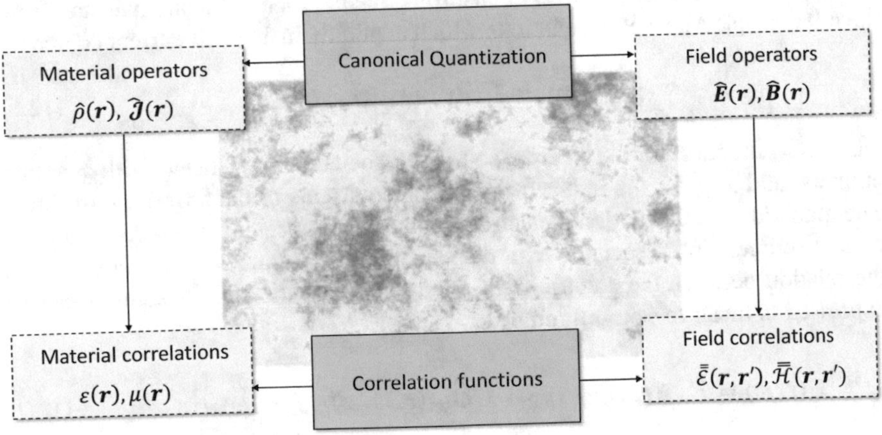

Fig. 12.2 The fluctuation–dissipation theorem relates the correlations in thermal equilibrium to the system's linear response. For the material part we can relate density–density correlations to the dielectric function and current–current correlations to the optical conductivity. For the light part we can relate the field correlations to the dyadic Green's function of classical electrodynamics

This theorem can be exploited in nano optics in various circumstances. In free space, the correlations of the electromagnetic field operators (the fluctuation part) can be related to the dyadic Green's function of classical electrodynamics (the dissipation part). For charged many-body systems and an external potential V_{ext}, the density–density correlations in thermal equilibrium (the fluctuation part) can be related to the induced charge density (the dissipation part). Through this, the density–density correlations can be related to the dielectric function. Similarly, the current–current correlations can be related to the optical conductivity. We can thus establish a direct link between the microscopic material description, based on correlation functions, and the macroscopic material description, based on permittivities and permeabilities. We will exploit this in later parts of this book for selected problems in the field of *quantum plasmonics*, such as nonlocal response functions or plasmon tunneling.

The connection between the material parameters ε, μ and the microscopic description in terms of density and current fluctuations is certainly an asset in nano optics and plasmonics. There exists, however, an alternative approach that allows bringing Maxwell's equations to the quantum domain without making contact to any microscopic description [120]. It is based on the multipolar Hamiltonian and has the polarization and magnetization operators as central objects, in close analogy to the macroscopic Maxwell's equations introduced in previous parts of this book. This approach is more popular when quantizing Maxwell's equations in presence of absorbing media, but since the multipolar Hamiltonian is not widely used in solid state physics we will stick to our microscopic description scheme and will only briefly comment on the phenomenological approach.

Step 3. Fluctuational Electrodynamics

Fluctuational electrodynamics in the quantum domain deals with the quantization of Maxwell's equations in presence of absorbing media, such as metals. We start from the microscopic Maxwell's equations which explicitly include all current sources

$$\hat{\boldsymbol{J}}(\boldsymbol{r}',\omega)=\hat{\boldsymbol{J}}_{\text{ext}}(\boldsymbol{r},\omega)+\hat{\boldsymbol{J}}_{\text{ind}}(\boldsymbol{r},\omega), \tag{12.6}$$

where $\hat{\boldsymbol{J}}_{\text{ext}}$ accounts for the external sources, associated, for instance, with quantum emitters, and $\hat{\boldsymbol{J}}_{\text{ind}}$ for the induced microscopic currents of the absorbing medium. The grand idea of fluctuational electrodynamics is to trace out in a second step the current sources $\hat{\boldsymbol{J}}_{\text{ind}}$ associated with the absorbing medium. As shown in Chap. 15, the relation between the electric field operator $\hat{\boldsymbol{E}}$ and the current sources can be expressed in linear response through

$$\hat{\boldsymbol{E}}(\boldsymbol{r},\omega)=\hat{\boldsymbol{E}}_{\text{inc}}(\boldsymbol{r},\omega)+i\mu_0\omega\int\bar{\bar{G}}_{\text{tot}}(\boldsymbol{r},\boldsymbol{r}',\omega)\cdot\hat{\boldsymbol{J}}(\boldsymbol{r}',\omega)\,d^3r', \tag{12.7}$$

where $\hat{\boldsymbol{E}}_{\text{inc}}$ is the field operator associated with an incoming radiation and G_{tot} is the total Green's function of classical electrodynamics. Note that this Green's function has to be computed for the photonic environment including all absorbing bodies. In order to bring fluctuational electrodynamics to work, we proceed as follows.

Observable. We start by specifying the observable we are interested in. Typical examples are the decay rate of a quantum emitter, the renormalization of its emission frequency—the so-called Lamb shift—or the Casimir–Polder force acting on the quantum emitter. In linear response, all these observables can be expressed in terms of field correlation functions.

Correlations. Using Eq. (12.7), the field fluctuations can be related to current fluctuations of the absorbing medium, and through Kubo's formalism the current fluctuations can be further related to the optical conductivity or the permittivity of the absorbing medium.

Evaluation. With this, the expressions for the observables can be written down in terms of (classical) Green's functions and (classical) permittivities, additionally decorated with distribution functions associated with the thermal photon occupation. These expressions can be finally evaluated along the same lines as discussed in previous parts of this book.

From the above discussion it is apparent that the quantization of Maxwell's equations in presence of absorbing media is different in comparison to other quantization schemes, which usually lead to operator expressions that are generally valid. Approximations are later introduced at the level of Schrödinger's equation. In contrast, fluctuational electrodynamics provides a prescription for obtaining working equations that include from the outset the approximation of a linear material response. Yet, the approach is extremely powerful, as will be shown in Chap. 15 for selected examples, and can be generally used for linear photonic environments. There exist generalization schemes for nonlinear material systems, but we will not discuss these modifications here.

12.2 The Quantum Optics Toolbox

In the final chapters of this book we investigate the situation where a generic few-level system, associated, for instance, with a fluorescent molecule or quantum dot, is embedded in a *linear* photonic environment, and the electromagnetic fields induce a *nonlinear* response of the system. The quantum optics toolbox provides a flexible machinery for describing such systems which additionally interact with their environment, so-called *open quantum systems*, and for computing their optical properties, as schematically shown in Fig. 12.3. In brief, the main ingredients of the toolbox can be summarized as follows.

Statistical Operator. A system interacting with its environment can no longer be described in terms of a wavefunction. Rather one has to introduce a statistical operator $\hat{\rho}$ to account for the fact that because of environment couplings the state of the system is known only with some probability.

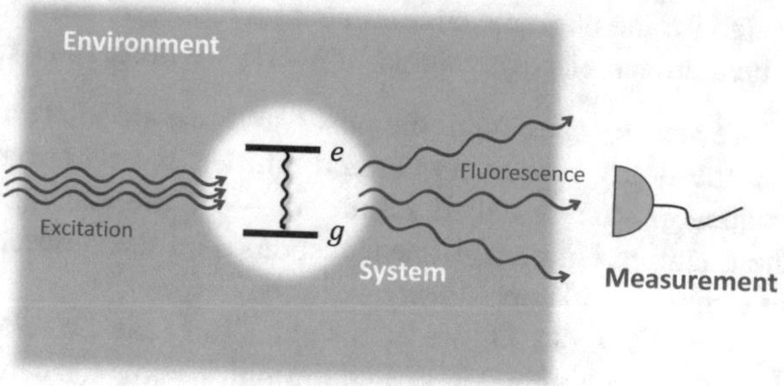

Fig. 12.3 Schematics of quantum optics toolbox. A few-level system, here a two-level system with a ground and excited state, is driven by a strong external light field, which induces a nonlinear response. The system additionally interacts with its environment as described in terms of dephasing and relaxation processes. The quantum optics toolbox provides a flexible machinery for describing the dynamics of open quantum systems, and for computing fluorescence spectra and photon correlations of the coherently driven system

Master Equation of Lindblad Form. A convenient way to describe the time evolution of the statistical operator is in terms of a master equation in Lindblad form, which accounts for both the *coherent* time evolution governed by external light fields and the *incoherent* time evolution attributed to environment couplings. Through the latter processes the system suffers dephasing and relaxation. Environment couplings are described through Lindblad operators, which depend on the initial and final states of a scattering process, as well as on the rate at which these scatterings occur.

Quantum Regression Theorem. One of the decay channels for optically excited systems is radiative decay, where the system decays by emitting a photon. In optical experiments it is precisely this emitted photon that is detected and which allows obtaining information about the system, either in the form of fluorescence spectra or photon correlations. With the quantum regression theorem we can compute photon correlation functions from the knowledge of the system dynamics alone, without accounting explicitly for the photon dynamics.

Although the above points do not look overly spectacular at first sight, it turns out that—on the contrary—the quantum optics toolbox is a real treasury which provides a machinery that is flexible, efficient, and robust, and which can be easily used once one understands its basic ingredients. There exists a lot of literature on the toolbox, mainly in textbooks on quantum optics, so we here keep the discussion rather short and mainly comment on how the toolbox can be used in the field of nano optics and plasmonics.

12.3 Summary of Book Chaps. 13–18

The contents of the chapters forming the second, "quantum" part of the book can be summarized as follows.

Chapter 13: Quantum Electrodynamics in a Nutshell. We start by discussing the quantization of Maxwell's equations and the material part using the canonical quantization procedure. The presentation is probably more pedantic and detailed than necessary; however, this has the advantage that it provides us with the full machinery applicable to the description of photons and light–matter interactions at the photon level.

Chapter 14: Correlation Functions. In this chapter we first introduce to correlation and spectral functions, and the cross-spectral density, without making contact to any specific physical system. The reason for doing so is that correlation functions are an extremely powerful tool in the field of nano optics and plasmonics: they allow us to relate field fluctuations in thermal equilibrium to the Green's function of classical electrodynamics, and provide a direct link between density–density correlations and current–current correlations with the dielectric functions and optical conductivities, respectively. We complement this chapter with selected applications in quantum plasmonics, including nonlocality and charge transfer plasmons, and present a quantum description of electron energy loss spectroscopy (EELS) of plasmonic nanoparticles.

Chapter 15: Thermal Effects in Nano Optics. This chapter is concerned with thermal effects in nano optics using the framework of fluctuational electrodynamics, which we develop in a general manner based on a microscopic material description. We show that this framework allows for a full quantum description of the decay rates and transition frequency renormalizations of quantum emitters situated in non-trivial photonic environments, as well as of Casimir–Polder forces acting on such emitters. We also discuss Casimir forces and heat transfer between macroscopic bodies.

Chapter 16: Two-Level Systems. The last chapters of this book are concerned with single quantum emitters, such as fluorescent molecules or quantum dots, which are embedded in a non-trivial photonic environment. The response of the environment is treated *linearly*, whereas the dynamics of the quantum emitter is described *nonlinearly*. We start by discussing generic two-level systems, where a simple pictorial representation is given by Bloch vectors. Relaxation and dephasing processes are introduced at the level of T_1 and T_2 times.

Chapter 17: Master Equation. We generalize the results of two-level systems to generic few-level systems, using a description in terms of a master equation of Lindblad form. Various solution schemes for the master equation are discussed, and we ponder on microscopic description schemes for relaxation and dephasing.

Chapter 18: Photon Noise. In this last chapter we introduce the quantum regression theorem, which allows computing fluorescence spectra and photon correlations from the system dynamics, without explicitly accounting for photons, and which brings the quantum optics toolbox to full glory.

With this, the reader should have acquired a fairly comprehensive picture of the theoretical and computational methods used in the field of nano optics and plasmonics and should be able to consult the scientific literature by him or herself without serious difficulties. It is fair to state that the selection of topics covered by this book is somewhat personal and selective, and many important topics have been left out. These include nonlinear material responses, ab-initio descriptions of quantum emitters and materials, or periodic structures and photonic crystals, to name just a few. So, there is still plenty of room to be discovered at the bottom of nano optics.

Chapter 13
Quantum Electrodynamics in a Nutshell

In this chapter we start by investigating charged particles interacting with electromagnetic fields as well as Maxwell's equations in the framework of the Lagrange formalism. Our primary motivation for doing so is that there exists a direct link between classical models in the Lagrange formalism and their quantized versions using the so-called *canonical quantization* procedure. With this we show how to describe light-matter coupling in quantum mechanics, submit the electromagnetic fields to canonical quantization, and introduce the concept of photons.

As we will discuss in the following chapters, in nano optics one can often trace out the photon degrees of freedom and use an object instead that should have become an old friend by now, the dyadic Green's function. So why bother spending a significant amount of time introducing the quantization of electromagnetic fields? First, in order to understand why the photon can be effectively removed from the theory one needs to understand why it is there in the first place. Secondly, the concept of photons is such a central theme in quantum optics and nano optics at the quantum level that it would be a shame to miss it. After all, generations of physicists have worked hard to come up with proper quantization procedures for field theories.

13.1 Preliminaries

Let us start with a short warmup in quantum physics. The main object in quantum mechanics is the wavefunction $|\psi(t)\rangle$ whose time evolution is governed by Schrödinger's equation

$$i\hbar\frac{d}{dt}\big|\psi(t)\big\rangle = \hat{H}(t)\big|\psi(t)\big\rangle. \tag{13.1}$$

U. Hohenester, *Nano and Quantum Optics*, Graduate Texts in Physics,
https://doi.org/10.1007/978-3-030-30504-8_13

Here $\hbar$ is the reduced Planck's constant, $\hat{H}(t)$ is the Hamilton operator that may depend on time, and we adopt the usual bra-ket notation throughout. Any observable can be associated with a Hermitian operator $\hat{A}$ whose expectation value is given by

$$\langle \hat{A} \rangle = \langle \psi(t) | \hat{A} | \psi(t) \rangle. \tag{13.2}$$

A powerful concept in quantum physics are unitary transformations, which can help to considerably simplify the theoretical approach without affecting the underlying physics. Let $\hat{\mathscr{U}}$ be a unitary operator with $\hat{\mathscr{U}}^\dagger \hat{\mathscr{U}} = \hat{\mathscr{U}} \hat{\mathscr{U}}^\dagger = \mathbb{1}$, where $\hat{\mathscr{U}}^\dagger$ denotes the Hermitian conjugate of the operator and $\mathbb{1}$ is the unit operator. Suppose that we transform the wavefunction with

$$|\psi(t)\rangle \longrightarrow |\psi'(t)\rangle = \hat{\mathscr{U}}^\dagger |\psi(t)\rangle,$$

and transform at the same time all operators according to

$$\hat{A} \longrightarrow \hat{A}' = \hat{\mathscr{U}}^\dagger \hat{A} \hat{\mathscr{U}}.$$

As can be easily seen, this transformation does not change the expectation value of $\hat{A}$,

$$\langle \psi'(t) | \hat{A}' | \psi'(t) \rangle = \Big(\langle \psi(t) | \hat{\mathscr{U}} \Big) \Big(\hat{\mathscr{U}}^\dagger \hat{A} \hat{\mathscr{U}} \Big) \Big(\hat{\mathscr{U}}^\dagger | \psi(t) \rangle \Big) = \langle \psi(t) | \hat{A} | \psi(t) \rangle. \tag{13.3}$$

In the above expression the products $\hat{\mathscr{U}} \hat{\mathscr{U}}^\dagger$ and $\hat{\mathscr{U}}^\dagger \hat{\mathscr{U}}$ give the unit operator. In a geometric interpretation, a unitary operator preserves the "lengths" and "angles" between vectors, and it can be considered as a type of rotation operator in an abstract vector space. It thus changes only the "coordinate system" without affecting the underlying physics.

Another type of unitary transformations is introduced by the time evolution operator $\hat{U}(t, t_0)$. This operator takes a wavefunction at time t_0 and propagates it to time t,

$$|\psi(t)\rangle = \hat{U}(t, t_0) |\psi(t_0)\rangle. \tag{13.4}$$

Inserting this expression into Schrödinger's equation gives

$$i\hbar \frac{d}{dt} \hat{U}(t, t_0) = \hat{H}(t) \hat{U}(t, t_0) \implies \hat{U}(t, t_0) = \mathbb{1} - \frac{i}{\hbar} \int_{t_0}^{t} \hat{H}(t') \hat{U}(t', t_0)\, dt', \tag{13.5}$$

where we have used $\hat{U}(t_0, t_0) = \mathbb{1}$. The solution of the integral equation on the right-hand side provides us with the time evolution operator. For a time-independent Hamiltonian $\hat{H}$ it can be solved readily and we arrive at the well-known result

$$\hat{U}(t, t_0) = \exp\left[-\frac{i}{\hbar}\hat{H}(t - t_0)\right]. \tag{13.6}$$

Note that the term in brackets has to be understood as the product between the time-independent Hamiltonian $\hat{H}$ and $t - t_0$. In the following we set t_0 equal to zero and denote the wavefunction at time zero by $|\psi_0\rangle$. We can now introduce the following "pictures" or representations.

In the **Schrödinger picture** we have time-dependent wavefunctions

$$\big|\psi_S(t)\big\rangle = \hat{U}(t, 0)\big|\psi_0\big\rangle \tag{13.7}$$

and time-independent operators $\hat{A}_S$ (unless the operators themselves depend explicitly on time). Thus, as time evolves only the wavefunction becomes modified but the operators remain unchanged. In contrast to that, in the **Heisenberg picture** the wavefunctions $|\psi_0\rangle$ are time independent and only the operators evolve in time according to

$$\hat{A}_H(t) = \hat{U}^\dagger(t, 0)\,\hat{A}_S\,\hat{U}(t, 0). \tag{13.8}$$

As can be easily verified, the expectation value of an operator $\hat{A}$ is the same in the Schrödinger and Heisenberg picture. Finally, taking the time derivative of an operator in the Heisenberg picture and using Eq. (13.5) for the time derivative of the time evolution operators leads us to the Heisenberg equation of motion for the operator $\hat{A}$

Heisenberg's Equation of Motion

$$i\hbar\frac{d}{dt}\hat{A}_H(t) = \hat{A}_H(t)\hat{H}(t) - \hat{H}(t)\hat{A}_H(t) = \left[\hat{A}_H(t), \hat{H}(t)\right], \tag{13.9}$$

with the usual abbreviation $[\hat{A}, \hat{B}] = \hat{A}\hat{B} - \hat{B}\hat{A}$ for the commutator.

There exists a third picture, the interaction picture or Dirac picture, which is highly useful for problems where the Hamiltonian can be separated into a part $\hat{H}_0$ that can be described exactly and a remainder $\hat{V}(t)$ that usually is treated as a perturbation,

$$\hat{H}(t) = \hat{H}_0 + \hat{V}(t).$$

We now introduce a time evolution operator $\hat{U}_0(t, 0)$ which is associated with the time evolution of $\hat{H}_0$ only,

$$i\hbar\frac{d}{dt}\hat{U}_0(t, 0) = \hat{H}_0\hat{U}_0(t, 0), \qquad \hat{U}_0(0, 0) = \mathbb{1}.$$

In the **interaction picture** we transform the wavefunctions and operators according to

$$\big|\psi_I(t)\big\rangle = \hat{U}_0^\dagger(t,0)\big|\psi_S(t)\big\rangle \tag{13.10a}$$

$$\hat{A}_I(t) = \hat{U}_0^\dagger(t,0)\hat{A}_S\hat{U}_0(t,0). \tag{13.10b}$$

One can readily show that the expectation value of an operator in the interaction picture is the same as in the Schrödinger and Heisenberg pictures. From the definition of the wavefunction in the interaction picture, Eq. (13.10a), we can define the time evolution operator in the interaction picture through

$$\hat{U}_I(t,0) = \hat{U}_0^\dagger(t,0)\hat{U}(t,0). \tag{13.11}$$

Taking the time derivative of $\hat{U}_I$ gives

$$i\hbar\frac{d}{dt}\hat{U}_I = i\hbar\left[\left(\frac{d}{dt}\hat{U}_0^\dagger\right)\hat{U} + \hat{U}_0^\dagger\left(\frac{d}{dt}\hat{U}\right)\right] = \left(-\hat{U}_0^\dagger\hat{H}_0\right)\hat{U} + \hat{U}_0^\dagger\left[\left(\hat{H}_0 + \hat{V}\right)\hat{U}\right],$$

where we have suppressed the time arguments of all operators. The terms with $\hat{H}_0$ cancel each other. Thus, we arrive at the defining equation for the time evolution operator in the interaction picture

Time Evolution Operator in Interaction Picture

$$i\hbar\frac{d}{dt}\hat{U}_I(t,0) = \hat{V}_I(t)\hat{U}_I(t,0), \qquad \hat{U}_I(0,0) = \mathbb{1}, \tag{13.12}$$

where the Hamiltonian $V_I(t)$ in the interaction picture is defined according to Eq. (13.10b). Equation (13.12) can be rewritten as an integral equation

$$\hat{U}_I(t,0) = \mathbb{1} - \frac{i}{\hbar}\int_0^t \hat{V}_I(t')\hat{U}_I(t',0)\,dt'. \tag{13.13}$$

From this expression the advantage of the interaction picture is particularly transparent. The part of the Hamiltonian which can be handled exactly, $\hat{H}_0$, has been completely absorbed into the wavefunctions and operators, and it is only the nontrivial part $\hat{V}$ of the Hamiltonian that governs the system's time evolution. Under broad circumstances the Volterra-type integral equation (13.12) can be solved iteratively, and we obtain from Eq. (13.13) in lowest order perturbation theory

$$\hat{U}_I(t,0) = \mathbb{1} - \frac{i}{\hbar}\int_0^t \hat{V}_I(t')\,dt' + \mathcal{O}\left(\hat{V}^2\right). \tag{13.14}$$

Similarly, the time evolution of an operator in the interaction picture can be computed from

$$\hat{U}_I^\dagger(t,0)\hat{A}_I(t)\hat{U}_I(t,0) = \hat{A}_I(t) - \frac{i}{\hbar}\int_0^t \left[\hat{A}_I(t), \hat{V}_I(t')\right] dt' + \mathcal{O}\left(\hat{V}^2\right). \tag{13.15}$$

These expressions will be used extensively in later parts of this book.

As a final comment, in this book we will often switch back and forth between the different pictures. For notational simplicity it is convenient to suppress the subscripts indicating the Schrödinger, Heisenberg, or interaction pictures. Unfortunately this introduces some danger of confusion, but we will always try to indicate carefully in which picture we work.

13.1.1 A First Glimpse of Quantum Electrodynamics

For the inpatient reader, we start by providing a first flavor of the quantum version of Maxwell's equations. All details will be worked out in later parts of the chapter. It will be convenient to work in the Coulomb gauge, see Sect. 13.3, where the scalar potential is given by the instantaneous Coulomb potential of electrostatics and the vector potential is entirely transverse.

A system of particles with charge q and mass m interacting with external electromagnetic fields is then described by the so-called minimal coupling Hamiltonian

$$\hat{H} = \sum_i \left[\frac{\left(\hat{\boldsymbol{\pi}}_i - q\boldsymbol{A}^\perp(\boldsymbol{r}_i,t)\right)^2}{2m} + qV(\boldsymbol{r}_i,t)\right] + \frac{1}{8\pi\varepsilon_0}\sum_{i\neq j}\frac{q^2}{|\boldsymbol{r}_i - \boldsymbol{r}_j|}. \tag{13.16}$$

Here $\hat{\boldsymbol{\pi}}_i$ is the canonical momentum of particle i, with the fundamental commutation relation $[r_{ik}, \hat{\pi}_{i'k'}] = i\hbar\delta_{ii'}\delta_{kk'}$, where i, i' label the different particles and k, k' the Cartesian coordinates. The term in brackets accounts for the kinetic energy and the coupling to the external electromagnetic fields, and the second term on the right-hand side for the instantaneous Coulomb interaction between the charged particles. It is important to realize that in quantum physics the basic quantities are the electromagnetic potentials, rather than the electromagnetic fields, although all observables must not depend on the chosen gauge. In Eq. (13.16) we have indicated with the superscript of $\boldsymbol{A}^\perp$ that the vector potential is transverse.

When considering quantized electromagnetic fields, the vector potential $\boldsymbol{A}^\perp$ has to be replaced by an operator $\hat{\boldsymbol{A}}^\perp$. In a plane wave basis this operator reads

The History of the Photon

From a historical perspective, quantized portions of radiation were first introduced by Planck in the context of black-body radiation and the famous radiation law named after him. They were later adopted by Einstein when describing the photoelectric effect. A short but comprehensive discussion of the photon history and the photoelectric effect is given by Roy Glauber [121].

The only response that the metals make to increasing the intensity of light lies in producing more photoelectrons. Einstein had a naively simple explanation for that. The light itself, he assumed, consists of localized energy packets and each possesses one quantum of energy. When light strikes the metal, each packet is absorbed by a single electron. That electron then flies off with a unique energy, an energy which is just the packet energy $h\nu$ minus whatever energy the electron needs to expend in order to escape the metal. [...]

It is worth pointing out a small shift in terminology that took place in the late 1920s. Once material particles were found to exhibit some of the wavelike behavior of light quanta, it seemed appropriate to acknowledge that the light quanta themselves might be elementary particles, and to call them "photons" as suggested by G. N. Lewis in 1926. They seemed every bit as discrete as material particles, even if their existence was more transitory, and they were at times freely created or annihilated.

$$\hat{\boldsymbol{A}}^{\perp}(\boldsymbol{r}) = \sum_{\boldsymbol{k},s} \left(\frac{\hbar}{2\Omega\varepsilon_0\omega}\right)^{\frac{1}{2}} \left(e^{i\boldsymbol{k}\cdot\boldsymbol{r}}\,\hat{a}_{\boldsymbol{k}s} + e^{-i\boldsymbol{k}\cdot\boldsymbol{r}}\,\hat{a}_{\boldsymbol{k}s}^{\dagger}\right)\boldsymbol{\epsilon}_{\boldsymbol{k}s}. \tag{13.17}$$

Here $\boldsymbol{k}$ is the photon wavevector, s labels the two orthogonal polarizations with polarization vectors $\boldsymbol{\epsilon}_{\boldsymbol{k}s}$, Ω is a quantization volume (which we may let approach infinity), and $\omega = kc$ is the angular frequency of light. The central objects of quantum electrodynamics are the operators $\hat{a}_{\boldsymbol{k}s}$, $\hat{a}_{\boldsymbol{k}s}^{\dagger}$, which annihilate and create photons, respectively. They fulfill the bosonic commutation relations

$$\left[\hat{a}_{\boldsymbol{k}s}, \hat{a}_{\boldsymbol{k}'s'}^{\dagger}\right] = \delta_{\boldsymbol{k}\boldsymbol{k}'}\delta_{ss'}, \quad \left[\hat{a}_{\boldsymbol{k}s}, \hat{a}_{\boldsymbol{k}'s'}\right] = \left[\hat{a}_{\boldsymbol{k}s}^{\dagger}, \hat{a}_{\boldsymbol{k}'s'}^{\dagger}\right] = 0. \tag{13.18}$$

With the vector potential operator of Eq. (13.17) the Hamiltonian for the electromagnetic fields can be written in the form

$$\hat{H}_{\text{em}} = \int \left[\frac{\varepsilon_0}{2}\hat{\boldsymbol{E}}^{\perp}(\boldsymbol{r})\cdot\hat{\boldsymbol{E}}^{\perp}(\boldsymbol{r}) + \frac{1}{2\mu_0}\hat{\boldsymbol{B}}(\boldsymbol{r})\cdot\hat{\boldsymbol{B}}(\boldsymbol{r})\right] d^3r, \tag{13.19}$$

where the transverse electric and magnetic fields are expressed in terms of the vector potential operator $\hat{\boldsymbol{A}}^{\perp}$. To be meaningful, the material and field operators must be additionally applied to a wavefunction.

For the quantized version of the interacting light-matter system we must combine the minimal coupling Hamiltonian of Eq. (13.16) with the field Hamiltonian of Eq. (13.19). When working in the Heisenberg picture, the time evolutions of the electric and magnetic field operators become

$$\varepsilon_0 \frac{\partial}{\partial t} \hat{\boldsymbol{E}}(\boldsymbol{r}, t) = \frac{1}{\mu_0} \nabla \times \hat{\boldsymbol{B}}(\boldsymbol{r}, t) - \hat{\boldsymbol{J}}(\boldsymbol{r}, t) \tag{13.20a}$$

$$\frac{\partial}{\partial t} \hat{\boldsymbol{B}}(\boldsymbol{r}, t) = -\nabla \times \hat{\boldsymbol{E}}(\boldsymbol{r}, t), \tag{13.20b}$$

where $\hat{\boldsymbol{J}}$ is the current operator associated with the charged particles. Additionally, the field operators are subject to the constraints

$$\nabla \cdot \hat{\boldsymbol{E}}(\boldsymbol{r}, t) = \frac{\hat{\rho}(\boldsymbol{r}, t)}{\varepsilon_0} \tag{13.20c}$$

$$\nabla \cdot \hat{\boldsymbol{B}}(\boldsymbol{r}, t) = 0, \tag{13.20d}$$

with the charge density operator $\hat{\rho}$. One can immediately identify these equations with those of Maxwell's theory in classical electrodynamics, with the main difference that they now have to be understood as operator equations. As we will show in later parts of this chapter, the propagation properties of photons are governed by those of classical waves, and it is primarily the noise properties which are different in a quantum approach. Additionally, the electromagnetic field operators $\hat{\boldsymbol{E}}$, $\hat{\boldsymbol{B}}$ do not commute in general, such that field measurements at different space-time points may influence each other.

In the remainder of this chapter we introduce the canonical quantization procedure, which allows us to start from a classical model and extract from it a quantum version. We apply this procedure to light-matter couplings and to Maxwell's equations, and discuss photon states, as well as the derivation of a multipolar Hamiltonian.

13.2 Canonical Quantization

We first recall the Lagrange formalism of classical mechanics, see also Fig. 13.1. For simplicity we consider a single particle only, the generalization for many particles follows along the same lines. Let $\boldsymbol{r}$ and $\boldsymbol{v} = \dot{\boldsymbol{r}}$ denote the position and velocity of the particle, which may differ from the usual Cartesian coordinates in case of additional constraints, although we shall not be interested in such problems here. For any possible trajectory $\boldsymbol{r}_{\text{trial}}(t)$ we can define an action

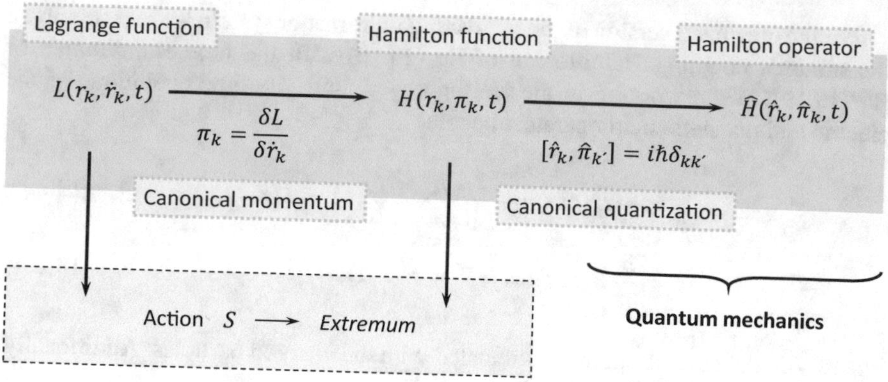

Fig. 13.1 Route from a classical description to a quantum description. In classical mechanics, one starts from a Lagrange function which depends on the generalized coordinates r_k and velocities $\dot{r}_k$, and performs a Legendre transformation using the canonical momenta π_k to arrive at the Hamilton function. Both the Lagrange and Hamilton function can be submitted to the action principle, where the system's equations of motions follow from the extremum principle of the action S. By replacing the coordinates r_k and canonical momenta π_k by operators, using the canonical quantization procedure, we end up with a quantized version of a classical model. This approach works for both the material degrees of freedom and for the electromagnetic fields, as will be discussed in more detail in later parts of this chapter

$$S = \int_{t_0}^{t_1} L\big(\boldsymbol{r}_{\text{trial}}(t), \dot{\boldsymbol{r}}_{\text{trial}}(t), t\big)\, dt, \tag{13.21}$$

where $L(\boldsymbol{r}, \dot{\boldsymbol{r}}, t)$ is the Lagrange function. The basic assumption of Lagrange's formalism is that this action S becomes an extremum for the true trajectory $\boldsymbol{r}(t)$ of the system

Lagrange Formalism

$$S \longrightarrow \text{Extremum for system's time evolution } \boldsymbol{r}(t). \tag{13.22}$$

To derive from this principle equations of motion for the particle, we proceed as follows.

Variations. We start by assuming that we already know the true solution $\boldsymbol{r}(t)$ and then consider small variations $\delta\boldsymbol{r}(t)$ around $\boldsymbol{r}(t)$, with the constraint of fixed positions at the initial and terminal times (see Fig. 13.2),

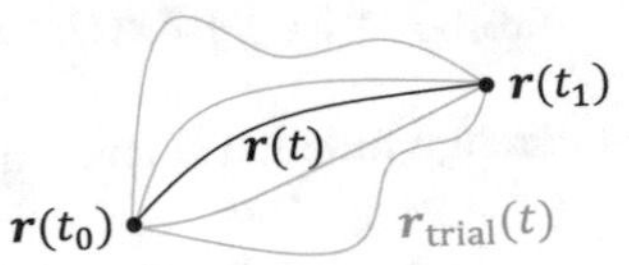

Fig. 13.2 Schematics for Lagrange formalism. For all possible trajectories $\boldsymbol{r}_{\text{trial}}(t)$ we fix the positions at the initial and final times t_0 and t_1. Within the Lagrange formalism, the action becomes an extremum for the true solution $\boldsymbol{r}(t)$

$$\delta\boldsymbol{r}(t_0) = \delta\boldsymbol{r}(t_1) = 0.$$

Linearization. A normal function $f(x)$ can be expanded around a position x_0 in a Taylor series

$$f(x) = f(x_0) + f'(x_0)\delta x + \frac{1}{2}f''(x_0)\delta x^2 + \mathcal{O}(\delta x^3),$$

with $\delta x = x - x_0$. At the extremum the first derivative of the function has to be zero,

$$f'(x_0) = 0.$$

Similarly, for the action integral of Eq. (13.21) we perform an expansion of the Lagrange function for small variations $\delta\boldsymbol{r}(t)$, and request that at the extremum the first-order contributions become zero.

The Lagrange formalism can be considered as an alternative to Newton's equations of motion, and has the appealing feature that it builds on a very general assumption, namely the minimization of S, although it gives in general no prescription of how to obtain L (similarly to Newton's equations of motion where the forces have to be guessed as well). In addition, the Lagrange formalism provides a clear route for the canonical quantization of a classical theory, as will be discussed further below.

Example of Harmonic Oscillator
For a simple but illustrative example of the Lagrange formalism we discuss the one-dimensional harmonic oscillator. The Lagrange function reads

$$L(x, \dot{x}) = T - V = \frac{1}{2}m\dot{x}^2 - \frac{1}{2}m\omega^2 x^2, \tag{13.23}$$

where T and V are the kinetic and potential energy, respectively, m is the mass of the particle, and ω the oscillator's frequency. We next investigate the behavior of the Lagrange function for trajectories close to the true trajectory $x(t)$ (which we want to compute, so far we only know that it minimizes S)

$$x_{\text{trial}}(t) = x(t) + \delta x(t).$$

For small variations δx the Lagrange function changes according to

$$\begin{aligned} L(x+\delta x, \dot{x}+\delta\dot{x}) &= \frac{1}{2}m\left\{\left(\dot{x}+\delta\dot{x}\right)^2 - \omega^2\left(x+\delta x\right)^2\right\} \\ &\approx L(x,\dot{x}) + m\left(\dot{x}\delta\dot{x}\right) - m\omega^2\left(x\delta x\right) + \mathcal{O}\left(\delta x^2\right). \end{aligned}$$

Here and in the following we neglect terms of order $(\delta x)^2$ and $(\delta\dot{x})^2$. Inserting the above expression into Eq. (13.21) then gives

$$\delta S = \int_{t_0}^{t_1} \left\{m\left(\dot{x}\delta\dot{x}\right) - m\omega^2\left(x\delta x\right)\right\} dt + \mathcal{O}(\delta x^2) = 0.$$

This equation has to be fulfilled for the true trajectory $x(t)$ and for small variations $\delta x(t)$ around it. The first term in brackets can be simplified by performing integration by parts

$$\int_{t_0}^{t_1} m\left(\dot{x}\delta\dot{x}\right) dt = m\left(\dot{x}\delta x\right)\Big|_{t_0}^{t_1} - \int_{t_0}^{t_1} m\left(\ddot{x}\delta x\right) dt.$$

If we keep the initial and final positions fixed, as we do here, the first term on the right-hand side becomes zero. Thus, the variation of the action integral becomes

$$\delta S = -\int_{t_0}^{t_1} \left(m\ddot{x} + m\omega^2 x\right)\delta x(t)\, dt = 0. \tag{13.24}$$

Since this expression must be fulfilled for any sufficiently small variation $\delta x(t)$, we conclude that the term in parentheses must be zero. This leads us to the equation of motion

$$\ddot{x}(t) = -\omega^2 x(t),$$

in accordance to Newton's equations of motion for the harmonic oscillator.

13.2.1 Euler–Lagrange Equations

The above procedure can be generalized to arbitrary potentials V and higher spatial dimensions, as well as to a larger number of particles, although we will not work out the many-particle case explicitly here. We start by considering trajectories which slightly deviate from the true solution $\boldsymbol{r}(t)$,

$$\boldsymbol{r}_{\text{trial}}(t) = \boldsymbol{r}(t) + \delta \boldsymbol{r}(t).$$

With $\delta \boldsymbol{r}(t_1) = \delta \boldsymbol{r}(t_2) = 0$ we assure again that all trial trajectories pass through the same initial and final positions. For sufficiently small deviations $\delta \boldsymbol{r}$, we can expand the Lagrange function around the true solution

$$\delta L = \sum_k \left(\frac{\partial L}{\partial \dot{r}_k} \delta \dot{r}_k + \frac{\partial L}{\partial r_k} \delta r_k \right) + \mathcal{O}(\delta r^2),$$

where the derivatives of the Lagrange function have to be understood as functional derivatives. At the extremum, the action S should not change for small modifications $\delta \boldsymbol{r}$, and we get from Eq. (13.21)

$$\delta S = \int_{t_0}^{t_1} \sum_k \left(\frac{\partial L}{\partial \dot{r}_k} \delta \dot{r}_k + \frac{\partial L}{\partial r_k} \delta r_k \right) dt = 0.$$

In the first term in parentheses we next perform integration by parts to shuffle the time derivative from $\delta \dot{r}_k$ to the preceding term. This gives

$$\sum_k \frac{\partial L}{\partial \dot{r}_k} \delta r_k \bigg|_{t_0}^{t_1} - \int_{t_0}^{t_1} \sum_k \left(\frac{d}{dt} \frac{\partial L}{\partial \dot{r}_k} - \frac{\partial L}{\partial r_k} \right) \delta r_k(t)\, dt = 0.$$

The first term on the left-hand side is zero because of the constraint $\delta r_k(t_0) = \delta r_k(t_1) = 0$. As the equality has to be fulfilled for arbitrary variations $\delta r_k(t)$, we conclude that the expression in parentheses must be zero. This leads us to the Euler–Lagrange equations

Euler–Lagrange Equations

$$\frac{d}{dt} \frac{\partial L}{\partial \dot{r}_k} - \frac{\partial L}{\partial r_k} = 0. \tag{13.25}$$

13.2.2 *Hamilton Formalism*

The Hamilton formalism is a variant of the Lagrange formalism which introduces the canonical momenta as basic quantities. It also provides a direct link to quantum mechanics, as will be discussed below. In the context of Hamilton dynamics one usually deals with different types of momenta.

Canonical Momentum. The canonical momentum is defined as[1]

$$\pi_k = \frac{\partial L}{\partial \dot{r}_k}, \tag{13.26}$$

and plays a central role in the Hamilton formalism and in the canonical quantization of classical models.

Kinetic Momentum. The kinetic momentum of a particle is defined as

$$\boldsymbol{\pi}_{\text{kin}} = m\dot{\boldsymbol{r}}. \tag{13.27}$$

In many cases of interest canonical and kinetic momenta are identical, but there are prominent examples where the two definitions differ. For instance, for particles in presence of electromagnetic fields the kinetic momentum accounts for the momentum of the charged particle only, whereas the canonical momentum accounts for both the momenta of the particle and the electromagnetic fields. This point will be discussed in more detail below.

The Hamilton function can be obtained from the Lagrange function through a Legendre transformation

$$H(r, p, t) = \sum_k \pi_k \dot{r}_k - L(r, \dot{r}, t). \tag{13.28}$$

H must be expressed in terms of the positions r_k and the canonical momenta π_k only, which means that velocities $\dot{r}_k$ have to be eliminated in favor of r_k and π_k throughout. In principle, the Hamilton formalism can be developed independently from an action principle similar to Eq. (13.21), without making contact with the Lagrange formalism. However, since we have L already at hand, we use Eq. (13.28) to explicitly compute H and the resulting equations of motion for r_k, π_k.

The total differential of the Lagrange function reads

$$dL = \sum_k \left(\frac{\partial L}{\partial \dot{r}_k} d\dot{r}_k + \frac{\partial L}{\partial r_k} dr_k \right) + \frac{\partial L}{\partial t} dt = \sum_k \left(\pi_k d\dot{r}_k + \dot{\pi}_k dr_k \right) + \frac{\partial L}{\partial t} dt.$$

We have used the Euler–Lagrange Eq. (13.25) to arrive at the final expression. Similarly, the total differential of H given in Eq. (13.28) is

$$dH = \sum_k \left(\pi_k d\dot{r}_k + \dot{r}_k d\pi_k \right) - dL = \sum_k \left(\dot{r}_k d\pi_k - \dot{\pi}_k dr_k \right) - \frac{\partial L}{\partial t} dt,$$

[1] We deviate here somewhat from the usual notation of p for the momenta, because the symbol is already reserved for the dipole moment, and use π instead.

where in the last step we have inserted the expression for dL to cancel the terms with $\pi_k d\dot{r}_k$. We can compare this with the total differential of the Hamiltonian

$$dH = \sum_k \left(\frac{\partial H}{\partial \pi_k} d\pi_k + \frac{\partial H}{\partial r_k} dr_k \right) + \frac{\partial H}{\partial t} dt,$$

and are led to Hamilton's equations of motion that describe the time evolution of r_k, π_k in terms of the Hamilton function H through

Hamilton's Equations of Motion

$$\dot{r}_k = \frac{\partial H}{\partial \pi_k}, \qquad \dot{\pi}_k = -\frac{\partial H}{\partial r_k}. \tag{13.29}$$

It is sometimes convenient to rewrite the above expressions. To this end we introduce the so-called **Poisson bracket**, which we define for two arbitrary functions u and v according to

$$\{u, v\} = \sum_k \left(\frac{\partial u}{\partial r_k}\frac{\partial v}{\partial \pi_k} - \frac{\partial v}{\partial r_k}\frac{\partial u}{\partial \pi_k} \right). \tag{13.30}$$

With this, Hamilton's equations of motion can be expressed through

$$\dot{r}_k = \{r_k, H\}, \qquad \dot{\pi}_k = \{\pi_k, H\},$$

as can be easily verified through explicit calculation. We also find the fundamental Poisson bracket of $\{r_k, \pi_{k'}\} = \delta_{kk'}$.

13.2.3 *Canonical Quantization*

There exists a remarkably simple prescription of how to submit a classical model to a quantization procedure for use within quantum mechanics, the so-called canonical quantization. Essentially it replaces the positions r_k and canonical momenta π_k by operators $\hat{r}_k$, $\hat{\pi}_k$, or q-numbers, as they were called by Paul Dirac [122]:

> q-numbers, of course, is just a name. Ordinary numbers, when one wants to make a distinction between them and these dynamical variables, may be called *c-numbers*. The reasons for the letters q and c are that q makes you think of quantum or queer, and c of classical or commuting. The essential difference between c-numbers and q-numbers is that c-numbers commute with everything, and q-numbers do not in general commute with q-numbers.

To go over from a classical theory to a quantum theory, within **canonical quantization** one has to make the replacement

$$\{u, v\} \longrightarrow \frac{1}{i\hbar}\left[\hat{u}, \hat{v}\right] = \frac{1}{i\hbar}\left(\hat{u}\hat{v} - \hat{v}\hat{u}\right), \tag{13.31}$$

where $\hat{u}$, $\hat{v}$ have now become queer or quantum operators, and the Poisson bracket has been replaced by a so-called commutator. Most importantly, the fundamental commutation relation between positions and momenta reads

Canonical Commutation Relation

$$\left[\hat{r}_k, \hat{\pi}_{k'}\right] = \hat{r}_k\hat{\pi}_{k'} - \hat{\pi}_{k'}\hat{r}_k = i\hbar\delta_{kk'}. \tag{13.32}$$

This quantization scheme will be used in the following for both the quantization of the material and electromagnetic field dynamics. In a real-space representation, the position operator becomes $\boldsymbol{r}$ (we will usually refrain from explicitly indicating that $\boldsymbol{r}$ is an operator) and the canonical momentum can be expressed as

$$\hat{\boldsymbol{\pi}} = -i\hbar\nabla. \tag{13.33}$$

It is surprising that nature has opted for the canonical quantization procedure to communicate between the classical and quantum world. Paul Dirac motivates this close connection between classical and quantum theories as follows [122].

> Let us consider now the connection between classical theory and quantum theory [...]. The Hamiltonian of the classical theory is a certain function of the classical dynamic variables and we set up the quantum Hamiltonian which is the same function of the quantum dynamic variables; but I would like you to note that this procedure of quantization is not uniquely determined. Given a classical Hamiltonian, we cannot in general say with certainty what the corresponding quantum Hamiltonian is. The classical Hamiltonian may involve a product of two factors that don't commute [...]. The classical theory doesn't tell us the order in which to put the two factors; we have to decide on the order when we go over to the quantum theory [...].
>
> There is no well-defined unique process for passing from classical theory to quantum theory. That means that when we set up a quantum theory we have to set it up to stand on its own feet, independent of the classical theory. The only value of the classical theory is to provide us with hints for getting a quantum theory; the quantum theory is then something that has to stand on its own right. If we were sufficiently clever to be able to think of a good quantum theory straight away, we could manage without the classical theory at all. But we're not that clever, and we have to get all the hints that we can to help us in setting up a good quantum theory.
>
> The classical theory can help us quite a lot. It provides us with suitable Hamiltonians with which we can start to work in the quantum theory. We can study them and see if they are good or not. When you've got one of the Hamiltonians and you've studied it, you might

find that you have to modify it; but still it's a good start. Without the classical theory you wouldn't have this start at all.

This is a beautiful motivation for canonical quantization. In our physics education we usually start with the classical theory, which we assume to exist on its own right. Canonical quantization then becomes a magic trick to bring us from the classical to the quantum realm. But this viewpoint is misleading. The classical theory does not exist without the quantum theory, and canonical quantization (hopefully) gives us the proper hints to uncover the quantum theory from the knowledge of its classical approximation.

Example of Harmonic Oscillator

We discuss canonical quantization for the one-dimensional harmonic oscillator with the Lagrange function

$$L = \frac{1}{2}m\left(\dot{x}^2 - \omega^2 x^2\right).$$

For the canonical momentum we immediately get

$$\pi = \frac{\partial L}{\partial \dot{x}} = m\dot{x},$$

which in this case is identical to the kinetic momentum. Upon insertion into Eq. (13.28), which relates the Lagrange to the Hamilton function, we arrive at

$$H = \frac{\pi^2}{2m} + \frac{1}{2}m\omega^2 x^2. \tag{13.34}$$

In quantum mechanics position and momentum become operators, with the fundamental commutation relations given by Eq. (13.32). The Hamilton operator directly follows from the classical Hamilton function for the harmonic oscillator and reads

$$\hat{H} = \frac{\hat{\pi}^2}{2m} + \frac{1}{2}m\omega^2 x^2. \tag{13.35}$$

Because the harmonic oscillator will be of importance for the quantization of Maxwell's equations, we continue to ponder on its eigenenergies and functions in the framework of quantum mechanics. It turns out to be convenient to introduce the operators

$$\hat{a} = \left(\frac{m\omega}{2\hbar}\right)^{\frac{1}{2}}\left[x + \frac{i}{m\omega}\hat{\pi}\right], \quad \hat{a}^\dagger = \left(\frac{m\omega}{2\hbar}\right)^{\frac{1}{2}}\left[x - \frac{i}{m\omega}\hat{\pi}\right]. \tag{13.36}$$

Using the canonical commutation relations of Eq. (13.32) it is easy to show that the operators fulfill the commutation relations

$$[\hat{a}, \hat{a}^\dagger] = 1, \quad [\hat{a}, \hat{a}] = [\hat{a}^\dagger, \hat{a}^\dagger] = 0, \tag{13.37}$$

and one can express the Hamilton operator in the form

$$\hat{H} = \hbar\omega\left(\hat{a}^\dagger\hat{a} + \frac{1}{2}\right).$$

As we will demonstrate next, with the operators $\hat{a}$, $\hat{a}^\dagger$ it is particularly easy to compute the eigenenergies and states of the harmonic oscillator. Suppose that we have already given one eigenstate $|n\rangle$ with energy E_n (we will discuss in a moment how this state actually can be computed). One can then show that also $\hat{a}^\dagger|n\rangle$ is an eigenstate,

$$\hat{H}\left(\hat{a}^\dagger|n\rangle\right) = \hbar\omega\left(\hat{a}^\dagger\hat{a}+\frac{1}{2}\right)\hat{a}^\dagger|n\rangle = \hbar\omega\left(\hat{a}^\dagger\left[\hat{a}^\dagger\hat{a}+1\right]+\frac{1}{2}\hat{a}^\dagger\right)|n\rangle = \hat{a}^\dagger\left(\hat{H}+\hbar\omega\right)|n\rangle,$$

where we have used the commutation relation of Eq. (13.37) to get $\hat{a}\hat{a}^\dagger = \hat{a}^\dagger\hat{a} + 1$. If we now exploit in the expression on the right-hand side that $|n\rangle$ is an eigenstate of $\hat{H}$, we end up with

$$\hat{H}\left(\hat{a}^\dagger|n\rangle\right) = \left(E_n + \hbar\omega\right)\left(\hat{a}^\dagger|n\rangle\right).$$

Thus, also $\hat{a}^\dagger|n\rangle$ is an eigenstate of the Hamilton operator, however, with the increased energy $E_n + \hbar\omega$. Along the same lines one can show that $\hat{a}|n\rangle$ is an (unnormalized) eigenstate

$$\hat{H}\left(\hat{a}|n\rangle\right) = \left(E_n - \hbar\omega\right)\left(\hat{a}|n\rangle\right),$$

but now with the reduced energy $E_n - \hbar\omega$. In words, the creation operator $\hat{a}^\dagger$ adds one excitation to the system and the annihilation operator $\hat{a}$ removes one excitation. All what remains to be done is to compute one eigenstate. Because the oscillator potential $\frac{1}{2}m\omega^2x^2$ is bounded from below, the system must possess a groundstate $|0\rangle$. Since there exists no state with a lower energy, we have

$$\hat{a}|0\rangle = 0 \implies \left(\frac{m\omega}{2\hbar}\right)^{\frac{1}{2}}\left[x + \frac{\hbar}{m\omega}\frac{d}{dx}\right]\psi_0(x) = 0,$$

where we have used the canonical momentum of Eq. (13.33) to arrive at the differential equation on the right-hand side. It is straightforward to show that the solution is given by

$$\psi_0(x) = \left(\frac{m\omega}{\pi\hbar}\right)^{\frac{1}{4}} \exp\left[-\frac{m\omega}{2\hbar}x^2\right], \quad E_0 = \frac{1}{2}\hbar\omega. \tag{13.38}$$

Starting from this groundstate, the normalized excited states of the harmonic oscillator can be obtained through application of the creation operators $\hat{a}^\dagger$,

Eigenstates of Harmonic Oscillator

$$|n\rangle = \frac{1}{\sqrt{n!}}\left(\hat{a}^\dagger\right)^n |0\rangle, \qquad E_n = \hbar\omega\left(n + \frac{1}{2}\right). \tag{13.39}$$

Figure 13.3 provides a graphical representation of how the creation and annihilation operators promote the system to a state of higher or lower energy, and hereby add or remove an energy quantum $\hbar\omega$. In later parts of this chapter we will use a similar procedure for the electromagnetic field Hamiltonian, which is quadratic in the generalized coordinates and velocities. Although there is no direct connection to a real physical oscillator, we will employ similar annihilation and creation operators, which remove or add a photon with energy $\hbar\omega$.

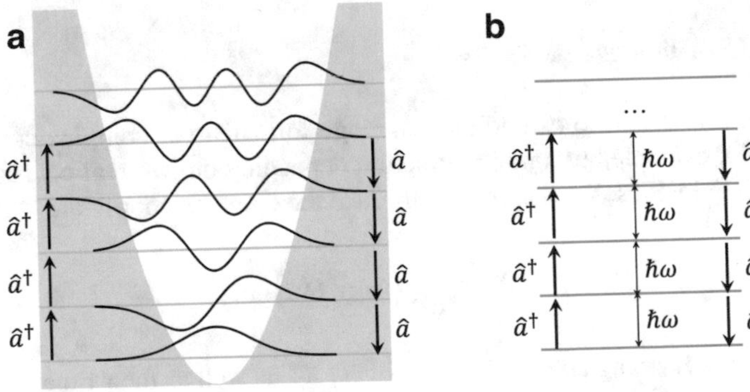

Fig. 13.3 Schematics for harmonic oscillator. In panel (**a**) the shaded area depicts the parabolic confinement potential. We additionally plot the eigenfunctions for the lowest eigenenergies $E_n = \hbar\omega(n + \frac{1}{2})$, which are offset for clarity. The creation operators $\hat{a}^\dagger$ promote the system to states of higher energy and add one excitation quantum $\hbar\omega$. Similarly, the annihilation operators $\hat{a}$ remove one excitation quantum and bring the system to a state of lower energy. (**b**) Later in this chapter we will use a similar scheme for the electromagnetic field Hamiltonian, which is quadratic in the generalized coordinates and velocities. Although there is no direct connection to an oscillator, we will employ similar annihilation and creation operators, which remove or add a photon with energy $\hbar\omega$

The solution scheme in terms of annihilation and creation operators can be also used in case of multiple oscillators

$$\hat{H} = \sum_{i=1}^{N} \left(\frac{\hat{\pi}_i^2}{2m_i} + \frac{1}{2} m\omega_i^2 x_i^2 \right).$$

For each oscillator we introduce the annihilation and creation operators $\hat{a}_i$, $\hat{a}_i^\dagger$ in accordance to Eq. (13.36), which are assumed to fulfill the commutation relations

$$[\hat{a}_i, \hat{a}_j^\dagger] = \delta_{ij}, \quad [\hat{a}_i, \hat{a}_j] = [\hat{a}_i^\dagger, \hat{a}_j^\dagger] = 0. \tag{13.40}$$

At this point, it is not obvious why we have chosen the operators for different oscillators to commute, but we will see later that with this choice we get a wavefunction that has proper bosonic symmetry properties. Finally, we define a groundstate $|0\rangle$ where all oscillators are in their respective states of lowest energy. Starting from $|0\rangle$ we can then obtain all excited states through

$$\big|n_1, n_2, \dots n_N\big\rangle = \big(n_1! n_2! \dots n_N!\big)^{-\frac{1}{2}} \left(\hat{a}_1^\dagger\right)^{n_1} \left(\hat{a}_2^\dagger\right)^{n_2} \dots \left(\hat{a}_N^\dagger\right)^{n_N} |0\rangle. \tag{13.41}$$

In later parts of this chapter we will show that these so-called Fock states play an important role in the quantization of electromagnetic fields.

Complex Coordinates

Sometimes it is useful to extend the Lagrange formalism to complex coordinates. Consider, for example, a Lagrange function depending on two real coordinates x_1, x_2 and their velocities. We may now introduce the complex coordinate

$$X = \frac{1}{\sqrt{2}}\big(x_1 + i x_2\big), \tag{13.42}$$

and rewrite the Lagrangian as a function of X, X^* as well as their time derivatives. It is easy to show that x_1, x_2 are related to X, X^* through

$$x_1 = \frac{1}{\sqrt{2}}\big(X + X^*\big), \quad x_2 = -\frac{i}{\sqrt{2}}\big(X - X^*\big).$$

In the following we consider X and X^* as two independent variables. With this we immediately obtain

$$\frac{\partial L}{\partial X} = \frac{\partial L}{\partial x_1}\frac{\partial x_1}{\partial X} + \frac{\partial L}{\partial x_2}\frac{\partial x_2}{\partial X} = \frac{1}{\sqrt{2}}\left(\frac{\partial L}{\partial x_1} - i\frac{\partial L}{\partial x_2}\right)$$

$$\frac{\partial L}{\partial X^*} = \frac{\partial L}{\partial x_1}\frac{\partial x_1}{\partial X^*} + \frac{\partial L}{\partial x_2}\frac{\partial x_2}{\partial X^*} = \frac{1}{\sqrt{2}}\left(\frac{\partial L}{\partial x_1} + i\frac{\partial L}{\partial x_2}\right),$$

and a similar expression for the velocities. In analogy to the coordinate transformation of Eq. (13.42), we define a complex momentum according to

$$\Pi = \frac{1}{\sqrt{2}}\left(\pi_1 + i\pi_2\right) = \left(\frac{\partial L}{\partial X^*}\right),$$

with a similar expression for Π^*. One can easily show that the following relation holds

$$\dot{x}_1\pi_1 + \dot{x}_2\pi_2 = \dot{X}\Pi^* + \dot{X}^*\Pi.$$

Thus, the relation between the Lagrange and Hamilton functions with complex arguments is given by

$$H = \dot{X}\Pi^* + \dot{X}^*\Pi - L, \tag{13.43}$$

where again the velocities $\dot{X}$, $\dot{X}^*$ have to be replaced in favor of the canonical momenta Π, Π^*. From the above expression it is obvious that H is an entirely real function. Some care has to be taken when submitting the above model to a canonical quantization. For the real fields we have

$$\left[\hat{x}_1, \hat{\pi}_1\right] = \left[\hat{x}_2, \hat{\pi}_2\right] = i\hbar, \quad \left[\hat{x}_1, \hat{\pi}_2\right] = \left[\hat{x}_2, \hat{\pi}_1\right] = 0.$$

With this, one readily finds

$$\left[\hat{X}, \hat{\Pi}\right] = \frac{1}{2}\left[\hat{x}_1 + i\hat{x}_2, \hat{\pi}_1 + i\hat{\pi}_2\right] = 0.$$

In contrast, the commutator with the Hermitian conjugate of $\hat{\Pi}$ becomes

$$\left[\hat{X}, \hat{\Pi}^\dagger\right] = \frac{1}{2}\left[\hat{x}_1 + i\hat{x}_2, \hat{\pi}_1 - i\hat{\pi}_2\right] = i\hbar, \tag{13.44}$$

which shows that the fundamental commutation relations have to be defined for $\hat{X}$, $\hat{\Pi}^\dagger$ rather than for $\hat{X}$, $\hat{\Pi}$.

13.3 Coulomb Gauge

In Sect. 2.2.1 we have discussed the concept of electromagnetic potentials, and have shown that they are not defined uniquely but can be submitted to *gauge transformations* which modify the potentials but not the electromagnetic fields. In this respect, gauge transformations can be used to facilitate the theoretical approach, however, without affecting the underlying physics.

Lorenz Gauge. In the first, "classical" part of this book we have exclusively adopted the Lorenz gauge. Here the scalar and vector potentials both fulfill wave equations, see Eq. (2.22), and the solutions can be expressed in terms of functions that propagate with the speed of light, as discussed at the beginning of the introduction Chap. 1. These solutions explicitly respect the relativistic principle that no information can propagate faster than the speed of light, which makes them highly attractive for most investigations.

Coulomb Gauge. The Coulomb gauge (or transverse gauge, as it is sometimes called) is an alternative formulation with

Coulomb Gauge

$$\nabla \cdot \boldsymbol{A} = 0. \tag{13.45}$$

From Eq. (2.45), which defines the decomposition of a vector function into its longitudinal and transverse parts, it is obvious that with this choice $\boldsymbol{A}$ is an entirely transverse function. Thus, the transverse electric and magnetic fields are solely determined by $\boldsymbol{A}$, whereas the longitudinal part is determined by the scalar potential V (by construction). Using Eq. (2.20), we find for the defining equation of the scalar potential

$$\nabla^2 V(\boldsymbol{r},t) = -\frac{\rho(\boldsymbol{r},t)}{\varepsilon_0} \implies V(\boldsymbol{r},t) = \frac{1}{4\pi\varepsilon_0}\int \frac{\rho(\boldsymbol{r}',t)}{|\boldsymbol{r}-\boldsymbol{r}'|}\,d^3r', \tag{13.46}$$

where we have written down the solution for an unbounded medium that has the same form as the Coulomb potential in electrostatics (which gives the gauge its name). Here, the scalar potential at some position $\boldsymbol{r}$ reacts *instantaneously* to the change of the charge distribution at some other position $\boldsymbol{r}'$, which appears to be in disagreement with the relativity principle. However, there is a solution to the problem, which is explained beautifully by Griffiths in his book on electrodynamics [1].

> There is a peculiar thing about the scalar potential in the Coulomb gauge: it is determined by the distribution of charge *right now*. If I move an electron in my laboratory, the potential V on the moon immediately records this change. That sounds particularly odd in the light of

special relativity, which allows no message to travel faster than the speed of light. The point is that V *by itself* is not a physically measurable quantity—all the man in the moon can measure $\boldsymbol{E}$, and that involves $\boldsymbol{A}$ as well. Somehow it is built into the vector potential, in the Coulomb gauge, that whereas V instantaneously reflects all changes in ρ, the combination $-\nabla V - (\partial \boldsymbol{A})/(\partial t)$ does *not*; $\boldsymbol{E}$ will change only after sufficient time has elapsed for the "news" to arrive.

Altogether, the Coulomb gauge appears somewhat odd, not only to the man in the moon but also to me, so I have avoided using it so far and have opted for the physically more transparent Lorenz gauge. However, in the realm of quantum mechanics and quantum electrodynamics the clear-cut separation of the electromagnetic fields into transverse and longitudinal parts directly related to $\boldsymbol{A}$ and V is so advantageous that **from here on we shall exclusively work in the Coulomb gauge**. In any case, for all results derived with electrodynamic potentials the physically measurable quantities must not depend on the chosen gauge. This principle is often denoted as *gauge invariance*.

As for the defining equation of the vector potential, we get from Eq. (2.20)

$$\left(\nabla^2 - \varepsilon_0\mu_0\frac{\partial^2}{\partial t^2}\right)\boldsymbol{A}(\boldsymbol{r},t) = -\mu_0\left[\boldsymbol{J}(\boldsymbol{r},t) - \varepsilon_0\nabla\left(\frac{\partial V(\boldsymbol{r},t)}{\partial t}\right)\right],$$

where we have used $\nabla\cdot\boldsymbol{A} = 0$. Together with the instantaneous Coulomb potential of Eq. (13.46) we then get for the term in brackets

$$\boldsymbol{J}(\boldsymbol{r},t) - \nabla\int\frac{1}{4\pi|\boldsymbol{r}-\boldsymbol{r}'|}\left[\frac{\partial\rho(\boldsymbol{r}',t)}{\partial t}\right]d^3r' = \boldsymbol{J}(\boldsymbol{r},t) + \nabla\int\frac{\nabla'\cdot\boldsymbol{J}(\boldsymbol{r}',t)}{4\pi|\boldsymbol{r}-\boldsymbol{r}'|}\,d^3r',$$

where we have employed the continuity equation to arrive at the second expression. Using the transverse delta function, see Eqs. (F.12) and (F.15), one can show that the term on the right-hand side equals the transverse current distribution $\boldsymbol{J}^\perp$. Note that the operation to make $\boldsymbol{J}$ transverse is nonlocal in space. We then get for the defining equation of the vector potential

$$\left(\nabla^2 - \mu_0\varepsilon_0\frac{\partial^2}{\partial t^2}\right)\boldsymbol{A}(\boldsymbol{r},t) = -\mu_0\boldsymbol{J}^\perp(\boldsymbol{r},t). \tag{13.47}$$

Dyadic Green's Function in Coulomb Gauge. In what follows, we show how to solve this differential equation using the scalar Green's function, and how to express the electric field in terms of V and $\boldsymbol{A}$. Upon Fourier transformation in time, Eq. (13.47) becomes

$$\left(\nabla^2 + k^2\right)\boldsymbol{A}(\boldsymbol{r}) = -\mu_0\boldsymbol{J}^\perp(\boldsymbol{r}) \implies \boldsymbol{A}(\boldsymbol{r}) = \frac{\mu_0}{4\pi}\int G(\boldsymbol{r},\boldsymbol{r}')\boldsymbol{J}^\perp(\boldsymbol{r}')\,d^3r',$$

with the scalar Green's function of Eq. (5.7). We have suppressed the dependence of the vector potential and current distribution on ω. The electric field can be obtained from

$$\boldsymbol{E}(\boldsymbol{r}) = i\mu_0\omega\int G(\boldsymbol{r},\boldsymbol{r}')\boldsymbol{J}^\perp(\boldsymbol{r}')\,d^3r' - \nabla\left(\frac{1}{4\pi\varepsilon_0}\int\frac{\rho(\boldsymbol{r}')}{|\boldsymbol{r}-\boldsymbol{r}'|}\,d^3r'\right).$$

Here the second term on the right-hand side reacts instantaneously to the charge distribution $\rho(\boldsymbol{r})$. However, as we will show next, this seemingly unphysical behavior is corrected by the vector potential that also has an instantaneous part associated with the nonlocal transversality operation. We can use Eq. (F.16) to exchange the transversality operation of the current distribution with the Green's function

$$\int G(\boldsymbol{r},\boldsymbol{r}')\boldsymbol{J}^{\perp}(\boldsymbol{r}')\,d^3r' = \int \bar{\bar{G}}(\boldsymbol{r},\boldsymbol{r}')\cdot\boldsymbol{J}(\boldsymbol{r}')\,d^3r' - \frac{1}{k^2}\nabla\int\frac{\nabla'\cdot\boldsymbol{J}(\boldsymbol{r}')}{4\pi|\boldsymbol{r}-\boldsymbol{r}'|}\,d^3r',$$

where $\bar{\bar{G}}$ is the dyadic Green's function of Eq. (5.19). When multiplying this expression with $i\mu_0\omega$, the second term on the right-hand side can be rewritten as

$$-\frac{i\mu_0\omega}{k^2}\nabla\int\frac{i\omega\rho(\boldsymbol{r}')}{4\pi|\boldsymbol{r}-\boldsymbol{r}'|}\,d^3r' = \nabla\int\frac{\rho(\boldsymbol{r}')}{4\pi\varepsilon|\boldsymbol{r}-\boldsymbol{r}'|}\,d^3r'.$$

This term is cancelled by the instantaneous Coulomb potential, which finally leads us to

$$\boldsymbol{E}(\boldsymbol{r}) = i\mu_0\omega\int\bar{\bar{G}}(\boldsymbol{r},\boldsymbol{r}')\cdot\boldsymbol{J}(\boldsymbol{r}')\,d^3r'.$$

This result has been previously derived in Chap. 5 using the Lorenz gauge. It is gratifying to see that the final result does not depend on the chosen gauge, in accordance to the principle of gauge invariance.

The Coulomb gauge is not a real beauty, but it has the appealing features that the longitudinal field components are solely governed by the scalar potential V, and the transverse field components solely by the vector potential $\boldsymbol{A}$. The price we have to pay is the physically not overly transparent fact that V reacts instantaneously to changes of the charge distribution ρ. However, since the transverse vector potential takes care that the electromagnetic fields fulfill the requirement of special relativity, namely that no information can travel faster than the speed of light, there is conceptually nothing wrong with the Coulomb gauge.

13.4 Canonical Quantization of Maxwell's Equations

In the following we set up a Lagrange formalism for electrodynamics and show how to obtain a quantum version of Maxwell's equations using the formalism of canonical quantization. This might be the right place to advertise the excellent book of Cohen-Tannoudji and co-authors on photons and atoms [123]. It is a remarkable textbook that explains the quantum theory of light in great detail and clarity. In contrast, we here present the topic in a rather condensed form, with emphasis on those aspects only which are needed for a basic understanding of the topic.

13.4.1 Lagrange Formalism

Within the framework of Maxwell's equation, the central objects of the Lagrange formalism are the scalar and vector potentials V and $\boldsymbol{A}$. A detailed discussion of the Lagrange formalism is given in Sect. 13.6. In brief, we find that the Lagrange function can be separated into three parts

$$L = L_{\mathrm{mat}} + L_{\mathrm{em}} + L_{\mathrm{int}}, \tag{13.48}$$

associated with the matter contribution L_{mat} for charged particles, the electromagnetic fields, L_{em}, and the light–matter interaction, L_{int}. For charged particles with mass m the Lagrange function consists of the kinetic part only,

$$L_{\mathrm{mat}} = \frac{1}{2}\sum\nolimits_i m v_i^2 . \tag{13.49a}$$

When dealing with particles that have an internal structure, such as molecules, we could add a potential term to account for the vibronic degrees of freedom, but for simplicity we will not consider such cases here. The Lagrange function for the electromagnetic fields reads [see also Eq. (13.105)]

$$L_{\mathrm{em}} = \int \left(\frac{\varepsilon_0}{2} \boldsymbol{E}\cdot\boldsymbol{E} - \frac{1}{2\mu_0}\boldsymbol{B}\cdot\boldsymbol{B} \right) d^3r, \tag{13.49b}$$

where the fields $\boldsymbol{E}$, $\boldsymbol{B}$ must be additionally expressed through the scalar and vector potentials V, $\boldsymbol{A}$. As explicitly worked out in Sect. 13.6.2, when submitting L_{em} together with the light–matter interaction L_{int} to the action principle of the Lagrange formalism we recover Maxwell's equations. The interaction of a single charged particle with electromagnetic fields is of the form [see also Eq. (13.101)]

$$L_{\mathrm{int}} = -qV(\boldsymbol{r},t) + q\boldsymbol{v}\cdot\boldsymbol{A}(\boldsymbol{r},t). \tag{13.49c}$$

In case of many particles we can introduce charge and current distributions, along the same lines as discussed for the transition from microscopic to macroscopic electrodynamics in Chap. 7, and arrive at the light–matter interaction

$$L_{\mathrm{int}} = \int \Big[-\rho(\boldsymbol{r},t)V(\boldsymbol{r},t) + \boldsymbol{J}(\boldsymbol{r},t)\cdot\boldsymbol{A}(\boldsymbol{r},t) \Big] d^3r. \tag{13.49d}$$

See Sect. 13.6.1 for further details. In what follows, we submit the Lagrange functions to canonical quantization, and proceed in three steps.

1. We start with the quantization of the matter degrees of freedom, and consider a single or many charged particles moving in presence of *classical* electromagnetic fields.

2. In a second step, we quantize Maxwell's equations in *free space* without any source terms, and introduce hereby the photon concept.
3. Finally, we combine the results derived in (1) and (2) to obtain the Hamiltonian for the quantized light-matter system.

13.4.2 Quantization of Matter Part

We start with the quantization of a system where one charged particle interacts with external electromagnetic fields. The fields are treated classically, but all of the following results prevail for a vector potential operator $\hat{\boldsymbol{A}}$ (which will be needed for the quantization of Maxwell's equations). From the Lagrange function $\frac{1}{2}mv^2 - qV + q\boldsymbol{v}\cdot\boldsymbol{A}$ for a charged particle coupled to the (classical) electromagnetic potentials we compute the canonical momentum

$$\pi_k = \frac{\partial}{\partial v_k}\Big(L_{\text{mat}} + L_{\text{int}}\Big) = mv_k + qA_k. \tag{13.50}$$

We next perform the Legendre transformation of Eq. (13.28) to arrive at

$$H = \boldsymbol{v}\cdot\big(m\boldsymbol{v} + q\boldsymbol{A}\big) - \frac{1}{2}mv^2 + qV - q\boldsymbol{v}\cdot\boldsymbol{A} = \frac{1}{2}mv^2 + qV.$$

With this, we get the Hamilton for a charged particle interacting with electromagnetic fields expressed in terms of the electromagnetic potentials V, $\boldsymbol{A}$,

$$H = \frac{\Big(\boldsymbol{\pi} - q\boldsymbol{A}(\boldsymbol{r},t)\Big)^2}{2m} + qV(\boldsymbol{r},t). \tag{13.51}$$

Using the canonical quantization procedure discussed in Sect. 13.2.3, see also Fig. 13.4, we are then led to the so-called minimal coupling Hamiltonian

Minimal Coupling Hamiltonian

$$\hat{H} = \frac{\Big(\hat{\boldsymbol{\pi}} - q\boldsymbol{A}(\boldsymbol{r},t)\Big)^2}{2m} + qV(\boldsymbol{r},t), \tag{13.52}$$

In Eq. (13.52) the **canonical momentum** operator

$$\hat{\boldsymbol{\pi}} = -i\hbar\nabla \tag{13.53}$$

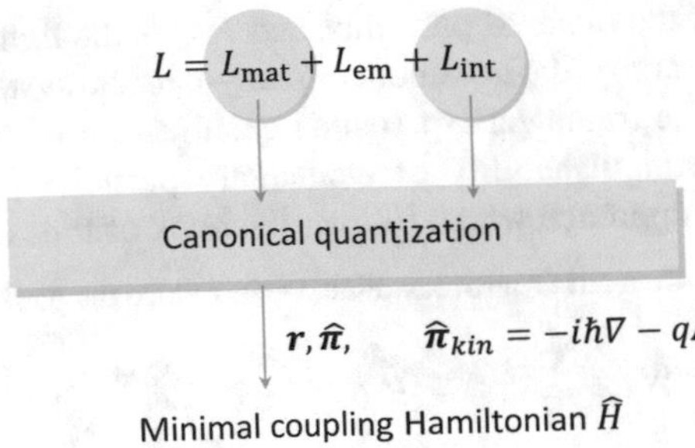

Fig. 13.4 Quantization of light-matter part. The Lagrangians for matter and light–matter interaction are submitted to canonical quantization, which yields the particle position $\boldsymbol{r}$ and canonical momentum $\hat{\boldsymbol{\pi}}$ operators as dynamic quantities, together with the minimal coupling Hamiltonian of Eq. (13.52). The external electromagnetic fields are treated on a classical level

accounts for the momentum of the charged particle in presence of the electromagnetic fields. In contrast, the **kinetic momentum** operator

$$\hat{\boldsymbol{\pi}}_{\rm kin} = -i\hbar\nabla - q\boldsymbol{A}(\boldsymbol{r}, t) \tag{13.54}$$

accounts for the momentum of the charged particle solely. We have already encountered these two momenta at the beginning of this chapter, as well as in the discussion of the Abraham–Minkowski controversy in Chap. 4. In the Coulomb gauge $\nabla \cdot \boldsymbol{A} = 0$ the momentum operator and the vector potential commute, and the minimal coupling Hamiltonian can be rewritten in the form

$$\hat{H} = \frac{\hat{\boldsymbol{\pi}}^2}{2m} - \frac{q}{m}\boldsymbol{A}(\boldsymbol{r}, t) \cdot \hat{\boldsymbol{\pi}} + qV(\boldsymbol{r}, t) + \frac{q^2 A^2(\boldsymbol{r})}{2m}. \tag{13.55}$$

In linear response the last term can be neglected, because it is quadratic in the vector potential, and we get a coupling of the form

$$\hat{H}_{\rm int} = -\frac{q}{m}\boldsymbol{A}(\boldsymbol{r}, t) \cdot \hat{\boldsymbol{\pi}} + qV(\boldsymbol{r}, t). \tag{13.56}$$

Light–Matter Interaction for Many-Particle Systems

We can generalize the approach to a system of many particles. For simplicity we assume that all particles have the same mass m and charge q, and write the Hamiltonian in the form

$$\hat{H}_0 + \hat{H}_{\rm int} = \sum_i \frac{\hat{\boldsymbol{\pi}}_i^2}{2m} + \sum_i \left[-\frac{q}{m}\boldsymbol{A}(\boldsymbol{r}_i) \cdot \hat{\boldsymbol{\pi}}_i + qV(\boldsymbol{r}_i) + \frac{q^2 A^2(\boldsymbol{r}_i)}{2m} \right]. \tag{13.57}$$

Here $\hat{H}_0$ accounts for the material part only, and $\hat{H}_{\rm int}$ is the light-matter coupling of Eq. (13.55). Quite generally, the treatment of many-particle systems is a complicated subject, and usually the framework of second quantization is employed to properly account for the indistinguishability of elementary particles [124, 125]. We here pursue a more simple approach which follows the book of Pines and Nozières [126].

Particle Density and Current Operators. First, we introduce the particle density operator

$$\hat{n}(\boldsymbol{r}) = \sum_i \delta(\boldsymbol{r} - \boldsymbol{r}_i) \iff \hat{n}_{\boldsymbol{k}} = \sum_i e^{-i\boldsymbol{k}\cdot\boldsymbol{r}_i}, \tag{13.58}$$

where $\hat{n}_{\boldsymbol{k}}$ is the Fourier transform of $\hat{n}(\boldsymbol{r})$. We next compute the time evolution of $\hat{n}(\boldsymbol{r})$ due to the system Hamiltonian $\hat{H}_0$ using Heisenberg's equations of motion,

$$i\hbar\frac{d\hat{n}_{\boldsymbol{k}}}{dt} = \left[\hat{n}_{\boldsymbol{k}}, \hat{H}_0\right] = \frac{1}{2m}\sum_i\left(\left[e^{-i\boldsymbol{k}\cdot\boldsymbol{r}_i}, \hat{\boldsymbol{\pi}}_i\right]\hat{\boldsymbol{\pi}}_i + \hat{\boldsymbol{\pi}}_i\left[e^{-i\boldsymbol{k}\cdot\boldsymbol{r}_i}, \hat{\boldsymbol{\pi}}_i\right]\right).$$

For notational simplicity we suppress the dependence of the operators on time throughout. With

$$[f(\boldsymbol{r}), \hat{\boldsymbol{\pi}}] = i\hbar\nabla f(\boldsymbol{r}),$$

which can be easily proven using $\hat{\boldsymbol{\pi}} = -i\hbar\nabla$, we then find

$$\frac{d\hat{n}_{\boldsymbol{k}}}{dt} = -\frac{i}{2m\hbar}\sum_i\left(e^{-i\boldsymbol{k}\cdot\boldsymbol{r}_i}\,\hbar\boldsymbol{k}\cdot\hat{\boldsymbol{\pi}}_i + \hbar\boldsymbol{k}\cdot\hat{\boldsymbol{\pi}}_i\,e^{-i\boldsymbol{k}\cdot\boldsymbol{r}_i}\right) = -i\boldsymbol{k}\cdot\hat{\mathbf{j}}_{\boldsymbol{k}}.$$

In the last expression we have introduced the particle current operator

$$\hat{\mathbf{j}}(\boldsymbol{r}) = \frac{1}{2m}\sum_i\left[\delta(\boldsymbol{r} - \boldsymbol{r}_i)\,\hat{\boldsymbol{\pi}}_i + \hat{\boldsymbol{\pi}}_i\,\delta(\boldsymbol{r} - \boldsymbol{r}_i)\right], \tag{13.59}$$

whose Fourier transform $\hat{\mathbf{j}}_{\boldsymbol{k}}$ is obtained by replacing $\delta(\boldsymbol{r}-\boldsymbol{r}_i)$ with $e^{-i\boldsymbol{k}\cdot\boldsymbol{r}_i}$. The symmetrized form is needed since $\boldsymbol{r}_i$ and $\hat{\boldsymbol{\pi}}_i$ do not commute. When considering in the time evolution of $\hat{n}(\boldsymbol{r})$ the system Hamiltonian $\hat{H}_0$ only, the particle density and current operators fulfill the continuity equation

$$\frac{d\hat{n}(\boldsymbol{r})}{dt} = -\nabla\cdot\hat{\mathbf{j}}(\boldsymbol{r}) \iff \frac{d\hat{n}_{\boldsymbol{k}}}{dt} = -i\boldsymbol{k}\cdot\hat{\mathbf{j}}_{\boldsymbol{k}}. \tag{13.60}$$

To additionally account for the light–matter interaction, we compute

$$\left[\hat{n}(\boldsymbol{r}), \hat{H}_{\rm int}\right] = -\frac{q}{m}\sum_i\left[\hat{n}(\boldsymbol{r}), \hat{\boldsymbol{\pi}}_i\right]\cdot\boldsymbol{A}(\boldsymbol{r}_i, t) = -\frac{iq\hbar}{m}\sum_i\nabla_i\hat{n}(\boldsymbol{r})\cdot\boldsymbol{A}(\boldsymbol{r}_i, t),$$

where we have used that $\hat{n}$ and A^2 commute. The last term can be rewritten using

$$\sum_i\nabla_i\hat{n}(\boldsymbol{r}) = \sum_{ij,\boldsymbol{k}}\nabla_i\left(e^{i\boldsymbol{k}\cdot(\boldsymbol{r}-\boldsymbol{r}_j)}\right) = \sum_{i,\boldsymbol{k}}\nabla_i\left(e^{i\boldsymbol{k}\cdot(\boldsymbol{r}-\boldsymbol{r}_i)}\right) = -\nabla\hat{n}(\boldsymbol{r}),$$

which allows us to pull out the nabla operator from the sum over all particles.

We thus obtain from Heisenberg's equations of motion for the density operator the modified continuity equation

$$\frac{d\hat{n}(\boldsymbol{r})}{dt} = -\nabla \cdot \left[\hat{\mathbf{j}}(\boldsymbol{r}) - \frac{q}{m}\hat{n}(\boldsymbol{r})\boldsymbol{A}(\boldsymbol{r},t)\right]. \tag{13.61}$$

The current operator on the right-hand side can be interpreted in terms of the kinetic momentum of Eq. (13.54), and is conveniently separated into paramagnetic and diamagnetic contributions defined through

$$\hat{\boldsymbol{J}}(\boldsymbol{r}) = \hat{\boldsymbol{J}}_{\text{para}}(\boldsymbol{r}) + \hat{\boldsymbol{J}}_{\text{dia}}(\boldsymbol{r}) = q\hat{\mathbf{j}}(\boldsymbol{r}) - \frac{q^2}{m}\hat{n}(\boldsymbol{r})\boldsymbol{A}(\boldsymbol{r},t). \tag{13.62}$$

With this, the light-matter coupling of a many-particle system can be written in the form

Light–Matter Interaction of Many-Particle System

$$\hat{H}_{\text{int}} = \int\left[-\left(q\hat{\mathbf{j}}(\boldsymbol{r})\right)\cdot\boldsymbol{A}(\boldsymbol{r},t) + \left(q\hat{n}(\boldsymbol{r})\right)V(\boldsymbol{r},t) + \frac{q^2\hat{n}(\boldsymbol{r})}{2m}A^2(\boldsymbol{r},t)\right]d^3r. \tag{13.63}$$

Note that within linear response it does not matter whether we take in the first term the paramagnetic or total current operator, since the two resulting contributions only differ by a term proportional to A^2, and we can also neglect the last term in brackets.

13.4.3 *Quantization of Maxwell's Equations*

For the quantization of Maxwell's equations we start from the Lagrange functions for the electromagnetic fields and the light–matter interaction, Eq. (13.49),

$$\begin{aligned} L_{\text{em}} + L_{\text{int}} &= \int\left[\frac{\varepsilon_0}{2}\left(-\nabla V - \dot{\boldsymbol{A}}\right)\cdot\left(-\nabla V - \dot{\boldsymbol{A}}\right)\right. \\ &\left. - \frac{1}{2\mu_0}\left(\nabla\times\boldsymbol{A}\right)\cdot\left(\nabla\times\boldsymbol{A}\right) + \boldsymbol{J}\cdot\boldsymbol{A} - \rho V\right]d^3r, \end{aligned} \tag{13.64}$$

where $\dot{\boldsymbol{A}}$ denotes the time derivative of the vector potential. When submitting this expression to the usual quantization procedure, we encounter the problem that the Lagrange function does not depend on the time derivative of the scalar potential and thus the canonical momentum of V is bound to zero.

A popular approach for overcoming this problem is to remove the scalar potential from the Lagrange function, a procedure that can be shown to be equivalent to the Coulomb gauge [123]. Details of this procedure are given in Sect. 13.6.3. It turns out to be convenient to work in Fourier space and to expand all quantities in a plane wave basis. Consider, for instance, the vector potential expansion

$$\boldsymbol{A}(\boldsymbol{r},t) = \Omega^{-1/2} \sum_{\boldsymbol{k}} e^{i\boldsymbol{k}\cdot\boldsymbol{r}} \boldsymbol{A}_{\boldsymbol{k}} = \Omega^{-1/2} \sum_{\boldsymbol{k}\in K+} \left(e^{i\boldsymbol{k}\cdot\boldsymbol{r}} \boldsymbol{A}_{\boldsymbol{k}} + e^{-i\boldsymbol{k}\cdot\boldsymbol{r}} \boldsymbol{A}_{\boldsymbol{k}}^* \right), \qquad (13.65)$$

with Ω being the quantization volume that we may send towards infinity at the end of the calculation. Because $\boldsymbol{A}(\boldsymbol{r},t)$ is a real function, the potential coefficients $\boldsymbol{A}_{\boldsymbol{k}}$ in wavenumber space are subject to the symmetry relation

$$\boldsymbol{A}_{\boldsymbol{k}} = \boldsymbol{A}_{-\boldsymbol{k}}^*.$$

We have used this in Eq. (13.65) by restricting the summation to half of the reciprocal space, denoted with K^+, and we relate the remaining wavevector components through the above symmetry relation. Eliminating redundant variables from the theory is important for submitting the model to canonical quantization, see also Fig. 13.5. With this, the Lagrange function of Eq. (13.64) is written in the form [Eq. (13.108)]

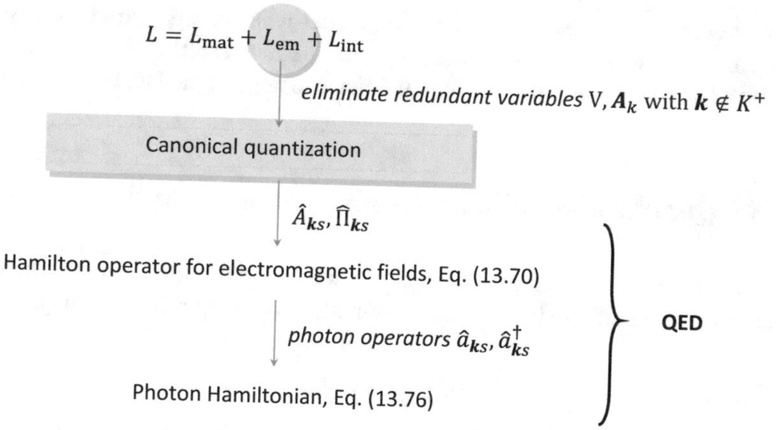

Fig. 13.5 Quantization of Maxwell's equations. We start from the Lagrange function of classical electrodynamics and remove the redundant variables: in the Coulomb gauge we express the scalar potential V in terms of the charge distribution, and in wavenumber space we only need one half-space $\boldsymbol{k} \in K^+$ because of the symmetry relation $\boldsymbol{A}_{\boldsymbol{k}} = \boldsymbol{A}_{-\boldsymbol{k}}^*$. With the transverse vector potential $A_{\boldsymbol{k}s}$ spanned by the two polarization vectors $\boldsymbol{\epsilon}_{\boldsymbol{k}s}$ and the generalized velocities, we submit the Lagrange function to the usual canonical quantization procedure. Finally, we introduce photon operators to rewrite the Hamiltonian in its final form of Eq. (13.76).

$$L_{\text{em}}^{\perp} + L_{\text{int}} = \sum_{k \in K^+} \left[\varepsilon_0 \left(\left| \dot{\boldsymbol{A}}_k \right|^2 - \omega^2 \left| \boldsymbol{A}_k \right|^2 \right) - \frac{\left| \rho_k \right|^2}{k^2 \varepsilon_0} + 2\text{Re}\left\{ \boldsymbol{J}_k \cdot \boldsymbol{A}_k^* \right\} \right], \tag{13.66}$$

where the first term in brackets accounts for the transverse electromagnetic fields, and the remaining two terms for the coupling of charged particles to the electromagnetic fields. Note that the scalar potential has been removed in favor of an instantaneous Coulomb interaction between the charged particles, see second term in brackets. We now exploit that $\boldsymbol{A}_k$ is a transverse vector function which can be spanned by two orthogonal polarization vectors $\boldsymbol{\epsilon}_{ks}$ perpendicular to $\boldsymbol{k}$, and we get

$$\boldsymbol{A}_k = \sum_s \boldsymbol{\epsilon}_{ks} \, A_{ks}, \tag{13.67}$$

with s labeling the different polarizations. The functions A_{ks} are related to the amplitudes of the vector potential components with wavevector $\boldsymbol{k}$ and polarization s. The Lagrange function for the transverse electromagnetic fields can now be expressed in the form

$$L_{\text{em}}^{\perp} = \sum_{k \in K^+} \sum_s \varepsilon_0 \left(\dot{A}_{ks}^* \dot{A}_{ks} - \omega^2 A_{ks}^* A_{ks} \right).$$

As discussed in the context of Eq. (13.43) for Lagrange functions of complex coordinates, we have to define the **canonical field momenta** via

$$\Pi_{ks} = \frac{\partial L_{\text{em}}^{\perp}}{\partial \dot{A}_{ks}^*} = \varepsilon_0 \dot{A}_{ks}, \tag{13.68}$$

and compute the Hamilton function according to

$$\begin{aligned} H_{\text{em}} &= \sum_{k \in K^+} \sum_s \left(\Pi_{ks} \dot{A}_{ks}^* + \Pi_{ks}^* \dot{A}_{ks} \right) - L_{\text{em}}^{\perp} \\ &= \sum_{k \in K^+} \sum_s \left(\frac{1}{\varepsilon_0} \Pi_{ks}^* \Pi_{ks} + \varepsilon_0 \omega^2 A_{ks}^* A_{ks} \right). \end{aligned}$$

We can now submit the field Hamiltonian to the usual canonical quantization procedure. We define the fundamental commutation relations

$$\left[\hat{A}_{ks}, \, \hat{\Pi}_{k's'}^{\dagger} \right] = \left[\hat{A}_{ks}^{\dagger}, \, \hat{\Pi}_{k's'} \right] = i\hbar \, \delta_{kk'} \delta_{ss'}, \tag{13.69}$$

where all other commutators are zero

$$\left[\hat{A}_{\boldsymbol{k}s},\,\hat{\Pi}_{\boldsymbol{k}'s'}\right]=\left[\hat{A}^\dagger_{\boldsymbol{k}s},\,\hat{\Pi}^\dagger_{\boldsymbol{k}'s'}\right]=0$$

$$\left[\hat{A}_{\boldsymbol{k}s},\,\hat{A}_{\boldsymbol{k}'s'}\right]=\left[\hat{A}_{\boldsymbol{k}s},\,\hat{A}^\dagger_{\boldsymbol{k}'s'}\right]=\left[\hat{\Pi}_{\boldsymbol{k}s},\,\hat{\Pi}_{\boldsymbol{k}'s'}\right]=\left[\hat{\Pi}_{\boldsymbol{k}s},\,\hat{\Pi}^\dagger_{\boldsymbol{k}'s'}\right]=0.$$

With this we obtain the Hamilton operator

$$\hat{H}_{\text{em}}=\frac{1}{2}\sum_{\boldsymbol{k},s}\left(\frac{1}{\varepsilon_0}\hat{\Pi}^\dagger_{\boldsymbol{k}s}\hat{\Pi}_{\boldsymbol{k}s}+\varepsilon_0\omega^2\hat{A}^\dagger_{\boldsymbol{k}s}\hat{A}_{\boldsymbol{k}s}\right). \tag{13.70}$$

After having quantized our classical model in reciprocal half-space, we can finally lift the summation restriction and introduce an additional factor of one half instead. With the field quantization and the Hamiltonian of Eq. (13.70) we are done in principle. All our cards are on the table now. What remains to be done is to investigate the consequences for the field quantization.

We should maybe pause here for a moment and reflect on the theory of quantum electrodynamics, which we have just developed. The field Hamiltonian of Eq. (13.70) accounts for excitations with a wave-particle duality inherent to quantum mechanics. The wave nature of light is well-known from classical Maxwell's theory, and, as we shall see later, the propagation of the elementary light quanta—**the photons**—is in complete analogy to classical light waves. The particle nature of light requests that we can only detect light in portions of $h\nu$, and we will show in a moment how this can be elegantly accounted for through photon creation and annihilation operators. Yet, it is fair to ask under which circumstances a quantum theory of light is needed and when we can use a classical approach. In the book of Mandel and Wolf [4] the authors reflect on this question.

> Up to now the electromagnetic field has been treated as a classical field, describable by c-number functions. The great success of classical electromagnetic theory in accounting for a variety of optical phenomena, particularly those connected with wave propagation, interference and diffraction, amply justifies the classical approach. [...] It might almost seem that there is little justification for going beyond the domain of classical wave theory in optics.
>
> On the other hand, it can be argued that optics lies well and truly in the quantum domain, in the sense that we often encounter situations in which very few quanta or photons are present. In the microwave region of the electromagnetic spectrum and at still longer wavelengths, the number of photons in each mode of the field is usually very large, and we are justified in treating the system classically. However, in the optical region the situation is usually just the opposite. For light produced by practically all sources other than lasers, the average number of photons is typically much less than unity. It might seem, therefore, that a classical description would be hopelessly inadequate, and that optics must always be regarded as a branch of quantum electrodynamics. [...]
>
> Why then does classical optics work so well in many cases? One answer to this question is that we rarely attempt to measure the non-classical features of light, like the wildly fluctuating phase of the wave. [...] For many purposes it suffices to deal with the light intensity, and this is frequently well described by classical wave optics. Indeed, classical optics is able to account for some phenomena even at extremely low light levels. However,

classical optics does not always work. There are some optical phenomena, usually, but not always, involving small quantum numbers, for which the field needs to be treated quantum mechanically, and quantum electrodynamics plays an essential role. The quantum theory of electromagnetic radiation is the most successful and all-embracing theory of optics, and to date none of the predictions has been contradicted by experiment.

The main theme of this reflection, namely that classical Maxwell's theory usually works much better than one would expect naively and only in some cases quantum corrections are needed, nicely summarizes what will be encountered in the following chapters. In many situations we can stay with a classical approach supplemented by certain quantum features.

We continue to combine the vector potential representation of Eq. (13.67) with the plane wave decomposition of Eq. (13.65), and obtain

$$\hat{\boldsymbol{A}}(\boldsymbol{r}) = \frac{1}{\sqrt{\Omega}}\sum_{\boldsymbol{k}\in K+}\sum_{s}\left(e^{i\boldsymbol{k}\cdot\boldsymbol{r}}\hat{A}_{\boldsymbol{k}s} + e^{-i\boldsymbol{k}\cdot\boldsymbol{r}}\hat{A}_{\boldsymbol{k}s}^{\dagger}\right)\boldsymbol{\epsilon}_{\boldsymbol{k}s}, \tag{13.71}$$

where we have assumed for simplicity that the polarization vector is real. The transverse components of the electric field can be obtained from

$$\boldsymbol{E}_{\boldsymbol{k}}^{\perp} = -\dot{\boldsymbol{A}}_{\boldsymbol{k}} \implies \hat{\boldsymbol{E}}_{\boldsymbol{k}} = -\frac{1}{\varepsilon_0}\sum_{s}\boldsymbol{\epsilon}_{\boldsymbol{k}s}\hat{\Pi}_{\boldsymbol{k}s}.$$

Proceeding in a similar fashion as for the vector potential, we get for the real-space representation of the transverse electric field operator

$$\hat{\boldsymbol{E}}^{\perp}(\boldsymbol{r}) = -\frac{1}{\varepsilon_0\sqrt{\Omega}}\sum_{\boldsymbol{k}\in K+}\sum_{s}\left(e^{i\boldsymbol{k}\cdot\boldsymbol{r}}\hat{\Pi}_{\boldsymbol{k}s} + e^{-i\boldsymbol{k}\cdot\boldsymbol{r}}\hat{\Pi}_{\boldsymbol{k}s}^{\dagger}\right)\boldsymbol{\epsilon}_{\boldsymbol{k}s}. \tag{13.72}$$

Additionally, the magnetic field operator is obtained from $\hat{\boldsymbol{B}}(\boldsymbol{r}) = \nabla\times\hat{\boldsymbol{A}}(\boldsymbol{r})$. With these operators one can rewrite the Hamiltonian of Eq. (13.70) in the intriguing form

Hamilton Operator for Free Electromagnetic Fields

$$\hat{H}_{\text{em}} = \int\left[\frac{\varepsilon_0}{2}\hat{\boldsymbol{E}}^{\perp}(\boldsymbol{r})\cdot\hat{\boldsymbol{E}}^{\perp}(\boldsymbol{r}) + \frac{1}{2\mu_0}\hat{\boldsymbol{B}}(\boldsymbol{r})\cdot\hat{\boldsymbol{B}}(\boldsymbol{r})\right]d^3r. \tag{13.73}$$

This result agrees with the classical result for the energy stored in the electromagnetic fields, however, the classical fields are now replaced by field operators.

13.4.4 Photons

The Hamiltonian $\hat{H}_{\mathrm{em}}$ for the electromagnetic fields is quadratic in both the vector potential and its conjugate momentum. As we will show in the following, we can rewrite this Hamiltonian using creation and destruction operators in complete analogy to the harmonic oscillator, although there is no "physical oscillator" involved in the electromagnetic case. Consider the operators

$$\hat{a}_{\boldsymbol{k}s} = \left(\frac{\varepsilon_0}{2\hbar\omega}\right)^{\frac{1}{2}} \left[\omega \hat{A}_{\boldsymbol{k}s} + \frac{i}{\varepsilon_0}\hat{\Pi}_{\boldsymbol{k}s}\right], \qquad \hat{a}^\dagger_{\boldsymbol{k}s} = \left(\frac{\varepsilon_0}{2\hbar\omega}\right)^{\frac{1}{2}} \left[\omega \hat{A}^\dagger_{\boldsymbol{k}s} - \frac{i}{\varepsilon_0}\hat{\Pi}^\dagger_{\boldsymbol{k}s}\right]. \tag{13.74}$$

Using the fundamental commutation relations of Eq. (13.69) one can readily prove the following expressions

$$\left[\hat{a}_{\boldsymbol{k}s}, \hat{a}^\dagger_{\boldsymbol{k}'s'}\right] = \delta_{\boldsymbol{k}\boldsymbol{k}'}\delta_{ss'}, \quad \left[\hat{a}_{\boldsymbol{k}s}, \hat{a}_{\boldsymbol{k}'s'}\right] = \left[\hat{a}^\dagger_{\boldsymbol{k}s}, \hat{a}^\dagger_{\boldsymbol{k}'s'}\right] = 0. \tag{13.75}$$

Furthermore we get

$$\hat{a}^\dagger_{\boldsymbol{k}s}\hat{a}_{\boldsymbol{k}s} = \left(\frac{\varepsilon_0}{2\hbar\omega}\right)\left[\frac{1}{\varepsilon_0^2}\hat{\Pi}^\dagger_{\boldsymbol{k}s}\hat{\Pi}_{\boldsymbol{k}s} + \omega^2\hat{A}^\dagger_{\boldsymbol{k}s}\hat{A}_{\boldsymbol{k}s} + \frac{i\omega}{\varepsilon_0}\left(\hat{A}^\dagger_{\boldsymbol{k}s}\hat{\Pi}_{\boldsymbol{k}s} - \hat{\Pi}^\dagger_{\boldsymbol{k}s}\hat{A}_{\boldsymbol{k}s}\right)\right]$$

$$\hat{a}_{\boldsymbol{k}s}\hat{a}^\dagger_{\boldsymbol{k}s} = \left(\frac{\varepsilon_0}{2\hbar\omega}\right)\left[\frac{1}{\varepsilon_0^2}\hat{\Pi}_{\boldsymbol{k}s}\hat{\Pi}^\dagger_{\boldsymbol{k}s} + \omega^2\hat{A}_{\boldsymbol{k}s}\hat{A}^\dagger_{\boldsymbol{k}s} + \frac{i\omega}{\varepsilon_0}\left(\hat{\Pi}_{\boldsymbol{k}s}\hat{A}^\dagger_{\boldsymbol{k}s} - \hat{A}_{\boldsymbol{k}s}\hat{\Pi}^\dagger_{\boldsymbol{k}s}\right)\right].$$

When adding the two terms and exploiting the fundamental commutation relations of Eq. (13.69), we find

$$\hat{a}^\dagger_{\boldsymbol{k}s}\hat{a}_{\boldsymbol{k}s} + \hat{a}_{\boldsymbol{k}s}\hat{a}^\dagger_{\boldsymbol{k}s} = \frac{1}{\hbar\omega}\left(\frac{1}{\varepsilon_0}\hat{\Pi}^\dagger_{\boldsymbol{k}s}\hat{\Pi}_{\boldsymbol{k}s} + \varepsilon_0\omega^2\,\hat{A}^\dagger_{\boldsymbol{k}s}\hat{A}_{\boldsymbol{k}s}\right) + \frac{i\omega}{\varepsilon_0}\big(\dots\big).$$

Using the symmetry relation $\hat{A}_{\boldsymbol{k},s} = \hat{A}^\dagger_{-\boldsymbol{k},s}$ and a similar one for the momentum operators, it can be shown that the last term in parentheses becomes zero when summed over the entire wavevector space. Thus, the field Hamiltonian of Eq. (13.70) can be written in the form

Photon Hamiltonian

$$\hat{H}_{\mathrm{em}} = \sum_{\boldsymbol{k},s}\hbar\omega\left(\hat{a}^\dagger_{\boldsymbol{k}s}\hat{a}_{\boldsymbol{k}s} + \frac{1}{2}\right), \tag{13.76}$$

with $\omega = ck$. It has the same form as the Hamiltonian describing a collection of harmonic oscillators, which we have previously discussed in this chapter. We can exploit the analogy by saying that $\hat{a}^\dagger_{\boldsymbol{k}s}$ creates a field excitation, namely a photon with wavenumber $\boldsymbol{k}$ and polarization s. Similarly the operator $\hat{a}_{\boldsymbol{k}s}$ removes one photon from the light field. The states on which these operators act are the so-called Fock states, which, in case of the photon interpretation, now consist of states with different photon numbers.

We can finally express the vector potential and the field operator in terms of $\hat{a}_{\boldsymbol{k}s}$, $\hat{a}^\dagger_{\boldsymbol{k}s}$. From the defining Eq. (13.74) for the photon operators we find

$$\hat{a}_{\boldsymbol{k},s} + \hat{a}^\dagger_{-\boldsymbol{k},s} = \left(\frac{2\varepsilon_0\omega}{\hbar}\right)^{\frac{1}{2}} \hat{A}_{\boldsymbol{k}s}, \quad \hat{a}_{\boldsymbol{k},s} - \hat{a}^\dagger_{-\boldsymbol{k},s} = i\left(\frac{2}{\hbar\varepsilon_0\omega}\right)^{\frac{1}{2}} \hat{\Pi}_{\boldsymbol{k}s}.$$

When inserting this into the expression of Eq. (13.71) for the vector potential operator we get

$$\hat{\boldsymbol{A}}(\boldsymbol{r}) = \sum_{\boldsymbol{k}\in K+}\sum_{s}\left(\frac{\hbar}{2\Omega\varepsilon_0\omega}\right)^{\frac{1}{2}}\left[e^{i\boldsymbol{k}\cdot\boldsymbol{r}}\left(\hat{a}_{\boldsymbol{k},s} + \hat{a}^\dagger_{-\boldsymbol{k},s}\right) + e^{-i\boldsymbol{k}\cdot\boldsymbol{r}}\left(\hat{a}^\dagger_{\boldsymbol{k},s} + \hat{a}_{-\boldsymbol{k},s}\right)\right]\boldsymbol{\epsilon}_{\boldsymbol{k}s}.$$

The terms with $\hat{a}_{-\boldsymbol{k},s}$, $\hat{a}^\dagger_{-\boldsymbol{k},s}$ can be described as the annihilation, respectively, creation of a photon with wavenumber $-\boldsymbol{k}$, in agreement to the corresponding plane wave exponentials. Thus, we can lift the restricted summation over half of the wavevector space and introduce an unrestricted summation instead, with the vector potential operator given by

$$\hat{\boldsymbol{A}}(\boldsymbol{r}) = \sum_{\boldsymbol{k},s}\left(\frac{\hbar}{2\Omega\varepsilon_0\omega}\right)^{\frac{1}{2}}\left[e^{i\boldsymbol{k}\cdot\boldsymbol{r}}\,\hat{a}_{\boldsymbol{k}s} + e^{-i\boldsymbol{k}\cdot\boldsymbol{r}}\,\hat{a}^\dagger_{\boldsymbol{k}s}\right]\boldsymbol{\epsilon}_{\boldsymbol{k}s}. \tag{13.77}$$

By a similar token we can derive the electric and magnetic field operators expressed in terms of the photon annihilation and creation operators through

Electric and Magnetic Field Operators

$$\hat{\boldsymbol{E}}^\perp(\boldsymbol{r}) = \sum_{\boldsymbol{k},s}\left(\frac{\hbar\omega}{2\Omega\varepsilon_0}\right)^{\frac{1}{2}} i\left[e^{i\boldsymbol{k}\cdot\boldsymbol{r}}\,\hat{a}_{\boldsymbol{k}s} - e^{-i\boldsymbol{k}\cdot\boldsymbol{r}}\,\hat{a}^\dagger_{\boldsymbol{k}s}\right]\boldsymbol{\epsilon}_{\boldsymbol{k}s}$$

$$\hat{\boldsymbol{B}}(\boldsymbol{r}) = \sum_{\boldsymbol{k},s}\left(\frac{\hbar}{2\Omega\varepsilon_0\omega}\right)^{\frac{1}{2}} i\left[e^{i\boldsymbol{k}\cdot\boldsymbol{r}}\,\hat{a}_{\boldsymbol{k}s} - e^{-i\boldsymbol{k}\cdot\boldsymbol{r}}\,\hat{a}^\dagger_{\boldsymbol{k}s}\right]\boldsymbol{k}\times\boldsymbol{\epsilon}_{\boldsymbol{k}s}. \tag{13.78}$$

Commutation Relations. Consider the commutator between the vector potential and the transverse electric field operators

$$\left[\hat{E}_i^\perp(\boldsymbol{r}'),\hat{A}_j(\boldsymbol{r})\right]=\frac{i\hbar}{2\Omega\varepsilon_0}\sum_{\boldsymbol{k},s}\left[e^{i\boldsymbol{k}\cdot(\boldsymbol{r}-\boldsymbol{r}')}+e^{-i\boldsymbol{k}\cdot(\boldsymbol{r}-\boldsymbol{r}')}\right]\left(\boldsymbol{\epsilon}_{\boldsymbol{k}s}\right)_i\left(\boldsymbol{\epsilon}_{\boldsymbol{k}s}\right)_j,$$

where we have used the fundamental commutation relations given in Eq. (13.75). We next exploit that the polarization vectors together with $\hat{\boldsymbol{k}}$ form a basis. Thus, the sum over the products of polarization vectors gives a projector on the directions perpendicular to $\hat{\boldsymbol{k}}$. We next perform the thermodynamic limit $\Omega\to\infty$ using the prescription of Eq. (10.11), and finally obtain

$$\left[\hat{E}_i^\perp(\boldsymbol{r}),\hat{A}_j(\boldsymbol{r}')\right]=\frac{i\hbar}{\varepsilon_0}\int_{-\infty}^{\infty}\left(\delta_{ij}-\hat{k}_i\hat{k}_j\right)e^{i\boldsymbol{k}\cdot\boldsymbol{r}}\frac{d^3k}{2\pi^3}=\frac{i\hbar}{\varepsilon_0}\delta_{ij}^\perp(\boldsymbol{r}-\boldsymbol{r}'),\tag{13.79}$$

with the transverse delta function defined in Eq. (F.12). By a similar token one finds

$$\left[\hat{B}_i(\boldsymbol{r}),\hat{A}_j(\boldsymbol{r}')\right]=0,$$

where we have exploited $(\boldsymbol{k}\times\boldsymbol{\epsilon})\cdot\boldsymbol{k}=0$. The other commutators between field operators will be investigated in the next chapter.

13.4.5 *Quantization of Coupled Light-Matter System*

When quantizing the coupled light-matter system, we start from the Lagrangian for charged particles together with Eq. (13.49a),

$$L=\frac{1}{2}\sum_i mv_i^2+\sum_{\boldsymbol{k}\in K^+}\left[\varepsilon_0\left(\left|\dot{\boldsymbol{A}}_{\boldsymbol{k}}\right|^2-\omega^2\left|\boldsymbol{A}_{\boldsymbol{k}}\right|^2\right)-\frac{\left|\rho_{\boldsymbol{k}}\right|^2}{k^2\varepsilon_0}+2\mathrm{Re}\left\{\boldsymbol{J}_{\boldsymbol{k}}\cdot\boldsymbol{A}_{\boldsymbol{k}}^*\right\}\right].$$

As explicitly worked out in Sect. 13.6.3, when going over to a real-space representation this Lagrange function can be rewritten in the form

$$L=\frac{1}{2}\sum_i mv_i^2+L_{\text{em}}^\perp-W_{\text{coul}}+\int\boldsymbol{J}(\boldsymbol{r})\cdot\boldsymbol{A}(\boldsymbol{r})\,d^3r,\tag{13.80}$$

with the instantaneous Coulomb interaction between the charged particles

$$W_{\text{coul}}=\frac{1}{2}\sum_{i\neq j}\frac{q^2}{4\pi\varepsilon_0|\boldsymbol{r}_i-\boldsymbol{r}_j|}.\tag{13.81}$$

For the Lagrange function of Eq. (13.80) the following two points can be easily shown.

1. The canonical quantization of the matter part can be performed in complete analogy to the discussion given in Sect. 13.4.2, however, using the vector potential operator $\hat{\boldsymbol{A}}$ rather than the classical vector potential $\boldsymbol{A}$.
2. The quantization of the transverse electromagnetic fields, described through the Lagrange function $L_{\text{em}}^{\perp}$, can be performed in complete analogy to the discussion given in Sect. 13.4.3.

Combining the results of these quantization procedures leads us to the total Hamiltonian for the coupled light-matter system

Hamiltonian for Interacting Light-Matter System

$$\hat{H} = \hat{H}_{\text{em}} + \sum_i \frac{\left[\hat{\boldsymbol{\pi}}_i - q\hat{\boldsymbol{A}}(\boldsymbol{r}_i)\right]^2}{2m} + W_{\text{coul}}. \tag{13.82}$$

The first term on the right-hand side is the photon Hamiltonian of Eq. (13.76). The second term is the minimal coupling Hamiltonian for a vector potential operator rather than a classical vector potential. The last term describes the instantaneous Coulomb coupling between the carriers forming the many-particle system. In principle, the Hamiltonian of Eq. (13.82) forms the starting point for a full quantum theory of interacting light-matter systems (Fig. 13.6).

If needed, we might now reintroduce a scalar potential associated with some external charge distribution. Suppose that the total charge distribution can be separated into two contributions

$$\rho(\boldsymbol{r}) = \rho^{(1)}(\boldsymbol{r}) + \rho^{(2)}(\boldsymbol{r}),$$

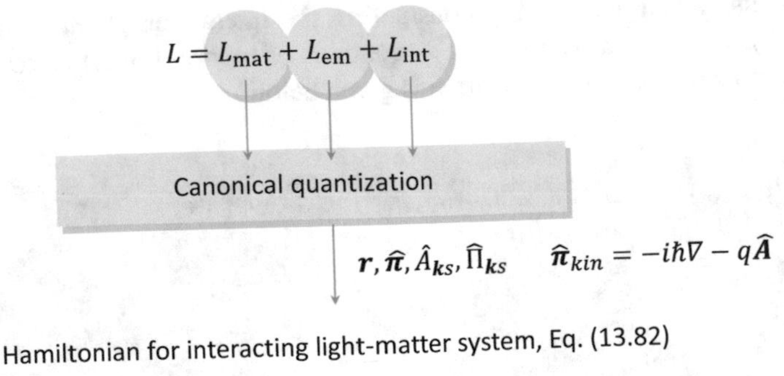

Fig. 13.6 Quantization of the coupled light-matter system. The dynamic variables are represented by $\boldsymbol{r}$ and the canonical momentum operators $\hat{\boldsymbol{\pi}}$ for the matter part, and the vector potential operator $\hat{\boldsymbol{A}}$ and its momentum $\hat{\boldsymbol{\Pi}}$ for the light fields

where the first part is for the carriers whose dynamics we would like to describe explicitly and the second one for the external charge distribution that can be controlled by some additional means. We can then split the instantaneous Coulomb potential of Eq. (13.81) into

$$W_{\text{coul}} = W_{\text{coul}}^{(1)} + W_{\text{coul}}^{(2)} + \sum_{i_1} q V_{\text{ext}}(\boldsymbol{r}_{i_1}), \tag{13.83}$$

where the summation of i_1 runs over the carriers of $\rho^{(1)}$ only, and V_{ext} is the scalar potential produced by the external charge distribution. The Hamiltonian for the photons and the first subsystem can then be written in the form

$$\hat{H} = \hat{H}_{\text{em}} + \sum_{i_1} \left(\frac{\left[\hat{\boldsymbol{\pi}}_{i_1} - q\hat{\boldsymbol{A}}(\boldsymbol{r}_{i_1})\right]^2}{2m} + q V_{\text{ext}}(\boldsymbol{r}_{i_1}) \right) + W_{\text{coul}}^{(1)}. \tag{13.84}$$

Through a similar procedure we might also include an external vector potential associated with an external current distribution, but we will not be interested in such problems here.

We conclude this section with the derivation of Maxwell's equations in operator form. Using the photon Hamiltonian of Eq. (13.76) it is easy to show that

$$\left[\hat{a}_{\boldsymbol{k}s}, \hat{H}_{\text{em}}\right] = \hbar\omega\, \hat{a}_{\boldsymbol{k}s}, \qquad \left[\hat{a}_{\boldsymbol{k}s}^\dagger, \hat{H}_{\text{em}}\right] = -\hbar\omega\, \hat{a}_{\boldsymbol{k}s}^\dagger$$

We can now use the electromagnetic field operators of Eq. (13.78) to prove the following relations

$$\frac{1}{i\hbar}\left[\hat{\boldsymbol{E}}^\perp(\boldsymbol{r}), \hat{H}_{\text{em}}\right] = \frac{1}{\mu_0\varepsilon_0}\nabla\times\hat{\boldsymbol{B}}(\boldsymbol{r}), \quad \frac{1}{i\hbar}\left[\hat{\boldsymbol{B}}(\boldsymbol{r}), \hat{H}_{\text{em}}\right] = -\nabla\times\hat{\boldsymbol{E}}^\perp(\boldsymbol{r}). \tag{13.85}$$

We have exploited the usual wave equation $\boldsymbol{k}\times\boldsymbol{k}\times\boldsymbol{\epsilon}_{\boldsymbol{k}s} + k^2\,\boldsymbol{\epsilon}_{\boldsymbol{k}s} = 0$ for deriving these results. When additionally considering the interaction Hamiltonian $\hat{H}_{\text{int}}$ of Eq. (13.63), we arrive at Heisenberg's equations of motion for the electromagnetic field operators (the detailed derivation is given below)

Heisenberg's Equations of Motion for Field Operators

$$\begin{aligned} \frac{\partial}{\partial t}\hat{\boldsymbol{B}}(\boldsymbol{r},t) &= -\nabla\times\hat{\boldsymbol{E}}(\boldsymbol{r},t) \\ \varepsilon_0\frac{\partial}{\partial t}\hat{\boldsymbol{E}}(\boldsymbol{r},t) &= \frac{1}{\mu_0}\nabla\times\hat{\boldsymbol{B}}(\boldsymbol{r},t) - \hat{\boldsymbol{J}}(\boldsymbol{r},t), \end{aligned} \tag{13.86}$$

where we have now explicitly indicated the time dependence of the operators. These are the curl equations of Maxwell's theory, however, with the classical fields replaced by field operators. To be meaningful, these operators must additionally act on a wavefunction.

Proof of Eq. (13.86). We work in the Heisenberg picture and suppress for notational simplicity the time dependence of all operators. First, we note that both $\hat{\boldsymbol{E}}^{\perp}$ and $\hat{\boldsymbol{B}}$ commute with an external scalar potential V_{ext} as well as the Coulomb term W_{coul}. With this and using that the commutator between $\hat{\boldsymbol{B}}$ and $\hat{\mathbf{A}}$ vanishes, we immediately find

$$\frac{1}{i\hbar}\left[\hat{\boldsymbol{B}}(\boldsymbol{r}),\,\hat{H}_{\text{int}}\right]=0,$$

where $\hat{H}_{\text{int}}$ is the Hamiltonian for the light–matter interaction. For the transverse electric field operator the calculation is somewhat longer. We start from

$$\frac{1}{i\hbar}\left[\hat{\boldsymbol{E}}^{\perp}(\boldsymbol{r}),\,\hat{H}_{\text{int}}\right]=\frac{1}{i\hbar}\int\left[\hat{\boldsymbol{E}}^{\perp}(\boldsymbol{r}),-\left(q\hat{\mathbf{j}}(\boldsymbol{r}')\right)\cdot\hat{\mathbf{A}}(\boldsymbol{r}')+\frac{q^2\hat{n}(\boldsymbol{r}')}{2m}\hat{A}^2(\boldsymbol{r}')\right]d^3r'.$$

Using Eq. (13.79) for the commutator between the electric field and the vector potential operators, we obtain

$$\frac{1}{i\hbar}\left[\hat{\boldsymbol{E}}^{\perp}(\boldsymbol{r}),\,\hat{H}_{\text{int}}\right]=\frac{1}{\varepsilon_0}\left(-q\hat{\mathbf{j}}^{\perp}(\boldsymbol{r})+\frac{q^2\hat{n}(\boldsymbol{r}')}{m}\hat{\mathbf{A}}(\boldsymbol{r})\right)=-\frac{1}{\varepsilon_0}\hat{\boldsymbol{J}}^{\perp}(\boldsymbol{r}),$$

with the transverse current operator $\hat{\boldsymbol{J}}^{\perp}$ defined in Eq. (13.62). We next additionally consider the longitudinal component of the electric field operator

$$\hat{\boldsymbol{E}}^{L}(\boldsymbol{r})=-\nabla\int\frac{q\hat{n}(\boldsymbol{r}')}{4\pi\varepsilon_0|\boldsymbol{r}-\boldsymbol{r}'|}\,d^3r'.$$

Taking the time derivative on both sides of the equation we then obtain together with the continuity equation of Eq. (13.61) the result

$$\frac{\partial}{\partial t}\hat{\boldsymbol{E}}^{L}(\boldsymbol{r})=\nabla\int\frac{\nabla'\cdot\hat{\boldsymbol{J}}(\boldsymbol{r}')}{4\pi\varepsilon_0|\boldsymbol{r}-\boldsymbol{r}'|}\,d^3r'=-\frac{1}{\varepsilon_0}\hat{\boldsymbol{J}}^{L}(\boldsymbol{r}),$$

where we have used the longitudinal delta function of Eq. (F.12). Putting together all results we arrive at Eq. (13.86).

Note that $\hat{\boldsymbol{E}}$ in Eq. (13.86) is the sum of transverse and longitudinal operators, and we have used $\nabla\times\hat{\boldsymbol{E}}^{L}=0$ to arrive at Ampere's law. Finally, the divergence parts of Maxwell's equations enter in the form of operator constraints. The scalar potential is a solution of Poisson's equation, see Eq. (13.46), which by construction fulfills Gauss' law. Similarly, with $\hat{\boldsymbol{B}}=\nabla\times\hat{\mathbf{A}}$ we always have $\nabla\cdot\hat{\boldsymbol{B}}=0$.

13.4.6 States of the Harmonic Oscillator

In quantum electrodynamics the photon states can be mapped onto those of a harmonic oscillator. Although in this book we will not be overly concerned with exotic photon states, in the following we briefly ponder on some of the most commonly used harmonic oscillator states with applications in quantum optics. A convenient way of plotting oscillator wavefunctions is using the Wigner quasiprobability distribution [127, 128]

Wigner Function

$$W(x,p)=\frac{1}{h}\int_{-\infty}^{\infty}e^{-ipy/\hbar}\,\psi\left(x+\frac{y}{2}\right)\psi^*\left(x-\frac{y}{2}\right)dy. \tag{13.87}$$

This function is real and has a number of remarkable properties. When integrated over all momenta p

$$\int_{-\infty}^{\infty}W(x,p)\,dp=\left|\psi(x)\right|^2$$

it gives the probability distribution in real space. Similarly, when integrated over all positions x

$$\int_{-\infty}^{\infty}W(x,p)\,dx=\left|\tilde\psi(p)\right|^2,\quad \tilde\psi(p)=h^{-\frac{1}{2}}\int_{-\infty}^{\infty}e^{-ipx/\hbar}\psi(x)\,dx$$

it gives the probability distribution in momentum space. Thus, $W(x,p)$ can be interpreted similarly to a classical (single-particle) phase-space distribution function, associated with the probability of finding a particle with momentum p at position x. There exist exotic quantum states with negative Wigner function values that cannot be interpreted classically.

We next employ the Wigner function to the states of the harmonic oscillator. We consider a Hamiltonian of the form

$$\hat H=\frac{\hat\pi^2}{2m}+\frac{1}{2}m\omega^2x^2,$$

where m is the mass and ω the angular frequency of the oscillator. Owing to the special properties of the harmonic oscillator, the time evolution of the Wigner function can be computed analytically [128, Eq. (48)]

$$W(x, p, t) = W\left(x \cos(\omega t) - \frac{p}{m\omega} \sin(\omega t), p \cos(\omega t) + m\omega x \sin(\omega t), 0\right), \tag{13.88}$$

In words, the time evolution of the Wigner function is obtained from the Wigner function at time zero by simply rotating it with angular frequency ω in phase space. We will come back to this observation in a moment.

Fock States. Figure 13.7 shows the Wigner function for Fock states with a given number of excitations. Panel (a) displays the groundstate which is a Gaussian distribution with equal widths $\Delta x = \Delta p$ and $\Delta x \Delta p = \frac{\hbar}{2}$. Panels (b, c) show the distributions for the first and second excited state, respectively, with an increasing number of nodes along the radial direction. Thus, the distributions have positive and negative values, corresponding to non-classical states. All Fock states have angular symmetry in phase space, in accordance to stationary states which exhibit no time evolution when the distribution rotates in phase space according to Eq. (13.88).

Coherent States. Coherent states are the most "classical" states of a harmonic oscillator which are obtained by displacing the groundstate wavefunction, see Fig. 13.8. As time goes on, these states rotate in phase space. When projected on either the position or momentum axis we obtain a motion reminiscent of a classical oscillator, however, with equal uncertainty $\Delta x = \Delta p$ and $\Delta x \Delta p = \frac{\hbar}{2}$. Coherent states are usually excellent approximations for laser light with a given phase and amplitude of the electric field (corresponding to the oscillator's momentum). A more detailed discussion of coherent states and coherent light can be found in the specialized literature on quantum optics, see, for instance, [4, 119, 127].

Squeezed States. Squeezed states are typically obtained from coherent states by reducing either Δx or Δp, while keeping $\Delta x \Delta p = \frac{\hbar}{2}$ constant. Figure 13.9 shows an example with reduced position fluctuations at time zero. As time

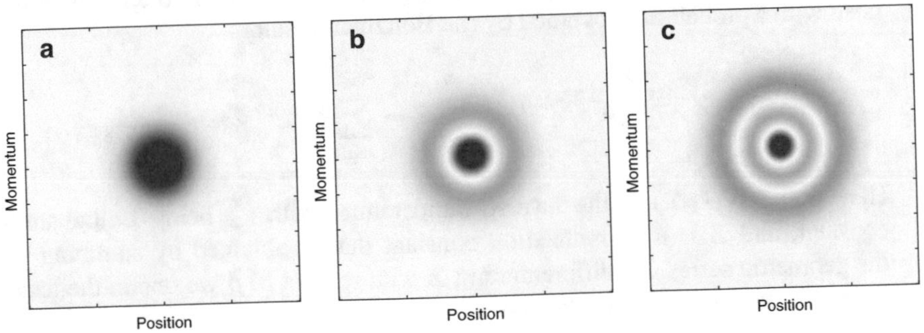

Fig. 13.7 Wigner distribution of Fock states. (**a**) Groundstate, (**b**) one-photon state, and (**c**) two-photon state. The groundstate has a Gaussian shape with equal widths Δx and Δp, which are subject to Heisenberg's uncertainty principle $\Delta x \Delta p = \frac{\hbar}{2}$. The excited states have an increasing number of nodes along the radial direction, see Fig. 13.3a for a plot of the wavefunctions

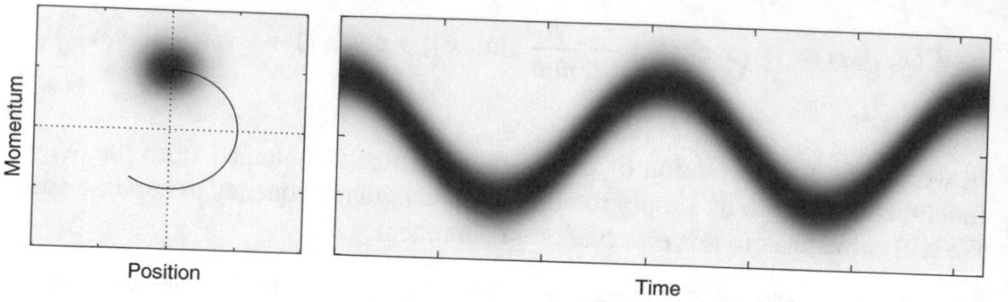

Fig. 13.8 Wigner distribution of coherent state. Coherent states are produced by displacing the groundstate of the harmonic oscillator in phase space, the ensuing time evolution is an oscillation of the displaced distribution with angular frequency ω. The left panel shows the projection of the clockwise oscillating distribution function on the momentum coordinate. Coherent states have equal position and momentum uncertainties Δx, Δp, and are often denoted as classical states

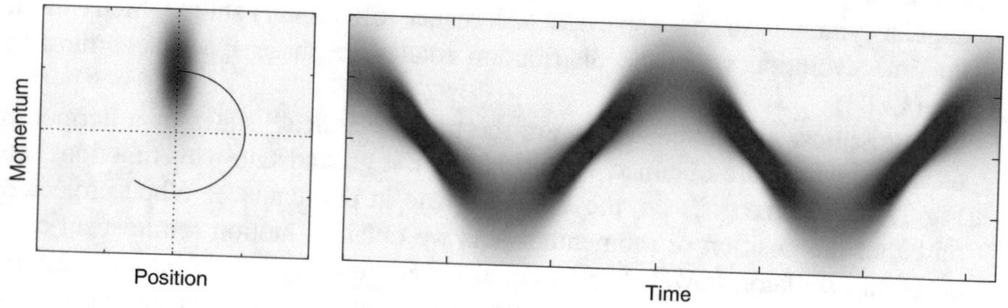

Fig. 13.9 Same as Fig. 13.8 but for squeezed states

goes on, the squeezed distribution rotates in phase space and the projection on either the position or momentum axis gives an amplitude that periodically sharpens and blurs. Squeezed states are conveniently created through nonlinear light–matter interactions, and are often employed for measurements with signal-to-noise ratios below the classical shot-noise limit [119, 127].

Thermal States. In thermal equilibrium the oscillator states are thermally populated with a probability governed by the Boltzmann factor

$$\bar{n}_{\rm th}(\omega) = Z^{-1}\sum_{n=0}^{\infty} e^{-\beta(n\hbar\omega)}n, \qquad Z = \sum_{n=0}^{\infty} e^{-\beta(n\hbar\omega)} = \frac{1}{1-e^{-\beta\hbar\omega}}.$$

Here $\beta = 1/(k_B T)$ is the inverse temperature, with k_B being Boltzmann's constant, and Z is a normalization constant that is obtained by summing up the geometric series. By differentiating Z with respect to β we obtain the usual Bose–Einstein distribution function

$$\bar{n}_{\rm th}(\omega) = \frac{1}{e^{\beta\hbar\omega}-1}. \tag{13.89}$$

Thermal photon states will be discussed in more detail in Chap. 15.

Entangled states. There exist other non-classical states formed by two or more photon modes. Consider, for instance, photon modes $\hat{a}_{\pm k,s}$ which propagate in the positive or negative $\boldsymbol{k}$ direction with polarizations $s_{1,2}$. The state

$$\psi = \frac{1}{\sqrt{2}}\left[\left(\hat{a}^\dagger_{+\boldsymbol{k},s_1}\right)\left(\hat{a}^\dagger_{-\boldsymbol{k},s_2}\right) - \left(\hat{a}^\dagger_{+\boldsymbol{k},s_2}\right)\left(\hat{a}^\dagger_{-\boldsymbol{k},s_1}\right)\right]|0\rangle$$

is then a polarization-entangled state with a number of most interesting properties [119, 127]. While such photon states play an important role in genuine quantum plasmonics experiments [129, 130], here we refrain from a more detailed discussion.

As a final comment, we note that the transition from quantum to classical electrodynamics is achieved most easily with **coherent states**. In general, these states are defined as [4, 119, 127]

$$|\alpha\rangle = \exp\left[\alpha\hat{a}^\dagger - \alpha^*\hat{a}\right]|0\rangle. \tag{13.90}$$

These states are eigenstates of the photon annihilation operator, $\hat{a}|\alpha\rangle = \alpha|\alpha\rangle$. Thus, for an electric field operator similar to Eq. (13.78),

$$\hat{\boldsymbol{E}}^\perp = i\mathscr{E}_0\left(e^{i\boldsymbol{k}\cdot\boldsymbol{r}}\,\hat{a} - e^{-i\boldsymbol{k}\cdot\boldsymbol{r}}\,\hat{a}^\dagger\right)\boldsymbol{\epsilon},$$

the expectation value $\langle\alpha|\hat{\boldsymbol{E}}^\perp|\alpha\rangle$ becomes a field which behaves almost classically, with "almost" referring to small (Poisson) field fluctuations around the mean value.

13.5 Multipolar Hamiltonian

In the remainder of this chapter we discuss a unitary transformation introduced by Power, Zienau, and Woolley [131, 132], which replaces the scalar and vector potential operators by the electromagnetic field operators. The resulting Hamiltonian is extremely useful in the context of the quantization of Maxwell's equations in matter, but in this book we will pursue a different approach. An excellent and detailed account of the topic can be found in Refs. [123, 133]. Consider the following unitary operator

Plasmon Entanglement

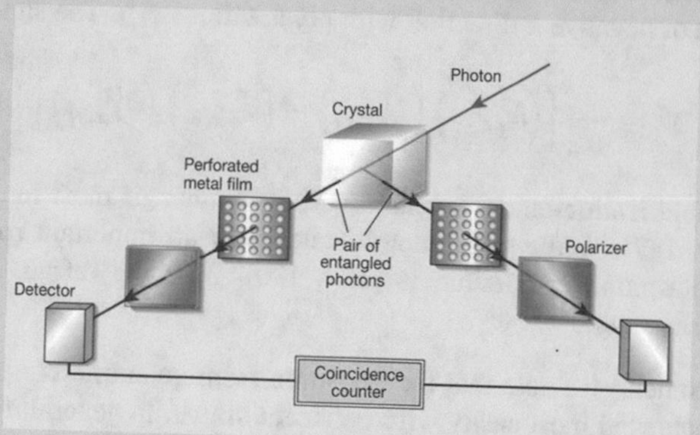

In a beautiful experiment, Altewischer and coworkers [129] sent entangled photons through perforated metal films, thereby transforming photons to surface plasmons, which couple from one side of the film to the other one, where they are again converted to photons. Surprisingly, they still observed a large degree of entanglement between the transmitted photons. In the abstract of their paper the authors write:

> *The state of a two-particle system is said to be entangled when its quantum-mechanical wavefunction cannot be factorized into two single-particle wavefunctions. This leads to one of the strongest counter-intuitive features of quantum mechanics, namely nonlocality. Experimental realization of quantum entanglement is relatively easy for photons; a starting photon can spontaneously split into a pair of entangled photons inside a nonlinear crystal. Here we investigate the effects of nanostructured metal optical elements on the properties of entangled photons. To this end, we place optically thick metal films perforated with a periodic array of sub-wavelength holes in the paths of the two entangled photons. Such arrays convert photons into surface-plasmon waves—optically excited compressive charge density waves—which tunnel through the holes before reradiating as photons at the far side. We address the question of whether the entanglement survives such a conversion process. Our coincidence counting measurements show that it does, so demonstrating that the surface plasmons have a true quantum nature.*

Figure taken from Bill Barnes, *"Survival of the entangled"*, Nature 418, 281–282 (2002).

$$\hat{\mathscr{U}} = e^{-i(q/\hbar)\boldsymbol{A}\cdot\boldsymbol{r}}\,.$$

We assume that the quantum-mechanical system under study is much smaller than the wavelength of the exciting light, such that we can approximately set $\boldsymbol{A}(\boldsymbol{r}) \approx \boldsymbol{A}(\boldsymbol{r}_0) = \boldsymbol{A}_0$, where $\boldsymbol{r}_0$ is the system's position and $\boldsymbol{A}_0$ is a constant (but time dependent) vector potential operator. Then,

$$\hat{\boldsymbol{\pi}} \longrightarrow \hat{\mathscr{U}}\,\hat{\boldsymbol{\pi}}\,\hat{\mathscr{U}}^\dagger = \left(e^{-i(q/\hbar)\boldsymbol{A}_0\cdot\boldsymbol{r}}\right)\hat{\boldsymbol{\pi}}\left(e^{i(q/\hbar)\boldsymbol{A}_0\cdot\boldsymbol{r}}\right) = \hat{\boldsymbol{\pi}} - \frac{iq}{\hbar}\Big[\boldsymbol{r}\cdot\boldsymbol{A}_0, \hat{\boldsymbol{\pi}}\Big],$$

where we have expanded $\hat{\mathscr{U}}$ in a Taylor series and have used that all higher-order terms are zero because the commutator $[\boldsymbol{r}\cdot\boldsymbol{A}_0, \hat{\boldsymbol{\pi}}] = i\hbar\boldsymbol{A}_0$ gives a constant vector that commutes with $\boldsymbol{A}_0\cdot\boldsymbol{r}$. We thus find

$$\hat{\boldsymbol{\pi}} \longrightarrow \hat{\boldsymbol{\pi}} + q\boldsymbol{A}_0.$$

The Schrödinger equation can be transformed correspondingly

$$i\hbar\hat{\mathscr{U}}\left[\frac{d}{dt}|\psi(t)\rangle\right] = i\hbar\frac{d}{dt}\left[\hat{\mathscr{U}}|\psi(t)\rangle\right] - i\hbar\frac{d\hat{\mathscr{U}}}{dt}|\psi(t)\rangle = \left[\hat{\mathscr{U}}\hat{H}\hat{\mathscr{U}}^\dagger\right]\hat{\mathscr{U}}|\psi(t)\rangle.$$

With

$$-i\hbar\frac{d\hat{\mathscr{U}}}{dt} = -i\hbar\left[-i\frac{q}{\hbar}\frac{d\boldsymbol{A}_0\cdot\boldsymbol{r}}{d}\right]\hat{\mathscr{U}} = q\left[\boldsymbol{r}\cdot\boldsymbol{E}_0^\perp\right]\hat{\mathscr{U}},$$

where $\boldsymbol{E}_0^\perp$ is the transverse electric field evaluated at the position of the atomic system, we are then led to the Hamiltonian in the electric-dipole approximation

Hamiltonian in the Electric-Dipole Approximation

$$\hat{H} = \left[\frac{\hat{\boldsymbol{\pi}}^2}{2m} + U(\boldsymbol{r})\right] - q\boldsymbol{r}\cdot\boldsymbol{E}_0^\perp. \tag{13.91}$$

For completeness, we have added a confinement potential $U(\boldsymbol{r})$ typically associated with the electrostatic Coulomb potentials of the nuclei or ions forming the molecule or nanostructure. Equation (13.91) states that the light-matter coupling for small quantum emitters can be described in terms of a product between the dipole operator $q\boldsymbol{r}$ and the electric field $\boldsymbol{E}_0^\perp$.

The Power–Zienau–Woolley transformation can be generalized to include higher multipole couplings. This is achieved by replacing the original unitary operator with

$$\hat{\mathscr{U}}(\boldsymbol{r}) = \exp\left[-\frac{iq}{\hbar}\int_{\boldsymbol{r}_0}^{\boldsymbol{r}}\boldsymbol{A}(\boldsymbol{r}')\cdot d\boldsymbol{r}'\right], \tag{13.92}$$

where the path integral is to be evaluated over the straight line between the field point $\boldsymbol{r}$ and the reference point for the charge distribution $\boldsymbol{r}_0$. The transformation introduced by $\hat{\mathscr{U}}$ is sometimes also referred to as the *Peierls substitution*. Woolley describes it as follows [132]:

> [...] the phase of the wavefunction is changed by an amount that would arise from the transfer of each particle in an aggregate from its position at $\boldsymbol{r}$ to the center of mass $\boldsymbol{r}_0$, in the presence of the transverse electromagnetic field described by the vector potential $\boldsymbol{A}(\boldsymbol{r})$. In other words, the transformation may be thought of as formally eliminating the finite extension of the molecular charge distribution, in favor of a superposition of point multipoles located at the center of mass $\boldsymbol{r}_0$, since the wavefunction is the probability amplitude that the components of the system will be found in some configuration at time t. Such an interpretation is consistent with Power and Zienau's contention that the transformation incorporates the molecular charge distribution and its fields into redefined modes of the radiation field.

Let us consider the transformation in more detail. We first introduce the polarization operator

$$\hat{\boldsymbol{P}}(\boldsymbol{r}') = q(\boldsymbol{r} - \boldsymbol{r}_0) \int_0^1 \delta\Big(\boldsymbol{r}' - \boldsymbol{r}_0 - \lambda[\boldsymbol{r} - \boldsymbol{r}_0]\Big)\, d\lambda, \tag{13.93}$$

which will be investigated more closely further below. The unitary transformation of Eq. (13.92) can thus be rewritten as

$$\hat{\mathscr{U}}(\boldsymbol{r}) = \exp\left[-\frac{i}{\hbar} \int \hat{\boldsymbol{P}}(\boldsymbol{r}') \cdot \boldsymbol{A}(\boldsymbol{r}')\, d^3r'\right]. \tag{13.94}$$

Proceeding in the same manner as for the constant vector potential, we find that the canonical momentum operator becomes

$$\hat{\boldsymbol{\pi}} \longrightarrow \hat{\mathscr{U}}\, \hat{\boldsymbol{\pi}}\, \hat{\mathscr{U}}^\dagger = \hat{\boldsymbol{\pi}} - \frac{i}{\hbar}\Big[\int \hat{\boldsymbol{P}}(\boldsymbol{r}') \cdot \boldsymbol{A}(\boldsymbol{r}')\, d^3r', \hat{\boldsymbol{\pi}}\Big] = \hat{\boldsymbol{\pi}} + \nabla \int \hat{\boldsymbol{P}}(\boldsymbol{r}') \cdot \boldsymbol{A}(\boldsymbol{r}')\, d^3r',$$

where we have used

$$[f(\boldsymbol{r}), \hat{\boldsymbol{\pi}}] = i\hbar \nabla f(\boldsymbol{r}).$$

The last term on the right-hand side can be rewritten. We set for a moment $\boldsymbol{r}_0 = 0$ and introduce the quantity

$$S = \int \boldsymbol{P}(\boldsymbol{r}') \cdot \boldsymbol{A}(\boldsymbol{r}')\, d^3r' = q \int_0^1 \boldsymbol{r} \cdot \boldsymbol{A}(\lambda \boldsymbol{r})\, d\lambda.$$

One can show that the following relation holds:

$$\frac{\partial S}{\partial r_i} = q \int_0^1 \left[A_i + r_j \frac{\partial A_j}{\partial r_i} \lambda\right] d\lambda = q \int_0^1 \Big[\boldsymbol{A} + \lambda \boldsymbol{r} \times (\nabla \times \boldsymbol{A}) + \lambda(\boldsymbol{r} \cdot \nabla)\boldsymbol{A}\Big]_i\, d\lambda.$$

The last term in brackets can be rewritten as $\lambda \frac{d\boldsymbol{A}}{d\lambda}$. Performing integration by parts and introducing again $\boldsymbol{r}_0$ then leads us to

$$\nabla \int \hat{\boldsymbol{P}}(\boldsymbol{r}') \cdot \boldsymbol{A}(\boldsymbol{r}')\, d^3r' = q \int_0^1 \boldsymbol{r} \times \boldsymbol{B}\Big(\boldsymbol{r}_0 + \lambda[\boldsymbol{r} - \boldsymbol{r}_0]\Big)\, d\lambda + q\boldsymbol{A}(\boldsymbol{r}), \tag{13.95}$$

with the magnetic field $\boldsymbol{B} = \nabla \times \boldsymbol{A}$. Putting together all results, we obtain after the unitary transformation of $\mathcal{U}$ instead of the minimal coupling Hamilton of Eq. (13.52) the multipolar Hamiltonian

Multipolar Hamiltonian (Single Particle)

$$\hat{H} = \frac{\left(\hat{\pi} + q\int_0^1 \boldsymbol{r} \times \boldsymbol{B}\, d\lambda\right)^2}{2m} - q\int_0^1 \boldsymbol{E}^\perp\, d\lambda + U(\boldsymbol{r}), \tag{13.96}$$

where the electromagnetic fields have to be evaluated at $\boldsymbol{r}_0 + \lambda(\boldsymbol{r} - \boldsymbol{r}_0)$. Note that the appearance of the transverse electric field term with $\boldsymbol{E}^\perp$ is for the same reasons as previously discussed for the case of a constant vector potential. By expanding $\boldsymbol{E}^\perp$, $\boldsymbol{B}$ in Taylor series around the center position $\boldsymbol{r}_0$ we then obtain the multipolar electric and magnetic interaction terms. We will not work this out explicitly here.

Polarization Operator. We now return to the polarization operator $\hat{\boldsymbol{P}}$ of Eq. (13.93), which we rewrite by expressing Dirac's delta function in wavenumber space

$$\hat{\boldsymbol{P}}(\boldsymbol{r}') = q\boldsymbol{r}\int_0^1 \left[\int_{-\infty}^{\infty} e^{i\boldsymbol{k}\cdot(\boldsymbol{r}'-\lambda\boldsymbol{r})} \frac{d^3k}{(2\pi)^3}\right] d\lambda,$$

where we have set for simplicity $\boldsymbol{r}_0 = 0$. For the divergence of $\hat{\boldsymbol{P}}$ we get

$$\begin{aligned}\nabla' \cdot \hat{\boldsymbol{P}}(\boldsymbol{r}') &= q\int_0^1 \left[\int_{-\infty}^{\infty} (i\boldsymbol{r}\cdot\boldsymbol{k}) e^{i\boldsymbol{k}\cdot(\boldsymbol{r}'-\lambda\boldsymbol{r})} \frac{d^3k}{(2\pi)^3}\right] d\lambda \\ &= -q\int_0^1 \frac{d}{d\lambda}\left[\int_{-\infty}^{\infty} e^{i\boldsymbol{k}\cdot(\boldsymbol{r}'-\lambda\boldsymbol{r})} \frac{d^3k}{(2\pi)^3}\right] d\lambda.\end{aligned}$$

Performing the λ integration and going back to the real-space representation for Dirac's delta function, we arrive at

$$-\nabla' \cdot \hat{\boldsymbol{P}}(\boldsymbol{r}') = q\Big[\delta(\boldsymbol{r}' - \boldsymbol{r}) - \delta(\boldsymbol{r}')\Big] = \rho(\boldsymbol{r}') - \rho_0(\boldsymbol{r}') \tag{13.97}$$

with $\rho(\boldsymbol{r}') = q\delta(\boldsymbol{r}' - \boldsymbol{r})$ and $\rho_0(\boldsymbol{r}') = q\delta(\boldsymbol{r}')$. The term on the right-hand side can be interpreted as a polarization charge distribution with two opposite point charges located at positions $\boldsymbol{r}$ and at the origin. ρ_0 is a reference charge distribution which is time independent. Along the same lines as for Eq. (13.60) we can derive the continuity equation

$$\frac{d\rho}{dt} = -\frac{i}{\hbar}\big[\rho, \hat{H}\big] = -\nabla \cdot \hat{\boldsymbol{J}},$$

with the charge current operator $\hat{\boldsymbol{J}} = q\hat{\mathbf{j}}$. We can use Eq. (13.97) to rewrite the above commutator according to

$$-\frac{i}{\hbar}[\rho, \hat{H}] = \frac{i}{\hbar}\big[-\nabla\cdot\hat{\boldsymbol{P}}, \hat{H}\big] = \frac{i}{\hbar}\nabla\cdot\big[\hat{\boldsymbol{P}}, \hat{H}\big] = -\nabla\cdot\frac{d\hat{\boldsymbol{P}}}{dt},$$

where we have used that $\hat{\boldsymbol{P}}$ depends on position and time only, and we thus can replace $\hat{H}$ with the kinetic energy part that commutes with the nabla operator. The above equations suggest the decomposition of the current operator into the contributions [123]

$$\hat{\boldsymbol{J}} = \frac{d\hat{\boldsymbol{P}}}{dt} + \nabla\times\hat{\boldsymbol{M}}. \tag{13.98}$$

The first term on the right-hand side is the polarization current and the second one a magnetization current, which is divergence free and will not be explicitly considered in the following (see Ref. [120] for a discussion of magnetization effects). Note that the above decomposition is completely along the lines of classical electrodynamics.

The scheme described so far can be extended to the case of many-particle systems. Without much ado, we state the multipolar Hamiltonian in the form

$$\hat{H} = \sum_i \frac{\hat{\boldsymbol{\pi}}_i^2}{2m} - \int \hat{\boldsymbol{P}}(\boldsymbol{r})\cdot\hat{\boldsymbol{E}}^{\perp}(\boldsymbol{r},t)\,d^3r + \int \hat{\rho}(\boldsymbol{r})V(\boldsymbol{r},t)\,d^3r,$$

with the polarization operator defined in analogy to Eq. (13.93),

$$\hat{\boldsymbol{P}}(\boldsymbol{r}') = \sum_i q(\boldsymbol{r}_i - \boldsymbol{r}_0)\int_0^1 \delta\Big(\boldsymbol{r}' - \boldsymbol{r}_0 - \lambda[\boldsymbol{r}_i - \boldsymbol{r}_0]\Big)\,d\lambda. \tag{13.99}$$

Here the sum extends over all particles of the system. The term with the scalar potential can be rewritten using Eq. (13.97) via

$$\int \hat{\rho}(\boldsymbol{r})V(\boldsymbol{r},t)\,d^3r = \int\Big[-\nabla\cdot\hat{\boldsymbol{P}}(\boldsymbol{r})\Big]V(\boldsymbol{r},t)\,d^3r = \int \hat{\boldsymbol{P}}(\boldsymbol{r})\cdot\Big[\nabla V(\boldsymbol{r},t)\Big]\,d^3r,$$

where we have performed integration by parts to arrive at the final term and have neglected the boundary term, which becomes zero for a localized charge distribution. Apparently, the term in brackets can be associated with the longitudinal part of the electric field. Putting together all results, we arrive at the light–matter interaction of the multipolar Hamiltonian

Multipolar Hamiltonian (Light–Matter Interaction)

$$\hat{H}_{\text{int}} = -\int \hat{\boldsymbol{P}}(\boldsymbol{r})\cdot\boldsymbol{E}(\boldsymbol{r},t)\,d^3r, \tag{13.100}$$

where $\hat{\boldsymbol{P}}$ is the polarization operator of Eq. (13.99) and $\boldsymbol{E}$ is a classical electric field. A similar expression can be obtained for the field operators. Equation (13.100) is the starting point for deriving a quantized version of Maxwell's equations in matter, as discussed in more length in [120, 134]. The strong point of this approach is that it allows carrying over the strategy of classical electrodynamics to separate charges and currents into free and bound contributions directly to the quantum realm. With this, one can develop a quantized version of Maxwell's equations in terms of permittivity and permeability functions only, without making any contact to a microscopic matter description. However, in solid state physics (and also plasmonics) one usually has a microscopic description already at hand, so it appears more natural to directly use it rather than taking the detour over polarizations and magnetizations. In the following chapters we will show how this can be done.

13.6 Details of Lagrange Formalism in Electrodynamics

In this section we show how to describe an interacting light-matter system within the framework of Lagrange functions. We start by considering a charged particle moving in presence of electromagnetic fields, and then ponder on the Lagrange formalism for Maxwell's equations.

13.6.1 *Lagrange Function for Charged Particle*

Consider a particle with mass m and charge q propagating in presence of the electromagnetic fields $\boldsymbol{E}$, $\boldsymbol{B}$. As we will proof through explicit calculation, Newton's equation of motion for the Lorentz force can be obtained from a Lagrange function containing the scalar and vector potentials V, $\boldsymbol{A}$ in the form

Lagrange Function for Point Charge and Electromagnetic Fields

$$L(\boldsymbol{r},\boldsymbol{v})=\frac{1}{2}mv^2-qV(\boldsymbol{r},t)+q\boldsymbol{v}\cdot\boldsymbol{A}(\boldsymbol{r},t). \tag{13.101}$$

Proof of Eq. (13.101). To show that this Lagrange function indeed leads to the Lorentz force, we first write it down in Cartesian components

$$L=\frac{1}{2}m\sum_j v_j^2-q\,V+q\sum_j v_jA_j,$$

where we have explicitly indicated the summation over j. The first functional derivative of the Euler–Lagrange Eq. (13.25) is evaluated to

$$\frac{d}{dt}\frac{\partial L}{\partial v_k} = \frac{d}{dt}\left(mv_k + qA_k\right) = m\dot{v}_k + q\frac{dA_k}{dt} = m\dot{v}_k + q\left(\sum_j \frac{\partial A_k}{\partial r_j}\frac{dr_j}{dt} + q\frac{\partial A_k}{\partial t}\right).$$

We have used that for the total time derivative of $\boldsymbol{A}$ we additionally have to consider the derivative with respect to the position $\boldsymbol{r}(t)$ where the vector potential is evaluated. As for the second functional derivative we get

$$\frac{\partial L}{\partial r_k} = -q\frac{\partial V}{\partial r_k} + q\sum_j v_j\frac{\partial A_j}{\partial r_k}.$$

Putting together the results, we get from the Euler–Lagrange Eq. (13.25)

$$m\dot{\boldsymbol{v}} = q\Big\{-\nabla V - \frac{\partial \boldsymbol{A}}{\partial t} + \nabla(\boldsymbol{v}\cdot\boldsymbol{A}) - (\boldsymbol{v}\cdot\nabla)\boldsymbol{A}\Big\}$$

The first two expressions in brackets can be identified as the electric field. For the other two terms we observe

$$\boldsymbol{v}\times\boldsymbol{B} = \boldsymbol{v}\times(\nabla\times\boldsymbol{A}) = \nabla(\boldsymbol{v}\cdot\boldsymbol{A}) - (\boldsymbol{v}\cdot\nabla)\boldsymbol{A}.$$

Thus, we recover from the Lagrangian of Eq. (13.101) Newton's equations of motion $m\dot{\boldsymbol{v}} = q(\boldsymbol{E} + \boldsymbol{v}\times\boldsymbol{B})$ for the Lorentz force, which completes our proof.

In case of multiple particles we can generalize the above Lagrange function according to

$$L = \sum_i \Big[\frac{1}{2}mv_i^2 - qV(\boldsymbol{r}_i, t) + q\boldsymbol{v}_i\cdot\boldsymbol{A}(\boldsymbol{r}_i, t)\Big].$$

We may now introduce the charge and current densities

$$\rho(\boldsymbol{r}) = q\sum_i \delta(\boldsymbol{r}-\boldsymbol{r}_i), \quad \boldsymbol{J}(\boldsymbol{r}) = q\sum_i \boldsymbol{v}_i\delta(\boldsymbol{r}-\boldsymbol{r}_i). \tag{13.102}$$

With this, the Lagrange function can be rewritten in the form

$$L = \sum_i \frac{1}{2}mv_i^2 + \int\Big[-\rho(\boldsymbol{r})V(\boldsymbol{r}, t) + \boldsymbol{J}(\boldsymbol{r}, t)\cdot\boldsymbol{A}(\boldsymbol{r}, t)\Big]d^3r. \tag{13.103}$$

13.6.2 Lagrange Function for Maxwell's Equations

In the context of Maxwell's equations, the central objects of the Lagrange formalism are the scalar and vector potential V, $\boldsymbol{A}$. With these potentials, the homogeneous

Maxwell's equations are automatically fulfilled. Being a field theory, we now have to deal with objects that depend on both space and time coordinates, contrary to $\boldsymbol{r}(t)$, $\dot{\boldsymbol{r}}(t)$ for point particles which depend on time only. Correspondingly, we introduce a **Lagrange density** $\mathscr{L}$ that depends on V, $\boldsymbol{A}$, as well as on their spatial and time derivatives. The action is related to $\mathscr{L}$ through

$$S = \int_{t_1}^{t_2} \int_{\Omega} \mathscr{L}\, d^3r dt \longrightarrow \text{Extremum.} \tag{13.104}$$

In the same spirit as in classical mechanics, we assume that the action has an extremum for the proper time evolution of the electromagnetic potentials. In our case, this procedure must therefore lead to the inhomogeneous Maxwell's equations. Again we assume that the potentials are fixed at the initial and final times. Additionally, we either consider an unbounded space where the potentials vanish at infinity, or fix the potentials at the boundary of the domain under consideration.

Before doing the actual calculations, let us give a brief summary of how to obtain the Euler–Lagrange equations. Suppose that we already know the proper solutions for V, $\boldsymbol{A}$, and consider small variations δV, $\delta \boldsymbol{A}$ around them. At the extremum, all contributions linear in the variations must then become zero. In analogy to the Lagrange formalism of classical mechanics, we use integration by parts to remove the spatial and time derivatives from δV, $\delta \boldsymbol{A}$. Without much ado, we claim that the Lagrange density for Maxwell's theory is given through

Lagrange Density for Maxwell's Theory

$$\mathscr{L} = \left(\frac{\varepsilon_0}{2}\boldsymbol{E}\cdot\boldsymbol{E} - \frac{1}{2\mu_0}\boldsymbol{B}\cdot\boldsymbol{B}\right) + \boldsymbol{J}\cdot\boldsymbol{A} - \rho V. \tag{13.105}$$

The fields $\boldsymbol{E}$, $\boldsymbol{B}$ must be additionally expressed through the scalar and vector potentials V, $\boldsymbol{A}$ which are the basic quantities of the Lagrange formalism for Maxwell's equations.

Proof of Eq. (13.105). We next prove that we indeed get the inhomogeneous Maxwell's equations from Eq. (13.105). Performing a variation in Eq. (13.104) we arrive at

$$\begin{aligned}
\delta S &= \int \left(\varepsilon_0 \boldsymbol{E}\cdot\delta\boldsymbol{E} - \frac{1}{\mu_0}\boldsymbol{B}\cdot\delta\boldsymbol{B} + \boldsymbol{J}\cdot\delta\boldsymbol{A} - \rho\delta V\right) d^3r dt \\
&= \int \left(\varepsilon_0 \boldsymbol{E}\cdot\left[-\nabla\delta V - \frac{\partial\delta\boldsymbol{A}}{\partial t}\right] - \frac{1}{\mu_0}\boldsymbol{B}\cdot\nabla\times\delta\boldsymbol{A} + \boldsymbol{J}\cdot\delta\boldsymbol{A} - \rho\delta V\right) d^3r dt,
\end{aligned}$$

where in the second line we have expressed $\delta\boldsymbol{E}$, $\delta\boldsymbol{B}$ in terms of the electromagnetic potentials. We perform integration by parts to remove the spatial and time derivatives from δV, $\delta\boldsymbol{A}$ and shuffle them over to the electromagnetic fields. We start with

$$-\varepsilon_0 \int \boldsymbol{E} \cdot \frac{\partial \delta \boldsymbol{A}}{\partial t}\, dt = \varepsilon_0 \int \frac{\partial \boldsymbol{E}}{\partial t} \cdot \delta \boldsymbol{A}\, dt + \text{b.t.},$$

where b.t. denotes the additional boundary term. This term can be ignored because the potential variations become zero at the boundary, for the reasons discussed above. Similarly, we get for the scalar potential

$$-\varepsilon_0 \int \boldsymbol{E} \cdot \nabla \delta V\, d^3r = -\varepsilon_0 \int \Big[\nabla \cdot (\boldsymbol{E} \delta V) - (\nabla \cdot \boldsymbol{E}) \delta V\Big] d^3r.$$

We can use Gauss' law to transform the first term in brackets to a boundary integral that can be ignored using the same reasoning as before. As for the magnetic field contribution, we find

$$-\frac{1}{\mu_0} \int \boldsymbol{B} \cdot \nabla \times \delta \boldsymbol{A}\, d^3r = \frac{1}{\mu_0} \int \Big[\nabla \cdot (\boldsymbol{B} \times \delta \boldsymbol{A}) - (\nabla \times \boldsymbol{B}) \cdot \delta \boldsymbol{A}\Big] d^3r,$$

where we have used the identity $\nabla \cdot \boldsymbol{B} \times \delta \boldsymbol{A} = (\nabla \times \boldsymbol{B}) \cdot \delta \boldsymbol{A} - \boldsymbol{B} \cdot \nabla \times \delta \boldsymbol{A}$. The first term in brackets can again be transformed to a boundary integral that becomes zero. Putting together all contributions, we get for the variation δS the expression

$$\delta S = \int \Big[\Big(\varepsilon_0 \nabla \cdot \boldsymbol{E} - \rho\Big) \delta V + \Big(\varepsilon_0 \frac{\partial \boldsymbol{E}}{\partial t} - \frac{1}{\mu_0} \nabla \times \boldsymbol{B} + \boldsymbol{J}\Big) \cdot \delta \boldsymbol{A}\Big] d^3r dt = 0.$$

Indeed, the first and second term in brackets represent Gauss' law and Ampere's law with Maxwell's displacement current, which completes our proof for the Lagrange density of Eq. (13.105).

13.6.3 Lagrange Function in Coulomb Gauge

We next express in the Lagrange density the fields $\boldsymbol{E}$, $\boldsymbol{B}$ in terms of the potentials V, $\boldsymbol{A}$, and arrive at

$$\begin{aligned} \mathscr{L} &= \frac{\varepsilon_0}{2}\Big(-\nabla V - \frac{\partial \boldsymbol{A}}{\partial t}\Big) \cdot \Big(-\nabla V - \frac{\partial \boldsymbol{A}}{\partial t}\Big) \\ &\quad - \frac{1}{2\mu_0}\Big(\nabla \times \boldsymbol{A}\Big) \cdot \Big(\nabla \times \boldsymbol{A}\Big) + \boldsymbol{J} \cdot \boldsymbol{A} - \rho V. \end{aligned} \tag{13.106}$$

Although this Lagrange density leads to the proper Maxwell's equations, it cannot be submitted directly to the canonical quantization procedure because it does not depend on the time derivative for V. As discussed at the beginning of Sect. 13.4.3, the problem can be overcome by formally eliminating the scalar potential from the Lagrange function. In our following discussion we start with electromagnetic potentials given in an arbitrary gauge and introduce the transverse vector potential $\boldsymbol{A}^\perp$ as the only dynamical quantity, a procedure that can be shown to be equivalent to the Coulomb gauge [123]. It turns out to be convenient to expand all quantities in a plane wave basis, see Eq. (13.65) for the expansion of the vector potential. Upon integration of the Lagrange density of Eq. (13.106) over the entire space we obtain

$$\begin{aligned} L_k &= \varepsilon_0\Big(-ikV_k - \dot{A}_k\Big)\cdot\Big(ikV_k^* - \dot{A}_k^*\Big) \\ &- \frac{1}{\mu_0}\Big(k\times A_k\Big)\cdot\Big(k\times A_k^*\Big) + 2\mathrm{Re}\Big(J_k\cdot A_k^* - \rho_k V_k^*\Big), \end{aligned} \tag{13.107}$$

where the total Lagrangian is obtained by summing the wavevectors over K^+. Because of the orthogonality of the plane wave basis, terms with $e^{\pm 2ik\cdot r}$ become zero when integrated over the entire space. In wavenumber space we find

$$E_k = -ikV_k - \dot{A}_k \implies -ik^2V_k = k\cdot E_k + k\cdot\dot{A}_k = -i\frac{\rho_k}{\varepsilon_0} + k\cdot\dot{A}_k,$$

with the obvious notation $\dot{A}_k$ for the time derivative. We have used Gauss' law to arrive at the last expression on the right-hand side. The electric field can thus be written in the form

$$E_k = \frac{1}{k^2}\left[-i\frac{k\rho_k}{\varepsilon_0} + k\left(k\cdot\dot{A}_k\right)\right] - \dot{A}_k = -i\frac{k\rho_k}{k^2\varepsilon_0} - \dot{A}_k^{\perp},$$

where we have now eliminated the scalar potential. In the last equation we have introduced the transverse part of the vector potential

$$A_i^{\perp} = \Big(\delta_{ij} - \frac{k_ik_j}{k^2}\Big)A_j,$$

where Einstein's sum convention has been adopted on the right-hand side. One can easily show that also the magnetic-field contribution of the Lagrange function can be expressed in terms of $A^{\perp}$ only,

$$\Big(k\times A\Big)\cdot\Big(k\times A^*\Big) = k^2\Big(\delta_{ij} - \frac{k_ik_j}{k^2}\Big)A_iA_j^* = k^2|A^{\perp}|^2.$$

Thus, the electromagnetic field terms of the Lagrange function can be written in the form

$$\varepsilon_0 E_k\cdot E_k^* - \frac{1}{\mu_0}B_k\cdot B_k^* = \frac{|\rho_k|^2}{k^2\varepsilon_0} + \varepsilon_0|\dot{A}_k^{\perp}|^2 - \frac{k^2}{\mu_0}|A_k^{\perp}|^2.$$

The source-term contributions read

$$2\mathrm{Re}\Big\{J_k\cdot A_k^* - \rho_kV_k^*\Big\} = 2\mathrm{Re}\left\{J_k\cdot A_k^* - \rho_k\left(\frac{\rho_k^*}{k^2\varepsilon_0} - \frac{ik}{k^2}\cdot\dot{A}_k^*\right)\right\}.$$

To simplify this term, we note that we can always add a total time derivative to L_k,

$$L_k \longrightarrow L_k + \frac{dF}{dt} = L_k + \frac{d}{dt}\,2\mathrm{Re}\left\{\rho_k\Big(-\frac{i\boldsymbol{k}}{k^2}\cdot \boldsymbol{A}_k^*\Big)\right\}.$$

This term only gives contributions to the action at the initial and final times, which do not contribute to δS since the potential variations of δV, $\delta \boldsymbol{A}$ are set to zero there. Thus, the source-term contribution becomes

$$2\mathrm{Re}\left\{\Big(\boldsymbol{J}_k - \dot{\rho}_k\frac{i\boldsymbol{k}}{k^2}\Big)\cdot \boldsymbol{A}_k^* - \frac{|\rho_k|^2}{k^2\varepsilon_0}\right\} = 2\mathrm{Re}\left\{\Big(\boldsymbol{J}_k - \frac{\boldsymbol{k}(\boldsymbol{k}\cdot \boldsymbol{J}_k)}{k^2}\Big)\cdot \boldsymbol{A}_k^* - \frac{|\rho_k|^2}{k^2\varepsilon_0}\right\},$$

where we have used the continuity equation $\dot{\rho}_k + i\boldsymbol{k}\cdot \boldsymbol{J}_k = 0$ to arrive at the last expression. Putting together all results, we get our final expression for the Lagrange function

Lagrange Function for Maxwell's Theory (Coulomb Gauge)

$$L = \sum_{k\in K^+}\left[\varepsilon_0\Big(|\dot{\boldsymbol{A}}_k^{\perp}|^2 - \omega^2|\boldsymbol{A}_k^{\perp}|^2\Big) - \frac{|\rho_k|^2}{k^2\varepsilon_0} + 2\mathrm{Re}\{\boldsymbol{J}_k^{\perp}\cdot \boldsymbol{A}_k^{\perp *}\}\right], \tag{13.108}$$

with the usual relation $k^2 = \varepsilon_0\mu_0\,\omega^2$. The first term in brackets can be associated with the transverse electromagnetic fields solely, whereas the other two terms account for interactions between matter and fields. This function only depends on the transverse vector potential, and can thus be submitted to a canonical quantization procedure. We may finally undo the plane wave decomposition and return to a real-space representation. All terms can be easily converted back, with the exception of the current contribution which must be handled with some care.

Transformation of the Current Term. In the $\boldsymbol{J}_k^{\perp}\cdot \boldsymbol{A}_k^{\perp *}$ term we can obviously replace the transverse current by the total current

$$\boldsymbol{J}_k^{\perp}\cdot \boldsymbol{A}_k^{\perp *} = \Big(\boldsymbol{J}_k^{\perp} + \boldsymbol{J}_k^{L}\Big)\cdot \boldsymbol{A}_k^{\perp *} = \boldsymbol{J}_k\cdot \boldsymbol{A}_k^{\perp *},$$

because $\boldsymbol{A}_k^{\perp *}$ is a transverse function and we automatically have $\boldsymbol{J}_k^{L}\cdot \boldsymbol{A}_k^{\perp *} = 0$. When returning to a real-space representation we get

$$\sum_{k\in K^+} 2\mathrm{Re}\{\boldsymbol{J}_k^{\perp}\cdot \boldsymbol{A}_k^{\perp *}\} \longrightarrow \int \boldsymbol{J}(\boldsymbol{r})\cdot \boldsymbol{A}^{\perp}(\boldsymbol{r})\,d^3r - \int \boldsymbol{J}^{L}(\boldsymbol{r})\cdot \boldsymbol{A}^{\perp}(\boldsymbol{r})\,d^3r,$$

where we have replaced in the first term the transverse current $J^\perp$ by the total one, and have subtracted the longitudinal component to correct for this. Quite generally, the transversality operation is nonlocal in space and it is not guaranteed that $J^L(r) \cdot A^\perp(r)$ vanishes. Only when integrated over the entire space one can use the longitudinal delta function δ^L of Eq. (F.14) to shuffle the derivatives of $\bar{\bar{\delta}}^L \cdot J$ over to $A^\perp$, hereby neglecting the additional boundary terms which vanish for a localized current distribution. With this, the second integral with $J^L(r) \cdot A^\perp(r)$ indeed becomes zero.

We finally add the kinetic energy contribution of the particles forming the charge distribution to the transformed Lagrangian of Eq. (13.108), and arrive at the total Lagrange function of the coupled light-matter system

$$
\begin{aligned}
L &= \int \left(\frac{\varepsilon_0}{2} \left| \dot{A}^\perp(r) \right|^2 - \frac{1}{2\mu_0} \left| \nabla \times A^\perp(r) \right|^2 \right) d^3r \\
&\quad + \frac{1}{2} \sum_i m v_i^2 - W_{\text{coul}} + \int J(r) \cdot A^\perp(r)\, d^3r.
\end{aligned} \tag{13.109}
$$

We have suppressed the dependence of all quantities on time and have introduced the Coulomb coupling W_{coul} of Eq. (13.81). In the main text we work exclusively in the Coulomb gauge and no longer indicate the transverse nature of the vector potential.

Exercises

Exercise 13.1 Consider the vector potential for light fields confined to a cavity formed by two mirrors at positions 0 and L,

$$
A(r, t) = \sqrt{\frac{2}{\Omega \varepsilon_0}}\, Q(t) \sin kz\, \hat{x},
$$

with $z \in [0, L]$. Here $Q(t)$ is a measure of the amplitude, Ω is the volume of the cavity, $k = \pi/L$ is a wavenumber, and the prefactor has been chosen for later convenience.

(a) Compute the corresponding electric and magnetic fields.
(b) Compute the Lagrange function using Eq. (13.105) and show that it has the form of a harmonic oscillator.
(c) Use the correspondence with the harmonic oscillator to submit the system to a canonical quantization, see Eq. (13.34) and discussion thereafter. Express the Hamiltonian in terms of field operators $\hat{a}$, $\hat{a}^\dagger$, and show that the electric field operator can be written in the form

$$\hat{\boldsymbol{E}}(\boldsymbol{r}) = i\left(\frac{\hbar\omega}{\Omega\varepsilon_0}\right)^{\frac{1}{2}} \sin kz \left(\hat{a} - \hat{a}^\dagger\right)\hat{\boldsymbol{x}}.$$

Exercise 13.2 Same as Exercise 13.1 but for a cavity sustaining two mode functions, with the vector potential

$$\boldsymbol{A}(\boldsymbol{r}, t) = \sqrt{\frac{2}{\Omega\varepsilon_0}}\left[Q_1(t)\sin kz + Q_2(t)\cos kz\right]\hat{\boldsymbol{x}}.$$

(a) Repeat the quantization of the previous exercise with the two mode functions.
(b) Introduce in $\boldsymbol{A}(\boldsymbol{r}, t)$ basis functions of the form $e^{\pm ikz}$, and perform the canonical quantization along the lines of Eq. (13.42) and discussion thereafter. Show that the electric field operator can be written in the form

$$\hat{\boldsymbol{E}}(\boldsymbol{r}) = \sum_{q=\pm k}\left(\frac{\hbar\omega}{2\Omega\varepsilon_0}\right)^{\frac{1}{2}} e^{iqz}\, i\left(\hat{a}_q - \hat{a}_q^\dagger\right)\hat{\boldsymbol{x}}.$$

Exercise 13.3 Compute the commutators

$$\left[\hat{E}_i^\perp(\boldsymbol{r}), \hat{E}_j^\perp(\boldsymbol{r}')\right], \quad \left[\hat{E}_i^\perp(\boldsymbol{r}), \hat{B}_j(\boldsymbol{r}')\right], \quad \left[\hat{B}_i(\boldsymbol{r}), \hat{B}_j(\boldsymbol{r}')\right].$$

The calculation can be performed along the same lines as the derivation of Eq. (13.79)

Exercise 13.4 Compute the mean value $\langle\hat{\boldsymbol{E}}^\perp(\boldsymbol{r})\rangle$ and the fluctuations $\langle\hat{\boldsymbol{E}}^\perp(\boldsymbol{r}) \cdot \hat{\boldsymbol{E}}^\perp(\boldsymbol{r})\rangle$ of the electric field operators at zero and finite temperature.

Exercise 13.5 Consider scalar and vector potentials V, $\boldsymbol{A}$ that do *not* fulfill the Coulomb gauge $\nabla \cdot \boldsymbol{A} = 0$. From Eq. (2.19), find a function $\lambda(\boldsymbol{r}, t)$ such that the transformed potentials fulfill the Coulomb gauge.

Exercise 13.6 For the scalar field $\phi(x, t)$, consider the action

$$S = \int\left[i\hbar\phi^*\partial_t\phi - \frac{\hbar^2}{2m}(\partial_x\phi^*)(\partial_x\phi) + \phi^* V(x)\phi\right]dxdt,$$

where ∂_t, ∂_x denote the derivatives with respect to t and x. Treat ϕ, ϕ^* as independent variable.

(a) Derive the equations of motion from the action principle $\delta S/\delta\phi = 0$, $\delta S/\delta\phi^* = 0$. Ignore all surface terms.
(b) What are the canonically conjugate momenta for ϕ and ϕ^*?
(c) Write down the Hamilton function.

Exercise 13.7 Consider the Lagrange function of Eq. (13.66) for a many particle system in presence of external electromagnetic fields

$$L = \frac{1}{2}\sum_i m v_i^2 - \frac{1}{2}\sum_{i \neq j} \frac{q^2}{4\pi\varepsilon_0 |\boldsymbol{r}_i - \boldsymbol{r}_j|} + \sum_i q\Big[- V_{\text{ext}}(\boldsymbol{r}_i, t) + \boldsymbol{v}_i \cdot \boldsymbol{A}_{\text{ext}}(\boldsymbol{r}_i, t)\Big].$$

Derive the classical Hamilton function, and obtain $\dot{\boldsymbol{r}}_i$, $\dot{\boldsymbol{v}}_i$ using Hamilton's equations of motion given in Eq. (13.29).

Exercise 13.8 A coherent photon state can be expressed in the form

$$\big|\alpha\big\rangle = e^{-\frac{1}{2}|\alpha|^2} \sum_{n=0}^{\infty} \frac{\alpha^n}{\sqrt{n!}} \big|n\big\rangle,$$

with $|n\rangle$ being the number state of Eq. (13.39).

(a) Show that $|\alpha\rangle$ is an eigenstate of $\hat{a}$.
(b) Compute the overlap $\langle\alpha|\alpha\rangle$.
(c) Compute the expectation value $\langle\alpha|\dots|\alpha\rangle$ for the operators $\hat{a}$, $\hat{a}^\dagger\hat{a}$, $\hat{a}^\dagger\hat{a}^\dagger\hat{a}\hat{a}$.

Exercise 13.9 Compute the photon correlation function $G^{(1)} = \langle\hat{a}^\dagger(t)\hat{a}(0)\rangle$ for the system in (a) the groundstate, (b) a thermal state, (c) a coherent state.

Exercise 13.10 Expand the multipolar Hamiltonian of Eq. (13.96) up to second order in $\boldsymbol{r} - \boldsymbol{r}_0$. Express the result in terms of electric and magnetic dipole and quadrupole moments.

Chapter 14
Correlation Functions

In this chapter we investigate correlation functions of the form

$$\big\langle \hat{u}(t)\hat{v}(0)\big\rangle_{\text{eq}},$$

where $\hat{u}(t)$, $\hat{v}(0)$ are two operators that act at different times on the system, and the brackets denote an average for the equilibrium system at either zero or finite temperature. Although at first sight this function appears to be of rather limited use, it will turn out that—on the contrary—correlation functions are a very powerful device and are heavily used in a variety of problems. Consider the example shown in Fig. 14.1 where the operator $\hat{v}(0)$ excites the system at time zero, the excitation propagates in absence of external couplings, and finally the operator $\hat{u}(t)$ measures some system property at a later time. This setup is reminiscent of Green's functions, previously discussed in Chap. 5, and indeed correlation functions take over many of the appealing features of Green's function to the quantum domain. Most importantly, we will find that the *linear* response of the system can be computed from the knowledge of the correlation function alone, without any necessity to describe the perturbed system explicitly.

We start this chapter with a rather general discussion of correlation functions, and derive many useful expressions without making contact to any specific physical system. Once developed, the full machinery of correlation functions will turn out to be extremely useful and applicable to various problems in nano optics. We will show that the correlation function of the electromagnetic field operators is closely related to the dyadic Green's function, whereas the permittivity and optical conductivity can be expressed in terms of density–density and current–current correlation functions. We complement this chapter with a few applications in quantum plasmonics, including nonlocality and charge transfer plasmons, and derive a quantum mechanical description of electron energy loss spectroscopy using Fermi's golden rule.

U. Hohenester, *Nano and Quantum Optics*, Graduate Texts in Physics,
https://doi.org/10.1007/978-3-030-30504-8_14

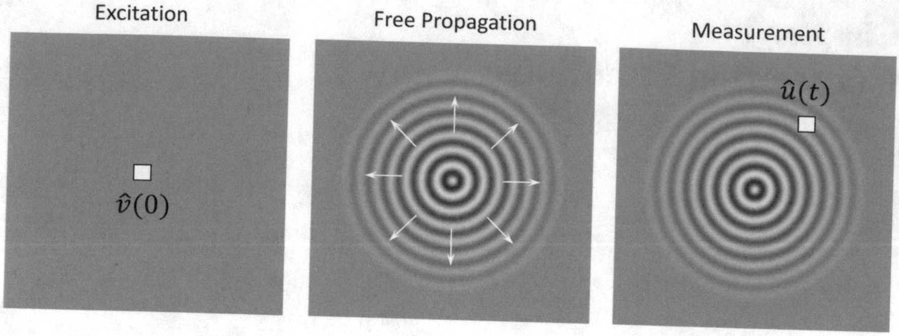

Fig. 14.1 Schematics of correlation function $\langle \hat{u}(t)\hat{v}(0)\rangle_{\text{eq}}$. An operator $\hat{v}(0)$ excites an equilibrium system at time zero. The perturbed system evolves freely in absence of any coupling to the outside world, and finally the expectation value of another operator $\hat{u}(t)$ is measured at a later time. Within Kubo's formalism, the linear response of a system to some external perturbation is computed from the correlation function alone, which can be evaluated in thermal equilibrium

14.1 Statistical Operator

In this chapter we compute wavefunction averages of operators and operator products. We consider a system described by the Hamiltonian $\hat{H}_0$ whose eigenenergies E_m and eigenstates $|m\rangle$ are defined through

$$\hat{H}_0\big|m\big\rangle = E_m\big|m\big\rangle. \tag{14.1}$$

The operator average can be performed in the usual manner at zero temperature

$$\big\langle \dots \big\rangle_{\text{eq}} = \big\langle 0\big| \dots \big|0\big\rangle, \tag{14.2}$$

where $|0\rangle$ denotes the system's groundstate and the dots in-between the brackets indicate the place for inserting the operators. We have also indicated that the expectation value is computed for the system's equilibrium, here its groundstate. At finite temperature we additionally have to sum over all states $|m\rangle$ which are populated with probability

$$p_m = Z^{-1}e^{-\beta E_m}, \quad Z = \sum_m e^{-\beta E_m}, \tag{14.3}$$

where Z is the partition function, which is needed for the normalization of the probability distribution, and $\beta = 1/(k_B T)$ is the inverse temperature with k_B being Boltzmann's constant. Instead of Eq. (14.2) we now have two kinds of averages, namely over wavefunctions and thermally populated states, and the operator expectation value becomes

$$\big\langle \dots \big\rangle_{\text{eq}} = Z^{-1} \sum_m e^{-\beta E_m} \big\langle m \big| \dots \big| m \big\rangle. \tag{14.4}$$

A convenient way to account for zero and finite temperature on the same footing is to introduce the equilibrium statistical operator

Equilibrium Statistical Operator

$$\hat{\rho}_{\text{eq}} = Z^{-1} \sum_m e^{-\beta E_m} \big| m \big\rangle \big\langle m \big|. \tag{14.5}$$

At zero temperature the statistical operator becomes

$$\hat{\rho}_{\text{eq}} = \big| 0 \big\rangle \big\langle 0 \big|.$$

The expectation value for operators can then be written in the compact form

$$\big\langle \dots \big\rangle_{\text{eq}} = \text{tr}\Big(\hat{\rho}_{\text{eq}} \dots \Big) = \sum_m \big\langle m \big| \hat{\rho}_{\text{eq}} \dots \big| m \big\rangle = Z^{-1} \sum_m e^{-\beta E_m} \big\langle m \big| \dots \big| m \big\rangle,$$

where we have used the definition of the trace as the sum over the expectation values for a complete basis set, for instance, the eigenstates of the Hamiltonian $\hat{H}_0$, such that we can exploit the orthogonality relation $\langle m|n\rangle = \delta_{mn}$ to arrive at the final expression on the right-hand side. Because the states $|m\rangle$ are eigenstates of the Hamiltonian, we can easily compute the statistical operator in the Heisenberg picture according to $\hat{H}_0$

$$e^{\frac{i}{\hbar}\hat{H}_0 t} \hat{\rho}_{\text{eq}} e^{-\frac{i}{\hbar}\hat{H}_0 t} = Z^{-1} \sum_m e^{-\beta E_m} \Big(e^{\frac{i}{\hbar} E_m t} \big| m \big\rangle \big\langle m \big| e^{-\frac{i}{\hbar} E_m t} \Big) = \hat{\rho}_{\text{eq}},$$

and find that the operator remains unchanged when going from the Schrödinger to the Heisenberg picture. The statistical operator in time-dependent systems will be investigated in more detail in Chap. 17.

Adiabatic Limit

Consider the correlation function for two operators $\hat{u}$ and $\hat{v}$

$$\big\langle \hat{u}(t)\hat{v}(t') \big\rangle_{\text{eq}} = Z^{-1} \sum_{m,n} e^{-\beta E_m} \big\langle m \big| \hat{u}(t) \big| n \big\rangle \big\langle n \big| \hat{v}(t') \big| m \big\rangle,$$

where we have inserted $\mathbb{1} = \sum_n |n\rangle\langle n|$ in-between the two operators. This representation in terms of a complete set of states is sometimes referred to as the **Lehmann representation**. Often it is hard or even impossible to actually compute the above expression, nevertheless one can obtain from it a number of useful relations. We first evaluate the time dependence of the matrix elements

$$\left\langle m\right|\left(e^{\frac{i}{\hbar}\hat{H}_0 t}\right)\hat{u}\left(e^{-\frac{i}{\hbar}\hat{H}_0 t}\right)\left|n\right\rangle = e^{-i\omega_{nm}t}\left\langle m\middle|\hat{u}\middle|n\right\rangle,$$

with the abbreviation $\hbar\omega_{nm} = E_n - E_m$. With this the correlation function can be worked out explicitly, and we arrive at

$$\left\langle \hat{u}(t)\hat{v}(t')\right\rangle_{\text{eq}} = Z^{-1}\sum_{m,n} e^{-\beta E_m}\left(e^{-i\omega_{nm}(t-t')}\right)\left\langle m\middle|\hat{u}\middle|n\right\rangle\!\left\langle n\middle|\hat{v}\middle|m\right\rangle.$$

Quite generally, when dealing with such expectation values for sufficiently large systems we can perform the **thermodynamic limit**, where we let the system size approach infinity. Or at least the size should be sufficiently large such that we can replace the summation over m and n by some integration over the system's degrees of freedom. Then, the correlation function decays fast as a function of time difference $t - t'$ because of destructive interference between the exponential terms oscillating with different frequencies.

Figure 14.2 shows a few representative examples. Panels (a,b) report Lorentzian lineshapes for the spectral function with different widths. The response in the time

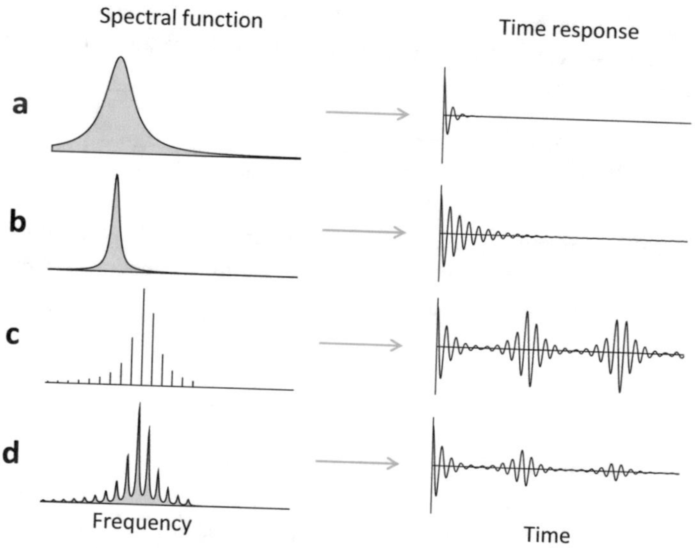

Fig. 14.2 Examples of spectral functions and of the corresponding time response. (**a**,**b**) Lorentzian lineshapes with different widths. (**c**) Series of delta peaks. (**d**) Series of broadened delta peaks

domain is an oscillation with an exponential damping. The larger the width of the Lorentzian, the faster the function decays in time. Similar behavior is found for other lineshapes, such as for instance Gaussians. Panel (c) shows a series of delta peaks, corresponding to the density of states of an isolated finite-size system. The time response exhibits a periodically returning pattern with no decay at all. In contrast, when a small broadening is added to the lineshape function, see panel (d), the temporal response function decays (although there are still a few recurrences observable). The last example refers to finite-size systems which are, however, sufficiently large such that even small line broadenings attributed to environment couplings suffice to lead to destructive interference and an overall time decay. For this reason, the thermodynamic limit can be often also safely applied to finite-size systems.

In many cases we would like to manipulate correlation functions prior to performing the thermodynamic limit. We then have to be careful about large time differences $t - t' \to \infty$ where the correlation function of the finite system continues to oscillate, in contrast to the correlation function of the system in the thermodynamic limit which has decayed long before because of destructive interference. In such cases one performs the **adiabatic limit** and adds a small damping term η to the correlation function,

$$\big\langle \hat{u}(t)\hat{v}(t')\big\rangle_{\text{eq}} = \lim_{\eta\to 0} Z^{-1}\sum_{m,n} e^{-\beta E_m}\left(e^{-i\omega_{nm}(t-t')}e^{-\eta(t-t')}\right)\big\langle m\big|\hat{u}\big|n\big\rangle\big\langle n\big|\hat{v}\big|m\big\rangle, \tag{14.6}$$

where we let η approach zero at the end of the calculation. With this additional damping term we can safely use correlation functions together with time integrals where the upper integration limit approaches infinity, as we will do further below. To make this point clear: the η contribution has nothing to do with any kind of physical damping, but is solely needed to use correlation functions in infinite time integrals *prior* to performing the thermodynamic limit.

14.2 Kubo Formalism

The Kubo formalism is a powerful device that allows computing the properties of a perturbed system in linear response using the properties of the unperturbed system only. See also Fig. 14.3. Suppose that the system's dynamics can be described in terms of a Hamiltonian $\hat{H}_0$ and a coupling to an external classical field $X(t)$ in the form

$$\hat{V}(t) = \hat{v}\, X(t). \tag{14.7}$$

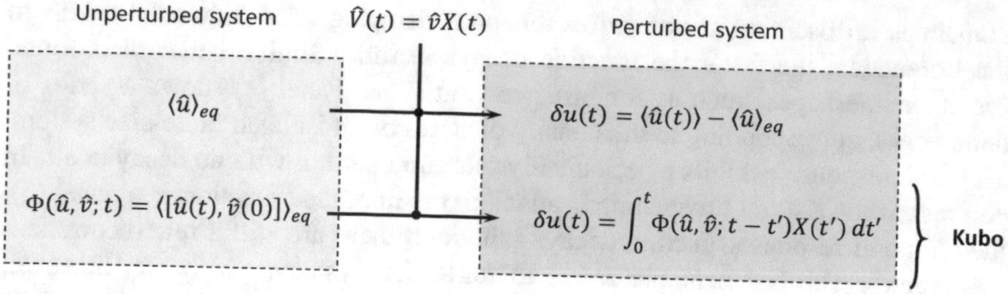

Fig. 14.3 Schematics of Kubo formalism. The change of the expectation value of an operator $\hat{u}$ in presence of a time-dependent Hamiltonian $\hat{V}(t) = \hat{v}X(t)$, where $\hat{v}$ is a system operator and $X(t)$ some external classical field, can be either computed from the time-dependent wavefunctions. Alternatively, in linear response one first computes the correlation function $\Phi(\hat{u}, \hat{v})$ for the *unperturbed* system, and then uses the Kubo formula of Eq. (14.8) to compute the expectation value from this correlation function and the external field $X(t)$ without explicitly accounting for the perturbed system

Here $\hat{v}$ is an operator accounting for the coupling to the external field. An example is the electric dipole interaction of Eq. (13.91). In the following we work in an interaction representation according to $\hat{H}_0$. In lowest order perturbation theory, the time evolution of some operator $\hat{u}$ can be computed from Eq. (13.15),

$$\hat{U}^\dagger(t,0)\,\hat{u}(t)\,\hat{U}(t,0) = \hat{u}(t) - \frac{i}{\hbar}\int_0^t \left[\hat{u}(t), \hat{v}(t')\right] X(t')\,dt' + \mathcal{O}\left(X^2\right),$$

where $\hat{U}$ is the time evolution operator of Eq. (13.12) in the interaction representation. We next introduce the abbreviation

$$\delta u(t) = \left\langle \hat{U}^\dagger(t,0)\,\hat{u}(t)\,\hat{U}(t,0) - \hat{u}(t)\right\rangle_{\text{eq}}.$$

With this we arrive at the Kubo formula for the expectation value of an operator $\hat{u}$

$$\delta u(t) = -\frac{i}{\hbar}\int_0^t \left\langle\left[\hat{u}(t), \hat{v}(t')\right]\right\rangle_{\text{eq}} X(t')\,dt'. \tag{14.8}$$

This expression allows computing the properties of the perturbed system in linear response using the correlation function of the unperturbed system only, which is highly beneficial for actual calculations. One usually introduces the correlation function

$$\Phi(\hat{u}, \hat{v}; t - t') = -\frac{i}{\hbar}\theta(t - t')\left\langle\left[\hat{u}(t), \hat{v}(t')\right]\right\rangle_{\text{eq}}, \tag{14.9}$$

which in thermal equilibrium depends on the time difference $t - t'$ only. θ is Heaviside's step function that accounts for the causal response. The Fourier transform of Kubo's formula becomes

$$\delta u(\omega) = \int_0^\infty e^{i\omega t}\left[\int_0^t \Phi(\hat{u}, \hat{v}; t - t')X(t')dt'\right]dt = \Phi(\hat{u}, \hat{v}; \omega)X(\omega),$$

where we have used that a convolution in time translates to a product in frequency space. We have introduced the Fourier transform of the correlation function

Correlation Function for Kubo Formalism

$$\Phi(\hat{u}, \hat{v}; \omega) = -\frac{i}{\hbar}\int_0^\infty e^{i\omega t}\Big\langle\Big[\hat{u}(t), \hat{v}(0)\Big]\Big\rangle_{\text{eq}} dt. \tag{14.10}$$

For notational clarity we keep the same symbols in the time and frequency domain, but are usually interested in the frequency representation only.

14.2.1 *Spectral Function*

It is often convenient to introduce the spectral function

Spectral Function

$$\rho(\hat{u}, \hat{v}; \omega) = \frac{1}{\hbar}\int_{-\infty}^\infty e^{i\omega t}\Big\langle\Big[\hat{u}(t), \hat{v}(0)\Big]\Big\rangle_{\text{eq}} dt, \tag{14.11}$$

which differs from the correlation function of Eq. (14.10) in the unlimited time integration, apart from a missing prefactor of $-i$. As we will show below, this function is often easier to compute and there exists a simple prescription of how to obtain from the spectral function the various correlation functions typically used within linear response theory. The spectral function in the time domain can be obtained through an inverse Fourier transform

$$\rho(\hat{u}, \hat{v}; t) = \int_{-\infty}^\infty e^{-i\omega' t}\rho(\hat{u}, \hat{v}; \omega')\,\frac{d\omega'}{2\pi},$$

where we have used ω' for a reason to become clear in a moment. Inserting this expression into Eq. (14.10) for the correlation function of Kubo's formalism gives

$$\Phi(\hat{u}, \hat{v}; \omega) = -i \lim_{\eta \to 0} \int_{-\infty}^{\infty} \left[\int_0^{\infty} e^{i(\omega - \omega' + i\eta)t} \, dt \right] \rho(\hat{u}, \hat{v}; \omega') \frac{d\omega'}{2\pi},$$

where we have added the small damping constant η in the spirit of the previously discussed adiabatic limit and to ensure that the integrand becomes zero at large time delays (recall that this is certainly the case for sufficiently large systems, because of destructive interference, but with the η term we are always on the safe side). Performing the time integral then leads us to the relation between the spectral function and the correlation function

$$\Phi(\hat{u}, \hat{v}; \omega) = \lim_{\eta \to 0} \int_{-\infty}^{\infty} \frac{\rho(\hat{u}, \hat{v}; \omega')}{\omega - \omega' + i\eta} \frac{d\omega'}{2\pi}, \tag{14.12}$$

where the integrand at time $t \to \infty$ has been neglected because of the η damping.

Symmetry Relations and Kramers-Kronig Relation. In thermal equilibrium the spectral function depends on the time difference between the operators $\hat{u}$ and $\hat{v}$ only, thus one finds from Eq. (14.11)

$$\rho(\hat{u}, \hat{v}; \omega) = \frac{1}{\hbar} \int_{-\infty}^{\infty} e^{i\omega t} \Big\langle \Big[\hat{u}(t), \hat{v}(0) \Big] \Big\rangle_{\text{eq}} dt = \frac{1}{\hbar} \int_{-\infty}^{\infty} e^{-i\omega t} \Big\langle \Big[\hat{u}(0), \hat{v}(t) \Big] \Big\rangle_{\text{eq}} dt. \tag{14.13}$$

This allows us to derive the following symmetry relations:

$$\rho(\hat{u}, \hat{v}; \omega) = -\rho(\hat{v}, \hat{u}; -\omega) = \rho^*(\hat{v}, \hat{u}; \omega), \tag{14.14}$$

where the last expression has been obtained by taking the complex conjugate of ρ and exploiting that $\hat{u}$, $\hat{v}$ are Hermitian operators.

In systems with time reversal symmetry one often has the situation that the correlation function is symmetric with respect to the exchange of operators $\hat{u}$ and $\hat{v}$ [43],

$$\Phi_{\text{sym}}(\hat{u}, \hat{v}; \omega) = \Phi_{\text{sym}}(\hat{v}, \hat{u}; \omega). \tag{14.15}$$

From Eq. (14.12) we then find

$$\begin{aligned} \Phi_{\text{sym}}(\hat{u}, \hat{v}; \omega) &= \frac{1}{2} \Big(\Phi(\hat{u}, \hat{v}; \omega) + \Phi(\hat{v}, \hat{u}; \omega) \Big) \\ &= \lim_{\eta \to 0} \frac{1}{2} \int_{-\infty}^{\infty} \frac{\rho(\hat{u}, \hat{v}; \omega') + \rho(\hat{v}, \hat{u}; \omega')}{\omega - \omega' + i\eta} \frac{d\omega'}{2\pi}. \end{aligned}$$

Using the symmetry relations for the spectral function it is easy to show that the symmetrized spectral function

$$\rho_{\text{sym}}(\hat{u}, \hat{v}; \omega) = \frac{1}{2}\Big(\rho(\hat{u}, \hat{v}; \omega) + \rho(\hat{v}, \hat{u}; \omega)\Big) = \rho^*_{\text{sym}}(\hat{u}, \hat{v}; \omega)$$

is entirely real. Inserting this function into Eq. (14.12) and taking on both sides of the equation the imaginary part leads us to

$$\Phi''_{\text{sym}}(\hat{u}, \hat{v}; \omega) = -\frac{1}{2}\rho_{\text{sym}}(\hat{u}, \hat{v}; \omega), \tag{14.16}$$

where we have used Eq. (F.6) to rewrite the frequency denominator in terms of a delta function and a principal value integral. Similarly, by taking on both sides of the equation the real part gives

$$\Phi'_{\text{sym}}(\hat{u}, \hat{v}; \omega) = -\mathcal{P}\int_{-\infty}^{\infty} \frac{\Phi''_{\text{sym}}(\hat{u}, \hat{v}; \omega')}{\omega - \omega'}\,\frac{d\omega'}{\pi}, \tag{14.17}$$

where $\mathcal{P}$ denotes the principal value integral. This shows that the real and imaginary parts of the symmetrized correlation function are related through an expression reminiscent of the Kramers–Kronig relations derived in Chap. 7. This is not accidental since our previous derivation has been based on the general assumptions of a linear and causal response. These assumptions obviously also apply to other types of correlation functions, for which corresponding Kramers–Kronig-like relations hold.

14.2.2 *Cross-Spectral Density*

Suppose that we would like to compute the correlation function for the operator product (rather than for the commutator)

$$\big\langle \hat{u}(t)\hat{v}(0)\big\rangle_{\text{eq}}.$$

As we will discuss in the following, in thermal equilibrium such functions can be related to the spectral function $\rho(\hat{u}, \hat{v}; t)$. As a preliminary step, let us consider the Fourier transform of the operator

$$\hat{u}(\omega) = \int_{-\infty}^{\infty} e^{i\omega t}\,\hat{u}(t)\,dt.$$

We then find

$$\big\langle \hat{u}(\omega)\hat{v}(\omega')\big\rangle_{\text{eq}} = \int_{-\infty}^{\infty} e^{i(\omega t + \omega' t')}\big\langle \hat{u}(t)\hat{v}(t')\big\rangle_{\text{eq}}\,dt\,dt'.$$

With the time coordinates $T = \frac{1}{2}(t + t')$ and $\tau = t - t'$ the above integral can be rewritten as

$$\langle \hat{u}(\omega)\hat{v}(\omega')\rangle_{\text{eq}} = \int_{-\infty}^{\infty} e^{i(\omega+\omega')T} e^{\frac{i}{2}(\omega-\omega')\tau} \Big\langle \hat{u}\big(T + \frac{1}{2}\tau\big)\hat{v}\big(T - \frac{1}{2}\tau\big)\Big\rangle_{\text{eq}} dT\, d\tau.$$

From this we are led to the following expression for the cross-spectral density

Cross-Spectral Density

$$\langle \hat{u}(\omega)\hat{v}(\omega')\rangle_{\text{eq}} = 2\pi\,\delta(\omega + \omega') \int_{-\infty}^{\infty} e^{i\omega t} \langle \hat{u}(t)\hat{v}(0)\rangle_{\text{eq}}\, dt. \tag{14.18}$$

To evaluate the integral on the right-hand side, we express the expectation value in thermal equilibrium $\langle \dots \rangle_{\text{eq}}$ using a complete basis to arrive at

$$\int_{-\infty}^{\infty} e^{i\omega t} \langle \hat{u}(t)\hat{v}(0)\rangle_{\text{eq}}\, dt = \int_{-\infty}^{\infty} Z^{-1} \sum_{m,n} e^{-\beta E_m} e^{i(\omega-\omega_{nm})t} \langle m|\hat{u}|n\rangle\langle n|\hat{v}|m\rangle\, dt.$$

Working out the time integration then leads us to

$$\int_{-\infty}^{\infty} e^{i\omega t} \langle \hat{u}(t)\hat{v}(0)\rangle_{\text{eq}}\, dt = Z^{-1} \sum_{m,n} e^{-\beta E_m} \langle m|\hat{u}|n\rangle\langle n|\hat{v}|m\rangle\, 2\pi\,\delta(\omega - \omega_{nm})$$

$$\int_{-\infty}^{\infty} e^{i\omega t} \langle \hat{v}(0)\hat{u}(t)\rangle_{\text{eq}}\, dt = Z^{-1} \sum_{m,n} e^{-\beta E_n} \langle n|\hat{v}|m\rangle\langle m|\hat{u}|n\rangle\, 2\pi\,\delta(\omega - \omega_{nm}),$$

where the expression in the second line has been obtained by reversing the order of operators and interchanging m with n. We can subtract the two expressions to get for the spectral function of Eq. (14.11)

$$\rho(\hat{u}, \hat{v}; \omega) = \frac{1}{\hbar Z} \sum_{m,n} e^{-\beta E_m} \langle m|\hat{u}|n\rangle\langle n|\hat{v}|m\rangle \Big(1 - e^{-\beta\hbar\omega}\Big)\, 2\pi\,\delta(\omega - \omega_{nm}). \tag{14.19}$$

We have used $E_n = E_m + \hbar\omega$ to pull out a common exponential for the thermal occupation of states. From this we observe that the cross-spectral density can be related to the spectral function through

$$\int_{-\infty}^{\infty} e^{i\omega t} \langle \hat{u}(t)\hat{v}(0)\rangle_{\text{eq}}\, dt = \frac{\hbar\rho(\hat{u}, \hat{v}; \omega)}{1 - e^{-\beta\hbar\omega}} = \hbar\rho(\hat{u}, \hat{v}; \omega)\Big(\bar{n}_{\text{th}}(\hbar\omega) + 1\Big),$$

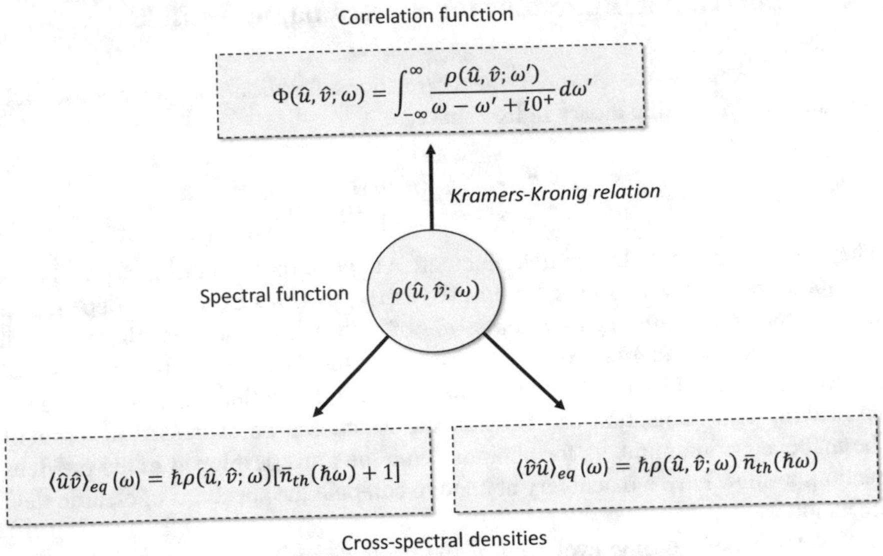

Fig. 14.4 Relation of spectral function to correlation function and cross-spectral densities. The spectral function of Eq. (14.11) is a measure of the fluctuations in the equilibrium system. It is related through the Kramers–Kronig-like expression of Eq. (14.12) to the correlation function, which allows computing within Kubo's formalism the linear response of the system using the fluctuations of the unperturbed system only. Often one is interested in the cross-spectral densities of Eq. (14.20), which are the expectation values of the operator products $\hat{u}\hat{v}$ and $\hat{v}\hat{u}$. In thermal equilibrium these densities can be expressed in terms of the spectral function and the Bose–Einstein distribution $\bar{n}_{\text{th}}$. Thus, all necessary correlation functions can be obtained from the knowledge of a single quantity, namely the spectral function $\rho(\hat{u},\,\hat{v};\,\omega)$

with the thermal Bose–Einstein distribution $\bar{n}_{\text{th}}(\hbar\omega)=1/(e^{\beta\hbar\omega}-1)$. We then arrive at our final expressions for the relations between the cross-spectral densities and the spectral function

Relation Between Cross-Spectral Density and Spectral Function

$$\int_{-\infty}^{\infty} e^{i\omega t}\big\langle\hat{u}(t)\hat{v}(0)\big\rangle_{\text{eq}}\,dt=\hbar\rho(\hat{u},\,\hat{v};\,\omega)\Big(\bar{n}_{\text{th}}(\hbar\omega)+1\Big)$$
$$\int_{-\infty}^{\infty} e^{i\omega t}\big\langle\hat{v}(0)\hat{u}(t)\big\rangle_{\text{eq}}\,dt=\hbar\rho(\hat{u},\,\hat{v};\,\omega)\,\bar{n}_{\text{th}}(\hbar\omega). \tag{14.20}$$

These relations are highly useful because they show that all possible correlation functions can be related to a single quantity, namely the spectral function. See also Fig. 14.4. In the following sections we give some representative examples.

14.3 Correlation Functions for Electromagnetic Fields

As a first example we consider the correlations function for the electric field operators of Maxwell's theory in free space,

$$\left\langle \hat{E}_i(\boldsymbol{r},t)\hat{E}_j(\boldsymbol{r}',t')\right\rangle_{\text{eq}}\,.$$

They describe how an electric field fluctuation at position $\boldsymbol{r}'$ and time t' propagates to another position $\boldsymbol{r}$ at a later time t. We expect that this expression depends on the one hand on the *propagation properties* of the electric fields, which, as we will show in a moment, can be expressed in terms of the Green's function of classical electrodynamics. On the other hand, the correlation function reflects how easily an electric field fluctuation can be induced, which can be expressed in terms of thermal occupation numbers for photons. With the tools developed in the previous section, we now have a machinery at hand to compute the spectrum of electric field fluctuations.

We start from the time evolution of the photon operators, using an interaction representation with respect to the photon Hamiltonian of Eq. (13.76),

$$i\hbar\frac{d}{dt}\hat{a}_{\boldsymbol{k}s}=\left[\hat{a}_{\boldsymbol{k}s},\hat{H}_{\text{em}}\right]=\omega\hat{a}_{\boldsymbol{k}s}\implies\hat{a}_{\boldsymbol{k}s}(t)=e^{-i\omega t}\hat{a}_{\boldsymbol{k}s}\,,$$

with the photon frequency $\omega=kc$. The vector potential operator of Eq. (13.77) can then be written in the form

$$\hat{\mathbf{A}}(\boldsymbol{r},t)=\sum_{\boldsymbol{k},s}\left(\frac{\hbar}{2\Omega\varepsilon_0\omega}\right)^{\frac{1}{2}}\left[e^{i(\boldsymbol{k}\cdot\boldsymbol{r}-\omega t)}\,\hat{a}_{\boldsymbol{k}s}+e^{-i(\boldsymbol{k}\cdot\boldsymbol{r}-\omega t)}\,\hat{a}^{\dagger}_{\boldsymbol{k}s}\right]\boldsymbol{\epsilon}_{\boldsymbol{k}s}\,.$$

It turns out to be convenient to introduce $\hat{\mathbf{A}}^{+}(\boldsymbol{r},t)$ for that part of $\hat{\mathbf{A}}(\boldsymbol{r},t)$ which contains the annihilation operators $\hat{a}_{\boldsymbol{k}s}$,

$$\hat{\mathbf{A}}^{+}(\boldsymbol{r},t)=\sum_{\boldsymbol{k},s}\left(\frac{\hbar}{2\Omega\varepsilon_0\omega}\right)^{\frac{1}{2}}e^{i(\boldsymbol{k}\cdot\boldsymbol{r}-i\omega)t}\,\hat{a}_{\boldsymbol{k}s}\,. \tag{14.21}$$

The plus symbol is a reminder that the field operator oscillates with positive frequencies according to $e^{-i\omega t}$; recall that this is the usual time dependence adopted in the physics literature. Similarly, $\hat{\mathbf{A}}^{-}(\boldsymbol{r},t)$ accounts for the contributions oscillating with negative frequencies $e^{-i(-\omega)t}=e^{i\omega t}$. We can thus split the vector potential operator into positive and negative frequency components

Positive and Negative Frequency Components

$$\hat{\boldsymbol{A}}(\boldsymbol{r},t) = \hat{\boldsymbol{A}}^{+}(\boldsymbol{r},t) + \hat{\boldsymbol{A}}^{-}(\boldsymbol{r},t). \tag{14.22}$$

Because $\hat{\boldsymbol{A}}^{-}$ is the Hermitian conjugate of $\hat{\boldsymbol{A}}^{+}$, the sum in Eq. (14.22) gives a real-valued operator. A corresponding splitting into positive and negative frequency components can be also performed for the electromagnetic field operators.

Field Commutators at Different Times. We next compute the commutator of the vector potential operators at different times t, t',

$$\left[\hat{A}_i(\boldsymbol{r},t), \hat{A}_j(\boldsymbol{r}',t')\right] = \left[\hat{A}_i^{+}(\boldsymbol{r},t), \hat{A}_j^{-}(\boldsymbol{r}',t')\right] + \left[\hat{A}_i^{-}(\boldsymbol{r},t), \hat{A}_j^{+}(\boldsymbol{r}',t')\right].$$

Here we have used that two annihilation or creation operators commute at all times. With the fundamental commutation relation $[\hat{a}_{\boldsymbol{k}s}, \hat{a}^{\dagger}_{\boldsymbol{k}'s'}] = \delta_{\boldsymbol{k}\boldsymbol{k}'}\delta_{ss'}$ we then get

$$\mathcal{I}_1 = \left[\hat{A}_i^{+}(\boldsymbol{r},t), \hat{A}_j^{-}(\boldsymbol{r}',t')\right] = \sum_{\boldsymbol{k},s} \frac{\hbar}{2\Omega\varepsilon_0\omega_k} e^{i\boldsymbol{k}\cdot(\boldsymbol{r}-\boldsymbol{r}')} e^{-i\omega_k(t-t')} \left(\epsilon_{+\boldsymbol{k}s}\right)_i \left(\epsilon_{+\boldsymbol{k}s}\right)_j$$

$$\mathcal{I}_2 = \left[\hat{A}_i^{-}(\boldsymbol{r},t), \hat{A}_j^{+}(\boldsymbol{r}',t')\right] = -\sum_{\boldsymbol{k},s} \frac{\hbar}{2\Omega\varepsilon_0\omega_k} e^{i\boldsymbol{k}\cdot(\boldsymbol{r}-\boldsymbol{r}')} e^{+i\omega_k(t-t')} \left(\epsilon_{-\boldsymbol{k}s}\right)_i \left(\epsilon_{-\boldsymbol{k}s}\right)_j.$$

Because $\boldsymbol{\epsilon}_{\boldsymbol{k}s}$ form together with $\boldsymbol{k}$ a basis, we readily observe that

$$\sum_s \left(\epsilon_{\boldsymbol{k}s}\right)_i \left(\epsilon_{\boldsymbol{k}s}\right)_j = \left(\delta_{ij} - \hat{k}_i\hat{k}_j\right)$$

is a projector on the transverse directions of $\boldsymbol{k}$. Performing the thermodynamic limit with $\sum_{\boldsymbol{k}} \to \Omega/(2\pi)^3 \int d^3k$, we are then led to

$$\mathcal{I}_{1,2} = \pm\frac{\hbar}{2\varepsilon_0 c}\int_{-\infty}^{\infty} \frac{1}{k}\left(\delta_{ij} - \hat{k}_i\hat{k}_j\right) e^{i\boldsymbol{k}\cdot(\boldsymbol{r}-\boldsymbol{r}')} e^{\mp ikc(t-t')} \frac{d^3k}{(2\pi)^3}.$$

We next introduce $\boldsymbol{R} = \boldsymbol{r} - \boldsymbol{r}'$ and $\tau = t - t'$ for the position and time differences, respectively, as well as spherical coordinates to rewrite the integral in the form

$$\mathcal{I}_{1,2} = \pm\frac{\hbar}{2\varepsilon_0 c}\int_0^{\infty} \frac{1}{k} e^{\mp ikc\tau} \left(\delta_{ij} + \frac{\partial_i\partial_j}{k^2}\right) \left[\oint e^{i\boldsymbol{k}\cdot\boldsymbol{R}} \frac{d\Omega}{(2\pi)^3}\right] k^2 dk.$$

The integral in brackets can be evaluated according to

$$\oint e^{i\boldsymbol{k}\cdot\boldsymbol{R}} \frac{d\Omega}{(2\pi)^3} = \frac{1}{4\pi^2}\int_{-1}^{1} e^{ikRu}\,du = \frac{1}{4i\pi^2 kR}\left[e^{ikR} - e^{-ikR}\right] = \frac{2}{\pi k}\mathrm{Im}[g(R)],$$

where we have introduced $u = \cos\theta$ for the polar angle integration together with the scalar Green's function $g(R) = e^{ikR}/(4\pi R)$ of Eq. (5.7). The commutation relations can thus be expressed with $\omega = ck$ as

$$\mathcal{I}_{1,2} = \pm 2\hbar\mu_0 \int_0^\infty e^{\mp i\omega\tau}\left(\delta_{ij} + \frac{\partial_i\partial_j}{k^2}\right)\text{Im}\big[g(R)\big]\,\frac{d\omega}{2\pi}. \tag{14.23}$$

Putting together all results, we are led to the commutation relations for the vector potential operators

$$\left[\hat{A}_i^\pm(\boldsymbol{r},t), \hat{A}_j^\mp(\boldsymbol{r}',t')\right] = \pm 2\hbar\mu_0 \int_0^\infty e^{\mp i\omega(t-t')}\,\text{Im}\big[\bar{\bar{G}}(\boldsymbol{r},\boldsymbol{r}')\big]_{ij}\,\frac{d\omega}{2\pi}, \tag{14.24}$$

with the dyadic Green's function of Eq. (5.19). The commutators with the electromagnetic field operators can be obtained by employing the usual relations

$$\hat{\boldsymbol{E}}^\pm(\boldsymbol{r},t) = -\frac{\partial}{\partial t}\hat{\boldsymbol{A}}^\pm(\boldsymbol{r},t), \quad \hat{\boldsymbol{B}}^\pm(\boldsymbol{r},t) = \nabla\times\hat{\boldsymbol{A}}^\pm(\boldsymbol{r},t)$$

between the vector potential operator and $\hat{\boldsymbol{E}}^\pm$, $\hat{\boldsymbol{B}}^\pm$. We here only state the results with the electric field operators

$$\left[\hat{E}_i^\pm(\boldsymbol{r},t), \hat{A}_j^\mp(\boldsymbol{r}',t')\right] = 2i\hbar\mu_0 \int_0^\infty e^{\mp i\omega(t-t')}\omega\,\text{Im}\big[\bar{\bar{G}}(\boldsymbol{r},\boldsymbol{r}',\omega)\big]_{ij}\,\frac{d\omega}{2\pi}$$
$$\left[\hat{E}_i^\pm(\boldsymbol{r},t), \hat{E}_j^\mp(\boldsymbol{r}',t')\right] = \pm 2\hbar\mu_0 \int_0^\infty e^{\mp i\omega(t-t')}\,\omega^2\text{Im}\big[\bar{\bar{G}}(\boldsymbol{r},\boldsymbol{r}')\big]_{ij}\,\frac{d\omega}{2\pi}, \tag{14.25}$$

and leave the relations with the magnetic field operators as an exercise. With the commutation relations of Eq. (14.25) we can readily compute the spectral function of Eq. (14.11) for two electric field operators,

$$\rho_{ij}^\pm(\boldsymbol{r},\boldsymbol{r}',\omega) = \frac{1}{\hbar}\int_{-\infty}^\infty e^{i\omega t}\left\langle\left[\hat{E}_i^\pm(\boldsymbol{r},t), \hat{E}_j^\mp(\boldsymbol{r}',0)\right]\right\rangle_{\text{eq}} dt. \tag{14.26}$$

We first note that the commutator gives a c-number, so we do not have to evaluate the average in thermal equilibrium explicitly. From Eq. (14.25) we get

$$\rho_{ij}^\pm(\boldsymbol{r},\boldsymbol{r}',\omega) = \pm 2\mu_0 \int_{-\infty}^\infty e^{i\omega\tau}\int_0^\infty e^{\mp i\omega'\tau}\omega'^2\text{Im}\left[\bar{\bar{G}}(\boldsymbol{r},\boldsymbol{r}',\omega')\right]_{ij}\frac{d\omega'}{2\pi}d\tau,$$

Working out the time integration, we express the spectral function for the electric field operators in the form

Spectral Function for Electric Field Operators

$$\rho^{\pm}_{ij}(\boldsymbol{r},\boldsymbol{r}',\omega) = \pm 2\mu_0\omega^2\,\mathrm{Im}\left[\bar{\bar{G}}(\boldsymbol{r},\boldsymbol{r}',\pm\omega)\right]_{ij}\theta(\pm\omega). \tag{14.27}$$

From the definition of the spectral function one can also verify that the following relation holds:

$$\rho^{-}_{ij}(\boldsymbol{r},\boldsymbol{r}',-\omega) = -\rho^{+}_{ij}(\boldsymbol{r},\boldsymbol{r}',\omega). \tag{14.28}$$

As shown in Fig. 14.5, from the spectral function of electric field operators we can compute the correlation function and the cross-spectral densities. Equally important, all of the electric field fluctuations can be expressed in terms of the dyadic

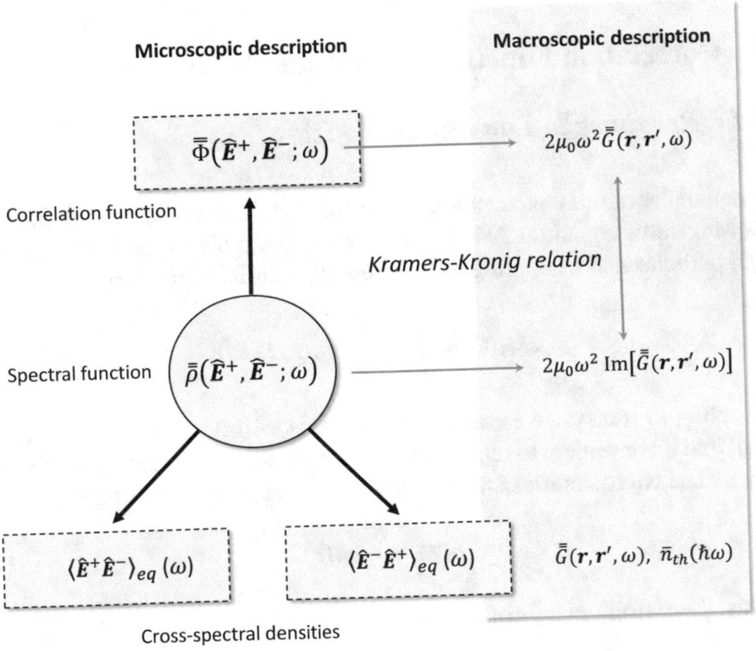

Fig. 14.5 Relation between spectral function $\bar{\bar{\rho}}$ of electric field operators in free space, and correlation function and cross-spectral densities. The spectral function is a measure of the field fluctuations in thermal equilibrium, which is related to the imaginary part of the dyadic Green's function of *classical* electrodynamics. The correlation function can be related through a Kramers–Kronig relation to the spectral function. The cross-spectral densities of electric field operators can be expressed in terms of the dyadic Green's function and of thermal occupation numbers $\bar{n}_{\mathrm{th}}$. As will be shown in Chap. 15, all of the above relations also hold for a non-trivial photonic environment formed by dielectric or metallic nanostructures

Green's function of classical electrodynamics together with the thermal occupation factors $\bar{n}_{\text{th}}(\hbar\omega)$. This provides us with a direct link between the microscopic and macroscopic field properties, expressed in terms of field fluctuations and the Green's function of classical Maxwell's theory.

Was it just for the field operators in free space, the above results would be interesting but not overly useful. Indeed, in quantum optics one usually survives without spectral and correlation functions, but keeps the electric field operator as the basic quantity. Things change considerably for a non-trivial photonic environment, which is the object of central interest in nano optics. As will be shown in Chap. 15, for such environments one can derive relations similar to those shown in Fig. 14.5, with the Green's function replaced by the total Green's function. The derivation is somewhat intricate and based on one important assumption, namely a *linear* optical response. We will show that the spectral function for the non-trivial photonic environment, as well as the corresponding correlation functions and cross-spectral densities play a central role in fluctuational electrodynamics, and will allow us to quantize Maxwell's equations in presence of absorbing media.

14.4 Correlation Functions for Coulomb Systems

14.4.1 Response to Longitudinal Fields

In the remainder of this chapter we focus on the correlation functions for the material part of Maxwell's equations. We start by applying Kubo's formalism to a system of charged particles and a coupling to an external potential of the form

$$\hat{H}_{\text{int}}(t) = \int \hat{\rho}(\boldsymbol{r})V_{\text{ext}}(\boldsymbol{r},t)\,d^3r.$$

The coupling to transverse electromagnetic fields will be presented further below. We shall find it convenient to replace the charge density of electrons $\hat{\rho}$ by the particle density $\hat{n}$. The two quantities are related via (e is the elementary charge)

$$\hat{\rho}(\boldsymbol{r}) = -e\hat{n}(\boldsymbol{r}). \tag{14.29}$$

With this, the Hamiltonian is of the form

$$\hat{H}_{\text{int}} = \int \hat{n}(\boldsymbol{r})\Big[-eV_{\text{ext}}(\boldsymbol{r},t)\Big]\,d^3r = \int \hat{n}(\boldsymbol{r})\,U_{\text{ext}}(\boldsymbol{r},t)\,d^3r, \tag{14.30}$$

where in the last expression we have introduced $U_{\text{ext}} = -eV_{\text{ext}}$ which has the dimension of an energy. Upon Fourier transformation we get for the induced particle density from Kubo's formula

$$\delta n(\boldsymbol{r},\omega)=\int\left[-\frac{i}{\hbar}\int_0^\infty e^{i\omega t}\Big\langle\big[\hat{n}(\boldsymbol{r},t),\hat{n}(\boldsymbol{r}',0)\big]\Big\rangle_{\rm eq}dt\right]\Big(-eV(\boldsymbol{r}',\omega)\Big)\,d^3r'. \tag{14.31}$$

Thus, the induced density can be related to the **density–density correlation function** of the equilibrium system. From this we can again establish a direct link between a microscopic and macroscopic material description, here expressed in terms of density fluctuations and the induced density.

There is an important point that has to be considered when using a linear response description for Coulomb systems. The alert reader may have noticed that in Eq. (14.31) we have introduced the *total* potential V rather than the external potential $V_{\rm ext}$, which would be the naive choice when straightforwardly applying Kubo's formula. However, in systems with charged carriers the response is *always* to the total field, which is the sum of the external and induced ones, the latter being caused by the induced particle density δn. See Fig. 14.6 for a graphical

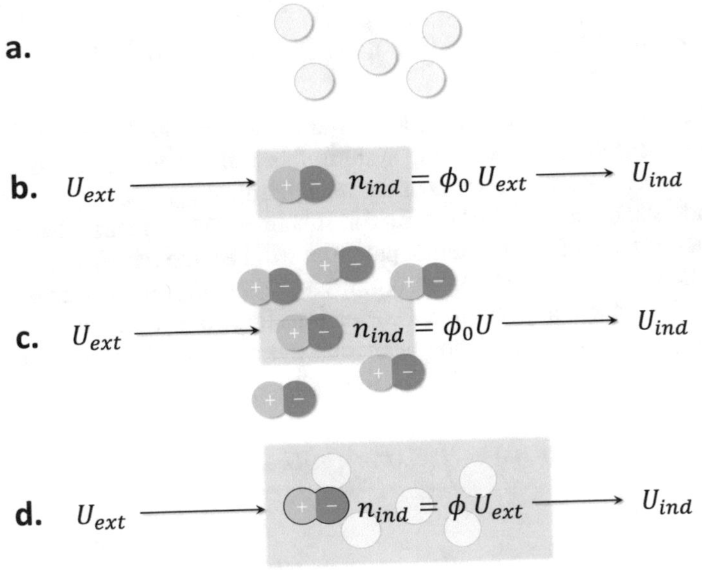

Fig. 14.6 Schematics of potential screening. (**a**) We consider a many-body system consisting of a polarizable medium, here represented by polarizable spheres. (**b**) When an external potential $U_{\rm ext}$ is applied, the system becomes polarized. According to Kubo's formalism, the induced density $n_{\rm ind}$ is given by the product of the density–density correlation function Φ_0 and the external potential $U_{\rm ext}$. The induced potential $U_{\rm ind} = U_0\, n_{\rm ind}$ can then be computed from the solution of Poisson's equation. (**c**) In a many-particle system the response of the system is due to the *total* potential $U = U_{\rm ext} + U_{\rm ind}$, which is the sum of the external potential and the induced potential originating from all polarization charges in the system. This leads to an equation for U that must be solved self-consistently. (**d**) Alternatively, we can introduce a density–density correlation function Φ for the *interacting* electron system, which already incorporates the effects of polarization charges such that the induced density can be directly computed from $n_{\rm ind} = \Phi U_{\rm ext}$. For details see text

representation. For this reason, δn should be related to the total potential V and not to V_{ext}. This choice is similar to Maxwell's equations in matter where we relate the polarization via $\boldsymbol{P} = \varepsilon_0 \chi_e \boldsymbol{E}$ to the *total* field $\boldsymbol{E}$, rather than to the dielectric displacement $\boldsymbol{D}$ that would correspond to the external potential V_{ext}.

In the Coulomb gauge we express the potential associated with the induced charge distribution $-e\delta n$ as

$$U_{\text{ind}}(\boldsymbol{r}, \omega) = -e \int \frac{(-e)\delta n(\boldsymbol{r}', \omega)}{4\pi\varepsilon_0 |\boldsymbol{r} - \boldsymbol{r}'|} d^3r' = \int U_0(\boldsymbol{r}, \boldsymbol{r}')\delta n(\boldsymbol{r}', \omega)\, d^3r',$$

where in the last equality we have introduced U_0 for the solution of Poisson's equation,

$$U_0(\boldsymbol{r}, \boldsymbol{r}') = \frac{e^2}{4\pi\varepsilon_0 |\boldsymbol{r} - \boldsymbol{r}'|}. \tag{14.32}$$

From Eq. (14.31) we then get (see also Fig. 14.7)

$$U_{\text{ind}}(\boldsymbol{r}, \omega) = \int U_0(\boldsymbol{r}, \boldsymbol{r}')\Phi_0\big(\hat{n}(\boldsymbol{r}'), \hat{n}(\boldsymbol{r}''); \omega\big)U(\boldsymbol{r}'', \omega)\, d^3r' d^3r'', \tag{14.33}$$

where the subscript zero on Φ_0 is a reminder that the correlation function has to be evaluated for the non-interacting system, this is in absence of mutual Coulomb interactions between the electrons of the many-particle system.

To understand what we are heading for, it is important to realize that as an effect of this induced potential the external potential will become screened. Further below we will show that the solution of Eq. (14.33) is particularly simple for homogeneous systems where one can perform a Fourier transform in space. However, we keep considering for a moment inhomogeneous systems and introduce an effective screening function $K(\boldsymbol{r}, \boldsymbol{r}', \omega)$ defined through

$$U(\boldsymbol{r}, \omega) = \int K(\boldsymbol{r}, \boldsymbol{r}', \omega)U_{\text{ext}}(\boldsymbol{r}', \omega)\, d^3r'. \tag{14.34}$$

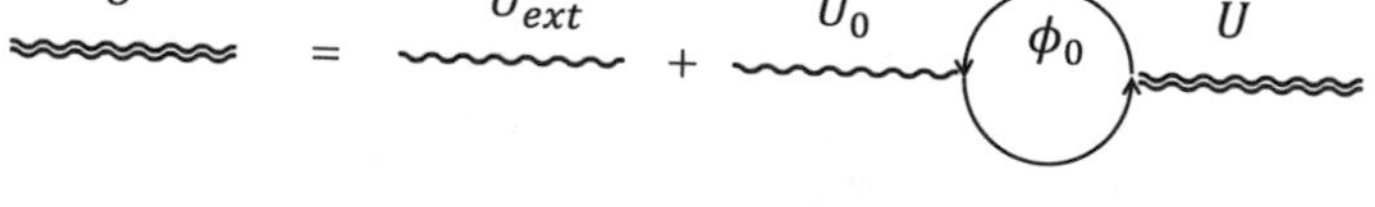

Fig. 14.7 Schematic representation of the screened Coulomb potential U. The screened Coulomb potential consists of the external Coulomb potential U_{ext} and an induced part. Φ_0 is the density–density correlation function, and $\delta n = \Phi_0 U$ is the density change caused by the *total* Coulomb potential. Through $U_0\delta n = U_0\Phi_0 U$ we obtain the induced polarization fields

If we associate U, U_{ext} with the fields $\boldsymbol{E}$, $\boldsymbol{D}$, we observe that K plays the role of an inverse permittivity ε^{-1}. Rewriting the left-hand side of Eq. (14.33) in the form $U - U_{\text{ext}}$ then leads us to

$$\int\left[\delta(\boldsymbol{r}-\boldsymbol{r}'')-\int U_0(\boldsymbol{r},\boldsymbol{r}')\Phi_0\big(\hat{n}(\boldsymbol{r}'),\hat{n}(\boldsymbol{r}'');\omega\big)\,d^3r'\right]U(\boldsymbol{r}'',\omega)\,d^3r'' = U_{\text{ext}}(\boldsymbol{r},\omega).$$

Comparison with Eq. (14.34) shows that the dielectric function for screening an external scalar potential is of the form

Dielectric Function for Screening of V_{ext}

$$K^{-1}(\boldsymbol{r},\boldsymbol{r}',\omega) = \delta(\boldsymbol{r}-\boldsymbol{r}') - \int U_0(\boldsymbol{r},\boldsymbol{r}'')\Phi_0\big(\hat{n}(\boldsymbol{r}''),\hat{n}(\boldsymbol{r}');\omega\big)\,d^3r''. \tag{14.35}$$

By multiplying this equation from the left-hand side with K and integrating over the entire space, we obtain an integral equation for the inverse dielectric function

$$K(\boldsymbol{r},\boldsymbol{r}',\omega) = \delta(\boldsymbol{r}-\boldsymbol{r}') + \int K(\boldsymbol{r},\boldsymbol{r}_1,\omega)U_0(\boldsymbol{r}_1,\boldsymbol{r}_2)\Phi_0\big(\hat{n}(\boldsymbol{r}_2),\hat{n}(\boldsymbol{r}');\omega\big)\,d^3r_1d^3r_2.$$

The solution of this integral equation then provides us with the inverse dielectric function.

Density-Density Correlation Function. Φ_0 accounts for the density response of the non-interacting system. It is sometimes convenient to introduce a density–density correlation function Φ for the *interacting* electron system, which accounts for the response to an external potential U_{ext} only and has already built in the Coulomb interactions between the electrons forming the many-particle system. In analogy to Eq. (14.31) we make the ansatz

$$\delta n(\boldsymbol{r},\omega) = \int \Phi\big(\hat{n}(\boldsymbol{r}),\hat{n}(\boldsymbol{r}');\omega\big)\,U_{\text{ext}}(\boldsymbol{r}',\omega)\,d^3r', \tag{14.36}$$

where we have removed the subscript zero from the density–density correlation function Φ. Similarly, instead of Eq. (14.33) we now obtain (Fig. 14.8)

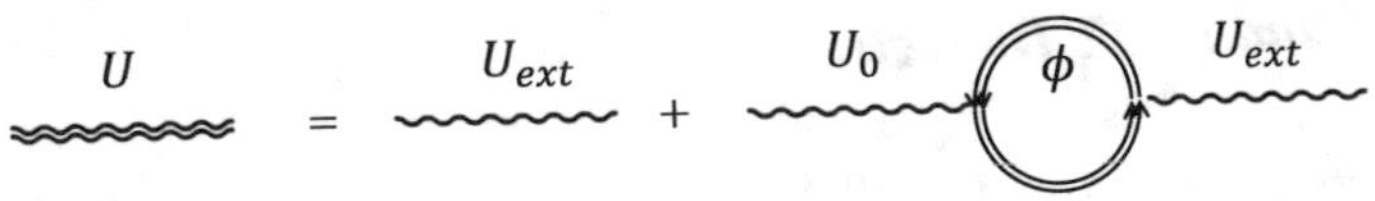

Fig. 14.8 Same as Fig. 14.7 but for density–density correlation function Φ of the *interacting* electron system. $\delta n = \Phi\, U_{\text{ext}}$ is the density change caused by the external Coulomb potential

$$U_{\text{ind}}(\boldsymbol{r},\omega)=\int U_0(\boldsymbol{r},\boldsymbol{r}')\Phi\big(\hat{n}(\boldsymbol{r}'),\hat{n}(\boldsymbol{r}'');\omega\big)U_{\text{ext}}(\boldsymbol{r}'',\omega)\,d^3r'd^3r''. \tag{14.37}$$

Here the potential on the right-hand side is the external potential (previously it has been the total potential), but the response function Φ is now for the interacting system rather than the non-interacting one. From this expression it is simple to establish a relation between Φ and Φ_0. In a short-hand notation, where the different contributions are assumed to be connected through a convolution in space, we get

$$U_{\text{ind}}=U_0\Phi U_{\text{ext}}=U_0\Phi_0 U=U_0\Phi_0\Big(U_{\text{ext}}+U_0\Phi U_{\text{ext}}\Big),$$

with the term in parentheses being precisely the total potential $U_{\text{ext}}+U_{\text{ind}}$. We thus get

$$\begin{aligned}\Phi\big(\hat{n}(\boldsymbol{r}),\hat{n}(\boldsymbol{r}');\omega\big)&=\Phi_0\big(\hat{n}(\boldsymbol{r}),\hat{n}(\boldsymbol{r}');\omega\big)\\&+\int\Phi_0\big(\hat{n}(\boldsymbol{r}),\hat{n}(\boldsymbol{r}_1);\omega\big)\,U_0(\boldsymbol{r}_1,\boldsymbol{r}_1')\,\Phi\big(\hat{n}(\boldsymbol{r}_1'),\hat{n}(\boldsymbol{r}');\omega\big)\,d^3r_1d^3r_1'.\end{aligned} \tag{14.38}$$

Suppose that the external potential is produced by a charge distribution $\rho_{\text{ext}}(\boldsymbol{r},\omega)=-en(\boldsymbol{r},\omega)$ through

$$U_{\text{ext}}(\boldsymbol{r},\omega)=\int U_0(\boldsymbol{r},\boldsymbol{r}')n_{\text{ext}}(\boldsymbol{r}',\omega)\,d^3r'.$$

Inserting this expression into Eq. (14.37) then gives

$$U_{\text{ind}}(\boldsymbol{r},\omega)=\int W_{\text{ind}}(\boldsymbol{r},\boldsymbol{r}',\omega)n_{\text{ext}}(\boldsymbol{r}',\omega)\,d^3r', \tag{14.39}$$

where we have introduced the screened Coulomb potential

$$W_{\text{ind}}(\boldsymbol{r},\boldsymbol{r}',\omega)=\int U_0(\boldsymbol{r},\boldsymbol{r}_1)\Phi\big(\hat{n}(\boldsymbol{r}_1),\hat{n}(\boldsymbol{r}_1');\omega\big)U_0(\boldsymbol{r}_1,\boldsymbol{r}')\,d^3r_1d^3r_1'. \tag{14.40}$$

It describes the response of an interacting electron system to a unit point charge located at position $\boldsymbol{r}'$. Apart from a prefactor e^2/ε_0 the screened Coulomb potential W_{ind} is thus identical to the "reflected" (induced) Green's function used in previous chapters. For local permittivities we can take over all techniques developed in the first "classical" part of this book. In case of nonlocal media we have to somewhat modify things, as will be discussed in Sect. 14.5.1.

14.4.2 Lindhard's Dielectric Function

The dielectric function of Eq. (14.35) can be solved most easily for a spatially homogeneous system. Here the convolution in real space becomes a product in wavenumber space, and we get

$$K^{-1}(\boldsymbol{q}, \omega) = 1 - \left(\frac{e^2}{\varepsilon_0 q^2}\right) \Phi^0_{nn}(\boldsymbol{q}, \omega),$$

where the term in parentheses is the Fourier transform of U_0 and $\Phi^0_{nn}(\boldsymbol{q}, \omega)$ is the Fourier transform of the density–density correlation function of the non-interacting system. The above expression is related to the permittivity of Maxwell's equations according to $\varepsilon(\boldsymbol{q}, \omega) = \varepsilon_0 K^{-1}(\boldsymbol{q}, \omega)$. Let us investigate in more detail the density–density correlation function Φ^0_{nn} that has to be computed from Eq. (14.10) using the Kubo formalism. We start from the Fourier transform of the density operator

$$\hat{n}_{\boldsymbol{q}} = \int e^{-i\boldsymbol{q}\cdot\boldsymbol{r}} \hat{n}(\boldsymbol{r})\, d^3 r, \quad \hat{n}(\boldsymbol{r}) = \sum_{\boldsymbol{q}} e^{i\boldsymbol{q}\cdot\boldsymbol{r}} \hat{n}_{\boldsymbol{q}}. \tag{14.41}$$

Inserting this into Kubo's formula gives

$$\Phi^0_{nn}(\boldsymbol{q}, \omega) = -\frac{i}{\hbar} \int e^{i\omega t} e^{-i\boldsymbol{q}\cdot(\boldsymbol{r}-\boldsymbol{r}')} \sum_{\boldsymbol{k},\boldsymbol{k}'} \int \left\langle \left[e^{i\boldsymbol{k}\cdot\boldsymbol{r}} \hat{n}_{\boldsymbol{k}}(t), e^{i\boldsymbol{k}'\cdot\boldsymbol{r}'} \hat{n}_{\boldsymbol{k}'} \right] \right\rangle_{\mathrm{eq}} dt\, d^3 r d^3 r',$$

which can be evaluated for a homogeneous system to

$$\Phi^0_{nn}(\boldsymbol{q}, \omega) = -\frac{i}{\hbar} \int_0^\infty e^{i\omega t} \left\langle \left[\hat{n}_{\boldsymbol{q}}(t), \hat{n}_{-\boldsymbol{q}} \right] \right\rangle_{\mathrm{eq}} dt. \tag{14.42}$$

In the following we assume that the eigenstates of the system Hamiltonian $\hat{H}_0$ are plane waves $|\boldsymbol{k}\rangle$ with energy $E_{\boldsymbol{k}}$, which are occupied with probability $f_0(E_{\boldsymbol{k}})$. Assuming momentum conservation, the density–density correlation function can be expressed as

$$\begin{aligned}\Phi^0_{nn}(\boldsymbol{q}, \omega) = -\frac{i}{\hbar} \sum_{\boldsymbol{k}} \int_0^\infty & e^{i(\omega+i\eta)t} \Big[f_0(E_{\boldsymbol{k}}) \langle \boldsymbol{k} | \hat{n}_{\boldsymbol{q}}(t) | \boldsymbol{k}+\boldsymbol{q} \rangle \langle \boldsymbol{k}+\boldsymbol{q} | \hat{n}_{-\boldsymbol{q}} | \boldsymbol{k} \rangle \\ & - f_0(E_{\boldsymbol{k}+\boldsymbol{q}}) \langle \boldsymbol{k}+\boldsymbol{q} | \hat{n}_{-\boldsymbol{q}} | \boldsymbol{k} \rangle \langle \boldsymbol{k} | \hat{n}_{\boldsymbol{q}}(t) | \boldsymbol{k}+\boldsymbol{q} \rangle \Big] dt.\end{aligned}$$

For plane waves the matrix elements become one and the time integration can be performed as previously discussed for the adiabatic limit. With this, we are led to the Lindhard's dielectric function

Lindhard's Dielectric Function

$$\varepsilon(\boldsymbol{q},\omega)/\varepsilon_0 = 1 - \left(\frac{e^2}{\varepsilon_0 q^2}\right)\lim_{\eta\to 0}\sum_{\boldsymbol{k}}\frac{f_0(E_{\boldsymbol{k}})-f_0(E_{\boldsymbol{k}+\boldsymbol{q}})}{\hbar\omega + i\eta + E_{\boldsymbol{k}} - E_{\boldsymbol{k}+\boldsymbol{q}}}. \tag{14.43}$$

This function has been previously used in Eq. (7.12) for a two-dimensional electron gas representative of graphene, where one additionally has to replace the term in parentheses with the two-dimensional Fourier transform of the Coulomb potential $e^2/(2\varepsilon_0 q)$. For a three-dimensional electron gas, representative for simple metals, the Lindhard's dielectric function can be computed analytically at zero temperature, as described in some detail in Refs. [124, 125]. We here only state the final result valid for $E_q \ll \hbar\omega$, $qv_F \ll \omega$, where v_F is the Fermi velocity, which reads [125, Eq. (5.5.9)]

$$\varepsilon'(q,\omega)/\varepsilon_0 = 1 - \frac{\omega_p^2}{\omega^2}\left\{1 + \frac{1}{\omega^2}\left[\frac{3}{5}(qv_F)^2 - E_q^2\right] + \mathcal{O}\left(\frac{1}{\omega^4}\right)\right\}. \tag{14.44}$$

Here ω_p is the plasma frequency. A similar dielectric function can be also obtained within the so-called hydrodynamic model, which describes the electrons in terms of density $n(\boldsymbol{r},t)$ and velocity $\boldsymbol{v}(\boldsymbol{r},t)$ distributions. The equation of motion is [135, 136]

$$m\left[\frac{\partial \boldsymbol{v}}{\partial t} + \boldsymbol{v}\cdot\nabla\boldsymbol{v}\right] = -m\gamma\boldsymbol{v} - \frac{\nabla p_{\text{deg}}}{n} - e\Big(\boldsymbol{E} + \boldsymbol{v}\times\boldsymbol{B}\Big), \tag{14.45}$$

where p_{deg} is the pressure of the degenerate electron gas, and γ is a damping constant. Within this model we obtain for the longitudinal dielectric function

$$\varepsilon^L(q,\omega)/\varepsilon_0 = 1 - \frac{\omega_p^2}{\omega^2 + i\gamma\omega - \beta^2 q^2}, \tag{14.46}$$

whereas the transverse permittivity is given by the usual Drude expression of Eq. (7.7). For a free electron gas we can set $\beta^2 = (3/5)v_F^2$, in agreement with Lindhard's expression of Eq. (14.44) for small q values. In many cases Eq. (14.46) captures the essential physics of nonlocal permittivities. We will return to such permittivity functions below in Sect. 14.5.1 when discussing effects of nonlocality in the context of quantum plasmonics.

14.4.3 *Response to Longitudinal and Transverse Fields*

In what follows we generalize our results for the response of a many-particle system to longitudinal and transverse electromagnetic fields. Our starting point is the many-body light–matter interaction of Eq. (13.63) expressed in terms of the current and density operators $\hat{\mathbf{j}}$, $\hat{n}$,

$$\hat{H}_{\text{int}} = -e \int \Big(-\hat{\mathbf{j}}(\boldsymbol{r},t)\cdot \boldsymbol{A}(\boldsymbol{r},t) + \hat{n}(\boldsymbol{r},t)V(\boldsymbol{r},t)\Big)\, d^3r.$$

As a result of the applied fields, a current will be induced in the system, which according to Eq. (13.62) consists of paramagnetic and diamagnetic contributions,

$$\boldsymbol{J}(\boldsymbol{r},t) = \Big\langle -e\hat{\mathbf{j}}(\boldsymbol{r},t) - \frac{e^2}{m}\hat{n}(\boldsymbol{r},t)\,\boldsymbol{A}(\boldsymbol{r},t)\Big\rangle. \tag{14.47}$$

From here on we work in an interaction representation with respect to the uncoupled light-matter system. We consider a classical vector potential, but our analysis would remain almost identical for $\boldsymbol{A}$ being replaced by an operator. In linear response the paramagnetic response (the first term in brackets) can be expressed in terms of a current–current correlation function. The diamagnetic response (the second term in brackets) depends already linearly on the vector potential, and we therefore do not have to consider any modifications of $\hat{n}$. Thus, we arrive at

$$\boldsymbol{J}(\boldsymbol{r},t) = \Big\langle -\frac{i}{\hbar}\int_0^t \Big[-e\hat{\mathbf{j}}(\boldsymbol{r},t),\, \hat{H}_{\text{int}}(t')\Big]\,dt' - \frac{e^2}{m}\hat{n}(\boldsymbol{r},t)\,\boldsymbol{A}(\boldsymbol{r},t)\Big\rangle_{\text{eq}},$$

where we have assumed that the current $\langle\hat{j}\rangle_{\text{eq}} = 0$ vanishes in the equilibrium system. As we will demonstrate below, the current response can be written in the form

$$\boldsymbol{J}(\boldsymbol{r},\omega) = \int \bar{\bar{\sigma}}(\boldsymbol{r},\boldsymbol{r}',\omega)\cdot \boldsymbol{E}(\boldsymbol{r}',\omega)\, d^3r', \tag{14.48}$$

with the nonlocal optical conductivity

Optical Conductivity

$$\bar{\bar{\sigma}}(\boldsymbol{r},\boldsymbol{r}',\omega) = \frac{i}{\omega}\Big[e^2\bar{\bar{\Phi}}_{\mathbf{jj}}(\boldsymbol{r},\boldsymbol{r}',\omega) + \frac{e^2 n_0(\boldsymbol{r})}{m}\,\delta(\boldsymbol{r}-\boldsymbol{r}')\mathbb{1}\Big]. \tag{14.49}$$

$n_0(\boldsymbol{r})$ is the equilibrium density of the system and $\bar{\bar{\Phi}}_{\mathbf{jj}}$ the **current–current correlation function**

$$\bar{\bar{\Phi}}_{\mathbf{jj}}(\boldsymbol{r},\boldsymbol{r}',\omega) = \Phi\big(\hat{\mathbf{j}}(\boldsymbol{r}),\hat{\mathbf{j}}(\boldsymbol{r}');\omega\big),$$

which accounts for the current response caused by the external electric field. Quite generally, the computation of the current–current correlation function can be fairly complicated, even for simple homogeneous systems [124–126]. The strength of the above approach is that it gives a prescription of how to proceed *in principle* provided that the current–current correlation is at hand. In this respect, the approach is similar to the dyadic Green's function approach of classical electrodynamics, which allows us, for instance, to express optical scattering rates for fluorescence or Raman scatterings in terms of this Green's function only. The question of how *to actually compute* the Green's function can then be tackled in a second step, but having a means for writing down a general solution can be highly beneficial in many cases.

Proof of Eq. (14.49) We start with the current response to the vector potential only, and correspondingly neglect in $\hat{H}_{\rm int}$ the term with the scalar potential. This leads us to

$$\boldsymbol{J}_1(\boldsymbol{r},t) = \left\langle \frac{ie^2}{\hbar}\int_0^t \Big[\hat{\mathbf{j}}(\boldsymbol{r},t),\int \hat{\mathbf{j}}(\boldsymbol{r}',t')\cdot\boldsymbol{A}(\boldsymbol{r}',t')\,d^3r'\Big]\,dt' - \frac{e^2}{m}\hat{n}(\boldsymbol{r},t)\,\boldsymbol{A}(\boldsymbol{r},t')\right\rangle_{\rm eq}.$$

Similarly, the current response to the scalar potential alone becomes

$$\boldsymbol{J}_2(\boldsymbol{r},t) = \left\langle \hat{\boldsymbol{J}}_2(\boldsymbol{r},t)\right\rangle_{\rm eq} = \left\langle -\frac{ie^2}{\hbar}\int_0^t \Big[\hat{\mathbf{j}}(\boldsymbol{r},t),\int \hat{n}(\boldsymbol{r}',t')V(\boldsymbol{r},t')\,d^3r'\Big]\,dt'\right\rangle_{\rm eq}.$$

We now perform integration by parts to rewrite the time integral in the form

$$\int_0^t \Big[\hat{\mathbf{j}}(\boldsymbol{r},t),\hat{n}(\boldsymbol{r}',t')V(\boldsymbol{r},t')\Big]\,dt' = \Big[\hat{\mathbf{j}}(\boldsymbol{r},t),\hat{n}(\boldsymbol{r}',t')\mathscr{V}(\boldsymbol{r}',t')\Big]\Big|_0^t - \int_0^t \Big[\hat{\mathbf{j}}(\boldsymbol{r},t),\frac{\partial\hat{n}(\boldsymbol{r}',t')}{\partial t'}\mathscr{V}(\boldsymbol{r}',t')\Big]\,dt',$$

where $\mathscr{V}$ is the scalar potential integrated in time. We next use the continuity equation of Eq. (13.61), where in linear response the diamagnetic current (which is proportional to the vector potential) can be neglected, to rewrite the last term in the form

$$-\Big[\hat{\mathbf{j}}(\boldsymbol{r},t),\frac{\partial\hat{n}(\boldsymbol{r}',t')}{\partial t'}\mathscr{V}(\boldsymbol{r}',t')\Big] \approx \Big[\hat{\mathbf{j}}(\boldsymbol{r},t),\nabla'\cdot\hat{\mathbf{j}}(\boldsymbol{r}',t')\mathscr{V}(\boldsymbol{r}',t')\Big].$$

Putting together all results and performing integration by parts, in order to pull over the derivative from $\hat{\mathbf{j}}$ to $\mathscr{V}$, then leads us to

$$\hat{\boldsymbol{J}}_2(\boldsymbol{r},t) = -\frac{ie^2}{\hbar}\int\bigg(\Big[\hat{\mathbf{j}}(\boldsymbol{r},t),\hat{n}(\boldsymbol{r}',t')\mathscr{V}(\boldsymbol{r}',t')\Big]\Big|_0^t + \int_0^t \Big[\hat{\mathbf{j}}(\boldsymbol{r},t),\hat{\mathbf{j}}(\boldsymbol{r}',t')\cdot\nabla'\mathscr{V}(\boldsymbol{r}',t')\Big]\,dt'\bigg)\,d^3r'.$$

In the first term in parentheses we assume that the commutator $[\hat{\mathbf{j}}(\boldsymbol{r},t),\hat{n}(\boldsymbol{r}',0)]$ can be neglected for sufficiently large time arguments t, using the same reasoning as for the adiabatic limit. With the exception of the initial transient regime around time zero this is expected to be an excellent approximation. For the equal time commutator we use Eqs. (13.58), (13.59) to get

$$\frac{1}{m}\sum_i \delta(\boldsymbol{r}-\boldsymbol{r}_i)\Big[\hat{\boldsymbol{\pi}}_i,\mathscr{V}(\boldsymbol{r}_i)\Big]=-\frac{i\hbar}{m}\sum_i \delta(\boldsymbol{r}-\boldsymbol{r}_i)\nabla_i\mathscr{V}(\boldsymbol{r}_i)=-\frac{i\hbar}{m}\hat{n}(\boldsymbol{r})\,\nabla\mathscr{V}(\boldsymbol{r}).$$

Thus, we obtain

$$\hat{J}_2(\boldsymbol{r},t)=-\frac{ie^2}{\hbar}\int_0^t\left[[\hat{\mathbf{j}}(\boldsymbol{r},t),\int\hat{\mathbf{j}}(\boldsymbol{r}',t')\cdot\nabla'\mathscr{V}(\boldsymbol{r}',t')\,d^3r'\right]dt'+\frac{e^2}{m}\hat{n}(\boldsymbol{r})\,\nabla\mathscr{V}(\boldsymbol{r}).$$

Comparison with the current $\boldsymbol{J}_1$ induced by the vector potential shows that $\boldsymbol{J}_2$ has the same form, but instead of $\boldsymbol{A}$ we now have $-\nabla\mathscr{V}$. Putting together all results we arrive at

$$\begin{aligned}\boldsymbol{J}(\boldsymbol{r},t)=&-\frac{e^2}{m}\Big\langle\hat{n}(\boldsymbol{r},t)\Big\rangle_{\text{eq}}\Big(\boldsymbol{A}(\boldsymbol{r},t)-\nabla'\mathscr{V}(\boldsymbol{r},t)\Big)\\&+\int\left\langle\frac{ie^2}{\hbar}\int_0^t\Big[\hat{\mathbf{j}}(\boldsymbol{r},t),\hat{\mathbf{j}}(\boldsymbol{r}',t')\cdot\Big(\boldsymbol{A}(\boldsymbol{r}',t')-\nabla'\mathscr{V}(\boldsymbol{r}',t')\Big)\Big]\,dt'\right\rangle_{\text{eq}}d^3r'.\end{aligned}\tag{14.50}$$

One readily observes that the combination of vector potential and time-integrated scalar potential is precisely the time-integrated electric field. In order to get our final expression of Eq. (14.49) we pull out the time-integrated field from the commutator and express the above equation in frequency space.

14.4.4 Fluctuation–Dissipation Theorem

We finally derive an important relation between the current fluctuations and the permittivity function, which will play an important role in the next chapter. We start with Maxwell's curl equations in matter

$$\nabla\times\boldsymbol{E}=i\omega\boldsymbol{B},\qquad\frac{1}{\mu_0}\nabla\times\boldsymbol{B}=\boldsymbol{J}_{\text{ext}}+\boldsymbol{J}_{\text{ind}}-i\omega\varepsilon_0\boldsymbol{E},$$

where for simplicity we have considered no magnetic response (which could be handled along the same lines). We now express the induced current in terms of the optical conductivity through

$$\boldsymbol{J}_{\text{ind}}(\boldsymbol{r},\omega)=\int\bar{\bar{\sigma}}(\boldsymbol{r},\boldsymbol{r}',\omega)\cdot\boldsymbol{E}(\boldsymbol{r}',\omega)\,d^3r'\implies\boldsymbol{J}_{\text{ind}}=\bar{\bar{\sigma}}\cdot\boldsymbol{E}.$$

Here we have introduced a short-hand notation where the nonlocal optical conductivity and the electric field are assumed to be connected through a convolution in space. The wave equation for the electric field can then be obtained as usually by taken the curl of Faraday's equation, which leads to

$$\nabla\times\nabla\times\boldsymbol{E}=i\omega\mu_0\boldsymbol{J}_{\text{ext}}+\omega^2\mu_0\left(\varepsilon_0\mathbb{1}+\frac{i\bar{\bar{\sigma}}}{\omega}\right)\cdot\boldsymbol{E}=i\omega\mu_0\boldsymbol{J}_{\text{ext}}+\omega^2\mu_0\bar{\bar{\varepsilon}}\cdot\boldsymbol{E}.$$

In the last expression we have introduced the nonlocal permittivity $\bar{\bar{\varepsilon}}$. In case of time reversal symmetry the dyadic current–current correlation function is symmetric [43]. Then also the nonlocal permittivity tensor is symmetric

$$\bar{\bar{\varepsilon}}(\boldsymbol{r},\boldsymbol{r}',\omega)=\delta(\boldsymbol{r}-\boldsymbol{r}')\left[\varepsilon_0-\frac{e^2n_0(\boldsymbol{r})}{m\omega^2}\right]\mathbb{1}-\frac{e^2}{\omega^2}\bar{\bar{\Phi}}_{\mathbf{jj}}(\boldsymbol{r},\boldsymbol{r}',\omega)=\bar{\bar{\varepsilon}}^T(\boldsymbol{r}',\boldsymbol{r},\omega),$$

where T denotes the matrix transpose and we have used Eq. (14.49) to relate the optical conductivity to the current–current correlation function. The above relation, together with Eqs. (14.16), (14.20), can be used to express the cross-spectral density of current operators through the imaginary part of the permittivity

Fluctuation-Dissipation Theorem

$$\frac{1}{\hbar}\int_{-\infty}^{\infty}e^{i\omega t}\Big\langle\hat{J}_i(\boldsymbol{r},t)\hat{J}_j(\boldsymbol{r}',0)\Big\rangle_{\text{eq}}dt=2\omega^2\,\text{Im}\Big[\bar{\bar{\varepsilon}}(\boldsymbol{r},\boldsymbol{r}',\omega)\Big]_{ij}\Big(\bar{n}_{\text{th}}(\hbar\omega)+1\Big).\tag{14.51}$$

This equation relates the current fluctuation spectrum on the left-hand side to the absorption losses (dissipation) on the right-hand side. Eq. (14.51) can be used for both positive and negative frequencies. The expression for negative frequencies can be rewritten by using Eq. (7.31), $\varepsilon(-\omega)=\varepsilon^*(\omega)$, together with

$$\bar{n}_{\text{th}}(-\hbar\omega)+1=\frac{e^{-\beta\hbar\omega}}{e^{-\beta\hbar\omega}-1}=-\bar{n}_{\text{th}}(\hbar\omega).$$

With this we get

$$\frac{1}{\hbar}\int_{-\infty}^{\infty}e^{-i\omega t}\Big\langle\hat{J}_i(\boldsymbol{r},t)\hat{J}_j(\boldsymbol{r}',0)\Big\rangle_{\text{eq}}dt=2\omega^2\,\text{Im}\Big[\bar{\bar{\varepsilon}}(\boldsymbol{r},\boldsymbol{r}',\omega)\Big]_{ij}\bar{n}_{\text{th}}(\hbar\omega).\tag{14.52}$$

The fluctuation–dissipation theorem of Eqs. (14.51) and (14.52) provides a direct link between the microscopic description of the material response in terms of correlation functions, and the material response functions of classical electrodynamics, see also Fig. 14.9.

Altogether, the correlation functions for electric and current field operators are made such that they work perfectly together, and relate to the dyadic Green's and

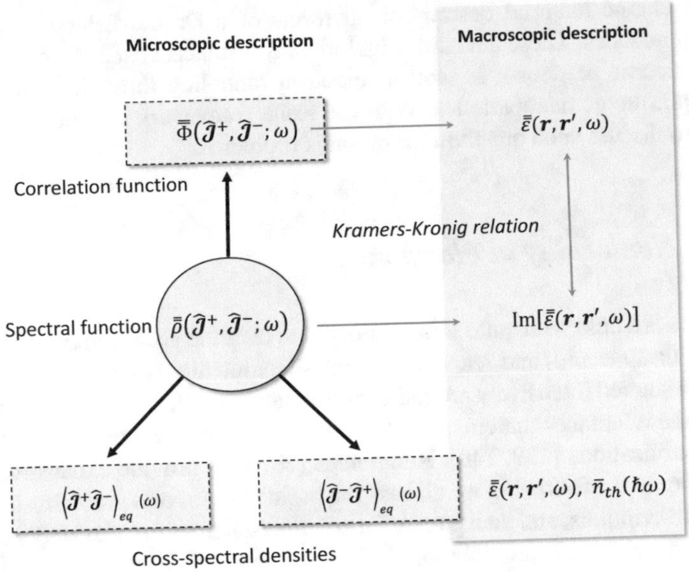

Fig. 14.9 Relation between spectral function $\bar{\bar{\rho}}$ of current operators, and correlation function and cross-spectral densities. The spectral function is a measure of the current fluctuations in thermal equilibrium, which is related to the imaginary part of the permittivity function of *classical* electrodynamics. The correlation function can be related through a Kramers–Kronig relation to the spectral function. The cross-spectral densities of current operators can be expressed through Eq. (14.51) in terms of the permittivity function and of thermal occupation numbers $\bar{n}_{\mathrm{th}}(\hbar\omega)$, thus establishing a direct link between a microscopic material description and the response functions of classical electrodynamics

permittivity functions of classical electrodynamics. This will be used in the next chapter when discussing the quantization of Maxwell's equations in presence of absorbing media. In the remainder of this chapter we discuss the effects of refined permittivity functions in the context of quantum plasmonics.

14.5 Quantum Plasmonics

For many years plasmonics has been a topic based on classical electrodynamics supplemented by a material description in terms of permittivities. While at the beginning of the twenty-first century more and more research areas went quantum, plasmonics remained classical for reasons highlighted at several places in this book: the concept of response functions (such as permittivities and permeabilities) as a "linker" between the electrodynamics and material worlds can easily accommodate for many types of quantum effects, and there is no urgent need to describe the quantum dynamics explicitly. However, at some point researchers had suffered too much and started to coin the terminology *quantum plasmonics* for deviations from the most classic material description in terms of a Drude dielectric function or related approaches. These deviations include, for instance, spatially nonlocal effects in the dielectric response, as well as electron tunneling through narrow gaps of coupled plasmonic nanoparticles. With the Kubo framework at hand, we are in the position to discuss such quantum plasmonics problems.

14.5.1 Nonlocality in Plasmonics

Plasmonics is a field with quite some history, and over the years certain subjects have appeared, disappeared, and reappeared again. Nonlocality is a prominent example. It has been studied intensively around 1980 for nanoparticles, see, for instance, [137, 138] and the references therein, and has come back to the agenda around 2012 with two key publications [139, 140]. References [73, 136] provide exhaustive reviews on the topic. In the following we discuss the basic features and problems inherent to nonlocal descriptions, and then ponder on solution schemes for Maxwell's equations within the quasistatic approximation.

Consider first the nonlocal dielectric function of Eq. (14.46) that is obtained within the so-called hydrodynamic model (from here on we skip the superscript L denoting the longitudinal character of the permittivity)

$$\varepsilon(q,\omega)/\varepsilon_0 = \kappa_b - \frac{\omega_p^2}{\omega^2 + i\gamma\omega - \beta^2 q^2}, \tag{14.53}$$

where ω_p is the plasma frequency, γ a damping constant, and β a factor related to the Fermi velocity. We have introduced in accordance to the Drude dielectric function of Eq. (7.7) an additional contribution κ_b associated with the screening of localized d-band electrons. In what follows, we exclusively consider the nonlocal permittivity given by Eq. (14.53), although most of our analysis can be easily carried over to more refined permittivity functions. For a linear but nonlocal response, the dielectric displacement $\boldsymbol{D}(\boldsymbol{r},\omega)$ at position $\boldsymbol{r}$ is related to the electric field $\boldsymbol{E}(\boldsymbol{r}',\omega)$ at position $\boldsymbol{r}'$ via

$$\boldsymbol{D}(\boldsymbol{r},\omega) = \int \varepsilon(\boldsymbol{r}-\boldsymbol{r}',\omega)\boldsymbol{E}(\boldsymbol{r}',\omega)\,d^3r'.$$

The electric field can then be expressed in terms of a scalar potential in the usual manner via $\boldsymbol{E}(\boldsymbol{r},\omega) = -\nabla V(\boldsymbol{r},\omega)$, and Gauss' law becomes

$$\nabla\cdot\boldsymbol{D}(\boldsymbol{r},\omega) = -\nabla\cdot\int \varepsilon(\boldsymbol{r}-\boldsymbol{r}',\omega)\nabla' V(\boldsymbol{r}',\omega)\,d^3r' = \rho(\boldsymbol{r},\omega), \tag{14.54}$$

with $\rho(\boldsymbol{r},\omega)$ being an external charge distribution. To solve this equation, it is convenient to introduce the Green's function $G_{\rm nl}(\boldsymbol{r}-\boldsymbol{r}',\omega)$ of the unbounded nonlocal medium, which is defined through

$$\nabla\cdot\int\varepsilon(\boldsymbol{r}-\boldsymbol{r}_1,\omega)\nabla_1 G_{\rm nl}(\boldsymbol{r}_1-\boldsymbol{r}',\omega)\,d^3r_1=-\delta(\boldsymbol{r}-\boldsymbol{r}'), \tag{14.55}$$

As shown below, the Green's function can be evaluated to

$$G_{\rm nl}(\boldsymbol{r},\omega)=\frac{1}{2\pi^2}\int_0^\infty\frac{1}{\varepsilon(q,\omega)}\,\frac{\sin qr}{qr}\,dq, \tag{14.56}$$

where $\varepsilon(q,\omega)$ is the Fourier transform of the real space permittivity $\varepsilon(\boldsymbol{r},\omega)$ and q a wavenumber.

Proof of Eq. (14.56) We first introduce the spatial Fourier transform and its inverse,

$$\tilde G_{\rm nl}(\boldsymbol{q},\omega)=\int e^{-i\boldsymbol{q}\cdot\boldsymbol{r}}\,G_{\rm nl}(\boldsymbol{r},\omega)\,d^3r \tag{14.57a}$$

$$G_{\rm nl}(\boldsymbol{r},\omega)=(2\pi)^{-3}\int e^{i\boldsymbol{q}\cdot\boldsymbol{r}}\;G_{\rm nl}(\boldsymbol{q},\omega)\,d^3q. \tag{14.57b}$$

The convolution of the permittivity and Green's function appearing in Eq. (14.55) can then be brought to the form

$$\nabla\cdot\int\varepsilon(\boldsymbol{r}-\boldsymbol{r}_1,\omega)\nabla G_{\rm nl}(\boldsymbol{r}_1-\boldsymbol{r}',\omega)\,d^3r_1=\int e^{i\boldsymbol{q}\cdot(\boldsymbol{r}-\boldsymbol{r}')}i\boldsymbol{q}\cdot\varepsilon(\boldsymbol{q},\omega)\,i\boldsymbol{q}\,\tilde G_{\rm nl}(\boldsymbol{q},\omega)\,\frac{d^3q}{(2\pi)^3}.$$

The terms $i\boldsymbol{q}$ stem from the spatial derivatives. Inserting this expression into Eq. (14.55) and using the wavevector decomposition for Dirac's delta function $\delta(\boldsymbol{r})=(2\pi)^{-3}\int e^{i\boldsymbol{q}\cdot\boldsymbol{r}}\,d^3q$ gives

$$\tilde G_{\rm nl}(\boldsymbol{q},\omega)=\frac{1}{q^2\varepsilon(q,\omega)}. \tag{14.58}$$

Thus, we get in spherical coordinates

$$G_{\rm nl}(\boldsymbol{r},\omega)=\frac{1}{4\pi^2}\int_0^\infty\left(\int_0^\pi e^{iqr\cos\theta}\,\sin\theta d\theta\right)\left(\frac{1}{q^2\varepsilon(q,\omega)}\right)q^2dq,$$

which finally leads us to Eq. (14.56).

For the hydrodynamic permittivity, the Green's function can be computed analytically. We start by stating the following integrals [141, Eqs. (3.721,3.725)]:

$$\int_0^\infty\frac{\sin(ax)}{x}\,dx=\frac{\pi}{2}\operatorname{sign}a \tag{14.59a}$$

$$\int_0^\infty\frac{\sin(ax)}{x(b^2+x^2)}\,dx=\frac{\pi}{2b^2}\left(1-e^{-ab}\right),\quad\text{for }\operatorname{Re}b>0,\ a>0. \tag{14.59b}$$

With the first integral we can compute the Green's function for a local permittivity $\varepsilon(0,\omega)$, alternatively we could undo the Fourier transform and use the Green's function result derived in previous chapters. The second integral can be proven by decomposing the sine into exponentials and performing complex integration as shown in Appendix A, but we here take a shortcut and simply use the results of Eq. (14.59). With the abbreviation

$$\omega_\gamma = \sqrt{\omega^2 + i\gamma\omega} \approx \omega + \frac{i}{2}\gamma,$$

the nonlocal dielectric function can be expressed as

$$\frac{\varepsilon(q,\omega)}{\varepsilon_0} = \kappa(q,\omega) = \kappa_b - \frac{\omega_p^2}{\omega_\gamma^2 - \beta^2 q^2} = \frac{\kappa(0,\omega)\omega_\gamma^2 - \beta^2 q^2}{\omega_\gamma^2 - \beta^2 q^2}.$$

We now introduce the wavenumber

$$Q = \frac{\omega_\gamma}{\beta}\sqrt{-\kappa(0,\omega)}, \tag{14.60}$$

where we have used that for plasmon excitations the dielectric function is always negative such that Q can be considered as almost real, with a small imaginary part associated with metal losses. With this, we obtain

$$\frac{1}{\kappa} = 1 + \left(\frac{1}{\kappa} - 1\right) = 1 + \frac{\omega_\gamma^2}{\beta^2}\left[\kappa(0,\omega) - 1\right]\left(\frac{1}{q^2 + Q^2}\right).$$

For the nonlocal Green's function of Eq. (14.56) we then find from the integrals of Eq. (14.59) the result

$$G_{\rm nl}(\boldsymbol{r},\omega) = \frac{1}{2\pi^2\varepsilon_0 r}\left[\frac{\pi}{2} + \frac{\pi\omega_\gamma^2[\kappa(0,\omega) - 1]}{2\beta^2 Q^2}\left(1 - e^{-Qr}\right)\right].$$

With a few trivial manipulations we are led to our final expression for the nonlocal Green's function within the hydrodynamic model

Green's Function for Nonlocal Hydrodynamic Permittivity

$$G_{\rm nl}(\boldsymbol{r},\omega) = \frac{1}{4\pi\varepsilon(0,\omega)r} + \left(\frac{1}{\varepsilon_0} - \frac{1}{\varepsilon(0,\omega)}\right)\frac{e^{-Qr}}{4\pi r}. \tag{14.61}$$

With the Drude parameters of Table 7.1 and a Fermi velocity of $v_F \approx 1.4 \times 10^6$ m/s for Au and Ag we obtain for a photon energy of $\hbar\omega = 1$ eV an inverse wavenumber of $Q^{-1} \approx 0.5$ nm, with an only small frequency dependence. Thus, nonlocal effects are expected to play a role when the particle geometry or field confinement vary significantly on a length scale of about 1 nm.

14.5.2 *Additional Boundary Conditions*

The consideration of a nonlocal response becomes more complicated in case of interfaces. Figure 14.10a shows the situation of a bulk medium where an electric field at position $\boldsymbol{r}$ induces a material response at position $\boldsymbol{r}'$ according to

$$\boldsymbol{P}(\boldsymbol{r},\omega) = \varepsilon_0 \int \chi_e(\boldsymbol{r}-\boldsymbol{r}',\omega)\boldsymbol{E}(\boldsymbol{r}',\omega)\, d^3r'. \tag{14.62}$$

Here $\boldsymbol{P}$ is the polarization and χ_e the electric susceptibility. Obviously, this description has to be modified in presence of interfaces separating the nonlocal metal from a local dielectric material, as shown in Fig. 14.10b. In our discussion we follow Halevi and Fuchs [138] and consider the wave equation in absence of external current distributions

$$\nabla \times \nabla \times \boldsymbol{E} - \mu_0\omega^2 \boldsymbol{D} = 0.$$

If we express the displacement in terms of the polarization $\boldsymbol{D} = \varepsilon_0\boldsymbol{E} + \boldsymbol{P}$ we get

$$\nabla^2\boldsymbol{E} - \nabla(\nabla\cdot\boldsymbol{E}) + \mu_0\varepsilon_0\omega^2\boldsymbol{E} = -\mu_0\omega^2\boldsymbol{P}. \tag{14.63}$$

Consider the situation depicted in Fig. 14.10b where a planar interface separates the region $z > 0$ filled with a local dielectric from the region $z < 0$ filled with a nonlocal metal. The electric field and polarization can be expressed in the form

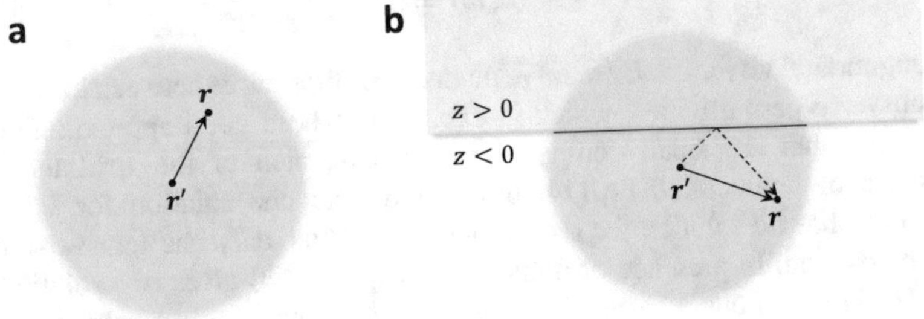

Fig. 14.10 For a nonlocal permittivity an excitation at position $\boldsymbol{r}'$ produces a response at position $\boldsymbol{r}$. The bulk response (**a**) is direct and depends on the relative position difference $\boldsymbol{r}-\boldsymbol{r}'$ only. The surface response (**b**) additionally contains an indirect part (dashed line). The shaded circle indicates the region with radius $\approx 1/Q$ where nonlocality plays a considerable role

$$E(r,\omega)=\tilde{E}(z,\omega)e^{ik_x x},\qquad P(r,\omega)=\tilde{P}(z,\omega)e^{ik_x x},$$

where we have used that, in accordance to our previous discussion of Chap. 8, the conservation of parallel electric field at the interface leads to the conservation of the parallel momentum k_x. The general linear response between the electric field and the polarization in the nonlinear material must have the form

$$\tilde{P}(z,\omega)=\varepsilon_0\int_{-\infty}^{0}\bar{\bar{\chi}}(z,z',\omega)\cdot\tilde{E}(z,\omega)\,dz, \tag{14.64}$$

where the susceptibility tensor additionally depends on the parallel momentum k_x. In the most simple approach we take the bulk susceptibility $\chi_e(z-z',\omega)$, which depends on the position difference $z-z'$ only, and consider the interface effects through

$$\chi_{ij}(z,z')=\Big(\chi_e(z-z',\omega)+U_i\,\chi_e(z+z',\omega)\Big)\delta_{ij}. \tag{14.65}$$

As schematically shown in Fig. 14.10b, the first term corresponds to the case that the excitation propagates directly from z' to z, whereas the second term accounts for a reflection of the excitation at the interface. The quantities U_i are, in general, complex numbers accounting for the reflection properties in a phenomenological manner. A particularly simple reflection model is due to Fuchs and Kliewer [142] who introduced with $U_{x,y}=1$ and $U_z=-1$ a *specular* reflection at the interface.

In Ref. [138] the authors continue to discuss the Fresnel coefficients for the interface between a local and nonlocal medium. Let us consider first a plane wave within an infinitely extended nonlocal medium. From the wave equation we find that the frequency and wavevector must satisfy the dispersion relation

$$\mu_0\varepsilon(k,\omega)\,\omega^2=k_x^2+k_z^2 \tag{14.66a}$$

for a transverse wave, $\boldsymbol{k}\cdot\boldsymbol{E}=0$, and

$$\varepsilon(k,\omega)=0 \tag{14.66b}$$

for a longitudinal wave, $\boldsymbol{k}\times\boldsymbol{E}=0$. We neglect the differences between longitudinal and transverse permittivities, which is expected to be a good approximation for wavevectors that are small compared to the extension of the Brillouin zone. In absence of nonlocality, Eq. (14.66a) would give one solution for k_z, while Eq. (14.66b) has no solution for real frequencies. Thus, only one transverse mode would be excited. In presence of nonlocality, Eq. (14.66a) gives two solutions for k_z and Eq. (14.66b) one additional solution. Thus, at the interface between a local and nonlocal medium, a wave impinging from the local side on the interface can excite *three* waves on the nonlocal side. Although we shall not work out the reflection and transmission coefficients here, for a detailed discussion see, for instance, Ref. [138], it is obvious that the usual boundary conditions for Maxwell's equations are insufficient to match the four waves at the interface (note that the continuity of the tangential fields has been already exploited in our wave ansatz).

We thus have to specify an **additional boundary condition** (ABC) that accounts for the properties of the nonlocal medium. The additional condition has to be extracted from the reflection properties at the interface, see Eq. (14.65), and depends on the reflection parameters U_i. In case of the Fuchs–Kliewer model of specular reflection one observes that the current flowing towards the interface is compensated by the current reflected at the interface. The polarization current is given by $\boldsymbol{J}_P = -i\omega\boldsymbol{P}$, and we thus express the additional boundary condition in the form

$$\lim_{z\to 0^-} P_z(z) = 0. \tag{14.67}$$

This form is valid if the nonlocal medium consists of free carriers only. It must be slightly modified in presence of additional bound electrons, as is the case for d-band electrons in transition metals whose dielectric response is described through the permittivity ε_b. We decompose the dielectric displacement into

$$\boldsymbol{D} = \varepsilon_0\boldsymbol{E} + \boldsymbol{P}_b + \boldsymbol{P}_f = \varepsilon_0\boldsymbol{E} + \big(\varepsilon_b - \varepsilon_0\big)\boldsymbol{E} + \boldsymbol{P}_f,$$

where $\boldsymbol{P}_b$ and $\boldsymbol{P}_f$ denote polarizations associated with bound and free electrons, respectively. For a general interface with an outer surface normal $\hat{\boldsymbol{n}}$ we then get for the additional boundary condition

$$\hat{\boldsymbol{n}}\cdot\boldsymbol{P}_f = \hat{\boldsymbol{n}}\cdot\Big(\boldsymbol{D} - \varepsilon_b\boldsymbol{E}\Big) = 0, \tag{14.68}$$

where $\boldsymbol{D}$, $\boldsymbol{E}$ must be evaluated in the nonlocal medium at the interface side.

Approach of Dasgupta and Fuchs for Spherical Nanoparticle

Maxwell's equations for nonlocal media can be solved analytically only for simple geometries, such as planar slabs, cylinders, or spheres. We here follow Dasgupta and Fuchs [137] and compute the optical response of a metallic nanosphere within the quasistatic approximation. Our analysis is closely related to the quasistatic Mie theory presented in Chap. 9, however, with a few important modifications to account for the nonlocal material.

The situation we have in mind consists of a metallic nanosphere (radius a) with a nonlocal permittivity ε_1, which is embedded in a local dielectric with permittivity ε_2, see Fig. 14.11. Because of nonlocality and as we wish to employ the nonlocal

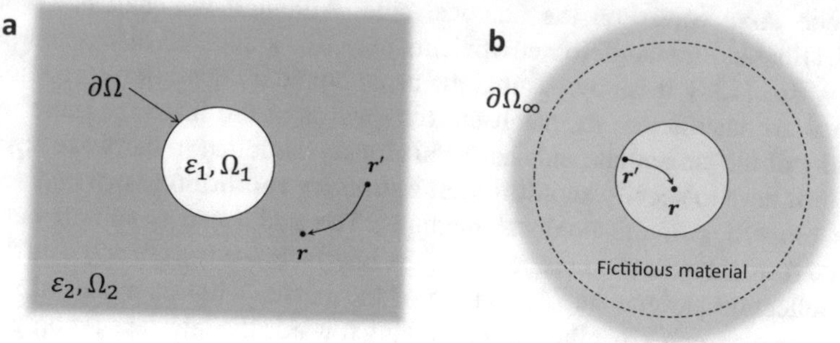

Fig. 14.11 Schematics of solution scheme of Dasgupta and Fuchs [137] for a metallic nanosphere (radius a, boundary $\partial\Omega$) with a nonlocal permittivity $\varepsilon_1(k,\omega)$, which is embedded in a local dielectric with permittivity ε_2. (**a**) In the outside region, Ω_2, Maxwell's equations in the quasistatic approximation are solved in complete analogy to the approach described in Chap. 9. (**b**) In the inside region, Ω_1, we extend the nonlocal material to the outside region, thereby introducing a fictitious material. The actual fields are assumed to coincide with those of the extended problem for $r < a$. To correct for the ad-hoc properties of the fictitious medium, we introduce so-called *additional boundary conditions* (ABC). The boundary $\partial\Omega_\infty$ far away from the sphere will be used later within the boundary integral method approach

Green's function of Eq. (14.56) for an unbounded medium, we extend for the solution of Maxwell's equations inside the sphere (1) the nonlocal medium to the outside region $r > a$, thereby introducing a fictitious medium in the region outside the sphere. (2) The fields inside the actual sphere are assumed to be identical to those of the extended medium in the $r < a$ region. (3) Possible errors caused by this procedure will be corrected at a later stage by employing the additional boundary condition of Eq. (14.68), such that the solutions are in agreement with the constraints of all boundary conditions.

As shown in a moment, the solution of the problem can be expressed in terms of the potentials V, V_D, V_E, which are determined by two parameters σ, p associated with a surface charge distribution and the induced dipole moment via

$$\begin{aligned}
r > a: &\quad \boldsymbol{E} = -\nabla V\,, &\dots\quad & V(r,\theta) = -E_0 r\cos\theta + \left(\frac{p}{4\pi\varepsilon_2 r^2}\right)\cos\theta\\
r < a: &\quad \boldsymbol{D} = -\nabla V_D, &\dots\quad & V_D(r,\theta) = \frac{1}{3}\sigma r\cos\theta\\
r < a: &\quad \boldsymbol{E} = -\nabla V_E, &\dots\quad & V_E(r,\theta) = a^2\sigma F(r)\cos\theta.
\end{aligned} \tag{14.69}$$

Here E_0 is the strength of the incoming electric field, which is assumed to be oriented along z, and $F(r)$ is a function defined as

$$F(r) = \frac{2}{\pi}\int_0^\infty \frac{j_1(ka)\,j_1(kr)}{\varepsilon_1(k,\omega)}\,dk. \tag{14.70}$$

Proof of Eq. (14.69) In the outside region with ε_2 we proceed analogously to the quasistatic Mie theory of Sect. 9.2.1, and make the ansatz

$$r > a: \qquad V(r,\theta) = -E_0 r\cos\theta + \left(\frac{p}{4\pi\varepsilon_2 r^2}\right)\cos\theta, \tag{14.71}$$

where the first term accounts for the electric field of the incoming wave and the second one for the induced dipole. By noting that $\cos\theta = P_1(\cos\theta)$ can be expressed in terms of Legendre polynomials, we can use the same reasoning as in Sect. 9.2.1 to discard other multipole contributions. Inside the sphere we introduce the potentials V_D, V_E. The potential for the dielectric displacement is defined through the equation

$$r < a: \qquad \nabla^2 V_D = -\sigma\delta(r-a)\cos\theta, \tag{14.72}$$

where σ represents a fictitious surface charge. Although the radial derivative of V_D is continuous at the actual sphere–dielectric interface, it is discontinuous at $r = a$ in our fictitious medium. To relate the dielectric displacement to the electric field, we perform a Fourier transformation to arrive at

$$-k^2\tilde V_D(\boldsymbol{k}) + a^2\sigma\left(\int e^{-i\boldsymbol{k}\cdot\boldsymbol{r}}\cos\theta\, d\Omega\right)_{r=a} = 0,$$

with $d\Omega$ denoting the solid angle integration. The integral can be computed by using Eq. (E.15)

$$e^{-i\boldsymbol{k}\cdot\boldsymbol{r}} = 4\pi\sum_{\ell,m}(-i)^\ell j_\ell(kr)\, Y^*_{\ell m}(\hat{\boldsymbol{r}})Y_{\ell m}(\hat{\boldsymbol{k}}),$$

where we have taken the complex conjugate of Eq. (E.15) to get an expansion for $e^{-i\boldsymbol{k}\cdot\boldsymbol{r}}$ rather than for $e^{i\boldsymbol{k}\cdot\boldsymbol{r}}$. We thus find

$$\int e^{-i\boldsymbol{k}\cdot\boldsymbol{r}}\cos\theta\, d\Omega = -4\pi i\; j_1(kr)P_1(\cos\theta_k).$$

From this we can compute the Fourier transforms

$$\tilde V_D(\boldsymbol{k}) = -\frac{4\pi i a^2 j_1(ka)}{k^2}\sigma\cos\theta_k \implies V_D(\boldsymbol{r}) = \frac{1}{3}\sigma r\cos\theta \tag{14.73}$$

$$\tilde V_E(\boldsymbol{k}) = -\frac{4\pi i a^2 j_1(ka)}{\varepsilon_1(k,\omega)k^2}\sigma\cos\theta_k \implies V_E(\boldsymbol{r}) = a^2\sigma F(r)\cos\theta.$$

In going back to real space, we use the integral

$$\frac{2}{\pi}\int_0^\infty j_1(ka)j_1(kr)\, dk = \frac{r}{3a^2}, \tag{14.74}$$

and introduce the function $F(r)$ defined in Eq. (14.70).

Solution without ABC. As a preliminary step, we compute the dipole moment by employing the usual boundary conditions (without the ABC) for the tangential electric field and the normal dielectric displacement

$$V_E(a^-) = V(a^+) \implies a^2\sigma F(a) = -E_0 a + \frac{p}{4\pi\varepsilon_2 a^2}$$

$$\left[\frac{\partial V_D(r)}{\partial r}\right]_{r=a^-} = \varepsilon_2\left[\frac{\partial V(r)}{\partial r}\right]_{r=a^+} \implies \frac{\sigma}{3} = -\varepsilon_2\left(E_0 + \frac{2p}{4\pi\varepsilon_2 a^3}\right),$$

where $a^\pm$ denote positions slightly outside or inside the sphere boundary. Solving for the dipole moment then leads us to

$$p = 4\pi\varepsilon_2\left(\frac{1 - 3\varepsilon_2 a F(a)}{1 + 6\varepsilon_2 a F(a)}\right) a^3 E_0. \tag{14.75}$$

In the local limit we find together with Eq. (14.74) that $\varepsilon_1(0,\omega)F(a) = 1/(3a)$, and we are thus led to the result of Eq. (9.9) for the quasistatic Mie theory.

Solution with ABC. So far we have not considered the additional boundary condition (ABC). Indeed, in general, we find

$$\left[\frac{\partial V_D}{\partial r} - \varepsilon_b\frac{\partial V_E}{\partial r}\right]_{r=a^-} \neq 0,$$

such that the ABC of Eq. (14.68) is not fulfilled. In Ref. [137] the authors suggest a procedure for overcoming this deficiency. They start by observing that in Eq. (14.69) one can always add a solution of the homogeneous Laplace equation to V_D, for instance, that of a constant electric field $-C\hat{z}$ in the fictitious medium,

$$V_D(r,\theta) = \frac{1}{3}\sigma r\cos\theta + Cr\cos\theta. \tag{14.76}$$

The field extends to infinity, however, this is not a problem since inside the nonlocal medium the "real" fields are those with $r < a$. The Fourier transform of a constant field is

$$\tilde{\boldsymbol{D}}(\boldsymbol{k}) = -\int e^{-i\boldsymbol{k}\cdot\boldsymbol{r}}\Big(-C\hat{z}\Big)\,d^3r = (2\pi)^3\delta(\boldsymbol{k})\Big(-C\hat{z}\Big).$$

Thus, the corresponding electric field becomes

$$\boldsymbol{E}(\boldsymbol{r}) = \int e^{i\boldsymbol{k}\cdot\boldsymbol{r}}\frac{\tilde{\boldsymbol{D}}(\boldsymbol{k})}{\varepsilon(k,\omega)}\frac{d^3k}{(2\pi)^3} = -\frac{C\hat{z}}{\varepsilon_2(0,\omega)}. \tag{14.77}$$

The potential for the electric field in Eq. (14.69) then has to be rewritten in the form

$$V_E(r,\theta) = a^2\sigma F(r)\cos\theta + \left(\frac{Cr}{\varepsilon_1(0,\omega)}\right)\cos\theta.$$

With this we arrive at the following set of boundary conditions

$$\begin{aligned}
&1^{\text{st}}\text{ b.c.:} && a^2\sigma F(a) + \frac{Ca}{\varepsilon_1(0,\omega)} = -E_0 a + \frac{p}{4\pi\varepsilon_2 a^2} \\
&2^{\text{nd}}\text{ b.c.:} && \frac{\sigma}{3} + C = -\varepsilon_2\left(E_0 + \frac{2p}{4\pi\varepsilon_2 a^3}\right) \\
&\text{ABC:} && \frac{\sigma}{3} + C - \varepsilon_b\left(a^2\sigma\left[\frac{dF(r)}{dr}\right]_{r=a^-} + \frac{C}{\varepsilon_1(0,\omega)}\right) = 0.
\end{aligned} \tag{14.78}$$

After a few elementary arrangements we get for the induced dipole moment of an optically excited nanosphere with a nonlocal permittivity the final result [137, Eq. (34)]

Dipole Moment of Nanosphere Including Nonlocality

$$p = 4\pi\varepsilon_2\left[\frac{1 - 3\varepsilon_2 aF(a) + K\big(\varepsilon_1(0,\omega) - \varepsilon_2\big)}{1 + 6\varepsilon_2 aF(a) + K\big(\varepsilon_1(0,\omega) + 2\varepsilon_2\big)}\right] a^3 E_0, \tag{14.79}$$

with the abbreviation

$$K = \Big(\varepsilon_1(0,\omega) - \varepsilon_b\Big)^{-1}\left(3\varepsilon_b a^2\left[\frac{dF}{dr}\right]_{r=a} - 1\right).$$

Figure 14.12 shows extinction spectra for a single nanosphere as computed without (dashed line) and with (solid line) nonlocal effects. One observes that because of nonlocality the dipolar surface plasmon peak is shifted to the blue, as an effect of the additional pressure term in Eq. (14.45). When comparing the results with and without the ABC, one observes that the inclusion of the ABC has a noticeable influence on the spectra. Quite generally one finds that nonlocal effects are important for small spheres only.

At first sight, the approach of Dasgupta and Fuchs [137] appears to be specifically tailored for spherical nanoparticles and one might wonder how to extend the approach to other nanoparticle geometries. In the following we briefly describe a few selected attempts for the generalization of the scheme.

Hydrodynamic model. In the hydrodynamic model one can relate the current through a velocity potential Ψ defined as

$$\boldsymbol{J} = -\nabla\Psi,$$

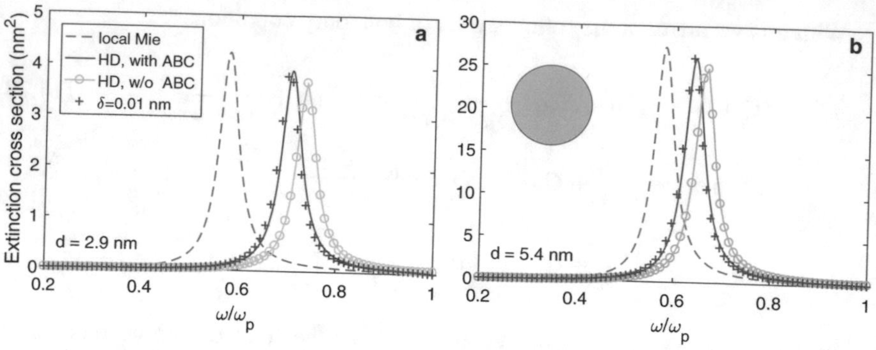

Fig. 14.12 Extinction cross section for gold nanosphere computed with and without nonlocal permittivities. (**a**) Results for sphere diameter of 2.9 nm and for local permittivity (dashed line), the hydrodynamic model with and without ABC (solid line, circle symbols), as computed from Eqs. (14.79) and (14.75). The cross symbols report results for the artificial coating layer of Luo et al. [144]. (**b**) same as panel (**a**) but for sphere diameter of 5.4 nm. For all simulations we use the Drude dielectric function of Ref. [144] representative of gold, with $\kappa_b = 1$, $\omega_p = 3.3$ eV, $\gamma = 0.165$ eV, and $\beta = 0.0036\,c$. We use $\varepsilon_2 = \varepsilon_0$

whose dynamics is governed by the linearized version of Eq. (14.45). This potential carries over many of the properties of the scalar potential V of Maxwell's theory, and one can thus treat Ψ, V on an equal footing. In Ref. [143] the authors derived within such an approach an expression similar to Eq. (14.79), as well as generalizations for spheres situated on substrates.

Artificial coating layer. In Ref. [144] the authors start by investigating the nonlocal optical response of selected geometries, such as stratified media, cylinders, and spheres, using the hydrodynamic model. From the analytic expressions they show that the effects of nonlocality can be mimicked through a surrogate system, consisting of (1) a nanoparticle with a *local* permittivity, which is (2) covered by a thin dielectric layer. To a good approximation, the permittivity $\varepsilon_{\text{layer}}$ of this artificial layer with thickness δ can be chosen according to [144]

$$\varepsilon_{\text{layer}} = \frac{\varepsilon_1(0,\omega)\varepsilon_2}{\varepsilon_1(0,\omega)-\varepsilon_2}\,Q\delta.$$

The apparent advantage of this approach is that one can use standard Maxwell's solvers for the simulation of nonlocal effects. The only modification needed is the inclusion of a sufficiently thin artificial layer. The cross symbols in Fig. 14.12 report simulation results for such an artificial layer model, with a geometry depicted in the inset of panel (b), which are in very good agreement with the predictions of Eq. (14.79).

Boundary element method. As in this book we have advocated so heavily for the boundary element method (BEM) approach, one might wonder whether nonlocality can be also considered there. A first step in this direction was made in Ref. [145], where, however, the ABC was not considered. In the following we show how to bring the BEM approach to full glory.

In the outside region we can relate the potential to a surface charge distribution in the usual manner

$$r \in \Omega_2 : \qquad V(r) = \oint_{\partial\Omega} G(r - s')\sigma_2(s')\,dS' + V_{\rm inc}(r),$$

where G is the Green's function $G(R) = 1/(4\pi R)$ for the Poisson equation and $V_{\rm inc}$ the external potential, associated, for instance, with a plane wave excitation. Inside the nanoparticle we make the following ansatz:

$$r \in \Omega_1 : \qquad V_D(r) = \oint_{\partial\Omega} G(r - s')\Big[\sigma_1(s') + \tilde\sigma_1(s')\Big]\,dS' \tag{14.80}$$

$$V_E(r) = \oint_{\partial\Omega} \Big[G_{\rm nl}(r - s', \omega)\sigma_1(s') + \frac{G(r - s')}{\varepsilon_1(0, \omega)}\tilde\sigma_1(s')\Big]\,dS',$$

with the nonlocal Green's function of Eq. (14.56). Let us discard for a moment the $\tilde\sigma_1$ contribution. We then observe that V_D is a solution of Eq. (14.72), and from the nonlocal Green's function definition of Eq. (14.55) we find that V_E is the electric field for the *same* surface charge distribution. Thus, the solutions of V_D, V_E are in complete agreement with Eq. (14.69) of the Dasgupta and Fuchs approach.

Up to here we have not invoked the ABC. As previously discussed, we can always add a solution of Laplace's equation to V_D and V_E. Figure 14.11b shows how this can be done. We attach a surface charge distribution to a boundary $\partial\Omega_\infty$ situated far away from the nanoparticle. As discussed in Chap. 5 in the context of the representation formula, the potential is completely determined when fixing its normal derivative (which is related to the surface charge) at the interface. For V_E we can proceed similarly and observe that for large distances $|r - r'|$ the nonlocal Green's function can be related to the local Green's function via

$$G_{\rm nl}(r - r', \omega) \underset{r \gg r'}{\longrightarrow} \frac{G(r - r')}{\varepsilon_1(0, \omega)}.$$

This asymptotic limit is particularly transparent for Eq. (14.61), but it can be also proven for other permittivity functions. Here comes the magic: for the solution of Poisson's equation it does not matter whether we attach $\tilde\sigma_1$ to a boundary at infinity or to the nanoparticle boundary, all that matters is that the potential value is given on some closed boundary. This is the essence of the representation formula. However, the representation formula does in general *not* hold for the electric field and a nonlocal permittivity (the reader may like to go back to Chap. 5 and check that in its derivation we have explicitly used Laplace's equation). For this reason we have taken the detour of a boundary at infinity, which allows us replacing the nonlocal Green's function by the local one, and to exploit then again the representation formula.

Equation (14.80) is equivalent to the Dasgupta and Fuchs ansatz including the homogeneous solution. We finally have to supplement the equations for V, V_D, V_E with the boundary conditions

$$V_E = V, \qquad \frac{\partial V_D}{\partial n} = \varepsilon_2 \frac{\partial V}{\partial n}, \qquad \frac{\partial V_D}{\partial n} - \varepsilon_b \frac{\partial V_E}{\partial n} = 0,$$

where all potentials and normal derivatives have to be evaluated at the particle boundary (V at the outside and V_D, V_E at the inside). The transformation from boundary integrals to boundary elements is along the same lines as discussed in Chap. 11, and we end up with a scheme that is not too different from that for local permittivities. An advantage of the BEM implementation with respect to other schemes is that it is not limited to hydrodynamic permittivities, but can be easily used with more sophisticated dielectric functions.

Figure 14.13 shows results for two coupled nanospheres. While in the local model the resonance position for the bonding dipole mode continuously shifts to the red when decreasing the gap distance, as an effect of the increasing coupling strength, in the nonlocal model the peak position saturates at a value of about 1000 nm. Such a behavior was observed experimentally in Ref. [140].

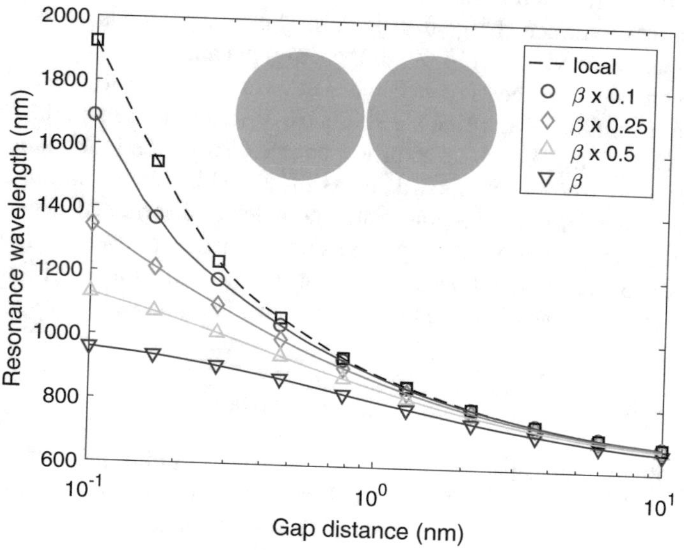

Fig. 14.13 Resonance wavelength for bonding mode of coupled spheres, with sphere diameters of 20 nm and the material parameters given in Fig. 14.12. In the local model the resonance continuously shifts with decreasing gap distance to the red, as an effect of the increasing coupling strength between the two spheres. When including nonlocality, one starts to see significant deviations for distance comparable to $Q^{-1} \approx 1$ nm, and for the smallest gap distances the peak position converges to a value of about 1000 nm. We also plot results for different β values, see Eq. (14.53)

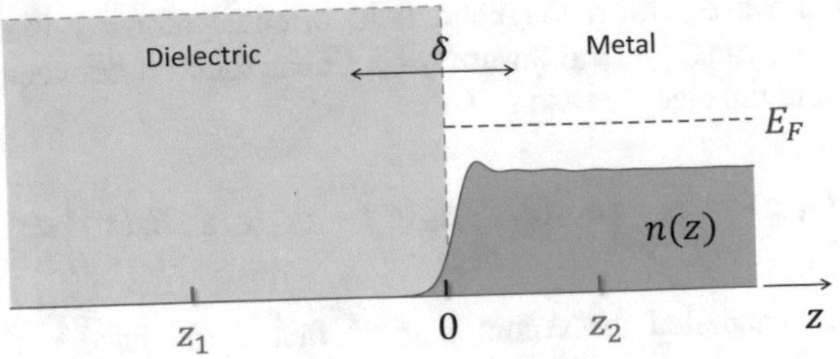

Fig. 14.14 Schematics of Feibelman parameters. $n(z)$ reports the groundstate electron density, E_F indicates the Fermi energy in the metal, and the gray shaded area reports the potential barrier of the dielectric. To account for quantum effects of the electron density in the metal, we introduce a description where we match the electromagnetic fields not directly at the interface, $z = 0$, but at positions z_1, z_2 which are located about a nanometer away. In the matching of the electromagnetic fields we then introduce the Feibelman parameters $d_\perp, d_\parallel$ which account for the actual form of the charge distribution

14.5.3 Feibelman Parameters

We conclude this section with a brief discussion regarding modifications of the dielectric response close to interfaces. A thorough discussion of the topic can be found in Refs. [146–148]. We consider a planar interface between a dielectric and a metal, which is excited by an incoming plane wave, as previously discussed in the context of the Fresnel coefficients in Chap. 8. So far we have only been concerned with situations where the material properties change abruptly at the interface. However, Fig. 14.14 shows a more realistic description where the material properties change smoothly around z, which is due to the quantum mechanical wavefunctions of the electrons. We are now seeking for a modified description of the reflection and transmission of electromagnetic waves at such interfaces. As we will discuss next, this can be done by (1) starting with the usual electromagnetic field description without any modifications due to the more complicated material response in the vicinity of the interface, and to (2) correct for these modifications through altered Fresnel coefficients.

Suppose that δ is the range where a refined microscopic description is needed. We then wish to match the fields at positions z_1 and z_2 that lie outside this δ range, as indicated in Fig. 14.14. The precise values of z_1, z_2 have in general no significant impact on the results, and we typically set $\delta \leq 1$ nm [146]. The central assumption of the following discussion is that the light wavelength λ is much larger than δ

$$\lambda \gg \delta .$$

Consider a TM wave with a magnetic field oriented along $\hat{\boldsymbol{y}}$ that impinges on the interface. The most general ansatz for a linear relation between the dielectric displacement and the electric field is

$$D_x(z) = \int_{-\infty}^{\infty}\Big[\varepsilon_{xx}(z,z')E_x(z') + \varepsilon_{xz}(z,z')E_z(z')\Big]\,dz',$$

where we have cancelled the common $e^{ik_x x}$ factor and have not indicated the dependence of all quantities on k_x.

Abrupt Interface For an abrupt interface the electromagnetic fields on the dielectric side with $z < 0$ can be expressed as

$$\begin{aligned} \boldsymbol{H}(z) &= H_0\Big[e^{ik_{1z}z} + R\,e^{-ik_{1z}z}\Big]\hat{\boldsymbol{y}} \\ \boldsymbol{E}(z) &= \frac{Z_1}{k_1}H_0\Big[\Big(e^{ik_{1z}z} - R\,e^{-ik_{1z}z}\Big)k_{1z}\hat{\boldsymbol{x}} - \Big(e^{ik_{1z}z} + R\,e^{-ik_{1z}z}\Big)k_x\hat{\boldsymbol{z}}\Big], \end{aligned} \tag{14.81a}$$

where R is the TM reflection coefficient and we have used Eq. (2.40) to relate the electric field to the magnetic field. Similarly, on the metal side with $z > 0$ we find by a similar token

$$\begin{aligned} \boldsymbol{H}(z) &= \quad H_0\Big[T e^{ik_{2z}z}\Big]\hat{\boldsymbol{y}} \\ \boldsymbol{E}(z) &= \frac{Z_2}{k_2}H_0\Big[T\,e^{ik_{2z}z}\Big]\Big(k_{2z}\hat{\boldsymbol{x}} - k_x\hat{\boldsymbol{z}}\Big), \end{aligned} \tag{14.81b}$$

where T is the TM transmission coefficient. For small arguments $k_z z$ the electromagnetic fields can be expanded according to

$$\begin{aligned} H_y(z) &\approx H_y(0^\pm) + \left[\frac{dH_y}{dz}\right]_{0^\pm} z \\ E_x(z) &\approx E_x(0^\pm) + \left[\frac{dE_x}{dz}\right]_{0^\pm} z. \end{aligned} \tag{14.81c}$$

The fields have to be evaluated at $z = 0^-$ on the dielectric side and at $z = 0^+$ on the metal side. The lowest-order contributions can be directly read off from the above expressions by replacing the exponentials with one, and the derivative terms can be simplified by using Maxwell's equations, as we will do in a moment.

For the usual boundary condition of Maxwell's equations the coefficients R, T would be given by the Fresnel coefficients of Eq. (8.26). However, we are here seeking for a refined description which accounts for modified boundary conditions. To match the electromagnetic fields at the interface, we start from Maxwell's equations and write down the curl parts for the fields of Eq. (14.81),

$$\begin{aligned} \frac{dH_y(z)}{dz} &= i\omega D_x(z) \\ ik_x E_z(x) - \frac{dE_x(z)}{dz} &= i\mu_0\omega H_y(z). \end{aligned} \tag{14.82}$$

Throughout we set the magnetic permeability to μ_0. We next integrate this set of equations from z_1 to z_2 to get

$$H_y(z_2) - H_y(z_1) = i\omega \int_{z_1}^{z_2} D_x(z)\,dz \tag{14.83}$$

$$E_x(z_2) - E_x(z_1) - ik_x \int_{z_1}^{z_2} E_z(z)\,dz = -i\mu_0\omega \int_{z_1}^{z_2} H_y(z)\,dz.$$

If we let z_1, z_2 approach zero, we recover the usual boundary conditions that the tangential fields of H_y, E_x are continuous at the interface. The integral contributions in the above equations thus provide the corrections.

Long Wavelength Expansion In the following we perform an expansion in powers of $k_z z$, and keep terms up to first order only. We start by decomposing D_x, E_z into

$$\begin{aligned} D_x(z) &= \varepsilon_1\theta(-z)\,E_x(0^-) + \varepsilon_2\theta(z)\,E_x(0^+) + \Delta D_x(z) \\ E_z(z) &= \quad \theta(-z)\,E_z(0^-) + \quad \theta(z)\,E_z(0^+) + \Delta E_z(z), \end{aligned} \tag{14.84}$$

where the first two terms on the right-hand side are the lowest order solutions of Eq. (14.81) and the last term is the remainder. A similar decomposition can be also made for the magnetic field H_y. Inserting these decompositions into Eq. (14.83) then gives

$$\begin{aligned} H_y(z_2) - H_y(z_1) &\approx i\omega\Big[\varepsilon_2 z_2 E_z(0^+) - \varepsilon_1 z_2 E_z^0(0^-) + \int_{z_1}^{z_2} \Delta D_x(z)\,dz\Big] \\ E_x(z_2) - E_x(z_1) &\approx ik_x\Big[\quad z_2 E_z(0^+) - \quad z_1 E_z(0^-) + \int_{z_1}^{z_2} \Delta E_z(z)\,dz\Big] \\ &\quad -i\mu_0\omega\Big[z_2 H_y(0^+) - z_1 H_y(0^-)\Big]. \end{aligned} \tag{14.85}$$

We have neglected the correction for H_y in the last term for the following reason. In contrast to D_x, E_z, which are discontinuous at the interface, H_y is continuous and thus the correction is of higher order in $k_z z$.

We next expand the terms on the left-hand side using Eq. (14.81c), and use that the terms proportional to z_1, z_2 cancel each other owing to Eq. (14.82),

$$\begin{aligned} \left[\frac{dH_y}{dz}\right]_{0+} z_2 - \left[\frac{dH_y}{dz}\right]_{0-} z_1 &= i\omega\Big[\varepsilon_2 z_2 E_z(0^+) - \varepsilon_1 z_1 E_z(0^-)\Big] \\ \left[\frac{dE_x}{dz}\right]_{0+} z_2 - \left[\frac{dE_x}{dz}\right]_{0-} z_1 &= ik_x\Big[\quad z_2 E_x(0^+) - \quad z_1 E_x(0^-)\Big] \\ &\quad -i\mu_0\omega\Big[z_2 H_y(0^+) - z_1 H_y(0^-)\Big]. \end{aligned}$$

With this we get instead of Eq. (14.85)

$$\begin{aligned} H_0\Big(T - 1 - R\Big) &= i\omega \int_{z_1}^{z_2} \Delta D_x(z)\,dz \\ \frac{Z_2 k_{2z}}{k_2} H_0\,T - \frac{Z_1 k_{1z}}{k_1} H_0\Big(1 - R\Big) &= ik_x \int_{z_1}^{z_2} \Delta E_z(z)\,dz. \end{aligned}$$

It is now convenient to introduce the *normalized* correction terms

$$\Delta\bar{D}_x = \frac{\Delta D_x}{\varepsilon_0 E_x(0^+)}, \qquad \Delta\bar{E}_z = \frac{\Delta E_z}{E_z(0^+)}, \tag{14.86}$$

where we have assumed that the corrections are proportional to the fields on the metal side. To the same order in $k_z z$ we could have also taken the fields on the dielectric side, but the above form appears to be better suited for the problem under study. With this, we rewrite the equations for R, T in the form

$$T - 1 - R = i\frac{\varepsilon_0 k_{2z} T}{\varepsilon_2}\int_{z_1}^{z_2} \Delta\bar{D}_x(z)\,dz$$

$$\frac{k_{2z}T}{\varepsilon_2\omega} - \frac{k_{1z}}{\varepsilon_1\omega}\Big(1 - R\Big) = \frac{ik_x^2 T}{\varepsilon_2\omega}\int_{z_1}^{z_2} \Delta\bar{E}_z(z)\,dz.$$

Solving for the reflection coefficient then gives [147, Eq. (14a)]

$$R = \frac{\varepsilon_2 k_{1z}\left(1 - \dfrac{i\varepsilon_0 k_{2z}}{\varepsilon_2}\displaystyle\int_{z_1}^{z_2}\Delta\bar{D}_x(z)\,dz\right) - \varepsilon_1 k_{2z}\left(1 + \dfrac{ik_x^2}{k_{2z}}\displaystyle\int_{z_1}^{z_2}\Delta\bar{E}_z(z)\,dz\right)}{\varepsilon_2 k_{1z}\left(1 - \dfrac{i\varepsilon_0 k_{2z}}{\varepsilon_2}\displaystyle\int_{z_1}^{z_2}\Delta\bar{D}_x(z)\,dz\right) + \varepsilon_1 k_{2z}\left(1 + \dfrac{ik_x^2}{k_{2z}}\displaystyle\int_{z_1}^{z_2}\Delta\bar{E}_z(z)\,dz\right)}.$$

We here follow Ref. [147] and introduce the effective wavenumbers

$$\tilde{k}_{1z} = \left[1 + ik_{2z}\left(1 - \frac{\varepsilon_0}{\varepsilon_2}\right)d_\parallel(\omega)\right]k_{1z}$$
$$\tilde{k}_{2z} = \left[1 + i\frac{k_x^2}{k_{2z}}\left(1 - \frac{\varepsilon_0}{\varepsilon_2}\right)d_\perp(\omega)\right]k_{2z}, \tag{14.87}$$

which allow us to rewrite the reflection coefficient in the form

$$R = \frac{\varepsilon_2\tilde{k}_{1z} - \varepsilon_1\tilde{k}_{2z}}{\varepsilon_2\tilde{k}_{1z} + \varepsilon_1\tilde{k}_{2z}}. \tag{14.88}$$

In the definitions of $\tilde{k}_{1z}$, $\tilde{k}_{2z}$ we have introduced the Feibelman parameters

Feibelman Parameters

$$d_\parallel(\omega) = \frac{\varepsilon_0}{\varepsilon_0 - \varepsilon_2}\int_{z_1}^{z_2}\Delta\bar{D}_x(z)\,dz$$
$$d_\perp(\omega) = -\frac{\varepsilon_0}{\varepsilon_0 - \varepsilon_2}\int_{z_1}^{z_2}\Delta\bar{E}_z(z)\,dz. \tag{14.89}$$

The neat thing about the Feibelman parameters is that they allow describing an in principle complicated problem, namely the microscopic electron response in the vicinity of interfaces, in terms of two parameters $d_\parallel$, $d_\perp$, which additionally may depend on the parallel wavevector component k_x.

In many cases the Feibelman parameters can be expressed even more explicitly with the induced charge distribution. We set the permittivity of the dielectric to ε_0 and start with Gauss' law

$$ik_x D_x(z) + \frac{dD_z(z)}{dz} \approx \frac{dD_z(z)}{dz} = \delta\rho(z),$$

where we have neglected the term proportional to k_x, which is expected to be much smaller. $\delta\rho$ is the induced charge distribution. Upon integration over z we get

$$\Big(\varepsilon_2 - \varepsilon_0\Big) E_z(0^+) \approx \int_{z_1}^{z_2} \delta\rho(z)\, dz,$$

with $D_z(0^-) \approx D_z(0^+)$. Through integration by parts we obtain

$$\int_{z_1}^{z_2} E_z(z)\, dz = zE_z(z)\Big|_{z_1}^{z_2} - \int_{z_1}^{z_2} z\frac{dE_z(z)}{dz}\, dz = -\int_{z_1}^{z_2} z\frac{d\Delta E_z(z)}{dz}\, dz$$
$$\approx -\left(\int_{z_1}^{z_2} z\frac{d\Delta\bar{E}_z(z)}{dz}\, dz\right) E_z(0^+) \approx -\varepsilon_0 \int_{z_1}^{z_2} z\delta\rho(z)\, dz,$$

where we have again employed Gauss' law in the last step. Combing the above equations finally leads us to the expression

$$d_\perp(\omega) \approx \left(\int_{z_1}^{z_2} z\delta\rho(z)\, dz\right) \Big/ \left(\int_{z_1}^{z_2} \delta\rho(z)\, dz\right). \tag{14.90}$$

Thus, $d_\perp$ is a measure of the position of the surface charge distribution. By a similar token one can show that $d_\parallel$ is a measure of the normal derivative of the induced current [146, 147].

Surface effects in the framework of Feibelman parameters have been studied for plasmonic nanoparticles in Ref. [149]. In Ref. [150] the authors used the Feibelman parameters to investigate whether nonlocal effects lead to a red or blue shift of surface plasmon resonances.

> The real part of the Feibelman parameter $d_\perp(\omega)$ gives the position of the centroid of the induced charge density at the interface and determines finite size effects in metal clusters as well as the dispersion of surface plasmons at flat surfaces. When measured from the jellium edge, it is positive for alkali metals, i.e., the screening charge is shifted into the vacuum because of the spill out of the conduction electrons outside the metal. [...] Importantly, for noble metals such as Au and Ag, finite size effects and surface screening lead to a blue shift of the dipole plasmon resonance. The difference between alkali and noble metals can be explained as due to the contribution of the localized d-electrons to the total screening in noble metals. When the d-electron contribution is accounted for, $\mathrm{Re}[d_\perp(\omega)]$ turns negative indicating that the screening charge is predominantly induced inside the metal.

Density Functional Theory

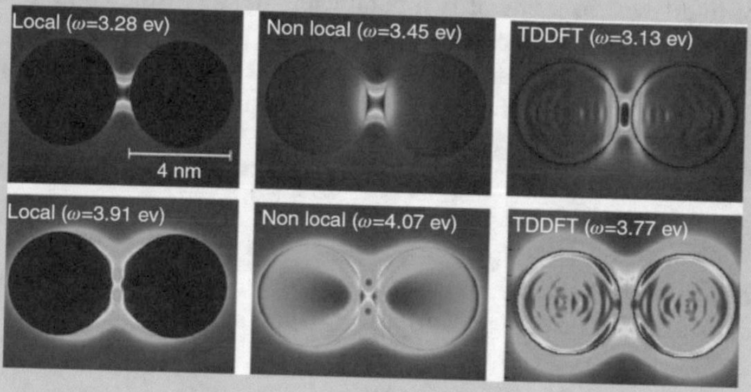

Recent years have seen strong efforts to employ first principles calculations for the theoretical modelling of plasmonic nanoparticles. The methods of choice are density functional theory (DFT) and time-dependent density functional theory (TDDFT). These methods are among *the* success stories of solid state theory, and are currently among the most applied simulation approaches in theoretical physics and chemistry. Part of the Nobel Prize in chemistry 1998 has been awarded to Walter Kohn "for his development of the density-functional theory." In this book we do not introduce these techniques, despite their importance and widespread use in many plasmonics studies, but refer the interested reader to the literature [73]. For a recent introduction to DFT and TDDFT see, for instance, Ref. [151].

The above image is taken from Ref. [152] and reports the field enhancement of the dipole (top row) and quadrupole (bottom row) resonances of a sodium nanowire dimer with diameters of 4 nm.

To summarize, the Feibelman parameters account for modifications of the boundary conditions of Maxwell's equations due to quantum effects of the electron charge density close to interfaces. Although the details of these modifications can be fairly involved, it often suffices to introduce two additional parameters (which, however, may still depend on frequency and on the parallel wavevector k_x) in order to describe the related physics properly. This suggests that there is still plenty of room between full ab-initio simulations and more phenomenological models based on effective material descriptions.

14.5.4 Charge Transfer Plasmons

As discussed in Chap. 9, when the gap distance between two metallic spheres is decreased the bonding plasmon mode increasingly shifts to the red. We have shown in Fig. 14.13 that this redshift undergoes a saturation for distances smaller than say a nanometer, owing to nonlocality. Another effect that occurs at small particle separations is quantum tunneling, where electrons tunnel through the gap region from one particle to the other one. This gives rise to a new type of plasmon resonance, which is usually denoted as charge transfer plasmons and which will be briefly discussed in this section. Quantum tunneling has received considerably interest in recent years, both from the experimental [153–155] as well as the theoretical side [73, 156, 157].

Figure 14.15 reports the confinement potential and the electron density for two metal slabs separated by a small gap in the sub-nanometer range. Electrons are confined in the metal regions, and it takes an energy larger than the metal work function, for *Au* and *Ag* in the range between 4 and 5 eV, to liberate the electron from an initial state located close to the Fermi energy. However, for energies smaller than the work function, electrons can penetrate into the classically forbidden gap region and tunnel through the gap. The electron wavefunction in the gap region has an evanescent character with an exponential decay, as can be inferred from the electron density profile depicted in the figure. Quite generally, a realistic density profile has to be obtained from sufficiently sophisticated solid state theories, such as density functional theory [151]. In the following we consider a considerably simpler approach based on a Jellium model. Here the ionic lattice of the metal is approximated by a homogeneous positive charge background, within which the electrons move as free particles, and the gap region is modelled as a potential barrier with some height $U_{\rm gap}$. The electron wavefunction is $e^{i\boldsymbol{k}_1\cdot\boldsymbol{r}}$ inside the metal and $e^{i\boldsymbol{k}_2\cdot\boldsymbol{r}}$ in the gap region. From energy conservation we find

$$E=\frac{\hbar^2}{2m}\left(k_x^2+k_{1z}^2\right)=\frac{\hbar^2}{2m}\left(k_x^2+k_{2z}^2\right)+U_{\rm gap}, \tag{14.91}$$

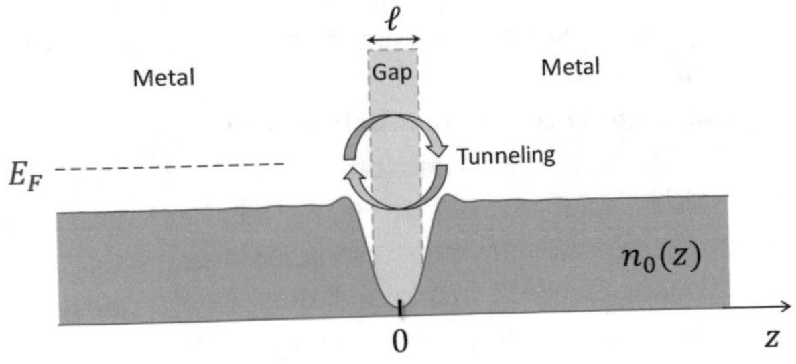

Fig. 14.15 Schematics of quantum tunneling between two metals that are separated by a gap in the sub-nanometer range (gap distance ℓ). $n_0(z)$ reports the groundstate electron density, E_F indicates the Fermi energy in the metal, and the gray shaded area the potential barrier of the gap

where m is the electron mass and we have assumed the conservation of parallel momentum at the metal–gap interface, which is a consequence of the wavefunction continuity [158]. For the electron energy being smaller than U_{gap}, the electron wavefunction has an evanescent character in the gap region. Using the continuity of the wavefunction derivative [158], we can then compute the amplitudes of the reflected and transmitted waves at the interface, in complete analogy to our discussion of the transfer matrix approach in Chap. 8. The only difference is the relation between k_{1z}, k_{2z} and k_x, E, which has to be computed from the parabolic dispersion of Eq. (14.91). The central quantity needed in the following analysis is the probability $T(\boldsymbol{k}, \ell)$ that an incoming electron with wavevector $\boldsymbol{k}$ is transferred through the gap region with distance ℓ. It can be computed either from the Jellium model description sketched above or from more sophisticated theoretical approaches which may include effects such as electron–electron interactions in the metal or image charges in the gap region [156, 157].

In Ref. [156] the authors suggested a quantum-corrected model based on classical electrodynamics, but using an artificial material in the gap region that has a constant conductivity $\sigma_{\text{gap}}(\ell)$ whose value is chosen such that it gives the proper tunnel current for an electric field applied along the gap direction. We can relate the electron transfer probability to the gap conductivity using [156, 157]

$$\sigma_{\text{gap}}(\ell) = \frac{\ell}{8\pi^3} \int_{-\infty}^{\infty} T(\boldsymbol{k}, \ell) f\big(E(\boldsymbol{k})\big)\, d^3k, \tag{14.92}$$

where f is the Fermi–Dirac distribution function for the metal electrons, which can be well approximated by a step function for zero temperature.

Figure 14.16 shows extinction spectra for coupled spheres computed without (gray lines) and with (red lines) consideration of tunneling. In case of tunneling, we use the quantum-corrected model with an artificial material representative for silver [157]. Without tunneling, the bonding dipole mode shifts continuously to the red when decreasing the gap distance (see dashed line), as previously discussed in Chap. 9. When considering quantum tunneling, (1) the bonding mode broadens and the peak position saturates, and (2) an additional charge transfer plasmon appears at photon energies below 1 eV. In the inset we report the physical nature of charge transfer plasmons, where electrons tunnel from one sphere to the other one, and a new dipolar plasmon mode is formed with a dipolar charge distribution that extends over both spheres.

A considerably different behavior is observed for the coupled cubes shown in Fig. 14.17. Here the extinction spectra look almost identical when neglecting or considering quantum tunneling. As has been previously discussed in the context of Eq. (9.36), the plasmon resonance energies depend on the ratio between the electrostatic energies inside and outside the particles. The cube configuration is reminiscent of a condensator, and correspondingly the electrostatic energy of the configuration saturates when decreasing the gap distance. Based on these arguments, the resonance frequency of the charge transfer plasmon is expected to be very similar

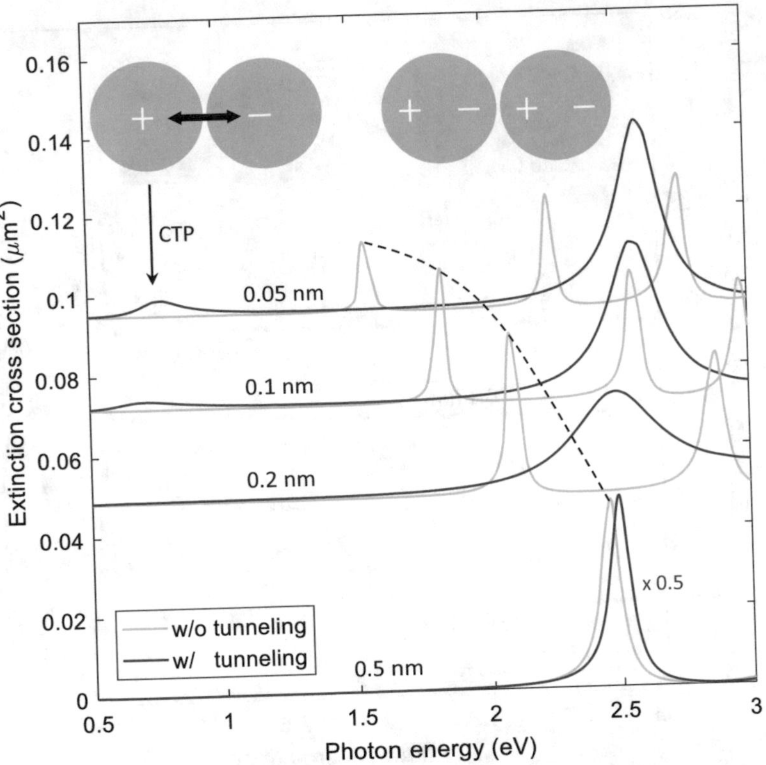

Fig. 14.16 Extinction cross sections for coupled spheres with varying gap distance (reported in the panels), without (gray lines) and with (red lines) consideration of quantum tunneling. The spheres have a diameter of 50 nm and we use Drude-type material and tunneling parameters representative for silver [157]. In absence of tunneling and when decreasing the gap distance, the bonding dipole peak continuously shifts to the red (the dashed line is a guide to the eye). Tunneling causes a distinctly different behavior for the smallest gap distances, say below 0.2 nm: the bonding mode broadens and the peak position saturates, and an additional charge transfer plasmon (CTP) appears at photon energies below 1 eV. As schematically shown in the inset by the black arrow, in the CTP electrons tunnel from one sphere to the other one and give rise to a novel, low-energy plasmon peak. The spectra are offset for clarity, and have been scaled by a factor of half for the largest gap distance of 0.5 nm as well as for all spectra without tunneling

to that of the bonding dipolar modes of particles with flat gap terminations, as schematically shown in the inset of the figure. Thus, the morphology of the gap has an important impact on the resonance energies of charge transfer plasmons [159]. Finally, the small peaks in the low-energy regions of the spectra are attributed to transverse cavity modes, which can be interpreted in terms of coupled surface plasmon modes within the gap region [159, 160]. Their physical origin has nothing to do with quantum tunneling, and correspondingly they show up in the spectra without and with consideration of tunneling.

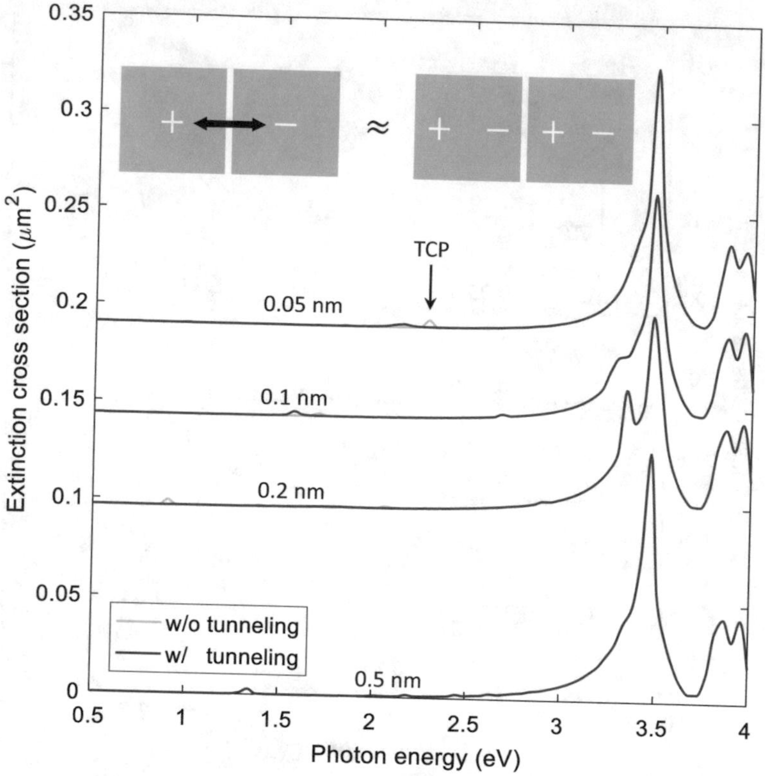

Fig. 14.17 Same as Fig. 14.16, but for silver cubes with a base length of 50 nm. The extinction spectra without (gray lines) and with (red lines) consideration of quantum tunneling are almost identical. This is attributed to the morphology of the gap, as discussed in Ref. [159] and in the text. The small peaks in the low-energy range are attributed to transverse cavity plasmons (TCPs) [159, 160]

14.6 Electron Energy Loss Spectroscopy Revisited

In the remainder of this chapter we present a general derivation of Fermi's golden rule, which is then used for a quantum description of electron energy loss spectroscopy (EELS) of plasmonic nanoparticles. We consider a generic setup where a system is coupled to a reservoir. For instance, in EELS the "system" consists of the swift electron, which is coupled to the "reservoir" formed by the many electrons of the plasmonic nanoparticle. The total Hamilton is of the form

$$\hat{H} = \hat{H}_0 + \hat{V},$$

where $\hat{H}_0$ accounts for the uncoupled system and reservoir, and $\hat{V}$ for the coupling. We adopt an interaction representation according to $\hat{H}_0$. The dynamics of the coupled system and reservoir will be investigated in more detail in Chap. 17, in the following we employ a perturbation theory in lowest order of $\hat{V}$. Suppose that the initial state is of the form $|i, r\rangle$, where i and r label the states of the system and reservoir, respectively. As an effect of the coupling, the system can be scattered to state $|f, r'\rangle$. The time evolution of the system and reservoir is according to

$$\hat{U}(t,0)\big|i,r\big\rangle,$$

where $\hat{U}$ is the time evolution operator of Eq. (13.12) in the interaction picture. For the moment we consider a single initial reservoir state only, but will introduce an average over the thermally populated reservoir states at the end. The probability of finding the system at a later time in the final state f is given by

$$P_f(t) = \sum_{r'} \Big\langle i,r\Big|\hat{U}^\dagger(t,0)\Big|f,r'\Big\rangle\Big\langle f,r'\Big|\hat{U}(t,0)\Big|i,r\Big\rangle,$$

where we have summed over all final reservoir states. For a monoexponential decay, the rate with which P_f increases is then proportional to the scattering rate $\Gamma_{i\to f}$,

$$\Gamma_{i\to f} = \frac{d}{dt}\sum_{r'} \Big\langle i,r\Big|\hat{U}^\dagger(t,0)\Big|f,r'\Big\rangle\Big\langle f,r'\Big|\hat{U}(t,0)\Big|i,r\Big\rangle.$$

To the lowest order in $\hat{V}$ we can express the time evolution operator in the form of Eq. (13.14), and are led together with Eq. (13.12) to

$$\Gamma_{i\to f} = \frac{1}{\hbar^2}\int_0^t \sum_{r'} \Big\langle i,r\Big|\hat{V}(t)\Big|f,r'\Big\rangle\Big\langle f,r'\Big|\hat{V}(t')\Big|i,r\Big\rangle dt' + \text{c.c.}, \tag{14.93}$$

with c.c. denoting the complex conjugate of the preceding term. We have used $\langle i, r|f, r'\rangle = 0$.

Evaluation of Fermi's Golden Rule. To arrive at our final expression we introduce two additional modifications. First, to account for the reservoir states which are thermally populated we introduce in analogy to Eq. (14.5) the statistical reservoir operator

$$\hat{\rho}_R = Z^{-1}\sum_r e^{-\beta E_r}\big|r\big\rangle\big\langle r\big|. \tag{14.94}$$

With this, the average over the reservoir degrees of freedom in Eq. (14.93) can be written in the form

$$\sum_{r,r'} p_r\big\langle r\big|\ldots\big|r'\big\rangle\big\langle r'\big|\ldots\big|r\big\rangle = \text{tr}_R\Big(\hat{\rho}_R\cdots\Big),$$

where we have exploited the completeness relation $\sum_{r'}|r'\rangle\langle r'| = \mathbb{1}$. Secondly, we assume that the reservoir is sufficiently large such that its spectrum is continuous. The integrand in

Eq. (14.93) is then assumed to decay fast as a function of time difference $t - t'$. We can thus introduce the adiabatic limit in complete accordance to our previous discussion

$$\int_0^t \Big\langle \dots \hat{V}(t) \dots \hat{V}(t') \dots \Big\rangle dt' \approx \lim_{\eta \to 0} \int_{-\infty}^0 e^{\eta t'} \Big\langle \dots \hat{V}(0) \dots \hat{V}(t') \dots \Big\rangle dt',$$

where we have assumed that the integrand depends on the time difference $t - t'$ only.

Putting together all results, we arrive at the scattering rate computed within Fermi's golden rule

Scattering Rate Within Fermi's Golden Rule

$$\Gamma_{i \to f} = \frac{1}{\hbar^2} \lim_{\eta \to 0} \int_{-\infty}^0 e^{\eta t} \, \mathrm{tr}_R \left(\hat{\rho}_R \big\langle i \big| \hat{V}(0) \big| f \big\rangle \big\langle f \big| \hat{V}(t) \big| i \big\rangle \right) dt + \text{c.c.}. \tag{14.95}$$

This expression can be interpreted such that at time t the system and reservoir interact, the perturbed system and reservoir propagate in time, and finally couple again at time zero. Owing to Heisenberg's uncertainty principle $\Delta E \Delta t \approx \hbar$ we observe that for a finite collision time Δt a final amount of energy ΔE can be exchanged in a scattering. See also Chap. 17 for a more detailed discussion.

14.6.1 Energy Loss of a Swift Electron

In Chap. 10 we have discussed electron energy loss spectroscopy (EELS) within a semiclassical framework. With the relation between the density–density correlation function and the permittivity of Eq. (14.34) we are now in the position to analyze the problem in the framework of quantum mechanics. Our discussion closely follows the original paper of Ritchie [161], see also Ref. [162] for further details, where the author investigates plasma losses by swift electrons in thin metallic films. This paper is the first one introducing the concept of *surface plasmons*, although the experimental situation was less convincing in 1957 when the paper appeared in Physical Review.

> The possibility occurs to one that these sub-plasma frequency losses may be identified with the low-lying losses observed by some experimenters using thin foils. It does not seem that the observed values of the losses are $1/\sqrt{2}$ times the "characteristic" losses observed in the same metals. However, it should be noted that thin metallic films may have a strongly granular structure. The strong variation of the grain structure with substrate composition, rate and amount of condensation, etc., of thin evaporated metallic films has been discussed by Heavens. In this reference are given electron micrographs which clearly show the transition from small grain size to a state in which the grains merge to form a nearly uniform films as the amount of material deposited is increased in a series of films. The

surface depolarization effect will certainly be larger for a small grain of average dimension a than for the semi-infinite plane foil of thickness a which was treated above. Thus one would expect the "lowered" losses in an actual foil to lie closer to the value $\hbar\omega_p/\sqrt{3}$, appropriate to a spherical grain, than to the value $\hbar\omega_p/\sqrt{2}$. This seems to be true for the low-lying losses which have been observed.

Apparently, the two types of excitation referred to are surface plasmons and particle plasmons, here in spherical particles, both of which have become workhorses in the research field of plasmonics. We next evaluate Fermi's golden rule for the energy loss probability of a swift electron in the aloof geometry, where the electron trajectory does not penetrate the plasmonic nanoparticle. See also Fig. 14.18. The coupling between the electron and the plasmonic nanoparticle is described in terms of the quasistatic approximation

$$\langle f|\hat{V}|i\rangle = \int \frac{e^2\psi_f^*(\boldsymbol{r})\psi_i(\boldsymbol{r})\,\hat{n}(\boldsymbol{r}')}{4\pi\varepsilon_0|\boldsymbol{r}-\boldsymbol{r}'|}\,d^3r d^3r' = \int \psi_f^*(\boldsymbol{r})\psi_i(\boldsymbol{r})U_0(\boldsymbol{r},\boldsymbol{r}')\hat{n}(\boldsymbol{r}')\,d^3r', \tag{14.96}$$

where $\psi(\boldsymbol{r})$ denotes the wavefunction of the swift electron, $\hat{n}(\boldsymbol{r})$ is the particle density operator for the plasmonic nanoparticle, and U_0 the Coulomb potential introduced in Eq. (14.32). Upon insertion into Eq. (14.95) the density operators can be combined to give

$$\int_{-\infty}^{0} e^{-i(\omega+i\eta)t}\Big\langle\hat{n}(\boldsymbol{r},0)\,\hat{n}(\boldsymbol{r}',t)\Big\rangle_{\text{eq}}dt = \int_0^{\infty} e^{i(\omega+i\eta)t}\Big\langle\hat{n}(\boldsymbol{r},t)\,\hat{n}(\boldsymbol{r}',0)\Big\rangle_{\text{eq}}dt,$$

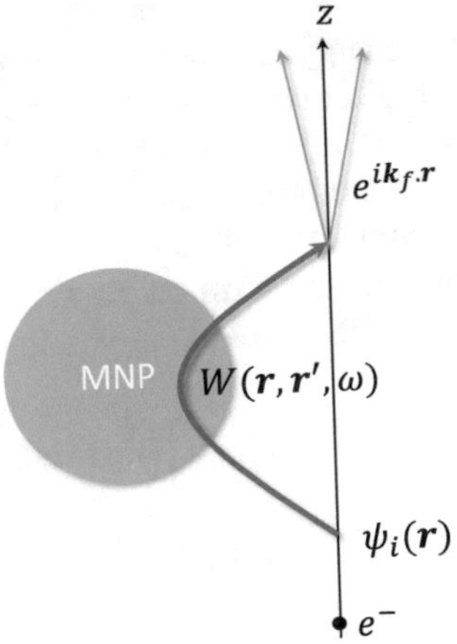

Fig. 14.18 Schematics of computation of electron energy loss probability. A swift electron interacts with a metallic nanoparticle and loses a tiny fraction of its kinetic energy. ψ_i denotes the initial wavefunction of the electron, the self-interaction process where the electron polarizes the nanoparticle and the polarization acts back on the electron is described through the screened Coulomb interaction W. The final wavefunction after the interaction is decomposed into plane waves with wavevector $\boldsymbol{k}_f$

with the abbreviation $\langle\dots\rangle_{\rm eq}=\mathrm{tr}_R(\hat\rho_R\dots)$. To arrive at the expression on the right-hand side we have used that the integral depends on the time difference between the operators only. Using Eqs. (14.12–14.17) together with Eq. (14.20), one can rewrite this in terms of the density–density correlation function via

$$\lim_{\eta\to 0}\int_0^\infty e^{i(\omega+i\eta)t}\Big\langle\hat n(\boldsymbol r,t)\,\hat n(\boldsymbol r',0)\Big\rangle_{\rm eq}dt=i\hbar\Phi\big(\hat n(\boldsymbol r),\hat n(\boldsymbol r');\omega\big)\Big(\bar n_{\rm th}(\hbar\omega)+1\Big).\tag{14.97}$$

Which correlation function, that for the interacting or non-interacting system? Since $\hat V$ includes the Coulomb coupling between the swift electron (an external charge distribution) and the metal electrons only, the mutual Coulomb couplings between the metal electrons are already included in the interaction representation. Thus, Φ has to be computed for the *interacting* system, as we have properly done in Eq. (14.97) by using Φ instead of Φ_0 for the non-interacting system. We can combine the density–density correlation function with the Coulomb potentials U_0 to arrive at

$$W_{\rm ind}(\boldsymbol r,\boldsymbol r',\omega)=\int U_0(\boldsymbol r,\boldsymbol r_1)\big[\Phi\big(\hat n(\boldsymbol r_1),\hat n(\boldsymbol r_1');\omega\big)\big]\,U_0(\boldsymbol r_1',\boldsymbol r')\,d^3r_1d^3r_1,$$

with the screened Coulomb potential of Eq. (14.40). Putting together all results we rewrite the scattering rate of Eq. (14.95) in the form

$$\Gamma=\frac{i}{\hbar}\int\psi_i^*(\boldsymbol r)\psi_f(\boldsymbol r)\,W_{\rm ind}(\boldsymbol r,\boldsymbol r',\omega)\Big(\bar n_{\rm th}(\hbar\omega)+1\Big)\,\psi_f^*(\boldsymbol r')\psi_i(\boldsymbol r')\,d^3rd^3r'+\text{c.c.},$$

where $\hbar\omega=E_f-E_i$ is the energy loss of the electron. E_i and E_f are the initial and final energies of the swift electron, respectively. For systems with time reversal symmetry the relation $W_{\rm ind}(\boldsymbol r,\boldsymbol r',\omega)=W_{\rm ind}(\boldsymbol r',\boldsymbol r,\omega)$ holds, as has been discussed in Chap. 7 in the context of the reciprocity theorem of optics for the full Maxwell's equations. Corresponding symmetry properties then also apply for the quasistatic limit. Thus, we can combine the first term on the right-hand side with the second one with the complex conjugate, to arrive at

$$\Gamma=\frac{2}{\hbar}\int\psi_i^*(\boldsymbol r)\psi_f(\boldsymbol r)\,\mathrm{Im}\Big[-W_{\rm ind}(\boldsymbol r,\boldsymbol r',\omega)\Big]\Big(\bar n_{\rm th}(\hbar\omega)+1\Big)\,\psi_f^*(\boldsymbol r')\psi_i(\boldsymbol r')\,d^3rd^3r'.$$

Evaluation of Scattering Rate We follow Ritchie's original work [161] and consider an initial electron wavepacket of the form

$$\psi_i(\boldsymbol r)=\frac{1}{\sqrt L}e^{ik_iz}\phi_\perp(\boldsymbol R)$$

with central wavenumber k_i that propagates along the z-direction, where L is the quantization length of the electron trajectory. $\phi_\perp(\boldsymbol R)$ is the wavefunction in the transverse directions $\boldsymbol R$, and we ignore dispersion along $\boldsymbol R$ [162]. The final states are expressed in terms of plane waves $\psi_f(\boldsymbol r)=e^{i\boldsymbol k_f\cdot\boldsymbol r}$, and we neglect in the energy conservation

$$\hbar\omega = E_i - E_f = \hbar v\left(k_i - k_{fz}\right) + \frac{\hbar^2 k_{f\rho}^2}{2m} \approx \hbar v\left(k_i - k_{fz}\right)$$

the recoil term $\hbar^2 k_{f\rho}^2/(2m)$, where k_{fz} and $k_{f\rho}$ are the z and transverse components of $\boldsymbol{k}_f$, respectively, and m is the free electron mass. The transition probability $\mathcal{P}(\hbar\omega)$ is obtained from the transition rate Γ by multiplying with the interaction time L/v, where v is the electron velocity. This gives

$$\mathcal{P}(\hbar\omega) = \frac{2}{v}\int \frac{d^3k_f}{(2\pi)^3}\left(\bar{n}_{\rm th}(\hbar\omega)+1\right)\delta\left(\hbar\omega - \hbar v\left[k_i - k_{fz}\right]\right)$$
$$\times \int e^{i\boldsymbol{k}_f\cdot\boldsymbol{r}}\phi_\perp^*(\boldsymbol{R})e^{-ik_iz}\,\mathrm{Im}\left[-W_{\rm ind}(\boldsymbol{r},\boldsymbol{r}',\omega)\right]e^{-i\boldsymbol{k}_f\cdot\boldsymbol{r}'}\phi_\perp(\boldsymbol{R}')e^{ik_iz'}\,d^3rd^3r'.$$

We have summed over all final states of the swift electron and have introduced a delta function for the loss energy $\hbar\omega$. The total loss probability is obtained by integrating over all loss energies $\hbar\omega$ and the z-component of k_{fz} is determined by Dirac's delta function. The integration over $\boldsymbol{k}_{f\perp}$ should only extend over the finite acceptance angle $\varphi_{\rm out}$ of the electron microscope, which has previously been introduced in Sect. 10.5.2 in the context of bulk losses. However, for kinetic electron energies in the 100 keV range and $\varphi_{\rm out} \gg 1$ mrad one can safely use the following approximation [162]

$$\int e^{i\boldsymbol{k}_f\cdot(\boldsymbol{R}-\boldsymbol{R}')}\,d^2k_f \approx (2\pi)^2\delta(\boldsymbol{R}-\boldsymbol{R}').$$

Thus, we are led to

$$\mathcal{P}(\hbar\omega) = \frac{1}{\pi\hbar v^2}\int |\phi_\perp(\boldsymbol{R})|^2\,e^{iq(z'-z)}\,\mathrm{Im}\left[-W_{\rm ind}(\boldsymbol{r},\boldsymbol{r}',\omega)\right]\left(\bar{n}_{\rm th}(\hbar\omega)+1\right)d^2R\,dz\,dz',$$

with $q = \omega/v$. If the transverse extension of the electron beam is strongly peaked around the impact parameter $\boldsymbol{R}_0$, we obtain the even more simple expression

$$\mathcal{P}(\boldsymbol{R}_0,\hbar\omega) = \frac{1}{\pi\hbar v^2}\int e^{iq(z'-z)}\,\mathrm{Im}\left[-W_{\rm ind}(\boldsymbol{R}_0,z,\boldsymbol{R}_0,z',\omega)\right]\left(\bar{n}_{\rm th}(\hbar\omega)+1\right)dz\,dz'. \tag{14.98}$$

We can now use the charge distribution $\rho(\boldsymbol{r},\omega)$ of the swift electron, Eq. (10.37), to rewrite

$$\frac{1}{v^2}\int e^{iq(z'-z)}\,W_{\rm ind}(\boldsymbol{R}_0,z,\boldsymbol{R}_0,z',\omega)\,dz\,dz' = \int \rho^*(\boldsymbol{r},\omega)V_{\rm ind}(\boldsymbol{r},\omega)\,d^3r,$$

where we have introduced in the spirit of Eq. (14.39) the induced potential

$$(-e)^2V_{\rm ind}(\boldsymbol{r},\omega) = \int W_{\rm ind}(\boldsymbol{r},\boldsymbol{r}',\omega)\rho(\boldsymbol{r}',\omega)\,d^3r'.$$

Putting together all results we arrive at the electron energy loss probability for a swift electron

Electron Energy Loss Probability

$$\mathcal{P}(\boldsymbol{R}_0,\hbar\omega) = -\frac{1}{\pi\hbar}\int \mathrm{Im}\left[\rho^*(\boldsymbol{r},\omega)V_{\rm ind}(\boldsymbol{r},\omega)\right]\left(\bar{n}_{\rm th}(\hbar\omega)+1\right)d^3r. \tag{14.99}$$

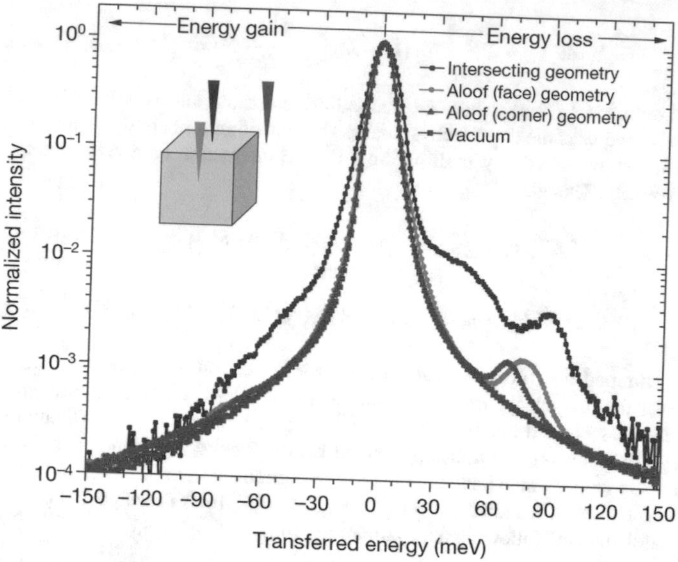

Fig. 14.19 Normalized electron energy loss spectra for an ionic nanocube, here a 150 nm MgO cube, acquired with the electron beam located in different positions (see inset): green, aloof geometry (beam located near a face); blue, aloof geometry (beam located near a corner); black, intersecting geometry. The spectrum in red is the ZLP spectra acquired in the vacuum. Curves were normalized so that the resonance peak maxima approximate a fractional scattering probability. Regions with positive and negative values of transferred energies are labeled energy loss and energy gain, respectively. Figure and caption adapted from Ref. [163]

This result is identical to the semiclassical result of Eq. (10.44), with the only exception of the thermal population factor. In the optical regime photon energies $\hbar\omega$ are in the eV range and thermal energies are usually much smaller, for instance, at room temperature we have approximately $k_B T \approx 25$ meV. Then the thermal enhancement can be ignored, and we end up with the semiclassical result.

Things somewhat change for non-plasmonic systems, such as ionic nanosystems to be described by a Lorentz-type dielectric function, Eq. (7.6), rather than a Drude-like function. Figure 14.19 shows electron energy loss spectra acquired for a MgO nanocube with a side length of 150 nm [163]. The resonances of the cube are similar to those of plasmonic nanoparticles, however, they are attributed to ion rather than electron motion. Because of the mass difference between ions and electrons, the resonance energies are shifted to the 100 meV range. On the energy loss side, indicated on top of the panel, we observe in blue and green losses to surface phonon polaritons that are located at the edges and faces of the nanocube, respectively, and which have very similar properties as the particle plasmons discussed in Chap. 9. In the theoretical approach we only have to modify the permittivity of the nanoparticle, everything else remains unchanged.

The reason why we here show Fig. 14.19 is that on the energy gain side, corresponding to negative transferred energies, the surface phonon peaks of the edge and face resonances are mirrored, however, with a significantly lower intensity. Here the swift electron *gains* energy by absorbing a thermally populated cube resonance. Indeed we can compute the corresponding probabilities from Eq. (14.99). We first note that

$$\mathrm{Im}\Big[-W_{\mathrm{ind}}(\boldsymbol{r},\boldsymbol{r}',-\omega)\Big]\Big(\bar{n}_{\mathrm{th}}(-\hbar\omega)+1\Big)=\mathrm{Im}\Big[W_{\mathrm{ind}}(\boldsymbol{r},\boldsymbol{r}',\omega)\Big]\Big(-\bar{n}_{\mathrm{th}}(\hbar\omega)\Big),$$

where we have used Eq. (14.14) to rewrite the first term. The relation in the second term can be easily verified through explicit calculation using $\bar{n}_{\mathrm{th}}(\hbar\omega)=1/(e^{\beta\hbar\omega}-1)$. We thus find for the electron energy gain probability

$$\mathcal{P}(\boldsymbol{R}_0,-\hbar\omega)=-\frac{1}{\pi\hbar}\int\mathrm{Im}\Big[\rho^*(\boldsymbol{r},\omega)V_{\mathrm{ind}}(\boldsymbol{r},\omega)\Big]\bar{n}_{\mathrm{th}}(\hbar\omega)\,d^3r,\tag{14.100}$$

where the energy gain $\hbar\omega$ is now considered to be a positive quantity. Thus, the thermal enhancement factors for energy loss and gain can be interpreted in terms of stimulated emission and absorption, respectively. For room temperature and a loss energy of 75 meV we then find $\bar{n}_{\mathrm{th}}:\bar{n}_{\mathrm{th}}+1\approx 1:20$, approximately in agreement with the situation depicted in the figure. In the next chapter we will discuss the impact of thermal effects in significantly more detail.

Exercises

Exercise 14.1 Consider a correlation function of the form

$$\mathcal{I}(t)=\int_{-\infty}^{\infty}e^{-i\omega t}\,\mathscr{D}(\omega)\,d\omega,$$

where $\mathscr{D}(\omega)$ is a Gaussian centered around ω_0 which has the width $\Delta\omega$. Evaluate the correlation function using the integral given in Exercise 1.4, and discuss the influence of $\Delta\omega$.

Exercise 14.2 Start from Eq. (14.24) to compute the commutation relation for magnetic field operators at different times.

Exercise 14.3 It is sometimes convenient to simplify the commutator with the electric field operators.

a. Show that Eq. (14.25) can be written in the form $\mathcal{I}_1+\mathcal{I}_2$, with

$$\mathcal{I}_{1,2}=\mp\frac{\hbar c}{\varepsilon_0}\left(\frac{\delta_{ij}}{c^2}\partial_t^2-\partial_i\partial_j\right)\int_0^{\infty}e^{\mp ikc\tau}\left[\frac{\sin kR}{4\pi^2R}\right]dk.$$

b. Work out the integrals to arrive at

$$\left[\hat{E}_i(\boldsymbol{r},t),\hat{E}_j(\boldsymbol{r}',t')\right]=\frac{i\hbar}{\varepsilon_0}\left(\frac{\delta_{ij}}{c^2}\partial_t^2-\partial_i\partial_j\right)\frac{1}{4\pi R}\left[\delta\left(\tau-\frac{R}{c}\right)-\delta\left(\tau+\frac{R}{c}\right)\right],$$

with the usual definition of $\boldsymbol{R}=\boldsymbol{r}-\boldsymbol{r}'$ and $\tau=t-t'$.

This result can be interpreted as follows: since the term in brackets is zero off the light cone connecting $\boldsymbol{r},t$ and $\boldsymbol{r}',t'$, electric fields at two different space-time points can be measured simultaneously only if the points cannot be connected by a light signal. If two space-time points can be connected by a light signal, the electric field measurements influence each other.

Exercise 14.4 Write down explicitly the Lindhard's dielectric function of Eq. (14.43) for electrons with a dispersion $E_k=\hbar^2k^2/(2m)$, and evaluate the angular integrals for the imaginary part of the permittivity.

Exercise 14.5 Use the two-dimensional Fourier transform $e^2/(2\varepsilon_0 q)$ of the Coulomb potential and a linear electron dispersion $E_k=\hbar v_f k$ to obtain Lindhard's dielectric function of Eq. (14.43) for a graphene-like material. Write down the imaginary part of the permittivity as an integral over a single variable. You might like to consult [39, 40] for a detailed analysis for the evaluation of the dielectric function.

Exercise 14.6 For a homogeneous system the spectral function for the density–density correlations can be written similarly to Eq. (14.42) in the form

$$\rho_{nn}(\boldsymbol{q},\omega)=\frac{1}{\hbar}\int_{-\infty}^{\infty}e^{i\omega t}\Big\langle\left[\hat{n}_{\boldsymbol{q}}(t),\hat{n}_{-\boldsymbol{q}}\right]\Big\rangle_{\text{eq}}\,dt.$$

a. Introduce a Lehmann representation to work out the time integration at zero temperature.
b. Show that the following relation holds:

$$\int_{-\infty}^{\infty}\omega\rho_{nn}(\boldsymbol{q},\omega)\,d\omega=const\times\Big\langle 0\Big|\left[[\hat{n}_{\boldsymbol{q}},\hat{H}],\hat{n}_{-\boldsymbol{q}}\right]\Big|0\Big\rangle,$$

where $\hat{H}$ is the many-body Hamiltonian of the system.
c. Work out the right-hand side of the above equation for $\hat{H}=\hat{\boldsymbol{\pi}}_{\boldsymbol{q}}^2/(2m)$. The resulting expression is the so-called f-sum rule.
d. Is the f-sum rule also valid at finite temperatures?

Exercise 14.7 Consider the interface between two materials with local but anisotropic material parameters $\bar{\bar{\varepsilon}}_1$, $\bar{\bar{\mu}}_1$ and $\bar{\bar{\varepsilon}}_2$, $\bar{\bar{\mu}}_2$. Derive the Fresnel coefficients along the lines discussed in Chap. 8.

Exercise 14.8 Consider the induced dipole moment of Eq. (14.79) for a nanosphere using nonlocality. Obtain in analogy to the quasistatic Mie theory of Sect. 9.2.1 a resonance condition for particle plasmons. Under which conditions is the plasmon resonance blue-shifted?

Exercise 14.9 Consider the modified reflection coefficient R of Eq. (14.88) containing the Feibelman parameters $d_\perp$, $d_\parallel$. Set $d_\parallel = 0$ and expand R in a Taylor series for small values of $d_\perp$. Discuss the difference between positive and negative $d_\perp$ values.

Chapter 15
Thermal Effects in Nano Optics

Thermal nearfield radiation and optical forces at the nanoscale have received considerable interest in recent years. Despite the importance of the topic, in this book we only provide a brief introduction and refer the interested reader to the rich literature on the subject, see, for instance, [134, 164–170] and the references therein. In Chap. 4 we have seen that electromagnetic fields $\boldsymbol{E}$, $\boldsymbol{H}$ transport energy and momentum, as described by the flow of the Poynting vector and Maxwell's stress tensor. Similarly, in thermal equilibrium energy and momentum can be transported through *field fluctuations*

$$\left\langle \hat{E}_i(\boldsymbol{r},\omega)\hat{E}_j(\boldsymbol{r}',\omega')\right\rangle_{\text{eq}}, \quad \left\langle \hat{E}_i(\boldsymbol{r},\omega)\hat{H}_j(\boldsymbol{r}',\omega')\right\rangle_{\text{eq}}, \quad \left\langle \hat{H}_i(\boldsymbol{r},\omega)\hat{H}_j(\boldsymbol{r}',\omega')\right\rangle_{\text{eq}},$$

which give rise to effects such as heat transfer or Casimir and Casimir–Polder forces. The systems we have in mind consist of one or several dielectric or metallic bodies, in the following synonymously referred to as absorbing bodies or materials, which are in thermal equilibrium. Alternatively, we will also consider quantum emitters, such as molecules or quantum dots, that interact with these absorbing bodies. The theoretical description within the so-called field of *fluctuational electrodynamics* combines several concepts discussed in previous chapters, and builds on a linear-response relation between the electromagnetic field operators and the current noise operators of the absorbing materials. The central steps of the theoretical description are depicted in Fig. 15.1 and can be summarized as follows.

Noise Current. One of the central equations of this chapter, Eq. (15.25), relates the electric field operator $\hat{\boldsymbol{E}}$ to the current operator $\hat{\boldsymbol{J}}$ of the absorbing material via

$$\hat{\boldsymbol{E}}(\boldsymbol{r},\omega) = \hat{\boldsymbol{E}}_{\text{inc}}(\boldsymbol{r},\omega) + i\mu_0\omega \int \bar{\bar{G}}_{\text{tot}}(\boldsymbol{r},\boldsymbol{r}',\omega)\cdot\hat{\boldsymbol{J}}(\boldsymbol{r}',\omega)\,d^3r'.$$

U. Hohenester, *Nano and Quantum Optics*, Graduate Texts in Physics,
https://doi.org/10.1007/978-3-030-30504-8_15

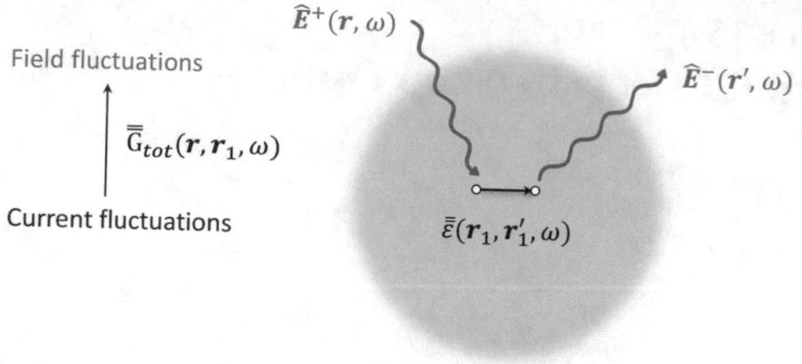

Fig. 15.1 Schematics of fluctuational electrodynamics. Current fluctuations in the absorbing medium (shaded area) can be related to the nonlocal permittivity $\bar{\bar{\varepsilon}}(\boldsymbol{r}_1,\boldsymbol{r}_1',\omega)$, and are transported through the dyadic Green's functions to the field positions $\boldsymbol{r}$ and $\boldsymbol{r}'$. By evaluating the expectation value of field operators in thermal equilibrium, we arrive at the cross-spectral density

Here $\hat{\boldsymbol{E}}_{\text{inc}}$ accounts for electric fields of different origin, such as incoming thermal radiation, and $\bar{\bar{G}}_{\text{tot}}$ is the total Green's function of classical electrodynamics. Importantly, this relation can be obtained in a general manner assuming a linear material response only.

Correlation Functions. The above relation between field and current operators can be used to express the field fluctuations in terms of current fluctuations. The latter can be evaluated using the fluctuation–dissipation theorem of Eq. (14.51),

$$\frac{1}{\hbar}\left\langle\hat{\boldsymbol{J}}(\boldsymbol{r},\omega)\otimes\hat{\boldsymbol{J}}^\dagger(\boldsymbol{r}',\omega')\right\rangle = 2\mu_0\omega^2\,\text{Im}\Big[\bar{\bar{\varepsilon}}(\boldsymbol{r},\boldsymbol{r}',\omega)\Big]\Big(\bar{n}_{\text{th}}(\hbar\omega)+1\Big)\,2\pi\,\delta(\omega-\omega'),$$

which relates the current correlations to the imaginary part of the permittivity together with a thermal occupation factor.

Green's Functions. Combining the relation between field and current operators with the fluctuation–dissipation theorem leads to expressions of the form

$$\int\bar{\bar{G}}_{\text{tot}}(\boldsymbol{r},\boldsymbol{r}_1,\omega)\cdot\text{Im}\Big[\bar{\bar{\varepsilon}}(\boldsymbol{r}_1,\boldsymbol{r}_1',\omega)\Big]\cdot\bar{\bar{G}}^*_{\text{tot}}(\boldsymbol{r}_1',\boldsymbol{r}',\omega)\,d^3r_1d^3r_1'.$$

As we will show below, the double integral can be considerably simplified using an expression reminiscent of the representation formula, previously derived in Chap. 5. With this, the field fluctuations and quantities related to them, for instance, those needed for the computation of radiative heat transfer or Casimir and Casimir–Polder forces, can be expressed in terms of classical Green's functions and thermal occupation factors only.

In the following we discuss the various steps of this approach in more detail.

15.1 Cross-Spectral Density and What We Can Do with It

The photonic environment can be often analyzed in terms of correlation functions for the electromagnetic fields. Consider the cross-spectral density of the electric fields [164, Eq. (23)]

Cross-Spectral Density

$$\mathscr{E}_{ij}(\boldsymbol{r},\boldsymbol{r}',\omega)=\frac{1}{\hbar}\int_{-\infty}^{\infty}e^{i\omega(t-t')}\Big\langle\hat{E}_i(\boldsymbol{r},t)\hat{E}_j(\boldsymbol{r}',t')\Big\rangle_{\rm eq}\,d(t-t'),\qquad(15.1)$$

where the average has to be performed for thermal equilibrium. It is sometimes convenient to perform a separate Fourier transform in t and t'. We first define the Fourier transform of the electric field operator

$$\hat{E}_i(\boldsymbol{r},\omega)=\int_{-\infty}^{\infty}e^{i\omega t}\,\hat{E}_i(\boldsymbol{r},t)\,dt.\qquad(15.2)$$

In terms of the field operators oscillating with positive and negative frequencies, Eq. (14.22), we can decompose the electric field operator of Eq. (15.2) into

$$\hat{E}_i(\boldsymbol{r},\omega)=\theta(\omega)\,\hat{E}_i^{+}(\boldsymbol{r},\omega)+\theta(-\omega)\,\hat{E}_i^{-}(\boldsymbol{r},\omega),$$

with Heaviside's step function $\theta(\pm\omega)$. The term $\hat{E}^+$ for positive frequencies can be associated with photon annihilations, whereas the second term $\hat{E}^-$ for negative frequencies with photon creations. By a similar token, we can decompose the cross-spectral density into positive and negative frequency components

$$\mathscr{E}_{ij}(\boldsymbol{r},\boldsymbol{r}',\omega)=\theta(\omega)\,\mathscr{E}_{ij}^{+}(\boldsymbol{r},\boldsymbol{r}',\omega)+\theta(-\omega)\,\mathscr{E}_{ij}^{-}(\boldsymbol{r},\boldsymbol{r}',\omega).$$

Because the cross-spectral density of Eq. (15.1) depends on the time difference $t-t'$ only, we get [164, Eq. (24)]

$$\hbar^{-1}\Big\langle\hat{E}_i(\boldsymbol{r},\omega)\hat{E}_j^{\dagger}(\boldsymbol{r}',\omega')\Big\rangle_{\rm eq}=2\pi\,\delta(\omega-\omega')\,\mathscr{E}_{ij}(\boldsymbol{r},\boldsymbol{r}',\omega).\qquad(15.3)$$

In deriving this expression we have used Eq. (14.18) together with

$$\hat{E}_i^{\dagger}(\boldsymbol{r},\omega)=\int_{-\infty}^{\infty}e^{-i\omega t}\,\hat{E}_i(\boldsymbol{r},t)\,dt.$$

Below we will also use the more compact notation

$$\hbar^{-1}\Big\langle\hat{\boldsymbol{E}}(\boldsymbol{r},\omega)\otimes\hat{\boldsymbol{E}}^{\dagger}(\boldsymbol{r}',\omega')\Big\rangle_{\text{eq}}=2\pi\,\delta(\omega-\omega')\,\bar{\bar{\mathscr{E}}}(\boldsymbol{r},\boldsymbol{r}',\omega),$$

where the product of electric field operators inside the brackets has to be understood as a dyadic one. We can use Maxwell's equations in operator form, Eq. (13.86), to relate the electric and magnetic field operators through

$$\nabla\times\hat{\boldsymbol{E}}(\boldsymbol{r},\omega)=i\omega\mu_0\hat{\boldsymbol{H}}(\boldsymbol{r},\omega).$$

With this, we obtain from the cross-spectral density for the electric fields also other expressions such as

$$\begin{aligned}
i\mu_0\omega\Big\langle\hat{\boldsymbol{H}}(\boldsymbol{r},\omega)\otimes\hat{\boldsymbol{E}}^{\dagger}(\boldsymbol{r}',\omega')\Big\rangle_{\text{eq}}&=\Big\langle\nabla\times\hat{\boldsymbol{E}}(\boldsymbol{r},\omega)\otimes\hat{\boldsymbol{E}}^{\dagger}(\boldsymbol{r}',\omega')\Big\rangle_{\text{eq}} \qquad (15.4)\\
-i\mu_0\omega\Big\langle\hat{\boldsymbol{E}}(\boldsymbol{r},\omega)\otimes\hat{\boldsymbol{H}}^{\dagger}(\boldsymbol{r}',\omega')\Big\rangle_{\text{eq}}&=\Big\langle\hat{\boldsymbol{E}}(\boldsymbol{r},\omega)\otimes\nabla'\times\hat{\boldsymbol{E}}^{\dagger}(\boldsymbol{r}',\omega')\Big\rangle_{\text{eq}}\\
\mu_0^2\omega^2\Big\langle\hat{\boldsymbol{H}}(\boldsymbol{r},\omega)\otimes\hat{\boldsymbol{H}}^{\dagger}(\boldsymbol{r}',\omega')\Big\rangle_{\text{eq}}&=\Big\langle\nabla\times\hat{\boldsymbol{E}}(\boldsymbol{r},\omega)\otimes\nabla'\times\hat{\boldsymbol{E}}^{\dagger}(\boldsymbol{r}',\omega')\Big\rangle_{\text{eq}}.
\end{aligned}$$

It turns out to be convenient to introduce for the curl applied to the second argument of a tensor the short-hand notation

$$\Big\langle\cdots\otimes\nabla'\times\hat{\boldsymbol{E}}^{\dagger}(\boldsymbol{r}',\omega)\Big\rangle_{\text{eq}}=\Big\langle\cdots\otimes\hat{\boldsymbol{E}}^{\dagger}(\boldsymbol{r}',\omega)\Big\rangle_{\text{eq}}\times\overleftarrow{\nabla}'. \qquad (15.5)$$

Then, the cross-spectral densities of Eq. (15.4) can be written in the compact form

$$\begin{aligned}
\frac{i\mu_0\omega}{\hbar}\Big\langle\hat{\boldsymbol{H}}(\boldsymbol{r},\omega)\otimes\hat{\boldsymbol{E}}^{\dagger}(\boldsymbol{r}',\omega')\Big\rangle&=2\pi\,\delta(\omega-\omega')\,\nabla\times\bar{\bar{\mathscr{E}}}(\boldsymbol{r},\boldsymbol{r}',\omega) \qquad (15.6)\\
-\frac{i\mu_0\omega}{\hbar}\Big\langle\hat{\boldsymbol{E}}(\boldsymbol{r},\omega)\otimes\hat{\boldsymbol{H}}^{\dagger}(\boldsymbol{r}',\omega')\Big\rangle&=2\pi\,\delta(\omega-\omega')\qquad\bar{\bar{\mathscr{E}}}(\boldsymbol{r},\boldsymbol{r}',\omega)\times\overleftarrow{\nabla}'\\
\frac{\mu_0^2\omega^2}{\hbar}\Big\langle\hat{\boldsymbol{H}}(\boldsymbol{r},\omega)\otimes\hat{\boldsymbol{H}}^{\dagger}(\boldsymbol{r}',\omega')\Big\rangle&=2\pi\,\delta(\omega-\omega')\,\nabla\times\bar{\bar{\mathscr{E}}}(\boldsymbol{r},\boldsymbol{r}',\omega)\times\overleftarrow{\nabla}'.
\end{aligned}$$

15.1.1 *Cross-Spectral Density in Free Space*

In free space, the cross-spectral density for the electric fields can be computed along the same lines as discussed for the Kubo formalism in Chap. 14. Using Eq. (14.20) we get

$$\bar{\bar{\mathscr{E}}}(\boldsymbol{r},\boldsymbol{r}',\omega)=\bar{\bar{\rho}}(\boldsymbol{r},\boldsymbol{r}',\omega)\big[\bar{n}_{\text{th}}(\hbar\omega)+1\big],$$

with the spectral function $\bar{\bar{\rho}}$ of electric field operators given in Eq. (14.11). Here $\bar{n}_{\text{th}}$ is the Bose–Einstein distribution function. For positive frequencies we find

$$\bar{\bar{\mathscr{E}}}^{+}(\boldsymbol{r},\boldsymbol{r}',\omega)=\bar{\bar{\rho}}^{+}(\boldsymbol{r},\boldsymbol{r}',\omega)\big[\bar{n}_{\text{th}}(\hbar\omega)+1\big].$$

For negative frequencies we get

$$\bar{\bar{\mathscr{E}}}^{-}(\boldsymbol{r},\boldsymbol{r}',\omega)=-\bar{\bar{\rho}}^{+}(\boldsymbol{r},\boldsymbol{r}',-\omega)\big[\bar{n}_{\text{th}}(\hbar\omega)+1\big],$$

where we have used Eq. (14.28) for the spectral function. Together with the relation $\bar{n}_{\text{th}}(\hbar\omega)+1=-\bar{n}_{\text{th}}(-\hbar\omega)$ we obtain

$$\bar{\bar{\mathscr{E}}}^{\pm}(\boldsymbol{r},\boldsymbol{r}',\omega)=\bar{\bar{\rho}}^{+}(\boldsymbol{r},\boldsymbol{r}',\pm\omega)\theta(\pm\omega)\left\{\begin{matrix}\bar{n}_{\text{th}}(+\hbar\omega)+1\\ \bar{n}_{\text{th}}(-\hbar\omega)\end{matrix}\right\}. \tag{15.7}$$

Through Eq. (14.11) we express the spectral function in terms of the dyadic Green's function,

$$\bar{\bar{\rho}}^{+}(\boldsymbol{r},\boldsymbol{r}',\omega)=2\mu_0\omega^2\,\text{Im}\left[\bar{\bar{G}}(\boldsymbol{r},\boldsymbol{r}',\omega)\right]\theta(\omega),$$

to finally arrive at the cross-spectral density of free space

$$\bar{\bar{\mathscr{E}}}^{\pm}(\boldsymbol{r},\boldsymbol{r}',\omega)=2\mu_0\omega^2\,\text{Im}\left[\bar{\bar{G}}(\boldsymbol{r},\boldsymbol{r}',\pm\omega)\right]\theta(\pm\omega)\left\{\begin{matrix}\bar{n}_{\text{th}}(+\hbar\omega)+1\\ \bar{n}_{\text{th}}(-\hbar\omega)\end{matrix}\right\}. \tag{15.8}$$

Here the propagation properties of the cross-spectral density are described by the Green's function of classical electrodynamics, in accordance to our previous discussion that in quantum electrodynamics the photon propagation properties are governed by those of light propagation in classical electrodynamics, and only the noise properties of light become altered in the quantum regime. In the above case, it is the Bose–Einstein factor that accounts for the thermal occupation of photon modes.

In later parts of this chapter we will show how to compute the cross-spectral density in case of a non-trivial photonic environment. Our analysis will build on a few novel concepts, such as a linear-response relation between electric field and current operators, but the final result will look very similar.

15.1.2 What We Can Do with Cross-Spectral Densities

Before entering the detailed discussion of the evaluation of the cross-spectral density, let us give some examples of what we can compute once this quantity is at hand.

Lifetime and Lamb Shift. Consider a quantum emitter, such as a molecule or quantum dot, which we shall denote as an "atom." It has a ground and excited state separated by an energy difference of $\hbar\omega$. Owing to the light–matter interaction, an excited atom can decay by emitting a photon. Using Fermi's golden rule of Eq. (14.95), we will show that in lowest order perturbation theory the corresponding rate can be computed from

$$\Gamma = \frac{1}{\hbar}\mathrm{Im}\left[\boldsymbol{p}\cdot\bar{\bar{\mathscr{E}}}^{+}(\boldsymbol{r}_0,\boldsymbol{r}_0,\omega)\cdot\boldsymbol{p}^*\right], \tag{15.9}$$

where $\boldsymbol{p}$ is the transition dipole moment of the atom and $\boldsymbol{r}_0$ the atom position. A further effect of the light–matter coupling is the modification of the atomic transition frequency by an amount of $\delta\omega(\boldsymbol{r}_0)$, which we shall refer to as the *Lamb shift*.

Casimir–Polder Force. In Chap. 4 we have seen that electromagnetic fields carry momentum that can exert forces on small polarizable particles. In the context of thermal fields and vacuum field fluctuations, these forces are denoted as *Casimir–Polder forces*, and can be computed from the spatial variation of the Lamb shift through [120, 134]

$$\boldsymbol{F}(\boldsymbol{r}_0) = -\nabla\Big(\hbar\delta\omega(\boldsymbol{r}_0)\Big). \tag{15.10}$$

When the center-of-mass motion of the atom can be described in a classical framework, the above result can be interpreted in terms of a classical force.

Casimir Force. In thermal equilibrium we can compute Maxwell's stress tensor due to vacuum and thermal field fluctuations from the cross-spectral densities of the electric and magnetic fields. When integrating Maxwell's tensor over a closed boundary, we obtain the forces exerted on macroscopic objects. In fluctuational electrodynamics they are known as *Casimir forces*.

Heat Transfer. The field fluctuations give rise to an energy flux that can be computed from the cross-spectral density between the electric and magnetic fields, and which can be interpreted in terms of Poynting's theorem. From this quantity one obtains detailed information about thermal radiation and heat flux at the nanoscale.

15.2 Noise Currents

In the following we analyze on an operator level the current response induced by external electromagnetic fields. Let $\hat{n}(\boldsymbol{r}, t), \hat{\mathbf{j}}(\boldsymbol{r}, t)$ be the particle and current density operators in an interaction picture with respect to the uncoupled light–matter system. The total current in presence of the external electromagnetic fields then consists of a paramagnetic and a diamagnetic term, see Eq. (13.62), and we get

$$\hat{J}_{\text{tot}}(\boldsymbol{r}, t) = \hat{U}_I^\dagger(t, 0)\left[-e\hat{\mathbf{j}}(\boldsymbol{r}, t) - \frac{e^2}{m}\hat{n}(\boldsymbol{r}, t)\hat{\mathbf{A}}(\boldsymbol{r}, t)\right]\hat{U}_I(t, 0).$$

Here $\hat{U}_I$ is the time evolution operator of Eq. (13.12) in the interaction picture, which accounts for the light–matter coupling of Eq. (13.63) that we have not considered so far. For a linear material response we proceed similarly to our previous discussion of Sect. 14.4.3 and get

$$\hat{J}_{\text{tot}}(\boldsymbol{r}, t) - \Big(-e\hat{\mathbf{j}}(\boldsymbol{r}, t)\Big) = -\frac{i}{\hbar}\int_0^t \Big[-e\hat{\mathbf{j}}(\boldsymbol{r}, t), \hat{H}_{\text{int}}(t')\Big]\, dt' - \frac{e^2}{m}\hat{n}(\boldsymbol{r}, t)\,\hat{\mathbf{A}}(\boldsymbol{r}, t). \tag{15.11}$$

The term on the left-hand side can be interpreted as the induced current operator, and the operator on the right-hand side is the paramagnetic (first term) and diamagnetic (second term) current operator. Equation (15.11) is an operator equation and we have to be careful when performing approximations on an operator level, because we should anticipate how the operator will be used at a later stage. In thermal equilibrium we have

$$\Big\langle\hat{\mathbf{j}}(\boldsymbol{r}, t)\Big\rangle_{\text{eq}} = 0, \qquad \Big\langle\hat{\mathbf{j}}(\boldsymbol{r}, t)\,\hat{\mathbf{j}}(\boldsymbol{r}', t')\Big\rangle_{\text{eq}} \neq 0.$$

In Sect. 14.4.3 we have been concerned with the expectation value of the current operator in presence of external, *classical* electromagnetic fields, and have discarded $\hat{\mathbf{j}}$ on the left-hand side of Eq. (15.11) because its expectation value vanishes in thermal equilibrium. In the context of the cross-spectral density we are concerned with current fluctuations, and thus must keep the current operator $\hat{\mathbf{j}}$ for later evaluations of fluctuation terms $\langle\hat{\mathbf{j}}\hat{\mathbf{j}}\rangle_{\text{eq}}$. The reader might like to go back to Sect. 14.4.3 to carefully verify that our proof of Eq. (14.49) can be also carried out on an operator level. Thus, we get in frequency space

$$\hat{J}_{\text{tot}}(\boldsymbol{r}, \omega) - \hat{J}(\boldsymbol{r}, \omega) = \int \hat{\bar{\bar{\sigma}}}(\boldsymbol{r}, \boldsymbol{r}', \omega) \cdot \hat{\boldsymbol{E}}(\boldsymbol{r}', \omega)\, d^3r',$$

where $\hat{J} = -e\hat{\mathbf{j}}$ is the current operator in the interaction picture, and we have introduced the nonlocal optical conductivity operator

$$\hat{\bar{\bar{\sigma}}}(\boldsymbol{r},\boldsymbol{r}',\omega) = \frac{ie^2}{\omega}\int_0^\infty e^{i\omega t}\left(-\frac{i}{\hbar}\int_0^t \left[\hat{\mathbf{j}}(\boldsymbol{r},t),\hat{\mathbf{j}}(\boldsymbol{r}',t')\right]dt' + \frac{\hat{n}(\boldsymbol{r},t)}{m}\delta(\boldsymbol{r}-\boldsymbol{r}')\mathbb{1}\right)dt. \tag{15.12}$$

When computing current correlation functions, we will typically evaluate expectation values of the form

$$\left\langle\left[\hat{\bar{\bar{\sigma}}}(\boldsymbol{r},\boldsymbol{r}',\omega)\right]\cdot\hat{\boldsymbol{E}}(\boldsymbol{r}',\omega)\dots\right\rangle_{\text{eq}} \longrightarrow \bar{\bar{\sigma}}(\boldsymbol{r},\boldsymbol{r}',\omega)\cdot\left\langle\hat{\boldsymbol{E}}(\boldsymbol{r}',\omega)\dots\right\rangle_{\text{eq}}.$$

As $\hat{\bar{\bar{\sigma}}}\cdot\hat{\boldsymbol{E}}$ already accounts for the material response in the lowest order of the electric field operator $\hat{\boldsymbol{E}}$, we can safely replace the conductivity tensor at the operator level by its equilibrium expectation value of Eq. (14.49). This has been done in the above equation on the right-hand side. Any modification caused by operators acting at earlier times (indicated with dots) would be at least of order $\hat{E}^2$, and can thus be ignored. In linear response the operator equation relating the current to the electric field can be written in the form

Generalized Ohm's Law in Operator Form

$$\hat{\boldsymbol{J}}_{\text{tot}}(\boldsymbol{r},\omega) = \int \bar{\bar{\sigma}}(\boldsymbol{r},\boldsymbol{r}',\omega)\cdot\hat{\boldsymbol{E}}(\boldsymbol{r}',\omega)\,d^3r' + \hat{\boldsymbol{J}}(\boldsymbol{r},\omega). \tag{15.13}$$

See Fig. 15.2 for a schematic representation. The first term on the right-hand side accounts for the current induced by the electric field, possibly also including dissipation, and the second term is a noise current that is needed to counteract these losses.

The separation into induced and noise currents could be also made without making contact with the microscopic material description introduced in the previous

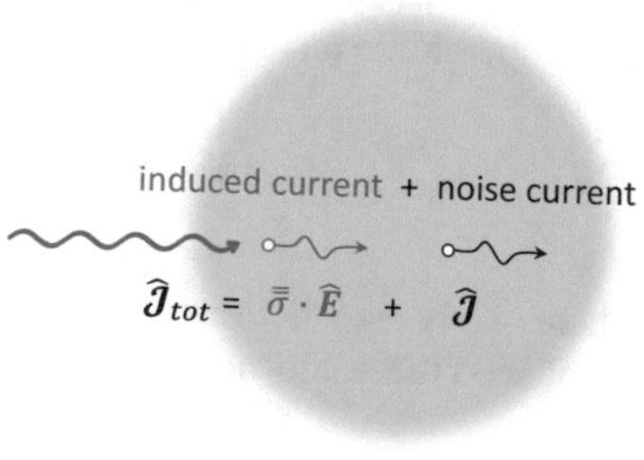

Fig. 15.2 Schematics for current sources. A current is either induced by an incoming photon, where the current and electric field operators are connected through the optical conductivity $\bar{\bar{\sigma}}$, or through a vacuum or thermal current fluctuation in the absorbing medium

chapter. Such an approach is presented, for instance, in the book of Vogel and Welsch [120] who developed a formalism that builds on generic permittivities and permeabilities solely. Generalizations to various kinds of refinements, including nonlocality and nonlinearity, have been developed by Scheel and coworkers [134].

Example of Coupled Oscillators

It is instructive to discuss the necessity of a noise term for a simple model system. We here follow the book of Scully and Zubairy [127] and consider the coupling of a harmonic oscillator, which we will denote as the "system," to a bath of harmonic oscillators denoted as the "reservoir." The system's harmonic oscillator has the frequency Ω and is described by the annihilation and creation operators $\hat{a}$, $\hat{a}^\dagger$. The reservoir consists of harmonic oscillators with frequency ω_k, which are described by the annihilation and creation operators $\hat{b}_k$, $\hat{b}_k^\dagger$. The total Hamiltonian is of the form

$$\hat{H}_0+\hat{V}=\left[\hbar\Omega\,\hat{a}^\dagger\hat{a}+\sum_k\hbar\omega_k\,\hat{b}_k^\dagger\hat{b}_k\right]+\hbar\sum_k g_k\left(\hat{a}\,\hat{b}_k^\dagger+\hat{b}_k\hat{a}^\dagger\right). \tag{15.14}$$

The second term on the right-hand side accounts for the system–reservoir coupling, with the coupling constant g_k, where either the excitation is transferred from the system to the reservoir (first term) or from the reservoir to the system (second term). Using an interaction representation with respect to the term in brackets, we rewrite the system–reservoir coupling in the form

$$\hat{V}(t)=\hbar\sum_k g_k\left(e^{-i(\Omega-\omega_k)t}\,\hat{a}\,\hat{b}_k^\dagger+e^{i(\Omega-\omega_k)t}\,\hat{b}_k\hat{a}^\dagger\right).$$

Heisenberg's equations of motion for the operators then read

$$\frac{d}{dt}\hat{a}(t)=-i\sum_k g_k e^{i(\Omega-\omega_k)t}\,b_k(t)$$

$$\frac{d}{dt}\hat{b}_k(t)=-ig_k e^{-i(\Omega-\omega_k)t}\,\hat{a}(t).$$

The second equation can be formally integrated to yield

$$\hat{b}_k(t)=\hat{b}_k(0)-ig_k\int_0^t e^{-i(\Omega-\omega_k)t'}\hat{a}(t')\,dt'.$$

The first term represents the free time evolution of the reservoir modes and the second term accounts for the system–reservoir coupling. Inserting this formal solution into the equation of motion for $\hat{a}$ then gives

$$\frac{d}{dt}\hat{a}(t) = -\sum_k g_k^2 \int_0^t e^{i(\Omega-\omega_k)(t-t')}\hat{a}(t')\,dt' + \left[-i\sum_k g_k\, e^{i(\Omega-\omega_k)t}\hat{b}_k(0)\right], \tag{15.15}$$

where the last term in brackets acts as a noise term that depends on the reservoir modes at time zero. Let us consider the first term on the right-hand side. For a sufficiently large reservoir the summation can be replaced by an integral, and the reservoir response function decays fast because of destructive interference, as discussed in the previous chapter. We then approximately find

$$\sum_k g_k^2 e^{i(\Omega-\omega_k)(t-t')} \approx g^2 \int_0^\infty e^{i(\Omega-\omega)(t-t')}\,d\omega \approx 2\pi g^2 \delta(t-t'), \tag{15.16}$$

where we have assumed that $g(\omega)$ changes slowly as a function of ω such that it can be replaced by a constant value. With this, the equation of motion for $\hat{a}(t)$ can be written in the form

$$\frac{d}{dt}\hat{a}(t) = -\frac{\Gamma}{2}\hat{a}(t) + \hat{\mathscr{F}}(t).$$

Here Γ is a scattering rate, whose value depends on the coupling constant g, and $\hat{\mathscr{F}}(t)$ is the noise operator given by the term in brackets of Eq. (15.15). If we would neglect $\hat{\mathscr{F}}(t)$, the system operator $\hat{a}$ would decay according to

$$\hat{a}(t) = e^{-(\Gamma/2)t}\hat{a}(0),$$

with a corresponding time evolution for $\hat{a}^\dagger(t)$. This decay is in clear contradiction to the probability conservation requested by quantum mechanics and the fact that the time evolution of the coupled system and reservoir is unitary. One thus must include the additional noise term $\hat{\mathscr{F}}(t)$ to correct for this. More specifically, we request that the noise operators $\hat{\mathscr{F}}(t)$ lead to the proper correlation functions

$$\begin{aligned}\left\langle \hat{\mathscr{F}}^\dagger(t)\hat{\mathscr{F}}(t')\right\rangle &= \left\langle \left[i\sum_k g_k\, e^{-i(\Omega-\omega_k)t}\hat{b}_k^\dagger\right]\left[-i\sum_{k'} g_{k'}\, e^{i(\Omega-\omega_{k'})t'}\hat{b}_{k'}\right]\right\rangle_{\text{eq}} \\ &\approx 2\pi g^2\,\bar{n}_{\text{th}}(\hbar\omega)\delta(t-t'),\end{aligned}$$

where we have used the same approximations as in Eq. (15.16) to arrive at the last term. The combination of scatterings and fluctuations, described by Γ and $\hat{\mathscr{F}}$, respectively, then leads to the desired probability conservation. There exists a huge amount of literature on these so-called Langevin noise operators, see, for instance, [119, 127] and the references therein, which will not be needed in our following discussion of current operators. They have already built in the proper noise properties because their dynamics is obtained from a microscopic description. Nevertheless, the simple harmonic oscillator example is illustrative to better understand the physics of open quantum systems in terms of dissipation and fluctuations, which come naturally together in open quantum systems.

15.2.1 Green's Function Solution

In Chap. 5 we have discussed the solution of differential equations using the concept of Green's functions. In the following we first extend this scheme to nonlocal differential operators $L(\boldsymbol{r},\boldsymbol{r}')$, and then generalize the approach to operator equations. Consider the linear differential equation

$$\int L(\boldsymbol{r},\boldsymbol{r}')f(\boldsymbol{r}')\,d^3r' = -Q(\boldsymbol{r}), \tag{15.17}$$

where $f(\boldsymbol{r}')$ is the solution we are seeking for and $Q(\boldsymbol{r})$ is a source term. To solve this equation, we introduce the Green's function defined through

$$\int L(\boldsymbol{r},\boldsymbol{r}')G(\boldsymbol{r}',\boldsymbol{r}_0)\,d^3r' = -\delta(\boldsymbol{r}-\boldsymbol{r}_0), \tag{15.18}$$

which describes the response of the system to a point-like source located at position $\boldsymbol{r}_0$. In addition, the Green's function is assumed to have built in the proper boundary conditions, such as outgoing waves at infinity. Then, the solution of the differential equation can be written in the form

$$f(\boldsymbol{r}) = \int G(\boldsymbol{r},\boldsymbol{r}')Q(\boldsymbol{r}')\,d^3r', \tag{15.19}$$

as can be easily verified by inserting this expression into Eq. (15.17) and using for the Green's function the defining expression of Eq. (15.18). Consider next the differential equation for an operator $\hat{F}(\boldsymbol{r})$,

$$\int L(\boldsymbol{r},\boldsymbol{r}')\hat{F}(\boldsymbol{r}')\,d^3r' = -\hat{Q}(\boldsymbol{r}), \tag{15.20}$$

where $\hat{Q}(\boldsymbol{r})$ is a source operator. If we assume that $\hat{F}(\boldsymbol{r})$ has the *same* boundary conditions as $f(\boldsymbol{r})$, such as outgoing waves in the case of the wave equation, we can use the *same* Green's function defined in Eq. (15.18) to express the operator solution in the form

$$\hat{F}(\boldsymbol{r}) = \int G(\boldsymbol{r},\boldsymbol{r}')\hat{Q}(\boldsymbol{r}')\,d^3r'. \tag{15.21}$$

This can be proven by inserting Eq. (15.21) into the differential equation of Eq. (15.20) and using the definition of the Green's function. In the following we employ the Green's function solution to the wave equation in operator form, and consider the situations that the current operator is associated with either an external current contribution, produced, for instance, by a quantum emitter, or the induced current of Eq. (15.13) consisting of dissipative and noise terms.

External Current Distribution. We start by considering an *external* current distribution $\hat{\boldsymbol{J}}_{\text{ext}}$ situated in free space. Maxwell's equations in operator form, Eq. (13.20), can be written in the frequency domain in terms of the usual wave equation

$$-\nabla\times\nabla\times\hat{\boldsymbol{E}}^{+}(\boldsymbol{r},\omega)+k^2\,\hat{\boldsymbol{E}}^{+}(\boldsymbol{r},\omega)=-i\mu_0\omega\hat{\boldsymbol{J}}^{+}_{\text{ext}}(\boldsymbol{r},\omega), \tag{15.22}$$

with $k^2=\varepsilon_0\mu_0\,\omega^2$. In analogy to the wave equation discussed in Sect. 5.3 we consider positive frequencies ω only. We can thus use the dyadic Green's function of free space, Eq. (5.19), to express the solution of Eq. (15.22) via

$$\hat{\boldsymbol{E}}^{+}(\boldsymbol{r},\omega)=\hat{\boldsymbol{E}}^{+}_{\text{inc}}(\boldsymbol{r},\omega)+i\mu_0\omega\int\bar{\bar{G}}(\boldsymbol{r},\boldsymbol{r}',\omega)\cdot\hat{\boldsymbol{J}}^{+}_{\text{ext}}(\boldsymbol{r}',\omega)\,d^3r', \tag{15.23}$$

where the incoming part $\hat{\boldsymbol{E}}^{+}_{\text{inc}}$ is a solution of the homogeneous Maxwell's equations. By construction, this solution fulfills the defining operator equation and has also built in the proper boundary conditions of outgoing waves at infinity. Note that the boundary conditions of outgoing waves only apply for electric field operators with positive frequencies, for $\hat{\boldsymbol{E}}^{-}$ we have to select ingoing waves instead. Alternatively, we can simply take the Hermitian conjugate of Eq. (15.23).

Induced Current Distribution. Consider next the case of Eq. (15.13) where $\hat{\boldsymbol{J}}$ is associated with the microscopic current sources of the absorbing medium, consisting of an induced and a noise term, respectively. See also Fig. 15.2. The wave equation then reads

$$-\nabla\times\nabla\times\hat{\boldsymbol{E}}^{+}(\boldsymbol{r})+k^2\,\hat{\boldsymbol{E}}^{+}(\boldsymbol{r})=-i\mu_0\omega\left[\int\bar{\bar{\sigma}}(\boldsymbol{r},\boldsymbol{r}')\cdot\hat{\boldsymbol{E}}^{+}(\boldsymbol{r}')\,d^3r'+\hat{\boldsymbol{\jmath}}^{+}(\boldsymbol{r})\right],$$

where for notational simplicity the ω dependence of all operators has been suppressed. This equation can be rewritten as

$$-\nabla\times\nabla\times\hat{\boldsymbol{E}}^{+}(\boldsymbol{r})+\mu_0\omega^2\int\left[\varepsilon_0\delta(\boldsymbol{r}-\boldsymbol{r}')\mathbb{1}+\frac{i\bar{\bar{\sigma}}(\boldsymbol{r},\boldsymbol{r}')}{\omega}\right]\cdot\hat{\boldsymbol{E}}^{+}(\boldsymbol{r}')\,d^3r'=-i\mu_0\omega\hat{\boldsymbol{\jmath}}^{+}(\boldsymbol{r}).$$

The term in brackets is the nonlocal permittivity tensor $\bar{\bar{\varepsilon}}(\boldsymbol{r},\boldsymbol{r}')$. To solve this equation, we introduce the total Green's function $\bar{\bar{G}}_{\text{tot}}$ defined through

$$-\nabla\times\nabla\times\bar{\bar{G}}_{\text{tot}}(\boldsymbol{r},\boldsymbol{r}',\omega) \tag{15.24}$$

$$+\mu_0\omega^2 \int \bar{\bar{\varepsilon}}(\boldsymbol{r}, \boldsymbol{r}_1, \omega) \cdot \bar{\bar{G}}_{\rm tot}(\boldsymbol{r}_1, \boldsymbol{r}', \omega)\, d^3 r_1 = -\delta(\boldsymbol{r} - \boldsymbol{r}')\mathbb{1}.$$

The electric field operator can then be written in the intriguing form

Relation Between Electric Field and Current Noise Operator

$$\hat{\boldsymbol{E}}^+(\boldsymbol{r}, \omega) = \hat{\boldsymbol{E}}^+_{\rm inc}(\boldsymbol{r}, \omega) + i\mu_0\omega \int \bar{\bar{G}}_{\rm tot}(\boldsymbol{r}, \boldsymbol{r}', \omega) \cdot \hat{\boldsymbol{J}}^+(\boldsymbol{r}', \omega)\, d^3 r'. \tag{15.25}$$

In comparison to the solution of Eq. (15.17) for the homogeneous medium this solution looks almost identical, with the only difference that we have replaced the Green's function by the total Green's function. However, the physical contents are more intricate. As previously discussed for the simple harmonic oscillator model, the Green's function may include losses associated with the imaginary part of the permittivity. To counteract such losses we have to introduce an additional current noise operator, which has appeared quite naturally in our approach building on the microscopic material dynamics.

We finally consider the situation where both external and induced current distributions are present. The electric field operator can then be expressed as

$$\hat{\boldsymbol{E}}^+(\boldsymbol{r}, \omega) = \hat{\boldsymbol{E}}^+_{\rm inc}(\boldsymbol{r}, \omega) + i\mu_0\omega \int \bar{\bar{G}}_{\rm tot}(\boldsymbol{r}, \boldsymbol{r}', \omega) \cdot \left[\hat{\boldsymbol{J}}^+_{\rm ext}(\boldsymbol{r}', \omega) + \hat{\boldsymbol{J}}^+(\boldsymbol{r}', \omega)\right] d^3 r', \tag{15.26}$$

where the first term in brackets accounts for the external current distribution, and the second one for the noise current distribution of the absorbing medium.

15.3 Cross-Spectral Density Revisited

We finally use the result of the previous section to compute for a non-trivial photonic environment the cross-spectral density

$$2\pi\delta(\omega - \omega')\bar{\bar{\mathscr{E}}}(\boldsymbol{r}, \boldsymbol{r}', \omega) = \left\langle \hat{\boldsymbol{E}}(\boldsymbol{r}, \omega) \otimes \hat{\boldsymbol{E}}^\dagger(\boldsymbol{r}', \omega') \right\rangle_{\rm eq}. \tag{15.27}$$

In case of positive frequencies ω the operator $\hat{\boldsymbol{E}}$ can be associated with the operator $\hat{\boldsymbol{E}}^+$ oscillating with positive frequencies. Then, the expression on the right-hand side can be evaluated using Eq. (15.25), and we get

$$\left\langle \left[\hat{\boldsymbol{E}}^{+}_{\rm inc}(\boldsymbol{r},\omega) + i\mu_0\omega \int \overline{\overline{G}}_{\rm tot}(\boldsymbol{r},\boldsymbol{r}_1,\omega)\cdot\hat{\boldsymbol{J}}^{+}(\boldsymbol{r}_1,\omega)\,d^3r_1 \right] \right.$$
$$\left. \otimes \left[\hat{\boldsymbol{E}}^{-}_{\rm inc}(\boldsymbol{r},\omega') - i\mu_0\omega' \int \overline{\overline{G}}^{*}_{\rm tot}(\boldsymbol{r}',\boldsymbol{r}'_1,\omega')\cdot\hat{\boldsymbol{J}}^{-}(\boldsymbol{r}'_1,\omega')\,d^3r'_1 \right] \right\rangle_{\rm eq}.$$

In thermal equilibrium the expectation value $\langle \hat{\boldsymbol{E}}_{\rm inc} \otimes \hat{\boldsymbol{J}} \rangle_{\rm eq}$ vanishes, because the fluctuations of the incoming thermal fields and the noise currents of the absorbing medium are uncorrelated. We can now use the fluctuation–dissipation theorem of Eq. (14.51) to relate the current fluctuations to the imaginary part of the permittivity,

$$\frac{1}{\hbar}\left\langle \hat{\boldsymbol{J}}^{+}(\boldsymbol{r},\omega)\otimes\hat{\boldsymbol{J}}^{-}(\boldsymbol{r}',\omega')\right\rangle = 2\omega^2\,{\rm Im}\Big[\overline{\overline{\varepsilon}}(\boldsymbol{r},\boldsymbol{r}',\omega)\Big]\Big(\bar{n}_{\rm th}(\hbar\omega)+1\Big)\,2\pi\,\delta(\omega-\omega').$$

We thus obtain from Eq. (15.27)

$$\begin{aligned}\overline{\overline{\mathscr{E}}}^{+}(\boldsymbol{r},\boldsymbol{r}',\omega) &= \overline{\overline{\mathscr{E}}}^{+}_{\rm inc}(\boldsymbol{r},\boldsymbol{r}',\omega) + 2\mu_0^2\omega^4\Big(\bar{n}_{\rm th}(\hbar\omega)+1\Big) \\ &\quad\times\int \overline{\overline{G}}_{\rm tot}(\boldsymbol{r},\boldsymbol{r}_1,\omega)\cdot{\rm Im}\Big[\overline{\overline{\varepsilon}}(\boldsymbol{r},\boldsymbol{r}',\omega)\Big]\cdot\overline{\overline{G}}^{*}_{\rm tot}(\boldsymbol{r}'_1,\boldsymbol{r}',\omega')\,d^3r_1d^3r'_1,\end{aligned} \tag{15.28}$$

where we have employed the reciprocity theorem of Sect. 7.4 to exchange the spatial coordinates in $\overline{\overline{G}}^{*}$. For the last term on the right-hand side we introduce for two dyadic functions $\overline{\overline{\mathscr{U}}}$, $\overline{\overline{\mathscr{V}}}$ the short-hand notation

$$\begin{aligned}&\mathbb{V}_{\Omega}\big[\overline{\overline{\mathscr{U}}},\overline{\overline{\mathscr{V}}}\big](\boldsymbol{r},\boldsymbol{r}') \\ &\quad=\mu_0\omega^2\int_{\Omega}\overline{\overline{\mathscr{U}}}(\boldsymbol{r},\boldsymbol{r}_1,\omega)\cdot{\rm Im}\Big[\overline{\overline{\varepsilon}}(\boldsymbol{r}_1,\boldsymbol{r}'_1,\omega)\Big]\cdot\overline{\overline{\mathscr{V}}}(\boldsymbol{r}'_1,\boldsymbol{r}',\omega)\,d^3r_1d^3r'_1,\end{aligned} \tag{15.29}$$

which can be associated with absorption losses in the absorbing medium. Equation (15.28) can then be written in the compact form

$$\begin{aligned}\overline{\overline{\mathscr{E}}}^{+}(\boldsymbol{r},\boldsymbol{r}',+\omega) &= \overline{\overline{\mathscr{E}}}^{+}_{\rm inc}(\boldsymbol{r},\boldsymbol{r}',+\omega) + 2\mu_0\omega^2\,\mathbb{V}_{\Omega}\Big[\overline{\overline{G}}_{\rm tot},\overline{\overline{G}}^{*}_{\rm tot}\Big](\boldsymbol{r},\boldsymbol{r}',\omega)\,\big(\bar{n}_{\rm th}(\hbar\omega)+1\big) \\ \overline{\overline{\mathscr{E}}}^{-}(\boldsymbol{r},\boldsymbol{r}',-\omega) &= \overline{\overline{\mathscr{E}}}^{-}_{\rm inc}(\boldsymbol{r},\boldsymbol{r}',-\omega) + 2\mu_0\omega^2\,\mathbb{V}_{\Omega}\Big[\overline{\overline{G}}^{*}_{\rm tot},\overline{\overline{G}}_{\rm tot}\Big](\boldsymbol{r},\boldsymbol{r}',\omega)\;\;\bar{n}_{\rm th}(\hbar\omega),\end{aligned} \tag{15.30}$$

where the cross-spectral density for negative frequencies has been obtained along the same lines as the expression for positive frequencies. Equation (15.30) is a remarkable result which expresses the photonic environment in thermal equilibrium in terms of the dyadic Green's functions of classical electrodynamics and the Bose–Einstein distribution functions.

15.3.1 Representation Formula for Green's Functions

As worked out explicitly in Sect. 15.7, the volume integrations $\mathbb{V}_\Omega$ can be considerably simplified using a Green's functions relation that is reminiscent of the representation formula for electromagnetic fields, Eq. (5.26),

Representation Formula for Green's Functions

$$\begin{aligned}\operatorname{Im}\Big[\bar{\bar{G}}_{\rm tot}(\boldsymbol{r},\boldsymbol{r}')\Big]\Theta_\Omega(\boldsymbol{r},\boldsymbol{r}') &= \mathbb{V}_\Omega\Big[\bar{\bar{G}}_{\rm tot},\bar{\bar{G}}^*_{\rm tot}\Big](\boldsymbol{r},\boldsymbol{r}') + \mathbb{B}_{\partial\Omega}\Big[\bar{\bar{G}}_{\rm tot},\bar{\bar{G}}^*_{\rm tot}\Big](\boldsymbol{r},\boldsymbol{r}')\\ \operatorname{Im}\Big[\bar{\bar{G}}_{\rm tot}(\boldsymbol{r},\boldsymbol{r}')\Big]\Theta_\Omega(\boldsymbol{r},\boldsymbol{r}') &= \mathbb{V}_\Omega\Big[\bar{\bar{G}}^*_{\rm tot},\bar{\bar{G}}_{\rm tot}\Big](\boldsymbol{r},\boldsymbol{r}') + \mathbb{B}_{\partial\Omega}\Big[\bar{\bar{G}}^*_{\rm tot},\bar{\bar{G}}_{\rm tot}\Big](\boldsymbol{r},\boldsymbol{r}').\end{aligned} \tag{15.31}$$

Here $\Theta_\Omega(\boldsymbol{r},\boldsymbol{r}')$ is one if $\boldsymbol{r}$, $\boldsymbol{r}'$ are located within the volume Ω and zero otherwise. We have introduced the boundary integral operator defined through

$$\begin{aligned}&\mathbb{B}_{\partial\Omega}\Big[\bar{\bar{\mathscr{U}}},\bar{\bar{\mathscr{V}}}\Big](\boldsymbol{r}_1,\boldsymbol{r}_2) \\ &= \frac{i}{2}\oint_{\partial\Omega}\Big\{\bar{\bar{\mathscr{U}}}(\boldsymbol{r}_1,\boldsymbol{r})\times\Big[\nabla\times\bar{\bar{\mathscr{V}}}(\boldsymbol{r},\boldsymbol{r}_2)\Big]+\Big[\nabla\times\bar{\bar{\mathscr{U}}}(\boldsymbol{r},\boldsymbol{r}')\Big]\times\bar{\bar{\mathscr{V}}}(\boldsymbol{r},\boldsymbol{r}_2)\Big\}\cdot\hat{\boldsymbol{n}}\,dS.\end{aligned} \tag{15.32}$$

Equation (15.31) can be interpreted such that the losses of the Green's function, associated with the imaginary part, can be separated into a volume term $\mathbb{V}$ associated with absorption losses and a boundary term $\mathbb{B}$ associated with scattering losses.

As an illustrative example, we consider the situation of an oscillating dipole situated in homogeneous space filled with a lossless material, whose permittivity correspondingly has no imaginary part. We then get from Eq. (15.31)

$$\operatorname{Im}\Big[\bar{\bar{G}}(\boldsymbol{r}_0,\boldsymbol{r}_0)\Big] = \mathbb{B}_{\partial\Omega}\Big[\bar{\bar{G}},\bar{\bar{G}}^*\Big](\boldsymbol{r}_0,\boldsymbol{r}_0). \tag{15.33}$$

where we have replaced the total Green's function by the Green's function $\bar{\bar{G}}$ for the unbounded homogeneous space, which we evaluate at the dipole position $\boldsymbol{r}_0$. Multiplying the equation from both sides with the dipole moment $\boldsymbol{p}$ and using the relation between the electric field and the dipole moment according to

$$\boldsymbol{E}(\boldsymbol{r}) = \mu_0\omega^2\bar{\bar{G}}(\boldsymbol{r},\boldsymbol{r}_0)\cdot\boldsymbol{p},$$

we rewrite Eq. (15.33) in the form

$$\mu_0^2\omega^4\, \boldsymbol{p}\cdot \mathrm{Im}\Big[\bar{\bar{G}}(\boldsymbol{r}_0,\boldsymbol{r}_0)\Big]\cdot \boldsymbol{p}^*$$
$$= \frac{i}{2}\oint_{\partial\Omega}\Big\{\big[\nabla\times \boldsymbol{E}(\boldsymbol{r})\big]\times \boldsymbol{E}^*(\boldsymbol{r})+\boldsymbol{E}(\boldsymbol{r})\times\big[\nabla\times \boldsymbol{E}^*(\boldsymbol{r})\big]\Big\}\cdot\hat{\boldsymbol{n}}dS.$$

Through Faraday's law we relate the curl of the electric field to the magnetic field, and are led to

$$\mu_0^2\omega^4\, \boldsymbol{p}^*\cdot \mathrm{Im}\Big[\bar{\bar{G}}(\boldsymbol{r}_0,\boldsymbol{r}_0)\Big]\cdot \boldsymbol{p}=\mu_0\omega\oint_{\partial\Omega}\mathrm{Re}\big[\boldsymbol{E}(\boldsymbol{r})\times \boldsymbol{H}^*(\boldsymbol{r})\big]\cdot\hat{\boldsymbol{n}}dS=2\mu_0\omega\, P_{\mathrm{sca}},$$

where we have used that the boundary integral on the right-hand side is just the scattering cross section of Eq. (4.22). Indeed, this is the expression of Eq. (10.2) previously derived for the averaged power dissipated by an oscillating dipole using a classical description scheme. This shows that the boundary term can be interpreted in terms of outgoing radiation produced by the current sources (here a single oscillating dipole) within a given volume.

15.3.2 Cross-Spectral Density for Absorbing Media

To obtain our final expression for the cross-spectral density, we start from the first line in Eq. (15.30) which we rewrite together with the representation formula in the form

$$\begin{aligned}\bar{\bar{\mathscr{E}}}^{+}(\boldsymbol{r},\boldsymbol{r}',\omega)&=\bar{\bar{\mathscr{E}}}^{+}_{\mathrm{inc}}(\boldsymbol{r},\boldsymbol{r}',\omega)\\&+2\mu_0\omega^2\left(\mathrm{Im}\Big[\bar{\bar{G}}_{\mathrm{tot}}(\boldsymbol{r},\boldsymbol{r}',\omega)\Big]-\mathbb{B}_{\partial\Omega}\Big[\bar{\bar{G}}_{\mathrm{tot}},\bar{\bar{G}}^*_{\mathrm{tot}}\Big](\boldsymbol{r},\boldsymbol{r}',\omega)\right)\big(\bar{n}_{\mathrm{th}}(\hbar\omega)+1\big).\end{aligned}\tag{15.34}$$

As we will argue next, in thermal equilibrium the incoming contribution and the boundary term cancel each other. We provide two derivations schematically sketched in Fig. 15.3. In both of them we are more specific how the thermal photons are generated. We here follow Ref. [171] and introduce an absorbing boundary shell that is situated far away (at "infinity") from all other absorbing bodies under consideration. The outside of the shell is a perfect mirror, such that no radiation escapes, whereas the inside is composed of a sufficiently thick material layer with small losses $\mathrm{Im}[\varepsilon_2]>0$. This absorbing material has two purposes: first, through the current fluctuations thermal radiation is created, and secondly because of material losses all incoming photons are absorbed (provided that the layer is thick enough). Because the shell is situated at infinity, the total Green's function $G_{\mathrm{tot}}(\boldsymbol{r},\boldsymbol{r}')$ is not noticeably modified by its presence, at least for positions $\boldsymbol{r}$, $\boldsymbol{r}'$ far away from the shell.

Derivation I. Suppose that we place the boundary $\partial\Omega=\partial\Omega_2$ used in the representation formula of Eq. (15.31) outside this shell, such that the boundary contribution $\mathbb{B}_{\partial\Omega}$ becomes zero. See Fig. 15.3a. We can also neglect the incoming

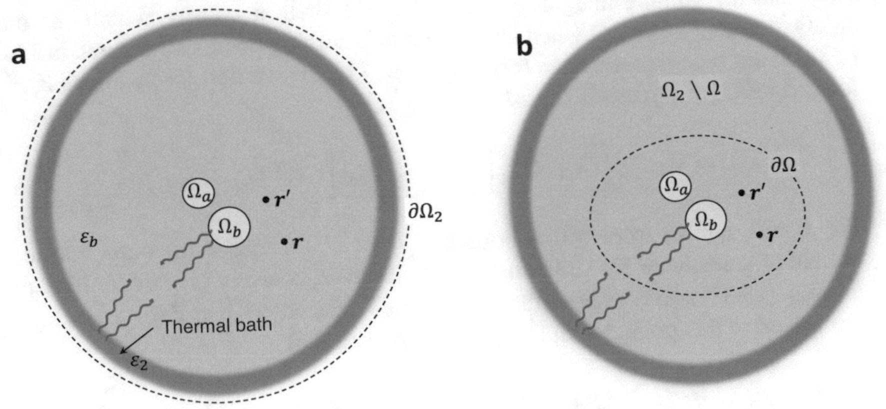

Fig. 15.3 Schematics of computation of cross-spectral density that correlates the electric fields at positions $\boldsymbol{r}$ and $\boldsymbol{r}'$ for absorbing bodies Ω_a and Ω_b in thermal equilibrium. We place far away from all other absorbing bodies a shell with a permittivity ε_2 that has a small imaginary part, such that it produces thermal radiation and absorbs incoming radiation. (**a**) When the boundary is placed outside this shell, there is no contribution from the boundary term $\mathbb{B}_{\partial\Omega_2}$. (**b**) When the boundary is placed away from the shell, we have contributions from the incoming thermal radiation and from the boundary term $\mathbb{B}_{\partial\Omega}$, which cancel each other. For details see text

cross-spectral density because all relevant current sources are located inside of Ω, and there is no incoming radiation from the outside. We then find from Eq. (15.31)

$$\mathrm{Im}\Big[\bar{\bar{G}}_{\mathrm{tot}}(\boldsymbol{r},\boldsymbol{r}')\Big] = \mathbb{V}_{\Omega}\Big[\bar{\bar{G}}_{\mathrm{tot}},\bar{\bar{G}}^*_{\mathrm{tot}}\Big](\boldsymbol{r},\boldsymbol{r}') + \mathbb{V}_{\Omega_2\setminus\Omega}\Big[\bar{\bar{G}}_{\mathrm{tot}},\bar{\bar{G}}^*_{\mathrm{tot}}\Big](\boldsymbol{r},\boldsymbol{r}'),$$

where the volume Ω encloses all absorbing bodies under study and $\Omega_2 \setminus \Omega$ is the volume of the shell at infinity, which produces thermal radiation and absorbs impinging photons. When we assume that G_{tot} is not noticeably influenced by the layer at infinity, which is an excellent approximation because the material with ε_2 is located sufficiently far away from all absorbing bodies and positions $\boldsymbol{r},\boldsymbol{r}'$ of interest, we get instead of Eq. (15.34) the more simple expression

$$\bar{\bar{\mathscr{E}}}^{+}(\boldsymbol{r},\boldsymbol{r}',\omega) = 2\mu_0\omega^2\,\mathrm{Im}\Big[\bar{\bar{G}}_{\mathrm{tot}}(\boldsymbol{r},\boldsymbol{r}',\omega)\Big]\big(\bar{n}_{\mathrm{th}}(\hbar\omega)+1\big), \tag{15.35}$$

where $\bar{\bar{G}}_{\mathrm{tot}}$ is the total Green's function in presence of the absorbing bodies.

Derivation II. Alternatively, we can consider the situation depicted in Fig. 15.3b where the boundary $\partial\Omega$ is located inside the absorbing shell at infinity. The incoming thermal radiation is produced by the current sources at infinity, and we obtain from Eq. (15.30)

$$\bar{\bar{\mathscr{E}}}^{+}_{\mathrm{inc}}(\boldsymbol{r},\boldsymbol{r}',\omega) = 2\mu_0\omega^2\,\mathbb{V}_{\Omega_2\setminus\Omega}\Big[\bar{\bar{G}}_{\mathrm{tot}},\bar{\bar{G}}^*_{\mathrm{tot}}\Big](\boldsymbol{r},\boldsymbol{r}',\omega)\big(\bar{n}_{\mathrm{th}}(\hbar\omega)+1\big).$$

Note that the volume integral extends over the shell at infinity, which can be safely extended to $\Omega_2 \setminus \Omega$ when all additional absorbing bodies are located inside of Ω. From the representation formula of Eq. (15.31) we find for positions $\boldsymbol{r}, \boldsymbol{r}'$ located inside of Ω (this is outside of $\Omega_2 \setminus \Omega$)

$$0 = \mathbb{V}_{\Omega_2\setminus\Omega}\Big[\bar{\bar{G}}_{\text{tot}}, \bar{\bar{G}}^*_{\text{tot}}\Big](\boldsymbol{r}, \boldsymbol{r}') - \mathbb{B}_{\partial\Omega}\Big[\bar{\bar{G}}_{\text{tot}}, \bar{\bar{G}}^*_{\text{tot}}\Big](\boldsymbol{r}, \boldsymbol{r}').$$

The minus sign in front of the boundary term $\mathbb{B}$ is because the outer surface normal of the volume $\Omega_2 \setminus \Omega$ points *into* volume Ω, contrary to our definition of $\hat{\boldsymbol{n}}$ in Eq. (15.32) as an *outer* surface normal of Ω. See also Chap. 5 for a thorough discussion of this point. From Eq. (15.34) we then find

$$\begin{aligned}\bar{\bar{\mathscr{E}}}^{+}(\boldsymbol{r}, \boldsymbol{r}', \omega) &= 2\mu_0\omega^2\, \mathbb{B}_{\partial\Omega}\Big[\bar{\bar{G}}_{\text{tot}}, \bar{\bar{G}}^*_{\text{tot}}\Big](\boldsymbol{r}, \boldsymbol{r}')\,\big(\bar{n}_{\text{th}}(\hbar\omega) + 1\big) \\ &+2\mu_0\omega^2\Big(\text{Im}\Big[\bar{\bar{G}}_{\text{tot}}(\boldsymbol{r}, \boldsymbol{r}', \omega)\Big] - \mathbb{B}_{\partial\Omega}\Big[\bar{\bar{G}}_{\text{tot}}, \bar{\bar{G}}^*_{\text{tot}}\Big](\boldsymbol{r}, \boldsymbol{r}', \omega)\,\big(\bar{n}_{\text{th}}(\hbar\omega) + 1\big)\Big),\end{aligned}$$

where the first term on the right-hand side is the incoming cross-spectral density $\mathscr{E}^{+}_{\text{inc}}$ expressed in terms of the boundary integral. We immediately observe that the two boundary terms cancel each other, and we again arrive at Eq. (15.35).

We can thus express the cross-spectral density in the intriguing form

Cross-Spectral Density in Thermal Equilibrium

$$\bar{\bar{\mathscr{E}}}^{\pm}(\boldsymbol{r}, \boldsymbol{r}', \omega) = 2\mu_0\omega^2\, \text{Im}\Big[\bar{\bar{G}}_{\text{tot}}(\boldsymbol{r}, \boldsymbol{r}', \pm\omega)\Big]\theta(\pm\omega)\begin{Bmatrix}\bar{n}_{\text{th}}(+\hbar\omega) + 1\\ \bar{n}_{\text{th}}(-\hbar\omega)\end{Bmatrix}. \tag{15.36}$$

The upper expression corresponds to positive frequencies, where the energy $\hbar\omega$ is transferred to the thermal photonic environment, and the lower one to the reversed process of energy gain from the photonic environment. Equation (15.36) is a quite remarkable result that expresses the cross-spectral density in terms of the dyadic Green's function, which is obtained within the framework of classical electrodynamics. The final result is almost identical to the free-space expression of Eq. (15.8), with the only difference that the free-space dyadic Green's function has been replaced by the total Green's function. The expression for $\mathscr{E}^{\pm}$ combines the photon propagation properties, which are accounted for by the Green's function of classical electrodynamics, with the noise properties of the field and current operators, as described by the Bose–Einstein distribution functions. Importantly, for electromagnetic couplings the linear response of a system in thermal equilibrium can be uniquely expressed in terms of cross-spectral densities. This might be the right place for a brief summary.

Correlation Functions. Correlation functions are an elegant means for describing the response properties of physical systems. In thermal equilibrium, all quantities, such as correlation functions or the cross-spectral density, can be expressed in terms of a single spectral function.

Linear Response. The key to our success has been the *linear* response we have investigated so far. It allows us to express the response of the perturbed system in terms of correlation functions only, as discussed in the context of Kubo's formalism, and to relate current sources to electric fields on an operator level, with the Green's dyadics accounting for the photon propagation properties. As we have discussed above, for absorbing materials we additionally have to introduce a noise term to counteract losses. Through the use of Green's functions we have effectively removed photons from our description. Their properties are still present in the form of the Green's function (accounting for photon propagation) and the thermal Bose–Einstein distribution functions, but there is no longer any need to describe the photon dynamics explicitly.

Green's Functions. As always in nano optics Green's functions play an important role, since they allow us to derive results *before* pondering on a specific system and because there exists a rich machinery for manipulating them. For instance, in the integral equation of Eq. (15.31) we have been able to convert the seemingly hard integration term of Eq. (15.28) into a relatively simple Green's function contribution.

15.4 Photonic Local Density of States Revisited

As a first application of the cross-spectral density, we consider a quantum emitter initially in an excited state e, which decays by emitting a photon as shown in Fig. 15.4. The quantum emitter can be an atom, molecule, or quantum dot. For simplicity, in the following we will refer to it as an "atom." We start from Eq. (13.91) accounting for the light–matter interaction in the dipole approximation

$$\hat{V} = -\hat{\boldsymbol{p}} \cdot \left[\hat{\boldsymbol{E}}^{+}(\boldsymbol{r}_0) + \hat{\boldsymbol{E}}^{-}(\boldsymbol{r}_0)\right], \tag{15.37}$$

but now introduce electric field operators which describe the annihilation and creation of photons. Here $\hat{\boldsymbol{p}}$ is the dipole operator and $\boldsymbol{r}_0$ is the position of the atom. We next introduce an approximation, the so-called rotating wave approximation (RWA), which we will investigate in more detail in the next chapter. In the same way as $\hat{\boldsymbol{E}}^{+}$ and $\hat{\boldsymbol{E}}^{-}$ account for the annihilation and creation of a light excitation (namely the photon), we can split the dipole operator $\hat{\boldsymbol{p}} = \hat{\boldsymbol{p}}^{+} + \hat{\boldsymbol{p}}^{-}$ into parts oscillating with positive and negative frequency components, respectively, which can be associated with the annihilation and creation of an atom excitation. In the rotating wave approximation we only keep the cross terms

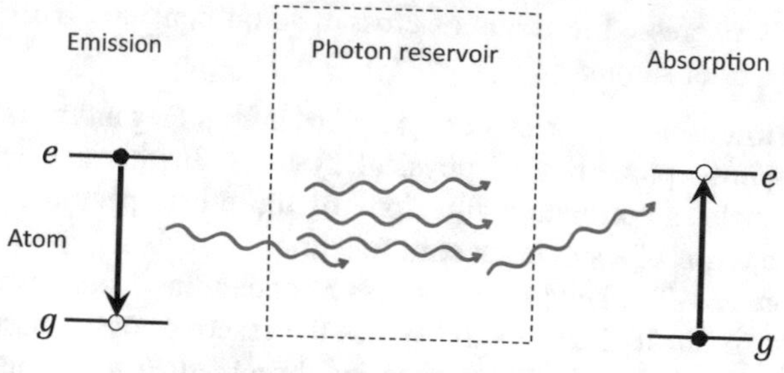

Fig. 15.4 Schematics of photon emission and absorption processes of atom. In case of emission, the atom decays from the excited state e to the groundstate g by emitting a photon. For absorption the atom is initially in the groundstate g and becomes promoted into the excited state e through absorption of a photon. The thermally populated photons in the reservoir enhance the scattering rates via the Bose–Einstein occupation factors

$$\hat{V} = -\left[\hat{\boldsymbol{p}}^- \cdot \hat{\boldsymbol{E}}^+(\boldsymbol{r}_0) + \hat{\boldsymbol{p}}^+ \cdot \hat{\boldsymbol{E}}^-(\boldsymbol{r}_0)\right], \tag{15.38}$$

where the first term accounts for the transfer of excitation from the light field to the atom, and the second one for the reversed process. Let $|g\rangle$, $|e\rangle$ denote the ground and excited state of the atom. The atom decay rate can be computed from Fermi's golden rule of Eq. (14.95), and we get

$$\Gamma^{\text{em}}_{e\to g} = \frac{1}{\hbar^2}\lim_{\eta\to 0}\int_0^\infty e^{-\eta t}\,\text{tr}_R\left(\hat{\rho}_R\,\langle e|\hat{V}(t)|g\rangle\langle g|\hat{V}(0)|e\rangle\right)dt + \text{c.c.}$$

$$\Gamma^{\text{abs}}_{g\to e} = \frac{1}{\hbar^2}\lim_{\eta\to 0}\int_0^\infty e^{-\eta t}\,\text{tr}_R\left(\hat{\rho}_R\,\langle g|\hat{V}(t)|e\rangle\langle e|\hat{V}(0)|g\rangle\right)dt + \text{c.c.},$$

where $\hat{\rho}_R$ is the statistical operator for the thermal photons ("the reservoir"). The expression in the first line accounts for photon emission, where the atom decays from the excited state to the groundstate, and the expression in the second line for photon absorption, where the atom is promoted from the ground to the excited state through absorption of a thermally populated photon. Let $\boldsymbol{p} = \langle e|\hat{\boldsymbol{p}}^-|g\rangle$ be the dipole moment. The scattering rates can then be rewritten in the form

$$\Gamma^{\text{em}}_{e\to g} = \frac{1}{\hbar^2}\lim_{\eta\to 0}\int_0^\infty e^{i(+\omega+i\eta)t}\,\boldsymbol{p}\cdot\left\langle\hat{\boldsymbol{E}}^+(\boldsymbol{r}_0,t)\otimes\hat{\boldsymbol{E}}^-(\boldsymbol{r}_0,0)\right\rangle_{\text{eq}}\cdot\boldsymbol{p}^*\,dt + \text{c.c.}$$

$$\Gamma^{\text{em}}_{e\to g} = \frac{1}{\hbar^2}\lim_{\eta\to 0}\int_0^\infty e^{i(-\omega+i\eta)t}\,\boldsymbol{p}^*\cdot\left\langle\hat{\boldsymbol{E}}^-(\boldsymbol{r}_0,t)\otimes\hat{\boldsymbol{E}}^+(\boldsymbol{r}_0,0)\right\rangle_{\text{eq}}\cdot\boldsymbol{p}\;dt + \text{c.c.},$$

where $\hbar\omega = E_e - E_g$ is the energy difference between the excited and groundstate of the atom. The product of the electric field operators has to be understood as a dyadic one, such that we have two scalar products between the dipole operators outside the brackets and the electric field operators inside the brackets. With the inverse Fourier transform

$$\frac{1}{\hbar}\left\langle \hat{\boldsymbol{E}}^{\pm}(\boldsymbol{r}_0,t)\otimes\hat{\boldsymbol{E}}^{\mp}(\boldsymbol{r}_0,0)\right\rangle_{\text{eq}}=\int_{-\infty}^{\infty}e^{-i\omega' t}\,\bar{\bar{\mathscr{E}}}^{\pm}(\boldsymbol{r}_0,\boldsymbol{r}_0,\omega')\,\frac{d\omega'}{2\pi}$$

one can easily derive the following expression

$$\frac{1}{\hbar}\int_0^{\infty}e^{i(\pm\omega+i\eta)t}\left\langle \hat{\boldsymbol{E}}^{\pm}(\boldsymbol{r}_0,t)\otimes\hat{\boldsymbol{E}}^{\mp}(\boldsymbol{r}_0,0)\right\rangle_{\text{eq}}dt=i\int_{-\infty}^{\infty}\frac{\bar{\bar{\mathscr{E}}}^{\pm}(\boldsymbol{r}_0,\boldsymbol{r}_0,\omega')}{\pm\omega-\omega'+i\eta}\,\frac{d\omega'}{2\pi}.$$

Additionally we find from the definition of the cross-spectral density, Eq. (15.1),

$$\left(\mathscr{E}^{+}_{ij}(\boldsymbol{r}_0,\boldsymbol{r}_0,\omega)\right)^*=\mathscr{E}^{+}_{ji}(\boldsymbol{r}_0,\boldsymbol{r}_0,\omega)\implies\left(\boldsymbol{p}\cdot\bar{\bar{\mathscr{E}}}^{+}\cdot\boldsymbol{p}^*\right)^*=\boldsymbol{p}\cdot\bar{\bar{\mathscr{E}}}^{+}\cdot\boldsymbol{p}^*$$

$$\left(\mathscr{E}^{-}_{ij}(\boldsymbol{r}_0,\boldsymbol{r}_0,\omega)\right)^*=\mathscr{E}^{-}_{ji}(\boldsymbol{r}_0,\boldsymbol{r}_0,\omega)\implies\left(\boldsymbol{p}^*\cdot\bar{\bar{\mathscr{E}}}^{-}\cdot\boldsymbol{p}\right)^*=\boldsymbol{p}^*\cdot\bar{\bar{\mathscr{E}}}^{-}\cdot\boldsymbol{p},$$

which shows that the terms on the right-hand sides are real-valued. We can use Eq. (F.6) to express the frequency denominators in terms of a Cauchy principal value integral and Dirac's delta function, where only the latter one contributes in combination with the complex conjugate. The Fermi's golden rule results for the decay rate of a quantum emitter then become together with Eq. (15.36) for the relation between the cross-spectral density and the dyadic Green's function

Decay Rate of Quantum Emitter

$$\begin{aligned}\hbar\Gamma^{\text{em}}_{e\to g}&=2\mu_0\omega^2\,\text{Im}\left[\boldsymbol{p}\cdot\bar{\bar{G}}_{\text{tot}}(\boldsymbol{r}_0,\boldsymbol{r}_0,\omega)\cdot\boldsymbol{p}^*\right]\left(\bar{n}_{\text{th}}(\hbar\omega)+1\right)\\ \hbar\Gamma^{\text{abs}}_{g\to e}&=2\mu_0\omega^2\,\text{Im}\left[\boldsymbol{p}^*\cdot\bar{\bar{G}}_{\text{tot}}(\boldsymbol{r}_0,\boldsymbol{r}_0,\omega)\cdot\boldsymbol{p}\right]\bar{n}_{\text{th}}(\hbar\omega).\end{aligned}\tag{15.39}$$

In the above equations $\hbar\omega$ is the transition energy of the atom, which is a positive quantity for both photon emission and absorption.

As a first example, we compute the decay rate of Eq. (15.39) for an atom in free space and at zero temperature. In this case we can use the result of Eq. (10.3) for the imaginary part of the Green's function, and get

$$2\omega^2\mu_0\,\text{Im}\left[\boldsymbol{p}\cdot\bar{\bar{G}}(\boldsymbol{r}_0,\boldsymbol{r}_0,\omega)\cdot\boldsymbol{p}^*\right]=2\omega^2\mu_0 p^2\left(\frac{k}{6\pi}\right),$$

We have used Eq. (10.3) to arrive at the last expression assuming a real dipole moment oriented along z, although we would get the same result also for different moments. Putting together all results, we obtain the Wigner–Weisskopf decay rate

$$\Gamma = \frac{\mu_0 \omega^3 p^2}{3\pi\hbar c}, \tag{15.40}$$

in agreement to Eq. (10.7) derived within a semiclassical framework.

Classical vs. Quantum Result In classical electrodynamics one uses for time-harmonic fields a complex notation where all quantities X have a time dependence of the form $e^{-i\omega t}$ and the physical quantity is obtained through the real space operation

$$X(t) \longrightarrow \text{Re}\left[e^{-i\omega t}X\right].$$

In quantum mechanics one introduces operators $\hat{X}^{\pm}$ oscillating with positive and negative frequency components, and the total operator is the sum

$$\hat{X} = \hat{X}^{+} + \hat{X}^{-}.$$

Thus, for a direct comparison between the classical and quantum results one should in principle rescale $\hat{X} \to \frac{1}{2}\hat{X}$. However, in this book we will keep the different definitions for classical and quantum physics, and will add a word of caution whenever needed.

Equation (15.39) is the same result as that derived for the classical Maxwell's equations, Eq. (10.15), with the only difference of the Bose–Einstein occupation numbers. For photon emission we find that the scattering rate is enhanced by a factor of $\bar{n}_{\text{th}} + 1$ because of the thermally populated photons which stimulate the emission process. For photon absorption the scattering rate depends on the photon occupation $\bar{n}_{\text{th}}$, and there exists no classical analog. Thus, **all results presented in Chap. 10 prevail** provided that the thermal photon populations are properly accounted for.

15.4.1 Lamb Shift

In the previous section we have discussed the atomic decay rate due to the photon emission or absorption, as computed from Fermi's golden rule. We next analyze the renormalization of the transition frequency of the atom as an effect of light–matter interaction, the so-called Lamb shift. Our starting point is the Rabi energy operator

$$\hbar\hat{\Omega}^{+} = \boldsymbol{p} \cdot \hat{\boldsymbol{E}}^{+}(\boldsymbol{r}_0), \tag{15.41}$$

and its Hermitian conjugate $\hbar\hat{\Omega}^{-}$. The light–matter interaction of Eq. (15.38) can then be expressed in the form

$$\hat{V}(t) = -\left(e^{i\omega t}\hbar\hat{\Omega}^{+}(t)\big|e\big\rangle\big\langle g\big| + e^{-i\omega t}\hbar\hat{\Omega}^{-}(t)\big|g\big\rangle\big\langle e\big|\right).$$

Throughout we use an interaction representation with respect to the uncoupled light and matter Hamiltonian $\hat{H}_0$. The first term in parentheses describes a process where a photon is destroyed and the system is promoted from the ground to the excited state, and the second term for the reversed process. Consider the polarization function

$$\mathscr{P}(t) = \langle \psi(t) | g \rangle \langle e | \psi(t) \rangle, \tag{15.42}$$

which accounts for the coherence between the ground and excited state of the atom. In the Schrödinger picture the time evolution due to the unperturbed system Hamiltonian would be given by

$$\dot{\mathscr{P}}_S(t) \approx -i\omega \mathscr{P}_S(t),$$

where ω is the transition frequency of the atom. In the interaction picture this fast time dependence is removed, and in absence of light–matter interactions the time evolution becomes $\dot{\mathscr{P}} \approx 0$. When considering light–matter interactions, we expect an approximate time evolution of the form

$$\dot{\mathscr{P}}(t) = -i\left(\delta\omega - \frac{i\Gamma}{2}\right)\mathscr{P}(t),$$

where $\delta\omega$ is the renormalization of the transition frequency and Γ the decay rate of the polarization. In what follows, we work out the time evolution of $\mathscr{P}$ in order to compute the parameters $\delta\omega$, Γ.

Wavefunction Solution In analogy to the derivation of Fermi's golden rule in Sect. 14.6.1 we consider basis states $|g, r\rangle$ and $|e, r\rangle$, where r denotes the reservoir (photon) degrees of freedom. A general state of the interacting system and reservoir can then be written in the form

$$|\psi(t)\rangle = \sum_r \Big(C_{gr}(t)|g, r\rangle + C_{er}(t)|e, r\rangle \Big), \tag{15.43}$$

where C_{gr}, C_{er} are the probability amplitudes. Inserting this wave function ansatz into Schrödinger's equation gives

$$\begin{aligned} i\hbar \sum_r \Big(\dot{C}_{gr}(t)|g, r\rangle + \dot{C}_{er}(t)|e, r\rangle \Big) \\ = -\sum_r \Big(\Big[e^{i\omega t}\hbar\hat{\Omega}^+(t) \Big] C_{gr}(t)|e, r\rangle + \Big[e^{-i\omega t}\hbar\hat{\Omega}^-(t) \Big] C_{er}(t)|g, r\rangle \Big). \end{aligned}$$

We next project from the left-hand side on the basis states, which leads to the equations of motion for the probability amplitudes

$$\begin{aligned} i\dot{C}_{gr}(t) &= -\sum_{r'} e^{-i\omega t}\,\Omega^-_{rr'}(t)\, C_{er'}(t) \\ i\dot{C}_{er}(t) &= -\sum_{r'} e^{+i\omega t}\,\Omega^+_{rr'}(t)\, C_{gr'}(t), \end{aligned}$$

with the matrix elements $\Omega^{\pm}_{rr'} = \langle r|\hat{\Omega}^{\pm}|r'\rangle$. We continue to formally integrate the above equations and insert the solutions into the original equations of motion. This gives

$$i\dot{C}_{gr}(t) = -\sum_{r'} e^{-i\omega t}\Omega^{-}_{rr'}(t)\left[C_{er'}(0) + i\int_0^t \sum_{r''} e^{+i\omega t'}\Omega^{+}_{r'r''}(t')\,C_{gr''}(t')\,dt'\right]$$

$$i\dot{C}_{er}(t) = -\sum_{r'} e^{+i\omega t}\Omega^{+}_{rr'}(t)\left[C_{gr'}(0) + i\int_0^t \sum_{r''} e^{-i\omega t'}\Omega^{-}_{r'r''}(t')\,C_{er''}(t')\,dt'\right].$$

In the following we assume that in the integrals the coefficients $C_{gr}(t')$, $C_{er}(t')$ change slowly as a function of time, such that they can be replaced by the coefficients at time t. This is the so-called Markov approximation, which is generally a good approximation because the integral kernels decay fast as function of time difference $t - t'$ owing to the destructive interference of reservoir modes; see also the discussion of the adiabatic limit in the previous chapter.

We additionally assume that at time zero the only non-vanishing coefficients $C_{gr} = \tilde{C}_g\delta_{rr_0}$, $C_{er} = \tilde{C}_e\delta_{rr_0}$ are for a single reservoir mode r_0, and we discard in the following all other coefficients, an approximation generally valid to the lowest order of the light–matter interaction. We then get

$$\begin{aligned} i\dot{\tilde{C}}_g(t)\delta_{rr_0} &= \left[-i\int_0^t e^{-i\omega(t-t')}\sum_{r'}\Omega^{-}_{r_0r'}(t)\Omega^{+}_{r'r_0}(t')\,dt'\right]\tilde{C}_g(t) \\ i\dot{\tilde{C}}_e(t)\delta_{rr_0} &= \left[-i\int_0^t e^{+i\omega(t-t')}\sum_{r'}\Omega^{+}_{r_0r'}(t)\Omega^{-}_{r'r_0}(t')\,dt'\right]\tilde{C}_e(t), \end{aligned} \tag{15.44}$$

where we have neglected the terms with the coefficients at the initial time because the matrix elements $\langle r_0|\hat{\Omega}^{\pm}|r_0\rangle$ are zero. The expression

$$\mathscr{P}(t) = \sum_r p_r\, C_{er}(t)C^*_{gr}(t) \tag{15.45}$$

corresponds to the polarization function of Eq. (15.42), however, with an additional summation over the reservoir states. When multiplying the terms in brackets of Eq. (15.44) with p_r and summing over r, we are led to

$$\int_0^\infty e^{i(-\omega+i\eta)t}\sum_{r,r'} p_r\Omega^{-}_{rr'}(t)\Omega^{+}_{r'r}(0)\,dt = \frac{i}{\hbar}\int_{-\infty}^{\infty}\frac{\boldsymbol{p}^*\cdot\bar{\bar{\mathscr{E}}}^{-}(\boldsymbol{r}_0,\boldsymbol{r}_0,\omega')\cdot\boldsymbol{p}}{-\omega-\omega'+i\eta}\frac{d\omega'}{2\pi}$$

$$\int_0^\infty e^{i(+\omega+i\eta)t}\sum_{r,r'} p_r\Omega^{+}_{rr'}(t)\Omega^{-}_{r'r}(0)\,dt = \frac{i}{\hbar}\int_{-\infty}^{\infty}\frac{\boldsymbol{p}\cdot\bar{\bar{\mathscr{E}}}^{+}(\boldsymbol{r}_0,\boldsymbol{r}_0,\omega')\cdot\boldsymbol{p}^*}{\omega-\omega'+i\eta}\frac{d\omega'}{2\pi},$$

where we have introduced the adiabatic limit which is valid for sufficiently large time delays. In arriving at the terms on the right-hand side we have used $\sum_{r'}|r'\rangle\langle r'| = \mathbb{1}$ and have evaluated the cross-spectral densities in accordance to our previous discussion of scattering rates. The time evolution of the polarization function of Eq. (15.45) can be finally brought to the desired form

$$\frac{d}{dt}\mathscr{P}(t) = -i\left(\delta\omega + \frac{i\Gamma}{2}\right)\mathscr{P}(t),$$

with the energy renormalization $\delta\omega$ and the decay rate Γ.

Putting together all results, we obtain for the frequency renormalization of the polarization function the result

$$\delta\omega - \frac{i\Gamma}{2} = \lim_{\eta\to 0}\frac{1}{\hbar}\int_{-\infty}^{\infty}\left[\frac{\boldsymbol{p}\cdot\bar{\bar{\mathscr{E}}}^{+}(\boldsymbol{r}_0,\boldsymbol{r}_0,\omega')\cdot\boldsymbol{p}^*}{\omega-\omega'+i\eta} + \frac{\boldsymbol{p}^*\cdot\bar{\bar{\mathscr{E}}}^{-}(\boldsymbol{r}_0,\boldsymbol{r}_0,\omega')\cdot\boldsymbol{p}}{\omega+\omega'+i\eta}\right]\frac{d\omega'}{2\pi}. \tag{15.46}$$

Comparison with Eq. (15.39) shows that the decay rate is the sum of the emission decay rate for the excited state, and the absorption decay rate for the groundstate. We finally use Eq. (15.36) to relate the cross-spectral densities to the dyadic Green's function, and arrive at

$$\delta\omega = \frac{\mu_0}{\pi\hbar}\mathcal{P}\int_0^{\infty}\boldsymbol{p}\cdot\mathrm{Im}\Big[\bar{\bar{G}}_{\rm tot}(\boldsymbol{r}_0,\boldsymbol{r}_0,\omega')\Big]\cdot\boldsymbol{p}^*\left(\frac{1+2\bar{n}_{\rm th}(\hbar\omega')}{\omega-\omega'}\right)\omega'^2 d\omega', \tag{15.47}$$

where $\mathcal{P}$ denotes Cauchy principal value integral, see Appendix F. The above expression is not ready for use yet, and must be adapted in two important ways. In the following we discuss the Lamb shift at zero and finite temperature separately. Consider first an atom embedded in free space, such that the Green's function can be replaced by the free-space function of Eq. (5.19). Using Eq. (10.3) we find

$$\boldsymbol{p}\cdot\mathrm{Im}\big[\bar{\bar{G}}(\boldsymbol{r}_0,\boldsymbol{r}_0,\omega')\big]\cdot\boldsymbol{p}^* = \frac{\omega p^2}{6\pi c},$$

which allows us to rewrite the Lamb shift in the form

$$\delta\omega = \frac{\mu_0 p^2}{6\pi^2\hbar n c}\mathcal{P}\int_0^{\infty}\frac{\omega'^3\,d\omega'}{\omega-\omega'}.$$

As can be immediately seen, this is a divergent integral. Within the field of quantum electrodynamics such divergent quantities appear at several places, and have to be removed using a renormalization approach. In this book we do not enter into this difficult topic. Rather we take the pragmatic viewpoint that the free-space frequency renormalization cannot be measured directly, and we thus assume that it is already included in the atomic transition frequency ω. As an effect of a non-trivial photonic environment, an additional Lamb shift occurs which can be computed by replacing the total Green's function in Eq. (15.47) by the reflected Green's function. This gives for the temperature-independent Lamb shift the final result

Lamb Shift (Temperature Independent)

$$\delta\omega^{(1)} = \frac{\mu_0}{\pi\hbar}\mathcal{P}\int_0^\infty \boldsymbol{p}\cdot \mathrm{Im}\Big[\bar{\bar{G}}_{\mathrm{refl}}(\boldsymbol{r}_0,\boldsymbol{r}_0,\omega')\Big]\cdot \boldsymbol{p}^*\left(\frac{1}{\omega-\omega'}\right)\omega'^2 d\omega'. \tag{15.48}$$

We next discuss the finite temperature case. As pointed out, for instance, in [172] and explicitly worked out in [134], at finite temperatures the rotating wave approximation, which only keeps the resonant terms where an excitation is transferred from the atom to the light fields and vice versa, is insufficient for computing the Lamb shift. A more detailed analysis including also the other non-resonant terms then gives for the **temperature-dependent Lamb shift** [134, 172]

$$\delta\omega^{(2)} = \frac{\mu_0}{\pi\hbar}\mathcal{P}\int_0^\infty \boldsymbol{p}\cdot \mathrm{Im}\Big[\bar{\bar{G}}_{\mathrm{tot}}(\boldsymbol{r}_0,\boldsymbol{r}_0,\omega')\Big]\cdot \boldsymbol{p}^*\left(\frac{\bar{n}_{\mathrm{th}}(\hbar\omega')}{\omega-\omega'}+\frac{\bar{n}_{\mathrm{th}}(\hbar\omega')}{\omega+\omega'}\right)\omega'^2 d\omega'. \tag{15.49}$$

For the temperature-independent Lamb shift we can use the Kramers–Kronig relation for the cross-spectral density, in accordance to Eq. (14.17) for general correlation functions, to rewrite Eq. (15.48) in the considerably simpler form

$$\delta\omega^{(1)} = -\frac{\mu_0\omega^2}{\hbar}\mathrm{Re}\Big[\boldsymbol{p}\cdot\bar{\bar{G}}_{\mathrm{refl}}(\boldsymbol{r}_0,\boldsymbol{r}_0,\omega)\cdot\boldsymbol{p}^*\Big]. \tag{15.50}$$

For a nanosphere we employ Mie's theory discussed in Appendix E to compute the scattering rate enhancement and Lamb's shift. Using Eq. (E.41) and its derivation, we get for a dipole located at $r_0\hat{z}$ and oriented along $\hat{z}$ the expressions (zero temperature)

$$\begin{aligned}\frac{\Gamma}{\Gamma_0} &= 1-\frac{3}{2}\mathrm{Re}\left\{\sum_{\ell=0}^{\infty}\ell(\ell+1)(2\ell+1)\,a_\ell\left[\frac{h_\ell^{(1)}(x)}{x}\right]^2_{x=kr_0}\right\} \qquad (15.51)\\ \frac{\delta\omega}{\Gamma_0} &= -\frac{3}{4}\mathrm{Im}\left\{\sum_{\ell=0}^{\infty}\ell(\ell+1)(2\ell+1)\,a_\ell\left[\frac{h_\ell^{(1)}(x)}{x}\right]^2_{x=kr_0}\right\}.\end{aligned}$$

Here Γ_0 is the Wigner–Weisskopf decay rate of free space, and a_ℓ are the Mie coefficients given in Eq. (E.22). Figure 15.5 reports $\delta\omega$, Γ for a silver nanosphere with a diameter of 40 nm, and for a dipole located 10 nm away from the sphere and with a dipole orientated along z. One observes that the Lamb shift can be either positive or negative, and can be enhanced by factors of up to hundreds. Figure 15.6 shows the Lamb shift as a function of distance between the dipole and the sphere, for the selected transition wavelengths indicated with dotted lines in Fig. 15.5.

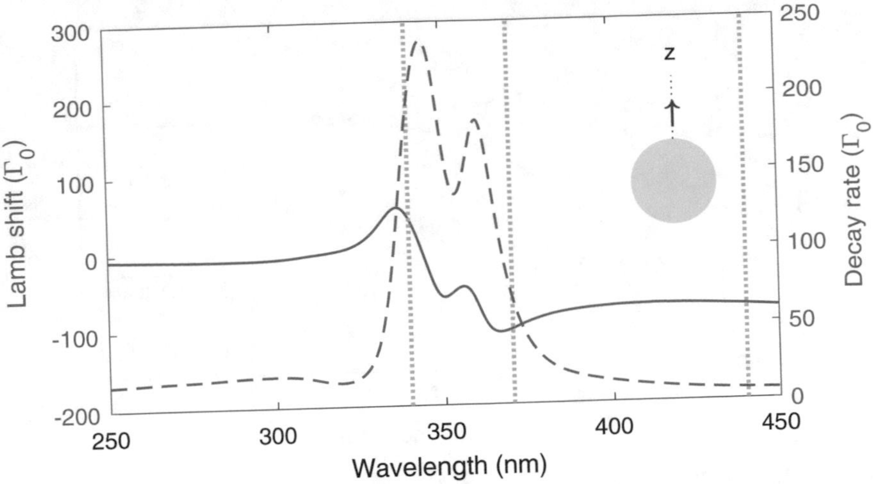

Fig. 15.5 Lamb shift $\delta\omega$ (solid line, left axis) and decay rate Γ (dashed line, right axis) for Ag nanosphere with 40 nm diameter and a dipole located 10 nm away from the sphere. Both $\delta\omega$, Γ are given in units of the Wigner–Weisskopf decay rate Γ_0. The dotted lines indicate the wavelengths where the distance-dependent Lamb shift is plotted in Fig. 15.6

15.5 Forces at the Nanoscale

In Chap. 4 we have discussed that electromagnetic fields carry momentum and can thus exert forces on dielectric and metallic bodies. We here show that such forces also exist in absence of external electromagnetic fields, where they are now mediated through field fluctuations. It is convenient to treat the forces exerted on small polarizable particles and larger bodies separately, and we will denote the forces as Casimir–Polder forces for small particles and Casimir forces for larger particles. Our analysis closely follows Refs. [120, 134].

15.5.1 Casimir–Polder Forces

When the atom moves sufficiently slowly, its dynamics can be approximately described by an effective Hamiltonian of the form

$$\hat{H}_{\text{atom}} = \frac{\hat{\boldsymbol{\pi}}^2}{2M} + \hbar\delta\omega(\boldsymbol{r}), \tag{15.52}$$

where M is the mass of the atom and $\delta\omega(\boldsymbol{r})$ the Lamb shift discussed above. In this model we neglect internal excitation processes caused by the atom motion, which can give rise to noncontact friction [173] for swift atoms. From Heisenberg's equations of motion we obtain for the force acting on the atom

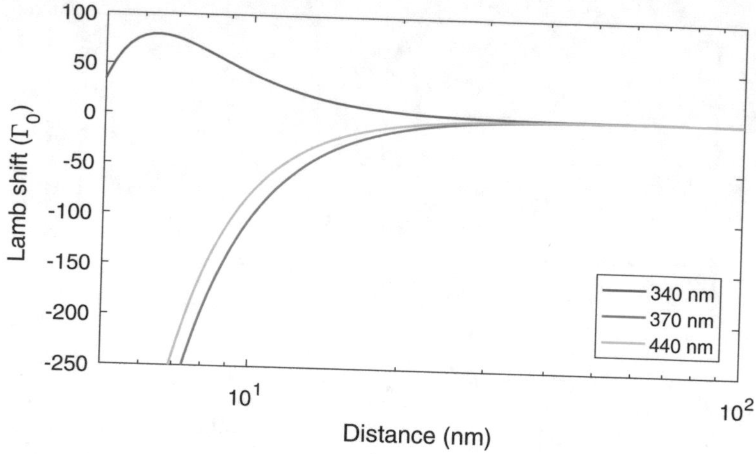

Fig. 15.6 Lamb shift when varying the distance between a *Ag* nanosphere and a dipole, with the same parameters as used in Fig. 15.5. The transition wavelengths for the dipole are indicated by dotted lines in Fig. 15.5

$$\hat{\boldsymbol{F}}(\boldsymbol{r}) = \frac{d}{dt}\hat{\boldsymbol{\pi}} = \frac{1}{i\hbar}\Big[\hat{\boldsymbol{\pi}}, \hat{H}_{\text{atom}}\Big] = -\nabla\Big(\hbar\delta\omega(\boldsymbol{r})\Big). \tag{15.53}$$

In many cases the atom motion can be treated classically, and $\boldsymbol{F}$ can be interpreted as a classical force acting on the atom. Apparently, the Casimir–Polder force of Eq. (15.53) is mediated by the electromagnetic field fluctuations and pushes the particle to regions of large, negative Lamb shifts.

Figure 15.6 shows the Lamb shift of a dipole in front of a metallic nanosphere as a function of dipole-sphere separation. One observes that for the investigated dipole transition wavelengths λ_{dip} the Casimir–Polder force is repulsive for $\lambda_{\text{dip}} =$ 340 nm, and attractive otherwise. Comparison with Fig. 15.5, which shows $\delta\omega$ as a function of resonance wavelengths, suggests the interpretation that the force is repulsive when $\hbar\omega_{\text{dip}}$ is on the high-energy side (blue detuning) of the plasmon resonances of the nanosphere, and becomes attractive when $\hbar\omega_{\text{dip}}$ is on the low-energy side (red detuning).

15.5.2 Casimir Forces

The Casimir forces between larger bodies have to be treated differently. We introduce two different description schemes, one based on vacuum fluctuations and a second, more general one based on Maxwell's stress tensor.

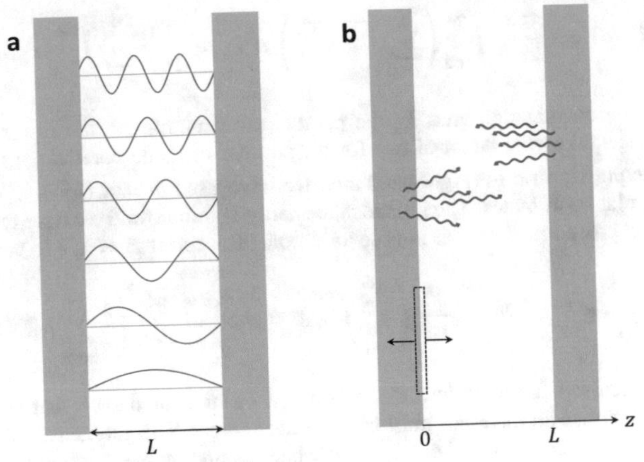

Fig. 15.7 Schematics for Casimir force. (**a**) Electromagnetic modes are confined between two perfectly conducting plates which are separated by a distance L. The photonic density of states is reduced with respect to its free-space value, which leads to an attractive force between the two plates. (**b**) The same effect can be also described differently in terms of (vacuum and thermal) radiation fluctuations which mediate a force. The dashed lines indicate a bounding box as needed for the computation of the force. For a discussion see text

Casimir Force Between Two Perfectly Conducting Plates

Figure 15.7 depicts a scenario where two perfectly conducting plates are separated by a distance L. When discussing the quantization of electromagnetic fields in Chap. 13 we have seen that the photon modes have an (infinite) groundstate energy

$$E_0 = \frac{1}{2}\sum_i \hbar\omega_i ,$$

where i labels the different photon modes. In general, this energy cannot be measured and is discarded from all calculations. However, in the case of field confinement, such as the one produced by the two conducting plates, the photonic density of states is reduced in the gap region, which leads to an attractive force between the plates. In the following we show how to compute this force.

Evaluation of Casimir Force First, we note that the modes perpendicular to the surface are discrete with wavenumbers

$$k_n = \frac{n\pi}{L},$$

where n is an integer. The groundstate energy can be computed from

$$E_0(d) = \frac{A\hbar c}{(2\pi)^2}\int_{-\infty}^{\infty}\left(\sum_{n=1}^{\infty}\sqrt{k_\parallel^2+k_n^2}\right)d^2k_\parallel = \frac{A\hbar}{2\pi c}\sum_{n=1}^{\infty}\int_{ck_n}^{\infty}\omega^2\,d\omega,$$

where A is the quantization area, $\boldsymbol{k}_\parallel$ the parallel momentum, and the factor of one half has been cancelled by the factor of two for the number of mode polarizations. In arriving at the last expression on the right-hand side we have used $\omega = c(k_\parallel^2 + k_n^2)^{\frac{1}{2}}$ and have introduced polar coordinates. The groundstate energy is infinite and we have to introduce a regularization procedure. We here impose an exponential cutoff $\Lambda = 1/\epsilon$ to get

$$\sum_{n=1}^{\infty}\int_{ck_n}^{\infty}e^{-\epsilon\omega}\omega^2\,d\omega = \frac{\partial^2}{\partial\epsilon^2}\left(\sum_{n=1}^{\infty}\int_{ck_n}^{\infty}e^{-\epsilon\omega}\,d\omega\right) = \frac{\partial^2}{\partial\epsilon^2}\left(\frac{1}{\epsilon}\sum_{n=1}^{\infty}e^{-\epsilon ck_n}\right).$$

As will be discussed further below, this cutoff can be motivated such that at sufficiently high photon energies all materials must become transparent. The geometric sum in the last expression can now be evaluated exactly, and when expanding the solution in powers of ϵ we get for the energy per unit area

$$\frac{E_0(d)}{A} \approx \left(\frac{3\hbar L}{\pi^2 c^2}\right)\Lambda^4 - \left(\frac{\hbar}{2\pi c}\right)\Lambda^3 - \frac{\pi^2}{720}\frac{\hbar c^2}{L^3}. \tag{15.54}$$

The first two terms are divergent in the limit $\Lambda \to \infty$ and will not be considered here. It can be shown that they do not contribute to the force between the plates.

The last term on the right-hand side is said to be the Casimir energy density [174], and it allows us computing the force per unit area via

$$F = -\frac{1}{A}\frac{dE(L)}{dL} = -\frac{\pi^2}{240}\frac{\hbar c}{L^4}. \tag{15.55}$$

Casimir Force Exerted by Field Fluctuations

The above derivation is along the lines of Casimir's original work [174], although he used a somewhat different reasoning to eliminate the divergent terms. These divergencies have been analyzed in great depth in the literature, see, for instance, [175, 176] and the references therein. In this section we provide an alternative derivation of this force, which is now mediated by the fluctuating vacuum or thermal fields, as schematically depicted in Fig. 15.7b. This approach is more in the spirit of fluctuational electrodynamics introduced in this chapter, and readers not overly happy with the above derivation will hopefully feel more comfortable with the following discussion.

We consider the situation depicted in Fig. 15.8 with two dielectric or metallic bodies situated in a background material with permittivity ε_b. Our goal is to evaluate the force acting on the body with volume Ω_b, which may have no sharp boundaries but a spatially inhomogeneous permittivity. In a first step we enclose the body in a volume Ω whose boundary is fully located in the background medium. We next use the total Lorentz force acting on the macroscopic charge distribution characterized by the charge density $\hat{\rho}$ and current density $\hat{\boldsymbol{J}}$, which has the form

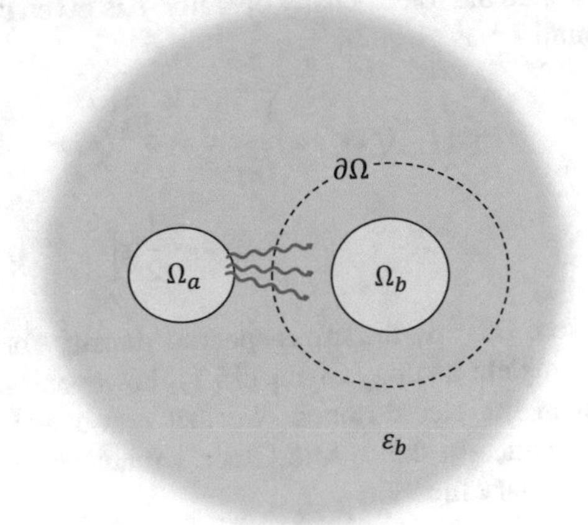

Fig. 15.8 Schematics for computation of Casimir force acting on a body with volume Ω_b. We enclose the body in a volume with boundary $\partial\Omega$ such that the boundary is fully located in the background material with permittivity ε_b. By evaluating the flux of Maxwell's stress tensor through $\partial\Omega$, we obtain the force acting on the dielectric or metallic body

$$\hat{\boldsymbol{F}} = \int_{\Omega} \Big[\hat{\rho}(\boldsymbol{r},t)\hat{\boldsymbol{E}}(\boldsymbol{r},t) + \hat{\boldsymbol{J}}(\boldsymbol{r},t) \times \hat{\boldsymbol{B}}(\boldsymbol{r},t)\Big]\, d^3r. \tag{15.56}$$

In Sect. 4.5 we have computed this force within the framework of classical electrodynamics. The reader might like to go back to this section to verify that the derivation can be carried out on an operator level without any noticeable modifications. We then get

$$\begin{aligned} \hat{\rho}(\boldsymbol{r},t)\hat{\boldsymbol{E}}(\boldsymbol{r},t) + \hat{\boldsymbol{J}}(\boldsymbol{r},t) \times \hat{\boldsymbol{B}}(\boldsymbol{r},t) = \nabla \cdot \lim_{\boldsymbol{r}'\to\boldsymbol{r}} \hat{\bar{\bar{T}}}(\boldsymbol{r},\boldsymbol{r}',t) \\ - \varepsilon_b \frac{d}{dt}\Big[\hat{\boldsymbol{E}}(\boldsymbol{r},t) \times \hat{\boldsymbol{B}}(\boldsymbol{r},t)\Big], \end{aligned} \tag{15.57}$$

where we have introduced the operator for Maxwell's stress tensor

$$\begin{aligned} \hat{T}_{ij}(\boldsymbol{r},\boldsymbol{r}',t) &= \varepsilon_b\, \hat{E}_i(\boldsymbol{r},t)\hat{E}_j(\boldsymbol{r}',t) + \frac{1}{\mu_0}\hat{B}_i(\boldsymbol{r},t)\hat{B}_j(\boldsymbol{r}',t) \\ &\quad - \frac{1}{2}\delta_{ij}\left(\varepsilon_b\, \hat{E}_k(\boldsymbol{r},t)\hat{E}_k(\boldsymbol{r}',t) + \frac{1}{\mu_0}\hat{B}_k(\boldsymbol{r},t)\hat{B}_k(\boldsymbol{r}',t)\right). \end{aligned}$$

In thermal equilibrium the time-derivative term on the right-hand side of Eq. (15.57) vanishes, and the expectation values of the field operator products can be related to the cross-spectral densities of Eq. (15.1). We then get

$$\Big\langle \hat{\boldsymbol{f}}(\boldsymbol{r})\Big\rangle_{\text{eq}} = \nabla \cdot \int_{-\infty}^{\infty} \lim_{\boldsymbol{r}'\to\boldsymbol{r}} \Big\langle \hat{\bar{\bar{T}}}^{+}(\boldsymbol{r},\boldsymbol{r}',\omega) + \hat{\bar{\bar{T}}}^{-}(\boldsymbol{r},\boldsymbol{r}',\omega)\Big\rangle_{\text{eq}} \frac{d\omega}{2\pi},$$

where the force density operator $\hat{f}$ is given through the left-hand side of Eq. (15.57), and $\hat{T}^{\pm}$ is given by

$$\left\langle \hat{\bar{\bar{T}}}^{\pm}(\boldsymbol{r},\boldsymbol{r}',\omega)\right\rangle_{\text{eq}} = \varepsilon_b \bar{\bar{\mathscr{E}}}^{\pm}(\boldsymbol{r},\boldsymbol{r}',\omega) + \mu_0 \bar{\bar{\mathscr{H}}}^{\pm}(\boldsymbol{r},\boldsymbol{r}',\omega) - \frac{1}{2}\text{tr}\Big[\varepsilon_b \bar{\bar{\mathscr{E}}}^{\pm}(\boldsymbol{r},\boldsymbol{r}',\omega) + \mu_0 \bar{\bar{\mathscr{H}}}^{\pm}(\boldsymbol{r},\boldsymbol{r}',\omega)\Big].$$

Here $\mathscr{H}^{\pm}$ is the cross-spectral density for magnetic fields, which is defined in complete analogy to Eq. (15.1), however, with the magnetic field operators rather than the electric ones. We can now use Eq. (15.36) to relate the cross-spectral densities to the dyadic Green's function. It turns out convenient to introduce the auxiliary function

$$\bar{\bar{\theta}}_{\text{tot}}(\boldsymbol{r},\boldsymbol{r}') = \frac{\hbar}{\pi}\int_0^\infty \big(1+2\bar{n}_{\text{th}}(\hbar\omega)\big) \times \text{Im}\Big[\mu_0\varepsilon_b\omega^2\,\bar{\bar{G}}_{\text{tot}}(\boldsymbol{r},\boldsymbol{r}',\omega) + \nabla\times\bar{\bar{G}}_{\text{tot}}(\boldsymbol{r},\boldsymbol{r}',\omega)\times\overleftarrow{\nabla}'\Big]\,d\omega, \tag{15.58}$$

where we have used Eq. (15.6) for the cross-spectral density of the magnetic fields. The force acting on the body enclosed within volume Ω can then be expressed as

$$\langle\hat{\boldsymbol{F}}\rangle_{\text{eq}} = \int_\Omega \nabla\cdot\lim_{\boldsymbol{r}'\to\boldsymbol{r}}\Big[\bar{\bar{\theta}}_{\text{tot}}(\boldsymbol{r},\boldsymbol{r}') - \frac{1}{2}\text{tr}\Big(\bar{\bar{\theta}}_{\text{tot}}(\boldsymbol{r},\boldsymbol{r}')\Big)\mathbb{1}\Big]\,d^3r.$$

So far we have avoided any singularities or infinities. However, for the limit $\boldsymbol{r}'\to\boldsymbol{r}$ in Eq. (15.58) it is apparent that the Green's function diverges. A suitable regularization procedure is obtained by splitting the Green's function into

$$\bar{\bar{G}}_{\text{tot}}(\boldsymbol{r},\boldsymbol{r}',\omega) = \bar{\bar{G}}(\boldsymbol{r},\boldsymbol{r}',\omega) + \bar{\bar{G}}_{\text{refl}}(\boldsymbol{r},\boldsymbol{r}',\omega),$$

where the first part is the Green's function of the homogeneous background, and the second "reflected" part is associated with the presence of the bodies. This decomposition has been discussed in some length in previous chapters. As no force is exerted on a single body embedded in the homogeneous background material, we can discard the (singular) contribution of $\bar{\bar{G}}$ and keep the reflected part only. With this, we find for the Casimir force acting on a body enclosed in volume Ω the expression

Casimir Force

$$\left\langle \hat{\boldsymbol{F}} \right\rangle_{\text{eq}} = \oint_{\partial\Omega} \lim_{\boldsymbol{r}'\to\boldsymbol{r}} \left[\bar{\bar{\theta}}_{\text{refl}}(\boldsymbol{r},\boldsymbol{r}') - \frac{1}{2}\text{tr}\left(\bar{\bar{\theta}}_{\text{refl}}(\boldsymbol{r},\boldsymbol{r}')\right)\mathbb{1} \right] \cdot \hat{\boldsymbol{n}}\, dS, \tag{15.59}$$

where we have used Gauss' law to convert the volume integral into a boundary integral. Equation (15.59) is a convenient equation for actual computation because its evaluation requires the knowledge for the classical Green's function only, albeit for the entire frequency range.

As a representative example we recompute the Casimir force between two perfectly conducting plates separated by a distance L, as depicted in Fig. 15.7. To be specific, we assume that the plates are located at positions 0 and L, respectively, and use the reflected Green's function of Eqs. (8.58), (8.59) for the slab structure. As detailed in exercise 15.8, the force per unit area in z-direction can be evaluated to

$$f_z = -\frac{\hbar}{2\pi^2} \int_0^\infty d\omega \big(1 + 2\bar{n}_{\text{th}}(\hbar\omega)\big) \int_0^\infty dk_\rho\, k_\rho k_z\, \text{Re}\left[\sum_\lambda \frac{e^{2ik_zL} R_1^\lambda R_2^\lambda}{1 - e^{2ik_zL} R_1^\lambda R_2^\lambda} \right], \tag{15.60}$$

where $R_{1,2}^\lambda$ are the reflection coefficients at the $z = 0$ and L interfaces, and the summation for λ runs over the transverse electric and magnetic modes. Remarkably, the force f_z is independent of z. This expression is usually attributed to Lifshitz [177]. In the following we compute it at zero temperature and consider frequency-independent reflection coefficients with $r_\lambda = R_1^\lambda R_2^\lambda$. We then get

$$f_z = -\frac{\hbar}{2\pi^2} \sum_\lambda \int_0^\infty \text{Re}\left[\sum_{n=1}^\infty r_\lambda^n \int_0^\infty k_\rho k_z e^{2ink_zL}\, dk_\rho \right] d\omega, \tag{15.61}$$

where we have expanded the denominator in a geometric series. We next extend the k_ρ integration into the complex plane, as discussed in Appendix B, and use the contour depicted in Fig. B.5. For the propagating modes we first integrate along the real axis from 0 to k and get for the term in brackets

$$\begin{aligned} &\text{Re}\left[\sum_{n=1}^\infty r_\lambda^n \int_0^k e^{2ink_zL} k_z^2\, dk_z \right] \\ &\quad = \frac{1}{8L^3}\, \text{Re}\left[\sum_{n=1}^\infty r_\lambda^n \left(-\frac{i\xi^2 e^{in\xi}}{n} + \frac{2\xi e^{in\xi}}{n^2} + \frac{2i}{n^3}\left(e^{in\xi} - 1\right) \right) \right], \end{aligned} \tag{15.62}$$

with $\xi = 2kL$. With the polylogarithm $\mathrm{Li}_m(z) = \sum_{n=1}^{\infty} \frac{z^n}{n^m}$ the term on the right-hand side can be rewritten as

$$\frac{1}{8L^3}\,\mathrm{Re}\Big[-i\xi^2\,\mathrm{Li}_1\left(r_\lambda e^{i\xi}\right) + 2\xi\,\mathrm{Li}_2\left(r_\lambda e^{i\xi}\right) + 2i\left(\mathrm{Li}_3\left(r_\lambda e^{i\xi}\right) - \mathrm{Li}_3(r_\lambda)\right)\Big].$$

For the evanescent modes we set $k_z = i\kappa$ to obtain in analogy to Eq. (15.62)

$$\mathrm{Re}\left[-i\sum_{n=1}^{\infty} r_\lambda^2 \int_0^\infty e^{-2n\kappa L}\kappa^2\,d\kappa\right] = \mathrm{Re}\left[\frac{2i}{8L^3}\,\mathrm{Li}_3(r_\lambda)\right],$$

which is cancelled by the corresponding term from the expression for the propagating modes. Putting together all results, we get from Eq. (15.61)

$$f_z = -\frac{\hbar c}{32\pi^2 L^4}\sum_\lambda \int_0^\infty \mathrm{Re}\Big[-i\xi^2\,\mathrm{Li}_1\left(r_\lambda e^{i\xi}\right) + 2\xi\,\mathrm{Li}_2\left(r_\lambda e^{i\xi}\right) + 2i\,\mathrm{Li}_3\left(r_\lambda e^{i\xi}\right)\Big]\,d\xi.$$

The integral can be performed analytically and we arrive at

$$f_z = -\frac{\hbar c}{32\pi^2 L^4}\sum_\lambda \mathrm{Re}\Big[-\xi^2\,\mathrm{Li}_2\left(r_\lambda e^{i\xi}\right) - 4i\xi\,\mathrm{Li}_3\left(r_\lambda e^{i\xi}\right) + 6\,\mathrm{Li}_4\left(r_\lambda e^{i\xi}\right)\Big]_0^\infty.$$

We finally assume that for the perfectly conducting plates the reflection coefficients are of the form $r_\lambda = e^{-\epsilon\omega}$, with $\epsilon \to 0$, such that the materials are perfect conductors at low frequencies but become transparent at large frequencies. Then, the integrated function vanishes at the upper integration limit,[1] and we arrive at

$$f_z = \frac{3\hbar c}{8\pi^2 L^4}\mathrm{Li}_4(1) = \frac{3\hbar c}{8\pi^2 L^4}\left(\frac{\pi^4}{90}\right) = \frac{\pi^2}{240}\frac{\hbar c}{L^4}. \tag{15.63}$$

This expression is in agreement to our previous result of Eq. (15.55). To understand the direction of the force, we consider the bounding box in Fig. 15.7b (dashed lines). Maxwell's stress tensor has the constant value of f_z in the gap region and is zero inside the plates, because the electromagnetic fields are completely screened there. Thus, the force per unit area acting on the plate at $z = 0$ can be computed from

$$F = \lim_{z\to 0^+}\left[\bar{\bar{\theta}}_{\mathrm{refl}}(\boldsymbol{r},\boldsymbol{r})\right]_{zz} - \lim_{z\to 0^-}\left[\bar{\bar{\theta}}_{\mathrm{refl}}(\boldsymbol{r},\boldsymbol{r})\right]_{zz} = f_z.$$

[1]In principle we should perform the ω integration for $r_\lambda = e^{-\epsilon\omega}$ and set $\epsilon \to 0$ at the end. However, as this procedure gives the same result as simply setting the function to zero at the upper integration limit, we here opt for the latter more simple approach.

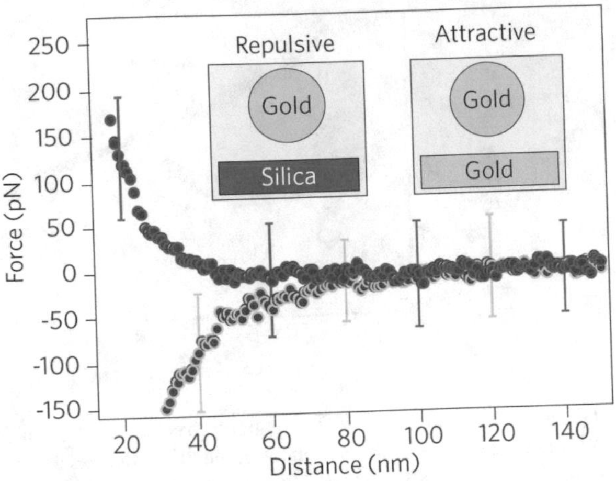

Fig. 15.9 Measured repulsive (blue) or attractive (yellow) Casimir force between a gold-coated (100 nm) polystyrene sphere and a silica (blue) or gold-coated (yellow) plate immersed in a supporting matrix. The circles represent the average force, averaged over 50 data sets, with corresponding error bars. Figure taken from [165]

By a similar token we find that the force acting on the plate at $z = L$ points in the negative z direction. This shows that the force between the two plates is attractive.

It is gratifying to see that both approaches for the Casimir force give the same result, despite the different physical nature of the underlying descriptions. As schematically shown in Fig. 15.7, the first approach builds on the reduced density of modes in the region between the plates and the resulting force associated with the reduction of groundstate energy of the quantized electromagnetic modes. The second approach uses the quantum and thermal field fluctuations to compute the force via Maxwell's stress tensor. In general, the agreement between the two approaches is only guaranteed for ideal conductors, whereas for bodies with arbitrary frequency-dependent permittivities one has to resort to Eq. (15.59), which provides an elegant and rather simple expression for computing the Casimir force once the Green's function of classical electrodynamics is at hand.

Figure 15.9 shows measured Casimir forces between a gold nanosphere and a dielectric or metallic substrate [165]. Depending on the material decomposition the Casimir force can be either repulsive or attractive. Figure 15.10 shows the measured force between two gold-coated spheres and the comparison with the theoretical predictions [178], which exhibit perfect agreement over the entire distance range.

15.6 Heat Transfer at the Nanoscale

We conclude this chapter with a brief discussion of heat transfer at the nanoscale, which, similarly to Casimir forces, is an active field of research with numerous interesting studies and applications. See, for instance, [164, 167, 173, 180] and the references therein. Figure 15.11 shows a seminal experiment where the authors investigated a thermal source, and showed that with a periodic microstructure

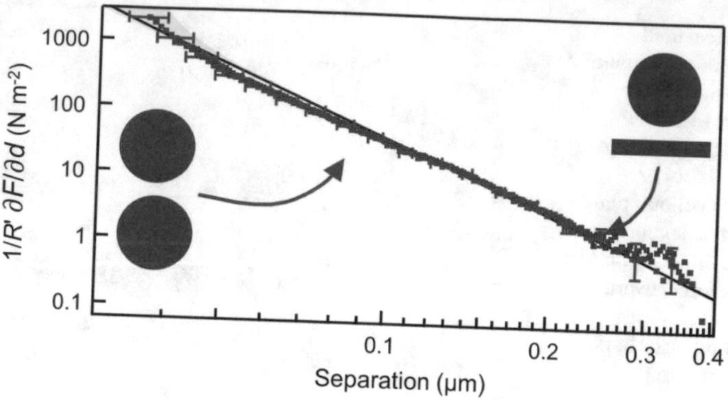

Fig. 15.10 Measurements of the spatial derivative of Casimir force for both sphere-plate (blue) and sphere-sphere (red) measurement geometries. Results are in agreement with calculated values of the Casimir force derivative for two gold spheres with a 4.9 nm rms perturbative roughness correction (black line). Gray shaded region shows the uncertainty in the roughness correction due to the uncertainty in the orientation of the spheres. Figure and caption taken from [178]

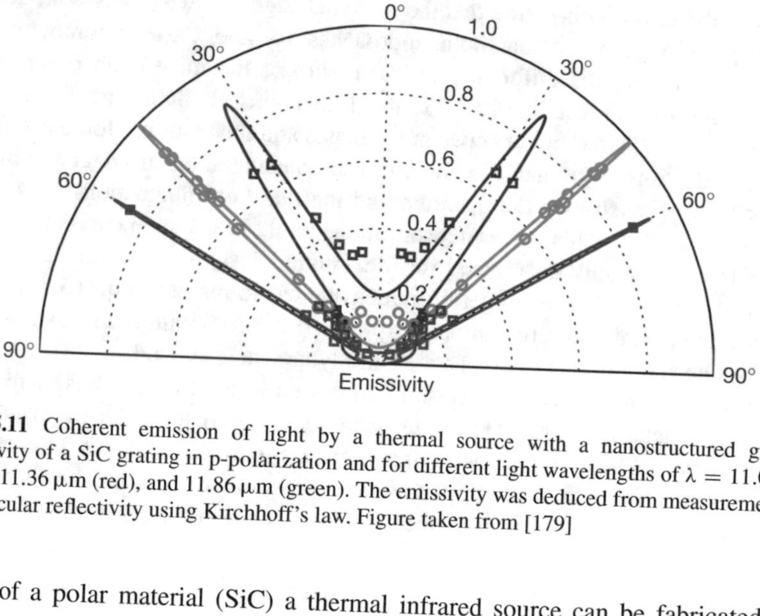

Fig. 15.11 Coherent emission of light by a thermal source with a nanostructured grating. Emissivity of a SiC grating in p-polarization and for different light wavelengths of $\lambda = 11.04\,\mu\text{m}$ (blue), $11.36\,\mu\text{m}$ (red), and $11.86\,\mu\text{m}$ (green). The emissivity was deduced from measurements of the specular reflectivity using Kirchhoff's law. Figure taken from [179]

made of a polar material (SiC) a thermal infrared source can be fabricated that is coherent over large distances (many wavelengths) and radiates in well-defined directions [179]. Quite generally, with the tools developed in this chapter we are in the position to compute the emitted radiation using the cross-spectral densities introduced at the beginning of this chapter.

Three Shades of Pink

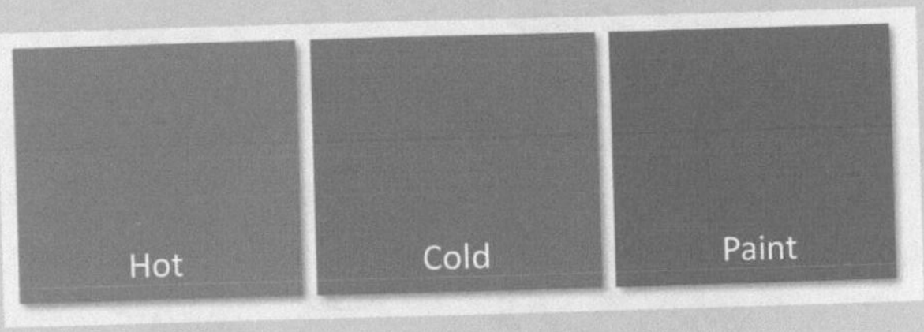

This has been a busy and theoretical chapter packed with (too) many equations, so it might be rewarding to see that with the tools developed one can also do practical things. Shanhui Fan and coworkers from Stanford University have shown over the last years that one can do marvelous things by modifying the emission and absorption properties of materials through nanostructuring. The above panels show three colors extracted from the samples presented in [181] reporting the "Photonic thermal management of colored objects." The temperature of an object depends on the influx and outflux of energy

$$P_{\rm net} = P_{\rm sun}^{\rm visible} + P_{\rm sun}^{\rm infrared} - \Big(P_{\rm rad} - P_{\rm atm}\Big),$$

where the first two quantities on the right-hand side correspond to the absorbed power from the sun, whereas $P_{\rm rad}$ is the total thermal radiation power by the object and $P_{\rm atm}$ is the absorbed thermal emission power from the atmosphere at ambient temperature. For objects with the approximately same color there is not too much room for improvement for $P_{\rm sun}^{\rm visible}$, but $P_{\rm sun}^{\rm infrared}$ and $P_{\rm rad}$ can be modified by nanostructuring the materials, in Fan's work by simply using stratified media. For rooftop temperature measurements at Stanford, California, the authors found for objects with approximately the same color extremely different temperatures, namely more than 80 °C for the hot sample, and about 60 °C and 40 °C for the paint and cold samples. Incidentally, the hot sample turned out to be even hotter than a black-painted sample.

In the following we investigate the heat transfer between two bodies held at different temperatures T_a and T_b, as shown in Fig. 15.12. We start from the Poynting vector

$$\Big\langle \hat{\boldsymbol{S}}(\boldsymbol{r})\Big\rangle_{\rm eq} = \frac{1}{2}{\rm Re}\Big\langle \hat{\boldsymbol{E}}^{+}(\boldsymbol{r}) \times \hat{\boldsymbol{H}}^{-}(\boldsymbol{r}) + \hat{\boldsymbol{E}}^{-}(\boldsymbol{r}) \times \hat{\boldsymbol{H}}^{+}(\boldsymbol{r})\Big\rangle_{\rm eq}, \tag{15.64}$$

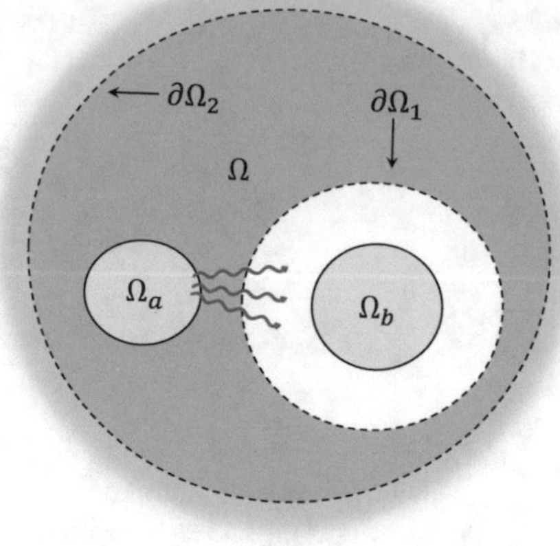

Fig. 15.12 Computation of heat transfer at the nanoscale. We consider two bodies Ω_a, Ω_b held at different temperatures T_a, T_b, and compute the radiation flowing between the bodies. We integrate over a volume Ω which includes body a and excludes body b. The outer boundary of Ω is denoted with $\partial\Omega_2$, and the boundary of the exclusion of Ω_b is denoted with $\partial\Omega_1$. By computing the flux of the Poynting vector from $a \to b$ and $b \to a$ we can evaluate the heat flux

which describes the energy flux density produced by the vacuum and thermal field fluctuations. We can use Eq. (15.6) to relate the Poynting vector to the cross-spectral density via

$$\left\langle \hat{\mathbf{S}}(\boldsymbol{r},\omega)\right\rangle_{\text{eq}} = \frac{1}{2}\lim_{\boldsymbol{r}'\to\boldsymbol{r}} \text{Re}\left[\frac{i}{\mu_0\omega}\left(\bar{\bar{\mathscr{E}}}^{+}(\boldsymbol{r},\boldsymbol{r}',\omega) - \bar{\bar{\mathscr{E}}}^{-}(\boldsymbol{r},\boldsymbol{r}',\omega)\right)\times\overleftarrow{\nabla}'\right].$$

Suppose that we are interested in the field fluctuations produced within a given volume Ω only. The corresponding cross-spectral densities can be obtained from Eq. (15.30),

$$\bar{\bar{\mathscr{E}}}^{+}(\boldsymbol{r},\boldsymbol{r}',+\omega)\Big|_{\Omega} = 2\mu_0\omega^2\,\mathbb{V}_{\Omega}\left[\bar{\bar{G}}_{\text{tot}},\bar{\bar{G}}^{*}_{\text{tot}}\right](\boldsymbol{r},\boldsymbol{r}',\omega)\big(\bar{n}_{\text{th}}(\hbar\omega)+1\big)$$

$$\bar{\bar{\mathscr{E}}}^{-}(\boldsymbol{r},\boldsymbol{r}',-\omega)\Big|_{\Omega} = 2\mu_0\omega^2\,\mathbb{V}_{\Omega}\left[\bar{\bar{G}}^{*}_{\text{tot}},\bar{\bar{G}}_{\text{tot}}\right](\boldsymbol{r},\boldsymbol{r}',\omega)\;\;\bar{n}_{\text{th}}(\hbar\omega).$$

Summing up these contributions and using the symmetry relation of Eq. (15.72) for the volume term, we can write the Poynting vector due to the current sources inside of Ω in the form

$$\left\langle \hat{\mathbf{S}}(\boldsymbol{r},\omega)\right\rangle_{\text{eq}}\Big|_{\Omega} = \omega\lim_{\boldsymbol{r}'\to\boldsymbol{r}} \text{Im}\left[\mathbb{V}_{\Omega}\left[\bar{\bar{G}}_{\text{tot}},\bar{\bar{G}}^{*}_{\text{tot}}\right](\boldsymbol{r},\boldsymbol{r}',\omega)\times\overleftarrow{\nabla}'\right]\big(1+2\bar{n}_{\text{th}}(\hbar\omega)\big). \tag{15.65}$$

The heat flux can now be computed from

$$Q_{a\to b} = -\oint_{\partial\Omega_1}\left[\int_0^\infty \left\langle \hat{\mathbf{S}}(\boldsymbol{r},\omega)\right\rangle_{\text{eq}}\Big|_{\Omega} d\omega\right]\cdot \hat{\boldsymbol{n}}_1 dS, \tag{15.66}$$

where the minus sign is a convention to make the quantity positive for temperatures T_a greater than T_b, see the absorption cross section in Chap. 4 for a related definition. In thermal equilibrium the heat flux $Q_{a\to b}$ must equal the flux $Q_{b\to a}$. When the bodies a and b are held at different temperatures, the net flux $\Delta Q_{a\to b}$ is carried by

$$\left\langle \Delta\hat{\mathbf{S}}(\boldsymbol{r},\omega)\right\rangle_{\text{eq}}\Big|_{\Omega} = \left\langle \hat{\mathbf{S}}(\boldsymbol{r},\omega)\right\rangle_{\text{eq}}\Big|_{\Omega}^{T_b} - \left\langle \hat{\mathbf{S}}(\boldsymbol{r},\omega)\right\rangle_{\text{eq}}\Big|_{\Omega}^{T_a}. \tag{15.67}$$

The net heat flux between two bodies a and b can then be computed from Poynting's theorem according to

Net Heat Flux Between Bodies *a* and *b*

$$\Delta Q_{a\to b} = -\oint_{\partial\Omega_1}\left[\int_0^\infty \left\langle \Delta\hat{\mathbf{S}}(\boldsymbol{r},\omega)\right\rangle_{\text{eq}}\Big|_{\Omega} d\omega\right]\cdot \hat{\boldsymbol{n}}_1 dS. \tag{15.68}$$

We may use Eq. (15.65) together with the representation formula for Green's functions, Eq. (15.31), to convert Eq. (15.68) into a double boundary integral. Related approaches are described in [182, 183], and are left as an exercise to the interested reader.

15.7 Details of Derivation of Representation Formula

In this section we show how to derive the representation formula for Green's functions of Eq. (15.31). We start from the Green's function definition of Eq. (15.24), which we rewrite in the form (we discard the subscript for the total Green's function)

$$\begin{aligned}
\left[-\nabla\times\nabla\times\bar{\bar{G}}(\boldsymbol{r},\boldsymbol{r}')\right]_{ij} + \mu_0\omega^2\int \varepsilon_{ik}(\boldsymbol{r},\boldsymbol{r}_1)G_{kj}(\boldsymbol{r}_1,\boldsymbol{r}')\,d^3r_1 &= -\delta(\boldsymbol{r}-\boldsymbol{r}')\delta_{ij}\\
\left[-\nabla'\times\nabla'\times\bar{\bar{G}}(\boldsymbol{r}',\boldsymbol{r})\right]_{ji} + \mu_0\omega^2\int \varepsilon_{jk}(\boldsymbol{r}',\boldsymbol{r}_1)G_{ki}(\boldsymbol{r}_1,\boldsymbol{r})\,d^3r_1 &= -\delta(\boldsymbol{r}-\boldsymbol{r}')\delta_{ij}.
\end{aligned} \tag{15.69}$$

The expression in the second line has been obtained by taking the transpose of the equation in the first line, and by exchanging the primed with the unprimed coordinates. We use the reciprocity relation of Eq. (7.35) for the dyadic Green's function together with a corresponding expression for the permittivity tensor $\varepsilon_{ij}(\boldsymbol{r},\boldsymbol{r}')=\varepsilon_{ji}(\boldsymbol{r}',\boldsymbol{r})$, to rewrite the second line as

$$\Big[-\nabla'\times\nabla'\times\bar{\bar{G}}(\boldsymbol{r},\boldsymbol{r}')\Big]_{ij}+\mu_0\omega^2\int G_{ik}(\boldsymbol{r},\boldsymbol{r}_1)\varepsilon_{kj}(\boldsymbol{r}_1,\boldsymbol{r}')\,d^3r_1=-\delta(\boldsymbol{r}-\boldsymbol{r}')\delta_{ij}.$$

Next, we multiply the second line of Eq. (15.69) from the right-hand side with $\bar{\bar{G}}^*$, and the complex conjugate of the second line from the left-hand side with $\bar{\bar{G}}$, to arrive at

$$\begin{aligned}-\delta(\boldsymbol{r}-\bar{\boldsymbol{r}})\bar{\bar{G}}^*(\bar{\boldsymbol{r}},\boldsymbol{r}')&=\Big[-\bar{\nabla}\times\bar{\nabla}\times\bar{\bar{G}}(\boldsymbol{r},\bar{\boldsymbol{r}})\Big]\cdot\bar{\bar{G}}^*(\bar{\boldsymbol{r}},\boldsymbol{r}')\\&\quad+\mu_0\omega^2\int_\Omega\bar{\bar{G}}(\boldsymbol{r},\bar{\boldsymbol{r}})\cdot\bar{\bar{\varepsilon}}(\bar{\boldsymbol{r}},\boldsymbol{r}_1)\cdot\bar{\bar{G}}^*(\boldsymbol{r}_1,\boldsymbol{r}')\,d^3r_1\\-\bar{\bar{G}}(\boldsymbol{r},\bar{\boldsymbol{r}})\delta(\bar{\boldsymbol{r}}-\boldsymbol{r}')&=\bar{\bar{G}}(\boldsymbol{r},\bar{\boldsymbol{r}})\cdot\Big[-\bar{\nabla}\times\bar{\nabla}\times\bar{\bar{G}}^*(\bar{\boldsymbol{r}},\boldsymbol{r}')\Big]\\&\quad+\mu_0\omega^2\int_\Omega\bar{\bar{G}}(\boldsymbol{r},\bar{\boldsymbol{r}})\cdot\bar{\bar{\varepsilon}}^*(\bar{\boldsymbol{r}},\boldsymbol{r}_1)\cdot\bar{\bar{G}}^*(\boldsymbol{r}_1,\boldsymbol{r}')\,d^3r_1.\end{aligned}\tag{15.70}$$

When subtracting the two expressions and integrating $\bar{\boldsymbol{r}}$ over volume Ω, we get

$$\begin{aligned}2i\,\mathrm{Im}\Big[\bar{\bar{G}}(\boldsymbol{r},\boldsymbol{r}')\Big]\Theta_\Omega(\boldsymbol{r},\boldsymbol{r}')&=2i\,\mathbb{V}_\Omega\big[\bar{\bar{G}},\bar{\bar{G}}^*\big](\boldsymbol{r},\boldsymbol{r}')+\int_\Omega\\&\times\Big\{\Big[-\bar{\nabla}\times\bar{\nabla}\times\bar{\bar{G}}(\boldsymbol{r},\bar{\boldsymbol{r}})\Big]\cdot\bar{\bar{G}}^*(\bar{\boldsymbol{r}},\boldsymbol{r}')+\bar{\bar{G}}(\boldsymbol{r},\bar{\boldsymbol{r}})\cdot\bar{\nabla}\times\bar{\nabla}\times\bar{\bar{G}}^*(\bar{\boldsymbol{r}},\boldsymbol{r}')\Big\}\,d^3\bar{r}.\end{aligned}\tag{15.71}$$

Here Θ_Ω is one or zero depending on whether $\boldsymbol{r}$, $\boldsymbol{r}'$ are both located inside or outside of volume Ω. We are not considering situations where $\boldsymbol{r}$, $\boldsymbol{r}'$ are located in different volumes. In the above equation we have introduced the volume integral of Eq. (15.29)

$$\begin{aligned}&\mathbb{V}_\Omega\Big[\bar{\bar{\mathscr{U}}},\bar{\bar{\mathscr{V}}}\Big](\boldsymbol{r},\boldsymbol{r}')\\&=\mu_0\omega^2\int_\Omega\bar{\bar{\mathscr{U}}}(\boldsymbol{r},\boldsymbol{r}_1,\omega)\cdot\mathrm{Im}\Big[\bar{\bar{\varepsilon}}(\boldsymbol{r}_1,\boldsymbol{r}_1',\omega)\Big]\cdot\bar{\bar{\mathscr{V}}}(\boldsymbol{r}_1',\boldsymbol{r}',\omega)\,d^3r_1d^3r_1',\end{aligned}$$

with $\bar{\bar{\mathscr{U}}}$, $\bar{\bar{\mathscr{V}}}$ being two arbitrary functions. Using the reciprocity theorem of optics and the definition of $\mathbb{V}$, one can easily derive the symmetry relations

$$\mathbb{V}_\Omega\Big[\bar{\bar{G}}_{\text{tot}}, \bar{\bar{G}}^*_{\text{tot}}\Big](\boldsymbol{r}, \boldsymbol{r}') = \mathbb{V}_\Omega\Big[\bar{\bar{G}}^*_{\text{tot}}, \bar{\bar{G}}_{\text{tot}}\Big](\boldsymbol{r}', \boldsymbol{r}) = \Big(\mathbb{V}_\Omega\Big[\bar{\bar{G}}^*_{\text{tot}}, \bar{\bar{G}}_{\text{tot}}\Big](\boldsymbol{r}, \boldsymbol{r}')\Big)^*, \tag{15.72}$$

where we have suppressed for simplicity the ω dependence.

In Sect. 5.5 we have derived for two vector functions $\boldsymbol{u}$, $\boldsymbol{v}$ the identity

$$\begin{aligned} &\int_\Omega \Big\{\Big[-\nabla\times\nabla\times\boldsymbol{u}(\boldsymbol{r})\Big]\cdot\boldsymbol{v}(\boldsymbol{r}) + \boldsymbol{u}(\boldsymbol{r})\cdot\nabla\times\nabla\times\boldsymbol{v}(\boldsymbol{r})\Big\}\, d^3r \\ &\qquad = -\oint_{\partial\Omega} \Big\{\big[\nabla\times\boldsymbol{u}(\boldsymbol{r})\big]\times\boldsymbol{v}(\boldsymbol{r}) + \boldsymbol{u}(\boldsymbol{r})\times\big[\nabla\times\boldsymbol{v}(\boldsymbol{r})\big]\Big\}\cdot\hat{\boldsymbol{n}}\,dS, \end{aligned} \tag{15.73}$$

which can be used to simplify the last term in Eq. (15.71). It turns out convenient to introduce in analogy to the volume integration $\mathbb{V}$ the abbreviation

$$\begin{aligned} \mathbb{B}_{\partial\Omega}\Big[\bar{\bar{\mathscr{U}}}, \bar{\bar{\mathscr{V}}}\Big](\boldsymbol{r}_1, \boldsymbol{r}_2) = \frac{i}{2}\oint_{\partial\Omega} \Big\{&\bar{\bar{\mathscr{U}}}(\boldsymbol{r}_1, \boldsymbol{r}) \times \Big[\nabla\times\bar{\bar{\mathscr{V}}}(\boldsymbol{r}, \boldsymbol{r}_2)\Big] \\ &+\Big[\nabla\times\bar{\bar{\mathscr{U}}}(\boldsymbol{r}, \boldsymbol{r}')\Big]\times\bar{\bar{\mathscr{V}}}(\boldsymbol{r}, \boldsymbol{r}_2)\Big\}\cdot\hat{\boldsymbol{n}}\,dS. \end{aligned}$$

Putting together all results, we are led to our final representation formula for Green's functions, Eq. (15.31), where the expression in the second line is obtained by taking the complex conjugate of the expression in the first line.

15.7.1 Evaluation of Boundary Term

One often wants to evaluate the boundary term $\mathbb{B}_{\partial\Omega}(\boldsymbol{r}, \boldsymbol{r}')$ for positions located *on* or close to the boundary $\partial\Omega$. When doing so, we have to be careful about the singular behavior of the Green's functions in the limit $\boldsymbol{r}_1, \boldsymbol{r}_2 \to \boldsymbol{r}$. Difficulties for such a limiting procedure can be avoided as follows. First, we assume that the boundary is located in a region with a constant background permittivity ε_b. We can then express the total Green's function as a sum of the Green's function $\bar{\bar{G}}_b$ for the unbounded background medium and a remainder, which we shall denote—in analogy to previous chapters—as the reflected Green's function,

$$\bar{\bar{G}}_{\text{tot}}(\boldsymbol{r}, \boldsymbol{r}') = \bar{\bar{G}}_b(\boldsymbol{r}, \boldsymbol{r}') + \bar{\bar{G}}_{\text{refl}}(\boldsymbol{r}, \boldsymbol{r}').$$

For $\boldsymbol{r}, \boldsymbol{r}'$ located inside the background region $\bar{\bar{G}}_b$ fulfills the defining equation

$$-\nabla\times\nabla\times\bar{\bar{G}}_b(\boldsymbol{r}, \boldsymbol{r}') + k_b^2\bar{\bar{G}}_b(\boldsymbol{r}, \boldsymbol{r}') = -\delta(\boldsymbol{r}-\boldsymbol{r}')\mathbb{1},$$

whereas the reflected Green's function fulfills the homogeneous wave equation

$$-\nabla \times \nabla \times \bar{\bar{G}}_{\rm refl}(\boldsymbol{r},\boldsymbol{r}') + k_b^2 \bar{\bar{G}}_{\rm refl}(\boldsymbol{r},\boldsymbol{r}') = 0.$$

For a non-absorbing medium with $\mathrm{Im}[\varepsilon_b] = 0$ we can use Eq. (15.33) to evaluate the highly singular part of the boundary term containing the product of free-space Green's functions. We then obtain

$$\begin{aligned} \mathbb{B}_{\partial\Omega}\Big[\bar{\bar{G}}_{\rm tot}, \bar{\bar{G}}^*_{\rm tot}\Big](\boldsymbol{r},\boldsymbol{r}') &= \mathbb{B}_{\partial\Omega}\Big[\bar{\bar{G}}_{\rm tot}, \bar{\bar{G}}^*_{\rm refl}\Big](\boldsymbol{r},\boldsymbol{r}') \\ &+ \mathbb{B}_{\partial\Omega}\Big[\bar{\bar{G}}_{\rm refl}, \bar{\bar{G}}^*_{\rm tot}\Big](\boldsymbol{r},\boldsymbol{r}') - \mathbb{B}_{\partial\Omega}\Big[\bar{\bar{G}}_{\rm refl}, \bar{\bar{G}}^*_{\rm refl}\Big](\boldsymbol{r},\boldsymbol{r}') + \mathrm{Im}\Big[\bar{\bar{G}}_b(\boldsymbol{r},\boldsymbol{r}')\Big]. \end{aligned} \tag{15.74}$$

The first two terms on the right-hand side can be evaluated in analogy to our discussion about the boundary integral and boundary element methods in Chaps. 9 and 11. The third term corrects for the double counting of the reflected Green's function term, and can be easily evaluated because all singular contributions have been removed from the reflected Green's function.

Exercises

Exercise 15.1 Repeat the calculations of Chap. 4 using field operators rather than classical fields for:

(a) the Poynting theorem of Eq. (4.14), and
(b) the optical forces of Eq. (4.36).

Exercise 15.2 Consider the light–matter interaction of Eq. (15.37) and the free-space electric field operator $\hat{\boldsymbol{E}}^{\perp}$ of Eq. (13.78). Work out explicitly the Fermi's golden rule expressions of Eq. (14.95) without using the cross-spectral density, and show that the final result agrees with the Wigner–Weisskopf decay rate of Eq. (15.40).

Exercise 15.3 Use from the NANOPT toolbox the file `demostrat03.m` to compute the lifetime reduction of a dipole located above a gold substrate.

(a) Rewrite the program such that it computes the scattering rate and Lamb shift of Eqs. (15.39), (15.50) at zero temperature.
(b) Investigate the dependence of the Lamb shift on the material. Compare Au, Ag, and a dielectric material with $\varepsilon : \varepsilon_0 = 10$ representative for silicon.
(c) Compute the enhancement of the scattering rate and the Lamb shift within the quasistatic approximation. Place a mirror dipole inside the substrate and approximate the reflected Green's function by the electric field of the mirror dipole. Compare with the exact results and identify the distance range where the quasistatic approximation works well.

Exercise 15.4 Use from the NANOPT toolbox the files `demodipmie01.m` and `demodipmie02.m` to compute the decay rate enhancement of a dipole located above a gold nanosphere. Rewrite the programs using the `lambshift` function of the `miesolver` class to compute the corresponding Lamb shifts. Use the programs to reproduce the results given in Figs. 15.5 and 15.6.

Exercise 15.5 Use from the NANOPT toolbox the files `demobem06.m` and `demobem07.m` to compute the lifetime reduction of a dipole located above a nanosphere using the quasistatic approximation.

(a) Modify the programs such that they can also compute the Lamb shift using Eq. (15.50). Compare for small spheres with the exact Mie results derived in Exercise 15.4.
(b) Compute the Lamb shift for nanoellipsoids with axis ratios of 1 : 2 and 1 : 3. Use transition frequencies at the dipole resonance, as well as red and blue shifted with respect to it.

Exercise 15.6 Consider Eq. (15.46) for the frequency renormalization and the decay rate of a quantum emitter.

(a) Express the cross-spectral densities in terms of the dyadic Green's function.
(b) Split the total Green's function into a free-space contribution G and a reflected part G_{refl}, and keep in the following the reflected part only.
(c) Use the Kramers–Kronig relation for the Green's function to establish a similar relation between the frequency renormalization and the decay rate, associated with the real and imaginary part of the Green's function, respectively.

Exercise 15.7 Use the results of Eq. (15.4) to compute the forces acting on a point dipole above a gold sphere with a diameter of 100 nm. Consider dipole-sphere separations of 10, 20, 50 nm, a dipole moment of 0.1 e nm, with e being the elementary charge, and compute the force in piconewton. At which transition frequency is the force largest?

Exercise 15.8 Derive Lifschitz's result of Eq. (15.60) for the force acting between two plates. Start from the reflected Green's function for a slab structure given in Eqs. (8.58), (8.59), which can be rewritten in the compact form

$$G_{ij}^{\text{refl}}(\boldsymbol{r},\boldsymbol{r}') = \frac{i}{8\pi^2}\int_{-\infty}^{\infty}\frac{e^{i\boldsymbol{k}_\parallel\cdot(\boldsymbol{r}-\boldsymbol{r}')}}{k_z}\sum_\lambda\left[\epsilon_i^{\lambda+}A^\lambda e^{ik_zz}+\epsilon_i^{\lambda-}B^\lambda e^{-ik_zz}\right]\epsilon_j^{\lambda\pm}\,dk_xdk_y.$$

The remaining calculation is somewhat technical and lengthy, and can be broken down into the following steps.

(a) Evaluate the Green's function elements $G_{xx}^{\text{refl}} = G_{yy}^{\text{refl}}$ and G_{zz}^{refl} in the limit $\boldsymbol{r}'\to\boldsymbol{r}$. Use Eq. (B.8) with

$$\left\langle \boldsymbol{\epsilon}^{TE}\boldsymbol{\epsilon}^{TE}\right\rangle = \frac{1}{2}\begin{pmatrix}1 & & \\ & 1 & \\ & & 0\end{pmatrix}, \quad \left\langle \boldsymbol{\epsilon}^{TM}\boldsymbol{\epsilon}^{TM}\right\rangle = \frac{1}{2k^2}\begin{pmatrix}\pm k_z^2 & & \\ & \pm k_z^2 & \\ & & 2k_\rho^2\end{pmatrix},$$

where the signs in the TM case depend on the combination of signs for the wavenumber arguments $\hat{\boldsymbol{k}}^{\pm}$ of the polarization vectors.

(b) For the evaluation of the magnetic Green's function, express the reflected Green's function given above in terms of $e^{i\boldsymbol{k}\cdot\boldsymbol{r}}, e^{-i\boldsymbol{k}'\cdot\boldsymbol{r}'}$, with $\boldsymbol{k} = \boldsymbol{k}^{\pm}$, $\boldsymbol{k}' = \boldsymbol{k}^{\pm}$. Use

$$\nabla \times \left(e^{i(\boldsymbol{k}\cdot\boldsymbol{r}-\boldsymbol{k}'\cdot\boldsymbol{r}')}\boldsymbol{\epsilon}^{\lambda}\boldsymbol{\epsilon}^{\lambda}\right) \times \overleftarrow{\nabla}' = \left(\boldsymbol{k}\times\boldsymbol{\epsilon}^{\lambda}\right)\left(\boldsymbol{k}'\times\boldsymbol{\epsilon}^{\lambda}\right)e^{i(\boldsymbol{k}\cdot\boldsymbol{r}-i\boldsymbol{k}'\cdot\boldsymbol{r}')},$$

and work out the cross products using Eq. (B.5) in order to compute the magnetic Green's function in the limit $\boldsymbol{r}' \to \boldsymbol{r}$.

(c) Consider the quantity

$$\bar{\bar{\theta}}(\boldsymbol{r},\boldsymbol{r},\omega) = \frac{\hbar}{\pi}\lim_{\boldsymbol{r}'\to\boldsymbol{r}}\text{Im}\left[k^2\bar{\bar{G}}_{\text{refl}}(\boldsymbol{r},\boldsymbol{r}',\omega) + \nabla\times\bar{\bar{G}}_{\text{refl}}(\boldsymbol{r},\boldsymbol{r}',\omega)\times\overleftarrow{\nabla}'\right].$$

which is defined in accordance to Eq. (15.58) in the limit $\boldsymbol{r}' \to \boldsymbol{r}$. Compute the force per unit area

$$f_z = \int\left[\frac{1}{2}\theta_{zz}(\boldsymbol{r},\boldsymbol{r},\omega) - \theta_{xx}(\boldsymbol{r},\boldsymbol{r},\omega)\right]d\omega,$$

to arrive at the final expression of Eq. (15.60).

Chapter 16
Two-Level Systems

In this book we have stayed so far with a linear material response. This chapter marks a change. We consider quantum emitters with a nonlinear dynamics. As a starting point, we introduce the most simple model of a two-level system, which will allow us to develop a pictorial representation of the system state in terms of a Bloch vector. The generalization to few-level systems will be discussed in the next chapter, where we will also present examples representative for nano optics and plasmonics.

16.1 Bloch Sphere

Two-level systems are the most simple quantum systems that can be brought into a superposition state. They have recently seen tremendous interest in the fields of quantum communication and quantum computation, because qubits, as these two-level systems are called there, exhibit practically all of the "exotic" behavior contained in quantum physics. In this book we stay on the old-school side and denote the states of the two-level system as g for the groundstate and e for the excited state. An arbitrary state of the system thus has the form

$$|\psi\rangle = C_g|g\rangle + C_e|e\rangle \longrightarrow \begin{pmatrix} C_e \\ C_g \end{pmatrix} = C_g \begin{pmatrix} 0 \\ 1 \end{pmatrix} + C_e \begin{pmatrix} 1 \\ 0 \end{pmatrix}. \tag{16.1}$$

Typical examples we have in mind are molecules or quantum dots, which can be promoted through photon absorption from the ground to the excited state as shown in Fig. 16.1. We will later generalize our results to a higher number of states, but most of the relevant physics is already present for two levels. Quite generally, the system is determined by two complex numbers C_g, C_e, or equivalently by four real numbers. If we additionally consider that the wavefunction is normalized and

U. Hohenester, *Nano and Quantum Optics*, Graduate Texts in Physics,
https://doi.org/10.1007/978-3-030-30504-8_16

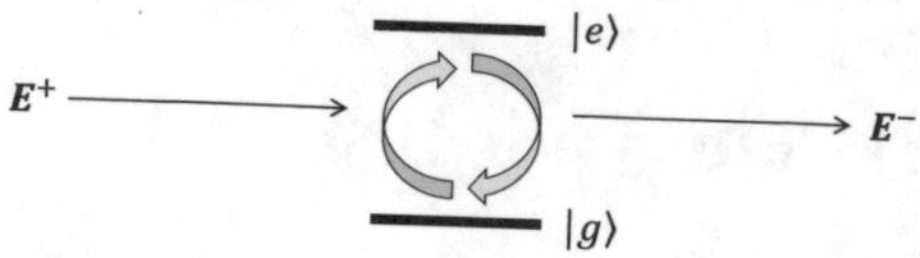

Fig. 16.1 Schematics of two-level system consisting of a groundstate g and an excited state e. Through photon absorption the system is promoted from the ground to the excited state, and, vice versa, through photon emission the system goes from the excited to the groundstate. $E^{\pm}$ are the electric field amplitudes associated with photon absorption and emission

uniquely defined only up to an irrelevant global phase, we find that the wavefunction can be characterized by two real numbers. A convenient parameterization is in terms of the angles ϕ, θ through

$$|\psi\rangle = \sin\left(\frac{\theta}{2}\right)|g\rangle + e^{-i\phi}\cos\left(\frac{\theta}{2}\right)|e\rangle. \tag{16.2}$$

With this we can visualize a state through a point on the Bloch sphere, as shown in Fig. 16.2. The groundstate is located at the south pole, the excited state at the north pole. We will also introduce a *Bloch vector* that points from the origin to the point on the sphere characterized by θ, ϕ. The Bloch vector language was originally introduced in the context of the electron spin, where its interpretation in terms of a three-dimensional vector becomes particularly obvious. However, the same language can be easily generalized to other two-level systems and turns out to be an extremely useful visualization tool in many situations.

16.1.1 Pauli Matrices

For two-level systems it is convenient to introduce the Pauli matrices

Pauli Matrices

$$\sigma_1 = \begin{pmatrix} 0 & 1 \\ 1 & 0 \end{pmatrix}, \quad \sigma_2 = \begin{pmatrix} 0 & -i \\ i & 0 \end{pmatrix}, \quad \sigma_3 = \begin{pmatrix} 1 & 0 \\ 0 & -1 \end{pmatrix}. \tag{16.3}$$

Together with the unit matrix

$$\mathbb{1} = \begin{pmatrix} 1 & 0 \\ 0 & 1 \end{pmatrix}$$

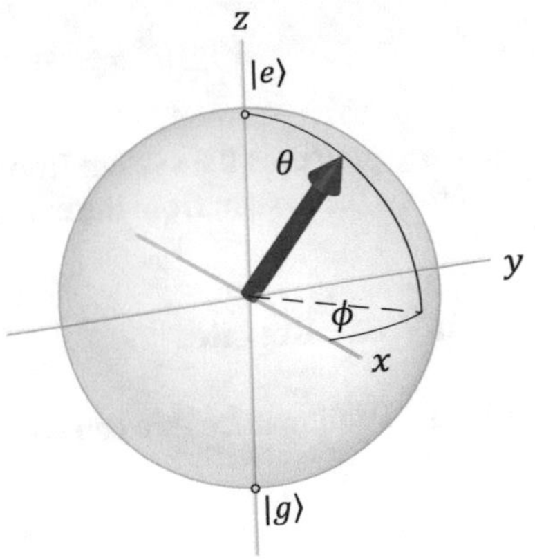

Fig. 16.2 Bloch sphere representation of a two-level system. A generic state of a two-level system corresponds to a point on the Bloch sphere characterized by two angles θ, ϕ. The groundstate is located at the south pole, the excited state at the north pole

the Pauli matrices form a complete basis for any two-level operator (see Exercise 16.1). This means that any operator $\hat{A}$ can be written in the form

$$\hat{A} = a_0 \mathbb{1} + \sum_{j=1}^{3} a_j \sigma_j = a_0 \mathbb{1} + \boldsymbol{a} \cdot \boldsymbol{\sigma}, \tag{16.4}$$

where in the last expression we have introduced a vector of matrices $\boldsymbol{\sigma} = \sigma_1 \hat{\boldsymbol{x}} + \sigma_2 \hat{\boldsymbol{y}} + \sigma_3 \hat{\boldsymbol{z}}$ together with the scalar product $\boldsymbol{a} \cdot \boldsymbol{\sigma}$, which has to be understood such that it gives in Eq. (16.4) the expression under the sum. The Pauli matrices are Hermitian

$$\sigma_j^\dagger = \sigma_j. \tag{16.5}$$

Thus, for a Hermitian operator $\hat{A} = \hat{A}^\dagger$ both the scalar a_0 and the vector $\boldsymbol{a}$ have to be real-valued. Other important properties of the Pauli matrices are that they are idempotent

$$\sigma_j^2 = \mathbb{1} \tag{16.6}$$

and fulfill the relation

$$\sigma_i \sigma_j = \delta_{ij} \mathbb{1} + i\, \epsilon_{ijk} \sigma_k, \tag{16.7}$$

with the totally anti-symmetric Levi-Civita tensor of Eq. (2.55). All of the above properties can be easily verified through explicit calculation. Below we will also use the operators

$$|e\rangle\langle g| \longrightarrow \sigma_+ = \frac{1}{2}(\sigma_1 + i\sigma_2) = \begin{pmatrix} 0 & 1 \\ 0 & 0 \end{pmatrix}$$

$$|g\rangle\langle e| \longrightarrow \sigma_- = \frac{1}{2}\left(\sigma_1 - i\sigma_2\right) = \begin{pmatrix} 0 & 0 \\ 1 & 0 \end{pmatrix}. \tag{16.8}$$

Here σ_+ promotes the system from the ground to the excited state, and conversely σ_- brings the system from the excited to the groundstate.

Bloch Vector Revisited

With the Pauli matrices we can write the Bloch vector $\boldsymbol{u}$ in the intriguing form

$$\boldsymbol{u} = \langle\psi|\boldsymbol{\sigma}|\psi\rangle = \begin{pmatrix} \cos\phi\sin\theta \\ \sin\phi\sin\theta \\ \cos\theta \end{pmatrix}, \tag{16.9}$$

where we have used the wavefunction of Eq. (16.2) to arrive at the last expression. To understand the physical meaning of the different components of the Bloch vector, we first evaluate the z-component for the wavefunction of Eq. (16.1),

$$u_z = \langle\sigma_3\rangle = |C_e|^2 - \left|C_g\right|^2,$$

which gives us the population difference between the excited and groundstate. This is -1 for the ground and $+1$ for the excited state, respectively, corresponding to states on the south and north pole of the Bloch sphere. Similarly we get

$$u_x = 2\,\mathrm{Re}\left\{C_g C_e^*\right\}, \quad u_y = 2\,\mathrm{Im}\left\{C_g C_e^*\right\}.$$

Thus, the x and y component are related to the superposition properties between the ground and excited state. As we will discuss further below, there also exists a direct relation to the polarization of Maxwell's equations.

16.2 Two-Level Dynamics

In the following we consider the dynamics of a two-level system in presence of a time-independent Hamiltonian

$$\hat{H} = E_0\mathbb{1} + \frac{1}{2}\hbar\boldsymbol{\Omega}\cdot\boldsymbol{\sigma}, \tag{16.10}$$

where the factor of $1/2$ has been chosen for later convenience. Because $\hat{H}$ is a Hermitian operator the vector $\boldsymbol{\Omega}$ must be real. Below we will also consider a time-dependent $\boldsymbol{\Omega}(t)$ vector, but for the moment we stay with the time-independent form and additionally set $E_0 = 0$. The time evolution operator for this Hamiltonian becomes

$$\hat{U}(t,0) = e^{-i\frac{1}{2}\boldsymbol{\Omega}\cdot\boldsymbol{\sigma}\,t} = \sum_{n=0}^{\infty} \frac{\left(-\frac{i}{2}t\right)^n}{n!} \left(\boldsymbol{\Omega}\cdot\boldsymbol{\sigma}\right)^n,$$

where in the last expression we have expanded the exponential in a Taylor series. We now use Eq. (16.7) to simplify the product of Pauli matrices according to

$$\left(\boldsymbol{\Omega}\cdot\boldsymbol{\sigma}\right)^2 = \Omega_i \Omega_j \left(\delta_{ij}\mathbb{1} + i\,\epsilon_{ijk}\sigma_k\right) = \Omega^2\,\mathbb{1},$$

where the term with the Levi-Civita tensor is zero because $\Omega_i\Omega_j$ is a symmetric matrix. We can thus rewrite the time evolution operator as

$$\hat{U}(t,0) = \sum_{n=0}^{\infty} \frac{\left(-\frac{i}{2}\Omega t\right)^{2n}}{(2n)!}\mathbb{1} + \sum_{n=0}^{\infty} \frac{\left(-\frac{i}{2}\Omega t\right)^{2n+1}}{(2n+1)!}\hat{\boldsymbol{\Omega}}\cdot\boldsymbol{\sigma},$$

with the unit vector $\hat{\boldsymbol{\Omega}}$. The time evolution operator for the time-independent Hamiltonian of Eq. (16.10) then reads

Time Evolution Operator for Two-Level System

$$\hat{U}(t,0) = \cos\left(\frac{\Omega t}{2}\right)\mathbb{1} - i\,\sin\left(\frac{\Omega t}{2}\right)\hat{\boldsymbol{\Omega}}\cdot\boldsymbol{\sigma}. \tag{16.11}$$

$\hat{U}(t,0)$ is a unitary matrix with $\hat{U}\hat{U}^\dagger = \mathbb{1}$, as can be easily checked in the above case through direct calculation. We also state the result for the unitary transformation of the Pauli matrix vector

$$e^{\frac{i}{2}\boldsymbol{\Omega}\cdot\boldsymbol{\sigma}t}\boldsymbol{\sigma}e^{-\frac{i}{2}\boldsymbol{\Omega}\cdot\boldsymbol{\sigma}t} = \cos(\Omega t)\,\boldsymbol{\sigma} + \sin(\Omega t)\hat{\boldsymbol{\Omega}}\times\boldsymbol{\sigma} + [1-\cos(\Omega t)]\,(\hat{\boldsymbol{\Omega}}\cdot\boldsymbol{\sigma})\hat{\boldsymbol{\Omega}}, \tag{16.12}$$

which can be obtained similarly to the above derivation.

16.2.1 *Optically Driven Two-Level System*

We next consider a Hamiltonian of the form

$$\hat{H}(t) = \sum_{i=e,g} E_i \, |i\rangle\langle i| + \hat{H}_{\rm op}(t), \tag{16.13}$$

where the first term accounts for the energies E_g, E_e of the ground and excited state, and

$$\hat{H}_{\rm op}(t) = -q\boldsymbol{r} \cdot \boldsymbol{E}(t)$$

is the light–matter coupling in the dipole approximation, see Eq. (13.91). $q\boldsymbol{r}$ is the dipole operator and $\boldsymbol{E}$ is the electric field at the position of the two-level system, which we model here as a classical field. In the basis of the two-level system the light–matter coupling reads

$$\hat{H}_{\rm op}(t) = -q\Big\{\langle e|\boldsymbol{r}|g\rangle\sigma_+ + \langle g|\boldsymbol{r}|e\rangle\sigma_-\Big\} \cdot \boldsymbol{E}(t),$$

where we have assumed that the diagonal elements $\langle g|\boldsymbol{r}|g\rangle$ and $\langle e|\boldsymbol{r}|e\rangle$ are both zero. The first term in brackets describes a dipole excitation from the ground to the excited state, and the second term the reversed de-excitation. Throughout the book we adopt the somewhat sloppy notation that the Pauli matrices $\sigma_{x,y,z}$ and the excitation and de-excitation operators $\sigma_\pm$ are written *without* the operator symbol. As for the electric field, we decompose it into components propagating with positive and negative frequencies

$$\boldsymbol{E}(t) = \boldsymbol{E}^+(t) + \boldsymbol{E}^-(t),$$

as discussed in more detail in Sect. 14.3. There we have also shown that $\boldsymbol{E}^+$ can be associated with photon absorption and $\boldsymbol{E}^-$ with photon emission, although in the following we will treat the electric fields on a classical level. In many cases of interest, this field decomposition allows for an important simplification of the light–matter Hamiltonian. Suppose that the light field oscillates with approximately the frequency corresponding to the two-level energy difference $\hbar\omega \approx E_e - E_g$, such that we can write

$$\boldsymbol{E}(t) = e^{-i\omega t}\,\boldsymbol{\mathcal{E}}^+(t) + e^{i\omega t}\,\boldsymbol{\mathcal{E}}^-(t), \tag{16.14}$$

with functions $\boldsymbol{\mathcal{E}}^\pm$ that change slowly as a function of time. In the interaction picture the light–matter coupling can then be expressed as

$$\begin{aligned}\hat{H}_{{\rm op},I}(t) = &-e^{+i(E_e-E_g-\omega)t}\Big[\boldsymbol{p}\;\cdot\boldsymbol{\mathcal{E}}^+(t)\Big]\sigma_+ - e^{+i(E_e-E_g+\omega)t}\Big[\boldsymbol{p}\;\cdot\boldsymbol{\mathcal{E}}^-(t)\Big]\sigma_+ \\ &-e^{-i(E_e-E_g-\omega)t}\Big[\boldsymbol{p}^*\cdot\boldsymbol{\mathcal{E}}^-(t)\Big]\sigma_- - e^{-i(E_e-E_g+\omega)t}\Big[\boldsymbol{p}^*\cdot\boldsymbol{\mathcal{E}}^+(t)\Big]\sigma_-,\end{aligned}$$

where we have introduced the dipole moment $\boldsymbol{p} = \langle e|q\boldsymbol{r}|g\rangle$. In the rotating wave approximation we only keep those terms which oscillate with a small frequency,

$$\hat{H}_{\text{op},I}(t) \approx -e^{+i(E_e-E_g-\omega)t}\Big[\boldsymbol{p}\cdot\boldsymbol{\mathcal{E}}^+(t)\Big]\sigma_+ - e^{-i(E_e-E_g-\omega)t}\Big[\boldsymbol{p}^*\cdot\boldsymbol{\mathcal{E}}^-(t)\Big]\sigma_-.$$

Transforming back to the Schrödinger picture, we get the light–matter Hamiltonian in the rotating wave approximation

Light–Matter Coupling in Rotating Wave Approximation

$$\hat{H}_{\text{op}}(t) = -\Big\{\hbar\Omega_R(t)\,\sigma_+ + \hbar\Omega_R^*(t)\,\sigma_-\Big\}. \tag{16.15}$$

Here we have introduced the Rabi energy

$$\hbar\Omega_R(t) = \big\langle e\big|q\boldsymbol{r}\big|g\big\rangle\cdot\boldsymbol{E}^+(t). \tag{16.16}$$

In this chapter we denote the Rabi energy with $\hbar\Omega_R$ and the Rabi frequency with Ω_R, to avoid any confusion with the $\boldsymbol{\Omega}$ vector introduced in Eq. (16.10), but we will drop the subscript in later chapters when there is no danger of confusion. In general, Ω_R is complex-valued, where the absolute value gives the light–matter interaction strength and the argument the phase relation between the dipole moment and the electric field vector $\boldsymbol{E}^+$.

The contributions which are neglected in the rotating wave approximation oscillate with the sum frequency $\omega + (E_e - E_g)/\hbar$, and are assumed to vary too rapidly to have any noticeable effect on the system's dynamics. The rotating wave approximation works extremely well for $\hbar\Omega_R \ll E_e - E_g$ which is usually well fulfilled for the systems we are interested in. Figure 16.3 shows calculations performed with and without the rotating wave approximation for a rather strong light–matter coupling of $\Omega_R : \omega = 1 : 10$ and for resonant excitation. Even for such a strong coupling the non-resonant terms play practically no role, which highlights the validity of the approximation.

Time-Harmonic Fields

For a time-harmonic light excitation the dynamics of a two-level system can be further simplified. We consider an electric field of the form

$$\boldsymbol{E}(t) = \frac{1}{2}\left(\boldsymbol{E}_0 e^{-i\omega t} + \boldsymbol{E}_0^* e^{i\omega t}\right),$$

where as usually the complex amplitude $\boldsymbol{E}_0$ accounts for both the strength and the phase of the light excitation. The energy difference between the ground and excited state of the two-level system is written in the form

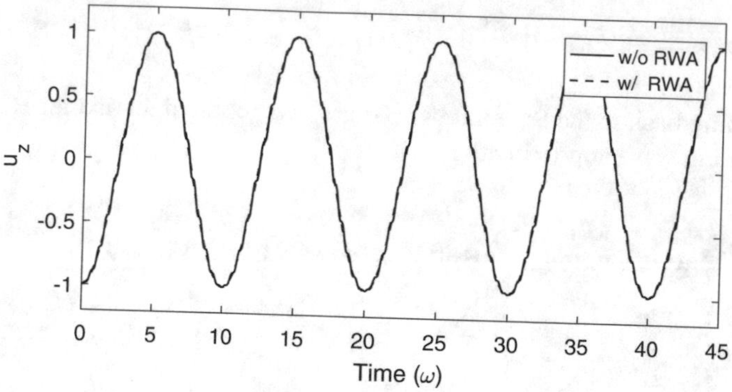

Fig. 16.3 Excitation of resonantly driven two-level system without (solid line) and with (dashed line) rotating wave approximation (RWA) for $\Omega_R : \omega = 1 : 10$. We plot the z-component of the Bloch vector as a function of time. Even for such a strong light–matter coupling the two results are almost indistinguishable

$$\hbar\omega + \hbar\Delta = E_e - E_g, \tag{16.17}$$

with ω being the frequency of the driving field and Δ the frequency detuning with respect to the two-level resonance. In terms of Pauli matrices the Hamiltonian can be expressed in the form

$$\hat{H}(t) = \frac{1}{2}\hbar\omega\sigma_3 + \frac{1}{2}\Big\{\hbar\Delta\sigma_3 - \Big(e^{-i\omega t}\hbar\Omega_R\,\sigma_+ + e^{i\omega t}\hbar\Omega_R^*\,\sigma_-\Big)\Big\} = \hat{H}_0 + \hat{V}(t),$$

with the complex-valued Rabi energy $\hbar\Omega_R = \langle e|q\boldsymbol{r}|g\rangle \cdot \boldsymbol{E}_0$. Next we introduce an interaction representation with respect to $\hat{H}_0 = \frac{1}{2}\hbar\omega\,\sigma_3$. As can be immediately seen, the time dependence of the Pauli matrices in the interaction representation cancels the time dependence of the electric fields, and we end up with a time-independent Hamiltonian for a harmonically driven two-level system

Hamiltonian for Harmonically Driven Two-Level System

$$\hat{V}_I = \frac{1}{2}\Big\{\hbar\Delta\sigma_3 - \hbar\Omega_R'\,\sigma_1 + \hbar\Omega_R''\,\sigma_2\Big\}. \tag{16.18}$$

Here Ω_R' and Ω_R'' are the real and imaginary part of the Rabi frequency. Remarkably, although describing a time-dependent problem the Hamiltonian in the interaction picture does not depend on time. It is important to realize that the fast time dependence of ω has been completely removed from the problem and the only relevant energies are the energy detuning $\hbar\Delta$ and the Rabi energy $\hbar\Omega_R$ accounting for the strength of the light–matter coupling.

From here on we work in the "rotating frame" that oscillates with the central light frequency ω, and use the interaction representation for $\hat{H}_0$, however, without indicating the interaction representation explicitly. The Hamiltonian of Eq. (16.18) can be rewritten in the form

$$\hat{V} = \frac{\hbar}{2}\Big(-\Omega_R'\hat{\mathbf{x}} + \Omega_R''\hat{\mathbf{y}} + \Delta\hat{\mathbf{z}}\Big)\cdot\boldsymbol{\sigma} = \frac{1}{2}\hbar\boldsymbol{\Omega}\cdot\boldsymbol{\sigma},$$

where we introduced in analogy to Eq. (16.10) the vector $\boldsymbol{\Omega} = (-\Omega_R', \Omega_R'', \Delta)$. The time evolution of the Bloch vector is obtained from

$$i\hbar\dot{u}_k = i\hbar\frac{d}{dt}\langle\psi|\sigma_k|\psi\rangle = \langle\psi|\Big[\sigma_k, \frac{1}{2}\hbar\boldsymbol{\Omega}\cdot\boldsymbol{\sigma}\Big]|\psi\rangle. \tag{16.19}$$

The last term can be simplified through

$$\Big[\sigma_k, \frac{1}{2}\hbar\boldsymbol{\Omega}\cdot\boldsymbol{\sigma}\Big] = \frac{1}{2}\hbar\Omega_\ell\Big[\sigma_k, \sigma_\ell\Big] = i\hbar\epsilon_{k\ell m}\Omega_\ell\sigma_m,$$

where we have used $[\sigma_k, \sigma_\ell] = 2i\epsilon_{k\ell m}\sigma_m$ which can be derived from the fundamental relation of Eq. (16.7) for the product of Pauli matrices. Thus, we end up with the time evolution of the Bloch vector for a harmonically driven two-level system

Time Evolution of Bloch Vector for Harmonic Driving

$$\frac{d\boldsymbol{u}}{dt} = \boldsymbol{\Omega}\times\boldsymbol{u}. \tag{16.20}$$

This can be interpreted as the evolution of a spin $\boldsymbol{u}$ in a fictitious magnetic field

$$\boldsymbol{\Omega} = -\Omega_R'\,\hat{\mathbf{x}} + \Omega_R''\,\hat{\mathbf{y}} + \Delta\,\hat{\mathbf{z}}.$$

The x and y components of this fictitious field are given by the real and imaginary part of the Rabi frequency, respectively, and the z-component by the detuning of the two-level resonance with respect to ω. The solution of the Bloch vector equation can be obtained from Eq. (16.12), and we get

$$\boldsymbol{u}(t) = \cos(\Omega t)\,\boldsymbol{u}_0 + \sin(\Omega t)\,\hat{\boldsymbol{\Omega}}\times\boldsymbol{u}_0 + [1-\cos(\Omega t)]\,(\hat{\boldsymbol{\Omega}}\cdot\boldsymbol{u}_0)\hat{\boldsymbol{\Omega}}, \tag{16.21}$$

where $\boldsymbol{u}_0$ is the Bloch vector at time zero. The above equation is also known as Rodrigues' rotation formula and describes the rotation of $\boldsymbol{u}_0$ around the rotation axis $\hat{\boldsymbol{\Omega}}$. Equation (16.21) can be written in matrix form

$$\boldsymbol{u} = \bar{\bar{R}}(\hat{\boldsymbol{\Omega}}, \Omega t) \cdot \boldsymbol{u}_0,$$

with the rotation matrix

$$R_{ij}(\hat{\boldsymbol{e}}, \theta) = \cos\theta\, \delta_{ij} + \sin\theta\, \epsilon_{ikj}\hat{e}_k + (1 - \cos\theta)\, \hat{e}_i \hat{e}_j.$$

Resonant Excitation and Free Time Evolution For later use we explicitly state two specific rotation matrices. First, for resonant excitation and for a real Rabi frequency we get $\hat{\boldsymbol{\Omega}} = -\hat{\boldsymbol{x}}$, such that $\bar{\bar{R}}$ becomes the rotation matrix around the x-axis

$$\bar{\bar{R}}(-\hat{\boldsymbol{x}}, \theta) = \begin{pmatrix} 1 & 0 & 0 \\ 0 & \cos\theta & \sin\theta \\ 0 & -\sin\theta & \cos\theta \end{pmatrix}. \tag{16.22}$$

Secondly, in absence of any driving the two-level system evolves in presence of the detuning term only, $\hat{\boldsymbol{\Omega}} = \hat{z}$, and $\bar{\bar{R}}$ becomes the rotation matrix around the z-axis

$$\bar{\bar{R}}(\hat{z}, \theta) = \begin{pmatrix} \cos\theta & -\sin\theta & 0 \\ \sin\theta & \cos\theta & 0 \\ 0 & 0 & 1 \end{pmatrix}. \tag{16.23}$$

Figure 16.4 shows trajectories of the Bloch vector for different $\boldsymbol{\Omega}$ vectors and for the system initially in the groundstate, corresponding to the south pole of the Bloch sphere. The left panel shows a large detuning, correspondingly $\boldsymbol{u}$ is not brought far away from the south pole upon excitation. Conversely, for zero detuning the Bloch vector oscillates between the south and north pole, as shown in the right panel.

The description of the system's dynamics in terms of a Bloch vector and a fictitious magnetic field is obviously borrowed from electron and nuclear spin resonance.

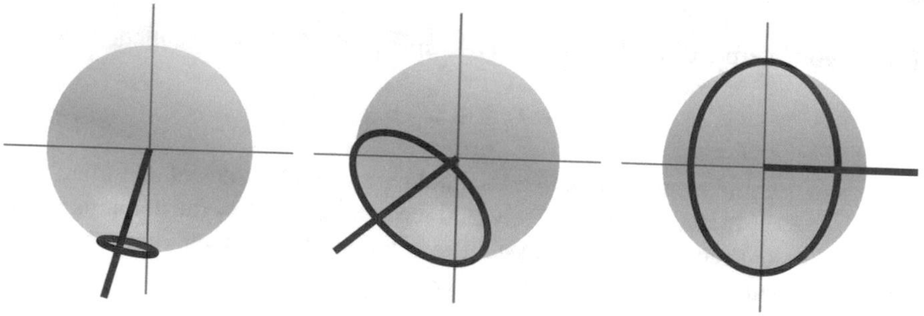

Fig. 16.4 Time evolution of Bloch vector for different $\boldsymbol{\Omega}$ vectors, as computed from Eq. (16.21), for the Bloch vector initially in groundstate (south pole). The blue lines report the direction of the $\boldsymbol{\Omega}$ vector

For electrons the coupling between the electron spin and a magnetic field is given through the Zeeman coupling

$$\hat{H} = -\hat{\boldsymbol{\mu}} \cdot \left(B_0 \hat{z} + \boldsymbol{B}_\perp \cos \omega t \right), \tag{16.24}$$

where B_0 is a static magnetic field applied along the z-direction, which energetically splits the two spin states of the electron, and $\boldsymbol{B}_\perp$ is an oscillating radio-frequency field oscillating with angular frequency ω, which induces transitions between the electron states. The magnetic moment operator is given by

$$\hat{\boldsymbol{\mu}} = \left(\frac{g\mu_B}{\hbar} \right) \boldsymbol{\sigma},$$

with the electron g-factor $g \approx 2$, the Bohr magneton μ_B, and the electron spin operator $\boldsymbol{\sigma}$. One can easily check that with this model one can introduce the rotating wave approximation in complete analogy to the optical case, and end up with an equation of motion similar to Eq. (16.20). However, the Bloch vector now points into the direction of the average electron spin, and the transverse components of the driving vector $\boldsymbol{\Omega}$ into the directions of the radio-frequency field $\boldsymbol{B}_\perp$. The longitudinal component Ω_z is again given by the detuning between the electron energy splitting and $\hbar\omega$ of the radio-frequency field. Despite the different physics underlying the two models of optical excitation of a two-level system and radio-frequency excitation of an electron spin situated in a static magnetic field, the description in terms of a Bloch vector and a Rabi vector for the driving field is universal, and can be applied to generic two-level systems.

16.3 Relaxation and Dephasing

Our discussion so far applies to ideal quantum-mechanical two-level systems that are isolated from the rest of the world. In general, systems interact with their environment and this interaction significantly influences the time evolution. It is common to distinguish between the following types of time evolutions:

Coherent Dynamics. The coherent time evolution is governed by the Schrödinger's equation, with a Hamiltonian that accounts for the isolated system and the coupling to the external fields through

$$i\hbar \frac{d}{dt} \big|\psi(t)\big\rangle = \left(\hat{H}_0 + \hat{H}_{\text{op}}(t) \right) \big|\psi(t)\big\rangle.$$

Incoherent Dynamics. The incoherent time evolution is due to environment couplings, which, in general, contain uncontrollable elements. Incoherent processes can be distinguished as follows:

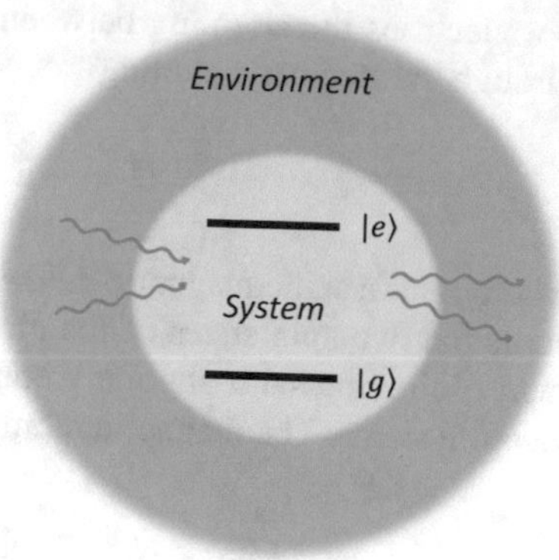

Fig. 16.5 Schematics of a two-level system interacting with its environment. This leads to scatterings between the levels (relaxation) and losses of the superposition properties (dephasing)

- *Relaxation* describes a process where the system undergoes a transition from an initial state to a final state, and energy is exchanged with the environment.
- *Dephasing* describes the process where the phase of superposition states is affected by environment couplings, typically without energy exchange with the environment. As an effect of such dephasing, the coherence properties of the system become diminished.

Figure 16.5 schematically depicts the situation of a two-level system interacting with its environment. We will discuss environment couplings in more detail in the next chapter, and here introduce only a simple and intuitive description scheme for two-level systems in terms of T_2, T_2^*, and T_1 times. Following the magnetic field language, one often calls T_2, T_2^* times which affect the transverse components u_x, u_y of the Bloch vector "transverse" relaxation times, and T_1 times which affect the z-component of the Bloch vector "longitudinal" relaxation times.

T_2 Dephasing Time

We start with the following simple model. Consider a two-level system that is resonantly driven by an external light field up to some random time t_r where it is assumed to interact with the environment. Suppose that the effect of this coupling is that the phase between the ground and excited state becomes scrambled up in the form

$$|\psi(t_r)\rangle = \sin\left(\frac{\theta}{2}\right)|g\rangle + e^{-i\phi}\cos\left(\frac{\theta}{2}\right)|e\rangle \longrightarrow \sin\left(\frac{\theta}{2}\right)|g\rangle + e^{-i\phi_r}\cos\left(\frac{\theta}{2}\right)|e\rangle,$$

where ϕ_r is some random phase. In terms of the Bloch vector this transition reads

$$\boldsymbol{u}(t_r) = \begin{pmatrix} u_\perp \cos\phi \\ u_\perp \sin\phi \\ u_z \end{pmatrix} \longrightarrow \begin{pmatrix} u_\perp \cos\phi_r \\ u_\perp \sin\phi_r \\ u_z \end{pmatrix}.$$

Here we have used the notation $u_\perp$ for the transverse component of the Bloch vector. After the scattering the system time evolution is again governed by the coherent dynamics described in the previous section, up to the point where it again picks up some random phase from the environment. Because the transverse component is scattered randomly, the time evolution differs from that of an identical system without environment couplings.

We next have to combine the above description with the way measurements are performed for quantum systems. In general, in experiments one either measures a large number of identical (or at least sufficiently similar) quantum systems, or one uses a single system and repeats the experiments many times. In the latter case the system has to be prepared in the same (or at least a sufficiently similar) state every time before the experiment is started. The result of the measurement is then an ensemble average, either for many systems or many repetitions of an experiment. For the averaged time evolution of the Bloch vector we obtain

$$\langle \boldsymbol{u}(t)\rangle = \frac{1}{N}\sum_{\mu=1}^{N}\boldsymbol{u}^{(\mu)}(t), \tag{16.25}$$

where N denotes the number of measurements and μ labels the individual measurement outcomes. Figure 16.6 shows results for a typical simulation of such dephasing

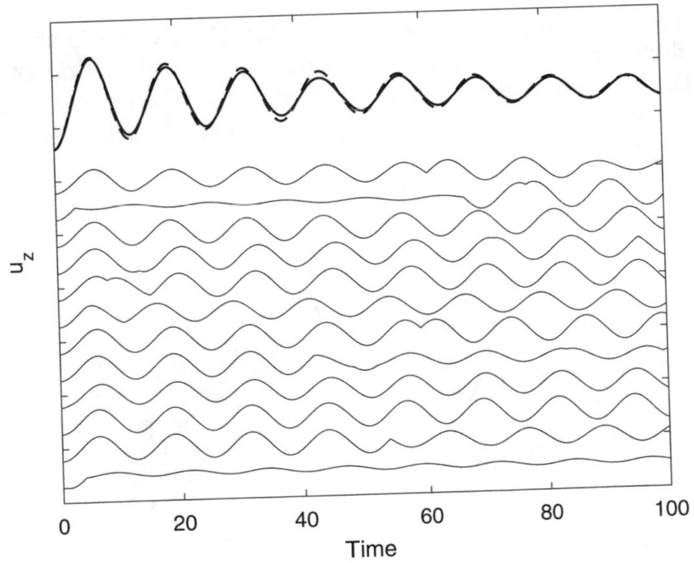

Fig. 16.6 Simulation of dephasing for an ensemble of 100 two-level systems that are resonantly driven by an external light field. Environment couplings are modelled through random phase jumps. The lower curves show individual trajectories of the z-component of the Bloch vector, the upper curves show the ensemble average (solid line) and an oscillation with an exponential damping (dashed line) for a damping constant of $2 \times T_2$. We use parameters of $\Omega = 0.5$ and $T_2 = 30$

for $N = 100$ two-level systems and for random phase jumps. The bottom part of the figure shows the trajectories of individual systems, which consist of a coherent time evolution interrupted by abrupt jumps of the phase. The scattering events occur at times randomly taken from an exponential $e^{-\Gamma t_r}$ distribution. The upper part shows the ensemble average which exhibits an exponential damping. In many cases of interest dephasing can then be modelled through a damping term in the Bloch equations

Pure Dephasing of Two-Level System With T_2 Time

$$\dot{\boldsymbol{u}} = \boldsymbol{\Omega} \times \boldsymbol{u} - \frac{1}{T_2}\boldsymbol{u}_\perp, \tag{16.26}$$

where $\boldsymbol{u}_\perp = \boldsymbol{u} - (\hat{z}\cdot\boldsymbol{u})\,\hat{z}$ is the transverse part of the Bloch vector. As shown by the thick solid line in Fig. 16.6, this phenomenological dephasing time model reproduces well the mean value of the simulation results. It is, however, important to realize that Eq. (16.26) does not rely on a specific model but provides a generic description in terms of a single time constant T_2 to be extracted from experiment or some other theoretical model. For this reason, the T_2 description has found widespread use in various fields of research.

Figure 16.7a reports the time evolution of the Bloch vector in absence of any external driving fields, and for an initial state located on the equator of the Bloch sphere. The u_z component remains constant while the Bloch vector rotates around the z-axis with a frequency determined by the detuning Δ, and decays towards the origin owing to dephasing losses.

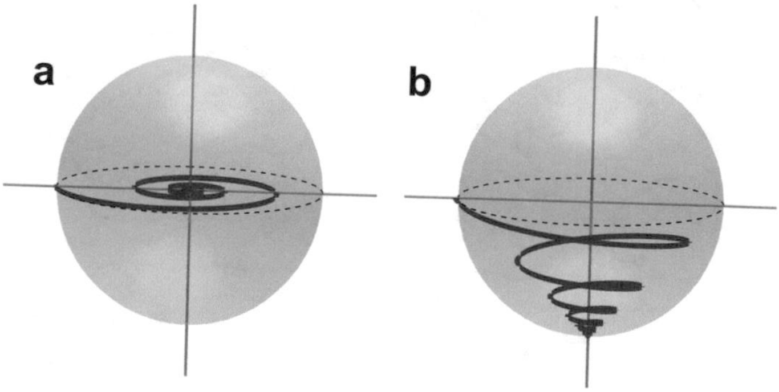

Fig. 16.7 Trajectories of Bloch vector for (**a**) pure dephasing, see Eq. (16.26), and (**b**) relaxation with T_1 only, see Eq. (16.28). In case of pure dephasing the u_z component does not change as a function of time, and only the transverse component $\boldsymbol{u}_\perp$ decays. For a pure relaxation the Bloch vector decays towards the south pole associated with the groundstate of the two-level system.

T_2^* Time and Inhomogeneous Broadening

Exponential damping can also occur for ensembles even in presence of a completely coherent time dynamics. Consider an ensemble of two-level systems with different detunings Δ_μ as is often the case for nanostructures, such as quantum dots, which are not identical but slightly differ in size or composition. But even for structurally identical molecules, environment effects, such as strain or field fluctuations, can lead to small modifications of the transition energies. The time-averaged Bloch vector then evolves according to

$$\langle \boldsymbol{u}(t) \rangle = \frac{1}{N} \sum_{\mu=1}^{N} \bar{\bar{R}} \left(\hat{\boldsymbol{\Omega}}_\mu, \Omega_\mu t \right) \cdot \boldsymbol{u}_0^{(\mu)} .$$

Thus, even when initially all two-level systems are synchronized, they will evolve with different pace according to the individual "clocks" determined by the detunings Δ_μ. The averaged response approximately decays exponentially with a time constant determined by the detuning distribution. Such dephasing caused by the inhomogeneous distribution of transition energies or transition dipole moments can be modelled through an effective dephasing time T_2^* via

Dephasing of Inhomogeneous Two-Level Ensemble With T_2^* Time

$$\dot{\boldsymbol{u}} = \boldsymbol{\Omega} \times \boldsymbol{u} - \frac{1}{T_2^*} \boldsymbol{u}_t . \tag{16.27}$$

Quite generally, T_2^* can contain contributions of both pure dephasing T_2 and ensemble effects, so the relation $T_2 \geq T_2^*$ must be always fulfilled. Figure 16.8a shows the evolution of an ensemble of Bloch vectors with an inhomogeneous broadening, and in absence of external driving fields.

Photon Echo There exists a beautiful technique which allows discriminating between pure dephasing and dephasing caused by inhomogeneous broadening, the photon echo. Suppose that initially all two-level systems are in the groundstate and at time zero a strong and short pulse with an area of $\pi/2$ brings them to $\boldsymbol{u}_0 = \hat{\boldsymbol{y}}$. The ensuing time evolution is then governed by

$$\langle \boldsymbol{u}(t) \rangle = \left\langle \begin{pmatrix} \cos \Delta t & -\sin \Delta t & 0 \\ \sin \Delta t & \cos \Delta t & 0 \\ 0 & 0 & 1 \end{pmatrix} \cdot \begin{pmatrix} 0 \\ 1 \\ 0 \end{pmatrix} \right\rangle = \left\langle \begin{pmatrix} -\sin \Delta t \\ \cos \Delta t \\ 0 \end{pmatrix} \right\rangle ,$$

where $\langle \dots \rangle$ denotes the average over the detuning ensemble and we have used the rotation matrix of Eq. (16.23) for the free time evolution in absence of any driving. Because of the different detunings Δ, the transverse component of the averaged Bloch vector decays in time. We next apply at time T a second, strong and short pulse with an area of π, see Eq. (16.22) for the corresponding rotation matrix, which changes the Bloch vector to

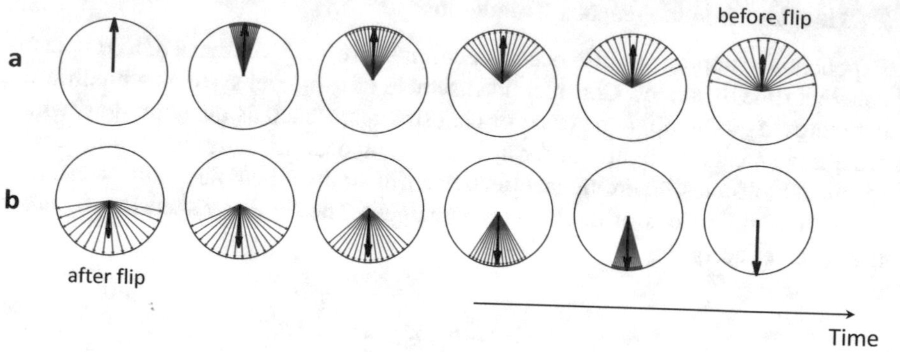

Fig. 16.8 (**a**) Schematics of time evolution of Bloch vectors in xy-plane for inhomogeneously broadened ensemble of two-level systems, in absence of a driving field. Each Bloch vector evolves with its own pace given by the detuning Δ_μ. The length of the averaged Bloch vector (black arrow) decreases as a function of time. (**b**) In photon echo experiments a π-pulse is applied after a time T, which flips the Bloch vectors at the x-axis. As time progresses, the Bloch vectors refocus again at time $2 \times T$, which gives rise to the so-called photon echo. For details see text

$$\langle \boldsymbol{u}(T) \rangle = \left\langle \begin{pmatrix} 1 & 0 & 0 \\ 0 & -1 & 0 \\ 0 & 0 & -1 \end{pmatrix} \cdot \begin{pmatrix} -\sin \Delta T \\ \cos \Delta T \\ 0 \end{pmatrix} \right\rangle = -\left\langle \begin{pmatrix} \sin \Delta T \\ \cos \Delta T \\ 0 \end{pmatrix} \right\rangle .$$

Computing the ensuing free time evolution by again applying the $\bar{\bar{R}}(\hat{z}, \Delta t)$ rotation matrix, we get

$$\langle \boldsymbol{u}(T+t) \rangle = -\left\langle \begin{pmatrix} \cos \Delta t \sin \Delta T - \sin \Delta t \cos \Delta T \\ \sin \Delta t \sin \Delta T + \cos \Delta t \cos \Delta T \\ 0 \end{pmatrix} \right\rangle .$$

Remarkably, after another waiting period T all Bloch vectors synchronize again at $\boldsymbol{u}(2T) = -\hat{\boldsymbol{y}}$, causing a so-called photon echo. See also Fig. 16.8 for a graphical representation. There exist a number of spectroscopy techniques based on the photon echo effect that allow measuring the genuine dephasing times T_2, which is often completely masked by inhomogeneous broadening for other spectroscopy techniques.

T_1 Relaxation Time

Energy exchange with the environment leads to scatterings between the states of the two-level system. Consider a scattering where the system decays from the excited state to the groundstate, with $\boldsymbol{u}_{\text{eq}} = -\hat{z}$ being the Bloch vector of the equilibrium state in absence of external excitations. It is convenient to describe such energy relaxation through a T_1 time and an additional term in the Bloch equation

Relaxation of Two-Level System With T_1 Time

$$\dot{\boldsymbol{u}} = \boldsymbol{\Omega} \times \boldsymbol{u} - \frac{1}{T_1}\Big(\boldsymbol{u} - \boldsymbol{u}_{\text{eq}}\Big) - \frac{1}{T_2^*}\boldsymbol{u}_\perp . \tag{16.28}$$

For completeness we have also included the dephasing contribution. Equation (16.28) often is a surprisingly good description for various systems consisting of only two effective relaxation and dephasing parameters, namely T_1 and T_2, T_2^*. In most cases of interest we have $T_2 \ll T_1$ which means that dephasing is more important than relaxation. Note, however, that also T_1 leads to a decay of the transverse component of the Bloch vector. We will come back to this in the next chapter.

Figure 16.7b shows the time evolution of the Bloch vector in absence of any driving fields, and for an initial state located on the equator of the Bloch sphere. The Bloch vector rotates around the z-axis, because of detuning, and decays towards the south pole, associated with the system's groundstate, because of relaxation processes.

16.4 Jaynes–Cummings Model

In the remainder of this chapter we discuss the so-called Jaynes–Cummings model [184] that can be solved analytically. It describes the interaction between a two-level system and a harmonic oscillator. For instance, in quantum optics this model is employed for an atom (the two-level atom) situated inside a cavity (the harmonic oscillator). The Hamiltonian reads

$$\hat{H} = \hbar\omega\Big(\hat{a}^\dagger\hat{a} + \frac{1}{2}\Big) + \sum_{i=g,e} E_i\,|i\rangle\langle i| + \frac{1}{2}\hbar\lambda\Big(\hat{a}^\dagger\,|g\rangle\langle e| + \hat{a}\,|e\rangle\langle g|\Big), \tag{16.29}$$

where $\hat{a}$, $\hat{a}^\dagger$ are the annihilation and creation operators for the harmonic oscillator, see Eq. (13.35) and discussion thereafter, and $|g\rangle$, $|e\rangle$ account for the ground and excited state of the two-level system. The last term on the right-hand side describes the interaction, with the coupling constant λ, where either the system is promoted from the excited state to the groundstate and the harmonic oscillator is excited (first term in parentheses), or the reversed process where an excitation is transferred from the harmonic oscillator to the two-level system (second term). This type of interaction is similar to the rotating wave approximation previously discussed in the context of Eq. (16.15). Owing to this special structure, the total number of excitations

$$\hat{N} = \hat{a}^\dagger \hat{a} + |e\rangle\langle e|$$

is conserved, where the two terms on the right-hand side measure the number of excitations for the harmonic oscillator and the two-level system, respectively, as can be easily shown through explicit calculation of $[\hat{N}, \hat{H}] = 0$. This conservation plays an important role in the solution of the Jaynes–Cummings model, as will be demonstrated in a moment.

First, we rewrite Eq. (16.29) in the form (apart from a constant term)

$$\hat{H}_0 + \hat{V} = \hbar\omega\left(\hat{a}^\dagger \hat{a} + \frac{1}{2}\sigma_3\right) + \frac{\hbar}{2}\left(\Delta\sigma_3 + \lambda\left[\hat{a}^\dagger \sigma_- + \hat{a}\sigma_+\right]\right),$$

with the detuning defined in analogy to Eq. (16.17). In the above expression we have also introduced the Pauli matrices defined at the beginning of this chapter. Using an interaction representation with respect to $\hat{H}_0$ we arrive at the Jaynes–Cummings Hamiltonian

Jaynes–Cummings Hamiltonian

$$\hat{V} = \frac{\hbar}{2}\left[\Delta\sigma_3 + \lambda\left(\hat{a}^\dagger \sigma_- + \hat{a}\sigma_+\right)\right]. \tag{16.30}$$

Consider a solution of the form

$$|\psi(t)\rangle = \sum_{n=0}^{\infty}\left(C_g^n(t)|g, n\rangle + C_e^n(t)|e, n\rangle\right), \tag{16.31}$$

where n labels the number of excitations of the harmonic oscillator, see Eq. (13.39), and C_g^n, C_e^n are the expansion coefficients of the wavefunction. Because the Jaynes–Cummings Hamiltonian conserves the number of excitations, it is easy to show that

- C_g^0 is not coupled to any other coefficient, and
- the other coefficients are coupled only pairwise in the form C_g^{n-1}, C_e^n.

We thus get from Eq. (16.30)

$$i\hbar\frac{d}{dt}\begin{pmatrix} C_e^n \\ C_g^{n-1} \end{pmatrix} = \frac{\hbar}{2}\left[\Delta\sigma_3 + \lambda\sqrt{n}\sigma_1\right]\begin{pmatrix} C_e^n \\ C_g^{n-1} \end{pmatrix},$$

which can be solved with the time evolution operator of Eq. (16.11) for a two-level system. We then get

$$\begin{pmatrix} C_e^n & (t) \\ C_g^{n-1}(t) \end{pmatrix} = \left[\cos\left(\frac{\Omega_n t}{2}\right)\mathbb{1} - i\sin\left(\frac{\Omega_n t}{2}\right)\frac{\lambda\sqrt{n}\,\sigma_1 + \Delta\sigma_3}{\Omega_n}\right]\begin{pmatrix} C_e^n & (0) \\ C_g^{n-1}(0) \end{pmatrix}, \tag{16.32}$$

with the effective frequency $\Omega_n = (\Delta^2 + \lambda^2 n)^{\frac{1}{2}}$. As a specific example, we consider the case of zero detuning and assume that at time zero the system is in the groundstate of the two-level system. It can then be readily shown that the probability of finding the system at a later time in the groundstate of the two-level system is given by

$$P_g(t) = \sum_{n=0}^{\infty}\left|C_g^n(t)\right|^2 = \frac{1}{2}\sum_{n=0}^{\infty}\mathscr{P}_n\left(1 + \cos\left[\sqrt{n}\,\lambda t\right]\right), \tag{16.33}$$

where $\mathscr{P}_n$ is the probability that the n'th excited state of the harmonic oscillator is populated at time zero. Suppose that initially only one state of the harmonic oscillator is populated, such that P_g oscillates with a single frequency $\lambda\sqrt{n}$. The situation is similar to the coherent driving discussed in Sect. 16.2.1, although now the excitation is transferred between the two-level system and the harmonic oscillator, rather than the classical light field. The oscillation frequency $\lambda\sqrt{n}$ increases with the excitation number n of the harmonic oscillator, owing to the enhancement of stimulated emission and absorption within the interaction processes.

Things change considerably for more complicated initial states of the harmonic oscillator. Figure 16.9 shows the situation where the harmonic oscillator is initially

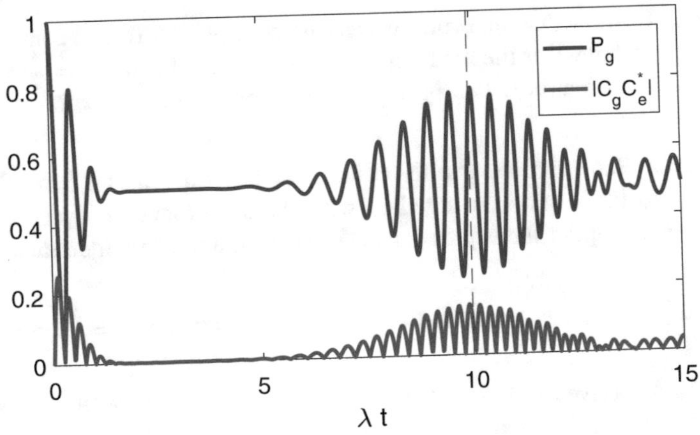

Fig. 16.9 Solution of Jaynes–Cummings model, Eq. (16.33), for the two-level system initially in its groundstate and the harmonic oscillator in a coherent state with a mean number $\bar{n} = 10$. The blue line reports the probability of finding the two-level system in its groundstate, and the magenta line shows the coherence properties of the two-level system as described by the transverse component $u_\perp$ of the Bloch vector

in a coherent state, see Eq. (13.90), with a mean number $\bar{n} = 10$. The probability $P_g(t)$ given by Eq. (16.33) is now the sum of oscillations with different frequencies which interfere destructively, similarly to the case of inhomogeneous broadening discussed above. Although the system's time evolution is governed by Schrödinger's equation and is thus completely coherent, the two-level subsystem behaves almost classically after an initial transient regime of $\lambda t \approx 2$: there is no coherence between the ground and excited state, which has previously been described by the transverse components of the Bloch vector, and the population of the ground and excited state can only be predicted with some probability. Because the system is finite, a recurrence of coherence occurs after some time, say around $\lambda t = 10$. It can be shown that this recurrence time scales according to [185]

$$T_{\text{revival}} \approx 2\pi \frac{\sqrt{\bar{n}}}{\lambda},$$

and thus becomes very large for almost "classical" oscillators with large $\bar{n}$ values and for small coupling constants λ. The dashed line in Fig. 16.9 shows that this estimate for the revival time nicely agrees with the exact results. The Jaynes–Cummings model can be considered as an example for open quantum systems, and shows that even for a completely coherent time evolution of the total system the dynamics of a subsystem may become incoherent, with a time evolution approximately described through relaxation and dephasing processes. This point will be discussed in more detail in the next chapter.

Exercises

Exercise 16.1 Show that the Pauli matrices together with $\mathbb{1}$ form a complete basis. You may like to consider the matrices $\sigma_\pm = \frac{1}{2}(\sigma_1 \pm i\sigma_2)$ and $P_\pm = \frac{1}{2}(\mathbb{1} \pm \sigma_3)$ to show that a 2×2 matrix with arbitrary matrix elements can be expressed in terms of these matrices.

Exercise 16.2 Show through explicit calculation that Eq. (16.9) is indeed fulfilled. Start with the Pauli matrices together with the wavefunction representation of Eq. (16.2), and simplify the expressions using the trigonometric identities

$$\sin\left(\frac{\theta}{2}\right)\cos\left(\frac{\theta}{2}\right) = \frac{1}{2}\sin\theta, \quad \cos^2\left(\frac{\theta}{2}\right) - \sin^2\left(\frac{\theta}{2}\right) = \frac{1}{2}\cos\theta.$$

Exercise 16.3 Derive from Eq. (16.21) the time evolution $\boldsymbol{u}(t)$ of the Bloch vector for the following parameters:

(a) Resonance excitation with $\Delta = 0$. Suppose that the external field Ω is turned off at time T, and show that the final Bloch vector only depends on the pulse area $\theta = \Omega T$.

(b) Strong off-resonance with $\Delta \gg \Omega$.

Exercise 16.4 Consider the situation where a system is excited optically by two resonant pulses of duration T that are temporally separated by a waiting period δt,

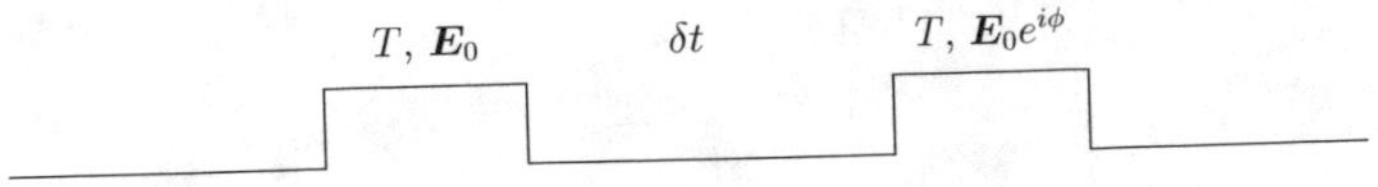

The field strength of the pulses is $\boldsymbol{E}_0$, and we consider an additional phase between the first and second pulse. Use the results of Exercise 16.3 to compute for the dipole moment $\boldsymbol{p}$ the final state of the system.

(a) For which phase ϕ is the excitation of the two-level system maximized?
(b) How should ϕ be chosen to bring the system back to the groundstate?
(c) Discuss qualitatively what happens in presence of dephasing and relaxation.

Exercise 16.5 Use from the NANOPT toolbox the file `demotwolevel01.m` to compute the time evolution of a coherently driven two-level system.

(a) What happens if the detuning is changed?
(b) Modify the program such that it computes the time evolution *with* the rotating wave approximation.
(c) Compare the simulations with and without the rotating wave approximation for various coupling strengths. How large must the coupling be chosen to observe noticeable differences?

Exercise 16.6 Compute from Eq. (16.28) the steady state $\dot{\boldsymbol{u}} = 0$ of a resonantly driven two-level system ($\Delta = 0$) in presence of dephasing and relaxation.

(a) Express the equation in matrix form, and solve it for the steady-state Bloch vector.
(b) Compute the population of the excited state as a function of the excitation strength Ω. Expand the function for small Ω values in a power series.

Exercise 16.7 Use from the NANOPT toolbox the file `demotwolevel04.m` to compute the time evolution of a coherently driven two-level system in presence of dephasing and losses. The Bloch vector trajectory is visualized on the Bloch sphere. Evaluate the influence of the parameters Δ, T_1, T_2, and give a physical interpretation.

Chapter 17
Master Equation

In the last chapter we have discussed the coherent and incoherent dynamics of a two-level system. We have extensively used the Bloch vector representation, which has allowed us to visualize the system's state and its dynamics in an intuitive manner. In this chapter we generalize the results to systems with more levels. Unfortunately, it turns out that we cannot easily extend the Bloch vector formalism to a higher number of states, and we thus have to resort to different description schemes. However, the modifications are not overly complicated and the tools we will develop are quite general and can be used under diverse circumstances.

17.1 Density Operator

Let $|\psi\rangle$ be a state of some system and $\hat{A}$ a Hermitian operator acting in the system's space. The expectation value of the operator

$$\langle\hat{A}\rangle = \langle\psi|\hat{A}|\psi\rangle$$

is an observable which provides information about the outcome of an experiment. Suppose that we perform the experiments for an ensemble of almost identical systems, or alternatively many times for the same system. Due to uncontrollable elements in the experiment, most likely caused by interactions with the environment, the state of the system $|\psi^{(\mu)}\rangle$ is only available with some probability p_μ. The expectation value

$$\langle\hat{A}\rangle = \sum_\mu p_\mu \left\langle\psi^{(\mu)}\middle|\hat{A}\middle|\psi^{(\mu)}\right\rangle \tag{17.1}$$

U. Hohenester, *Nano and Quantum Optics*, Graduate Texts in Physics,
https://doi.org/10.1007/978-3-030-30504-8_17

then contains two kinds of averages, one for the quantum-mechanical operator and one due to our uncertainty of the state of the system. The formalism we will develop next is made such that we can deal with these averaging procedures on the same footing. As a preliminary step, we consider a complete basis

$$\sum_i |i\rangle\langle i| = \mathbb{1},$$

where i labels the different states of the basis. The trace of an operator is then defined as

$$\mathrm{tr}\{\hat{A}\} = \sum_i \langle i|\hat{A}|i\rangle.$$

When dealing with the trace over operator products one can cyclically permute the operators under the trace

$$\mathrm{tr}\{\hat{A}_1\hat{A}_2\ldots\hat{A}_n\} = \mathrm{tr}\{\hat{A}_n\hat{A}_1\ldots\hat{A}_{n-1}\}, \tag{17.2}$$

as explicitly worked out in Exercise 17.1. We next expand in Eq. (17.1) the states $|\psi^{(\mu)}\rangle$ in the basis $|i\rangle$,

$$\begin{aligned}\langle\hat{A}\rangle &= \sum_\mu p_\mu \sum_{ij} \langle\psi^{(\mu)}|i\rangle\langle i|\hat{A}|j\rangle\langle j|\psi^{(\mu)}\rangle \\ &= \sum_\mu p_\mu \sum_{ij} \langle j|\psi^{(\mu)}\rangle\langle\psi^{(\mu)}|i\rangle\langle i|\hat{A}|j\rangle,\end{aligned}$$

where in the last expression we have exchanged the order of the different terms. To deal with such expectation values, it turns out to be convenient to introduce the density operator

Density Operator

$$\hat{\rho} = \sum_\mu p_\mu\, |\psi^{(\mu)}\rangle\langle\psi^{(\mu)}|. \tag{17.3}$$

With this operator the expectation value of an operator can be simply expressed as

$$\langle\hat{A}\rangle = \mathrm{tr}\{\hat{\rho}\hat{A}\}, \tag{17.4}$$

as can be easily verified by expanding the trace in terms of a complete basis. The above definition extends that of Eq. (14.5) for systems in thermal equilibrium. The density operator is constructed such that it provides two kinds of averaging.

Quantum Mechanical Average. A quantum mechanical averaging which has to be interpreted in the usual sense that for a pure state $|\psi^{(\mu)}\rangle$ a single measurement gives with some probability one of the eigenvalues of the operator $\hat{A}$, and a number of measurements is needed to obtain the mean value.

Statistical Average. When the state of the system is not known with certainty before the measurement, as is usually the case for systems interacting with their environment, one has to perform an additional averaging to account for the probabilities p_μ of the system being in state $|\psi^{(\mu)}\rangle$.

The density operator $\hat{\rho}$ fulfills a few important relations. We first evaluate the trace over $\hat{\rho}$ in terms of the complete basis

$$\mathrm{tr}\left\{\hat{\rho}\right\} = \sum_\mu \sum_i p_\mu \langle i|\psi^{(\mu)}\rangle\langle\psi^{(\mu)}|i\rangle = \sum_\mu \sum_i p_\mu \langle\psi^{(\mu)}|i\rangle\langle i|\psi^{(\mu)}\rangle.$$

Removing again the basis we find

$$\mathrm{tr}\left\{\hat{\rho}\right\} = \sum_\mu p_\mu \langle\psi^{(\mu)}|\psi^{(\mu)}\rangle = \sum_\mu p_\mu = 1,$$

where we have used that the wavefunctions are normalized and the sum over all probabilities has to be one. In principle, a density operator can be also computed for a pure state

$$\hat{\rho}_{\text{pure}} = |\psi\rangle\langle\psi|. \tag{17.5}$$

For pure states the density operator becomes a projector with

$$\hat{\rho}^2_{\text{pure}} = |\psi\rangle\langle\psi|\psi\rangle\langle\psi| = \hat{\rho}_{\text{pure}}.$$

We are thus led to the following properties that any density operator must fulfill

Properties of Density Operator

$$\mathrm{tr}\{\hat{\rho}\} = 1, \quad \mathrm{tr}\{\hat{\rho}^2\} \le 1, \tag{17.6}$$

where the equal sign of the last expression holds for pure states only.

Example of Two-Level System

The density operator of a two-level system can be written in the form

$$\hat{\rho} = \frac{1}{2}\big(\mathbb{1} + \boldsymbol{u}\cdot\boldsymbol{\sigma}\big). \tag{17.7}$$

To get the coefficient for the $\mathbb{1}$ term we note that $\mathrm{tr}\{\mathbb{1}\} = 2$. In contrast, the trace of the Pauli matrices is zero. The different components of the Bloch vector $\boldsymbol{u}$ can then be obtained from

$$u_k = \mathrm{tr}\{\hat{\rho}\sigma_k\} = \frac{1}{2}\mathrm{tr}\Big\{\sigma_k + \sum_j u_j\sigma_j\sigma_k\Big\} = \frac{1}{2}\mathrm{tr}\Big\{\sum_j u_j\Big(\delta_{jk}\mathbb{1} + i\epsilon_{jkl}\sigma_\ell\Big)\Big\},$$

and using again the trace properties of $\mathbb{1}$ and the Pauli matrices. The purity statement relates to

$$\mathrm{tr}\left\{\hat{\rho}^2\right\} = \frac{1}{4}\mathrm{tr}\Big\{\mathbb{1} + u_iu_j\big(\delta_{ij}\mathbb{1} + i\epsilon_{ijk}\sigma_k\big) + 2\boldsymbol{u}\cdot\boldsymbol{\sigma}\Big\} \le 1,$$

where we have used the result of Eq. (17.7). If we use that the trace of $\mathbb{1}$ and σ_k are two and zero, respectively, we find

$$\boldsymbol{u}\cdot\boldsymbol{u} \le 1. \tag{17.8}$$

Indeed, in the last section we have found that for a coherent time evolution the length of the Bloch vector is one and $\boldsymbol{u}$ is located *on* the Bloch sphere. Upon consideration of the T_1 and T_2 relaxation and dephasing times, the length of the Bloch vector becomes reduced an $\boldsymbol{u}$ is located *inside* the Bloch sphere. This is a direct consequence of the purity statement $\mathrm{tr}\,\hat{\rho}^2 \le 1$.

17.1.1 Time Evolution of Density Operator

The coherent time evolution of the density operator can be obtained by computing its time derivative using Schrödinger's equation for the bra and ket,

$$i\hbar\frac{d\hat{\rho}}{dt} = \sum_\mu p_\mu\left(\hat{H}(t)|\psi^{(\mu)}\rangle\langle\psi^{(\mu)}| - |\psi^{(\mu)}\rangle\langle\psi^{(\mu)}|\hat{H}(t)\right),$$

where the bra derivative acquires a negative sign. We next pull out the Hamilton from the sum to arrive at the von-Neumann equation, which gives the time derivative of the density operator according to

von-Neumann Equation for Time Evolution of $\hat{\rho}$

$$i\hbar\frac{d\hat{\rho}}{dt} = \left[\hat{H}(t), \hat{\rho}\right]. \tag{17.9}$$

The von-Neumann equation has to be used whenever dealing with density operators, and is the analog of Schrödinger's equation for states. We will discuss further below that Eq. (17.9) has to be modified to account for incoherent environment couplings, in close analogy to the dephasing and relaxation terms introduced in the Bloch equations of Eqs. (16.26) and (16.28). The von-Neumann equation in the above form can be easily solved using the time evolution operator of Eq. (13.5), and we get

$$\hat{\rho}(t) = \hat{U}(t,0)\,\hat{\rho}_0\,\hat{U}^\dagger(t,0), \tag{17.10}$$

where $\hat{\rho}_0$ is the density operator at time zero.

Example of Two-Level System

The time evolution of the Bloch vector can be obtained from

$$i\hbar\dot{u}_k = \text{tr}\left\{i\hbar\frac{d\hat{\rho}}{dt}\sigma_k\right\} = \text{tr}\left\{\left[\hat{H}(t), \hat{\rho}\right]\sigma_k\right\}.$$

We can perform a cyclic permutation under the trace to swap the commutator from the density operator to σ_k through

$$i\hbar\dot{u}_k = \text{tr}\left\{\left(\hat{H}\hat{\rho} - \hat{\rho}\hat{H}\right)\sigma_k\right\} = \text{tr}\left\{\hat{\rho}\left(\sigma_k\hat{H} - \hat{H}\sigma_k\right)\right\} = \left\langle\left[\sigma_k, \hat{H}\right]\right\rangle.$$

Comparison with Eq. (16.19) shows that this is precisely the expression previously evaluated for the two-level system in order to derive the time evolution of the Bloch vector. The above derivation, however, is slightly more general because it not only applies to pure states but also to mixtures, where an additional ensemble average is needed.

Optically Driven Few-Level System

The description of an optically driven few-level system is very similar to the case of a two-level system, however, we have to be somewhat more careful with the notation which slightly complicates our discussion without adding too much novel physics. We consider a few-level system with a basis $|i\rangle$ and a Hamiltonian of the form

$$\hat{H}(t) = \sum_i E_i \, |i\rangle\langle i| + \hat{H}_{\rm op}(t), \tag{17.11}$$

where the first term accounts for the energies E_i. The light–matter interaction in the dipole approximation reads

$$\hat{H}_{\rm op}(t) = -\sum_{ij} \Big[\big\langle i \big| q\boldsymbol{r} \big| j \big\rangle \cdot \boldsymbol{E}(t) \Big] |i\rangle\langle j| .$$

In the following we split the electric field $\boldsymbol{E} = \boldsymbol{E}^+ + \boldsymbol{E}^-$ again into contributions oscillating with positive and negative frequencies, respectively. Similarly, we introduce a transition operator with the following properties

$$\sigma_{ij} = |i\rangle\langle j| = \begin{cases} \sigma_{ij}^+ & \text{for } E_i < E_j \\ \sigma_{ij}^- & \text{for } E_i > E_j . \end{cases} \tag{17.12}$$

Using an interaction picture, $\sigma_{ij}^\pm$ then oscillate with positive and negative frequencies, respectively,[1] as discussed, for instance, in Sect. 16.2.1. In the rotating wave approximation, previously discussed in Eq. (16.15) for a two-level system, we keep solely terms varying slowly in time

$$\hat{H}_{\rm op}(t) \approx -\sum_{ij}{}' \Big(\hbar\Omega_{ij}(t)\, \sigma_{ij}^- + \hbar\Omega_{ji}^*(t)\, \sigma_{ji}^+ \Big) . \tag{17.13}$$

The prime in the sum is a reminder that only pairs with $E_i < E_j$ have to be considered, and we have introduced the (complex-valued) Rabi energies

$$\hbar\Omega_{ij}(t) = \big\langle i \big| q\boldsymbol{r} \big| j \big\rangle \cdot \boldsymbol{E}^+(t) .$$

We next consider time-harmonic electromagnetic fields oscillating with frequency ω. Our main assumption for the following discussion is that $\hbar\omega$ provides a characteristic energy scale for the system, such that the excited states have energies of the order $E_i \approx n_i\, \hbar\omega$, where n_i is an integer number, and optical transitions only connect states that differ by a single quantum $\hbar\omega$. Quantum dots with their excitonic and biexcitonic states are typical examples, see Fig. 17.1. If the above assumption does not apply we have to keep the full time dependence in the light-matter coupling of Eq. (17.13). Otherwise we split the total Hamiltonian into the contributions

[1] Note that with this definition of positive and negative frequency components $\sigma^\pm$ de-excite and excite the system, respectively, contrary to the Pauli matrices $\sigma_\pm$ for excitation and de-excitation of the two-level system.

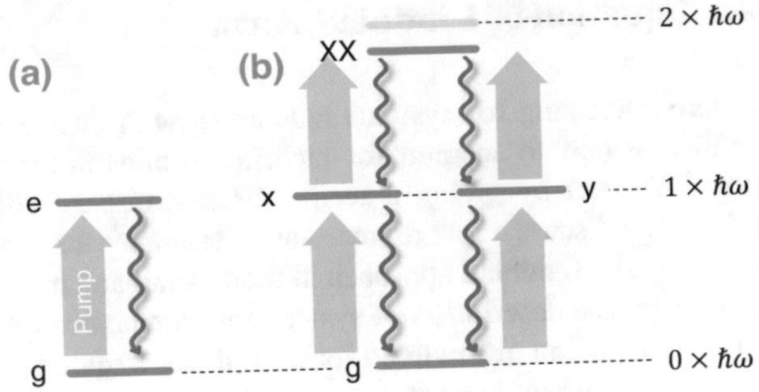

Fig. 17.1 Typical few-level systems. (**a**) Two-level system with ground and excited states g and e, which are separated by an energy distance of approximately $\hbar\omega$. (**b**) Quantum-dot level scheme with groundstate g, single-exciton states x, y, and biexciton state XX, whose energy is reduced with respect to twice the exciton energy by the so-called biexciton binding energy. For sufficiently weak excitation fields one can only add single excitations to the system, and there is no direct transition from the groundstate to XX

$$\hat{H}(t) = \sum_i n_i \hbar\omega\sigma_{ii} + \left[\sum_i \hbar\Delta_i\sigma_{ii} - \sum_{ij}{}' \left(\hbar\Omega_{ij}(t)\sigma_{ij}^- + \hbar\Omega_{ji}^*(t)\sigma_{ji}^+\right)\right], \tag{17.14}$$

where we have introduced the detunings

$$\hbar\Delta_i = E_i - n_i\,\hbar\omega.$$

We next employ an interaction representation with respect to the first term in Eq. (17.14). In the same manner as previously discussed for the two-level system, this interaction representation removes the time-dependent factors in the Rabi energies and we end up with a time-independent Hamiltonian for a harmonically driven few-level system

Hamiltonian for Harmonically Driven Few-Level System

$$\hat{V}_I = \sum_i \hbar\Delta_i\sigma_{ii} - \sum_{ij}{}' \left(\hbar\Omega_{ij}\,\sigma_{ij}^- + \hbar\Omega_{ji}^*\,\sigma_{ji}^+\right). \tag{17.15}$$

17.2 Master Equation of Lindblad Form

Life becomes more interesting for systems interacting with their environment. In this section we discuss how to account for such interactions in the framework of density operators. We start by giving a recipe for environment couplings using Lindblad operators, and discuss a few representative examples. In later parts of this chapter we will justify the Lindblad approach in more detail and make contact with microscopic models for the description of system-environment couplings. Without much ado, the master equation in Lindblad form for the description of the coherent and incoherent dynamics of few-level systems has the form

Master Equation of Lindblad Form

$$\frac{d\hat{\rho}}{dt} = -\frac{i}{\hbar}\left(\hat{H}_{\text{eff}}\hat{\rho}(t) - \hat{\rho}\hat{H}_{\text{eff}}^{\dagger}(t)\right) + \sum_k \hat{L}_k\hat{\rho}\hat{L}_k^{\dagger}. \tag{17.16}$$

Here we have introduced the Lindblad operators $\hat{L}_k$ and the effective Hamiltonian

$$\hat{H}_{\text{eff}}(t) = \hat{H}(t) - \frac{i\hbar}{2}\sum_k \hat{L}_k^{\dagger}\hat{L}_k,$$

which differs from the usual Hamiltonian in that it is a non-Hermitian operator, and thus the time evolution due to $\hat{H}_{\text{eff}}$ alone would not preserve the norm of the wavefunction. The Lindblad operators $\hat{L}_k$ are typically of the form

$$\hat{L}_k = \sqrt{\Gamma_k}\,\sigma_{fi}$$

and describe the scattering from an initial state i to a final state f, with the scattering rate Γ_k. See Fig. 17.2 for a simple example. Equation (17.16) can be rewritten in the form

$$\frac{d\hat{\rho}}{dt} = -\frac{i}{\hbar}\left[\hat{H},\hat{\rho}\right] + \sum_k \left\{\hat{L}_k\hat{\rho}\hat{L}_k^{\dagger} - \frac{1}{2}\left(\hat{L}_k^{\dagger}\hat{L}_k\hat{\rho} + \hat{\rho}\hat{L}_k^{\dagger}\hat{L}_k\right)\right\}. \tag{17.17}$$

By taking the trace on both sides of the equation, we find upon cyclic permutations that the term on the right-hand side becomes zero. Thus, the master equation of Lindblad form has the important property that it preserves the trace of the density operator. The Lindblad equation is the most general form of a master equation with a Markovian time evolution and trace preservation.

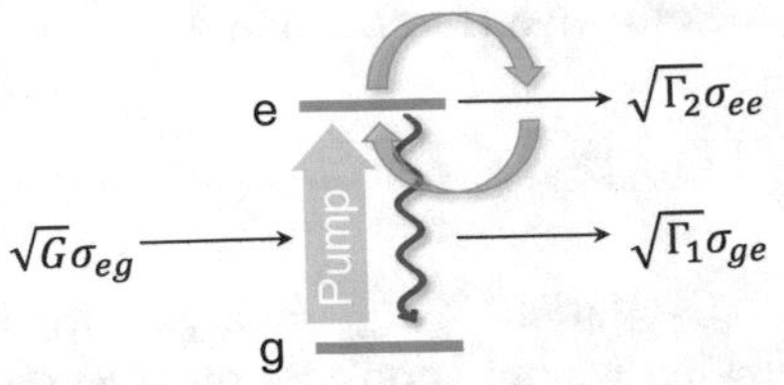

Fig. 17.2 Example for Lindblad operators to describe the incoherent dynamics of a two-level system. G is the generation rate for the incoherent pumping from the ground to the excited state, alternatively one can also use a coherent driving field in accordance to the discussion of the previous section. Γ_1 is the radiative decay rate from the excited to the groundstate, and Γ_2 is the dephasing rate for the excited state

Markovian. In a Markovian time evolution the system's dynamics can be described in terms of $\hat{\rho}(t)$ at a single time solely. As we will discuss below, for a system interacting with its environment the dynamics depends in general on the past of the system when the environmental degrees of freedom are traced out from the theory, so the Markovian time evolution is only an approximation.

Trace Preservation. The solutions of the master equation must be such that the trace of $\hat{\rho}$ remains unity and all diagonal elements are positive (the operator is thus positive definite), in accordance to the requirements of quantum mechanics and statistical physics that interpret the diagonal elements as probabilities for finding the system in the corresponding states.

Example of Two-Level System

We next apply the Lindblad formalism to the two-level system of the previous chapter, see also Fig. 17.2. From here on it turns out convenient to work in the state basis $|g\rangle$, $|e\rangle$, and to change to the Bloch vector description only at the end of the calculation.

Relaxation. Relaxation is a transition from the excited to the groundstate with rate Γ_1,

$$\hat{L}_1 = \sqrt{\Gamma_1}\,\sigma_{ge}. \tag{17.18}$$

Note that, as always in quantum mechanics, operators should be read from the right to the left, so that σ_{ge} describes the transition from $e \to g$.

Pure Dephasing. Pure dephasing only affects the excited state with rate Γ_2, and scatters from $e \to e$

$$\hat{L}_2 = \sqrt{\Gamma_2}\,\sigma_{ee}. \tag{17.19}$$

Nevertheless, as we will show below, this operator has an effect on the density operator and leads to the dephasing previously discussed for the two-level system.

We start by computing the effective Hamiltonian

$$\hat{H}_{\text{eff}} = \hat{H}_0 - \frac{i\hbar}{2}\Big(\Gamma_1 \sigma_{eg}\sigma_{ge} + \Gamma_2 \sigma_{ee}\sigma_{ee}\Big) = \hat{H}_0 - \frac{i\hbar}{2}(\Gamma_1 + \Gamma_2)\sigma_{ee}.$$

We have used the obvious relation $\sigma_{ij}\sigma_{kl} = \delta_{jk}\sigma_{il}$ for the transition operators $\sigma_{ij} = |i\rangle\langle j|$. In the following we only consider the incoherent part of the effective Hamiltonian to express in Eq. (17.17) the terms in brackets for the two-level system as

$$\left(\frac{d\hat{\rho}}{dt}\right)_{\text{incoh}} = \Gamma_1\Big(\sigma_{ge}\,\hat{\rho}\,\sigma_{eg}\Big) + \Gamma_2\Big(\sigma_{ee}\hat{\rho}\,\sigma_{ee}\Big) - \frac{1}{2}(\Gamma_1 + \Gamma_2)\,\Big(\sigma_{ee}\,\hat{\rho} + \hat{\rho}\,\sigma_{ee}\Big).$$

This operator equation can be brought to matrix form by projecting it from the left- and right-hand side on the basis states,

$$\begin{aligned}
\big(\dot{\rho}_{ee}\big)_{\text{incoh}} &= \Big\langle e\Big|\left(\frac{d\hat{\rho}}{dt}\right)_{\text{incoh}}\Big|e\Big\rangle = \Gamma_2\,\rho_{ee} - (\Gamma_1 + \Gamma_2)\,\rho_{ee} \\
\big(\dot{\rho}_{gg}\big)_{\text{incoh}} &= \Big\langle g\Big|\left(\frac{d\hat{\rho}}{dt}\right)_{\text{incoh}}\Big|g\Big\rangle = \Gamma_1\,\rho_{ee} \\
\big(\dot{\rho}_{ge}\big)_{\text{incoh}} &= \Big\langle g\Big|\left(\frac{d\hat{\rho}}{dt}\right)_{\text{incoh}}\Big|e\Big\rangle = -\frac{1}{2}\,(\Gamma_1 + \Gamma_2)\,\rho_{ge}.
\end{aligned}$$

We finally use the relation

$$\boldsymbol{u} = 2\,\text{Re}\big\{\rho_{ge}\big\}\,\hat{\boldsymbol{x}} + 2\,\text{Im}\big\{\rho_{ge}\big\}\,\hat{\boldsymbol{y}} + \big(\rho_{ee} - \rho_{gg}\big)\,\hat{\boldsymbol{z}}$$

between the density matrix elements and the Bloch vector to obtain for the incoherent part of the time evolution

$$\begin{aligned}
\big(\dot{u}_1\big)_{\text{incoh}} &= 2\,\text{Re}\Big\{\big(\dot{\rho}_{ge}\big)_{\text{incoh}}\Big\} = -\frac{1}{2}\,(\Gamma_1 + \Gamma_2)\,u_1 \\
\big(\dot{u}_2\big)_{\text{incoh}} &= 2\,\text{Im}\Big\{\big(\dot{\rho}_{ge}\big)_{\text{incoh}}\Big\} = -\frac{1}{2}\,(\Gamma_1 + \Gamma_2)\,u_2 \\
\big(\dot{u}_3\big)_{\text{incoh}} &= \big(\dot{\rho}_{ee} - \dot{\rho}_{gg}\big)_{\text{incoh}} = -2\Gamma_1\rho_{ee}.
\end{aligned} \tag{17.20}$$

The last term can be rewritten using the trace relation $\rho_{gg} + \rho_{ee} = 1$ to give

$$2\rho_{ee} = \rho_{ee} + \big(1 - \rho_{gg}\big) = u_3 + 1.$$

Comparison with Eq. (16.28) shows that the two results are identical when the following assignments are made:

$$\frac{1}{T_1} = \Gamma_1, \quad \frac{2}{T_2} = \Gamma_1 + \Gamma_2. \tag{17.21}$$

Thus, the relaxation and dephasing time description for a two-level system can be derived within a Lindblad approach for the master equation.

17.3 Solving the Master Equation of Lindblad Form

The central equation of this chapter is the master equation, which can be written in compact form

$$i\hbar \frac{d\hat{\rho}}{dt} = \mathbb{L}(t)\, \hat{\rho}. \tag{17.22}$$

Here $\mathbb{L}$ is the Liouville operator, sometimes also called a super-operator because it acts on an operator itself. For the von-Neumann equation $\mathbb{L}$ is defined through

$$\mathbb{L}(t)\hat{\rho} = \left[\hat{H}(t), \hat{\rho}\right]. \tag{17.23}$$

For the master equation of Lindblad form, $\mathbb{L}$ has to be modified such that it gives the right-hand side of Eq. (17.16). Expanding $\mathbb{L}$ and the density operator in some complete basis allows us to rewrite the master equation in the form

$$i\hbar \dot{\rho}_{ij} = \mathbb{L}_{ij,kl}\, \rho_{kl}. \tag{17.24}$$

From this expression it is obvious that the solution of the equation is not particularly more complicated than that of say the Schrödinger's equation. Numerically one often relates the pairs ij of indices to a single super-index $[\![ij]\!]$, which can be done in terms of simple book keeping. Then, the master equation

$$i\hbar \dot{\rho}_{[ij]} = \mathbb{L}_{[\![ij]\!][\![kl]\!]}\, \rho_{[\![kl]\!]}$$

becomes a vector equation with $\mathbb{L}$ being a matrix. We next briefly discuss possible solution schemes for the Liouville equation of Eq. (17.22).

Differential Equation. Equation (17.22) is a time-dependent differential equation, which, in many cases, can be solved directly in the time domain. In particular when using the rotating wave approximation, the fast time dependencies have been removed from the Hamiltonian and the system of differential equations can be solved efficiently on a computer, using, for instance, a Runge–Kutta scheme [8].

Eigenmode Decomposition. For time-harmonic fields one can obtain the Hamiltonian of Eq. (17.15) in the interaction picture, which does not depend on time. In this case one can seek for the eigenmodes

$$\mathbb{L}\,\mathbb{X} = \mathbb{X}\,\Lambda, \quad \tilde{\mathbb{X}}\,\mathbb{L} = \Lambda\,\tilde{\mathbb{X}}. \tag{17.25}$$

In general $\mathbb{L}$ is neither a symmetric nor Hermitian matrix, so we must compute both the left and right matrix of eigenvectors $\tilde{\mathbb{X}}$, $\mathbb{X}$, respectively. Λ is a matrix with the eigenvalues on its diagonals. As shown in Exercise 17.2, the left and right eigenvectors form a biorthogonal set with

$$\tilde{\mathbb{X}}\mathbb{X} = \mathbb{X}\tilde{\mathbb{X}} = \mathbb{1},$$

where $\mathbb{1}$ is the unit matrix. The Liouville equation of Eq. (17.22) can then be solved in the form

$$\exp\left(-\frac{i}{\hbar}\mathbb{L}t\right)\mathbb{X}\tilde{\mathbb{X}} = \mathbb{X}\exp\left(-\frac{i}{\hbar}\Lambda t\right)\tilde{\mathbb{X}}.$$

Because Λ is a diagonal matrix, the exponential function can be easily evaluated. This allows us to solve the master equation for the density operator in terms of its eigenmodes through

Eigenmode Solution of Master Equation

$$\rho_{ij}(t) = \left[\mathbb{X}\exp\left(-\frac{i}{\hbar}\Lambda t\right)\tilde{\mathbb{X}}\right]_{ij,kl}\rho_{kl}(0). \tag{17.26}$$

In case of an optically driven system subject to dephasing and relaxation, the above expression even allows to obtain the steady state of the system starting from the equilibrium groundstate through

$$\rho_{ij}^{\infty} = \lim_{t\to\infty}\left[\mathbb{X}\exp\left(-\frac{i}{\hbar}\Lambda t\right)\tilde{\mathbb{X}}\right]_{ij,kl}\rho_{kl}^{\mathrm{eq}}.$$

We will return to this expression in the next chapter when discussing photon correlation measurements.

The master equation of Lindblad form can be also solved by a stochastic method, which is known as the "unraveling of the master equation" [186] and has found widespread use in many fields of research [187]. It consists of two parts, a solution of the Schrödinger's equation for the non-Hermitian Hamiltonian $\hat{H}_{\mathrm{eff}}$ and a stochastic part, associated with the scattering term of the master equation, where the wavefunction "jumps" to one of the possible scattering state.

Stochastic Unraveling The density operator is a statistical mixture of state vectors

$$\hat{\rho} = \sum_{\mu} p_{\mu} |\psi^{(\mu)}\rangle\langle\psi^{(\mu)}|,$$

where the summation over μ results from the statistical average over the various pure states. For simplicity, in the following we restrict ourselves to a single state vector $|\psi\rangle$. The general case can be obtained in a similar fashion. Inserting the projector $|\psi\rangle\langle\psi|$ into the master equation of Lindblad form, Eq. (17.16), we get

$$\frac{d}{dt}|\psi\rangle\langle\psi| = -\frac{i}{\hbar}\left(\hat{H}_{\text{eff}}(t)\,|\psi\rangle\langle\psi| - |\psi\rangle\langle\psi|\,\hat{H}^{\dagger}_{\text{eff}}(t)\right) + \sum_k \hat{L}_k\,|\psi\rangle\langle\psi|\,\hat{L}^{\dagger}_k. \tag{17.27}$$

The first term on the right-hand side can be interpreted as a non-Hermitian, Schrödinger-like evolution under the influence of $\hat{H}_{\text{eff}}$, which can be solved using the (non-unitary) time evolution operator defined through

$$i\hbar\frac{d}{dt}\hat{U}_{\text{eff}}(t,t') = \hat{H}_{\text{eff}}(t)\hat{U}_{\text{eff}}(t,t'), \quad \hat{U}_{\text{eff}}(t,t) = \mathbb{1}. \tag{17.28}$$

In contrast, the second term describes a time evolution where $|\psi\rangle$ is projected—or jumps—to one of the possible states $\hat{L}_k|\psi\rangle$. For sufficiently small time intervals δt the time evolution according to $\hat{H}_{\text{eff}}$ is given by

$$\big|\psi(t+\delta t)\big\rangle = \hat{U}_{\text{eff}}(t+\delta t,t)\big|\psi(t)\big\rangle \approx \left(1 - \frac{i}{\hbar}\hat{H}_{\text{eff}}\,\delta t\right)\big|\psi(t)\big\rangle.$$

Note that $\hat{H}_{\text{eff}}$ is non-Hermitian and consequently the wavefunction at later time is not normalized. To the lowest order in δt the decrease of norm δp is given by

$$\delta p = \frac{i\delta t}{\hbar}\left\langle\psi(t)\middle|\hat{H}_{\text{eff}}(t) - H^{\dagger}_{\text{eff}}(t)\middle|\psi(t)\right\rangle = \delta t\sum_k\left\langle\psi(t)\middle|\hat{L}^{\dagger}_k\hat{L}_k\middle|\psi(t)\right\rangle = \sum_k \delta p_k. \tag{17.29}$$

The full master equation evolution has to preserve the norm. This missing norm δp is brought in by the states $\hat{L}_k|\psi(t)\rangle$ to which the system is scattered with probability δp_k. The time evolution of the density operator can thus be decomposed into

$$\hat{\rho}(t+\delta t) \approx \left[\hat{U}_{\text{eff}}(t+\delta t,t)\,\hat{\rho}\,\hat{U}^{\dagger}_{\text{eff}}(t+\delta t,t)\right] + \delta t\left[\sum_k \hat{L}_k\hat{\rho}\hat{L}^{\dagger}_k\right]. \tag{17.30}$$

The terms in brackets on the right-hand side are often referred to as conditional density operators [186]. The first term describes the propagation of the system in presence of a coherent time evolution and out-scatterings, where the system is scattered away from its current state, and the second one for in-scatterings that are needed to preserve the trace of $\hat{\rho}$. This stochastic solution has the advantage that we have to consider wavefunctions only, which depend on a single state index, contrary to the density matrices which depend on two state indices. For a more detailed discussion the interested readers are referred to the literature [187].

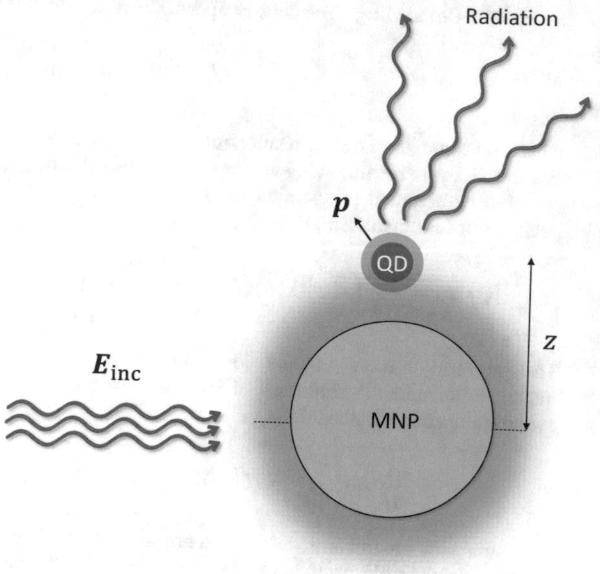

Fig. 17.3 Quantum dot (QD) coupled to metallic nanoparticle (MNP). The system is excited by an incoming laser field, where both the excitation and relaxation processes of the quantum dot are modified by the presence of the metallic nanoparticle

17.3.1 Quantum Dot Coupled to Metallic Nanosphere

The master equation of Lindblad form can be combined with non-trivial photonic environments, as discussed in previous chapters. We here follow the work of Artuso and Bryant [188] and consider a quantum dot coupled to a metallic nanosphere, as schematically shown in Fig. 17.3. The quantum dot is treated as a two-level system with a dipole moment $\boldsymbol{p}$, so we can adopt most of the results derived for generic two-level systems. The light-matter coupling is described by the Rabi energy

$$\hbar\Omega = \boldsymbol{p}\cdot\Big(\boldsymbol{E}^{+}_{\rm inc}(\boldsymbol{r}_{\rm dot}) + \boldsymbol{E}^{+}_{\rm refl}(\boldsymbol{r}_{\rm dot})\Big), \tag{17.31}$$

where $\boldsymbol{E}^{+}_{\rm inc}$, $\boldsymbol{E}^{+}_{\rm refl}$ account for the incoming and reflected light at the position $\boldsymbol{r}_{\rm dot}$ of the quantum dot, respectively. Before pondering on the solution of the master equation for the coupled dot-nanosphere system, let us analyze the problem from a more general perspective.

Photonic Environment. In what follows we consider a *linear* response for the metallic nanoparticle, which we describe through a reflected Green's function or some approximation of it. The dot dynamics will be treated in a *nonlinear* fashion by using the master equation of Lindblad form. Note that this separation into a linear photonic environment and a nonlinear dynamics for the quantum emitter located inside this photonic environment is of general nature, and can be immediately applied to other setups. The total Green's function properly accounts for the propagation of the electromagnetic fields towards and away from the quantum emitter.

Coherent Dynamics. As schematically shown in Fig. 17.3 and explicitly stated in Eq. (17.31), the excitation of the quantum dot is modified by the presence of the metallic nanoparticle. Thus, the polarized nanosphere modifies the excitation channel for the quantum dot, and conversely the transition dipole moment of the quantum dot modifies the polarization of the nanosphere. This leads to a coupling that must be accounted for in a self-consistent manner.

Incoherent Dynamics. Finally, the scattering and dephasing channels of the quantum dot may become modified within a non-trivial photonic environment, as has been discussed in Sect. 15.4 for photon emission and absorption. This leads to modified scattering and dephasing rates Γ_1, Γ_2, which enter the Lindblad operators of the master equation approach.

In the following we introduce in accordance to Ref. [188] a simplified description scheme where the electromagnetic response of the quantum dot and nanoparticle is modelled in terms of point dipoles $\boldsymbol{P}_{\rm dot}$, $\boldsymbol{P}_{\rm sph}$, which are separated by a distance z. The induced dipole of the quantum dot can be expressed as

$$\boldsymbol{P}_{\rm dot} = \boldsymbol{p}\,\rho_{ge}(t).$$

For the induced dipole of the sphere we find

$$\boldsymbol{P}_{\rm sph} \approx \left[4\pi\varepsilon_b\left(\frac{\varepsilon_b-\varepsilon_{\rm sph}}{2\varepsilon_b+\varepsilon_{\rm sph}}\right)a^3\right]\left(\boldsymbol{E}_{\rm inc}+\frac{1}{4\pi\varepsilon_b}\frac{3(\boldsymbol{P}_{\rm dot}\cdot\hat{\boldsymbol{z}})\hat{\boldsymbol{z}}-\boldsymbol{P}_{\rm dot}}{z^3}\right),$$

where the term in brackets is the polarizability $\alpha_{\rm sph}$ of the sphere with radius a, Eq. (9.9), and the term in parentheses is the total electric field at the sphere origin. $\varepsilon_{\rm sph}, \varepsilon_b$ are the permittivities of the metallic nanosphere and the embedding medium, respectively. Similarly, we get for the total electric field at the dot position

$$\boldsymbol{E}(\boldsymbol{r}_{\rm dot}) \approx \boldsymbol{E}_{\rm inc}+\frac{1}{4\pi\varepsilon_b}\left(\frac{3(\boldsymbol{P}_{\rm sph}\cdot\hat{\boldsymbol{z}})\hat{\boldsymbol{z}}-\boldsymbol{P}_{\rm sph}}{z^3}\right).$$

With the Hamiltonian of Eq. (17.15)

$$\hat{V} = \hbar\Delta\,\sigma_{ee} - \left(\hbar\Omega\,\sigma_{eg} + \hbar\Omega^*\,\sigma_{ge}\right) \tag{17.32}$$

and the Lindblad operators for relaxation and dephasing, we then obtain the equations of motion for the density matrix elements

$$\begin{aligned}
\dot\rho_{ee} &= -2\,\mathrm{Im}\left(\Omega^*\rho_{ge}\right) - \Gamma_1\rho_{ee}\\
\dot\rho_{gg} &= +2\,\mathrm{Im}\left(\Omega^*\rho_{ge}\right) + \Gamma_1\rho_{ee}\\
\dot\rho_{ge} &= -i\Delta\,\rho_{ge} - \frac{1}{2}(\Gamma_1+\Gamma_2)\rho_{ge} + i\Omega^*\left(\rho_{ee}-\rho_{gg}\right),
\end{aligned} \tag{17.33}$$

which are sometimes denoted as the optical Bloch equations. One of the first two equations is redundant because of the constraint $\rho_{gg} + \rho_{ee} = 1$ that must be fulfilled at all times.

The driving field Ω has two contributions associated with direct excitation, which becomes modified in presence of the nanosphere, and a self-interaction. In the latter case the dot dipole polarizes the nanosphere, and the induced sphere dipole acts back on the dot dipole. This self-interaction leads to a complicated dot dynamics, which has been investigated in some detail in Ref. [188] where the authors showed that depending on the excitation and coupling parameters the system exhibits a variety of different physical phenomena, such as Fano lineshapes or exciton-induced transparency, see Fig. 17.4 for selected examples.

17.3.2 *Lasers and Spasers*

The laser—the acronym stands for Light Amplification by Stimulated Emission of Radiation—has revolutionized the field of optics [189]. There exist tons of literature on lasers, and it would be presumptuous to assume that we can add something overly useful to the subject here. Yet, with the tools developed so far we are in the position to understand the basic theory of the laser, so why not taking a moment to show how to proceed in principal.

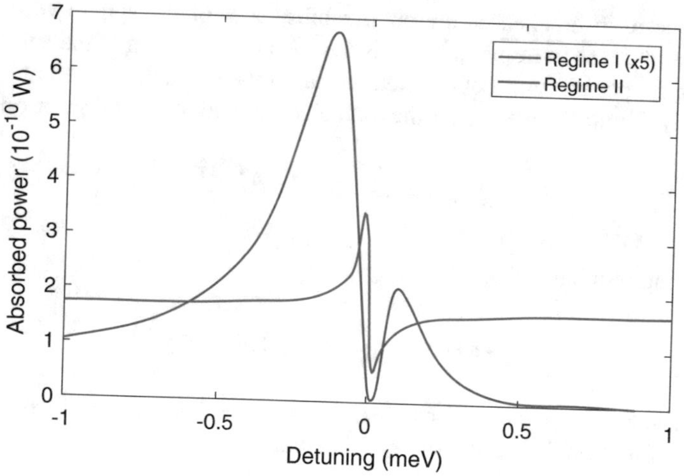

Fig. 17.4 Absorbed power of a driven quantum dot coupled to a metallic nanosphere, as a function of detuning ($\hbar\omega = 2.5\,\text{eV}$). We take the same simulation parameters as in Ref. [188] representative for a gold nanosphere ($a = 13\,\text{nm}$), finding a Fano lineshape (Regime I) for the smaller quantum dot dipole moment of $p = 0.25\,e\,\text{nm}$ and an exciton-induced transparency (Regime II) for the larger dipole moment $p = 2\,e\,\text{nm}$. The results of Regime I have been scaled by a factor of five for better visibility

We start with the optical Bloch equations of Eq. (17.33) and assume that Ω is associated with the external driving field only, with no self-interaction modifications, as would be the case for an isolated quantum emitter (from here on denoted as an "atom"). In the stationary state the time derivatives in Eq. (17.33) have to be set zero, and we obtain three linear equations for the unknowns ρ'_{ge}, ρ''_{ge}, ρ_{ee}, which can be solved to give

$$\rho_{ee}^{\infty} = \frac{\Gamma\Omega^2}{\Delta^2\Gamma_1 + \frac{1}{4}\Gamma^2\Gamma_1 + 2\Gamma\Omega^2},$$

where $\Gamma = \Gamma_1 + \Gamma_2$ and we have assumed that Ω is real. Most importantly, from the above equation we observe that even for large Ω values, associated with strong driving fields, the occupation of the excited state is bound to values smaller or equal than 1/2. As the prerequisite for lasing is population inversion, such that stimulated emission becomes larger than absorption, the active medium for lasers in general has to be made with quantum emitters that have three or more levels.

Figure 17.5 shows the most simple three-level diagram consisting of a ground and excited state g, e, together with an auxiliary state that becomes populated through optical or electrical pumping. We assume that there is a fast decay from the auxiliary to the excited state, where the atom remains until it becomes depopulated either through spontaneous or stimulated emission. When stimulated emission wins over the spontaneous one, we achieve light amplification through stimulated emission, which forms the basis of the laser. Although we could easily describe the above three-level system with the master equation approach of Lindblad form, we here adopt a somewhat simpler description by effectively removing the auxiliary atom

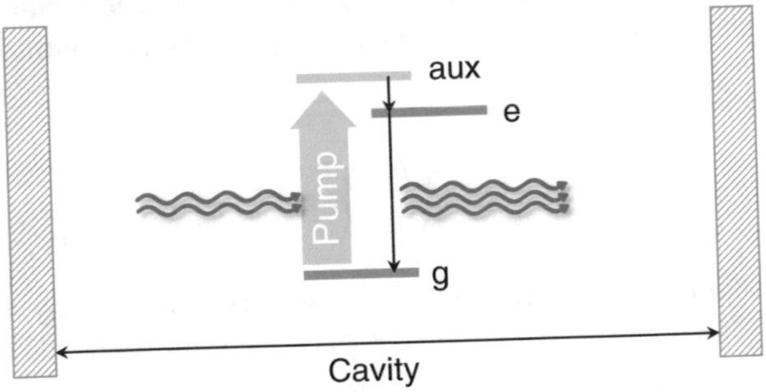

Fig. 17.5 Schematics of laser. The light field is confined in a cavity, for instance, by using mirrors. The active medium is composed of effective three-level systems, consisting of a ground and excited state g, e, together with an auxiliary state. The atoms are pumped from the ground to the auxiliary state, where they decay fast into e, thus obtaining population inversion between e and g. When an incoming photon hits the atom it undergoes a stimulated emission, hereby leading to an amplification of the light field within the cavity

state and considering incoherent pumping of the exciting state, with the Lindblad operator shown in Fig. 17.2. In order to describe the laser process we introduce the following ingredients.

Cavity Field. For the cavity we consider a single mode $\boldsymbol{u}(\boldsymbol{r})$, which is assumed to be a solution of the wave equation for the electric field inside the cavity, and express the electric field operator according to exercise 13.1 in the form

$$\hat{\boldsymbol{E}}(\boldsymbol{r}) = i\left(\frac{\hbar\omega}{\varepsilon_0}\right)^{\frac{1}{2}} \boldsymbol{u}(\boldsymbol{r})\left(\hat{a} - \hat{a}^\dagger\right). \tag{17.34}$$

Here ω is the frequency of the cavity mode, and $\hat{a}$, $\hat{a}^\dagger$ are the annihilation and creation operators for cavity photons.

Active Medium. The active medium is described as an ensemble of N two-level system, with density matrices of the form $\rho^{(\mu)}$, where μ labels the different atoms. In principle, the different atoms could have a detuning distribution $\Delta^{(\mu)}$, but for simplicity we here consider only identical atoms.

Hamiltonian. The Hamiltonian accounting for the coupled system of cavity and atoms can be expressed as

$$\hat{H} = \hbar\omega\left[\sum_\mu \sigma_{ii}^{(\mu)} + \left(\hat{a}^\dagger\hat{a} + \frac{1}{2}\right)\right] + \left[\hbar\Delta\sum_\mu \sigma_{ii}^{(\mu)} - \hbar\Omega\sum_\mu i\left(\hat{a}\sigma_{eg}^{(\mu)} - \hat{a}^\dagger\sigma_{ge}^{(\mu)}\right)\right], \tag{17.35}$$

where the last term in brackets accounts for the interaction, and the remaining terms for the uncoupled atoms and cavity photons. We have introduced the Rabi energy

$$\hbar\Omega = \left(\frac{\hbar\omega}{\varepsilon_0}\right)^{\frac{1}{2}} \boldsymbol{u}(\boldsymbol{r})\cdot\boldsymbol{p},$$

which we assume to be real-valued. Throughout we employ an interaction representation with respect to the first term on the right-hand side of Eq. (17.35).

We here follow Refs. [192, 194] and consider a semiclassical description, where the atoms are described in terms of their density matrices ρ and the electric field of the laser in terms of the expectation value $a = \langle\hat{a}\rangle$. The optical Bloch equations, Eq. (17.33), then become

$$\dot{\rho}_{ee} = -2\,\mathrm{Re}\big(\Omega a^*\rho_{ge}\big) - \Gamma_1\rho_{ee} + G\big(1 - \rho_{ee}\big) \tag{17.36a}$$

$$\dot{\rho}_{ge} = -i\Delta\,\rho_{ge} - \frac{1}{2}\Gamma\rho_{ge} + \Omega a\left(2\rho_{ee} - 1\right), \tag{17.36b}$$

Mark Stockman's Spaser

SPASER is the acronym for Surface Plasmon Amplification by Stimulated Emission of Radiation, which was suggested by David Bergman and Mark Stockman back in 2003 as follows [190]:

> We make a step towards quantum nanoplasmonics: surface plasmon fields of a nanosystem are quantized and their stimulated emission is considered. We introduce a quantum generator for surface plasmon quanta and consider the phenomenon of surface plasmon amplification by stimulated emission of radiation (spaser). Spaser generates temporally coherent high-intensity fields of selected surface plasmon modes that can be strongly localized on the nanoscale, including dark modes that do not couple to far-zone electromagnetic fields.

For many years Mark has been among the leading theoreticians in the field of plasmonics, equally gifted with physical intuition and the ability to carry out back-of-the-envelope calculations in order to estimate the feasibility of such proposals. It is always instructive and illustrative to read his papers. Also his sketchy figures are quite legendary, see, for instance, Fig. 1 of Ref. [190] for a beautiful example.

Mark has strongly advertised his SPASER at various occasions, but despite strong efforts the experimental realization has been more difficult than expected. The main reason is probably the strong metal losses which make it hard to reach the conditions necessary for stimulated emission. One of the first realizations has been presented by Noginov and coworkers [191] who used a gold core surrounded by a silica shell with organic dye molecules, but the experiment has received some criticism because it could not be repeated by others [192]. A more promising approach has been presented by Oulton and coworkers [193] who used a semiconductor CdS nanowire separated from a silver surface by a thin insulating gap. For a review of this and other experiments see, for instance, Ref. [192].

with $\Gamma = \Gamma_1 + \Gamma_2 + G$ and G being the generation rate, see Fig. 17.2, and we have not explicitly indicated the μ dependence of the density matrices. From Heisenberg's equations of motion for $\hat{a}$ we get

$$\dot{a} = -\frac{\gamma}{2}a + N\Omega\rho_{ge}, \tag{17.36c}$$

where γ accounts for cavity losses. The set of Eqs. (17.36) is analyzed in some detail in Ref. [192], in the following we only investigate the steady state for $\Delta = 0$. From Eq. (17.36b) we get

$$\rho_{ge} = \frac{2\Omega a}{\Gamma}\big(2\rho_{ee} - 1\big).$$

Inserting this into the expression for the cavity field gives

$$\left[\frac{2N\Omega^2}{\Gamma}(2\rho_{ee}-1)-\frac{\gamma}{2}\right]a=0 \implies \frac{4N\Omega^2}{\gamma\Gamma}\Big(2\rho_{ee}-1\Big)=1,$$

which has either the trivial solution $a = 0$ or the term in brackets must be zero. The latter condition leads to the expression on the right-hand side, which shows that lasing can only be achieved through population inversion. In addition, the following laser condition must be fulfilled [192, Eq. (39)]

$$4N\Omega^2 \geq \gamma\Gamma. \tag{17.37}$$

This implies that the atom-cavity couplings must be sufficiently strong in comparison to the atom and cavity losses. When this condition is met, the threshold pumping rate G and the mean number of photons in the cavity a can be computed from Eq. (17.36a).

17.4 Environment Couplings

In this section we make contact with a microscopic description of environment couplings. Quite generally, this is a difficult topic and there exists a lot of specialized literature, see, for instance, the book of Breuer and Petruccione [187]. Our main purpose here is to show how to derive generalized master equations *in principle*, and which approximations are typically involved. We start with the mother of all transport equations, the Boltzmann's equation. Although relying on classical mechanics, its derivation provides a blueprint for basically all other transport equations including those used in the field of quantum mechanics.

17.4.1 Boltzmann's Equation

Ludwig Boltzmann spent almost twenty years at the University of Graz, where I teach and work, and each day I pass by his memorial plaque. It is said that his time in Graz was among the most happy and productive periods of his life. He had a house on a hill located somewhat outside of Graz, close to the place where I grew up, and rumor says that he even owned a cow to enjoy a farmer's life. For sure he formulated his celebrated second law of thermodynamics during his period in Graz, which is nowadays known as

$$S = k_B \ln W, \tag{17.38}$$

with k_B being the Boltzmann's constant. This equation relates the entropy S with the thermodynamic probability W, and states that large systems tend towards the more probable configurations. For very large systems, with typically around 10^{23} particles as is usually the case for gases or solids, "probable" then means (almost) certain.

The second law of thermodynamics fundamentally differs from other theories in the sense that it does not make contact with the system dynamics, but only relies on probability reasonings. While this general approach makes it so powerful—indeed it can be applied to all fundamental theories including classical and quantum mechanics, as well as electrodynamics—one sometimes feels a little betrayed by Eq. (17.38) because it does not give any information on *how* a system approaches equilibrium.

Boltzmann was aware of this problem and backed the second law with a transport equation that should describe the approach of a system towards equilibrium. Initially he thought that this equation would be equally fundamental as the second law, but he soon had to realize that several approximations are needed to obtain a meaningful approach. The central object of Boltzmann's transport equation is the distribution function $f(\boldsymbol{r}, \boldsymbol{v}, t)$ which gives the density of particles at position $\boldsymbol{r}$ with velocity $\boldsymbol{v}$. For particles moving in presence of an external force $\boldsymbol{F}$ but without internal collisions, one gets within a small time interval dt

$$f\left(\boldsymbol{r} + \boldsymbol{v}\,dt, \boldsymbol{v} + \frac{\boldsymbol{F}}{m}\,dt, t + dt\right) d^3r d^3p - f\left(\boldsymbol{r}, \boldsymbol{v}, t\right) d^3r d^3p = 0.$$

The above expression states that the number of particles within a phase space volume element does not change upon diffusion $\boldsymbol{r} + \boldsymbol{v}\,dt$ and drift $\boldsymbol{v} + \frac{\boldsymbol{F}}{m}\,dt$ when using a reference frame that moves along with the particles. Expanding the first term on the left-hand side in a Taylor series and dividing by dt, we arrive at

$$\frac{df}{dt} = \left(\nabla_r f\right) \cdot \boldsymbol{v} + \left(\nabla_v f\right) \cdot \frac{\boldsymbol{F}}{m} + \frac{\partial f}{\partial t} = 0.$$

When we consider interactions between particles, the distribution function can change as an effect of collisions caused by these interactions

Boltzmann's Equation with Collision Integral

$$\left(\nabla_r f\right) \cdot \boldsymbol{v} + \left(\nabla_v f\right) \cdot \frac{\boldsymbol{F}}{m} + \frac{\partial f}{\partial t} = \left(\frac{\partial f}{\partial t}\right)_{\text{coll}}. \tag{17.39}$$

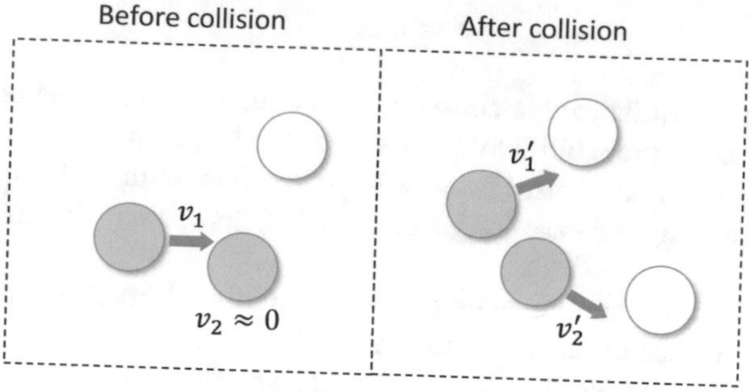

Fig. 17.6 Schematics of Boltzmann's famous Stoßzahl ansatz for an ensemble of hard spheres. Before the collision the two gray particles with velocities $\boldsymbol{v}_1$, $\boldsymbol{v}_2$ are uncorrelated, but become correlated *after* the collision through the differential cross section that gives the probability for the two particles having velocities $\boldsymbol{v}_1'$, $\boldsymbol{v}_2'$ after completion of the scattering process. Thus, even when starting with an uncorrelated ensemble of spheres, they become correlated as an effect of scatterings. The Stoßzahl ansatz neglects these correlations using the molecular chaos hypotheses, which is described in more detail in the text

The term on the right-hand side is the collision integral that accounts for the effect of particle interactions. To render this equation useful for applications we must specify this term, and here the dilemma starts.

Quite generally, even when all particles are uncorrelated at time zero, as a result of scattering they become correlated. Let us assume that the interparticle forces are sufficiently short-ranged such that the effect of the scattering can be described in terms of a differential cross section

$$\sigma_{\text{coll}}(\boldsymbol{v}_1, \boldsymbol{v}_2 \to \boldsymbol{v}_1', \boldsymbol{v}_2'), \tag{17.40}$$

which accounts for the probability of a scattering from the initial states $\boldsymbol{v}_1$, $\boldsymbol{v}_2$ to the final states $\boldsymbol{v}_1'$, $\boldsymbol{v}_2'$. An idealization are hard spheres which interact only during contact, such that the duration time of a scattering approaches zero and the transition occurs instantaneously. See Fig. 17.6 for a schematic drawing. As an effect of such scatterings, the two scattering partners become correlated after the collision,

$$f(\boldsymbol{r}, \boldsymbol{v}_1, t) f(\boldsymbol{r}, \boldsymbol{v}_2, t) \underset{\sigma_{\text{coll}}}{\longrightarrow} f_2(\boldsymbol{r}, \boldsymbol{v}_1', \boldsymbol{r}, \boldsymbol{v}_2', t).$$

Here we have assumed that the collisions take place at position $\boldsymbol{r}$, and have introduced the two-particle distribution function $f_2(\boldsymbol{r}_1, \boldsymbol{v}_1, \boldsymbol{r}_2, \boldsymbol{v}_2, t)$ that gives the probability that one particle is at position $\boldsymbol{r}_1$ with velocity $\boldsymbol{v}_1$ while another particle is located at position $\boldsymbol{r}_2$ with velocity $\boldsymbol{v}_2$. Obviously, the next collision correlates three particles, to be described by a three-particle distribution function f_3 and so on.

To truncate this infinite hierarchy, Boltzmann factorized the two-particle distribution function into a product of two one-particle functions

Factorization of Two-Particle Correlations

$$f_2(\boldsymbol{r}_1, \boldsymbol{v}_1, \boldsymbol{r}_2, \boldsymbol{v}_2, t) \approx f(\boldsymbol{r}_1, \boldsymbol{v}_1, t) f(\boldsymbol{r}_2, \boldsymbol{v}_2, t). \tag{17.41}$$

His motivation for this factorization is formulated in the so-called molecular chaos. It states that in a system with a sufficiently large number of particles the sharing of correlations among an ever increasing number of particles has the net effect that correlations built up in the past become diluted in the sea of many-particle correlations, and are no longer of importance for the ensuing time evolution. This approximation turns out to work extremely well for all systems of interest, yet it is an ad-hoc assumption that can be motivated but not proven. Even worse, the factorization per se introduces an irreversibility in the dynamics, as it breaks time reversal symmetry, which is a somewhat unpleasant feature for an equation that was meant to demonstrate the transition from unlikely to probable states. As a result, Boltzmann was seriously attacked by his contemporaries for this, and he did not receive the recognition during lifetime that he would have deserved.

Quite generally, this is all I wanted to tell you about Boltzmann's transport equation, but since we already started the derivation we should also bring it to an end. The differential cross section is defined as

$$\sigma_{\text{coll}} = \frac{\text{number of scattered particles}}{\text{incident flux}}.$$

The incident flux of a particle with velocity $\boldsymbol{v}_1$ on a second particle with velocity $\boldsymbol{v}_2$ is $|\boldsymbol{v}_1 - \boldsymbol{v}_2|$. The collision integral then consists of two contributions, a negative contribution for particles that are scattered out from a given phase space element located around $\boldsymbol{r}$, $\boldsymbol{v}_1$, and a positive contribution for particles that are scattered in

$$\begin{aligned} &\left(\frac{\partial f(\boldsymbol{r}, \boldsymbol{v}_1, t)}{\partial t}\right)_{\text{coll}} \\ &= -\int \sigma_{\text{coll}}(\boldsymbol{v}_1, \boldsymbol{v}_2 \to \boldsymbol{v}_1', \boldsymbol{v}_2') |\boldsymbol{v}_1 - \boldsymbol{v}_2| f(\boldsymbol{r}, \boldsymbol{v}_1, t) f(\boldsymbol{r}, \boldsymbol{v}_2, t) d^3v_1' d^3v_2 d^3v_2' \\ &\quad + \int \sigma_{\text{coll}}(\boldsymbol{v}_1', \boldsymbol{v}_2' \to \boldsymbol{v}_1, \boldsymbol{v}_2) |\boldsymbol{v}_1' - \boldsymbol{v}_2'| f(\boldsymbol{r}, \boldsymbol{v}_1', t) f(\boldsymbol{r}, \boldsymbol{v}_2', t) d^3v_1' d^3v_2 d^3v_2'. \end{aligned} \tag{17.42}$$

This expression is the famous Stoßzahl ansatz of Boltzmann. It can be further simplified by exploiting time reversal symmetry and momentum conservation for the scattering cross section, but we keep Eq. (17.42) as our final expression. The two central approximations entering Boltzmann's equation can be summarized as follows.

Scatterings. Particle interactions are described in terms of scattering probabilities. The ideal case is hard spheres where the scattering is instantaneous. If this is not the case, scattering processes should be at last short enough that one scattering is completed before the next one takes place.

Neglect of Correlations. As a result of such scatterings the particles become correlated. We neglect these correlations in the spirit of the "molecular chaos" assumption, which states that in a sufficiently large system correlations are transferred to more and more particles, and the probability that a correlation produced in a past scattering between two particles will act back on one of these particles in the future is so small that it can be safely neglected.

There have been many attempts to go beyond these approximations, but then similar assumptions have to be made somewhere else, for instance, by neglecting three-body correlations rather than two-body correlations. Boltzmann's transport equation is a great tool that works (almost too) well in most cases of interest, but it builds on assumptions that cannot be proven from first principles. This has been and will always be its glory and misery.

17.4.2 Nakajima–Zwanzig Equation

We next derive a Boltzmann-like equation for the master equation using quantum physics rather than classical mechanics. We assume that the problem under consideration can be divided into two parts, one for the system S we are interested in and a remainder for the "reservoir" R with which the system is assumed to weakly interact. Let $\hat{w}$ be the density operator that accounts for both the system and reservoir degrees of freedom. We assume that the system and reservoir density operators can be obtained by tracing over the reservoir and system degrees of freedom, respectively,

$$\hat{\rho}_S(t) = \mathrm{tr}_R\left\{\hat{w}(t)\right\}, \quad \hat{\rho}_R(t) = \mathrm{tr}_S\left\{\hat{w}(t)\right\}. \tag{17.43}$$

Below we will return to our previous notation $\hat{\rho} = \hat{\rho}_S$. We additionally introduce projection operators $\mathbb{P}$, $\mathbb{Q}$, with $\mathbb{P}+\mathbb{Q} = \mathbb{1}$, which project onto subspaces where the system and reservoir are uncorrelated

$$\mathbb{P}\hat{w}(t) = \hat{\rho}_S(t) \otimes \hat{\rho}_R(t), \tag{17.44}$$

and onto the complement $\mathbb{Q}\hat{w}(t)$. See Fig. 17.7 for a schematics. In the above expression "$\otimes$" denotes a direct product between the system and reservoir subspaces. Using Eqs. (17.43) and (17.44) one can readily show that

$$\mathrm{tr}_R\left\{\mathbb{Q}\hat{w}\right\} = \mathrm{tr}_S\left\{\mathbb{Q}\hat{w}\right\} = 0.$$

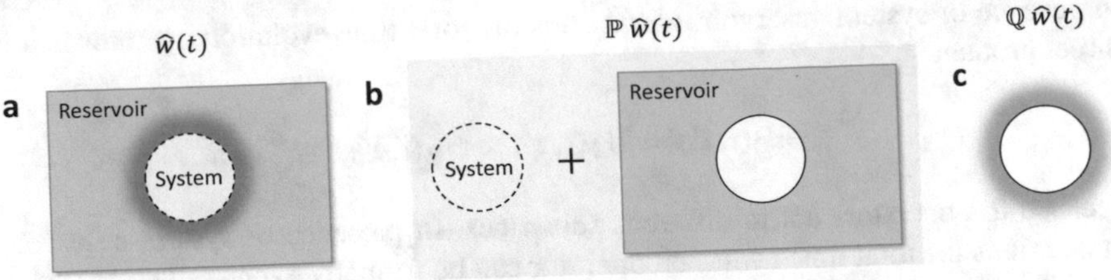

Fig. 17.7 Schematics of density and projection operators for Nakajima–Zwanzig approach. (**a**) The total density operator $\hat{w}(t)$ accounts for the interacting system and reservoir. (**b**) The (super)operator $\mathbb{P}$ projects on the subspace where the system and reservoir are uncoupled. By tracing over the reservoir degrees of freedom, one obtains the system operator $\hat{\rho}_S$. (**c**) The (super)operator $\mathbb{Q}$ projects on the complementary subspace accounting for the correlations between system and reservoir

Thus $\mathrm{tr}_R\{\hat{w}\} = \mathrm{tr}_R\{\mathbb{P}\hat{w}\} = \hat{\rho}_S$, and the master equation for the system can be obtained by tracing over the reservoir degrees of freedom via

$$i\hbar\frac{d\hat{\rho}_S}{dt} = i\hbar\,\mathrm{tr}_R\left\{\frac{d\hat{w}}{dt}\right\} = \mathrm{tr}_R\left\{\mathbb{P}\Big(\mathbb{L}_S(t) + \mathbb{L}_R + \mathbb{V}\Big)\hat{w}\right\}.$$

Here $\mathbb{L}_S, \mathbb{L}_R$ are the Liouville operators for the system and reservoir, respectively, and $\mathbb{V}$ accounts for the system-reservoir couplings. We have assumed that only the system part depends explicitly on time, although we will not further exploit this point. The important observation about the above equation is that we can trace out the reservoir in the time-derivative term only, on the right-hand side the coupling $\mathbb{V}$ leads to a correlation between system and reservoir. The expression can be simplified by observing that the projection operator $\mathbb{P}$ commutes with $\mathbb{L}_S, \mathbb{L}_R$ which account for dynamics in the uncoupled system and reservoir subspaces, respectively. With this we get

$$\begin{aligned} i\hbar\frac{d\hat{\rho}_S}{dt} &= \mathrm{tr}_R\left\{\mathbb{P}\Big(\mathbb{L}_S(t) + \mathbb{L}_R + \mathbb{V}\Big)(\mathbb{P}+\mathbb{Q})\hat{w}\right\} \\ &= \mathrm{tr}_R\Big\{\big(\mathbb{P}\mathbb{L}_S\mathbb{P}\big)\hat{\rho}_S + \big(\mathbb{P}\mathbb{V}\mathbb{Q}\big)\hat{\rho}_R\Big\}, \end{aligned} \tag{17.45}$$

where we have used that $\mathbb{P}\mathbb{L}_{S,R}\mathbb{Q} = 0$ and $\mathbb{P}\mathbb{V}\mathbb{P} = 0$. The latter expression is zero because $\mathbb{V}$ couples from the uncorrelated to the correlated subspace, and thus only terms of the form $\mathbb{P}\mathbb{V}\mathbb{Q}$ and $\mathbb{P}\mathbb{V}\mathbb{Q}$ give a non-zero contribution. We next introduce time evolution operators for the system and reservoir, respectively,

$$i\hbar\frac{d}{dt}\mathbb{U}_S(t,t') = \mathbb{L}_S(t)\mathbb{U}_S(t,t'), \quad \mathbb{U}_S(t,t) = \mathbb{1}$$

$$i\hbar\frac{d}{dt}\mathbb{U}_R(t,t') = \mathbb{L}_R\mathbb{U}_R(t,t'), \qquad \mathbb{U}_R(t,t) = \mathbb{1}.$$

In absence of system–reservoir interactions the total time evolution operator is the direct product

$$\mathbb{U}_0(t,t') = \mathbb{U}_S(t,t') \otimes \mathbb{U}_R(t,t'),$$

because the operators act in different subspaces. In presence of system–reservoir interactions the total time evolution operator can be formally expressed as

$$\mathbb{U}(t,t') = \mathbb{U}_0(t,t') - \frac{i}{\hbar}\int_{t'}^{t} \mathbb{U}_0(t,\tau)\,\mathbb{V}\,\mathbb{U}(\tau,t')\,d\tau. \tag{17.46}$$

Indeed, by taking the time derivative on both sides of the equation we are led to

$$\begin{aligned} i\hbar\frac{d}{dt}\mathbb{U}(t,t') &= \Big(\mathbb{L}_S(t) + \mathbb{L}_R\Big)\mathbb{U}_0(t,t') + \mathbb{U}_0(t,t)\mathbb{V}\,\mathbb{U}(\tau,t') \\ &\quad - \Big(\mathbb{L}_S(t) + \mathbb{L}_R\Big)\frac{i}{\hbar}\int_{t'}^{t} \mathbb{U}_0(t,\tau)\,\mathbb{V}\,\mathbb{U}(\tau,t')\,d\tau. \end{aligned}$$

Expressing via Eq. (17.46) the integral term on the right-hand side as $\mathbb{U}(t,t') - \mathbb{U}_0(t,t')$ we get

$$i\hbar\frac{d}{dt}\mathbb{U}(t,t') = \Big(\mathbb{L}_S(t) + \mathbb{L}_R + \mathbb{V}\Big)\mathbb{U}(t,t'),$$

which shows that $\mathbb{U}(t,t')$ fulfills the equation of motion for the time evolution operator together with the boundary condition $\mathbb{U}(t,t) = \mathbb{1}$. We can now obtain a formal solution for $\hat{\rho}_R$ by applying the time evolution operator to the density operator, $\mathbb{U}(t,0)\hat{w}(0)$, and acting with $\mathbb{Q}$ from the left-hand side on Eq. (17.46). This gives

$$\mathbb{Q}\hat{w}(t) = \mathbb{Q}\mathbb{U}_0(t,0)\hat{w}(0) - \frac{i}{\hbar}\int_{t_0}^{t} \mathbb{Q}\mathbb{U}_0(t,\tau)\Big[\mathbb{Q}\mathbb{V}\mathbb{P}\Big]\mathbb{U}(\tau,t_0)\hat{w}(t_0)\,d\tau,$$

where t_0 is the initial time of our problem. For clarity, we have decorated the interaction term $\mathbb{V}$ in brackets with projection operators. The $\mathbb{Q}$ projector on the left-hand side can be inserted because $\mathbb{U}_0$ does not couple between the different subspaces, and thus $\mathbb{Q}\mathbb{U}_0 = \mathbb{Q}\mathbb{U}_0\mathbb{Q}$. The $\mathbb{P}$ projector on the right-hand side can be inserted because $\mathbb{V}$ couples between the different subspaces, and thus $\mathbb{Q}\mathbb{V} = \mathbb{Q}\mathbb{V}\mathbb{P}$.

Inserting the above expression into the reduced master equation for the system density operator, Eq. (17.45), we finally obtain the Nakajima–Zwanzig equation that describes the time evolution of $\hat{\rho}_S$ in presence of additional system–reservoir interactions [195]

Nakajima–Zwanzig Equation

$$i\hbar\frac{d\hat{\rho}_S}{dt} = \mathbb{L}_S(t)\hat{\rho}_S + \mathrm{tr}_R\left\{\mathbb{PVQ}\,\hat{w}(t) - \frac{i}{\hbar}\int_{t_0}^{t}\Big[\mathbb{PVU}_0(t,t')\mathbb{V}\Big]\hat{\rho}_S(t')\otimes\hat{\rho}_R(t')\,dt'\right\}. \tag{17.47}$$

Eq. (17.47) is a formal solution of the master equation that was derived without introducing any approximation. In this sense it is very helpful for drawing general conclusions but yet worthless from a practical point of view, although we will discuss approximation strategies in a moment. The first interesting point about the Nakajima–Zwanzig equation is that it is nonlocal in time and extends into the system's past, whereas the master equation of the total density operator, which served us as the starting expression, is local in time. The transition from local to nonlocal in time is due to the fact that we have formally traced out the degrees of freedom of the reservoir. The price we pay for eliminating these degrees is that we have to consider the system's past explicitly. The different contributions on the right-hand side of Eq. (17.47) can be interpreted as follows.

System Degrees of Freedom. The first term with the Liouvillian $\mathbb{L}_S$ accounts for interactions in the system subspace only. This part has been previously assigned to coherent system interactions and could be described in terms of a Schrödinger's equation alone.

Reservoir Correlations. The second term $\mathbb{PVQ}w(t)$ accounts for possible initial correlations between the system and reservoir, and will be neglected in the spirit of Boltzmann's molecular chaos assumption.

System-Reservoir Scatterings. The last term accounts for scatterings between the system and the environment. Working through the expression in brackets from the right to left (see also Fig. 17.8), we find that initially the system interacts with the reservoir through $\mathbb{V}$, the induced fluctuation propagates in the system and reservoir subspaces independently, and finally the fluctuation acts back to the system through another system–reservoir interaction $\mathbb{V}$. As will be discussed in the following, this buildup and back-action of correlations can be approximately described in terms of a scattering cross section, which often allows us to reduce the Nakajima–Zwanzig equation to a master equation of Lindblad form.

Master Equation Revisited

Although the derivation of the Nakajima–Zwanzig equation has been general, we take the luxury and re-derive it in lowest order perturbation theory without using

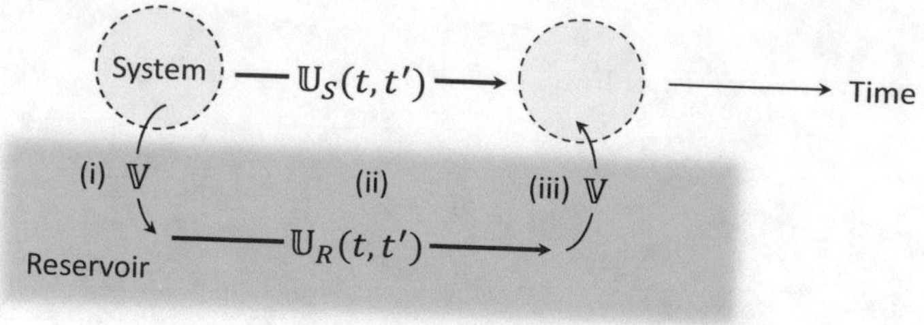

Fig. 17.8 Schematics of system–reservoir interaction within the Nakajima–Zwanzig approach. Instead of describing the interacting system and reservoir explicitly, we trace out the reservoir degrees of freedom and end up with the Nakajima–Zwanzig equation, Eq. (17.47), for the system density operator $\hat{\rho}_S$ only. System–reservoir interactions are described in terms of scattering-type processes: (i) at time t' the system and reservoir interact, as described by the (super)operator $\mathbb{V}$ that induces a system-reservoir correlation, (ii) the correlation propagates in the isolated subspaces of system and reservoir, and (iii) finally at time t the correlation acts back on the system subspace

super-operators. We start by introducing the usual time evolution operators for the system and reservoir, respectively, which are defined through

$$i\hbar\frac{d}{dt}\hat{U}_S(t,t') = \hat{H}_S(t)\hat{U}_S(t,t'), \quad \hat{U}_S(t,t) = \mathbb{1}_S$$

$$i\hbar\frac{d}{dt}\hat{U}_R(t,t') = \hat{H}_R(t)\hat{U}_R(t,t'), \quad \hat{U}_R(t,t) = \mathbb{1}_R.$$

The von-Neumann equation of the total density operator $\hat{w}(t)$ can be written in the form

$$i\hbar\frac{d\hat{w}(t)}{dt} = \Big[\hat{H}_S(t) + \hat{H}_R + \hat{V}, \hat{w}(t)\Big],$$

where $\hat{V}$ is the coupling between the system and reservoir. We next introduce an interaction representation for the density operator

$$\hat{w}_I(t) = \hat{U}_0^\dagger(t,0)\hat{w}(t)\hat{U}_0(t,0), \quad \hat{U}_0(t,0) = \hat{U}_S(t,0)\otimes\hat{U}_R(t,0),$$

and a corresponding transformation for the interaction term $\hat{V}_I(t)$. The von-Neumann equation for $\hat{w}_I(t)$ then becomes

$$i\hbar\frac{d\hat{w}_I(t)}{dt} = \Big[\hat{V}_I(t), \hat{w}_I(t)\Big], \tag{17.48}$$

as can be readily shown by taking the time derivative on both sides of the equation and using the defining equations for $\hat{U}_S$, $\hat{U}_R$. Note that the time derivative of $\hat{w}_I$ is now only caused by the system–reservoir interaction. To get the time evolution for the system part only, we trace out in Eq. (17.48) the reservoir degrees of freedom

$$i\hbar\frac{d\hat{\rho}_I(t)}{dt} = \mathrm{tr}_R\left\{\left[\hat{V}_I(t), \hat{w}_I(t)\right]\right\}, \tag{17.49}$$

where we have switched back to $\hat{\rho}$ (without subscript S) for the system density operator. As in the derivation of the Nakajima–Zwanzig equation, the tracing procedure can be only safely performed for the time-derivative term. We next integrate Eq. (17.48) in time

$$\hat{w}_I(t) = \hat{w}_I(t_0) - \frac{i}{\hbar}\int_{t_0}^{t}\left[\hat{V}_I(\tau), \hat{w}_I(t')\right]dt',$$

and insert the resulting expression into Eq. (17.49). At the same time we introduce a few additional approximations.

- In the spirit of the molecular chaos we ignore the $\hat{w}_I(t_0)$ term in the system's dynamics.
- We assume that before the first system–reservoir interaction the system and the reservoir are uncorrelated, such that

$$\hat{w}_I(\tau) \approx \hat{\rho}_I(t') \otimes \hat{\rho}_{R,I}(t').$$

- We neglect the time dependence of the reservoir density operator and set

$$\hat{\rho}_I(t') \approx \hat{\rho}_I(t) + \mathcal{O}(\hat{V}).$$

  Note that this replacement is correct up to the lowest order in $\hat{V}$.

With these replacements we are finally led to the master equation, which is correct up to second order in the system-reservoir Hamiltonian $\hat{V}$ and describes the time evolution of $\hat{\rho}_I$ in presence of environment couplings

Master Equation for System Coupled to Reservoir

$$\left(\frac{d\hat{\rho}_I(t)}{dt}\right)_{\text{incoh}} = -\frac{1}{\hbar^2}\int_{t_0}^{t}\mathrm{tr}_R\left\{\left[\hat{V}_I(t), \left[\hat{V}_I(t'), \hat{\rho}_I(t)\otimes\hat{\rho}_R\right]\right]\right\}dt'. \tag{17.50}$$

This equation has the same physical contents as the Nakajima–Zwanzig equation previously derived, however, it is more practical for actual implementations.

17.4.3 Fermi's Golden Rule

As an example, in the following we consider the situation where a two-level system is coupled to an ensemble of harmonic oscillators

$$\hat{H} = \frac{1}{2}\hbar\omega\,\sigma_3 + \sum_\lambda \hbar\omega_\lambda\,\hat{a}_\lambda^\dagger\hat{a}_\lambda + i\sum_\lambda g_\lambda\left(\hat{a}_\lambda^\dagger\,\sigma_- - \hat{a}_\lambda\,\sigma_+\right). \tag{17.51}$$

This model is sometimes referred to as the Caldeira–Leggett model [196]. $\hbar\omega$ is the energy splitting between ground and excited state of the two-level system, $\hbar\omega_\lambda$ are the energies of the harmonic oscillators which are described by the bosonic field operator $\hat{a}_\lambda$, and g_λ are the system-oscillator coupling constants which are assumed to be real. The last term on the right-hand side defines the system–environment interaction $\hat{V}$ consisting of terms where energy is transferred from the two-level system to the environment and vice versa. $\sigma_\pm$ are the usual raising and lowering operators for the two-level system, previously introduced in Eq. (16.8).

If we insert $\hat{V}$ in the interaction representation into Eq. (17.50) we obtain after some straightforward calculation

$$\begin{aligned}
\left(\frac{\partial\hat{\rho}_I(t)}{\partial t}\right)_{\text{incoh}} = -\int_{t_0}^{t}\sum_{\lambda\lambda'} g_\lambda g_{\lambda'}\Big[\\
&\left\langle\hat{a}_\lambda(t)\hat{a}_{\lambda'}^\dagger(t')\right\rangle\sigma_+(t)\,\sigma_-(t')\,\hat{\rho}_I(t) + \left\langle\hat{a}_{\lambda'}(t')\hat{a}_\lambda^\dagger(t)\right\rangle\hat{\rho}_I(t)\,\sigma_+(t')\,\sigma_-(t)\\
&-\left\langle\hat{a}_\lambda(t)\hat{a}_{\lambda'}^\dagger(\tau)\right\rangle\sigma_-(t')\,\hat{\rho}_I(t)\,\sigma_+(t) - \left\langle\hat{a}_{\lambda'}(t')\hat{a}_\lambda^\dagger(t)\right\rangle\sigma_-(t)\,\hat{\rho}_I(t)\,\sigma_+(t')\\
&+\left\langle\hat{a}_\lambda^\dagger(t)\hat{a}_{\lambda'}(t')\right\rangle\sigma_-(t)\,\sigma_+(t')\,\hat{\rho}_I(t) + \left\langle\hat{a}_{\lambda'}^\dagger(t')a_\lambda(t)\right\rangle\hat{\rho}_I(t)\,\sigma_-(t')\,\sigma_+(t)\\
&-\left\langle\hat{a}_\lambda^\dagger(t)\hat{a}_{\lambda'}(t')\right\rangle\sigma_+(t')\,\hat{\rho}_I(t)\,\sigma_-(t) - \left\langle\hat{a}_{\lambda'}^\dagger(t)\hat{a}_\lambda(t)\right\rangle\sigma_+(t)\,\hat{\rho}_I(t)\,\sigma_-(t')\Big]\,dt'.
\end{aligned} \tag{17.52}$$

The terms with $\hat{\rho}_I$ and $\sigma_\pm$ describe the effects of environment couplings on the system. The terms in brackets $\langle\ldots\rangle = \mathrm{tr}_R\{\hat{\rho}_R\ldots\}$ have been derived by use of cyclic permutation under the trace and describe the propagation of excitations in the environment. In thermal equilibrium they can be simplified using

$$\left\langle a_\lambda^\dagger a_{\lambda'}\right\rangle = \bar{n}(\hbar\omega_\lambda)\,\delta_{\lambda\lambda'},$$

with $\bar{n}$ being the Bose–Einstein distribution function. Note that in Eq. (17.52) we have not considered terms with two annihilation or creation operators because their expectation values would vanish in thermal equilibrium. The expressions in the second and third line of Eq. (17.52) describe emission processes, where energy is transferred from the system to the environment, and those in the fourth and fifth line absorption processes. With $\Delta_\lambda = \omega_\lambda - \omega$ we can rewrite the terms in the brackets as

$$
\begin{aligned}
(\bar{n}_\lambda + 1) & \Big[e^{-i\Delta_\lambda(t-t')} \left(\sigma_+\sigma_-\hat{\rho}_I - \sigma_-\hat{\rho}_I\sigma_+\right) + e^{i\Delta_\lambda(t-t')} \left(\hat{\rho}_I\sigma_+\sigma_- - \sigma_-\hat{\rho}_I\sigma_+\right) \Big] \\
+\bar{n}_\lambda & \Big[e^{i\Delta_\lambda(t-t')} \left(\sigma_-\sigma_+\hat{\rho}_I - \sigma_+\hat{\rho}_I\sigma_-\right) + e^{-i\Delta_\lambda(t-t')} \left(\hat{\rho}_I\sigma_-\sigma_+ - \sigma_+\hat{\rho}_I\sigma_-\right) \Big],
\end{aligned}
\tag{17.53}
$$

where we have not explicitly indicated the time dependence of $\hat{\rho}_I(t)$ and have introduced the short-hand notation $\bar{n}_\lambda$ for the thermal distribution function. The remaining integrals are then of the form

$$
\mathcal{I} = \sum_\lambda \int_{t_0}^{t} \mathcal{F}_\lambda \, e^{\pm i\Delta_\lambda(t-t')} \, dt' \approx \lim_{\eta\to 0} \sum_\lambda \int_{-\infty}^{t} \mathcal{F}_\lambda e^{(\pm i\Delta_\lambda - \eta)(t-t')} \, dt',
\tag{17.54}
$$

where $\mathcal{F}_\lambda$ is a term consisting of the coupling constants and the thermal distribution functions. In the thermodynamic limit the integrands decay fast as a function of $t-t'$, as discussed in more length in Chap. 14, which allows us to introduce the adiabatic limit which is worked out on the right-hand side. The above integral provides two time scales, which can be analyzed in the light of Boltzmann's theory.

Memory Time. The decay of the integrand gives the memory time of the system, which can be interpreted as the time it takes for a collision to be completed. Similarly to the hard sphere limit of Boltzmann's equation, one can hope that this time is much shorter than the time between successive scatterings. This is the case for most systems of interest, but in principle the assumption has to be justified for each problem individually.

Scattering Time. The value of the integral $\mathcal{I}$ is related to the scattering rate, this is how often a scattering occurs. The corresponding time scale is not related to the memory time associated with the time of one collision, as can be easily seen by considering a scaling of $\mathcal{F}$ which only affects $\mathcal{I}$ but not the decay time of the integrand. Expressed in terms of hard spheres, the collision time is governed by the interaction range whereas the time between successive scatterings is determined by the velocity distribution and the concentration of spheres.

Finally, working out the integral of Eq. (17.54) we are led to

$$
\mathcal{I} \approx \pm i \lim_{\eta\to 0} \sum_\lambda \frac{\mathcal{F}_\lambda}{\Delta_\lambda \pm i\eta} \approx \sum_\lambda \mathcal{F}_\lambda \, \pi\delta(\Delta_\lambda),
$$

where we have used Eq. (F.6) to arrive at the last expression, and have ignored the energy renormalization of the two-level system associated with Cauchy's principal part. Working out the expressions given in Eq. (17.53), we find that the resulting integrals can be related to two scattering rates

Fermi's Golden Rule Result for Scattering Rates

$$\Gamma_{\text{abs}} = \frac{2\pi}{\hbar}\sum_\lambda g_\lambda^2\, \bar{n}(\hbar\omega_\lambda)\, \delta\big(\hbar\omega_\lambda - \hbar\omega_0\big)$$

$$\Gamma_{\text{em}} = \frac{2\pi}{\hbar}\sum_\lambda g_\lambda^2 \big[\bar{n}(\hbar\omega_\lambda) + 1\big]\, \delta\big(\hbar\omega_\lambda - \hbar\omega_0\big). \tag{17.55}$$

These scattering rates are associated with absorption and emission processes, and could have been alternatively derived using Fermi's golden rule. However, it is reassuring to see that our more detailed approach comes up with the same result. Finally, Eq. (17.52) can be brought to the form of a master equation in Lindblad form, Eq. (17.16), with the Lindblad operators accounting for absorption and emission through

$$\hat{L}_{\text{abs}} = \sqrt{\Gamma_{\text{abs}}}\,\sigma_+, \quad \hat{L}_{\text{em}} = \sqrt{\Gamma_{\text{em}}}\,\sigma_-.$$

Résumé The purpose of the chapter has been to show how to describe the quantum dynamics of open few-level systems coupled to the environment. We have started by giving a simple recipe for describing such couplings using the master equation of Lindblad form, and have then backed our working equations by a microscopic description in terms of a Boltzmann-like transport equation. The latter part may give the impression that such a formal derivation has to be carried out for every system individually. In fact, this is not the case.

The nice thing about the master equation of Lindblad form is that one can guess the Lindblad operators for the scatterings using physical intuition. We will do so in the next chapter, and often the experimental results provide a direct hint to the scattering channels and rates. In this sense the approach should be considered as being of semi-quantitative nature, although often the results are surprisingly good. If the problem under consideration forces one to dig deeper into the description of environment couplings, using, for instance, a microscopic description, the tools developed in this chapter will help as a first starting point, but a closer look to the specialized literature will be certainly needed.

Exercises

Exercise 17.1 Rewrite Eq. (17.2) in terms of matrix elements

$$\mathrm{tr}\{\hat{A}_1\hat{A}_2\ldots\hat{A}_n\} = \sum_{i_1}\sum_{i_2}\cdots\sum_{i_n}\langle i_1|\hat{A}_1|i_2\rangle\langle i_2|\hat{A}_2|i_3\rangle\ldots\langle i_n|\hat{A}_n|i_1\rangle,$$

to show that the operators can be permuted cyclically under the trace.

Exercise 17.2 Consider the eigenvalue equations

$$\bar{\bar{A}}\cdot\boldsymbol{x}_k^R = \lambda^R\boldsymbol{x}_k^R, \quad \left(\boldsymbol{x}_k^L\right)^T\cdot\bar{\bar{A}} = \lambda_k^L\left(\boldsymbol{x}_k^L\right)^T.$$

(a) Show that the second equation can be written in the form $\bar{\bar{A}}^T\boldsymbol{x}_k^L = \lambda_k^L\boldsymbol{x}_k^L$.

(b) Using the determinant properties, argue why $\bar{\bar{A}}$ and $\bar{\bar{A}}^T$ have the same eigenvalues. From this it immediately follows that the left and right eigenvalues are identical.

(c) Show that the eigenvalue equations can be rewritten in the form

$$\bar{\bar{A}}\cdot\bar{\bar{X}}^R = \bar{\bar{X}}^R\cdot\bar{\bar{\Lambda}}, \qquad \bar{\bar{X}}^L\cdot\bar{\bar{A}} = \bar{\bar{\Lambda}}\cdot\bar{\bar{X}}^L,$$

where $\bar{\bar{\Lambda}}$ is a matrix with the eigenvalues on the diagonal, and $\bar{\bar{X}}^L$, $\bar{\bar{X}}^R$ are matrices formed by the eigenvectors.

(d) Multiply the first equation from the left-hand side with $\bar{\bar{X}}^L$ and the second equation from the right-hand side with $\bar{\bar{X}}^R$, to get upon subtraction of the two matrices $[\bar{\bar{X}}^L\bar{\bar{X}}^R, \bar{\bar{\Lambda}}] = 0$. From this follows that

$$\bar{\bar{X}}^L\bar{\bar{X}}^R = \bar{\bar{X}}^R\bar{\bar{X}}^L = \mathbb{1}.$$

Exercise 17.3 Consider a two-level system with $\Delta = 0$ and two Lindblad operators of the form $L_1 = \sqrt{\Gamma_1}\sigma_{ge}$, $L_2 = \sqrt{\Gamma_2}\sigma_{eg}$.

(a) Compute the steady state of the system. This can be done most easily by setting $\dot{\rho} = 0$ in the master equation.

(b) Compute the occupation of the states in thermal equilibrium.

(c) Determine which relation Γ_1, Γ_2 must have such that the steady state of the two-level system is in thermal equilibrium.

Exercise 17.4 Use from the NANOPT toolbox the file `democascade01.m` for a cascade structure, as shown in Fig. 17.1b. It consists of three states 0, 1, 2, which are optically coupled via $0 \to 1$ and $1 \to 2$. Correspondingly, we introduce Lindblad operators for the radiative decay from $1 \to 0$, $2 \to 1$, and detunings Δ_1, Δ_2 subject to $\Delta_2 = \Delta_1 - 1$. Find the frequency of the driving field for which the population of state 2 becomes largest, and interpret the result.

Exercise 17.5 Show explicitly how to derive from Eq. (17.52) and the scattering rates of Eq. (17.55) a master equation of Lindblad form.

Exercise 17.6 Repeat the derivation of Fermi's golden rule of Sect. 17.4.3, however, for a light–matter interaction of the form

$$\hat{V}_I(t) = -\left(e^{i\omega t}\,\hat{\Omega}_I^+(t)\sigma_{eg} + e^{-i\omega t}\,\hat{\Omega}_I^-(t)\sigma_{ge}\right),$$

with the Rabi energy operator $\hbar\hat{\Omega}^+ = \boldsymbol{p}\cdot\hat{\boldsymbol{E}}^+(\boldsymbol{r}_0)$ and a corresponding expression for $\hbar\hat{\Omega}^-$. We have adopted an interaction representation with respect to the uncoupled light-matter system, with $\hbar\omega$ being the energy difference of the two-level system.

(a) Derive an expression similarly to Eq. (17.52) and express the expectation values of the electric field operators in terms of the cross-spectral densities of Eq. (15.1).
(b) Introduce in the time integrals the adiabatic limit and evaluate the integrals.
(c) Show that the resulting expressions are identical to the photon scattering rates of Eq. (15.39) and to the Lamb shift of Eq. (15.47).

Chapter 18
Photon Noise

In the last chapters we have shown how to describe the coherent and incoherent dynamics of optically driven few-level systems using a master equation approach of Lindblad form. We next approach the problem typically encountered in experiment, namely acquiring information about the dynamics of this few-level system. The situation we have in mind is sketched in Fig. 18.1 and consists of an optically driven few-level system, where the excited state suffers losses due to dephasing and photon scatterings. In the latter process a photon is emitted and the system is promoted to its groundstate. Importantly, in experiment one observes precisely this emitted photon which thus provides us with the information about the system's dynamics. Describing the emission and detection of the photon is the topic of this chapter.

In Chap. 13 we have seen that within quantum electrodynamics the electromagnetic fields are described by operators, which can be related to the photon creation and annihilation operators and which must additionally act on a wavefunction. So what remains to be done is to describe how the photon is created in the decay of the few-level system, follow the photon all its way from the quantum emitter to the detector, and finally compute the detection probability, which can then be directly compared with experiment. Fortunately there is a shortcut to this approach. If the propagation of the photon from its source (the quantum emitter) to the detector can be described in terms of the linear Maxwell's equations, we can use a good old friend, the Green's dyadics introduced in Chap. 5, to relate the electric field to the current source. In doing so, we end up with current correlation functions that have to be evaluated in the system's subspace only.

The quantum regression theorem, which we will derive in the following, accomplishes the task of computing such correlation functions in a relatively simple manner, and will bring the quantum optics toolbox to full glory. As emphasized at several places in this book—and here again—the quantum optics toolbox is made to be used and will unveil its true beauty only to those who are willing to work with it. We conclude this final chapter with a few selected examples in the field of nano optics.

U. Hohenester, *Nano and Quantum Optics*, Graduate Texts in Physics,
https://doi.org/10.1007/978-3-030-30504-8_18

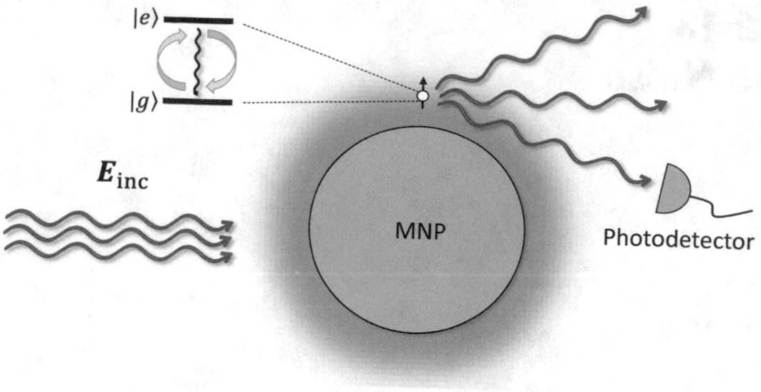

Fig. 18.1 Optically driven two-level system subject to dephasing and photon emissions. We consider a generic two-level system, described in terms of a point dipole, that is placed in a non-trivial photonic environment, here close to a metallic nanosphere. The system is driven coherently by an incoming light field, and its incoherent dynamics is described in terms of Lindblad operators. Most importantly, through radiative decay photons are emitted which are observed by a photodetector. In this chapter we discuss how to describe such photon detection experiments theoretically using the quantum optics toolbox

18.1 Photon Detectors and Spectrometers

Figure 18.1 schematically depicts a setup where a two-level system—in the following denoted as the "atom"—is located in a non-trivial photonic environment, here formed by a metallic nanosphere. The atom is driven coherently by an incoming light field, where the excitation becomes modified by the nearby nanoparticle, and its dynamics is described by a master equation with Lindblad operators accounting for photon emission, dephasing, and possibly other interaction channels. Through radiative decay a photon is emitted, which is detected in an optical experiment ultimately in a photodetector. Our starting point for the following discussion is Eq. (15.18):

$$\begin{aligned}\hat{\boldsymbol{E}}^{+}(\boldsymbol{r},\omega) &= \hat{\boldsymbol{E}}^{+}_{\rm inc}(\boldsymbol{r},\omega)\\ &\quad + i\mu_0\omega\int \bar{\bar{G}}_{\rm tot}(\boldsymbol{r},\boldsymbol{r}',\omega)\cdot\Big[\hat{\boldsymbol{J}}^{+}_{\rm atom}(\boldsymbol{r}',\omega)+\hat{\boldsymbol{J}}^{+}_{\rm ind}(\boldsymbol{r}',\omega)\Big]\,d^3r',\end{aligned} \tag{18.1}$$

where we have separated the field operator into a part associated with the incoming radiation and another one associated with the current sources of the atom and the absorbing materials in the photontic environment. A few comments are at place.

Linearity. In Eq. (18.1) we implicitly assume that the photon propagation can be described in terms of the linear Maxwell's equations, where the electric field operator can be related to the current operators via the dyadic Green's function of classical electrodynamics. Note that the dynamics of the atom can nonetheless be nonlinear.

Light Excitation. Throughout this chapter we assume that the incoming fields can be treated classically, that is we replace the operator $\hat{\boldsymbol{E}}_{\text{inc}}$ with a classical field that can be incorporated into the master equation along the lines described in the previous chapters. Note that in principle nothing hinders us to consider for the excitation also non-classical light fields.

Current Sources. In Eq. (18.1) we have separated the current sources into a part associated with the atom

$$\hat{\boldsymbol{J}}^{+}_{\text{atom}}(\boldsymbol{r},\omega) = -i\omega\boldsymbol{p}\,\delta(\boldsymbol{r}-\boldsymbol{r}_{\text{atom}})\,\sigma^{+}_{ge}\,, \tag{18.2}$$

where $\boldsymbol{r}_{\text{atom}}$ is the atom position and $\boldsymbol{p}$ the transition dipole moment. The induced part $\hat{\boldsymbol{J}}^{+}_{\text{ind}}$ is due to noise currents of the absorbing materials of the photonic environment, which are responsible for thermal radiation, as previously discussed in Chap. 15. They will be neglected for simplicity from here on, but could be included without too much additional efforts.

Retardation. Performing a Fourier transform in Eq. (18.1) we are led to

$$\begin{aligned}\hat{\boldsymbol{E}}^{+}(\boldsymbol{r},t) &= \hat{\boldsymbol{E}}^{+}_{\text{inc}}(\boldsymbol{r},t)\\ &\quad -\mu_0\int_{-\infty}^{\infty}\int_{-\infty}^{t}\bar{\bar{G}}_{\text{tot}}(\boldsymbol{r},\boldsymbol{r}',t-t')\cdot\left[\frac{\partial}{\partial t'}\hat{\boldsymbol{J}}^{+}_{\text{atom}}(\boldsymbol{r}',t')\right]dt'\,d^3r',\end{aligned} \tag{18.3}$$

where we have used that a product in frequency space translates to a convolution in time. From the above expression it is apparent that it takes a while for the photon to propagate from the atom to the detector. We will discuss in a moment how this can be considered in a simple manner.

18.1.1 Photodetector

The light intensity at a given position $\boldsymbol{r}$ can be related to the light intensity operator [4, Eq. (12.2–9)]

$$\hat{I}(\boldsymbol{r},t) = \hat{\boldsymbol{E}}^{-}(\boldsymbol{r},t)\hat{\boldsymbol{E}}^{+}(\boldsymbol{r},t)\,. \tag{18.4}$$

When expressed in terms of photon creation and annihilation operators this intensity is related to the number $\hat{a}^\dagger\hat{a}$ of photons. The detection probability for a photon is obtained by integration over the solid angle $\partial\Omega_{\text{det}}$ of the detector and taking the expectation value of the light fields

$$P^{(1)}_{\text{det}}(t) = \text{const}\times\oint_{\partial\Omega_{\text{det}}}\left\langle\hat{I}(s,t)\right\rangle dS\,, \tag{18.5}$$

where the constant factor depends on the detection efficiency, which can be lumped into a prefactor without significance for our following discussion.

Normal Ordering We will be also interested in two-photon intensities at equal or different times

$$\hat{I}(\boldsymbol{r},t_1)\hat{I}(\boldsymbol{r},t_2)=\hat{\boldsymbol{E}}^-(\boldsymbol{r},t_1)\hat{\boldsymbol{E}}^+(\boldsymbol{r},t_1)\,\hat{\boldsymbol{E}}^-(\boldsymbol{r},t_2)\hat{\boldsymbol{E}}^+(\boldsymbol{r},t_2)\,.$$

When dealing with such expressions we have to be careful about the fact that the electric field operators $\hat{\boldsymbol{E}}^+$ and $\hat{\boldsymbol{E}}^-$ do not commute in general, which can lead to spurious contributions. Expressed in photon creation and annihilation operators we would get

$$\hat{a}^\dagger\hat{a}\,\hat{a}^\dagger\hat{a}=\hat{a}^\dagger\hat{a}^\dagger\hat{a}\hat{a}+\hat{a}^\dagger\hat{a}\,,$$

where we have used the fundamental commutation relation $[\hat{a},\hat{a}^\dagger]=1$ to exchange the annihilation and creation operators on the left-hand side. Obviously, the second term on the right-hand side would give a two-photon intensity even for a single-photon state. This is obviously a wrong result that must be discarded when computing photon detection probabilities. To this end, it is convenient to introduce the concept of normal ordering [4]

$$:\hat{a}^\dagger\hat{a}\,\hat{a}^\dagger\hat{a}:=\hat{a}^\dagger\hat{a}^\dagger\hat{a}\hat{a}\,, \tag{18.6}$$

where the operators in between the $:(\dots):$ expression are rearranged such that all creation operators are on the left-hand side of all annihilation operators. For the electric field operators we correspondingly get

$$:\hat{I}(\boldsymbol{r}_1,t_1)\hat{I}(\boldsymbol{r}_2,t_2):=\hat{\boldsymbol{E}}^-(\boldsymbol{r}_1,t_1)\hat{\boldsymbol{E}}^-(\boldsymbol{r}_2,t_2)\,\hat{\boldsymbol{E}}^+(\boldsymbol{r}_2,t_2)\hat{\boldsymbol{E}}^+(\boldsymbol{r}_1,t_1)\,. \tag{18.7}$$

18.1.2 Spectrometer

In a spectrometer the incoming light is decomposed into its different frequency components, and finally the light intensity for a given frequency (more precisely for a small frequency range) is measured. We are not concerned with the question of how the spectrometer performs this decomposition in experiment, but simply assume that the field operators exiting the spectrometer can be written in the form

$$\hat{\boldsymbol{E}}^+_\omega(\boldsymbol{r},t)=\int_0^\infty e^{i\omega\tau}\,\hat{\boldsymbol{E}}^+(\boldsymbol{r},t-\tau)\,d\tau\,. \tag{18.8}$$

From here on we suppress the spatial dependence of $\hat{\boldsymbol{E}}^\pm$ and assume that the system is measured under steady-state conditions, that is the spectra should not depend on time. Outside the spectrometer we place an array of photodetectors or a charged-coupled device (CCD) camera, where the time-averaged intensity of one of the detectors is

$$S(\omega)\propto\lim_{T\to\infty}\frac{1}{T}\int_0^T\left\langle\hat{\boldsymbol{E}}^-_\omega(t)\cdot\hat{\boldsymbol{E}}^+_\omega(t)\right\rangle dt=\left\langle\!\left\langle\hat{\boldsymbol{E}}^-_\omega(t)\cdot\hat{\boldsymbol{E}}^+_\omega(t)\right\rangle\!\right\rangle,$$

with the additional brackets indicating time average. Inserting Eq. (18.8) into the above expression gives

$$S(\omega) \propto \int_0^\infty e^{i\omega(\tau'-\tau)} \left\langle\!\left\langle \hat{\boldsymbol{E}}^-(t-\tau') \cdot \hat{\boldsymbol{E}}^+(t-\tau) \right\rangle\!\right\rangle d\tau d\tau' .$$

The term in brackets depends on contributions where $\hat{\boldsymbol{E}}^-$ is either at a later or earlier time than $\hat{\boldsymbol{E}}^+$, and we can rewrite the integral in the form[1]

$$S(\omega) \propto \int_{-\infty}^\infty e^{i\omega\tau} \left\langle\!\left\langle \hat{\boldsymbol{E}}^-(t) \cdot \hat{\boldsymbol{E}}^+(t+\tau) \right\rangle\!\right\rangle d\tau .$$

We finally use the relation

$$\left\{ e^{i\omega\tau} \left\langle\!\left\langle \hat{\boldsymbol{E}}^-(t) \cdot \hat{\boldsymbol{E}}^+(t+\tau) \right\rangle\!\right\rangle \right\}^* = e^{-i\omega\tau} \left\langle\!\left\langle \hat{\boldsymbol{E}}^-(t+\tau) \cdot \hat{\boldsymbol{E}}^+(t) \right\rangle\!\right\rangle$$

to arrive at

$$S(\omega) \propto \mathrm{Re} \left\{ \int_0^\infty e^{i\omega\tau} \left\langle \hat{\boldsymbol{E}}^-(0) \cdot \hat{\boldsymbol{E}}^+(\tau) \right\rangle d\tau \right\} . \tag{18.9}$$

Here we have removed again the time average and have inserted a specific time zero for the first operator, which is of no importance under the steady-state conditions we are interested in. We next use Eq. (18.1) to relate the electric field and current operators, and to express the spectrum of an optically driven atom in the form

Spectrum of Optically Driven Atom

$$S(\omega) \propto \left(\mu^2 \omega^4 \oint_{\partial\Omega_{\text{det}}} \left| \bar{\bar{G}}(\boldsymbol{s}, \boldsymbol{r}_{\text{atom}}, \omega) \cdot \boldsymbol{p} \right|^2 dS \right)$$
$$\times \mathrm{Re} \left\{ \int_0^\infty e^{i\omega\tau} \left\langle \sigma^-(0) \sigma^+(\tau) \right\rangle d\tau \right\} . \tag{18.10}$$

[1] To arrive safely at the final result, one should rather use field operators smeared over a small frequency window

$$\hat{\boldsymbol{E}}_\omega^+(t) = \int \left\{ \int_0^\infty e^{i\omega'\tau} \hat{\boldsymbol{E}}^+(t-\tau)\, d\tau \right\} f_{\delta\omega}(\omega-\omega') d\omega' ,$$

where $f_{\delta\omega}$ is a function peaked at frequency ω with some small width $\delta\omega$.

The first term in parentheses describes the photonic part and accounts for the propagation of the emitted photons from the atom to the spectrometer. The second part accounts for the atom dynamics, and expresses the fluorescence spectrum in terms of current correlation functions to be evaluated under the nonequilibrium conditions of the optically driven atom. We will show below how such correlation functions can be evaluated conveniently when using a master equation approach. Equation (18.10) can be readily generalized to few-level systems by introducing an additional summation over the different decay channels and transition dipole moments. We shall find it convenient to rewrite Eq. (18.10) in the more compact form

$$S(\omega) \propto \Big[N(\boldsymbol{r}_{\text{atom}}, \omega)\Big] \text{Re} \left\{ \int_0^\infty e^{i\omega\tau} \Big\langle \sigma^-(0)\sigma^+(\tau) \Big\rangle \, d\tau \right\}, \tag{18.11}$$

where $N(\boldsymbol{r}, \omega)$ is the photonic part to be directly read off from the term in parentheses of Eq. (18.10).

18.1.3 Photon Correlations

Useful information about the dynamics of a single few-level system can be also obtained by measuring the photon correlations emitted by the few-level system

$$G^{(2)}(t, \tau) \propto \Big\langle : \hat{I}(t+\tau)\hat{I}(t) : \Big\rangle . \tag{18.12}$$

Here we have introduced normal ordering of Eq. (18.7) and the intensity operator for the detector

$$\hat{I}(t) = \oint_{\partial\Omega_{\text{det}}} \hat{I}(s, t)\, dS .$$

The photon correlation gives information about the probability of measuring a photon at time $t + \tau$ *conditional* to a previous measurement of another photon at time t. A typical experimental setup is depicted in Fig. 18.2 and consists of a semi-transparent mirror M and two photodetectors D_1, D_2. In principle, a single photodetector would suffice for the measurement, however, in general the dead time of a detector—this is the time it takes for the detector to recover after a photon measurement and to be ready to measure a second photon—is very long, and photon correlations with short τ times can usually be only measured with two detectors.

At time t one photon impinges on the mirror M and is randomly reflected or transmitted (with equal probabilities) to detector $D1$ or $D2$, where it is detected. A second photon at later time $t + \tau$ impinges on the same mirror and is again reflected or transmitted to one of the detectors. In case of incidence on the second detector

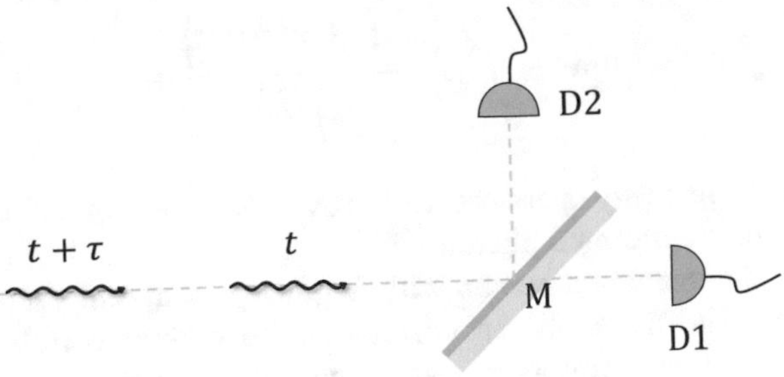

Fig. 18.2 Schematics of photon correlation measurements. A photon impinges at time t on a semi-transparent mirror M, and becomes transmitted or reflected with equal probability to detector $D1$ or $D2$. After detection, it takes a long time for the detector to recover. When a second photon arrives at a later time $t+\tau$, it has a fifty percent probability to be detected by the second photodetector. By collecting events where both detectors click, one can measure two-photon correlations

the photon arrival is measured, and the time delay τ is recorded to be accumulated for the correlation function $G^{(2)}$. In case of incidence on the same detector, which is still recovering from the first detection, no second photon is recorded, and nothing is added to the correlation function.

Under steady-state excitation the two-photon correlation of Eq. (18.12) does not depend on t, and it is convenient to introduce a normalized two-photon correlation defined as

Two-Photon Correlation Function (Normalized)

$$g^{(2)}(\tau) = \lim_{T\to\infty} \frac{1}{T} \int_0^T \frac{\left\langle : \hat{I}(t+\tau)\hat{I}(t) : \right\rangle}{\left\langle \hat{I} \right\rangle^2} dt \,. \tag{18.13}$$

Note that the above quantity can be also defined for negative delay times τ since the order of photon detections, that is which photon pushes the start and stop signals for the correlation measurement, is irrelevant.

In some cases it can be also useful to correlate photons emitted from different transitions of a few-level system. If the photons are emitted at different frequencies ω_a, ω_b and filtering for these frequency windows does not introduce a noticeable time delay (see also discussion below), we can introduce another correlation function

$$g_{ab}^{(2)}(\tau) = \lim_{T\to\infty} \frac{1}{T} \int_0^T \frac{\left\langle : \hat{I}_a(t+\tau)\hat{I}_b(t) : \right\rangle}{\left\langle \hat{I}_a \right\rangle \left\langle \hat{I}_b \right\rangle} dt \,. \tag{18.14}$$

Here $\hat{I}_a$, $\hat{I}_b$ are the intensity operators for the two transitions. Equation (18.14) gives the probability of measuring a photon with frequency ω_a and time delay τ subject to a second measurement of a photon with frequency ω_b at time "zero."

In order to relate these photon correlations to current correlations, we first introduce the transition operators $\sigma_{a,b}$ and dipole elements $\boldsymbol{p}_{a,b}$ associated with the corresponding atom transitions. For photon correlations of a single transition, for instance, in case of a two-level system, the operators and dipole moments can be chosen identical. We now make the following assumptions.

Atom Dynamics. The relevant time scales for the atom dynamics are assumed to be given by T_1, T_2 associated with relaxation and dephasing. For typical fluorophores or quantum dots, these times are in the range of pico to nano second.

Photonic Environment. We assume that the Green's function of the photonic environment changes slowly in the frequency range of $1/T_{1,2}$ centered around the transition frequency $\omega_{a,b}$. This assumption excludes highly structured environments, including for instance guided modes in stratified media, but certainly works for photonic environments with plasmon peak widths in the range of several tenth of eV.

Under the above assumptions, we can approximate in Eq. (18.1) the frequency of the Green's function by the central frequency $\omega_{a,b}$, and perform a Fourier transform in order to get

$$\hat{\boldsymbol{E}}^+(\boldsymbol{r},t) \approx \hat{\boldsymbol{E}}^+_{\text{inc}}(\boldsymbol{r},t) + i\mu_0\omega_{a,b} \int \bar{\bar{G}}_{\text{tot}}(\boldsymbol{r},\boldsymbol{r}',\omega_{a,b}) \cdot \hat{\boldsymbol{J}}^+_{\text{atom}}(\boldsymbol{r}',t-\delta t)\, d^3r' \,, \tag{18.15}$$

where δt is the time it takes for the photon to propagate from the atom to the detector. Equation (18.15) is a good approximation for non-dispersive media where the time envelope of the photon wavepacket is not modified during propagation, and propagation effects can be accounted for through a propagation time δt together with the Green's function evaluated at the central frequency. If the assumptions stated above are not valid, we have to resort to Eq. (18.3) which somewhat complicates our analysis, however, without adding anything particularly new. The two-photon correlation of Eq. (18.12) can then be rewritten as

$$G_{ab}^{(2)}(t,\tau) \propto \Big[N(\boldsymbol{r}_{\text{atom}},\omega_a)N(\boldsymbol{r}_{\text{atom}},\omega_b)\Big]\Big\langle \sigma_a^-(t)\sigma_b^-(t+\tau)\sigma_b^+(t+\tau)\sigma_a^+(t)\Big\rangle \,. \tag{18.16}$$

The first term in brackets accounts for the photon propagation and the second term characterizes the two-time photon correlation function, where we have discarded the constant time delay δt between photon emission and detection.

18.2 Quantum Regression Theorem

Before bringing the quantum optics toolbox to work, we have to add a final piece to the puzzle. In this section we derive the so-called *quantum regression theorem* which gives us the prescription of how to compute multi-time correlation functions of the form

$$\mathcal{I} = \left\langle \hat{A}_1^-(t_1)\hat{A}_2^-(t_2)\dots\hat{A}_n^-(t_n)\hat{A}_n^+(t_n)\dots\hat{A}_2^+(t_2)\hat{A}_1^+(t_1)\right\rangle . \tag{18.17}$$

They consist of "twin" operators $\hat{A}_i^\pm(t_i)$ acting in the subspace of the few-level system and propagating with positive and negative frequency components, which can be associated with photon annihilation and creation operators given in normal order—this is annihilation operators to the right and creation operators to the left. When at a specific time only an annihilation or creation operator is present, such as for the calculation of the fluorescence spectra, the other twin operator should be replaced by the unit operator $\mathbb{1}$. In the above expression the operators are given in the Heisenberg picture and we assume $t_1 < t_2 < \cdots < t_n$. We start by investigating the expression

$$\tilde{\mathcal{I}} = \left\langle \hat{A}_1^-(t_1)\hat{B}(t_2)\hat{A}_1^+(t_1)\right\rangle = \mathrm{tr}\left\{\left[\hat{A}_1^+(t_1)\hat{w}\,\hat{A}_1^-(t_1)\right]\hat{B}(t_2)\right\},$$

where we have lumped all operators at times later than t_1 into $\hat{B}(t_2)$ and have used a cyclic permutation under the trace to arrive at the last expression. $\hat{w}$ is the statistical operator of the total system comprising the atom and the photonic environment. The computation of a multi-time correlation function then follows through induction. The quantum regression theorem is quite general and builds on two assumptions.

Initial Conditions. We assume that initially the system and the photonic reservoir are uncorrelated, such that $\hat{w} = \hat{\rho}_S \otimes \hat{\rho}_R$.

Markovian Dynamics. We consider a Markovian time evolution where the system interacts with the reservoir in terms of scatterings, which are so fast that the system–reservoir correlations are only important during the scatterings but the density operator factorizes otherwise. Let us consider the time evolution (super)operator $\mathbb{U}$ for the system

$$\hat{\rho}_S(t) = \mathrm{tr}_R\left\{\hat{U}(t,t_0)\left[\hat{\rho}_s \otimes \hat{\rho}_R\right]\hat{U}^\dagger(t,t_0)\right\} = \mathbb{U}\left[\hat{\rho}_S\right](t,t_0), \tag{18.18}$$

which propagates $\hat{\rho}_S$ from time t_0 to t. One can think of the Markovian time evolution in terms of a propagation of $\hat{w}$ through small time steps

$$\hat{w}(t+\delta t) \approx \left\{\mathbb{U}\left[\hat{\rho}_S(t)\right](t+\delta t,t)\right\} \otimes \hat{\rho}_R ,$$

where δt is long enough that a scattering can be completed but short enough that $\hat{\rho}_S$ does not change noticeably (and its time evolution can be described in terms of a Markovian master equation). Thus, within δt the system and reservoir interact, which leads to relaxation and dephasing in the system, but at the end of δt the operator $\hat{w}$ factorizes again in the spirit of the molecular chaos assumption discussed in the previous chapter. Quite generally, in the quantum regression theorem we only require that such factorization can be performed after the measurement operators $\hat{A}^{\pm}$ act on the system, but since the measurement times are arbitrary it is better to stay with the above, more stringent assumption.

We next consider the expectation value of operators

$$\begin{aligned}\tilde{\mathcal{I}} &= \mathrm{tr}\left\{\left[\hat{U}^\dagger(t_1,t_0)\hat{A}_1^+\hat{U}(t_1,t_0)\right]\left[\hat{\rho}_S\otimes\hat{\rho}_R\right]\left[\hat{U}^\dagger(t_1,t_0)\hat{A}_1^-\hat{U}(t_1,t_0)\right]\hat{B}(t_2)\right\}\\ &= \mathrm{tr}\left\{\left[\hat{A}_1^+\hat{\rho}_S(t_1)\hat{A}_1^-\otimes\hat{\rho}_R\right]\left[\hat{U}(t_1,t_0)\hat{U}^\dagger(t_2,t_0)\,\hat{B}\,\hat{U}(t_2,t_0)\hat{U}^\dagger(t_1,t_0)\right]\right\}.\end{aligned}$$

In the first line we have transformed the operators $\hat{A}^{\pm}$ from the Heisenberg to the Schrödinger picture using the time evolution operators $\hat{U}$. In the second line we have propagated $\hat{\rho}_S$ using $\hat{U}$, and have used that $\hat{A}_1$ only acts in the system subspace and can thus be exchanged with the density operator of the reservoir. We have also transformed $\hat{B}$ from the Heisenberg to the Schrödinger picture, and have shuffled the operators cyclically under the trace. We now exploit the unitarity of $\hat{U}$ together with the group property to combine two successive time propagations through a single one,

$$\hat{U}(t_2,t_0)\hat{U}^\dagger(t_1,t_0)=\hat{U}(t_2,t_0)\hat{U}(t_0,t_1)=\hat{U}(t_2,t_1)\,.$$

Thus we find for the expectation value

$$\mathrm{tr}\left\{\hat{U}(t_2,t_1)\left[\hat{A}_1^+\hat{\rho}_S\hat{A}_1^-\otimes\hat{\rho}_R\right]\hat{U}^\dagger(t_2,t_1)\hat{B}\right\}=\mathrm{tr}_S\left\{\mathbb{U}\left[\hat{A}_1^+\hat{\rho}_S(t_1)\hat{A}_1^-\right](t_2,t_1)\,\hat{B}\right\}.$$

The important point here is that the fluctuation $\hat{A}_1^+\rho_S\hat{A}_1^-$ propagates in the same way as the density operator $\hat{\rho}_S$ in Eq. (18.18). Thus, if we have a machinery at hand to compute the time evolution of $\hat{\rho}_S$, for instance, the master equation of Lindblad form, we can immediately use the *same* machinery to compute multi-time correlation functions.

This is the essence of the quantum regression theorem which we now formulate in a general form. Suppose that the time evolution of the density operator can be written in the form (we skip again the subscript S)

$$i\hbar\frac{d\hat{\rho}}{dt}=\mathbb{L}(t)\hat{\rho}\,.$$

Here $\mathbb{L}$ is the Liouville operator governing the dynamic equation, in its most general form the master equation of Lindblad form. The time-propagated density operator can be formally written in the form

$$\hat{\rho}(t) = \mathbb{U}\big[\hat{\rho}(t_0)\big](t, t_0)\,,$$

with the time evolution operator introduced in Eq. (18.18). A multi-time correlation function of the form given in Eq. (18.17) can then be computed using the quantum regression theorem

Quantum Regression Theorem

$$\begin{aligned}
\delta\hat{\rho}_1 &= \hat{A}_1^+ \left\{\mathbb{U}\big[\hat{\rho}(t_0)\big](t_1, t_0)\right\} \hat{A}_1^- \\
\delta\hat{\rho}_i &= \hat{A}_i^+ \left\{\mathbb{U}\big[\delta\hat{\rho}_{i-1}(t_{i-1})\big](t_i, t_{i-1})\right\} \hat{A}_i^-\,, \quad i = 2, \dots n \\
\mathcal{I} &= \mathrm{tr}\big\{\,\delta\hat{\rho}_n\,\big\}\,.
\end{aligned} \tag{18.19}$$

As schematically depicted in Fig. 18.3, we first propagate the system density operator to time t_1, and act with the measurement operators $\hat{A}_i^\pm$ on $\hat{\rho}$ to get a fluctuation $\delta\hat{\rho}_1$ in the system. The quantum regression theorem states that the fluctuations propagate in the same manner as the system density operator, and thus the same differential equation can be used for its time propagation. We then propagate the fluctuation forward in time till the next pair of operators $\hat{A}_i^\pm$ acts on $\delta\hat{\rho}$, and so on till the last operator pair has acted. The expectation value $\mathcal{I}$ is finally given by the trace over the last fluctuation $\delta\hat{\rho}_n$.

18.3 Photon Correlations and Fluorescence Spectra

The master equation of Lindblad form and the quantum regression theorem are made to be used. Up to this point we have spent a considerable amount of time deriving the basic ideas and concepts, and readers not familiar with the topic might get the impression that this is a beautiful approach but too difficult to be actually used. This is not the case! As we will demonstrate in the following for a few selected examples, once the machinery is at hand the remaining calculations appear almost too simple. And also in numerical implementations the crucial steps usually only require a few lines of codes.

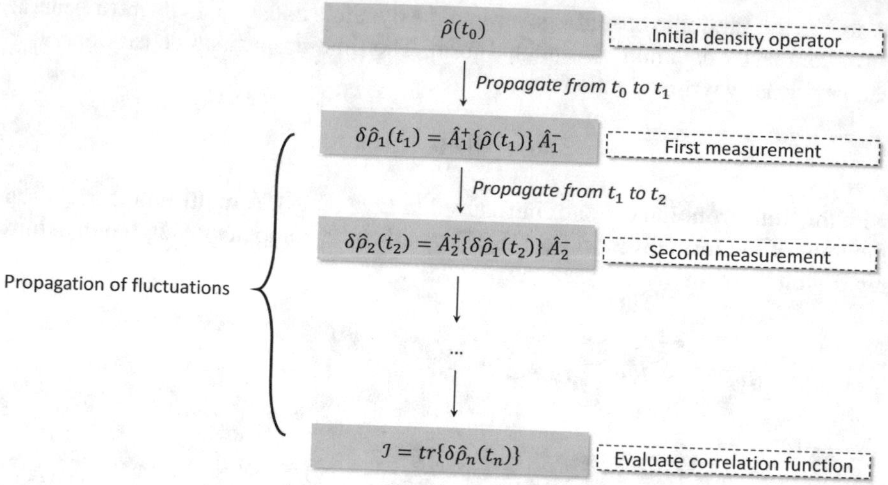

Fig. 18.3 Schematics of quantum regression theorem. We start with the initial density operator $\hat{\rho}(t_0)$ that is propagated in time to t_1 when the first pair of measurement operators acts on the system. By applying the twin operators from both sides on $\hat{\rho}(t_1)$ we induce a fluctuation in the system, as described by the possibly non-Hermitian fluctuation operator $\delta\hat{\rho}_1(t_1)$. The fluctuation is propagated forward in time using the *same* differential equation as for the density operator, till the time t_2 when the second pair of operators acts on the system. This procedure is repeated until we arrive at the final pair of operators at time t_n. The multi-time correlation function is finally obtained by taking the trace over $\delta\hat{\rho}_n$

18.3.1 Incoherently Driven Two-Level System

We start by considering a two-level system with the Hamiltonian

$$\hat{H} = \hbar\Delta\,\sigma_{ee}\,,$$

together with the Lindblad operators

$$\hat{L}_1 = \sqrt{\Gamma_1}\sigma_{ge}\,, \quad \hat{L}_2 = \sqrt{\Gamma_2}\sigma_{ee}\,, \quad \hat{L}_3 = \sqrt{G}\sigma_{eg}\,.$$

The first two operators account for relaxation due to photon emission and dephasing, respectively, and have been discussed in some length in Sect. 17.2. $\hat{L}_3$ describes the excitation of the two-level system from the ground to the excited state with a generation rate G through incoherent pumping. Note that the frequency ω of the exciting light field does not play any role for this simplified model, and correspondingly the value of the detuning Δ is of no particular significance. For clarity we keep Δ in the following rather than setting it equal to zero. The master equation of Lindblad form can be immediately obtained in the form

$$\dot{\rho}_{gg} = -G\rho_{gg} + \Gamma_1\rho_{ee}$$
$$\dot{\rho}_{ee} = \quad G\rho_{gg} - \Gamma_1\rho_{ee}$$
$$\dot{\rho}_{eg} = \big(-i\Delta - \frac{1}{2}\Gamma_{\text{tot}}\big)\rho_{eg}\,, \tag{18.20}$$

where we have introduced the abbreviation $\Gamma_{\text{tot}} = \Gamma_1 + \Gamma_2 + G$. Because of the incoherent pumping, the population elements ρ_{gg}, ρ_{ee} are not coupled to the polarization term ρ_{eg}, and their equations of motion have the form of simple rate equations. In the stationary state we have $\dot{\rho}_{gg}^{\infty} = \dot{\rho}_{ee}^{\infty} = 0$ and we thus get

$$G(1 - \rho_{ee}^{\infty}) = \Gamma_1\rho_{ee}^{\infty} \implies \rho_{ee}^{\infty} = \frac{G}{G + \Gamma_1}\,. \tag{18.21}$$

We have used the trace relation $\rho_{gg}^{\infty} + \rho_{ee}^{\infty} = 1$ which must be fulfilled at all times. The excited state population in Eq. (18.21) depends on the ratio between the generation rate G and the sum of generation and relaxation rates $G + \Gamma_1$, as one could have easily guessed based on simple reasoning.

The few-level operator accounting for optical transitions propagating with a positive frequency is $\hat{A}^+ = \sigma_{ge}$. With this we get for the (unnormalized) fluorescence intensity in the steady state

$$\left\langle \hat{A}^-\hat{A}^+\right\rangle = \text{tr}\left\{\sigma_{ge}^+\hat{\rho}^{\infty}\sigma_{eg}^-\right\} = \rho_{ee}^{\infty}\,.$$

We can now tackle the problem of computing the photon correlations and fluorescence spectra of the incoherently driven two-level system.

Photon Correlations

For the calculation of the two-photon correlation function of Eq. (18.13) we assume that the first photon detection occurs at time zero. Following the prescription of the quantum regression theorem of Eq. (18.19), the fluctuation after photon detection can be computed from

$$\delta\hat{\rho}_1(0) = \sigma_{ge}^+\hat{\rho}^{\infty}\sigma_{eg}^- = \rho_{ee}^{\infty}\,\sigma_{gg}\,. \tag{18.22}$$

For clarity we have added both the superscripts for positive and negative oscillation frequencies as well as the subscripts for the state transitions to the σ matrices. This result can be easily interpreted. The magnitude of the fluctuation depends on the excited state population, which relates to the probability of emitting a photon (and correspondingly the probability of detecting the first photon). Directly after photon emission the system is in the groundstate σ_{gg}, in agreement with the von Neumann measurement postulate. From the quantum regression theorem we then find for the (unnormalized) probability of detecting a second photon at time t the expression

$$\mathcal{I} = \mathrm{tr}\Big\{\sigma_{ge}^{+}\Big[\delta\hat{\rho}_1(t)\Big]\sigma_{eg}^{-}\Big\} = \delta\rho_{1,ee}(t)\,.$$

The above expression states that the probability for detecting a second photon is related to the excited state population of the fluctuation $\delta\hat{\rho}_1(t)$, subject to the condition that at time zero $\delta\hat{\rho}_1(0)$ is in the groundstate. From the master equation of Lindblad form, Eq. (18.20), we obtain for the buildup of the excited state population $\rho_{ee}(t)$ for a system in the groundstate at time zero

$$\rho_{ee}(t) = \Big(1 - e^{-(G+\Gamma_1)t}\Big)\,\rho_{ee}^{\infty}\,.$$

To get the fluctuation term we have to multiply this result with the magnitude of the fluctuation at time zero, $\delta\rho_{1,ee}(t) = \rho_{ee}^{\infty}\,\rho_{ee}(t)$. The normalized two-photon correlation is then obtained from Eq. (18.13), and we are led to our final result

$$g^{(2)}(t) = \frac{\big\langle :\hat{I}(t)\hat{I}(0): \big\rangle}{\big\langle \hat{I}\big\rangle^2} = 1 - e^{-(G+\Gamma_1)t}\,, \tag{18.23}$$

with the intensity operator $\hat{I} = \hat{A}^{-}\hat{A}^{+}$. It gives the probability of detecting a photon at time t subject to the condition that another photon has been detected at time zero, and the expression is normalized such that it approaches one at sufficiently long delay times where the two detections are uncorrelated. The physical meaning of Eq. (18.23) is that after the photo detection the system is in the groundstate, and it has to be pumped again into the excited state to emit a second photon, see also Fig. 18.4. The recovery time is governed by the generation and recombination rates G and Γ_1, respectively. For this reason, the probability of detecting two photons at the same time is zero, which is referred to as "photon antibunching" and constitutes a clear fingerprint for single two-level systems. In the figure we plot $g^{(2)}(t)$ also for negative delay times, as is often done in experiment because when using two photodetectors for the correlation measurement, see Fig. 18.2, a positive delay means that detector $D1$ records a photon before detector $D2$, while for negative delay times the detection order is reversed.

The two-photon correlation function can be also computed for a coherently driven two-level system, see Eq. (16.28) for the optical Bloch equation including relaxation and dephasing times T_1 and T_2, respectively. The (unnormalized) two-photon correlation function then corresponds to the solution of Eq. (16.28) for the system in the groundstate at time zero, this is when the first photon is detected. The result for resonant excitation can be found for instance in Ref. [197] and reads

$$g^{(2)}(t) = 1 - \left[\cos(\Omega_\Gamma t) + \frac{T_1^{-1} + T_2^{-1}}{2\Omega_\Gamma}\sin(\Omega_\Gamma t)\right] e^{-\frac{1}{2}(T_1^{-1}+T_2^{-1})t}\,, \tag{18.24}$$

where the effective Rabi frequency is defined as

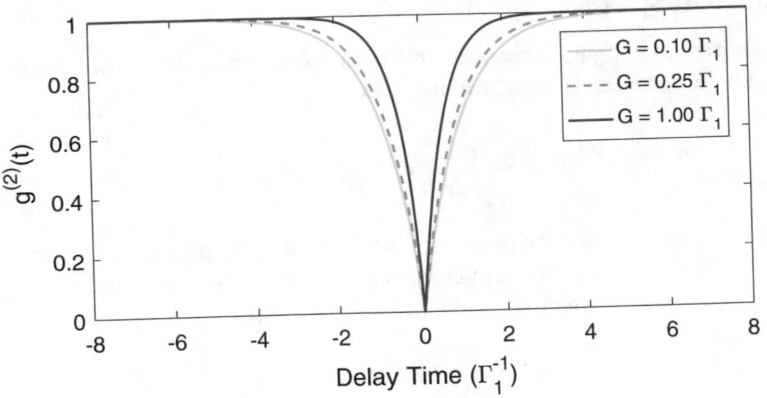

Fig. 18.4 Two-photon correlations for incoherently driven two-level system, as computed from Eq. (18.23). We compare different generation rates and keep the radiative decay rate Γ_1 fixed. With increasing G the excited state of the two-level system is faster populated, as reflected by the faster increase of $g^{(2)}$

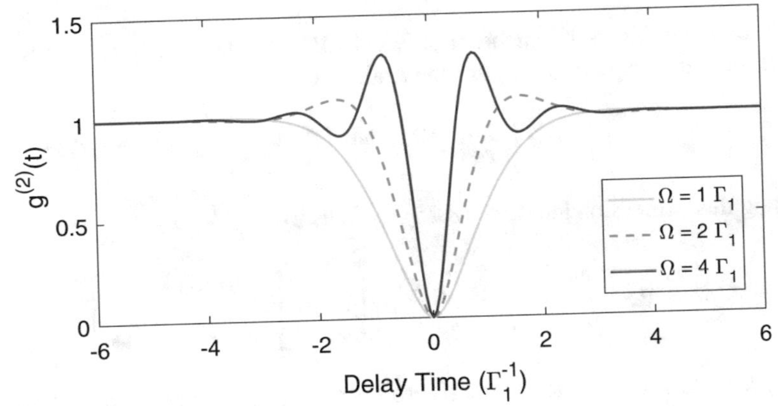

Fig. 18.5 Two-photon correlations for coherently driven two-level system, as computed from Eq. (18.24). We compare different Rabi frequencies Ω and keep the radiative decay rate Γ_1 fixed. The dephasing time is $T_2 = \frac{1}{2}T_1$. With increasing Ω the excited state of the two-level system is faster populated and we observe the onset of Rabi oscillations

$$\Omega_\Gamma = \sqrt{\Omega^2 - \frac{1}{4}(T_1^{-1} - T_2^{-1})^2}\,.$$

Figure 18.5 shows results for different Rabi frequencies Ω. One observes that with increasing Ω the excited state becomes faster populated, and we additionally observe the onset of Rabi oscillations at the highest Ω values.

Fluorescence Spectra

We continue to compute the spectrum of an incoherently driven two-level system using the result of Eq. (18.10), which reads

$$S(\omega) \propto \mathrm{Re}\left\{\int_0^\infty e^{i\omega t}\Big\langle\sigma_{eg}^-(0)\sigma_{ge}^+(t)\Big\rangle\,dt\right\}, \tag{18.25}$$

where we have neglected the photonic part describing the propagation of the photon from the atom to the detector. By acting at time zero with σ^- on the density operator and replacing the missing twin operator by the unit operator, we get

$$\delta\hat{\rho}_1(0) = \mathbb{1}\,\hat{\rho}^\infty\,\sigma_{eg}^- = \rho_{ee}^\infty\,\sigma_{eg}^-\,.$$

The time evolution of the fluctuation operator is determined by the last expression in Eq. (18.20), which can be solved to give

$$\delta\hat{\rho}_1(t) = \rho_{ee}^\infty\,e^{-i(\Delta-\frac{i}{2}\Gamma_{\rm tot})t}\sigma_{eg}^-\,.$$

For the quantum regression theorem of Eq. (18.19) we additionally have to compute the action of the operator $\sigma^+(t)$ at time t, and we get

$$\mathrm{tr}\Big\{\sigma_{ge}^+\delta\hat{\rho}_1(t)\,\mathbb{1}\Big\} = \mathrm{tr}\left\{\sigma_{ge}^+\left[\rho_{ee}^\infty\,e^{-i(\Delta-\frac{i}{2}\Gamma_{\rm tot})t}\sigma_{eg}^-\right]\right\} = e^{-i(\Delta-\frac{i}{2}\Gamma_{\rm tot})t}\,.$$

Inserting this expression into Eq. (18.25) leads us to

$$S(\omega) \propto \mathrm{Re}\left\{\int_0^\infty e^{i(\omega-\Delta+\frac{i}{2}\Gamma_{\rm tot})t}\,dt\right\} = \mathrm{Im}\left[\frac{e^{i(\omega-\Delta+\frac{i}{2}\Gamma_{\rm tot})t}}{\omega-\Delta+\frac{i}{2}\Gamma_{\rm tot}}\right]_0^\infty\,.$$

The upper integration limit becomes zero because of the damping term with $\Gamma_{\rm tot}$, and we finally obtain for the fluorescence spectrum the usual Lorentzian profile

$$S(\omega) \propto \frac{\left(\frac{\Gamma_{\rm tot}}{2}\right)}{(\omega-\Delta)^2+\left(\frac{\Gamma_{\rm tot}}{2}\right)^2}\,. \tag{18.26}$$

Figure 18.6 shows the fluorescence spectra for $\Delta = 0$ and for different pure dephasing rates Γ_2, as computed from Eq. (18.26). The spectra consist of a single Lorentzian peak, associated with photon emission for the decay of the two-level system from e to g, and the peak broadens with increasing Γ_2.

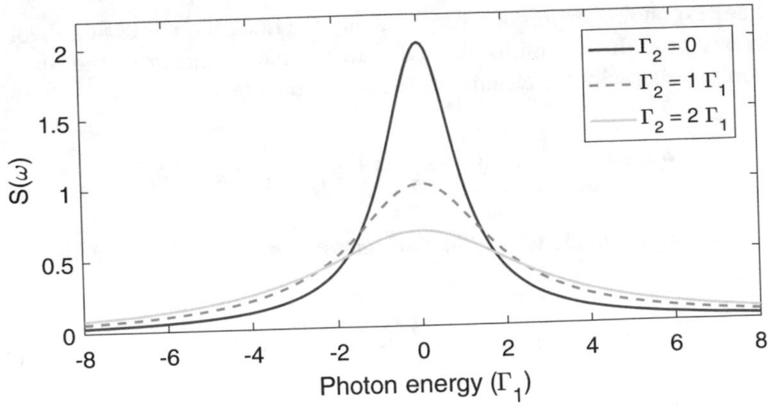

Fig. 18.6 Fluorescence spectra for incoherently driven two-level system and for different pure dephasing rates Γ_2, as computed from Eq. (18.26). $\hbar\omega = 0$ corresponds to the transition energy of the two-level system

18.3.2 *Quantum Regression Theorem and Eigenmodes*

The evaluation of the quantum regression theorem becomes particularly simple when using the eigenmodes for the Lindblad equation. Previously we have shown in Eq. (17.26) that the density matrix can be propagated in time through

$$\rho_{ij}(t) = \left[\mathbb{X}\exp\left(-\frac{i}{\hbar}\Lambda t\right)\tilde{\mathbb{X}}\right]_{ij,kl}\rho_{kl}(0)\,.$$

Here Λ is a matrix with the eigenvalues on the diagonal, and $\mathbb{X}$, $\tilde{\mathbb{X}}$ are the corresponding right and left eigenvectors. When propagating the initial density operator for sufficiently long times, we can use the above expression to compute the steady-state density matrix ρ_{ij}^{∞}.

Photon Correlations

Let σ_a^+ and σ_b^+ be the operators for transitions between few-level states which both propagate with positive frequencies. The frequencies ω_a, ω_b can be the same but could also differ, for instance, when considering different transitions in a single or two independent quantum systems. In the following we evaluate the probability of detecting a photon associated with the transition operator σ_a^+ at time t prior to a photon detection at time zero associated with the transition operator σ_b^+, see also Eq. (18.14). We shall find it convenient to speak about photons a and b, respectively. The intensity for photon a is

$$I_a = \mathrm{tr}\left\{\sigma_a^+\hat{\rho}^{\infty}\sigma_a^-\right\},$$

with a corresponding expression for I_b. Using the quantum regression theorem we can compute the fluctuation associated with the measurement of the first photon detection together with the ensuing time evolution from

$$\delta\rho_{1,ij}(t) = \left[\mathbb{X} \exp\left(-\frac{i}{\hbar}\Lambda t\right)\tilde{\mathbb{X}}\right]_{ij,kl} \left\langle k\middle|\sigma_b^+\hat{\rho}^\infty\sigma_b^-\middle|l\right\rangle .$$

From this we immediately obtain the two-photon correlation function

$$g^{(2)}(t) = \frac{\mathrm{tr}\left\{\sigma_a^+\,\delta\hat{\rho}_1(t)\,\sigma_a^-\right\}}{I_a I_b} . \tag{18.27}$$

Fluorescence Spectra

In a similar fashion we can compute the fluorescence spectra for a generic few-level system. Let σ^+ denote the operator for optically allowed transitions propagating with positive frequencies. Working out the expression for the spectrum of Eq. (18.10) by use of the quantum regression theorem gives

$$S(\omega) \propto \mathrm{Re}\left\{\int_0^\infty \sum_i \sigma_{ij}^+ \left[\mathbb{X} \exp\left(-\frac{i}{\hbar}\Lambda t\right)\tilde{\mathbb{X}}\right]_{ji,kl} \left\langle k\middle|\hat{\rho}^\infty\sigma^-\middle|l\right\rangle dt\right\} .$$

The term $\hat{\rho}^\infty\sigma^-$ accounts for the creation of the fluctuation at time zero, the term in brackets propagates the fluctuation forward in time, and finally at time t the second operator σ^+ acts on the fluctuation. Note that we have explicitly worked out the trace by introducing the sum over i. We can finally perform the time integration, just as discussed before for the incoherently driven two-level system, and arrive after some simple calculations at

$$S(\omega) = -\mathrm{Im}\left\{\sum_i \sigma_{ij}^+ \left[\mathbb{X}\,\frac{1}{\omega - \hbar^{-1}\Lambda}\,\tilde{\mathbb{X}}\right]_{ji,kl} \left\langle k\middle|\hat{\rho}^\infty\sigma^-\middle|l\right\rangle\right\} . \tag{18.28}$$

Λ is a diagonal matrix that can be easily inverted.

Figure 18.7 shows fluorescence spectra for a coherently driven two-level system as a function of photon energy $\hbar\omega$ and Rabi energy $\hbar\Omega$. With increasing Rabi energy the peaks broaden and finally split into three separate peaks, the so-called Mollow spectrum, which become energetically further separated with increasing $\hbar\Omega$. The peak splitting in the frequency domain can be related to the onset of Rabi oscillations in the time domain, and depends on the relative importance of coherent driving and incoherent loss channels.

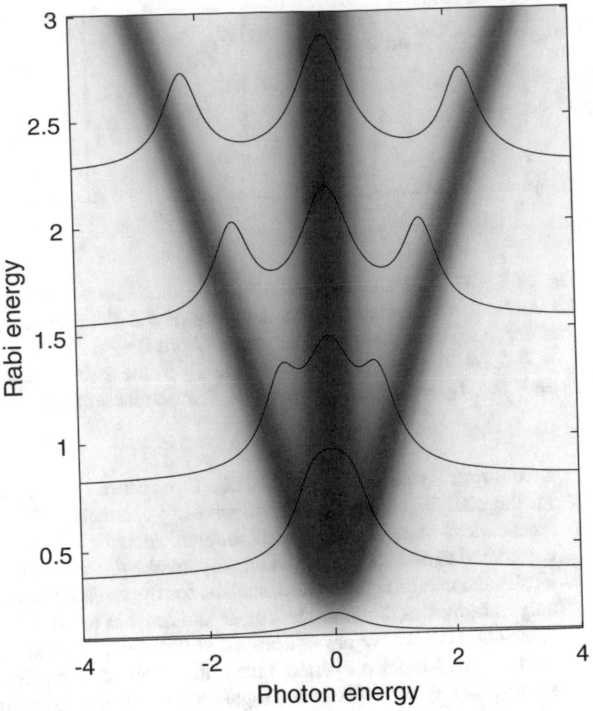

Fig. 18.7 Fluorescence spectra of coherently driven two-level system for different Rabi energies $\hbar\Omega$. We use $T_1 = 5$ and $T_2 = 2$ in arbitrary units. The density plot shows the spectra for different values of $\hbar\omega$ and $\hbar\Omega$, the solid lines shows spectra for selected $\hbar\Omega$ values which are offset for clarity. With increasing $\hbar\Omega$ the fluorescence peak broadens and splits into three peaks, the so-called Mollow spectrum, which become energetically further separated with increasing $\hbar\Omega$

18.3.3 Three-Level System

As another representative example we consider the three-level system depicted in the inset of Fig. 18.8. The groundstate $|0\rangle$ is separated through an energy gap $\hbar\omega_a$ from the first excited state $|1\rangle$, which is again separated through an energy gap $\hbar\omega_b = \hbar\omega_a - \hbar\Delta$ from the second excited state $|2\rangle$. Starting from the groundstate we first populate state $|1\rangle$, which can then be promoted to the second excited state $|2\rangle$. This level scheme mimics biexciton states in semiconductor quantum dots, where the groundstate corresponds to the empty dot, the state $|1\rangle$ to a dot populated by a single electron-hole pair (exciton), and the state $|2\rangle$ to a dot populated by two electron-hole pairs (biexciton), whose energy is reduced because of Coulomb correlations among the photoexcited carriers. To describe the above system, we use an interaction representation with respect to the central energy $\hbar\omega_a$. For the coherent and incoherent dynamics we introduce

- the Hamiltonian for the unperturbed states $\hat{H} = -\hbar\Delta\,|2\rangle\langle 2|$,
- the Lindblad operators for pumping $\sqrt{G}|1\rangle\langle 0|$, $\sqrt{G}|2\rangle\langle 1|$,
- the Lindblad operators for radiative decay $\sqrt{\Gamma_1}|0\rangle\langle 1|$, $\sqrt{\Gamma_1}|1\rangle\langle 2|$, and
- the Lindblad operators for dephasing $\sqrt{\Gamma_2}|1\rangle\langle 1|$, $\sqrt{\Gamma_2}|2\rangle\langle 2|$.

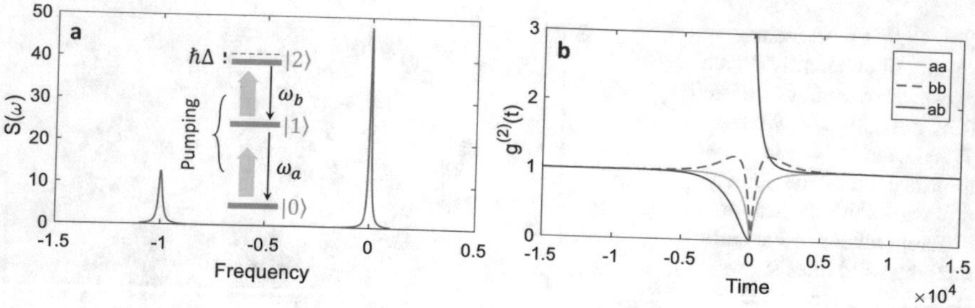

Fig. 18.8 (**a**) Fluorescence spectra and (**b**) photon correlations for three-level system depicted in the inset, where the state 2 has a detuning of $\hbar\Delta$ with respect to the transition energy $\hbar\omega_a$. We consider incoherent pumping with rate G from $0 \to 1$, $1 \to 2$, radiative decay with rate Γ_1 from $1 \to 0$, $2 \to 1$, and dephasing with rate Γ_2 in the states 1, 2. In the simulations we use $\Delta = 1$, $\Gamma_1 = 1/1000$, $\Gamma_2 = 1/100$, and $G = 1/2000$. For details see text

Incoherent Dynamics and Transition Operators When choosing the Lindblad operators for the pumping, relaxation, and dephasing channels it is important to introduce for each independent channel a *separate* Lindblad operator L_k. Transition channels are generally considered to be independent when they invoke *different* initial and final states. Usually this identification is straightforward, such as for the three-level system under consideration, and only in rare cases things are less clear and one has to go back to the microscopic scattering description (which we have discussed at the end of the previous chapter) in order to figure out how the Lindblad operators have to be properly chosen. For the computation of the fluorescence spectra we can add together the corresponding transition dipole operators

$$\sigma^+ = \sigma_a^+ + \sigma_b^+ ,$$

where $\sigma_{a,b}^+$ are associated with the transitions from $1 \to 0$, $2 \to 1$. The reason is that before detecting the photon (which passes through the spectrometer) we have no knowledge in which transition the photon has been created, correspondingly we keep—as always in quantum mechanics—all choices open by taking a linear combination. The master equation in Lindblad form then properly accounts for the incoherent loss channels and all the rest. This scheme has to be adapted when the emitted photons can be distinguished in some sense, for instance through their polarization, in which case we have to introduce different dipole operators for the different polarizations.

The situation is different for the computation of photon correlations in presence of spectral filtering. It is first important to guarantee that the spectral separation does not significantly affect the time resolution, which can be achieved for relatively broad frequency windows. If this can be done, we must use *separate* transition operators σ_a, σ_b for the different transitions, as the filter process has increased our knowledge about the state of the photon.

Figure 18.8a shows the fluorescence spectrum for the incoherently excited three-level system. It consists of two peaks, the first one centered at zero detuning (corresponding to the $1 \to 0$ transition) and another one at the detuning of state 2 (corresponding to the $2 \to 1$ transition. From the heights of the peaks we can infer the relative population of the two excited states. Panel (b) shows the different photon correlation functions. The behavior of the correlations between a photons is similar to the previously discussed two-level system, but the recovery of state 1 is slower because we now have the additional depletion channel of pumping from $1 \to 2$. The correlations for the b photons shows a dip at time zero (photon antibunching) together with a faster recovery in comparison to the a photons, because of the different population characteristics of states 1 and 2.

Things change considerably for the correlation between b and a photons. Once a b photon has been observed there is a considerably higher probability of detecting another a photon. This is no surprise for the cascade decay under consideration: once we have detected a b photon we know with certainty that the system is in the 1 state, owing to the von Neumann measurement postulate, so the probability of detecting a second a photon is high. This increased two-photon correlation is usually denoted as *photon bunching*. In contrast, for negative time delays the a photon is detected prior to the b one, thus for the detection of the b photon the system has to be pumped all the way from the groundstate to state 2. Figure 18.9 shows the experimental observation of such photon correlation measurements in a quantum dot structure [198].

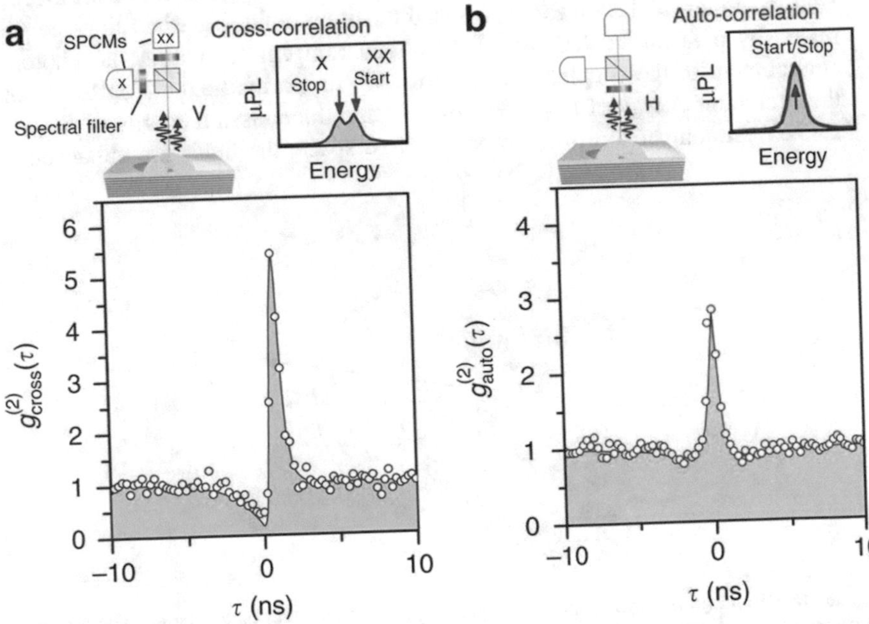

Fig. 18.9 Two-photon correlations for the case that the photons from the biexciton and exciton decay (**a**) can be distinguished, (**b**) cannot be distinguished. Image taken from Ref. [198]

18.4 Molecule Interacting with Metallic Nanospheres

We conclude this chapter with the example depicted in Fig. 18.10 where a quantum emitter is located in the gap region between two silver nanospheres. The level structure is depicted on the right-hand side of the figure and consists of a groundstate 1 that is optically coupled to state 2, from which the system relaxes to state 1 where it decays radiatively. Through the two excited states 1, 2 the excitation and decay channels are decoupled, which will allow us to compute the fluorescence spectra more easily. However, the formalism presented in the following needs to be adapted only slightly to account for the excitation and decay of a genuine two-level system. In what follows, we discuss the ingredients of our model and show that the theoretical analysis can be broken down into three steps, associated with the photonic environment, the specifications for the coherent and incoherent dynamics, and the solution of the master equation.

Photonic Environment. Due to the presence of the metallic nanoparticles, the excitation and decay channels of the quantum emitter become modified, as has been already discussed in Chap. 10 in the context of surface-enhanced Raman scattering (SERS). The red lobes in Fig. 18.10 report the modulus of the electric field in the far-field zone for an isolated quantum emitter (dashed lines, results scaled by a factor of 200) and for the quantum emitter within the photonic environment of the coupled nanospheres. We report the field strengths at both the excitation energy of $\hbar\omega_{01} = 2.925\,\text{eV}$ and the emission energy of $\hbar\omega_{20} = 2.85\,\text{eV}$. In comparison to the dipole in free space, the fields are enhanced by

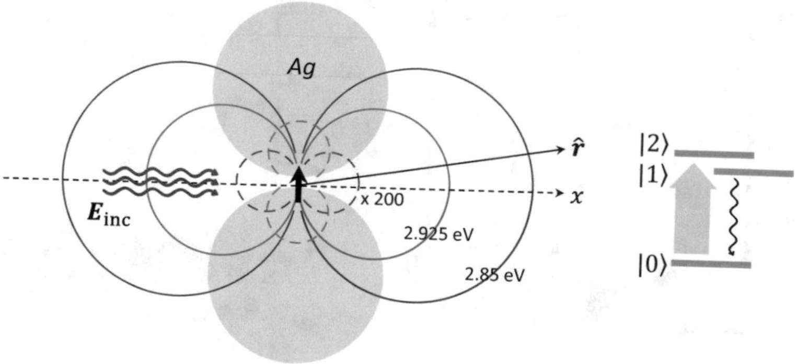

Fig. 18.10 Dipole (black arrow) located in gap region of two silver nanospheres. The red lobes report the emission pattern of the dipole at the dipole resonance of 2.85 eV (red) and at a detuned energy of 2.925 eV (light red). The dashed lines report the emission pattern of a dipole in free space for orientation along z (red) and x (gray), which are magnified by a factor of 200. For sphere diameters of 40 nm and a gap distance of 2.5 nm the electric field in the far-field zone is enhanced by a factor of approximately 750. In the simulations we use the level diagram reported on the right-hand side, with a groundstate 0, a state 2 to which the system is pumped, and a state 1 from which emission occurs. The decay channel from $2 \to 1$ is not shown

factors of approximately $f(\omega_{10}) \approx 800$ and $f(\omega_{02}) \approx 500$. In the following we assume that the quantum emitter is excited by a light field propagating along x and polarized along z, such that we can use the same enhancement factors also for excitation. Figure 18.11 reports the (a) extinction spectra and the (b) enhancement of the radiative and total decay rates, which have been discussed in some length in previous parts of this book. In the spectra we observe for a light polarization along z a pronounced peak centered around 2.85 eV, which is attributed to the bonding mode of the sphere dimer. This mode resonance is also visible in the radiative and total decay rates of the quantum emitter depicted in panel (b), which show enhancements in the range of up to millions.

Coherent and Incoherent Dynamics. Once the photonic environment has been specified, we have to set up the Hamiltonian and Lindblad operators for the three-level system. If we assume that the frequency of the exciting light is tuned to the ω_{02} transition, the Hamiltonian accounting for the coherent dynamics is of the form

$$\hat{H} = \Big(E_1 - \hbar\omega_{20}\Big)|1\rangle\langle 1| - \Big[f(\omega_{20})\hbar\Omega\Big]\Big(|1\rangle\langle 0| + |2\rangle\langle 0| + \text{h.c.}\Big), \qquad (18.29)$$

where h.c. denotes the Hermitian conjugate of the preceding term, and we have assumed that the Rabi frequency Ω is real and the dipole moments for the two transitions are identical. We consider Lindblad operators

$$L_{\text{tot},1} = \big[\Gamma(\omega_{10})\big]^{1/2}|0\rangle\langle 1|, \quad L_{\text{tot},2} = \big[\Gamma(\omega_{20})\big]^{1/2}|0\rangle\langle 2|$$

accounting for radiative and non-radiative decay, as well as dephasing with rate Γ_2, and an internal relaxation between states 1 and 2,

$$L_{12} = \big[\Gamma_{2\to 1}\big]^{1/2}|1\rangle\langle 2|.$$

Solution of Master Equation. Once the Hamiltonian and the Lindblad operators have been specified, we can solve the master equation of Lindblad form along

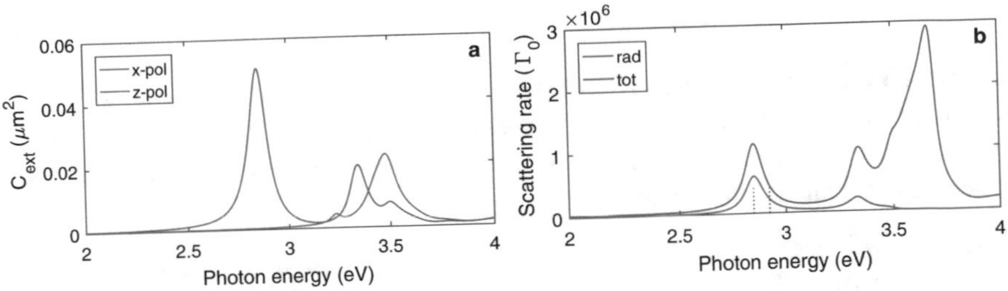

Fig. 18.11 (**a**) Extinction cross section and (**b**) enhancement of radiative and total decay rate for dipole oriented along z and placed in between two silver nanospheres, as depicted in Fig. 18.10. The dotted lines in panel (**b**) report the energies of the different molecule transitions considered in our calculations

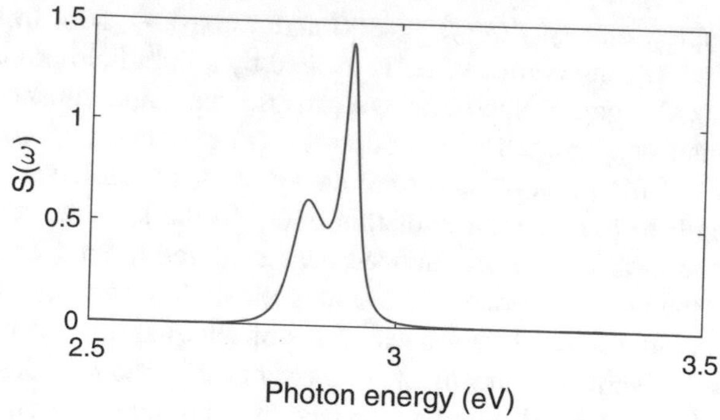

Fig. 18.12 Calculated fluorescence spectrum for optically driven quantum emitter placed in the gap region of two metallic nanospheres. For the emitter in free space we use a scattering rate of $\Gamma_0 = 0.1\,\text{ns}^{-1}$. The dephasing rates are set to $\Gamma_2 = 1\,\text{ps}^{-1}$, the interlevel scattering rate is $\Gamma_{2\to 1} = 10\,\Gamma_0$, and the Rabi energy for the excitation is $10\,\mu\text{eV}$

the same lines as previously discussed. The fluorescence spectra of Eq. (18.10) have to be only slightly modified to account for the emission enhancement according to

$$S(\omega) \propto \omega^4 f^2(\omega)\,\text{Re}\left\{\int_0^\infty e^{i\omega\tau}\Big\langle \sigma^-(0)\sigma^+(\tau)\Big\rangle\, d\tau\right\}. \tag{18.30}$$

Figure 18.12 shows the fluorescence spectra as computed with this model and for the parameters given in the caption. It consists of two peak, associated with the $1 \to 0$ and $2 \to 0$ decay. Note that around the energy of the driving laser field we should also consider light scattered directly by the nanosphere. In Ref. [199, 200] the authors generalized the above scheme to molecules with vibronic excitations, and showed that within such an approach one can develop a model suitable for the description of surface-enhanced Raman scattering (SERS).

Exercises

Exercise 18.1 Consider Eq. (18.24) for the two-photon correlations $g^{(2)}$ of a coherently driven two-level system in presence of relaxation and dephasing. Under which conditions are Rabi oscillations visible in $g^{(2)}$? Discuss separately the cases where Ω_Γ is real and imaginary.

Exercise 18.2 Use from the NANOPT toolbox the file `demotwolevel05.m` to compute the two-photon correlations $g^{(2)}$ of an optically driven two-level system.

(a) Investigate the influence of the light–matter coupling, denoted in the program with g rather than Ω, which depends on the field amplitude of the exciting light field. When do you observe the onset of Rabi oscillations?
(b) Investigate the influence of dephasing, and the interplay of dephasing and g. How does it affect the Rabi oscillations observable in $g^{(2)}$? Compare with the findings of exercise 18.1.
(c) Use the file `demotwolevel04.m` to visualize the time evolution of the fluctuation operator for two-photon correlations on the Bloch sphere. How does the vector become modified through the photon measurements?

Exercise 18.3 Start from the expression of Eq. (18.11) for the fluorescence spectrum, and use for a time-independent Liouville operator $\mathbb{L}$ the eigenvectors $\mathbb{X}$, $\tilde{\mathbb{X}}$ and eigenvalues Λ to arrive at Eq. (18.28).

Exercise 18.4 Use from the NANOPT toolbox the file `demotwolevel06.m` to compute the fluorescence spectra of an optically driven two-level system.

(a) Investigate the influence of the light–matter coupling, and determine the condition under which peak splitting occurs.
(b) Determine for sufficiently strong field strengths the dependence of the peak positions on the light–matter coupling.

Exercise 18.5 Use from the NANOPT toolbox the file `demothreelevel01.m` for the three-level system depicted in the inset of Fig. 18.8.

(a) Replace the coherent excitation with an incoherent pumping.
(b) Modify the program such that it computes the fluorescence spectrum. You may like to use `demotwolevel06.m` as a template.
(c) Modify the program such that it computes the photon correlations using *separate* detection operators for the decays of $2 \to 1$ and $1 \to 0$. Compute all two-photon correlations, and discuss the conditions for observing photon bunching or antibunching.

Appendix A
Complex Analysis

A.1 Cauchy's Theorem

Consider a complex function $f(z)$ that is *analytic* in a given region of complex space, which means that the function can be expanded in a Taylor series around any point z_0, and within a given neighborhood of z_0 the function value does not depend on how the point is approached,

$$\lim_{z \to a} f(z) = f(a)\,.$$

Analytic functions are smooth, that is, infinitely differentiable. They have the remarkable property that any integration in complex plane along a closed path gives zero,

Cauchy's Theorem

$$\oint_C f(z)\,dz = 0\,. \tag{A.1}$$

To proof this relation, we first use $z = x + iy$ and decompose the function into real and imaginary parts,

$$f(x, y) = u(x, y) + iv(x, y)\,.$$

The derivative of the function can then be computed from

U. Hohenester, *Nano and Quantum Optics*, Graduate Texts in Physics,
https://doi.org/10.1007/978-3-030-30504-8

$$f'(z) = \lim_{\eta\to 0} \frac{f(z+\eta)-f(z)}{\eta}$$
$$= \lim_{\eta\to 0} \frac{u(x+\eta,y)+iv(x+\eta,y)-u(x,y)-iv(x,y)}{\eta} = \frac{\partial u}{\partial x} + i\frac{\partial v}{\partial x}\,.$$

However, for an analytic function we can also take the derivative along a different direction, say along the imaginary axis

$$f'(z) = \lim_{\eta\to 0} \frac{f(z+i\eta)-f(z)}{i\eta}$$
$$= -i\lim_{\eta\to 0} \frac{u(x,y+\eta)+iv(x,y+\eta)-u(x,y)-iv(x,y)}{\eta} = -i\frac{\partial u}{\partial y} + \frac{\partial v}{\partial y}\,.$$

Because the function is assumed to be analytical, both expressions must be identical. A comparison leads to the so-called Cauchy–Riemann equations for analytic functions

$$\frac{\partial u}{\partial x} = \frac{\partial v}{\partial y}\,,\quad \frac{\partial v}{\partial x} = -\frac{\partial u}{\partial y}\,. \tag{A.2}$$

We now return to Eq. (A.1) and split the integrand $f = u + iv$ as well as the differential $dz = x + i\,dy$ into their real and imaginary parts,

$$\oint_C (u+iv)(dx+i\,dy) = \oint_C (u\,dx - v\,dy) + i\oint_C (v\,dx + u\,dy)\,.$$

This expression can be rewritten using Green's theorem[1]

$$\oint_C (u\,dx + v\,dy) = \int \left(\frac{\partial v}{\partial x} - \frac{\partial u}{\partial y}\right) dxdy\,, \tag{A.3}$$

and we obtain for the real and imaginary parts

$$\oint_C (u\,dx - v\,dy) = \int \left(-\frac{\partial v}{\partial x} - \frac{\partial u}{\partial y}\right) dxdy = 0$$
$$\oint_C (v\,dx + u\,dy) = \int \left(\frac{\partial u}{\partial x} - \frac{\partial v}{\partial y}\right) dxdy = 0\,.$$

We have used the Cauchy–Riemann equations of Eq. (A.2) to evaluate the integrals to zero. This completes our proof of the Cauchy's theorem.

---

[1] Green's theorem can be derived from Stokes' theorem of Eq. (2.13) using

$$\boldsymbol{F} = u(x,y)\,\hat{\boldsymbol{x}} + v(x,y)\,\hat{\boldsymbol{y}}\,,\quad d\boldsymbol{\ell} = \hat{\boldsymbol{x}}\,dx + \hat{\boldsymbol{y}}\,dy\,.$$

A.2 Residue Theorem

Consider a complex integral of the form

$$\oint_C \left[\frac{f(z)}{z-z_0}\right] dz = \oint_C g(z)\, dz\,. \tag{A.4}$$

Here $f(z)$ is an analytic function and C is a contour that encloses the critical point z_0, as schematically shown in Fig. A.1. In complex analysis the critical point z_0 is called a pole, and the function $g(z)$ is analytic everywhere except at z_0. We now deform the integration path such that it approaches the critical point z_0 along A, it moves around z_0 along a circle with radius $r \to 0$, and finally goes back along A to the contour C. The two contributions along the integration path A cancel each other because of the opposite integration directions. As for the circle, we introduce polar coordinates

$$z = z_0 + re^{i\phi}\,, \qquad dz = ir^{i\phi}\, d\phi\,.$$

Thus, we get for the integration path along the circle

$$-\int_0^{2\pi} \left[\frac{f\left(z_0 + re^{i\phi}\right)}{\left(z_0 + re^{i\phi}\right) - z_0}\right] \left(ire^{i\phi} d\phi\right) = -f(z_0) \int_0^{2\pi} \frac{ire^{i\phi}}{ire^{i\phi}}\, d\phi \underset{r\to 0}{\longrightarrow} -2\pi i\, f(z_0).$$

We have pulled out the function $f(z)$ from the integral because it is assumed to change only slowly in the vicinity of z_0. The negative sign of the integral is due to the fact that we move around z_0 in a clockwise direction. We then obtain for the integration path along C and the small circle around z_0

$$\oint_C \left[\frac{f(z)}{z-z_0}\right] dz - 2\pi i f(z_0) = 0\,.$$

The above expression is zero because of Eq. (A.1) and the fact that the combined integration path now excludes the pole, and thus the integrand is analytic within the entire integration domain. With this, we find the most simple form of the residue theorem

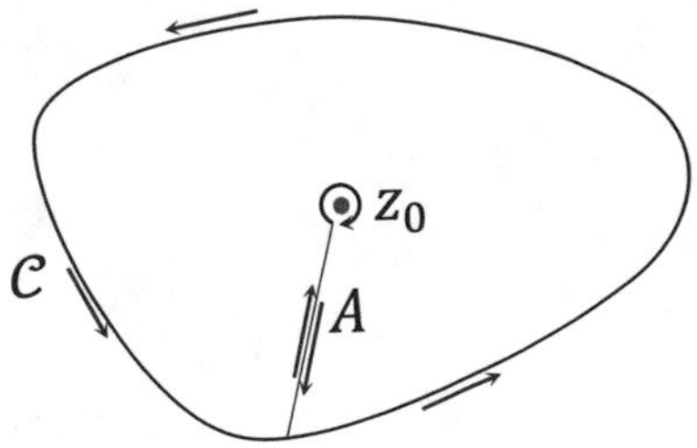

Fig. A.1 Schematics for residue theorem. A complex function $g(z)$ with a pole at z_0 is integrated along a contour C that encloses the critical point z_0

Residue Theorem

$$\oint_C \left[\frac{f(z)}{z - z_0}\right] dz = 2\pi i f(z_0)\,. \tag{A.5}$$

The above result can be easily generalized for a larger number of poles.

Appendix B
Spectral Green's Function

In this appendix we show how to decompose the Green's functions into plane waves. It turns out to be convenient to consider a decomposition using complex wavenumbers such that the Green's function becomes a function with complex arguments. For such functions there exists an important theorem of complex analysis, the so-called Cauchy's theorem, which states that the contour integral of an analytic function is zero. This theorem is sketched in Appendix A and plays an important role in the following discussion.

B.1 Spectral Decomposition of Scalar Green's Function

We start by decomposing the Green's function of the Helmholtz equation, Eq. (5.5), into plane waves. The defining equation reads

$$\left(\nabla^2 + k_1^2\right) g(\boldsymbol{r}) = -\delta(\boldsymbol{r}),$$

where k_1 is a wavenumber. We use the subscript on k_1 to distinguish it from the wavevector $\boldsymbol{k}$ of the Fourier transforms for the Green's function and Dirac's delta function

$$g(\boldsymbol{r}) = (2\pi)^{-3} \int_{-\infty}^{\infty} e^{i\boldsymbol{k}\cdot\boldsymbol{r}}\, \tilde{g}(\boldsymbol{k})\, d^3k$$

$$\delta(\boldsymbol{r}) = (2\pi)^{-3} \int_{-\infty}^{\infty} e^{i\boldsymbol{k}\cdot\boldsymbol{r}}\, d^3k\,.$$

Inserting these expressions into the defining equation for the Green's function we obtain

U. Hohenester, *Nano and Quantum Optics*, Graduate Texts in Physics,
https://doi.org/10.1007/978-3-030-30504-8

$$(2\pi)^{-3}\int_{-\infty}^{\infty} e^{i\boldsymbol{k}\cdot\boldsymbol{r}}\Big[(k_1^2-k^2)\tilde{g}(\boldsymbol{k})+1\Big]\,d^3k=0\,.$$

Because the equation has to be fulfilled for all values of $\boldsymbol{r}$ we are led to $\tilde{g}(\boldsymbol{k}) = 1/(k^2-k_1^2)$ and, in turn,

$$g(\boldsymbol{r})=(2\pi)^{-3}\int_{-\infty}^{\infty}\frac{e^{i\boldsymbol{k}\cdot\boldsymbol{r}}}{k^2-k_1^2}\,d^3k\,. \tag{B.1}$$

For a lossy material with $\varepsilon''>0$ we have

$$k_1^2=\varepsilon_1\mu_1\,\omega^2\longrightarrow \mathrm{Im}\{k_1\}>0$$

and the integrand of Eq. (B.1) is well-defined for all values of k. To additionally consider lossless materials we (1) add a small loss term $i\eta$ to the wavenumber, and (2) let $\eta\to 0$ approach zero at the end of the calculation. With this procedure we ensure that the Fourier transform only includes outgoing waves, similar to our discussion of Chap. 5 about the boundary condition of Green's functions. We next rewrite the Green's function decomposition in Cartesian coordinates[1]

$$g(x,y,z)=(2\pi)^{-3}\int_{-\infty}^{\infty}\frac{e^{i(k_xx+k_yy+k_zz)}}{k_z^2-(k_{1z}+i\eta)^2}\,dk_xdk_ydk_z\,,$$

with $k_{1z}=\sqrt{k_1^2-k_x^2-k_y^2}$. To evaluate the k_z part of the integral we employ Cauchy's theorem of Eq. (A.1). Let us consider the case of $z>0$ first. For a complex wavenumber $k_z=k_z'+ik_z''$ we have

$$e^{i(k_z'+ik_z'')z}=e^{ik_z'z}e^{-k_z''z}\underset{z\to\infty}{\longrightarrow}0\,.$$

Thus, we can add to the integration contour a semicircle in the upper complex k_z-plane, as depicted in Fig. B.1, whose contribution becomes zero in the limit $R\to\infty$. The contour integral along the real k_z axis and back along the semicircle would be zero if the integrand was an analytic function. However, the pole at $k_{1z}+i\eta$ has to be treated with care and we deform the integration path as follows:

- We move along the line indicated with A from the real axis to the pole.
- We move along a circle $k_z=k_{1z}+i\eta+re^{i\phi}$ around the pole and set $r\to 0$.
- We finally move again along line A back to the real axis.

[1]Note that if k_1 has a small and positive imaginary part, then also $k_{1z}=\sqrt{k_1^2-k_x^2-k_y^2}$ has a small and positive imaginary part.

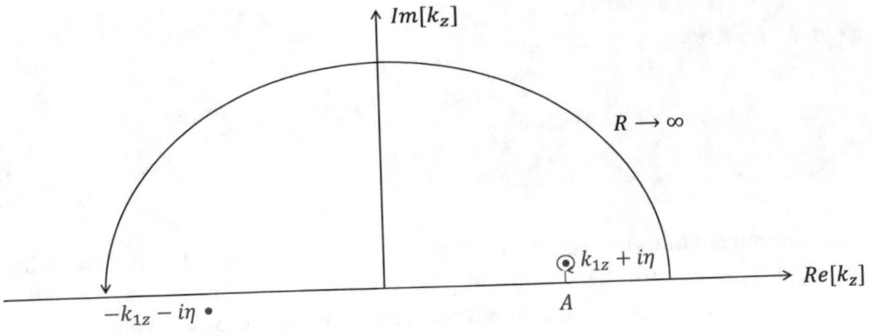

Fig. B.1 Path of integration for Green's function in lossy material. The poles of the integrand are located at $\pm(k_{1z}+i\eta)$

The two contributions along A cancel each other because of the opposite integration directions. For the small circle around the pole (the so-called residuum) we get

$$\mathcal{R} = \oint \frac{(e^{ik_z z})\,dk_z}{(k_z - k_{1z} - i\eta)(k_z + k_{1z} + i\eta)}$$
$$= \left(e^{i(k_{1z}+i\eta)z}\right)\lim_{r\to 0}\int_{\pi}^{-\pi}\frac{ire^{i\phi}\,d\phi}{(re^{i\phi})(2k_{1z}+2i\eta+re^{i\phi})} = -2\pi i\,\frac{e^{i(k_{1z}+i\eta)z}}{2(k_{1z}+i\eta)}\,.$$

When going from the first to the second line we have pulled out the exponential from the integral, because for $r\to 0$ it varies slowly as a function of ϕ and can be safely evaluated at the pole value $k_{1z}+i\eta$. Additionally, we have set $dk_z = ire^{i\phi}\,d\phi$ in polar coordinates. Thus, we obtain for the entire integration path

$$\int_{-\infty}^{\infty}\frac{e^{ik_z z}\,dk_z}{k_z^2-(k_{1z}+i\eta)^2} + \mathcal{R} = 0 \implies \int_{-\infty}^{\infty}\frac{e^{ik_z z}\,dk_z}{k_z^2-(k_{1z}+i\eta)^2} = i\pi\,\frac{e^{ik_{1z}z}}{k_{1z}}\,,$$

where we have set $\eta\to 0$ to arrive at the final expression. The Green's function then becomes

$$g(x,y,z) = \frac{i}{8\pi^2}\int_{-\infty}^{\infty}\frac{e^{i(k_x x+k_y y+k_{1z}|z|)}}{k_{1z}}\,dk_x dk_y\,.$$

Note that the above expression can be also used for negative z values. In this case we have to close the integration path in the lower half of the complex plane. With this, we are finally led to the Weyl identity for the expansion of a spherical wave in terms of plane waves

Weyl Identity

$$\frac{e^{ikr}}{r} = \frac{i}{2\pi}\int_{-\infty}^{\infty} \frac{e^{i(k_x x + k_y y + k_{1z}|z|)}}{k_z}\, dk_x dk_y\,. \tag{B.2}$$

The k_z component of the wavevector has to be computed from the dispersion relation $k_1^2 = k_x^2 + k_y^2 + k_z^2$, and for lossless media we implicitly assume that a small loss term in $k_z + i\eta$ is present. From the integral identity for the Bessel function

$$\int_0^{2\pi} e^{ix\cos\phi}\, d\phi = 2\pi\ J_0(x)$$

we can rewrite the Weyl identity by introducing polar coordinates (ϕ, k_ρ) for the (k_x, k_y) coordinates, and we arrive at the Sommerfeld identity

$$\frac{e^{ikr}}{r} = i\int_0^{\infty} \frac{k_\rho}{k_z} J_0(k_\rho \rho) e^{ik_z|z|}\, dk_\rho\,. \tag{B.3}$$

B.2 Spectral Representation of Dyadic Green's Function

We next show how to decompose the Green's dyadics into plane waves. Using the Green's dyadics definition of Eq. (5.19), together with the plane wave decomposition for the scalar Green's function of Eq. (B.1), we are led to

$$g_{ij}(\boldsymbol{r}) = \left(\delta_{ij} + \frac{\partial_i \partial_j}{k_1^2}\right)\frac{e^{ik_1 r}}{4\pi r} = \frac{1}{8\pi^3 k_1^2}\int_{-\infty}^{\infty} e^{i\boldsymbol{k}\cdot\boldsymbol{r}}\left(\frac{k_1^2\delta_{ij} - k_i k_j}{k^2 - k_1^2}\right) d^3k\,.$$

Before performing the k_z integration analytically using the complex integration procedure discussed above, we note that in the limit $k_z \to \infty$ the term in parentheses on the right-hand side becomes $-\hat{z}_i\hat{z}_j$ and the integral is not well behaved for g_{zz}. We thus add and subtract the term

$$g_{ij}(\boldsymbol{r}) = \frac{1}{8\pi^3 k_1^2}\int_{-\infty}^{\infty} e^{i\boldsymbol{k}\cdot\boldsymbol{r}}\left(\frac{k_1^2\delta_{ij} - k_i k_j}{k^2 - k_1^2} + \hat{z}_i\hat{z}_j\right) d^3k - \frac{\hat{z}_i\hat{z}_j}{8\pi^3 k_1^2}\int_{-\infty}^{\infty} e^{i\boldsymbol{k}\cdot\boldsymbol{r}} d^3k\,. \tag{B.4}$$

The second integral gives $-\hat{z}_i\hat{z}_j k_1^{-2}\delta(\boldsymbol{r})$, and the first one is now well behaved in the limit $k_z \to \infty$ and can be evaluated using complex integration as previously discussed in the context of the Weyl identity, Eq. (B.2). We have to distinguish the following cases:

- For $z > 0$ we close the semicircle in the upper complex half-space and evaluate the residuum at $k_{1z}^{+} = k_{1z} + i\eta$.
- For $z < 0$ we close the semicircle in the lower complex half-space and evaluate the residuum at $k_{1z}^{-} = -k_{1z} - i\eta$.

At the poles $k_{1z}^{\pm}$ the term in parentheses of Eq. (B.4) becomes

$$k_1^2\delta_{ij} - k_i k_j + \hat{z}_i\hat{z}_j\left(k^2 - k_1^2\right) \longrightarrow k_1^2\delta_{ij} - k_{1i}^{\pm}k_{1j}^{\pm}\,,$$

where we have introduced the wavenumber

$$\boldsymbol{k}_1^{\pm} = k_x\hat{\boldsymbol{x}} + k_y\hat{\boldsymbol{y}} \pm k_{1z}\hat{\boldsymbol{z}}\,,\quad k_{1z} = \sqrt{k_1^2 - k_x^2 - k_y^2} + i\eta\,.$$

Putting together the results we are led to

$$g_{ij}(\boldsymbol{r}) = \frac{i}{8\pi^2}\int_{-\infty}^{\infty} e^{i\boldsymbol{k}_1^{\pm}\cdot\boldsymbol{r}}\left(\frac{\delta_{ij} - \hat{k}_{1i}^{\pm}\hat{k}_{1j}^{\pm}}{k_{1z}}\right)dk_x dk_y - \frac{\hat{z}_i\hat{z}_j}{k_1^2}\delta(\boldsymbol{r})\,.$$

When acting on a vector, the matrix $\delta_{ij} - \hat{k}_{1i}^{\pm}\hat{k}_{1j}^{\pm}$ projects on the directions perpendicular to $\hat{\boldsymbol{k}}_1^{\pm}$, as discussed for instance in Sect. 2.5.

Decomposition into TE and TM Modes It turns out to be convenient to introduce a triad formed by $\hat{\boldsymbol{k}}_1^{\pm}$ and the following vectors

$$\boldsymbol{\epsilon}^{TE}(\hat{\boldsymbol{k}}_1^{\pm}) = \frac{\hat{\boldsymbol{k}}_1^{\pm}\times\hat{\boldsymbol{z}}}{|\hat{\boldsymbol{k}}_1^{\pm}\times\hat{\boldsymbol{z}}|} = \frac{1}{k_\rho}\left(k_y\hat{\boldsymbol{x}} - k_x\hat{\boldsymbol{y}}\right) \tag{B.5a}$$

$$\boldsymbol{\epsilon}^{TM}(\hat{\boldsymbol{k}}_1^{\pm}) = \hat{\boldsymbol{k}}_1^{\pm}\times\boldsymbol{\epsilon}^{TE}(\hat{\boldsymbol{k}}_1^{\pm}) = \pm\frac{k_{1z}}{k_1 k_\rho}\left(k_x\hat{\boldsymbol{x}} + k_y\hat{\boldsymbol{y}}\right) - \frac{k_\rho}{k_1}\hat{\boldsymbol{z}}\,, \tag{B.5b}$$

with $k_\rho = (k_x^2 + k_y^2)^{\frac{1}{2}}$. When considering reflections and transmissions at the interfaces of a stratified medium, as done in Chap. 8, this decomposition proves particularly useful because the two basis vectors can be associated with TE and TM fields. We thus get

$$\begin{aligned} g_{ij}(\boldsymbol{r}) &= -\frac{\hat{z}_i\hat{z}_j}{k_1^2}\delta(\boldsymbol{r}) \\ &+ \frac{i}{8\pi^2}\int_{-\infty}^{\infty}\frac{e^{i\boldsymbol{k}_1^{\pm}\cdot\boldsymbol{r}}}{k_{1z}}\left\{\epsilon_i^{TE}(\hat{\boldsymbol{k}}_1^{\pm})\epsilon_j^{TE}(\hat{\boldsymbol{k}}_1^{\pm}) + \epsilon_i^{TM}(\hat{\boldsymbol{k}}_1^{\pm})\epsilon_j^{TM}(\hat{\boldsymbol{k}}_1^{\pm})\right\}dk_x dk_y\,. \end{aligned} \tag{B.6}$$

In what follows it turns out to be convenient to consider the more general case of $\epsilon_i^{TE}(\boldsymbol{k}_1)\epsilon_j^{TE}(\boldsymbol{k}_2)$ and a similar expression for the TM fields, where $\boldsymbol{k}_1$ and $\boldsymbol{k}_2$ have the same parallel wavevectors but can differ regarding their z-components. Such combinations are needed for the description of reflected and transmitted waves in the case of stratified media. We then get

$$\left[\epsilon_i^{TE}(\boldsymbol{k}_1)\,\epsilon_j^{TE}(\boldsymbol{k}_2)\right] = \frac{1}{2}\begin{bmatrix} 1-\cos 2\phi & -\sin 2\phi & 0 \\ -\sin 2\phi & 1+\cos 2\phi & 0 \\ 0 & 0 & 0 \end{bmatrix}_{ij}$$

$$\left[\epsilon_i^{TM}(\boldsymbol{k}_1)\epsilon_j^{TM}(\boldsymbol{k}_2)\right] = \frac{1}{2k_1k_2} \tag{B.7}$$

$$\times\begin{bmatrix} k_{1z}k_{2z}(1+\cos 2\phi) & k_{1z}k_{2z}\sin 2\phi & -2k_{1z}k_\rho\cos\phi \\ k_{1z}k_{2z}\sin 2\phi & k_{1z}k_{2z}(1-\cos 2\phi) & -k_{1z}k_\rho\sin\phi \\ -2k_\rho k_{2z}\cos\phi & -2k_\rho k_{2z}\sin\phi & 2k_\rho^2 \end{bmatrix}_{ij},$$

where we have introduced polar coordinates (ϕ, k_ρ) for the parallel components of the wavevector (k_x, k_y). Transforming to cylinder coordinates (φ, ρ, z) for the position $\boldsymbol{r}$, the integration over the azimuthal angle

$$\langle\ldots\rangle = \frac{1}{2\pi}\int_0^{2\pi} e^{ik_\rho\rho\,\cos(\phi-\varphi)}\left[\ldots\right]d\phi$$

can be performed analytically using Eq. (3.21). With this we arrive at

$$\left\langle\epsilon_i^{TE}(\boldsymbol{k}_1)\epsilon_j^{TE}(\boldsymbol{k}_2)\right\rangle = \frac{1}{2}\begin{bmatrix} J_0+J_2\cos 2\varphi & J_2\sin 2\varphi & 0 \\ J_2\sin 2\varphi & J_0-J_2\cos 2\varphi & 0 \\ 0 & 0 & 0 \end{bmatrix}_{ij}$$

$$\left\langle\epsilon_i^{TM}(\boldsymbol{k}_1)\epsilon_j^{TM}(\boldsymbol{k}_2)\right\rangle = \frac{1}{2k_1k_2} \tag{B.8}$$

$$\times\begin{bmatrix} k_{1z}k_{2z}(J_0-J_2\cos 2\varphi) & -k_{1z}k_{2z}J_2\sin 2\varphi & -2ik_\rho k_{1z}J_1\cos\varphi \\ -k_{1z}k_{2z}J_2\sin 2\varphi & k_{1z}k_{2z}(J_0+J_2\cos 2\varphi) & -2ik_\rho k_{1z}J_1\sin\varphi \\ -2ik_\rho k_{2z}J_1\cos\varphi & -2ik_\rho k_{2z}J_1\sin\varphi & 2k_\rho^2 J_0 \end{bmatrix}_{ij},$$

where for notational simplicity we have suppressed the arguments of the Bessel functions $J_n(k_\rho\rho)$.

Putting together all results, we are led to the Weyl decomposition of the dyadic Green's function

Weyl Decomposition of Dyadic Green's Function

$$G_{ij}(\boldsymbol{r},\boldsymbol{r}') = -\frac{\hat{z}_i\hat{z}_j}{k_1^2}\delta(\boldsymbol{r}-\boldsymbol{r}') \tag{B.9}$$

$$+\frac{i}{4\pi}\int_0^\infty \frac{e^{ik_{1z}|z-z'|}}{k_{1z}}\left\{\left\langle\epsilon_i^{TE}(\boldsymbol{k}_1^\pm)\epsilon_j^{TE}(\boldsymbol{k}_1^\mp)\right\rangle + \left\langle\epsilon_i^{TM}(\boldsymbol{k}_1^\pm)\epsilon_j^{TM}(\boldsymbol{k}_1^\mp)\right\rangle\right\} k_\rho dk_\rho\,.$$

B.3 Sommerfeld Integration Path

The integral of Eq. (B.9) has to be evaluated under the prescription that $k_{1z}+i\eta$ has a small imaginary part. However, for numerical evaluation, which we will need for stratified media, one has to be careful about $k_\rho = k_1$, where k_{1z} in the denominator becomes very small, as well as for large k_ρ arguments where the integrand decays very slowly for small $|z-z'|$ values. As we will discuss in the following, to avoid any difficulties with these limits we can

- express the k_ρ integration of Eq. (B.9) as a complex contour integration,
- and deform the contour such that the integration path stays sufficiently far away from all critical points or regions.

First, we observe from Eq. (B.9) and from the matrices in Eq. (B.8) that the integrands are of the form $k_\rho J_0(k_\rho\rho)$, $k_\rho^2 J_1(k_\rho\rho)$, and $k_\rho J_2(k_\rho\rho)$ multiplied with functions that depend on k_{1z} only. Using the relation

$$J_n(x) = \frac{1}{2}\left[H_n^{(1)}(x) + H_n^{(2)}(x)\right] \tag{B.10}$$

between the Bessel and Hankel functions, together with

$$H_n^{(1)}(-z) = -e^{-i\pi n} H_n^{(2)}(z)\,,$$

one can easily show that Eq. (B.9) can be expressed as

$$G_{ij}(\boldsymbol{r},\boldsymbol{r}') = -\frac{\hat{z}_i\hat{z}_j}{k_1^2}\delta(\boldsymbol{r}-\boldsymbol{r}') + \frac{i}{8\pi}\int_{-\infty}^{\infty}\frac{e^{ik_{1z}|z-z'|}}{k_{1z}}\left\{J_n \longrightarrow H_n^{(1)}\right\} k_\rho dk_\rho\,, \tag{B.11}$$

where the term in curly brackets is identical to the one in Eq. (B.9) with the only difference that all Bessel functions are replaced by Hankel functions. From the asymptotic form of the Hankel functions

$$H_n^{(1)}(k_\rho\rho) \underset{x\to\infty}{\longrightarrow} \sqrt{\frac{2}{\pi k_\rho \rho}}\, e^{i[k_\rho\rho - \frac{\pi}{2}(n+\frac{1}{2})]} \tag{B.12}$$

we then immediately observe that we can add a semicircle in the upper k_ρ plane, similarly to the integration path shown in Fig. B.1, whose contribution becomes zero in the limit $R\to\infty$. To evaluate Eq. (B.11) we then proceed as follows:

- The integration path in Eq. (B.11) is replaced by a contour along the real axis and back over a semi-circle in the upper half of the complex plane.
- Because of Cauchy's theorem, we can continue to further deform the path provided that we keep all poles inside the integration contour. As we will discuss

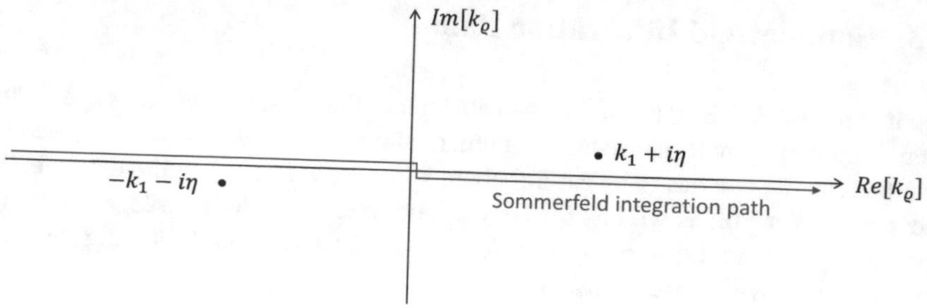

Fig. B.2 Sommerfeld integration path for the evaluation of the integral of Eq. (B.11) for the dyadic Green's function

in the following, we additionally have to be careful about branch points and cuts in the complex plane originating from multi-valued functions such as the square root.

A viable path is the Sommerfeld integration path shown in Fig. B.2 which goes slightly above the real k_ρ axis for negative k_ρ values, and slightly below the real k_ρ axis for positive k_ρ values. In the following we discuss the reasons for this choice.

Riemann Sheets and Branch Cuts

Consider the square root function $f(z) = \sqrt{z}$ which is a double-valued function because of the sign ambiguity of the square root. In the complex plane a complex number can be expressed as

$$z = re^{i\phi},$$

where r is the modulus and ϕ the phase of z. Correspondingly, the square root has the solution

$$\sqrt{z} = \sqrt{r}e^{i\frac{\phi}{2}}$$

and $\sqrt{z}$ is a function with a periodicity of 4π. Suppose that we vary ϕ in the range $(-\pi, \pi)$. In the z-plane shown in Fig. B.3b, we start at position B and then move anticlockwise around a circle until we end up at position A. In the complex $\sqrt{z}$ space shown in Fig. B.3a, the corresponding path is a semi-circle from B to A which is located in the sector for positive Re$\sqrt{z}$ values. When ϕ further increases we get in the z-plane the same sequence of complex numbers, but in the $\sqrt{z}$ plane we now move in the sector with negative Re$\sqrt{z}$ values. It is convenient to assign the two sectors with the top and bottom *Riemann sheets*. The cut between the two sheets is given by Re$\sqrt{z} = 0$ and is called a *branch cut*. It starts at the branch point $z = 0$ and ends at $z \to \infty$. When modifying an integration contour in the complex plane we have to be careful about such branch cuts. In general one tries to avoid crossing branch cuts.

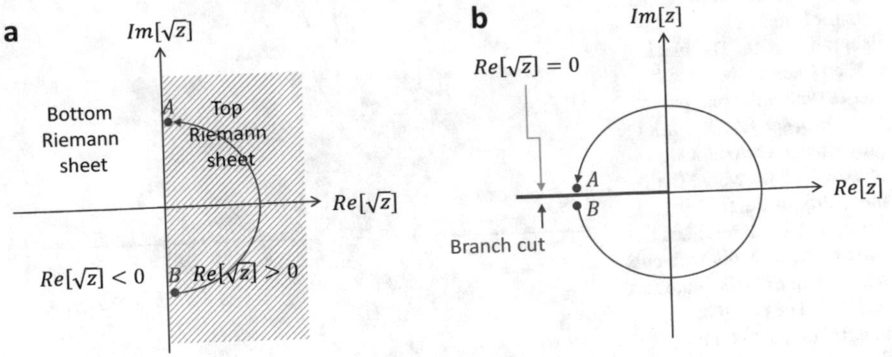

Fig. B.3 The complex plane for (**a**) the square root function $\sqrt{z}$ and (**b**) the corresponding z values. The half-plane $\mathrm{Re}\sqrt{z} > 0$ is mapped onto the upper Riemann sheet, the half-plane $\mathrm{Re}\sqrt{z} < 0$ is mapped onto the lower Riemann sheet

Branch Cuts for Stratified Media

We now return to the integral of Eq. (B.11). Our plan is to deform the integration path subject to the conditions that all poles are located inside the original and deformed paths, and we avoid crossing branch cuts. For the deformed path it is important that $\mathrm{Im}(k_z) > 0$ throughout. We first decompose both k_1 as well as k_ρ into real and imaginary parts,

$$k_z = \sqrt{k_1^2 - k_\rho^2} = \left[k_1'^2 - k_1''^2 + 2ik_1'k_1'' - k_\rho'^2 + k_\rho''^2 - 2ik_\rho'k_\rho''\right]^{1/2}. \tag{B.13}$$

In order for k_z to be real, or $\mathrm{Im}(k_z) = 0$, we need that

$$k_\rho'k_\rho'' = k_1'k_1'' \tag{B.14a}$$

$$k_\rho'^2 - k_\rho''^2 \le k_1'^2 - k_1''^2 . \tag{B.14b}$$

Figure B.4 shows how to locate those k_ρ values where k_z is real. Equation (B.14a) defines hyperbolas in the first and third quadrant, respectively, which asymptotically approach the x and y axes, whereas Eq. (B.14b) defines the k_ρ values bounded by the hyperbolas which asymptotically approach the $y = \pm x$ lines. Thus, the real k_z values are located on the hyperbola branches indicated by the dashed lines. The Sommerfeld integration branch navigates around all critical points, and from

$$k_z = \sqrt{k_1^2 - \left(|k_\rho| - i\eta\right)^2}$$

we immediately observe that k_z has a small positive imaginary part throughout. For stratified media and the computation of the reflected Green's function, see Eqs. (8.51) and (8.52), we additionally have to consider the poles of the generalized

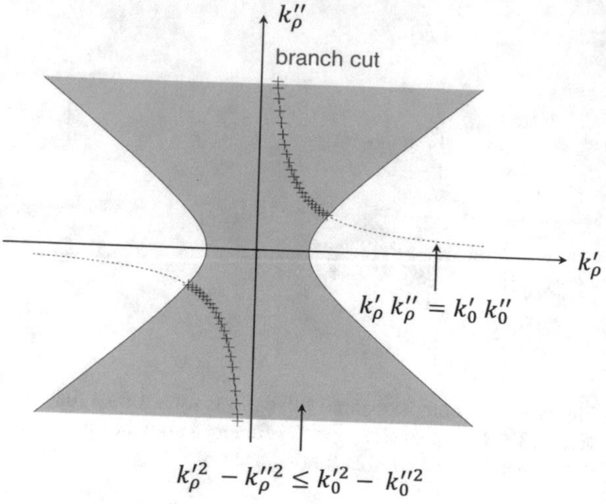

Fig. B.4 Location of branch cuts for Sommerfeld integration path. The black dashed lines show those k_ρ values where the real part of k_z is zero, see Eq. (B.14a), the gray shaded area indicate those k_ρ values where the inequality of Eq. (B.14b) holds. The thick red lines thus correspond to the branch cuts where both Eqs. (B.14a,b) are fulfilled. The complex integration path can be deformed provided that it does not cross the branch cuts

reflection and transmission coefficients. These poles are associated with guided modes, either in one of the layers of the stratified medium or bound to an interface, as discussed in Sect. 8.1 for surface plasmons. All poles have an imaginary part that is greater or equal than zero, such that the radial waves in Eq. (B.12) become asymptotically damped. As regarding the real parts of the modes they are all located within $[0, k'_{\max}]$, where $k'_{\max}$ is the maximal wavenumber of the materials forming the layer structure. A more detailed discussion of the poles and branch cuts can be found in Ref. [20].

Numerical Integration

When evaluating integrals of the form of Eq. (B.11) numerically we can further deform the integration path in the complex plane, provided that the deformed path does not cross any branch cuts and does not exclude any pole. A vast amount of literature is available on the subject, and if one computes Green's functions for stratified media on a daily basis it makes sense to dig deeper into the subject. Some techniques are discussed for instance by Chew [20], such as the method of the steepest descent, although the discussion given there is by no means exhaustive.

In the field of plasmonics and nanophotonics Paulus and coworkers [201] have suggested a clear and simple prescription of how to choose the path in the complex plane, which has found widespread use in the community. The suggested integration path is shown in Fig. B.5 and follows direction (a) for sufficiently large $|z - z'|$ values, and direction (b) if $|z - z'|$ is substantially smaller than ρ. In the figure $k'_{\max}$ is the maximum of the real parts of the wavenumbers in the different materials of the stratified medium.

A, A′. We first integrate along a semi-ellipse from the origin to $k'_{\max} + k_0$, where k_0 has been added as a safety margin. Along this path we stay sufficiently far

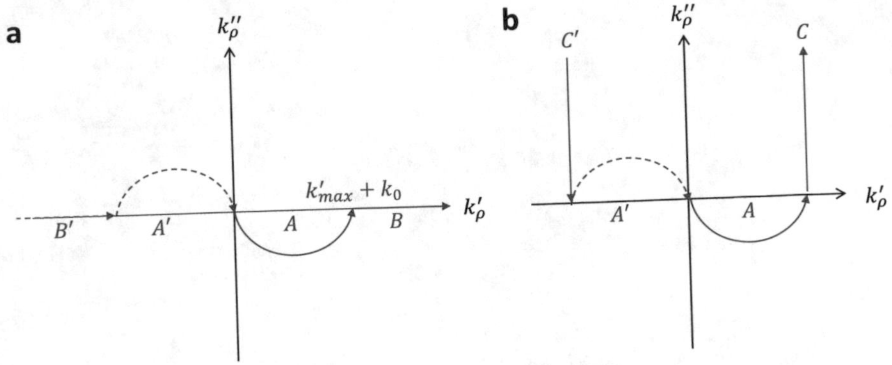

Fig. B.5 Integration path of Paulus et al. [201]. We first integrate in the complex plane along the semi-ellipses A, A' from the origin to the points $\pm(k'_{\max} + k_0)$ on the real axis, where $k_{\max}$ is the maximum of the real parts of the wavenumbers in the different materials of a stratified medium, and we have added the free-space wavenumber as a safety margin. (**a**) For sufficiently large $|z - z'|$ values the exponential factor in Eq. (B.11) leads to exponential damping for large real k_ρ values (and correspondingly large imaginary k_z values), and we thus integrate along the real axis, see B, B'. (**b**) For small $|z - z'|$ values we deform the integration path into the complex plane, see lines C, C' where the Hankel function $H_n^{(1)}$ for large imaginary values guarantees a fast decay of the integrand. Note that the integration paths A, A' and B, B' can be combined using Eq. (B.10) such that the sum of Hankel functions gives a Bessel function

away from the real k_ρ axis where the poles of lossless materials are located. Paulus et al. suggest an axis ratio of $1 : 1/1000$ for the semi-ellipse, although in many cases also more moderate ratios such as $1 : 1/10$ will do the job.

B, B'. When $|z - z'|$ is sufficiently large, the integrand becomes exponentially damped for large k'_ρ values because of the corresponding large, imaginary k_z values, see Eq. (B.13). For this reason we integrate along the real axis from $k'_{\max} + k_0$ to infinity. Note that we can combine for the paths A, A' and B, B' the Hankel functions for positive and negative arguments using Eq. (B.10) to get a single Bessel function, which is then integrated along paths A and B only.

C, C'. When $|z - z'|$ is significantly smaller than ρ, say by a factor of ten, we exploit the asymptotic form of Eq. (B.12) for the Hankel function to achieve faster convergence along the integration paths C, C'. We can safely deform the integration path in this way because all poles remain inside the integration contour and no branch cuts are crossed.

Appendix C
Spherical Wave Equation

In this appendix we show how to solve the scalar wave equation

$$\left(\nabla^2 + k^2\right)\psi(\boldsymbol{r}) = 0$$

in spherical coordinates, with k being a wavenumber. In doing so, we will introduce a number of special functions, namely Legendre polynomials, spherical harmonics, and spherical Bessel and Hankel functions. Everything presented here can be found in much more detail in other textbooks, and we give the main results for completeness only. Our discussion closely follows the book of Jackson [2] and we will refer to the corresponding equations whenever possible to facilitate a direct comparison. In spherical coordinates the wave equation reads

$$\frac{1}{r^2}\frac{\partial}{\partial r}\left(r^2\frac{\partial\psi}{\partial r}\right) + \frac{1}{r^2\sin\theta}\frac{\partial}{\partial\theta}\left(\sin\theta\frac{\partial\psi}{\partial\theta}\right) + \frac{1}{r^2\sin^2\theta}\frac{\partial^2\psi}{\partial\phi^2} + k^2\psi = 0\,. \tag{C.1}$$

Because of the spherical symmetry, we can make a product ansatz for the solution

$$\psi(r,\theta,\phi) = R(r)P(\theta)Q(\phi)\,.$$

Inserting this ansatz into Eq. (C.1), dividing by ψ, and multiplying with $r^2\sin^2\theta$ then gives

$$\left[\frac{\sin^2\theta}{R}\frac{d}{dr}\left(r^2\frac{dR}{dr}\right) + \frac{\sin\theta}{P}\frac{d}{d\theta}\left(\sin\theta\frac{dP}{d\theta}\right) + k^2r^2\sin^2\theta\right] + \frac{1}{Q}\frac{d^2Q}{d\phi^2} = 0\,. \tag{C.2}$$

This equation must be fulfilled for arbitrary values of r, θ, ϕ, which can only be achieved if the two terms on the left-hand side are constants. Consequently,

U. Hohenester, *Nano and Quantum Optics*, Graduate Texts in Physics,
https://doi.org/10.1007/978-3-030-30504-8

$$\frac{1}{Q}\frac{d^2Q}{d\phi^2} = -m^2\,, \tag{C.3}$$

with some constant m. This has the solution

$$Q(\phi) = e^{\pm im\phi}\,. \tag{C.4}$$

Because the function $\psi(r, \theta, \phi)$ must be periodic in ϕ, we observe that m has to be an integer. We then obtain from Eq. (C.2)

$$\left[\frac{1}{R}\frac{d}{dr}\left(r^2\frac{dR}{dr}\right) + k^2 r^2\right] + \left[\frac{1}{\sin\theta P}\frac{d}{d\theta}\left(\sin\theta\frac{dP}{d\theta}\right) - \frac{m^2}{\sin^2\theta}\right] = 0\,.$$

To fulfill this equation for arbitrary r, θ values the two terms in brackets must again be constants. This leads us to the following equations.

Associated Legendre Polynomials. For the polar angle part we get

$$\frac{1}{\sin\theta}\frac{d}{d\theta}\left(\sin\theta\frac{dP(\theta)}{d\theta}\right) + \left[\ell(\ell+1) - \frac{m^2}{\sin^2\theta}\right]P(\theta) = 0\,, \tag{C.5}$$

where the constant has been written in the form $\ell(\ell + 1)$. As will be discussed below, the solutions of this equation are given by the associated Legendre polynomials $P_{\ell,m}(\theta, \phi)$.

Spherical Bessel Functions. The radial part becomes

$$\frac{1}{r^2}\frac{d}{dr}\left(r^2\frac{dR(r)}{dr}\right) + \left[k^2 - \frac{\ell(\ell+1)}{r^2}\right]R(r) = 0\,. \tag{C.6}$$

The solutions $R(r) = f_\ell(kr)$ of this equation are linear combinations of spherical Bessel and Hankel functions.

It turns out to be convenient to combine the functions $Q(\phi)$ and $P(\theta)$ to the so-called spherical harmonics $Y_{\ell m}(\theta, \phi)$ which provide a complete set of functions for the angular degrees of freedom. With this the solution of the wave equation can be written as a linear combination of these fundamental solutions in the form [2, Eq. (9.80)]

Solution of Spherical Wave Equation

$$\psi(r, \theta, \phi) = \sum_{\ell=0}^{\infty}\sum_{m=-\ell}^{\ell} f_\ell(kr)Y_{\ell m}(\theta, \phi)\,, \tag{C.7}$$

where ℓ and m are the spherical degree and order, respectively.

C.1 Legendre Polynomials

We start with Eq. (C.5) for the Legendre polynomials, which for $m = 0$ are solutions of

$$\frac{d}{dx}\left[(1-x^2)\frac{dP_\ell(x)}{dx}\right] + \ell(\ell+1)P_\ell(x) = 0\,, \tag{C.8}$$

where we have introduced $x = \cos\theta$. Throughout we assume that $\theta \in [0, \pi]$ and $x \in [-1, 1]$, such that we can always take the positive sign for $\sin\theta = \sqrt{1-x^2}$. The solution can be represented by a power series of the form [2, Eq. (3.11)]

$$P_\ell(x) = \sum_{j=0}^{\infty} a_j x^j\,.$$

In order to remain finite for all values for x, the series must truncate for some j value which can only be achieved if ℓ is an integer. Using this requirement one can show that the Legendre polynomials $P_\ell(x)$ can be expressed in terms of the so-called Rodrigues' formula [2, Eq. (3.16)]

Rodrigues' Formula for Legendre Polynomials

$$P_\ell(x) = \frac{1}{2^\ell \ell!}\frac{d^\ell}{dx^\ell}\left(x^2-1\right)^\ell\,. \tag{C.9}$$

The Legendre polynomials are normalized such that $P_\ell(1) = 1$. They are even functions for even values of ℓ, and odd functions for odd values of ℓ. More explicitly, the first few Legendre polynomials read [2, Eq. (3.15)]

$$P_0(x) = 1\,, \quad P_1(x) = x\,, \quad P_2(x) = \frac{1}{2}(3x^2-1)\,, \quad P_3(x) = \frac{1}{2}(5x^2-3x)\,. \tag{C.10}$$

The Legendre polynomials are orthogonal to each other [2, Eq. (3.21)]

$$\int_{-1}^{1} P_{\ell'}(x)P_\ell(x)\,dx = \frac{2}{2\ell+1}\delta_{\ell'\ell}\,, \tag{C.11}$$

and form a complete set, such that any function in the interval $x \in [-1, 1]$ can be expanded in terms of these polynomials. There exist various recurrence formulas, such as [2, Eq. (3.29)]

$$\begin{aligned}(\ell+1)P_{\ell+1}(x) - (2\ell+1)xP_\ell(x) + \ell P_{\ell-1}(x) &= 0\\ \frac{dP_\ell(x)}{dx} - x\frac{dP_\ell(x)}{dx} - (\ell+1)P_\ell(x) &= 0\,,\end{aligned} \tag{C.12}$$

which can be used to compute the Legendre polynomials and their derivatives numerically once two starting values are known.

Associated Legendre Polynomials

The defining equation for the associated Legendre polynomials $P_{\ell m}(x)$ for arbitrary m values, Eq. (C.5), reads

$$\frac{d}{dx}\left[(1-x^2)\frac{dP_\ell^m(x)}{dx}\right]+\left[\ell(\ell+1)-\frac{m^2}{1-x^2}\right]P_\ell^m(x)=0\,. \tag{C.13}$$

For positive m values the associated Legendre polynomials can be computed from [2, Eq. (3.49)]

Associated Legendre Polynomials for $m > 0$

$$P_\ell^m(x)=(-1)^m(1-x^2)^{m/2}\frac{d^m}{dx^m}P_\ell(x)\,, \tag{C.14}$$

whereas the Legendre polynomials for negative m values are given by [2, Eq. (3.51)]

$$P_\ell^{-m}(x)=(-1)^m\frac{(\ell-m)!}{(\ell+m)!}P_\ell^m(x)\,. \tag{C.15}$$

For fixed values of m the associated Legendre polynomials form a complete set of functions which are orthogonal to each other [2, Eq. (3.52)]

$$\int_{-1}^{1}P_{\ell'}^m(x)P_\ell^m(x)\,dx=\frac{2}{2\ell+1}\frac{(\ell+m)!}{(\ell-m)!}\delta_{\ell'\ell}\,. \tag{C.16}$$

C.2 Spherical Harmonics

It turns out to be convenient to combine the solutions $Q(\phi)$ and $P(\theta)$ to the so-called spherical harmonics [2, Eq. (3.53)]

Spherical Harmonics

$$Y_{\ell m}(\theta,\phi)=\sqrt{\frac{2\ell+1}{4\pi}\frac{(\ell-m)!}{(\ell+m)!}}\,P_\ell^m(\cos\theta)e^{im\phi}\,. \tag{C.17}$$

Sometimes we will use the alternative notation

$$Y_{\ell m}(\hat{\boldsymbol{r}})\,, \quad \hat{\boldsymbol{r}} = \cos\phi\sin\theta\,\hat{\boldsymbol{x}} + \sin\phi\sin\theta\,\hat{\boldsymbol{y}} + \cos\theta\,\hat{\boldsymbol{z}}\,, \tag{C.18}$$

where $\hat{\boldsymbol{r}}$ is a unit vector defined by the polar and azimuthal angles θ and ϕ, respectively. From Eq. (C.15) one can readily show that [2, Eq. (3.54)]

$$Y^*_{\ell m}(\theta,\phi) = (-1)^m Y_{\ell,-m}(\theta,\phi)\,. \tag{C.19}$$

The spherical harmonics form a complete and orthonormal set of functions for the angular degrees of freedom with [2, Eq. (3.55)]

$$\int_0^{2\pi} d\phi \int_0^{\pi} \sin\theta d\theta\, Y^*_{\ell' m'}(\theta,\phi) Y_{\ell m}(\theta,\phi) = \delta_{\ell'\ell}\delta_{m'm}\,. \tag{C.20}$$

The completeness relation is [2, Eq. (3.56)]

$$\sum_{\ell=0}^{\infty}\sum_{m=-\ell}^{\ell} Y^*_{\ell m}(\theta',\phi') Y_{\ell m}(\theta,\phi) = \delta(\phi-\phi')\delta(\cos\theta-\cos\theta')\,. \tag{C.21}$$

A few selected spherical harmonics read

$$\ell = 0 \qquad Y_{00} = \frac{1}{\sqrt{4\pi}}$$

$$\ell = 1 \quad \begin{cases} Y_{11} = -\sqrt{\dfrac{3}{8\pi}}\sin\theta e^{i\phi} \\ Y_{10} = \sqrt{\dfrac{3}{4\pi}}\cos\theta \end{cases}$$

$$\ell = 2 \quad \begin{cases} Y_{22} = \sqrt{\dfrac{5}{32\pi}}\sin^2\theta e^{2i\phi} \\ Y_{21} = -\sqrt{\dfrac{15}{8\pi}}\sin\cos\theta e^{i\phi} \\ Y_{20} = \sqrt{\dfrac{5}{4\pi}}(\dfrac{3}{2}\cos^2\theta - \dfrac{1}{2})\,. \end{cases}$$

For $\theta = 0$ one finds

$$Y_{\ell m}(\hat{\boldsymbol{z}}) = \sqrt{\frac{2\ell+1}{4\pi}}\,\delta_{m0}\,. \tag{C.22}$$

Figure C.1 shows a few selected spherical harmonics on the units sphere. The function with $\ell = 0$ is a constant, and the functions with $\ell = 1, 2$ have one or two nodes in either the polar or azimuthal directions. Sometimes one uses an alternative

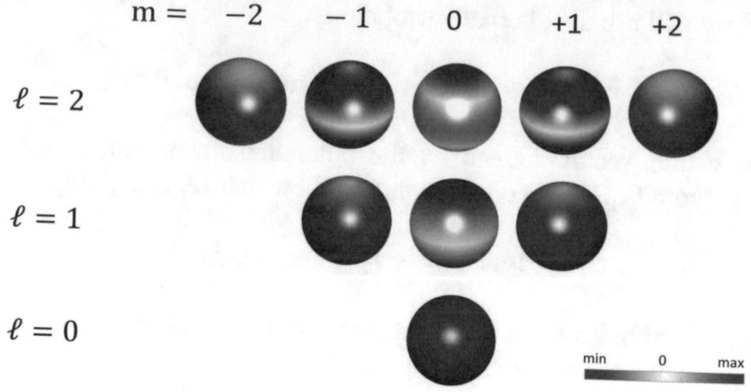

Fig. C.1 Visualization of vector spherical harmonics $Y_{\ell m}(\theta,\phi)$ for different angular degrees and orders ℓ and m, respectively. We plot $e^{-im\phi}Y_{\ell m}(\theta,\phi)$ on the unit sphere using the colomap shown at the bottom of the figure

visualization of $Y_{\ell m}(\theta,\phi)$ by deforming the radius according to the absolute value of the spherical harmonics, as shown in Fig. C.2 at the example of Y_{20} and in Fig. C.3a for the spherical harmonics of lowest degree. By taking the linear combinations

$$\begin{cases} \dfrac{i}{\sqrt{2}}\Big(Y_{\ell m}-(-1)^m Y_{\ell,-m}\Big) & \text{for } m<0 \\ \dfrac{1}{\sqrt{2}}\Big(Y_{\ell m}+(-1)^m Y_{\ell,-m}\Big) & \text{for } m>0 \end{cases} \tag{C.23}$$

and using Eq. (C.19) one can define a set of real-valued functions, which are shown in Fig. C.3b.

C.3 Spherical Bessel and Hankel Functions

We finally address the radial part of the spherical wave equation, see Eq. (C.6),

$$\frac{1}{r^2}\frac{d}{dr}\left(r^2\frac{df_\ell(r)}{dr}\right)+\left[k^2-\frac{\ell(\ell+1)}{r^2}\right]f_\ell(r)=0\,. \tag{C.24}$$

The solutions are given by linear combinations of spherical Bessel and Hankel functions j_ℓ and $h_\ell^{(1)}$, respectively [2, Eq. (9.84)]

Fig. C.2 Alternative visualization of vector spherical harmonics. For each θ, ϕ value we scale the radius from one to the absolute value of $Y_{\ell m}(\theta, \phi)$, and finally end up with the plot on the right

Solution of Radial Part of Scalar Wave Equation

$$f_\ell(kr) = A_{\ell m}\, j_\ell(kr) + B_{\ell m}\, h_\ell^{(1)}(kr)\,, \tag{C.25}$$

where $A_{\ell m}$, $B_{\ell m}$ are arbitrary constants. It is customary to define the spherical Bessel and Hankel functions $j_\ell(x), n_\ell(x), h_\ell^{(1,2)}(x)$, which are related to the Bessel functions $J_\ell(x)$, $N_\ell(x)$ via [2, Eq. (9.85)]

$$\begin{aligned} j_\ell(x) &= \sqrt{\frac{\pi}{2x}}\, J_{\ell+\frac{1}{2}}(x) \\ n_\ell(x) &= \sqrt{\frac{\pi}{2x}}\, N_{\ell+\frac{1}{2}}(x) \\ h_\ell^{(1,2)}(x) &= j_\ell(x) \pm i n_\ell(x)\,. \end{aligned} \tag{C.26}$$

The spherical Bessel functions can be obtained from [2, Eq. (9.86)]

$$\begin{aligned} j_\ell(x) &= \quad (-x)^\ell \left(\frac{1}{x}\frac{d}{dx}\right)^\ell \left(\frac{\sin x}{x}\right) \\ n_\ell(x) &= -(-x)^\ell \left(\frac{1}{x}\frac{d}{dx}\right)^\ell \left(\frac{\cos x}{x}\right)\,. \end{aligned} \tag{C.27}$$

The first few Bessel and Hankel functions read

$$j_0(x) = \frac{\sin x}{x}\,, \quad j_1(x) = \frac{\sin x}{x^2} - \frac{\cos x}{x}\,, \quad j_2(x) = \left(\frac{3}{x^2} - \frac{1}{x}\right)\sin x - \frac{3\cos x}{x}\,,$$

$$h_0^{(1)}(x) = \frac{e^{ix}}{x}\,, \quad h_1^{(1)}(x) = -\frac{e^{ix}}{x}\left(1 + \frac{1}{x}\right)\,, \quad h_2^{(1)}(x) = \frac{ie^{ix}}{x}\left(1 + \frac{3i}{x} - \frac{3}{x^2}\right)\,.$$

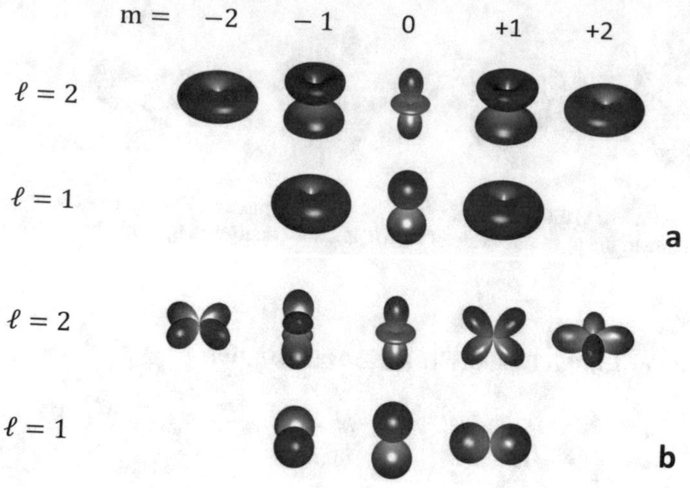

Fig. C.3 Visualization of (**a**) $e^{-im\phi}Y_{\ell m}$ and (**b**) the real-valued representation of Eq. (C.23) for the lowest angular degrees using the deformation procedure shown in Fig. C.2

For small x values one can derive the series expansions [2, Eq. (9.88)]

$$\begin{aligned} j_\ell(x) &\rightarrow \frac{x^\ell}{(2\ell+1)!!}\left[1-\frac{x^2}{2(2\ell+3)}+\ldots\right] \\ n_\ell(x) &\rightarrow -\frac{(2\ell-1)!!}{x^{\ell+1}}\left[1-\frac{x^2}{2(1-2\ell)}+\ldots\right], \end{aligned} \tag{C.28}$$

and for large arguments we get the asymptotic expansions [2, Eq. (9.89)]

$$\begin{aligned} j_\ell(x) &\rightarrow \frac{1}{x}\sin\left(x-\frac{\ell\pi}{2}\right) \\ n_\ell(x) &\rightarrow -\frac{1}{x}\cos\left(x-\frac{\ell\pi}{2}\right) \\ h_\ell^{(1)}(x) &\rightarrow (-i)^{\ell+1}\frac{e^{ix}}{x}. \end{aligned} \tag{C.29}$$

From these expansions we can draw the following general conclusions about the coefficients $A_{\ell m}$, $B_{\ell m}$ for the spherical wave equation in Eq. (C.25).

Bessel Functions. When using the spherical wave solution in a range $r \in [0, r_{\max}]$ we have to ensure that $f_\ell(kr)$ remains finite at the origin. For this reason we set

$$f_\ell(kr) = A_{\ell m} j_\ell(kr) \qquad \text{for } r \in [0, r_{\max}] \,.$$

Hankel Functions. When using the spherical wave solution in a range $r \in [r_{\min}, \infty)$ we have to ensure that for large kr values $f_\ell(kr)$ becomes an outgoing wave. For this reason we set

$$f_\ell(kr) = B_{\ell m} h_\ell^{(1)}(kr) \qquad \text{for } r \in [r_{\min}, \infty) \,.$$

The spherical Bessel functions satisfy the recursion formulas [2, Eq. (9.90)]

$$\begin{aligned}
\frac{2\ell+1}{x} z_\ell(x) &= z_{\ell-1}(x) + z_{\ell+1}(x) \\
z'_\ell(x) &= \frac{1}{2\ell+1}\Big[\ell z_{\ell-1}(x) - (\ell+1) z_{\ell+1}(x)\Big] \\
\frac{d}{dx}\Big[x z_\ell(x)\Big] &= x z_{\ell-1}(x) - \ell z_\ell(x) \,,
\end{aligned} \tag{C.30}$$

where $z_\ell(x)$ is any of the functions $j_\ell(x)$, $n_\ell(x)$, $h_\ell^{(1,2)}(x)$. Numerically one should start for the Bessel functions from $j_0(x)$, $j_1(x)$ to compute the higher-order Bessel functions using an upward scheme, and from $h_\ell^{(1)}(x)$, $h_{\ell-1}^{(1)}(x)$ to compute the lower-order Hankel functions in a downward scheme. Otherwise an iterative solution may become numerically instable for large spherical degrees.

We finally give the expression for the decomposition of the Green's function of Eq. (5.7) in terms of spherical functions [2, Eq. (9.98)]

Spherical Green's Function Expansion

$$\frac{e^{ik|\boldsymbol{r}-\boldsymbol{r}'|}}{4\pi|\boldsymbol{r}-\boldsymbol{r}'|} = ik \sum_{\ell=0}^{\infty} \sum_{m=-\ell}^{\ell} j_\ell(kr_<) h_\ell^{(1)}(kr_>)\, Y_{\ell m}^*(\hat{\boldsymbol{r}}') Y_{\ell m}(\hat{\boldsymbol{r}}) \,. \tag{C.31}$$

Here $r_<$ is the smaller value of r, r' and $r_>$ the larger value.

Appendix D
Vector Spherical Harmonics

In this appendix we discuss the solution of the vectorial wave equation

$$-\nabla \times \nabla \times \boldsymbol{F}(\boldsymbol{r}) + k^2 \boldsymbol{F}(\boldsymbol{r}) = 0 \tag{D.1}$$

in terms of longitudinal and transverse basis functions. For spherical coordinates this will lead us to the so-called vector spherical harmonics, to be introduced further below. We start our discussion with a detour and ponder on the solution of the wave equation in Cartesian coordinates.

Wave Equation in Cartesian Coordinates

In Sect. 2.5 we have shown that any vector field can be decomposed into its longitudinal and transverse parts. For a plane wave with wavenumber $\boldsymbol{k}$ the longitudinal and transverse parts can be expressed as

$$\boldsymbol{F}_k^L = \left(\hat{\boldsymbol{k}} \cdot \boldsymbol{F}_k\right) \hat{\boldsymbol{k}}\,, \quad \boldsymbol{F}_k^\perp = \boldsymbol{F}_k - \boldsymbol{F}_k^L\,.$$

For special geometries such as stratified media one can further use Eq. (2.43) to decompose the transverse fields into TE and TM modes. The field $\boldsymbol{F}_k$ can thus be spanned by the following (unnormalized) vectors:

$$\boldsymbol{k}\,, \quad \boldsymbol{k} \times \hat{z}\,, \quad \boldsymbol{k} \times \left(\boldsymbol{k} \times \hat{z}\right).$$

Projection on $\boldsymbol{k}$ gives the longitudinal component $\boldsymbol{F}_k^L$, whereas projection on the other two vectors gives the transverse components with TE and TM character, respectively, if $\boldsymbol{F}_k$ represents an electric field, and TM and TE character for a magnetic field.

U. Hohenester, *Nano and Quantum Optics*, Graduate Texts in Physics,
https://doi.org/10.1007/978-3-030-30504-8

Vector WaveFunctions

We can investigate this decomposition procedure from a slightly different perspective. Consider a scalar potential $\psi(\boldsymbol{r})$ which fulfills the Helmholtz equation

$$\left(\nabla^2 + k^2\right)\psi(\boldsymbol{r}) = 0\,.$$

From such a potential we can derive the following set of vector functions:

$$\boldsymbol{L}_\psi(\boldsymbol{r}) = \nabla\psi(\boldsymbol{r})\,, \quad \boldsymbol{M}_\psi(\boldsymbol{r}) = \nabla\times\boldsymbol{c}\psi(\boldsymbol{r})\,, \quad \boldsymbol{N}_\psi(\boldsymbol{r}) = \frac{1}{k}\nabla\times\boldsymbol{M}_\psi(\boldsymbol{r})\,, \tag{D.2}$$

where $\boldsymbol{c}$ is a pilot vector to be specified in a moment. By construction $\boldsymbol{L}_\psi$ is a longitudinal vector function, whereas both $\boldsymbol{M}_\psi$ and $\boldsymbol{N}_\psi$ are transverse vector functions, sometimes referred to as solenoidal vector functions, which both fulfill the wave equation. To show this, we insert $\boldsymbol{M}_\psi$ into the wave equation and get

$$\nabla\Big(\nabla\cdot\nabla\times\boldsymbol{c}\psi(\boldsymbol{r})\Big) - \nabla^2\Big(\nabla\times\boldsymbol{c}\psi(\boldsymbol{r})\Big) + k^2\nabla\times\boldsymbol{c}\psi(\boldsymbol{r}) = 0\,.$$

The first term vanishes, because the divergence of a curl is always zero, and thus $\boldsymbol{M}_\psi$ indeed fulfills Eq. (D.1). By the same token we can show that also $\boldsymbol{N}_\psi$ fulfills the wave equation. It can be readily verified that $\boldsymbol{M}_\psi$ and $\boldsymbol{N}_\psi$ are related to each other via

$$k\boldsymbol{N}_\psi = \nabla\times\boldsymbol{M}_\psi\,, \quad k\boldsymbol{M}_\psi = \nabla\times\boldsymbol{N}_\psi\,.$$

Thus, if $\boldsymbol{M}_\psi$ represents an electric field then $\boldsymbol{N}_\psi$ is a magnetic field, and vice versa. From the above discussion it follows that any vector function can be expressed in terms of three scalar functions $u(\boldsymbol{r})$, $v(\boldsymbol{r})$, $w(\boldsymbol{r})$ as

Decompositon into Longitudinal and Transverse Basis Functions

$$\boldsymbol{F}(\boldsymbol{r}) = \boldsymbol{L}_w(\boldsymbol{r}) + \boldsymbol{M}_u(\boldsymbol{r}) + \boldsymbol{N}_v(\boldsymbol{r})\,. \tag{D.3}$$

Similarly, any transverse vector function can be written in the form

$$\boldsymbol{F}^\perp(\boldsymbol{r}) = \boldsymbol{M}_u(\boldsymbol{r}) + \boldsymbol{N}_v(\boldsymbol{r})\,. \tag{D.4}$$

If we consider for the generating potentials $\psi(\boldsymbol{r}) = e^{i\boldsymbol{k}\cdot\boldsymbol{r}}$ plane waves and for the pilot vector $\boldsymbol{c} = \hat{\boldsymbol{z}}$, we observe that the vector functions become

$$\boldsymbol{L}(\boldsymbol{r}) = i\boldsymbol{k}e^{i\boldsymbol{k}\cdot\boldsymbol{r}}\,, \quad \boldsymbol{M}(\boldsymbol{r}) = i\boldsymbol{k}\times\hat{\boldsymbol{z}}e^{i\boldsymbol{k}\cdot\boldsymbol{r}}\,, \quad \boldsymbol{N}(\boldsymbol{r}) = -\frac{1}{k^2}\boldsymbol{k}\times\left(\boldsymbol{k}\times\hat{\boldsymbol{z}}e^{i\boldsymbol{k}\cdot\boldsymbol{r}}\right).$$

This corresponds to the basis considered above.

Spherical Wave Equation

The procedure just outlined can be also applied to the wave equation of Eq. (C.7) in spherical coordinates

$$\psi(r,\theta,\phi)=\sum_{\ell m}\Big[A_{lm}\,j_\ell(kr)+B_{lm}h_\ell^{(1)}(kr)\Big]\,Y_{\ell m}(\theta,\phi)\,,$$

where ℓ and m are the spherical degree and order, respectively, A_{lm}, B_{lm} are coefficients to be determined for the problem under consideration, j_ℓ, $h_\ell^{(1)}$ are the spherical Bessel and Hankel function of first order, respectively, and $Y_{\ell m}$ are the spherical harmonics. For a detailed discussion of the solution and the special functions see Appendix C. We now set the pilot vector to $\boldsymbol{c}=\boldsymbol{r}$ such that

$$\boldsymbol{M}_\psi(\boldsymbol{r})=\nabla\times\boldsymbol{r}\psi(\boldsymbol{r})=\nabla\times\boldsymbol{r}\,f_\ell(r)Y_{lm}(\theta,\phi)\,, \tag{D.5}$$

where f_ℓ is some linear combination of spherical Bessel and Hankel functions. In the above expression we use $\nabla\times\boldsymbol{r}=-\boldsymbol{r}\times\nabla$ to get

$$\boldsymbol{M}_\psi(\boldsymbol{r})=-i\left(\frac{1}{i}\boldsymbol{r}\times\nabla\right)f_\ell(r)Y_{lm}(\theta,\phi)=-i\hat{\boldsymbol{L}}\,f_\ell(r)Y_{lm}(\theta,\phi)\,, \tag{D.6}$$

where we have introduced the angular momentum operator

$$\hat{\boldsymbol{L}}=-i\boldsymbol{r}\times\nabla \tag{D.7}$$

known from quantum mechanics (without $\hbar$). This operator is introduced on purely formal grounds and has nothing to do with quantum effects. However, $\hat{\boldsymbol{L}}$ will allow us to adopt several results well-known from quantum mechanics.

D.1 Vector Spherical Harmonics

It turns out to be convenient to introduce the *vector spherical harmonics* for the decomposition of transverse electromagnetic fields in spherical coordinates [2, Eq. (9.119)]

Vector Spherical Harmonics

$$\boldsymbol{X}_{\ell m}(\theta,\phi)=\frac{1}{\sqrt{\ell(\ell+1)}}\,\hat{\boldsymbol{L}}\,Y_{\ell m}(\theta,\phi)\,. \tag{D.8}$$

Note that in the literature there exist several slightly different definitions of vector spherical harmonics and we here follow the book of Jackson [2]. From the representation [202]

$$\hat{\boldsymbol{L}} = i\left(\hat{\boldsymbol{\theta}}\frac{1}{\sin\theta}\frac{\partial}{\partial\phi} - \hat{\boldsymbol{\phi}}\frac{\partial}{\partial\theta}\right) \tag{D.9}$$

one can then obtain the vector spherical harmonics, where $\hat{\boldsymbol{\theta}}$, $\hat{\boldsymbol{\phi}}$ are the unit vectors in the polar and azimuthal direction, respectively. With this we get

$$\ell = 0 \qquad \boldsymbol{X}_{00} = 0$$

$$\ell = 1 \quad \begin{cases} \boldsymbol{X}_{11} = \sqrt{\dfrac{3}{16\pi}}e^{i\phi}\left(\hat{\boldsymbol{\theta}} + i\cos\theta\,\hat{\boldsymbol{\phi}}\right) \\ \boldsymbol{X}_{10} = i\sqrt{\dfrac{3}{8\pi}}\sin\theta\,\hat{\boldsymbol{\phi}} \end{cases}$$

$$\ell = 2 \quad \begin{cases} \boldsymbol{X}_{22} = -\sqrt{\dfrac{5}{16\pi}}\sin\theta e^{2i\phi}\left(\hat{\boldsymbol{\theta}} + i\cos\theta\,\hat{\boldsymbol{\phi}}\right) \\ \boldsymbol{X}_{21} = \sqrt{\dfrac{5}{16\pi}}e^{i\phi}\left(\cos\theta\,\hat{\boldsymbol{\theta}} + i\cos 2\theta\,\hat{\boldsymbol{\phi}}\right) \\ \boldsymbol{X}_{20} = i\sqrt{\dfrac{15}{32\pi}}\sin 2\theta\,\hat{\boldsymbol{\phi}}\,. \end{cases}$$

From Eq. (C.19) we find

$$\boldsymbol{X}^*_{\ell m}(\theta,\phi) = (-1)^{m+1}\boldsymbol{X}_{\ell,-m}(\theta,\phi)\,. \tag{D.10}$$

By using the linear combinations of Eq. (C.23) we obtain a real-valued representation for the vector spherical harmonics, which is shown for a few angular degrees and orders in Fig. D.1. The $\boldsymbol{X}_{\ell m}$ are vector functions that are tangential to the unit sphere and have ℓ nodes in the polar and azimuthal directions. From Eq. (C.22) one can show that

$$\boldsymbol{X}_{\ell,m}(\hat{\boldsymbol{z}}) = \sqrt{\frac{2\ell+1}{16\pi}}\boldsymbol{\epsilon}_\pm\,\delta_{m,\pm 1}\,, \quad \boldsymbol{\epsilon}_\pm = \hat{\boldsymbol{x}} \pm i\,\hat{\boldsymbol{y}}\,. \tag{D.11}$$

D.2 Orthogonality Relations

The vector spherical harmonics can be used to construct a complete basis for the solution of the wave equation in spherical coordinates. This basis consists of the three vector functions

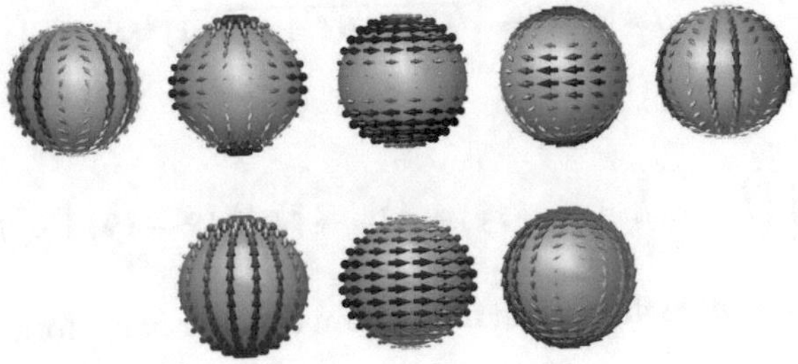

Fig. D.1 Vector spherical harmonics for different angular degrees ℓ. We form linear combinations and plot for $m \leq 0$ the imaginary part of $\boldsymbol{X}_{\ell m}$, and for $m > 0$ the real part. The sizes of the vectors correspond to the norm of $\boldsymbol{X}_{\ell m}$

Basis Functions for Spherical Wave Equation

$$\hat{\boldsymbol{r}}\, h_\ell(kr) Y_{\ell m}\,, \quad g_\ell(kr)\boldsymbol{X}_{\ell m}\,, \quad \nabla \times f_\ell(kr)\boldsymbol{X}_{\ell m}\,, \tag{D.12}$$

where f_ℓ, g_ℓ, and h_ℓ are spherical Bessel or Hankel functions (or linear combinations thereof). The first vector function describes a longitudinal vector field, which will not be needed in our following analysis, whereas the other two functions are transverse. All functions are orthogonal to each other, as will be demonstrated in a moment. Before doing so we recapitulate the angular momentum algebra usually employed in quantum mechanics [202], which will allow us to derive a number of useful expressions in a particularly simple manner.

Angular Momentum Algebra

In the following we introduce in analogy to quantum mechanics the momentum operator $\hat{\boldsymbol{\pi}} = -i\nabla$ (without $\hbar$) and write the angular momentum operator in the form $\hat{\boldsymbol{L}} = \boldsymbol{r} \times \hat{\boldsymbol{\pi}}$. Let

$$Y_{\ell m}(\theta, \phi) = \big\langle \theta, \phi \big| \ell, m \big\rangle$$

denote the spherical harmonics using the bra-ket formalism of quantum mechanics. The spherical harmonics are eigenfunctions of $\hat{L}_z$ and $\hat{L}^2$,

$$\hat{L}_z \big|\ell, m\big\rangle = m\big|\ell, m\big\rangle\,, \quad \hat{L}^2\big|\ell, m\big\rangle = \ell(\ell+1)\big|\ell, m\big\rangle\,. \tag{D.13}$$

It is often useful to introduce the operators [2, Eq. (9.102)]

$$\hat{L}_\pm = \hat{L}_x \pm i\hat{L}_y\,, \tag{D.14}$$

which, upon acting on the angular momentum eigenstates, give [2, Eq. (9.104)]

$$\hat{L}_{\pm}|\ell, m\rangle = \sqrt{(\ell \mp 1)(\ell \pm m + 1)}\,|\ell, m \pm 1\rangle. \tag{D.15}$$

Using

$$\frac{1}{2}\Big\{\left(\hat{L}_x + i\hat{L}_y\right)(\hat{\boldsymbol{x}} - i\hat{\boldsymbol{y}}) + \left(\hat{L}_x - i\hat{L}_y\right)(\hat{\boldsymbol{x}} + i\hat{\boldsymbol{y}})\Big\} = \hat{L}_x\hat{\boldsymbol{x}} + \hat{L}_y\hat{\boldsymbol{y}}$$

one can decompose the angular momentum operator in the form

$$\hat{\boldsymbol{L}} = \frac{1}{2}\Big\{\hat{L}_+\boldsymbol{\epsilon}_+^* + \hat{L}_-\boldsymbol{\epsilon}_-^*\Big\} + \hat{L}_z\hat{\boldsymbol{z}}, \tag{D.16}$$

where $\boldsymbol{\epsilon}_\pm = \hat{\boldsymbol{x}} \pm i\,\hat{\boldsymbol{y}}$. With this we can compute the vector spherical harmonics from

$$\sqrt{\ell(\ell+1)}\boldsymbol{X}_{\ell m}(\theta, \phi) = \frac{1}{2}\Big\{\hat{L}_+\boldsymbol{\epsilon}_+^* + \hat{L}_-\boldsymbol{\epsilon}_-^*\Big\}Y_{\ell m}(\theta, \phi) + m\,Y_{\ell m}(\theta, \phi)\,\hat{\boldsymbol{z}}, \tag{D.17}$$

where the first term on the right-hand side has to be evaluated using Eq. (D.15). From the fundamental commutation relation $[r_m, \hat{\pi}_n] = i\delta_{mn}$ one can derive a number of useful relations [2, Eq. (9.105)]

$$\hat{\boldsymbol{L}}\nabla^2 = \nabla^2\hat{\boldsymbol{L}}, \quad \hat{\boldsymbol{L}} \times \hat{\boldsymbol{L}} = i\hat{\boldsymbol{L}}, \quad \nabla^2 = \frac{1}{r}\frac{\partial^2}{\partial r^2} - \frac{\hat{L}^2}{r^2}, \tag{D.18}$$

which we will use in the following.

Derivation of the Orthogonality Relations

We next show that the three functions

$$\big\langle\theta, \phi\big|\boldsymbol{r}\big|\ell, m\big\rangle, \quad \big\langle\theta, \phi\big|\hat{\boldsymbol{L}}\big|\ell, m\big\rangle, \quad \big\langle\theta, \phi\big|\boldsymbol{r} \times \hat{\boldsymbol{L}}\big|\ell, m\big\rangle \tag{D.19}$$

span an (unnormalized) basis. In doing so we exploit the orthogonality relation of Eq. (C.20) for the spherical harmonics

$$\begin{aligned}\langle\ell', m'|\ell, m\rangle &= \langle\ell', m'|\left[\oint |\theta, \phi\rangle\langle\theta, \phi|\,d\Omega\right]|\ell, m\rangle \\ &= \oint Y_{\ell' m'}^*(\theta, \phi)Y_{\ell m}(\theta, \phi)\,d\Omega = \delta_{\ell'\ell}\delta_{m'm},\end{aligned}$$

where we have inserted in the first line the unit operator to express the spherical harmonics in spherical coordinates, and have used $d\Omega$ for the integration over the unit sphere. Using the identities

$$\boldsymbol{r} \cdot \hat{\boldsymbol{L}} = \boldsymbol{r} \cdot \left(\boldsymbol{r} \times \hat{\boldsymbol{L}}\right) = \hat{\boldsymbol{L}} \cdot \left(\boldsymbol{r} \times \hat{\boldsymbol{L}}\right) = 0,$$

which can be easily verified using the properties of the operators, one can show that the functions defined in Eq. (D.19) are orthogonal to each other

$$\begin{aligned}\langle \ell', m' | \boldsymbol{r} \cdot \hat{\boldsymbol{L}} | \ell, m \rangle &= 0 \\ \langle \ell', m' | \boldsymbol{r} \cdot \boldsymbol{r} \times \hat{\boldsymbol{L}} | \ell, m \rangle &= 0 \\ \langle \ell', m' | \hat{\boldsymbol{L}} \cdot \boldsymbol{r} \times \hat{\boldsymbol{L}} | \ell, m \rangle &= 0 \,. \end{aligned} \tag{D.20}$$

In a similar fashion we find[1]

$$\begin{aligned}\langle \ell', m' | \boldsymbol{r} \cdot \boldsymbol{r} | \ell, m \rangle &= r^2 \, \delta_{\ell'\ell} \delta_{m'm} \\ \langle \ell', m' | \hat{\boldsymbol{L}} \cdot \hat{\boldsymbol{L}} | \ell, m \rangle &= \ell(\ell+1) \, \delta_{\ell'\ell} \delta_{m'm} \\ \langle \ell', m' | (\boldsymbol{r} \times \hat{\boldsymbol{L}}) \cdot (\boldsymbol{r} \times \hat{\boldsymbol{L}}) | \ell, m \rangle &= r^2 \ell(\ell+1) \, \delta_{\ell'\ell} \delta_{m'm} \,. \end{aligned} \tag{D.21}$$

Using these expressions we obtain the orthogonality relations

Orthogonality Relations for Vector Spherical Harmonics I

$$\oint \boldsymbol{X}^*_{\ell'm'}(\theta, \phi) \cdot \Big[g_\ell(r) \boldsymbol{X}_{\ell m}(\theta, \phi) \Big] \, d\Omega = g_\ell(r) \, \delta_{\ell'\ell} \delta_{m'm} \tag{D.22a}$$

$$\oint \boldsymbol{X}^*_{\ell'm'}(\theta, \phi) \cdot \Big[\nabla \times f_\ell(r) \boldsymbol{X}_{\ell m}(\theta, \phi) \Big] \, d\Omega = 0 \,. \tag{D.22b}$$

Eq. (D.22a) can be easily proven from the second equality of Eq. (D.21). To prove the second relation we employ the decomposition of the momentum operator derived in exercise D.3 to arrive at [2, Eq. (10.60)]

$$\hat{\boldsymbol{\pi}} \times f_\ell(r) \hat{\boldsymbol{L}} = -\frac{i}{r^2} \frac{d}{dr} \Big[r f_\ell(r) \Big] \boldsymbol{r} \times \hat{\boldsymbol{L}} + \frac{f_\ell(r)}{r^2} \boldsymbol{r} \hat{L}^2 \,. \tag{D.23}$$

Using the orthogonality relations of Eq. (D.20) we find

$$\Big\langle \ell', m' \Big| \hat{\boldsymbol{L}} \cdot \left\{ -\frac{i}{r^2} \frac{d}{dr} \Big[r f_\ell(r) \Big] \boldsymbol{r} \times \hat{\boldsymbol{L}} + \frac{f_\ell(r)}{r^2} \boldsymbol{r} \hat{L}^2 \right\} \Big| \ell, m \Big\rangle = 0 \,,$$

[1] For the last expression we use $[r_i, \hat{L}_j] = i \epsilon_{ijk} r_k$ together with

$$(\boldsymbol{r} \times \hat{\boldsymbol{L}}) \cdot (\boldsymbol{r} \times \hat{\boldsymbol{L}}) = \sum_{ij} \left(r_i \hat{L}_j r_i \hat{L}_j - r_i \hat{L}_j r_j \hat{L}_i \right) .$$

which proves Eq. (D.22b). Along the same lines we can exploit the orthogonality relations between the basis states of Eq. (D.19) to obtain the second set of orthogonality relations

Orthogonality Relations for Vector Spherical Harmonics II

$$\oint \boldsymbol{r}\times\boldsymbol{X}^*_{\ell'm'}(\theta,\phi)\cdot\Big[g_\ell(r)\boldsymbol{X}_{\ell m}(\theta,\phi)\Big]\,d\Omega = 0 \tag{D.24a}$$

$$\oint \boldsymbol{r}\times\boldsymbol{X}^*_{\ell'm'}(\theta,\phi)\cdot\Big[\nabla\times f_\ell(r)\boldsymbol{X}_{\ell m}(\theta,\phi)\Big]\,d\Omega = -i\left[\frac{d}{dr}rf_\ell(r)\right]\delta_{\ell\ell'}\delta_{mm'}\,, \tag{D.24b}$$

which will be used below in the context of Mie theory. In principle, we could derive the orthogonality relations for the longitudinal vector function $\hat{\boldsymbol{r}}h_\ell(kr)Y_{\ell m}$ along the same lines. However, it is often sufficient to consider the transverse vector functions only, and we thus leave the orthogonality relations for the longitudinal function as an exercise to the interested reader.

Exercises

Exercise D.1 Show that $\boldsymbol{N}_\psi(\boldsymbol{r})$ defined in Eq. (D.2) fulfills the wave equation of Eq. (D.1).

Exercise D.2 Show that $\boldsymbol{M}_\psi$ and $\boldsymbol{N}_\psi$ defined in Eq. (D.2) are related to each other via $k\boldsymbol{M}_\psi = \nabla\times\boldsymbol{N}_\psi$. Start from the definition $\boldsymbol{N}_\psi = \frac{1}{k}\nabla\times\boldsymbol{M}_\psi$ and use that $\nabla\cdot\boldsymbol{M}_\psi = 0$.

Exercise D.3 Consider the momentum operator $\hat{\boldsymbol{\pi}} = -i\nabla$. Use the fundamental commutation relation $[r_m,\hat{\pi}_n] = i\delta_{mn}$ to prove the decomposition

$$\hat{\boldsymbol{\pi}} = -\frac{i\boldsymbol{r}}{r}\frac{\partial}{\partial r} - \frac{1}{r^2}\boldsymbol{r}\times\hat{\boldsymbol{L}}\,.$$

Exercise D.4 Derive through detailed calculation the orthogonality relations of Eq. (D.24)

Exercise D.5 Compute for $\ell = 1$ the basis functions $\nabla\times f_\ell(kr)\boldsymbol{X}_{\ell m}$ using Eq. (D.23).

Exercise D.6 Prove the orthogonality relations of Eq. (D.24).

Appendix E
Mie Theory

In this appendix we show how to solve Maxwell's equations for a spherical particle. The approach is usually referred to as "Mie theory" in honor of Gustav Mie's original work on this topic [74]. Mie theory is a feast of special functions, namely spherical harmonics, vector spherical harmonics, as well as spherical Bessel and Hankel functions, and its derivation is somewhat intricate. Yet, it is one of the few problems in electrodynamics that can be solved analytically and Mie solutions have found widespread use in various fields of research, so it might be worth looking to the problem in slightly more detail.

E.1 Multipole Expansion of Electromagnetic Fields

In Appendix D we have shown that any transverse vector function can be expanded in terms of the basis functions

$$\boldsymbol{M}_f(\boldsymbol{r}) = \nabla \times \boldsymbol{r} f_\ell(kr) Y_{\ell m}\,, \quad \boldsymbol{N}_g(\boldsymbol{r}) = \frac{1}{k} \nabla \times \nabla \times \boldsymbol{r} g_\ell(kr) Y_{\ell m}\,,$$

where f_ℓ and g_ℓ are spherical Bessel or Hankel functions. The two vector functions are related through

$$k\boldsymbol{M}_f = \nabla \times \boldsymbol{N}_f\,, \quad k\boldsymbol{N}_f = \nabla \times \boldsymbol{M}_f\,. \tag{E.1}$$

In the following we apply this decomposition to electromagnetic fields. Apart from an unimportant prefactor the vector function $\boldsymbol{M}_g$ for the electric field can be expressed as

$$\text{(electric field)} \qquad \boldsymbol{M}_g(\boldsymbol{r}) = g_\ell(r) \boldsymbol{X}_{\ell m}(\theta, \phi)\,, \tag{E.2}$$

U. Hohenester, *Nano and Quantum Optics*, Graduate Texts in Physics,
https://doi.org/10.1007/978-3-030-30504-8

where $\boldsymbol{X}_{\ell m}$ is the vector spherical harmonic defined in Eq. (D.8). Through Faraday's law $\boldsymbol{M}_g$ is related to a magnetic field of the form

$$ikZ\,\boldsymbol{H}(\boldsymbol{r}) = \nabla \times \boldsymbol{M}_g(\boldsymbol{r}) = k\boldsymbol{N}_g(\boldsymbol{r})\,,$$

with Z being the impedance. Similarly, we can express the vector function $\boldsymbol{M}_f$ for the magnetic field in the form

$$\text{(magnetic field)} \qquad Z^{-1}\boldsymbol{M}_f(\boldsymbol{r}) = f_\ell(r)\boldsymbol{X}_{\ell m}(\theta,\phi)\,, \tag{E.3}$$

which is related to an electric of the form

$$-ikZ^{-1}\boldsymbol{E}(\boldsymbol{r}) = Z^{-1}\nabla \times \boldsymbol{M}_f(\boldsymbol{r}) = Z^{-1}k\,\boldsymbol{N}_f(\boldsymbol{r})\,.$$

Putting together the electric field components $\boldsymbol{M}_g(\boldsymbol{r})+\boldsymbol{N}_f(\boldsymbol{r})$ and the magnetic field components $\boldsymbol{M}_f(\boldsymbol{r})+\boldsymbol{N}_g(\boldsymbol{r})$ we arrive at the decomposition of the electromagnetic fields in terms of vector spherical harmonics

Multipole Expansion of Electromagnetic Fields

$$\begin{aligned}\boldsymbol{E}(\boldsymbol{r}) &= Z\sum_{\ell,m}\left[b_{\ell m}g_\ell(kr)\boldsymbol{X}_{\ell m}(\theta,\phi) + \frac{i}{k}a_{\ell m}\nabla\times f_\ell(kr)\boldsymbol{X}_{\ell m}(\theta,\phi)\right]\\ \boldsymbol{H}(\boldsymbol{r}) &= \sum_{\ell,m}\left[a_{\ell m}f_\ell(kr)\boldsymbol{X}_{\ell m}(\theta,\phi) - \frac{i}{k}b_{\ell m}\nabla\times g_\ell(kr)\boldsymbol{X}_{\ell m}(\theta,\phi)\right].\end{aligned} \tag{E.4}$$

This expression provides a general decomposition for transverse electromagnetic fields where the coefficients $a_{\ell m}$, $b_{\ell m}$ and the radial functions f_ℓ, h_ℓ have to be determined for the problem under study.

Multipole Coefficients

Suppose that we know the (transverse) electromagnetic fields $\boldsymbol{E}$, $\boldsymbol{H}$ and want to compute the corresponding expansion coefficients $a_{\ell m}$, $b_{\ell m}$. Our starting point is the multipole expansion for the electric field of Eq. (E.4)

$$\boldsymbol{E}(\boldsymbol{r}) = Z\sum_{\ell,m}\left[b_{\ell m}g_\ell(kr)\boldsymbol{X}_{\ell m}(\theta,\phi) + \frac{i}{k}a_{\ell m}\nabla\times f_\ell(kr)\boldsymbol{X}_{\ell m}(\theta,\phi)\right].$$

We multiply both sides from the left-hand side with $\boldsymbol{r}$ and note that $\boldsymbol{r}\cdot\hat{\boldsymbol{L}} = 0$ such that the first term in brackets becomes zero. The second term can be simplified using a cyclic permutation in the triple product via

$$i\boldsymbol{r}\cdot\nabla\times f_\ell(kr)\boldsymbol{X}_{\ell m}=\Big(i\boldsymbol{r}\times\nabla\Big)\cdot f_\ell(kr)\boldsymbol{X}_{\ell m}=-\hat{\boldsymbol{L}}\cdot f_\ell(kr)\boldsymbol{X}_{\ell m}\,.$$

Thus, we get

$$\boldsymbol{r}\cdot\boldsymbol{E}=-Z\sum_{\ell,m}\frac{\sqrt{\ell(\ell+1)}}{k}\,a_{\ell m}\,f_\ell(kr)Y_{\ell,m}(\theta,\phi)\,,$$

where we have used the definition of Eq. (D.8) for the vector spherical harmonics. If we multiply the above equation with a spherical harmonic of given degree and order, we get upon integration over all angles the multipole coefficients expressed in terms of the transverse electromagnetic fields

Multipole Expansion Coefficients

$$a_{\ell m}\,f_\ell(kr)=-\frac{Z^{-1}k}{\sqrt{\ell(\ell+1)}}\oint Y^*_{\ell,m}(\theta,\phi)\Big[\boldsymbol{r}\cdot\boldsymbol{E}(\boldsymbol{r})\Big]\,d\Omega$$

$$b_{\ell m}\,g_\ell(kr)=\frac{k}{\sqrt{\ell(\ell+1)}}\oint Y^*_{\ell,m}(\theta,\phi)\Big[\boldsymbol{r}\cdot\boldsymbol{H}(\boldsymbol{r})\Big]\,d\Omega\,. \tag{E.5}$$

The second expression for $b_{\ell m}$ can be derived by the same token from the multipole expansion of the magnetic field in Eq. (E.4).

E.2 Mie Coefficients

We next consider the problem of a spherical nanoparticle with radius R and with homogeneous material properties ε_1, μ_1 inside the sphere and ε_2, μ_2 outside the sphere, see Fig. E.1. We decompose the electric fields $\boldsymbol{E}_1(\boldsymbol{r})$, $\boldsymbol{E}_2(\boldsymbol{r})$ inside and outside the sphere in the form

$$\boldsymbol{E}_1(\boldsymbol{r})=\boldsymbol{E}_1^{\rm sca}(\boldsymbol{r})\,,\quad \boldsymbol{E}_2(\boldsymbol{r})=\boldsymbol{E}_2^{\rm inc}(\boldsymbol{r})+\boldsymbol{E}_2^{\rm sca}(\boldsymbol{r})\,,$$

with a corresponding expression for the magnetic fields. Here $\boldsymbol{E}^{\rm inc}(\boldsymbol{r})$ is the incoming field, for instance associated with a plane wave excitation or an oscillating dipole, and $\boldsymbol{E}_{1,2}^{\rm sca}(\boldsymbol{r})$ are the scattered fields which describe the response of the spherical nanoparticle. The **incoming electromagnetic fields** can be expressed in terms of the multipole expansion of Eq. (E.4) in the form

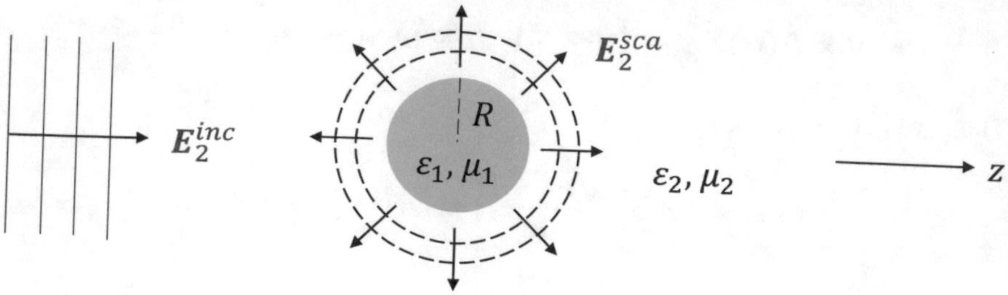

Fig. E.1 Schematics of Mie problem. A spherical particle with radius R and material properties ε_1, μ_1 is embedded in a medium with ε_2, μ_2. The particle is excited by an incoming field $\boldsymbol{E}_2^{\rm inc}$, here a plane wave, and the response of the sphere is described by the scattered fields $\boldsymbol{E}_2^{\rm sca}$ and $\boldsymbol{E}_1^{\rm sca}$ outside and inside the spherical particle, respectively. Within Mie theory these scattered fields are described in terms of the so-called Mie coefficients

$$\begin{aligned}\boldsymbol{E}_2^{\rm inc} &= Z_2\sum_{\ell,m}\left[b_{\ell m}^{\rm inc}g_\ell(k_2r)\boldsymbol{X}_{\ell m}(\theta,\phi)+\frac{i}{k_2}a_{\ell m}^{\rm inc}\nabla\times f_\ell(k_2r)\boldsymbol{X}_{\ell m}(\theta,\phi)\right]\\ \boldsymbol{H}_2^{\rm inc} &= \sum_{\ell,m}\left[a_{\ell m}^{\rm inc}f_\ell(k_2r)\boldsymbol{X}_{\ell m}(\theta,\phi)-\frac{i}{k_2}b_{\ell m}^{\rm inc}\nabla\times g_\ell(k_2r)\boldsymbol{X}_{\ell m}(\theta,\phi)\right],\end{aligned}\tag{E.6}$$

where the coefficients $a_{\ell m}^{\rm inc}$, $b_{\ell m}^{\rm inc}$ as well as the combination of spherical Bessel and Hankel functions $f_\ell(k_2r)$, $g_\ell(k_2r)$ have to be determined for each type of excitation separately, as will be discussed further below. k_2 and Z_2 are the wavenumber and impedance of the embedding medium, respectively. The **scattered fields outside the sphere** become[1]

$$\begin{aligned}\boldsymbol{E}_2^{\rm sca} &= -Z_2\sum_{\ell,m}\left[b_{\ell m}h_\ell^{(1)}(k_2r)\boldsymbol{X}_{\ell m}(\theta,\phi)+\frac{i}{k_2}a_{\ell m}\nabla\times h_\ell^{(1)}(k_2r)\boldsymbol{X}_{\ell m}(\theta,\phi)\right]\\ \boldsymbol{H}_2^{\rm sca} &= -\sum_{\ell,m}\left[a_{\ell m}h_\ell^{(1)}(k_2r)\boldsymbol{X}_{\ell m}(\theta,\phi)-\frac{i}{k_2}b_{\ell m}\nabla\times h_\ell^{(1)}(k_2r)\boldsymbol{X}_{\ell m}(\theta,\phi)\right].\end{aligned}\tag{E.7}$$

Here we have replaced f_ℓ, g_ℓ by the spherical Hankel function $h_\ell^{(1)}$ of the first kind because they possess for large arguments the proper boundary conditions of outgoing waves, see Eq. (C.29). Similarly, inside the particle we replace f_ℓ, g_ℓ by the spherical Bessel function j_ℓ which remain finite at the origin. This leads us to the **scattered fields inside the sphere**

[1]The negative sign in front of the summation for the scattered fields is chosen for convenience to get the same Mie coefficients as those derived by Bohren and Huffman [60].

$$
\begin{aligned}
\boldsymbol{E}_1^{\text{sca}} &= Z_1 \sum_{\ell,m} \left[d_{\ell m} j_\ell(k_1 r) \boldsymbol{X}_{\ell m}(\theta,\phi) + \frac{i}{k_1} c_{\ell m} \nabla \times j_\ell(k_1 r) \boldsymbol{X}_{\ell m}(\theta,\phi) \right] \\
\boldsymbol{H}_1^{\text{sca}} &= \sum_{\ell,m} \left[c_{\ell m} j_\ell(k_1 r) \boldsymbol{X}_{\ell m}(\theta,\phi) - \frac{i}{k_1} d_{\ell m} \nabla \times j_\ell(k_1 r) \boldsymbol{X}_{\ell m}(\theta,\phi) \right] .
\end{aligned}
\tag{E.8}
$$

To compute the unknown coefficients $a_{\ell m}$, $b_{\ell m}$ at the particle outside and $c_{\ell m}$, $d_{\ell m}$ at the particle inside, we need to match the electromagnetic fields at the particle boundary. This matching procedure is facilitated by the fact that in the multipole expansion of Eqs. (E.6–E.8) the electromagnetic fields are already purely tangential. Thus, we get the boundary conditions

$$
\boldsymbol{E}_1^{\text{sca}}\Big|_{r=R} = \Big[\boldsymbol{E}_2^{\text{inc}} + \boldsymbol{E}_2^{\text{sca}}\Big]_{r=R}, \quad \boldsymbol{H}_1^{\text{sca}}\Big|_{r=R} = \Big[\boldsymbol{H}_2^{\text{inc}} + \boldsymbol{H}_2^{\text{sca}}\Big]_{r=R}.
$$

We first multiply the above equations with $\boldsymbol{X}_{\ell m}^*$, integrate over all angles, and use the orthogonality relations of Eq. (D.22). This leads us to

$$
\begin{aligned}
\frac{Z_1}{Z_2} d_{\ell m} j_\ell(k_1 R) &= b_{\ell m}^{\text{inc}} g_\ell(k_2 R) - b_{\ell m} h_\ell^{(1)}(k_2 R) \\
c_{\ell m} j_\ell(k_1 R) &= a_{\ell m}^{\text{inc}} f_\ell(k_2 R) - a_{\ell m} h_\ell^{(1)}(k_2 R) .
\end{aligned}
\tag{E.9}
$$

Similarly, we multiply the boundary conditions with $\boldsymbol{r} \times \boldsymbol{X}_{\ell m}^*$, integrate over all angles, and use the orthogonality relations of Eq. (D.24) to arrive at

$$
\begin{aligned}
\frac{Z_1}{Z_2}\frac{c_{\ell m}}{k_1}\left[\frac{d}{dr} r j_\ell(k_1 r)\right]_{r=R} &= \frac{a_{\ell m}^{\text{inc}}}{k_2}\left[\frac{d}{dr} r f_\ell(k_2 r)\right]_{r=R} - \frac{a_{\ell m}}{k_2}\left[\frac{d}{dr} r h_\ell^{(1)}(k_2 r)\right]_{r=R} \\
\frac{d_{\ell m}}{k_1}\left[\frac{d}{dr} r j_\ell(k_1 r)\right]_{r=R} &= \frac{b_{\ell m}^{\text{inc}}}{k_2}\left[\frac{d}{dr} r g_\ell(k_2 r)\right]_{r=R} - \frac{b_{\ell m}}{k_2}\left[\frac{d}{dr} r h_\ell^{(1)}(k_2 r)\right]_{r=R} .
\end{aligned}
\tag{E.10}
$$

We next introduce the abbreviations $x_1 = k_1 R$, $x_2 = k_2 R$, and the Ricatti–Bessel functions and their derivatives

Riccati-Bessel Functions and Derivatives

$$
\begin{aligned}
\psi_\ell(x) &= x j_\ell(x) , \quad & \psi_\ell'(x) &= \frac{d}{dx}\Big[x j_\ell(x)\Big] \\
\xi_\ell(x) &= x h_\ell^{(1)}(x) , \quad & \xi_\ell'(x) &= \frac{d}{dx}\Big[x h_\ell^{(1)}(x)\Big] .
\end{aligned}
\tag{E.11}
$$

We additionally introduce the functions $F_\ell(x) = xf_\ell(x)$, $G_\ell(x) = xg_\ell(x)$ and their derivatives. The Mie coefficients for the fields at the sphere outside can then be expressed in the form

$$\begin{aligned} a_{\ell m} &= \left[\frac{Z_2\psi_\ell(x_1)F_\ell'(x_2) - Z_1\psi_\ell'(x_1)F_\ell(x_2)}{Z_2\psi_\ell(x_1)\xi_\ell'(x_2) - Z_1\psi_\ell'(x_1)\xi_\ell(x_2)}\right] a_{\ell m}^{\rm inc} \\ b_{\ell m} &= \left[\frac{Z_2\psi_\ell'(x_1)G_\ell(x_2) - Z_1\psi_\ell(x_1)G_\ell'(x_2)}{Z_2\psi_\ell'(x_1)\xi_\ell(x_2) - Z_1\psi_\ell(x_1)\xi_\ell'(x_2)}\right] b_{\ell m}^{\rm inc}\,. \end{aligned} \tag{E.12}$$

Similarly, the field coefficients inside the sphere are given by

$$\begin{aligned} c_{\ell m} &= \frac{k_1}{k_2}\left[\frac{Z_1\xi_\ell'(x_2)F_\ell(x_2) - Z_1\xi_\ell(x_2)F_\ell'(x_2)}{Z_2\psi_\ell(x_1)\xi_\ell'(x_2) - Z_1\psi_\ell'(x_1)\xi_\ell(x_2)}\right] a_{\ell m}^{\rm inc} \\ d_{\ell m} &= \frac{k_1}{k_2}\left[\frac{Z_1\psi_\ell(x_2)G_\ell'(x_2) - Z_1\psi_\ell(x_2)G_\ell(x_2)}{Z_2\psi_\ell'(x_1)\xi_\ell(x_2) - Z_1\psi_\ell(x_1)\xi_\ell'(x_2)}\right] b_{\ell m}^{\rm inc}\,. \end{aligned} \tag{E.13}$$

Thus, the solution of Maxwell's equations can be captured by four coefficients, which are conveniently referred to as the *Mie coefficients*.

E.3 Plane Wave Excitation

Suppose that the sphere is excited by an incoming plane wave, as depicted in Fig. E.1. Because the incoming fields are purely transverse, $\nabla\cdot\boldsymbol{E}^{\rm inc} = \nabla\cdot\boldsymbol{H}^{\rm inc} = 0$, also the scattered must fulfill $\nabla\cdot\boldsymbol{E}^{\rm sca} = \nabla\cdot\boldsymbol{H}^{\rm sca} = 0$. This can be only achieved when the longitudinal parts of the scattered fields are zero, and we thus have to consider the transverse vector functions only.

Expansion Coefficients for Plane Wave Excitation

We first show how to compute the coefficients $a_{\ell m}^{\rm inc}$, $b_{\ell m}^{\rm inc}$ for a plane wave excitation. Our starting point is Eq. (C.31) for the decomposition of a spherical wave in terms of spherical harmonics

$$\frac{e^{ikR}}{4\pi R} = ik\sum_{\ell,m} j_\ell(kr_<)h_\ell^{(1)}(kr_>)\,Y_{\ell m}^*(\hat{\boldsymbol{r}}')Y_{\ell m}(\hat{\boldsymbol{r}})\,, \tag{E.14}$$

where $\boldsymbol{R} = \boldsymbol{r} - \boldsymbol{r}'$, $r_<$ is the smaller value of r and r', and $r_>$ the larger one. For large values of r' and for $r' \gg r$ we can approximate the expression on the left-hand side using

$$\frac{e^{ikR}}{4\pi R} \xrightarrow[r'\to\infty]{} \left[\frac{e^{ikr'}}{4\pi r'}\right] e^{-ik\hat{\boldsymbol{r}}'\cdot\boldsymbol{r}}\,,$$

as discussed in more detail in Sect. 5.3.1 (to compare the results one additionally has to exchange $\boldsymbol{r}$ and $\boldsymbol{r}'$). Using the asymptotic form of Eq. (C.29) for the spherical Hankel function, inserting the expansions for large arguments into Eq. (E.14), and taking the complex conjugate on both sides of the equation we are led to

$$e^{i\boldsymbol{k}\cdot\boldsymbol{r}} = 4\pi \sum_{\ell,m} i^\ell j_\ell(kr)\, Y^*_{\ell m}(\hat{\boldsymbol{r}}) Y_{\ell m}(\hat{\boldsymbol{k}}) \,, \tag{E.15}$$

with $\boldsymbol{k} = k\hat{\boldsymbol{r}}'$. In the following we assume that the incoming wave propagates along the z-direction, $\theta' = 0$, and use the addition theorem for spherical harmonics [2, Eq. (3.62)]

$$P_\ell(\cos\theta) = \frac{4\pi}{2\ell+1} \sum_{m=-\ell}^{\ell} Y^*_{\ell m}(\theta,\phi) Y_{\ell m}(\hat{\boldsymbol{z}}) \,.$$

With this, we are led to the expansion of a plane wave in terms of spherical waves

Spherical Wave Expansion of Plane Wave I

$$e^{ikz} = \sum_\ell i^\ell \sqrt{4\pi(2\ell+1)}\, j_\ell(kr) Y_{\ell,0}(\theta,\phi) \,. \tag{E.16}$$

In the following we consider a circularly polarized plane wave propagating along z. We introduce the polarization vectors $\boldsymbol{\epsilon}_\pm = \hat{\boldsymbol{x}} \pm i\hat{\boldsymbol{y}}$ for helicity $\pm$. The electromagnetic fields can then be expressed as

$$\boldsymbol{E} = \boldsymbol{\epsilon}_\pm E_0 e^{ikz} \,, \quad Z\boldsymbol{H} = \hat{\boldsymbol{z}} \times \boldsymbol{E} = \mp i \boldsymbol{\epsilon}_\pm E_0 e^{ikz} \,, \tag{E.17}$$

where E_0 is the electric field amplitude of the incoming wave. A wave with linear polarization can be expressed as a linear combination of the two helicity states. We next multiply the multipole expansion of the electromagnetic fields given by Eq. (E.4) from both sides with $\boldsymbol{X}^*_{\ell m}$, integrate over all angles, and use the orthogonality relations of the vector spherical harmonics to get

$$\begin{aligned}
\oint \boldsymbol{X}^*_{\ell m} \cdot \boldsymbol{E}(\boldsymbol{r})\, d\Omega &= Z\, b^\pm_{lm} g_l(kr) = \oint \boldsymbol{X}^*_{\ell m} \cdot \Big[\quad \boldsymbol{\epsilon}_\pm E_0 e^{ikz} \Big]\, d\Omega \\
Z \oint \boldsymbol{X}^*_{\ell m} \cdot \boldsymbol{H}(\boldsymbol{r})\, d\Omega &= Z\, a^\pm_{lm} f_l(kr) = \oint \boldsymbol{X}^*_{\ell m} \cdot \Big[\mp i \boldsymbol{\epsilon}_\pm E_0 e^{ikz} \Big]\, d\Omega \,.
\end{aligned} \tag{E.18}$$

To evaluate the integrals on the right-hand side we first note that

$$\sqrt{\ell(\ell+1)}\, \boldsymbol{\epsilon}^*_\pm \cdot \boldsymbol{X}_{\ell m} = \hat{L}_\mp Y_{\ell,m} \,, \tag{E.19}$$

with the operators $\hat{L}_\pm = \hat{L}_x \pm i\hat{L}_y$ introduced in Eq. (D.14). To compute $\hat{L}_\mp Y_{\ell,m}$ we use Eq. (D.15) and arrive at

$$\oint \left[\boldsymbol{\epsilon}_\pm^* \cdot \boldsymbol{X}_{\ell m}\right]^* Y_{\ell,0}\, d\Omega = \sqrt{\frac{(\ell \pm m)(\ell \mp m + 1)}{\ell(\ell+1)}}\,\delta_{m\mp 1,0} = \delta_{m,\pm 1}\,.$$

Thus, if we insert the spherical wave expansion of Eq. (E.16) into Eq. (E.18) we get

$$Z\, b_{lm}^{\pm} g_l(kr) = i^\ell \sqrt{4\pi(2\ell+1)}\,\delta_{m,\pm 1}\, j_\ell(kr)\,, \tag{E.20}$$

as well as $a_{\ell,m}^{\pm} = \mp i b_{\ell,m}^{\pm}$. Putting together all results, we are led to the expansion of an incoming plane wave with helicity $\pm$ that is propagating along the z-direction in terms of vector spherical harmonics [2, Eq. (10.55)]

Spherical Wave Expansion of Plane Wave II

$$\begin{aligned} \boldsymbol{E} &= E_0 \sum_\ell i^\ell \sqrt{4\pi(2\ell+1)}\left[\quad j_\ell(kr)\boldsymbol{X}_{\ell,\pm 1} \pm \frac{1}{k}\nabla\times j_\ell(kr)\boldsymbol{X}_{\ell,\pm 1}\right] \\ Z\boldsymbol{H} &= E_0 \sum_\ell i^\ell \sqrt{4\pi(2\ell+1)}\left[\mp i j_\ell(kr)\boldsymbol{X}_{\ell,\pm 1} - \frac{i}{k}\nabla\times j_\ell(kr)\boldsymbol{X}_{\ell,\pm 1}\right]. \end{aligned} \tag{E.21}$$

Plane Wave Excitation of Spherical Particle

We next use Eq. (E.7) to express the scattered electromagnetic fields outside the metallic nanoparticle in terms of Mie coefficients, with the incoming coefficients $a_{\ell m}^{\mathrm{inc}}$, $b_{\ell m}^{\mathrm{inc}}$ given through Eq. (E.21). The calculation of the fields inside the sphere is left as exercise to the interested reader. We first rewrite the terms in brackets of Eq. (E.12) for $F_\ell(x) = G_\ell(x) = \psi_\ell(x)$ in the form

Mie Coefficients for Plane-Wave Excitation

$$\begin{aligned} a_\ell &= \frac{Z_2\psi_\ell(x_1)\psi_\ell'(x_2) - Z_1\psi_\ell'(x_1)\psi_\ell(x_2)}{Z_2\psi_\ell(x_1)\xi_\ell'(x_2) - Z_1\psi_\ell'(x_1)\xi_\ell(x_2)} \\ b_\ell &= \frac{Z_2\psi_\ell'(x_1)\psi_\ell(x_2) - Z_1\psi_\ell(x_1)\psi_\ell'(x_2)}{Z_2\psi_\ell'(x_1)\xi_\ell(x_2) - Z_1\psi_\ell(x_1)\xi_\ell'(x_2)}, \end{aligned} \tag{E.22}$$

with the Riccati-Bessel functions ψ_ℓ, ξ_ℓ given in Eq. (E.11). k_1, k_2 are the wavenumbers inside and outside the sphere, and Z_1, Z_2 are the corresponding impedances.

In addition, we use the abbreviations $x_1 = k_1 R$, $x_2 = k_2 R$, with R being the sphere radius. With these Mie coefficients the electromagnetic fields outside the sphere can be calculated from Eq. (E.7), and are given in the form

$$\begin{aligned}
\boldsymbol{E}_2^{\text{sca}} &= -E_0 \sum_\ell i^\ell \sqrt{4\pi(2\ell+1)} \Big[b_\ell h_\ell^{(1)}(k_2 r) \boldsymbol{X}_{\ell,\pm 1} \\
&\quad \pm \frac{a_\ell}{k_2} \nabla \times h_\ell^{(1)}(k_2 r) \boldsymbol{X}_{\ell,\pm 1} \Big] \\
Z_2 \boldsymbol{H}_2^{\text{sca}} &= -E_0 \sum_\ell i^\ell \sqrt{4\pi(2\ell+1)} \Big[\mp i a_\ell h_\ell^{(1)}(k_2 r) \boldsymbol{X}_{\ell,\pm 1} \\
&\quad - \frac{i b_\ell}{k_2} \nabla \times h_\ell^{(1)}(k_2 r) \boldsymbol{X}_{\ell,\pm 1} \Big] .
\end{aligned} \tag{E.23}$$

Extinction Cross Section

To compute the extinction cross section, we start from the optical theorem of Eq. (4.27) which expresses the extinction power by

$$P_{\text{ext}} = \frac{2\pi}{k_2} Z_2^{-1} \, \text{Im} \Big[E_0^* \boldsymbol{\epsilon}_\pm^* \cdot \boldsymbol{F}_2^{\text{sca}}(\hat{\boldsymbol{z}}) \Big] .$$

Here $\boldsymbol{F}_2^{\text{sca}}(\hat{\boldsymbol{z}})$ is the far-field amplitude of the scattered electric field in the direction of $\hat{\boldsymbol{z}}$. Using the asymptotic form of Eq. (C.29) for the spherical Hankel functions we get from Eq. (E.23) the far-field amplitude

$$\boldsymbol{F}_2^{\text{sca}}(\hat{\boldsymbol{z}}) = \frac{i E_0}{k_2} \sum_\ell \sqrt{4\pi(2\ell+1)} \Big[b_\ell \boldsymbol{X}_{\ell,\pm 1} \pm \frac{a_\ell}{k_2} (i k_2 \hat{\boldsymbol{z}}) \times \boldsymbol{X}_{\ell,\pm 1} \Big] .$$

Multiplication of the above expression with $\boldsymbol{\epsilon}_\pm^*$ gives

$$\boldsymbol{\epsilon}_\pm^* \cdot \boldsymbol{F}_2^{\text{sca}}(\hat{\boldsymbol{z}}) = \frac{i E_0}{k_2} \sum_\ell \sqrt{4\pi(2\ell+1)} \Big[b_\ell \, \boldsymbol{\epsilon}_\pm^* \cdot \boldsymbol{X}_{\ell,\pm 1} \pm i a_\ell \, \boldsymbol{\epsilon}_\pm^* \cdot \hat{\boldsymbol{z}} \times \boldsymbol{X}_{\ell,\pm 1} \Big] .$$

The second term in brackets can be rewritten using cyclic permutation of a triple product via

$$\boldsymbol{\epsilon}_\pm^* \cdot \hat{\boldsymbol{z}} \times \boldsymbol{X}_{\ell,\pm 1} = \boldsymbol{\epsilon}_\pm^* \times \hat{\boldsymbol{z}} \cdot \boldsymbol{X}_{\ell,\pm 1} = \mp i \boldsymbol{\epsilon}_\pm^* \cdot \boldsymbol{X}_{\ell,\pm 1} . \tag{E.24}$$

Using Eq. (E.19) we then obtain

$$\boldsymbol{\epsilon}_\pm^* \cdot \boldsymbol{F}_2^{\text{sca}}(\hat{\boldsymbol{z}}) = \frac{i E_0}{k_2} \sum_\ell \sqrt{4\pi(2\ell+1)} \, (a_\ell + b_\ell) \left[\hat{L}_\mp \frac{Y_{\ell,\pm 1}}{\sqrt{\ell(\ell+1)}} \right] ,$$

where the term in brackets becomes $Y_{\ell 0}$ and has to be evaluated for the angles corresponding to the propagation direction of the incoming plane wave, in our case $\theta = 0$. We can thus use $Y_{\ell,0}(\hat{z}) = \sqrt{\frac{2\ell+1}{4\pi}}$ to express the extinction power in the form

$$P_{\text{ext}} = \frac{2\pi}{k_2^2} Z_2^{-1} |E_0|^2 \sum_\ell (2\ell+1)\text{Re}\Big[a_\ell + b_\ell\Big]. \tag{E.25}$$

The intensity of the incoming plane wave is $I_{\text{inc}} = \frac{1}{2} Z_2^{-1} \left|\sqrt{2}E_0\right|^2$, where the factor of $\sqrt{2}$ is introduced because the polarization vectors $\boldsymbol{\epsilon}_\pm = \hat{\boldsymbol{x}} \pm i\hat{\boldsymbol{y}}$ are not normalized. The ratio $P_{\text{ext}} : I_{\text{inc}}$ then gives the extinction cross section for a spherical particle excited by an incoming plane wave

Extinction Cross Section (Mie Theory)

$$C_{\text{ext}} = \frac{2\pi}{k_2^2} \sum_\ell (2\ell+1)\text{Re}\Big[a_\ell + b_\ell\Big]. \tag{E.26}$$

Scattering Cross Section

Consider the scattered fields of Eq. (E.7) at the particle outside. Far away from the particle we can use the asymptotic form of Eq. (C.29) for the Hankel functions to express the electromagnetic fields as

$$\begin{aligned} \boldsymbol{H}_2^{\text{sca}} &\to \frac{e^{ik_2 r}}{k_2 r} \sum_{\ell,m} (-i)^{l+1} \Big[a_{\ell,m} \boldsymbol{X}_{\ell,m} + b_{\ell,m} \hat{\boldsymbol{k}}_2 \times \boldsymbol{X}_{\ell,m}\Big] \\ \boldsymbol{E}_2^{\text{sca}} &\to Z_2 \boldsymbol{H}_2^{\text{sca}} \times \hat{\boldsymbol{k}}_2 . \end{aligned} \tag{E.27}$$

The time-averaged power radiated per unit solid angle by the scatterer can be obtained from the Poynting vector $\frac{1}{2}\text{Re}(\boldsymbol{E} \times \boldsymbol{H}^*) \cdot \hat{\boldsymbol{k}}_2$ projected on the propagation direction, which gives

$$\begin{aligned} \frac{dP_{\text{sca}}}{d\Omega} &= \frac{1}{2}\text{Re}\Big[r^2 \hat{\boldsymbol{k}}_2 \cdot \boldsymbol{E}_2^{\text{sca}} \times \boldsymbol{H}_2^{\text{sca}*}\Big] \\ &= \frac{Z_2}{2k_2^2} \left| \sum_{\ell,m} (-i)^{l+1} \Big[a_{\ell,m} \boldsymbol{X}_{\ell,m} \times \hat{\boldsymbol{k}}_2 + b_{\ell,m} \boldsymbol{X}_{\ell,m}\Big] \right|^2 . \end{aligned} \tag{E.28}$$

The total radiated power can be obtained by integrating this expression over all angles. In doing so, one readily observes that the interference terms do not contribute because of the orthogonality of the vector spherical harmonics, and the total radiated power is just an incoherent sum of the different multipole contributions

$$P_{\text{sca}} = \frac{Z_2}{2k_2^2} \sum_{\ell,m} \left(\left|a_{\ell,m}\right|^2 + \left|b_{\ell,m}\right|^2 \right) . \tag{E.29}$$

The above expression is general and can be used for any type of scattered fields written in the form of Eq. (E.7). For a plane wave excitation the coefficients $a_{\ell,m}^{\text{inc}}$, $b_{\ell,m}^{\text{inc}}$ are given by Eq. (E.20) and we get

$$P_{\text{sca}} = \frac{2\pi}{k_2^2} Z_2^{-1} \sum_{\ell,m} (2\ell+1) \left(|a_\ell|^2 + |b_\ell|^2 \right) ,$$

with the Mie coefficients of Eq. (E.22). Dividing by the intensity I_{inc} of the incoming plane wave then gives the scattering cross section for a spherical nanoparticle excited by an incoming plane wave

Scattering Cross Section (Mie Theory)

$$C_{\text{sca}} = \frac{2\pi}{k_2^2} \sum_{\ell} (2\ell+1) \left(\left|a_\ell\right|^2 + \left|b_\ell\right|^2 \right) . \tag{E.30}$$

E.4 Dipole Excitation

We next consider the situation shown in Fig. E.2 of an oscillating dipole with dipole moment $\boldsymbol{p}$ located at position r_0 outside a sphere [203, 204]. The "incoming"

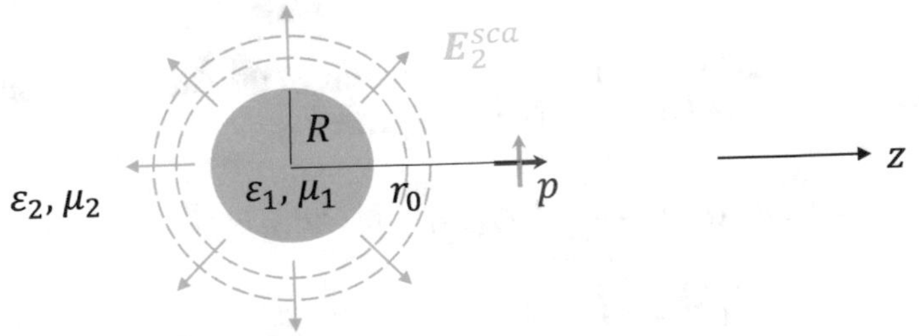

Fig. E.2 Schematics of dipole excitation of spherical nanoparticle. A dipole with dipole moment $\boldsymbol{p}$ (oriented along $\boldsymbol{\epsilon}_\pm$ or $\hat{z}$) is located at position $r_0\hat{z}$ and oscillates with frequency ω. The scattered fields act back on the dipole and modify its radiative and non-radiative properties electric field $\boldsymbol{E}_2^{\rm inc}$ has both transverse and longitudinal components, the latter being determined from $\varepsilon_2 \nabla \cdot \boldsymbol{E}_2^{\rm inc} = \rho$, where ρ is the charge distribution of the dipole. For the scattered fields we find from $\varepsilon \nabla \cdot (\boldsymbol{E}_2^{\rm inc} + \boldsymbol{E}_2^{\rm sca}) = \rho$ that they are transverse because $\nabla \cdot \boldsymbol{E}_2^{\rm sca} = 0$ must be fulfilled in the entire space. It thus suffices to consider in the following the transverse components of the electromagnetic fields only.

Multipole Expansion Coefficients for Current Source

We start from Eq. (E.5) for the calculation of the multipole expansion coefficients in terms of the transverse fields $\boldsymbol{E}$, $\boldsymbol{H}$. We introduce the transverse electric field

$$\boldsymbol{E}^{\perp} = \boldsymbol{E} + \frac{i}{\omega\varepsilon} \boldsymbol{J} \,. \tag{E.31}$$

Together with the continuity equation $i\omega\rho = \nabla \cdot \boldsymbol{J}$ one can easily show that with this we indeed get $\nabla \cdot \boldsymbol{E}^{\perp} = 0$. The curl equations of Maxwell's equations can then be written in the form

$$\nabla \times \boldsymbol{E}^{\perp} = i\omega\mu \boldsymbol{H} + \frac{i}{\omega\varepsilon} \nabla \times \boldsymbol{J} \,, \quad \nabla \times \boldsymbol{H} = -i\omega\varepsilon \boldsymbol{E}^{\perp} \,.$$

Applying the curl to both sides of the equation and using $\nabla \cdot \boldsymbol{E}^{\perp} = \nabla \cdot \boldsymbol{H} = 0$ we obtain the wave equations

$$\begin{aligned} \left(\nabla^2 + k^2\right) \boldsymbol{E}^{\perp} &= -\frac{i}{\omega\varepsilon} \nabla \times \nabla \times \boldsymbol{J} \\ \left(\nabla^2 + k^2\right) \boldsymbol{H} \quad &= -\nabla \times \boldsymbol{J} \,. \end{aligned}$$

We next multiply both sides of the equations from the left-hand side with $\boldsymbol{r}$ and use the general vector identity $\nabla^2(\boldsymbol{r} \cdot \boldsymbol{A}) = \boldsymbol{r}(\nabla^2 \boldsymbol{A}) + 2\nabla \cdot \boldsymbol{A}$ to get[2]

$$\begin{aligned} \left(\nabla^2 + k^2\right) \boldsymbol{r} \cdot \boldsymbol{E}^{\perp} &= \frac{1}{\omega\varepsilon} \hat{\boldsymbol{L}} \cdot \nabla \times \boldsymbol{J} \\ \left(\nabla^2 + k^2\right) \boldsymbol{r} \cdot \boldsymbol{H} \quad &= -i \hat{\boldsymbol{L}} \cdot \boldsymbol{J} \,. \end{aligned}$$

The above wave equations can be solved by means of the Green's function for the Helmholtz equation, see Eq. (5.7), and we get

$$\begin{aligned} \boldsymbol{r} \cdot \boldsymbol{E}^{\perp}(\boldsymbol{r}) &= -\frac{1}{\omega\varepsilon} \int G(\boldsymbol{r}, \boldsymbol{r}') \, \hat{\boldsymbol{L}}' \cdot \nabla' \times \boldsymbol{J}(\boldsymbol{r}') \, d^3 r' \\ \boldsymbol{r} \cdot \boldsymbol{H}(\boldsymbol{r}) &= i \int G(\boldsymbol{r}, \boldsymbol{r}') \, \hat{\boldsymbol{L}}' \cdot \boldsymbol{J}(\boldsymbol{r}') \, d^3 r' \,. \end{aligned}$$

[2] In the first term we use cyclic permutation of a triple product to rewrite $\boldsymbol{r} \cdot \nabla \times \nabla \times \boldsymbol{J} = \boldsymbol{r} \times \nabla \cdot \nabla \times \boldsymbol{J}$.

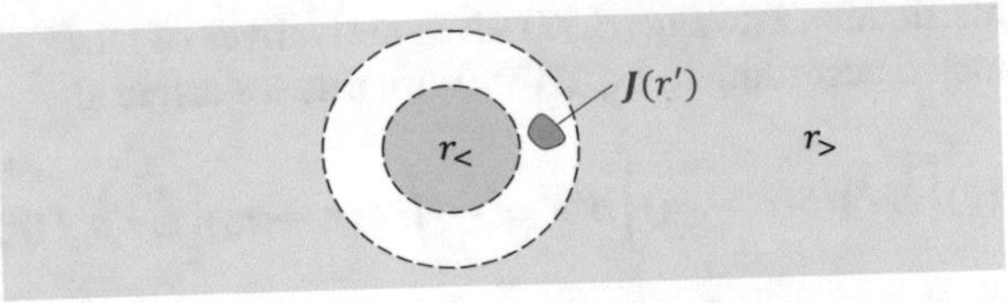

Fig. E.3 Multipole expansion for current distribution $\boldsymbol{J}(\boldsymbol{r}')$. We assume that the point $\boldsymbol{r}$ where the fields are computed is either located inside a sphere with a radius smaller than all r' values of the source, see region denoted with $r_<$, or outside a sphere with a radius larger than all r' values, see region denoted with $r_>$

We next rewrite the Green's function in terms of spherical harmonics using Eq. (C.31). In doing so we assume that the current distribution is located within some region Ω' and chose for $\boldsymbol{r}$ a value that is located inside or outside the sphere shell including the entire source, see $r_<$, $r_>$ regions shown in Fig. E.3. Together with [2, Eq. (9.164)]

$$\oint Y_{\ell,m}(\theta,\phi)G(\boldsymbol{r},\boldsymbol{r}')\,d\Omega = ik\left\{\begin{array}{c} h_\ell^{(1)}(kr_>)j_\ell(kr') \\ j_\ell(kr_<)h_\ell^{(1)}(kr') \end{array}\right\} Y^*_{\ell,m}(\theta',\phi')\,,$$

which directly follows from Eq. (C.31), we can then write the multipole expansion coefficients for an arbitrary current source in the form [2, Eq. (9.165)]

$$\begin{aligned} a_{\ell m}^{\text{inc}} &= \frac{ik}{\sqrt{\ell(\ell+1)}}\int f_\ell(kr)Y^*_{\ell,m}(\theta,\phi)\,\hat{\boldsymbol{L}}\cdot\nabla\times\boldsymbol{J}(\boldsymbol{r})\,d^3r \\ b_{\ell m}^{\text{inc}} &= -\frac{k^2}{\sqrt{\ell(\ell+1)}}\int f_\ell(kr)Y^*_{\ell,m}(\theta,\phi)\,\hat{\boldsymbol{L}}\cdot\boldsymbol{J}(\boldsymbol{r})\,d^3r\,. \end{aligned} \tag{E.32}$$

Note that in the above expressions we have changed the integration variable from $\boldsymbol{r}'$ to $\boldsymbol{r}$. In the evaluation of the multipole coefficients we must distinguish the following two cases.

Case $\boldsymbol{r_>}$. When the observation point is outside the sphere shell including the source, corresponding to region $r_>$, we must use $f_\ell(kr) = j_\ell(kr)$.

Case $\boldsymbol{r_<}$. When the observation point is inside the sphere shell including the source, corresponding to region $r_<$, we must use $f_\ell(kr) = h_\ell^{(1)}(kr)$.

Multipole Expansion Coefficients for Point Dipole

We next consider for the current distribution the expression of Eq. (6.1) for a point dipole, which reads

$$\boldsymbol{J} = -i\omega\boldsymbol{p}\,\delta(\boldsymbol{r}-\boldsymbol{r}_0)\,.$$

Inserting this distribution into Eq. (E.32) gives derivatives of Dirac's delta function, which are evaluated according to Eq. (F.3). With this we arrive at

$$\int f(\boldsymbol{r})\Big[\hat{\boldsymbol{L}}\cdot\boldsymbol{p}\,\delta(\boldsymbol{r}-\boldsymbol{r}_0)\Big]d^3r = -\int \delta(\boldsymbol{r}-\boldsymbol{r}_0)\Big[\hat{\boldsymbol{L}}\cdot\boldsymbol{p}\,f(\boldsymbol{r})\Big]d^3r\,,$$
$$\int f(\boldsymbol{r})\Big[\hat{\boldsymbol{L}}\cdot\nabla\times\boldsymbol{p}\,\delta(\boldsymbol{r}-\boldsymbol{r}_0)\Big]d^3r = \int \delta(\boldsymbol{r}-\boldsymbol{r}_0)\Big[\hat{\boldsymbol{L}}\cdot\nabla\times\boldsymbol{p}\,f(\boldsymbol{r})\Big]d^3r\,.$$

From these results we get after some simple manipulations

$$a_{\ell m}^{\rm inc} = \frac{\omega k}{\sqrt{\ell(\ell+1)}}\Big[\hat{\boldsymbol{L}}\cdot\nabla\times\boldsymbol{p}\,f_\ell(kr)Y_{\ell,m}^*\Big]_{\boldsymbol{r}=\boldsymbol{r}_0}$$
$$b_{\ell m}^{\rm inc} = \frac{i\omega k^2}{\sqrt{\ell(\ell+1)}}\Big[\hat{\boldsymbol{L}}\cdot\boldsymbol{p}\,f_\ell(kr)Y_{\ell,m}^*\Big]_{\boldsymbol{r}=\boldsymbol{r}_0}\,.$$

Using the decomposition of the nabla operator given in exercise D.3 one can show that

$$\hat{\boldsymbol{L}}\cdot\nabla\times\boldsymbol{p} = \boldsymbol{p}\cdot\hat{\boldsymbol{L}}\times\nabla = -\frac{i\boldsymbol{p}}{r^2}\cdot\left(i\boldsymbol{r}\times\hat{\boldsymbol{L}}\left[r\frac{\partial}{\partial r}+1\right]+\boldsymbol{r}\hat{L}^2\right)\,.$$

Thus, we can express the multipole expansion coefficients for a point dipole with dipole moment $\boldsymbol{p}$ located at position $\boldsymbol{r}_0$ in the form [204, Eq. (16)]

Multipole Expansion Coefficients for Point Dipole

$$a_{\ell m}^{\rm inc} = -\frac{ik\omega}{r^2}\Big[\sqrt{\ell(\ell+1)}\boldsymbol{p}\cdot\boldsymbol{r}\,f_\ell(kr)Y_{\ell,m}^* + i\boldsymbol{p}\cdot\boldsymbol{r}\times[xf_\ell(x)]'_{x=kr}\boldsymbol{X}_{\ell,m}^*\Big]_{\boldsymbol{r}=\boldsymbol{r}_0}$$
$$b_{\ell m}^{\rm inc} = i\omega k^2\Big[\boldsymbol{p}\cdot f_\ell(kr)\boldsymbol{X}_{\ell,m}^*\Big]_{\boldsymbol{r}=\boldsymbol{r}_0}\,. \tag{E.33}$$

Here the prime $[xf_\ell(x)]'$ denotes differentiation with respect to x. In the following we consider the situation where the dipole is located outside the sphere and on the z-axis, such that $\boldsymbol{r}_0 = r_0\hat{\boldsymbol{z}}$, and use Eqs. (C.22) and (D.11)

$$Y_{\ell,m}(\hat{\boldsymbol{z}}) = \sqrt{\frac{2\ell+1}{4\pi}}\,\delta_{m,0}\,,\quad \boldsymbol{X}_{\ell,m}(\hat{\boldsymbol{z}}) = \sqrt{\frac{2\ell+1}{16\pi}}\,\boldsymbol{\epsilon}_\pm\,\delta_{m,\pm1}\,. \tag{E.34}$$

We separately treat the cases of a dipole oriented parallel and perpendicular to the z-axis. For a dipole orientation $\boldsymbol{p} = p\,\hat{\boldsymbol{z}}$ we get from Eq. (E.33)

$$a_{\ell m}^{\rm inc} = -ip\,\omega k^2\sqrt{\frac{\ell(\ell+1)(2\ell+1)}{4\pi}}\,\frac{f_\ell(kr_0)}{kr_0}\delta_{m,0}\,,\quad b_{\ell m}^{\rm inc} = 0\,. \tag{E.35}$$

Similarly, for a dipole $\boldsymbol{p} = p\,\boldsymbol{\epsilon}_\pm$ we get

$$a_{\ell m}^{\rm inc} = \pm i p\,\omega k^2 \sqrt{\frac{2\ell+1}{4\pi}}\,\frac{[x f_\ell(x)]'_{x=kr_0}}{kr_0}\,\delta_{m,\pm 1}$$

$$b_{\ell m}^{\rm inc} = \quad i p\,\omega k^2 \sqrt{\frac{2\ell+1}{4\pi}}\,f_\ell(kr_0)\,\delta_{m,\pm 1}\,, \tag{E.36}$$

where we have used $\hat{z} \times \boldsymbol{\epsilon}_\pm = \mp i \boldsymbol{\epsilon}_\pm$. Note that the vector $\boldsymbol{\epsilon}_\pm$ is not normalized and we have to account for that in the evaluation of the radiated and dissipated powers.

Radiated Power of Oscillating Dipole

We start with a consistency test and compute the radiated power P_0 of the oscillating dipole alone. The problem has been investigated in Chap. 10 where we have obtained in Eq. (10.4) the expression

$$P_0 = \frac{\mu\omega^4 p^2}{12\pi c}\,.$$

We now use the multipole expansion coefficient of Eq. (E.35) to demonstrate that we get the same result within Mie theory. For the scattered power we use Eq. (E.29) to get for a dipole oriented along z the result

$$P_{\rm sca} = \frac{Z}{2k^2}\sum_\ell |a_\ell|^2 = \frac{Zp^2\omega^2k^4}{8\pi k^2}\sum_\ell \ell(\ell+1)(2\ell+1)\left|\frac{j_\ell(kr_0)}{kr_0}\right|^2\,.$$

Note that here and in the following we suppress the subscript 2 for the outer medium. As in the above expression we evaluate the fields far away from the sphere, corresponding to region $r_>$, we have to use the spherical Bessel function j_ℓ in the expansion coefficient. Dividing the scattered power by P_0 and expanding j_ℓ in a power series for small arguments of $x = kr_0$, see Eq. (C.29), we get

$$\frac{P_{\rm sca}}{P_0} = \frac{3}{2}\sum_\ell \ell(\ell+1)(2\ell+1)\left|\frac{x^\ell}{x(2\ell+1)!!} + \mathcal{O}(x^{\ell+1})\right|^2 \to 1\,,$$

where the last limit corresponds to $x \to 0$. Thus, for a dipole oriented along z and located at the origin we indeed get the proper result. A similar analysis can be also applied to the dipole orientations $\boldsymbol{\epsilon}_\pm$.

Enhancement of Radiated Power

Putting together the results for an oscillating dipole oriented along $\hat{z}$ we get for the enhancement of the radiated power the result [203, Eq. (18)]

Enhancement of Radiated Power for Dipole Orientation $\hat{z}$

$$\frac{P^z_{\rm sca}}{P_0}=\frac{3}{2}\sum_{\ell=0}^{\infty}\ell(\ell+1)(2\ell+1)\left|\frac{j_\ell(x)+a_\ell h^{(1)}_\ell(x)}{x}\right|^2_{x=kr_0}. \tag{E.37}$$

The term $a_\ell h^{(1)}_\ell$ corresponds to the scattered far-fields, Eq. (E.7), where a_ℓ is the Mie coefficient of Eq. (E.22). For the computation of the expansion coefficient we have used in Eq. (E.35) the solution $f_\ell = h^{(1)}_\ell$ for the region $r_<$ as the dipole is assumed to be located outside the sphere. Similarly, for an oscillating dipole oriented along $\boldsymbol{\epsilon}_\pm = \hat{\boldsymbol{x}} \pm i\hat{\boldsymbol{y}}$ we get for the enhancement of the radiated power the result [203, Eq. (20)]

Enhancement of Radiated Power for Dipole Orientation $\boldsymbol{\epsilon}_\pm$

$$\frac{P^\pm_{\rm sca}}{P_0}=\frac{3}{4}\sum_{\ell=0}^{\infty}(2\ell+1)\left\{\left|j_\ell(x)+a_\ell h^{(1)}_\ell(x)\right|^2+\left|\frac{\psi'_\ell(x)+b_\ell\xi'_\ell(x)}{x}\right|^2\right\}_{x=kr_0}, \tag{E.38}$$

where $\psi(x)$, $\xi(x)$ are the Riccati-Bessel functions given in Eq. (E.11). Note that we have introduced in the front of the sum an additional factor 1/2 because the dipole vector $\boldsymbol{\epsilon}_\pm$ is not normalized, and correspondingly P_0 has to be multiplied by a factor of two.

Enhancement of Dissipated Power

The enhancement for the total dissipated power can be computed from Eq. (10.5),

$$\frac{P}{P_0}=1+\frac{6\pi}{k}\frac{1}{\mu\omega^2p^2}\,\mathrm{Im}\Big\{\boldsymbol{p}^*\cdot\boldsymbol{E}^{\rm sca}(\boldsymbol{r}_0)\Big\}, \tag{E.39}$$

where we have related the reflected Green's function to the scattered (induced) electric field. Our starting point is Eq. (E.7) for this scattered field

$$\boldsymbol{E}^{\rm sca}=-Z\sum_{\ell,m}\left[b_{\ell m}h^{(1)}_\ell(kr)\boldsymbol{X}_{\ell m}+\frac{i}{k}a_{\ell m}\nabla\times h^{(1)}_\ell(kr)\boldsymbol{X}_{\ell m}\right]. \tag{E.40}$$

We first consider a dipole orientation along $\hat{z}$ and use Eq. (D.23) to bring the term in brackets to the form

$$\Big[b_{\ell m}h_\ell^{(1)}(kr)\boldsymbol{X}_{\ell m}+\frac{a_{\ell m}}{kr^2}\Big(i\xi_\ell'(kr)\boldsymbol{r}\times\boldsymbol{X}_{\ell m}-\sqrt{\ell(\ell+1)}h_\ell^{(1)}(kr)\,\boldsymbol{r}Y_{\ell m}\Big)\Big]\,.$$

Thus, we get

$$\begin{aligned}\hat{z}\cdot\boldsymbol{E}^{\rm sca}(r_0\hat{z})&=Z\sum_{\ell,m}\sqrt{\ell(\ell+1)}\,a_{\ell m}\left[\frac{h_\ell^{(1)}(kr_0)}{kr_0}\right]Y_{\ell m}(\hat{z})\\&=-iZp\omega k^2\sum_\ell\sqrt{\ell(\ell+1)}\sqrt{\frac{\ell(\ell+1)(2\ell+1)}{4\pi}}a_\ell\left[\frac{h_\ell^{(1)}(kr_0)}{kr_0}\right]^2\sqrt{\frac{2\ell+1}{4\pi}}\,,\end{aligned}$$

where we have explicitly written in the second line the expressions for the Mie coefficient and the spherical harmonics, Eq. (E.34). Inserting the electric field into Eq. (E.39) we obtain the enhancement of the total dissipated power for a dipole oriented along $\hat{z}$ [203, Eq. (17)][3]

Enhancement of Dissipated Power for Dipole Orientation $\hat{z}$

$$\frac{P_{\rm tot}^z}{P_0}=1-\frac{3}{2}{\rm Re}\left\{\sum_{\ell=0}^{\infty}\ell(\ell+1)(2\ell+1)\,a_\ell\left[\frac{h_\ell^{(1)}(x)}{x}\right]^2_{x=kr_0}\right\}.\qquad\text{(E.41)}$$

Similarly, for a dipole orientation $\boldsymbol{\epsilon}_\pm$ the scattered field is of the form

$$\boldsymbol{\epsilon}_\pm^*\cdot\boldsymbol{E}^{\rm sca}=-Z\sum_{\ell,m}\Big[\Big(b_{\ell m}h_\ell^{(1)}(kr)\Big)\boldsymbol{\epsilon}_\pm^*\cdot\boldsymbol{X}_{\ell m}+\Big(\frac{a_{\ell m}}{kr^2}\xi_\ell'(kr)\Big)\boldsymbol{\epsilon}_\pm^*\cdot\boldsymbol{r}\times\boldsymbol{X}_{\ell m}\Big]\,.$$

At the dipole position $r_0\hat{z}$ we get for the first term in brackets

$$\left(ip\,\omega k^2\sqrt{\frac{2\ell+1}{4\pi}}\,h_\ell^{(1)}(kr_0)\,b_\ell h_\ell^{(1)}(kr_0)\right)\sqrt{\frac{2\ell+1}{4\pi}}\,\delta_{m,\pm1}\,,$$

where we have used Eq. (E.36) for the expansion coefficients $b_{\ell m}$. For the second term we use Eq. (E.24) to simplify the triple product, and obtain after some simple manipulations

$$\left(\pm ip\,\omega k^2\sqrt{\frac{2\ell+1}{4\pi}}\frac{\xi_\ell'(kr_0)}{kr_0}\,ia_\ell\frac{\xi_\ell'(kr_0)}{kr_0}\right)\left(\mp i\sqrt{\frac{2\ell+1}{4\pi}}\right)\delta_{m,\pm1}\,,$$

[3] Note that the different signs in Eq. (E.41) and in Eq. (17) of Ref. [203] are due to the different definitions of Mie coefficients.

with the Ricatti–Bessel function $\xi_\ell(x)$ of Eq. (E.11). Putting together all results we finally obtain the enhancement of the total dissipated power for a dipole oriented along $\boldsymbol{\epsilon}_\pm$ [203, Eq. (19)]

Enhancement of Dissipated Power for Dipole Orientation $\boldsymbol{\epsilon}_\pm$

$$\frac{P_{\rm tot}^{\pm}}{P_0} = 1 - \frac{3}{4}\mathrm{Re}\left\{\sum_{\ell=0}^{\infty}(2\ell+1)\left(a_\ell\left[\frac{\xi_\ell'(x)}{x}\right]^2_{x=kr_0} + b_\ell\left[h_\ell^{(1)}(x)\right]^2_{x=kr_0}\right)\right\}. \tag{E.42}$$

The second term has again been multiplied by a factor of 1/2 because the dipole moment $\boldsymbol{\epsilon}_\pm$ is not normalized.

Appendix F
Dirac's Delta Function

Dirac's delta function is defined through

Dirac's Delta Function

$$\int_{x_0}^{x_1} \delta(x-a) f(x)\, dx = \begin{cases} f(a) & \text{if } a \in (x_0, x_1) \\ 0 & \text{else}\,. \end{cases} \tag{F.1}$$

Expressed in words, the integral with Dirac's delta function gives the function value of $f(a)$ if a is located within the integration limits, and zero otherwise. In fact, such a behavior cannot be achieved with a normal function but rather with a distribution defined through some limiting procedure such as

$$\delta(x) = \frac{1}{\pi} \lim_{\eta \to 0} \frac{\eta}{x^2+\eta^2} = \frac{1}{2\sqrt{\pi}} \lim_{\eta \to 0} \eta^{-\frac{1}{2}} \exp\left(-\frac{x^2}{4\eta}\right) . \tag{F.2}$$

We also assume that the function $f(x)$ in Eq. (F.1) is sufficiently well behaved. In this case we can shuffle derivatives of Dirac's delta function over to the function via

$$\int_{-\infty}^{\infty} f(x) \left[\frac{d^n}{dx^n}\delta(x-a)\right] dx = (-1)^n \left[\frac{d^n f(x)}{dx^n}\right]_{x=a} . \tag{F.3}$$

Another useful relation one can obtain from Eq. (F.2) is

$$\delta\Big(g(x)\Big) = \sum_{i=1}^{n} \frac{\delta(x-x_i)}{|g'(x_i)|} , \tag{F.4}$$

U. Hohenester, *Nano and Quantum Optics*, Graduate Texts in Physics,
https://doi.org/10.1007/978-3-030-30504-8

where $g(x)$ is assumed to possess only simple zero points x_i and the summation runs over all x_i, with $g(x_i) = 0$ and $g'(x)$ denoting the derivative of g with respect to x. From this expression we immediately find

$$\delta(ax) = \frac{1}{|a|}\delta(x)\,. \tag{F.5}$$

From the definition of the Dirac's delta function it follows that

$$\lim_{\eta\to 0}\left[\frac{1}{x \pm i\eta}\right] = \mathcal{P}\left(\frac{1}{x}\right) \mp i\pi\delta(x) \tag{F.6}$$

where $\mathcal{P}$ denotes Cauchy's principal value

$$\mathcal{P}\left(\frac{1}{x}\right) = \lim_{\eta\to 0}\frac{x}{x^2 + \eta^2}\,. \tag{F.7}$$

When using Eq. (F.6) under an integral we arrive at the important relation

$$\lim_{\eta\to 0}\int_{-\infty}^{\infty}\frac{f(x)}{x - a \pm i\eta}\,dx = \mathcal{P}\int_{-\infty}^{\infty}\frac{f(x)}{x - a}\,dx \mp i\pi f(a)\,, \tag{F.8}$$

where the principal value integral can be written in the form

$$\mathcal{P}\int_{-\infty}^{\infty}\frac{f(x)}{x - a}\,dx = \lim_{\eta\to 0}\left[\int_{-\infty}^{a-\eta}\frac{f(x)}{x - a}\,dx + \int_{a+\eta}^{\infty}\frac{f(x)}{x - a}\,dx\right]. \tag{F.9}$$

The Fourier transform of Dirac's delta function reads

$$\delta(x) = \frac{1}{2\pi}\int_{-\infty}^{\infty} e^{ikx}\,dk\,. \tag{F.10}$$

Dirac's delta function can be defined also for vectors

$$\delta^{(3)}(\boldsymbol{r} - \boldsymbol{a}) = \delta(x - a_x)\delta(y - a_y)\delta(z - a_z)\,. \tag{F.11}$$

In this book we adopt throughout the somewhat sloppy notation $\delta(\boldsymbol{r} - \boldsymbol{a})$ instead of the more correct form of $\delta^{(3)}(\boldsymbol{r} - \boldsymbol{a})$.

F.1 Transverse and Longitudinal Delta Function

In the context of vector functions it is convenient to introduce the transverse delta function

Transverse Delta Function

$$\delta_{ij}^{\perp}(\boldsymbol{r}-\boldsymbol{r}') = \int_{-\infty}^{\infty} e^{i\boldsymbol{k}\cdot(\boldsymbol{r}-\boldsymbol{r}')}\left(\delta_{ij} - \hat{k}_i\hat{k}_j\right)\frac{d^3k}{(2\pi)^3}\,, \tag{F.12}$$

where $\hat{\boldsymbol{k}}$ is the unit vector of $\boldsymbol{k}$. Applying $\delta^{\perp}$ to some arbitrary vector function $\boldsymbol{F}(\boldsymbol{r})$ gives

$$F_i^{\perp}(\boldsymbol{r}) = \int \delta_{ij}^{\perp}(\boldsymbol{r}-\boldsymbol{r}')F_j(\boldsymbol{r}')\,d^3r' = \int_{-\infty}^{\infty} e^{i\boldsymbol{k}\cdot(\boldsymbol{r}-\boldsymbol{r}')}\left(\delta_{ij} - \hat{k}_i\hat{k}_j\right)F_j(\boldsymbol{k})\,\frac{d^3k}{(2\pi)^3}\,.$$

When going from the first to the second expression we have used that a convolution in real space becomes a product in wavenumber space. Thus, the transverse delta-function projects on the transverse directions of $\boldsymbol{F}(\boldsymbol{r})$. We can rewrite Eq. (F.12) in a slightly different form. We start by working out the two terms in parentheses,

$$\delta_{ij}^{\perp}(\boldsymbol{r}-\boldsymbol{r}') = \delta_{ij}\delta(\boldsymbol{r}-\boldsymbol{r}') + \partial_i\partial_j\left(\int_{-\infty}^{\infty} e^{i\boldsymbol{k}\cdot(\boldsymbol{r}-\boldsymbol{r}')}\frac{1}{k^2}\frac{d^3k}{(2\pi)^3}\right)\,.$$

The integral on the right-hand side is the Fourier transform of the Coulomb potential $1/(4\pi|\boldsymbol{r}-\boldsymbol{r}'|)$. Thus, we can rewrite the transverse delta function in a real-space representation

$$\delta_{ij}^{\perp}(\boldsymbol{r}-\boldsymbol{r}') = \delta_{ij}\delta(\boldsymbol{r}-\boldsymbol{r}') + \partial_i\partial_j\left(\frac{1}{4\pi|\boldsymbol{r}-\boldsymbol{r}'|}\right)\,. \tag{F.13}$$

This relation suggests introducing the longitudinal delta function

Longitudinal Delta Function

$$\delta_{ij}^{L}(\boldsymbol{r}-\boldsymbol{r}') = -\partial_i\partial_j\left(\frac{1}{4\pi|\boldsymbol{r}-\boldsymbol{r}'|}\right)\,. \tag{F.14}$$

When applying the longitudinal delta function to some vector function we get

$$F_i^{L}(\boldsymbol{r}) = \partial_i\int\left[\partial_j'\frac{1}{4\pi|\boldsymbol{r}-\boldsymbol{r}'|}\right]F_j(\boldsymbol{r}')\,d^3r' = -\partial_i\int\frac{\partial_j'F_j(\boldsymbol{r}')}{4\pi|\boldsymbol{r}-\boldsymbol{r}'|}\,d^3r'\,,$$

where we have performed integration by parts to shuffle the derivative from the $1/|\boldsymbol{r}-\boldsymbol{r}'|$ term to $\boldsymbol{F}(\boldsymbol{r}')$. We have also neglected the additional boundary term which arises from the partial integration, a neglect that only works for a localized vector function $\boldsymbol{F}(\boldsymbol{r})$ becoming zero for large values of $\boldsymbol{r}$. We thus find

$$\boldsymbol{F}^L(\boldsymbol{r}) = -\nabla \int \frac{\nabla' \cdot \boldsymbol{F}(\boldsymbol{r}')}{4\pi |\boldsymbol{r}-\boldsymbol{r}'|}\, d^3r' . \tag{F.15}$$

With the transverse and longitudinal delta functions we immediately get

$$\delta_{ij}\delta(\boldsymbol{r}-\boldsymbol{r}') = \delta_{ij}^{\perp}(\boldsymbol{r}-\boldsymbol{r}') + \delta_{ij}^{L}(\boldsymbol{r}-\boldsymbol{r}') .$$

Thus, by applying $\delta^{\perp}$ and δ^{L} to some vector function $\boldsymbol{F}$, one can decompose it into its transverse and longitudinal parts. Note that the corresponding operations are nonlocal in space.

Green's Function Acting on Transverse Vector Function We conclude this appendix with the derivation of a useful relation for the Green's function acting on a transverse vector function. Consider first the expression

$$\int G(\boldsymbol{r},\boldsymbol{r}')\boldsymbol{F}^{\perp}(\boldsymbol{r}')\, d^3r' = \int \left[\frac{e^{ik|\boldsymbol{r}-\boldsymbol{r}'|}}{4\pi|\boldsymbol{r}-\boldsymbol{r}'|}\right] \boldsymbol{F}^{\perp}(\boldsymbol{r}')\, d^3r' ,$$

where $G(\boldsymbol{r},\boldsymbol{r}')$ is the scalar Green's function of Eq. (5.7), given by the term in brackets on right-hand side, and k is a wavenumber. Together with Eq. (F.13) we are led to

$$\int G(\boldsymbol{r},\boldsymbol{r}')F_i(\boldsymbol{r}')\, d^3r' + \int G(\boldsymbol{r},\boldsymbol{r}')\, \partial_i'\partial_j' \left[\frac{F_j(\boldsymbol{r}'')}{4\pi|\boldsymbol{r}'-\boldsymbol{r}''|}\right] d^3r'' d^3r' = \mathcal{I}_1 + \mathcal{I}_2 ,$$

with $\mathcal{I}_{1,2}$ denoting the first and second term on the left-hand side, respectively. The second term can be rewritten using the same procedure as in the derivation of Eq. (F.15), and we get

$$\mathcal{I}_2 = \int G(\boldsymbol{r}-\boldsymbol{r}')\,\nabla' \left(\frac{\nabla''\cdot \boldsymbol{F}(\boldsymbol{r}'')}{4\pi|\boldsymbol{r}'-\boldsymbol{r}''|}\right) d^3r'' d^3r' .$$

We next use the defining equation for the scalar Green's function to arrive at

$$\left(\nabla'^2 + k^2\right) G(\boldsymbol{r},\boldsymbol{r}') = -\delta(\boldsymbol{r}-\boldsymbol{r}') \implies G(\boldsymbol{r},\boldsymbol{r}') = -\frac{1}{k^2}\left[\delta(\boldsymbol{r}-\boldsymbol{r}') + \nabla'^2 G(\boldsymbol{r},\boldsymbol{r}')\right],$$

which finally leads us to

$$\mathcal{I}_2 = -\frac{1}{k^2}\left[\nabla \int \frac{\nabla'\cdot \boldsymbol{F}(\boldsymbol{r}')}{4\pi|\boldsymbol{r}-\boldsymbol{r}'|}\, d^3r' + \int G(\boldsymbol{r},\boldsymbol{r}')\nabla'\nabla'^2 \left(\frac{\nabla''\cdot \boldsymbol{F}(\boldsymbol{r}'')}{4\pi|\boldsymbol{r}'-\boldsymbol{r}''|}\right) d^3r'' d^3r'\right] .$$

In the second term we have performed integration by parts to shuffle the Laplacian ∇'^2 from the Green's function to the second term, and have again neglected all boundary terms. We have also used that the derivatives of the Laplacian and the nabla operator commute. With

$$\nabla'^2 \left(\frac{1}{4\pi|\boldsymbol{r}'-\boldsymbol{r}''|}\right) = -\delta(\boldsymbol{r}-\boldsymbol{r}') .$$

we can rewrite $\mathcal{I}_2$ in the form

$$\mathcal{I}_2 = -\frac{1}{k^2}\left[\nabla\int\frac{\nabla'\cdot\boldsymbol{F}(\boldsymbol{r}')}{4\pi|\boldsymbol{r}-\boldsymbol{r}'|}\,d^3r - \int G(\boldsymbol{r},\boldsymbol{r}')\nabla'\Big(\nabla'\cdot\boldsymbol{F}(\boldsymbol{r}')\Big)\,d^3r'\right].$$

We finally perform integration by parts in the second term in brackets in order to shuffle the derivatives from the vector function $\boldsymbol{F}$ to the scalar Green's function, ignoring again all boundary terms.

Putting together all results, we can rewrite the product of the scalar Green's function with a transverse vector function in the form

Integral of Green's Function and Transverse Vector Function

$$\int G(\boldsymbol{r},\boldsymbol{r}')\boldsymbol{F}^{\perp}(\boldsymbol{r}')\,d^3r' \tag{F.16}$$
$$= \int\left(1+\frac{\nabla\nabla}{k^2}\right)G(\boldsymbol{r},\boldsymbol{r}')\cdot\boldsymbol{F}(\boldsymbol{r}')\,d^3r' - \frac{1}{k^2}\nabla\int\frac{\nabla'\cdot\boldsymbol{F}(\boldsymbol{r}')}{4\pi|\boldsymbol{r}-\boldsymbol{r}'|}\,d^3r'.$$

The product of nabla operators in the first term on the right-hand side has to be understood as a dyadic product, in the same way as previously used for the dyadic Green's function, see Eq. (5.19). Note that the second term on the right-hand side is proportional to the longitudinal component of the vector function, $\boldsymbol{F}^L(\boldsymbol{r})$. Equation (F.16) is particularly useful when working in the Coulomb gauge, where the Coulomb potential is instantaneous and the vector potential transverse. See Chap. 13 for the use of Eq. (F.16).

References

1. D.J. Griffiths, *Introduction to Electrodynamics* (Pearson, San Francisco, 2008)
2. J.D. Jackson, *Classical Electrodynamics* (Wiley, New York, 1999)
3. B. Mahon, How Maxwell's equations came to light. Nat. Photonics **9**, 2–4 (2015)
4. L. Mandel, E. Wolf, *Optical Coherence and Quantum Optics* (Cambridge University Press, Cambridge, 1995)
5. B. Richards, E. Wolf, Electromagnetic simulation in optical systems II. Structure of the image field in an aplanatic system. Proc. R. Soc. Lond. Ser. A **253**, 358 (1959)
6. L. Novotny, B. Hecht, *Principles of Nano-Optics* (Cambridge University Press, Cambridge, 2012)
7. J. Dongarra, F. Sullivan, Guest editors introduction to the top 10 algorithms. Comput. Sci. Eng. **2**, 22 (2000)
8. W.H. Press, S.A. Teukolsky, W.T. Vetterling, B.P. Flannery, *Numerical Recipes in C++: The Art of Scientific Computing*, 2nd edn. (Cambridge University Press, Cambridge, 2002)
9. P.H. Jones, O.M. Marago, G. Volpe, *Optical Tweezers* (Cambridge University Press, Cambridge, 2015)
10. A. Gennerich (ed.), *Optical Tweezers* (Springer, Berlin, 2017)
11. O.M. Marago, P.H. Jones, P.G. Gucciardi, G. Volpe, A.C. Ferrari, Optical trapping and manipulation of nanostructures. Nat. Nanotechnol. **8**, 807 (2013)
12. S. Chu, Nobel lecture: the manipulation of neutral particles. Rev. Mod. Phys. **70**, 685–706 (1998)
13. F.M. Fazal, S.M. Block, Optical tweezers study life under tension. Nat. Photonics **5**, 318 (2011)
14. R.N.C. Pfeifer, T.A. Nieminen, N.R. Heckenberg, H. Rubinsztein-Dunlop, Colloquium: momentum of an electromagnetic wave in dielectric media. Rev. Mod. Phys. **79**, 1197–1216 (2007)
15. S.M. Barnett, Resolution of the Abraham-Minkowski dilemma. Phys. Rev. Lett. **104**, 070401 (2010)
16. A.M. Yao, M.J. Padgett, Orbital angular momentum: origins, behavior, and applications. Adv. Optics Photonics **3**, 161–204 (2011)
17. M.J. Padgett, Orbital angular momentum 25 years on. Opt. Express **25**, 11265 (2017)
18. K.T. Gahagan, G.A. Swartzlander, Simultaneous trapping of low-index and high-index nanoparticles observed with an optical-vortex trap. J. Opt. Soc. Am. B **16**, 533 (1999)
19. L. Challis, F. Sheard, The Green of the Green functions. Phy. Today **41** (2003)
20. W.C. Chew, *Waves and Fields in Inhomogeneous Media* (IEEE Press, Picsatoway, 1995)

U. Hohenester, *Nano and Quantum Optics*, Graduate Texts in Physics,
https://doi.org/10.1007/978-3-030-30504-8

21. J.A. Stratton, L.J. Chu, Diffraction theory of electromagnetic waves. Phys. Rev. **56**, 99–107 (1939)
22. E. Abbe, Beiträge zur Theorie des Mikroskops und der mikroskopischen Wahrnehmung. Archiv Mikroskop Anat. **9**, 413 (1873)
23. B. Hecht, B. Sick, U.P. Wild, V. Deckert, R. Zenobi, O.J.F. Martin, D.W. Pohl, Scanning near-field optical microscopy with aperture probes: fundamentals and applications. J. Chem. Phys. **112**, 7761 (2000)
24. M.A. Paesler, P.J. Moyer, *Near-Field Optics: Theory, Instrumentation, and Applications* (Wiley, New York, 1996)
25. H.A. Bethe, Theory of diffraction by small holes. Phys. Rev. **66**, 163 (1944)
26. C.J. Bouwkamp, On Bethe's theory of diffraction by small holes. Philips Res. Rep. **5**, 321 (1950)
27. H.F. Hess, E. Betzig, T.D. Harris, L.N. Pfeiffer, K.W. West, Near-field spectroscopy of the quantum constituents of a luminescent system. Science **264**, 1740 (1994)
28. E. Betzig, G.H. Patterson, R. Sougrat, O.W. Lindwasser, S. Olenych, J.S. Bonifacino, M.W. Davidson, J. Lippincott-Schwartz, H.F. Hess, Imaging intracellular fluorescent proteins at nanometer resolution. Science **313**, 1642–1645 (2006)
29. M.J. Rust, M. Bates, X. Zhuang, Sub diffraction-limit imaging by stochastic optical reconstruction microscopy (STORM). Nat. Methods **3**, 793–796 (2006)
30. S.W. Hell, J. Wichmann, Breaking the diffraction resolution limit by stimulated emission: stimulated-emission-depletion fluorescence microscopy. Op. Lett. **19**, 780–782 (1994)
31. P. Tinnefeld, C. Eggeling, S.W. Hell (eds.), *Far-Field Optical Nanoscopy* (Springer, Berlin, 2015)
32. R.E. Thompson, D.R. Larson, W.W. Webb, Precise nanometer localization analysis for individual fluorescent probes. Biophys. J. **82**, 2775–2783 (2002)
33. F. Göttfert, C.A. Wurm, V. Mueller, S. Berning, V.C. Cordes, A. Honigmann, S.W. Hell, Coaligned dual-channel STED nanoscopy and molecular diffusion analysis at 20 nm resolution. Biophys. J. **105**, L01–L03 (2013)
34. P.B. Johnson, R.W. Christy, Optical constants of the noble metals. Phys. Rev. B **6**, 4370 (1972)
35. E.D. Palik, *Handbook of Optical Constants of Solids* (Academic, San Diego, 1985)
36. N.W. Ashcroft, N.D. Mermin, *Solid State Physics* (Saunders, Fort Worth, 1976)
37. A.H. Castro Neto, F. Guinea, N.M.R. Peres, K.S. Novoselov, A.K. Geim, The electronic properties of graphene. Rev. Mod. Phys. **81**, 109 (2009)
38. F.J. Garcia de Abajo, Graphene plasmonics: challenges and opportunities. ACS Photonics **1**, 135 (2014)
39. B. Wunsch, T. Stauber, F. Sols, F. Guinea, Dynamical polarization of graphene at finite doping. New J. Phys. **8**, 318 (2006)
40. E.H. Hwang, S. Das Sarma, Dielectric function, screening, and plasmons in 2d graphene. Phys. Rev. B **75**, 205418 (2007)
41. J.B. Pendry, A.J. Holden, D.J. Robbins, W.J. Stewart, Magnetism from conductors, and enhanced non-linear phenomena. IEEE Trans. Microwave Theory Tech. **47**, 2075 (1999)
42. C.M. Soukoulis, M. Wegener, Past achievements and future challenges in the development of three-dimensional photonic metamaterials. Nat. Photonics **5**, 523 (2011)
43. R.J. Potton, Reciprocity in optics. Rep. Prog. Phys. **67**, 717 (2004)
44. H. Atwater, The promise of plasmonics. Sci. Am. **296**(4), 56 (2007)
45. J. Heber, News feature: surfing the wave. Nature **461**, 720 (2009)
46. A. Otto, Excitation of nonradiative surface plasma waves in silver by the method of frustrated total reflection. Z. Phys. **216**(4), 398–410 (1968)
47. E. Kretschmann, Die Bestimmung optischer Konstanten von Metallen durch Anregung von Oberflächenplasmaschwingungen. Z. Phys. **241**, 313 (1971)
48. T.W. Ebbesen, H.J. Lezec, H.F. Ghaemi, T. Thio, P.A. Wolff, Extraordinary optical transmission through sub-wavelength hole arrays. Nature **391**, 667–669 (1998)
49. S. Xiao, X. Zhu, B.-H. Li, N.A. Mortensen, Graphene-plasmon polaritons: from fundamental properties to potential applications. Front. Phys. **11**, 117801 (2016).

50. J. Chen, M. Badioli, P. Alonso-Gonzalez, S. Thongrattanasiri, F. Huth, J. Osmond, M. Spasenovic, A. Centeno, A. Pesquera, P. Godignon, A. Z. Elorza, N. Camara, F.J. Garcia de Abajo, R. Hillenbrand, F. Koppens, Optical nano-imaging of gate-tunable graphene plasmons. Nature **487**, 77 (2012)
51. Z. Fei, A.S. Rodin, G.O. Andreev, W. Bao, A.S. McLeod, M. Wagner, L.M. Zhang, Z. Zhao, G. Dominguez M. Thiemens, M.M. Fogler, A.H. Castro Neto, C.N. Lau, F. Keilmann, D.N. Basov, Gate-tuning of graphene plasmons revealed by infrared nano-imaging. Nature **487**, 82 (2012)
52. M.A. Cooper, Optical biosensors in drug discovery. Nat. Rev. Drug Discov. **1**, 515 (2002)
53. V.G. Veselago, The electrodynamics of substances with simultaneously negative values of ε and μ. Sov. Phys. Uspekhi **56**, 509 (1964)
54. J.B. Pendry, Negative refraction makes a perfect lens. Phys. Rev. Lett. **85**, 3966 (2000)
55. K.Y. Bliokh, Y.P. Bliokh, V. Freilikher, S. Savel'ev, F. Nori, Colloquium: unusual resonators: plasmonics, metamaterials, and random media. Rev. Mod. Phys. **80**, 1201–1213 (2008)
56. J.B. Pendry, D. Schurig, D.R. Smith, Controlling electromagnetic fields. Science **312**, 1780 (2006)
57. U. Leonhardt, Optical conformal mapping. Science **312**, 1777 (2006)
58. D. Schurig, J.J. Mock, B.J. Justice, S.A. Cummer, J.B. Pendry, A.F. Starr, D.R. Smith, Metamaterial electromagnetic cloak at microwave frequencies. Science **314**, 977 (2006)
59. N. Fang, H. Lee, C. Sun, X. Zhang, Subdiffraction-limited optical imaging with a silver superlens. Science **308**, 534 (2005)
60. C.F. Bohren, D.R. Huffman, *Absorption and Scattering of Light* (Wiley, New York, 1983)
61. F.J. García de Abajo, J. Aizpurua, Numerical simulation of electron energy loss near inhomogeneous dielectrics. Phys. Rev. B **56**, 15873 (1997)
62. G. Boudarham, M. Kociak, Modal decompositions of the local electromagnetic density of states and spatially resolved electron energy loss probability in terms of geometric modes. Phys. Rev. B **85**, 245447 (2012)
63. F.-P. Schmidt, H. Ditlbacher, U. Hohenester, A. Hohenau, F. Hofer, J.R. Krenn, Dark plasmonic breathing modes in silver nanodisks. Nano Lett. **12**, 5780 (2012)
64. M.I. Stockman, Nanoplasmonics: past, present, and glimpse into future. Opt. Express **19**, 22029 (2011)
65. I.D. Mayergoyz, D.R. Fredkin, Z. Zhang, Electrostatic (plasmon) resonances in nanoparticles. Phys. Rev. B **72**, 155412 (2005)
66. P. Zijlstra, P.M. Paulo, M. Orrit, Optical detection of single non-absorbing molecules using the surface plasmon resonance of a gold nanorod. Nat. Nanotechnol. **7**, 379 (2012)
67. J. Becker, A. Trügler, A. Jakab, U. Hohenester, C. Sönnichsen, The optimal aspect ratio of gold nanorods for plasmonic bio-sensing. Plasmonics **5**, 161 (2010)
68. E. Prodan, C. Radloff, N.J. Halas, P. Nordlander, Hybridization model for the plasmon response of complex nanostructures. Science **302**, 419 (2003)
69. A. Aubry, D. Yuan Lei, A.I. Fernandez-Dominguez, Y. Sonnefraud, S.A. Maier, J.B. Pendry, Plasmonic light-harvesting devices over the whole visible spectrum. Nano Lett. **10**, 2574 (2010)
70. R.C. McPhedran, W.T. Perrins, Electrostatic and optical resonances of cylinder pairs. Appl. Phys. **24**, 311 (1981)
71. A. Aubry, D. Yuan Lei, S.A. Maier, J.B. Pendry, Conformal transformation applied to plasmonics beyond the quasistatic limit. Phys. Rev. B **82**, 205109 (2010)
72. D.Y. Lei, A. Aubry, S.A. Maier, J.B. Pendry, Broadband nano-focusing of light using kissing nanowires. New. J. Phys. **12**, 093030 (2010)
73. W. Zhu, R. Esteban, A.G. Borisov, J.J. Baumberg, P. Nordlander, H.J. Lezec, J. Aizpurua, K.B. Crozier, Quantum mechanical effects in plasmonic structures with subnanometre gaps. Nat. Commun. **7**, 11495 (2016)
74. G. Mie, Beiträge zur Optik trüber Medien, speziell kolloidaler Metallösungen. Ann. Phys. **330**, 377 (1908)

75. Y. Chang, R. Harrington, A surface formulation for characteristic modes of material bodies. IEEE Trans. Antennas Propag. **25**(6), 789–795 (1977)
76. A.J. Poggio, E.K. Miller, Chapter 4: integral equation solutions of three-dimensional scattering problems, in *Computer Techniques for Electromagnetics*, ed. by R. Mittra. International Series of Monographs in Electrical Engineering (Pergamon, 1973), pp. 159–264
77. T.K. Wu, L.L. Tsai, Scattering from arbitrarily-shaped lossy dielectric bodies of revolution. Radio Sci. **12**(5), 709–718 (1977)
78. P.T. Leung, S.Y. Liu, K. Young, Completeness and orthogonality of quasinormal modes in leaky optical cavities. Phys. Rev. A **49**, 3057 (1994)
79. C. Sauvan, J.P. Hugonin, I.S. Maksymov, P. Lalanne, Theory of the spontaneous optical emission of nanosize photonic and plasmon resonators. Phys. Rev. Lett. **110**, 237401 (2013)
80. F. Ouyang, M. Isaacson, Surface plasmon excitation of objects with arbitrary shape and dielectric constant. Philos. Mag. B **60**, 481 (1989)
81. J. Petersen, J. Volz, A. Rauschenbeutel, Chiral nanophotonic waveguide interface based on spin-orbit interaction of light. Science **346**, 67 (2014)
82. E.M. Purcell, H.C. Torry, R.V. Pound, Resonance absorption by nuclear magnetic moments in a solid. Phys. Rev. **69**, 37 (1946)
83. R. Carminati, J.J. Greffet, C. Henkel, J.M. Vigoureux, Radiative and non-radiative decay of a single molecule close to a metallic nanoparticle. Opt. Commun. **216**, 368 (2006)
84. P. Anger, P. Bharadwaj, L. Novotny, Enhancement and quenching of single-molecule fluorescence. Phys. Rev. Lett. **96**, 113002 (2006)
85. A. Hörl, G. Haberfehlner, A. Trügler, F. Schmidt, U. Hohenester, G. Kothleitner, Tomographic reconstruction of the photonic environment of plasmonic nanoparticles. Nat. Commun. **8**, 37 (2017)
86. K. Joulain, R. Carminati, J.-P. Mulet, J.-J. Greffet, Definition and measurement of the local density of electromagnetic states close to an interface. Phys. Rev. B **68**, 245405 (2003)
87. K.H. Drexhage, Influence of a dielectric interface on fluorescence decay time. J. Lumin. **12**, 693 (1970)
88. R.R. Chance, A. Prock, R. Silbey, *Molecular Fluorescence and Energy Transfer Near Interface*, vol. 37 (Wiley, New York, 1978).
89. E.C. Le Ru, P.G. Etchegoin, *Principles of Surface Enhanced Raman Spectroscopy* (Elsevier, Amsterdam, 2009)
90. S. Nie, S.R. Emory, Probing single molecules and single nanoparticles by surface enhanced raman scattering. Science **275**, 1102 (1997)
91. M. Fleischmann, P.J. Hendra, A.J. McQuillan, Raman spectra of pyridine adsorbed at a silver electrode. Chem. Phys. Lett. **26**, 163 (1974)
92. K. Kneipp, M. Moskovits, M. Kneipp (eds.), *Surface Enhanced Raman Scattering* (Springer, Berlin, 2008)
93. T. Förster, Energiewanderung und Fluoreszenz. Naturwissenschaften **33**, 166 (1946)
94. P. Andrew, W.L. Barnes, Energy transfer across a metal film mediated by surface plasmon polaritons. Science **306**, 1002 (2004)
95. J.I. Gersten, A. Nitzan, Accelerated energy transfer between molecules near a solid particle. Chem. Phys. Lett. **104**, 31 (1984)
96. C. Cherqui, N. Thakkar, G. Li, J.P. Camden, D.J. Masiello, Characterizing localized surface plasmons using electron energy-loss spectroscopy. Annu. Rev. Phys. Chem. **67**, 331 (2015)
97. C.J. Powell, J.B. Swan, Origin of the characteristic electron energy losses in aluminum. Phys. Rev. **115**, 869 (1959)
98. M. Bosman, V.J. Keast, M. Watanabe, A.I. Maaroof, M.B. Cortie, Mapping surface plasmons at the nanometre scale with an electron beam. Nanotechnology **18**, 165505 (2007)
99. J. Nelayah, M. Kociak, O. Stephan, F.J. García de Abajo, M. Tence, L. Henrard, D. Taverna, I. Pastoriza-Santos, L. M. Liz-Martin, C. Colliex, Mapping surface plasmons on a single metallic nanoparticle. Nat. Phys. **3**, 348 (2007)
100. F.J. García de Abajo, Optical excitations in electron microscopy. Rev. Mod. Phys. **82**, 209 (2010)

101. C. Colliex, M. Kociak, O. Stephan, Electron energy loss spectroscopy imaging of surface plasmons at the nanoscale. Ultramicroscopy **162**, A1 (2016)
102. U.S. Inan, R.A. Marshall, *Numerical Electromagnetics* (Cambridge University Press, Cambridge, 2011)
103. A. Taflove, S.C. Hagness, *Computational electrodynamics* (Artech House, Boston, 2005)
104. K.S. Yee, Numerical solution of initial boundary value problems involving Maxwell's equations in isotropic media. IEEE Trans. Antennas Propag. **14**, 302 (1966)
105. A. Taflove, M.E. Browdin, Numerical solution of steady-state electromagnetic scattering problems using the time-dependent Maxwell's equations. IEEE Trans. Microwave Theory Tech. **23**, 623 (1975)
106. A. Taflove, M.E. Browdin, Computation of the electromagnetic fields and induced temperatures within a model of the microwave-irradiated human eye. IEEE Trans. Microwave Theory Tech. **23**, 888 (1975)
107. J. Berenger, A perfectly matched layer for the absorption of electromagnetic waves. J. Comput. Phys. **114**, 185 (1994)
108. R. Fuchs, S.H. Liu, Sum rule for the polarizability of small particles. Phys. Rev. B **14**, 5521 (1976)
109. F.J. García de Abajo, A. Howie, Retarded field calculation of electron energy loss in inhomogeneous dielectrics. Phys. Rev. B **65**, 115418 (2002)
110. U. Hohenester, A. Trügler, MNPBEM—a Matlab Toolbox for the simulation of plasmonic nanoparticles. Comp. Phys. Commun. **183**, 370 (2012)
111. A.M. Kern, O.J.F. Martin, Surface integral formulation for 3D simulations of plasmonic and high permittivity nanostructures. J. Opt. Soc. Am. A **26**, 732 (2009)
112. P. Arcioni, M. Bressan, L. Perregrini, On the evaluation of the double surface integrals arising in the application of the boundary integral method to 3d problems. IEEE Trans. Microwave Theory Tech. **45**, 436 (1997)
113. D.J. Taylor, Accurate and efficient numerical integration of weakly singulars integrals in Galerkin EFIE solutions. IEEE Trans. Antennas Propag. **51**, 2543 (2003)
114. S. Sarraf, E. Lopez, G. Rios Rodriguez, J. D'Elia, Validation of a Galerkin technique on a boundary integral equation for creeping flow around a torus. Comp. Appl. Math. **33**, 63 (2014)
115. J.S. Hesthaven, T. Warburton, High-order/spectral methods on unstructured grids I. time-domain solution of Maxwell's equations. J. Comput. Phys. **181**, 186 (2002)
116. J.S. Hesthaven, High-order accurate methods in time-domain computational electromagnetics: a review. Adv. Imaging Electron Phys. **127**, 59–123 (2003)
117. J.C. Nedelec, Mixed finite elements in R3. Numer. Math. **35**, 315 (1980)
118. K. Busch, M. König, J. Niegemann, Discontinuous Galerkin method in nanophotonics. Laser Photonics Rev. **5**, 773–809 (2011)
119. D.F. Walls, G.J. Millburn, *Quantum Optics* (Springer, Berlin, 1995)
120. W. Vogel, D. Welsch, *Quantum Optics* (Wiley, Weinheim, 2006)
121. R. Glauber, Nobel lecture: one hundred years of light quanta. Rev. Mod. Phys. **78**, 1267 (2006)
122. P.A.M. Dirac, *Lectures on Quantum Field Theory* (Academic Press, New York, 1966)
123. C. Cohen-Tannoudji, J. Dupont-Roc, G. Grynberg, *Photons and Atoms* (Wiley, New York, 1989)
124. A.L. Fetter, J.D. Walecka, *Quantum Theory of Many-Particle Systems* (McGraw-Hill, New York, 1971)
125. G.D. Mahan, *Many-Particle Physics* (Plenum, New York, 1981)
126. D. Pines, P. Nozieres, *The Theory of Quantum Liquids* (Benjamin, New York, 1966)
127. M.O. Scully, M.S. Zubairy, *Quantum Optics* (Cambridge University Press, Cambridge, 1997)
128. W.B. Case, Wigner functions and Weyl transforms for pedestrians. Am. J. Phys. **76**, 937 (2008)
129. E. Altewischer, M.P. van Exter, J.P. Woerdman, Plasmon-assisted transmission of entangled photons. Nature **418**, 304 (2002)

130. S.I. Bozhevolnyi, L. Martin-Moreno, F. Garcia-Vidal (eds.), *Quantum Plasmonics*. Springer Series in Solid State Sciences (Springer, Berlin, 2017)
131. E. A. Power, S. Zienau, H.S. Massey, Coulomb gauge in non-relativistic quantum electrodynamics and the shape of spectral lines. Philos. Trans. R. Soc. Lond. Ser. A **251**, 457 (1959)
132. R.G. Woolley, Molecular quantum electrodynamics. Proc. R. Soc. Lond. Ser. A **321**, 557 (1971)
133. D.L. Adrews, G.A. Jones, A. Salam, R. Woolley, Perspective: Quantum Hamiltonians for optical interactions. J. Chem. Phys. **148**, 040901 (2018)
134. S. Scheel, S.Y. Buhmann, Macroscopic quantum electrodynamics—concepts and applications. Acta Phys. Slovaca **58**, 675 (2008)
135. A. Eguiluz, J.J. Quinn, Hydrodynamic model for surface plasmons in metals and degenerate semiconductors. Phys. Rev. B **14**, 1347–1361 (1976)
136. S. Raza, S.I. Bozhevolnyi, M. Wubs, N.A. Mortensen, Nonlocal optical response in metallic nanostructures. J. Phys. Condens. Matter **27**, 183204 (2015)
137. B.B. Dasgupta, R. Fuchs, Polarizability of a small sphere including nonlocal effects. Phys. Rev. B **24**, 554 (1981)
138. P. Halevi, R. Fuchs, Gerneralised additional boundary conditions for non-local dielectrics: I. Reflectivity. J. Phys. C Solid State Phys. **17**, 3869 (1984)
139. J.A. Scholl, A.L. Koh, J.A. Dionne, Quantum plasmon resonances of individual metallic nanoparticles. Nature **483**, 421 (2012)
140. C. Ciraci, R.T. Hill, Y. Urzhumov, A.I. Fernandez-Dominguez, S.A. Maier, J.B. Pendry, A. Chilkoti, D.R. Smith, Probing the ultimate limits of plasmonic enhancement. Science **337**, 1072 (2012)
141. I.S. Gradshteyn, I.M. Ryzhik, *Table of Integrals, Series, and Products* (Academic Press, San Diego, 2000)
142. R. Fuchs, K.L. Kliewer, Surface plasmon in a semi-infinite free-electron gas. Phys. Rev. B **3**, 2270–2278 (1971)
143. S. Raza, W. Yan, N. Stenger, M. Wubs, N.A. Mortensen, Blueshift of the surface plasmon resonance shift in silver nanoparticles. Opt. Express **21**, 27344 (2013)
144. Y. Luo, A.I. Fernandez-Dominguez, A. Wiener, S.A. Maier, J.B. Pendry, Surface plasmons and nonlocality: a simple model. Phys. Rev. Lett. **111**, 093901 (2013)
145. A. Trügler, U. Hohenester, F.J. Garcia de Abajo, Plasmonics simulations including nonlocal effects using a boundary element method approach. Int. J. Mod. Phys. B **31**, 1740007 (2017)
146. P.J. Feibelman, Surface electromagnetic fields. Prog. Surf. Sci. **12**, 287 (1982)
147. P. Apell, A simple derivation of the surface contribution to the reflectivity of a metal, and its use in the van der Waals interaction. Phys. Scr. **24**, 795 (1981)
148. J.M. Pitarke, V.M. Silkin, E.V. Chulkov, P.M. Echenique, Theory of surface plasmons and surface-plasmon polaritons. Rep. Prog. Phys. **70**(1), 1–87 (2007)
149. T. Christensen, W. Yan, A.-P. Jauho, M. Soljacic, N.A. Mortensen, Quantum corrections in nanoplasmonics: shape, scale, and material. Phys. Rev. Lett. **118**, 157402 (2017)
150. T.V. Teperik, P. Nordlander, J. Aizpurua, A.G. Borisov, Robust subnanometric plasmon ruler by rescaling of the nonlocal optical response. Phys. Rev. Lett. **110**, 263901 (2013)
151. R.M. Martin, L. Reining, D.M. Ceperly, *Interacting Electrons* (Cambridge University Press, Cambridge, 2016)
152. A. Varas, P. Garcia-Gonzalez, J. Feist, F.J. Garcia-Vidal, A. Rubio, Quantum plasmonics: from jellium models to ab initio calculations. Nanophotonics **5**, 409 (2016)
153. K.J. Savage, M.M. Hawkeye, R. Esteband, A.G. Borisov, J. Aizpurua, J.J. Baumberg, Revealing the quantum regime in tunneling plasmonics. Nature **491**, 574 (2012)
154. J.A. Scholl, A. Garcia-Etxarri, A. Leen Koh, J.A. Dionne, Observation of quantum tunneling between two plasmonic nanoparticles. Nano Lett. **13**, 564 (2013)
155. S.F. Tan, L. Wu, J.K.W. Yang, P. Bai, M. Bosman, C.A. Nijhuis, Quantum plasmon resonances controlled by molecular tunnel junctions. Science **343**, 1496 (2014)
156. R. Esteban, A.G. Borisov, P. Nordlander, J. Aizpurua, Bridging quantum and classical plasmonics with a quantum-corrected model. Nat. Commun. **3**, 825 (2012)

157. R. Esteban, A. Zugarramurdi, P. Zhang, P. Nordlander, F. J. Garcia-Vidal, A.G. Borisov, J. Aizpurua, A classical treatment of optical tunneling in plasmonic gaps: extending the quantum corrected model to practical situations. Faraday Discuss. **178**, 151 (2015)
158. A. Messiah, *Quantum Mechanics* (North-Holland, Amsterdam, 1965)
159. R. Esteban, G. Aguirregabiria, A.G. Borisov, Y.M. Wang, P. Nordlander, G.W. Bryant, J. Aizpurua, The morphology of narrow gaps modifies the plasmonic response. ACS Photonics **2**, 295 (2015)
160. D. Knebl, A. Hörl, A. Trügler, J. Kern, J.R. Krenn, P. Puschnig, U. Hohenester, Gap plasmonics of silver nanocube dimers. Phys. Rev. B **93**, 081405 (2016)
161. R.H. Ritchie, Plasma losses by fast electrons in thin films. Phys. Rev. **106**, 874 (1957)
162. R.H. Ritchie, A. Howie, Inelastic scattering probabilities in scanning transmission electron microscopy. Philos. Mag. A **5**, 753 (1988)
163. M.J. Lagos, A. Trügler, U. Hohenester, P.E. Batson, Mapping vibrational surface and bulk modes in a single nanocube. Nature **543**, 533 (2017)
164. K. Joulain, J.-P. Mulet, F. Marquier, R. Carminati, J.-J. Greffet, Surface electromagnetic waves thermally excited: radiative heat transfer, coherence properties, and casimir forces revisited in the near field. Surf. Sci. Rep. **57**, 59 (2005)
165. A.W. Rodriguez, F. Capasso, S.G. Johnson, The Casimir effect in microstructured geometries. Nature Photonics **5**, 211 (2011)
166. E. Rousseau, A. Siria, G. Jourdan, S. Volz, F. Comin, J. Chevrier, J.J. Greffet, Radiative heat transfer at the nanoscale. Nature Photonics **3**, 514 (2009)
167. S.-A. Biehs, P. Ben-Andallah, F.S. Rosa, Nanoscale radiative heat transfer and its applications, in *Infrared Radiation* (InTech, London, 2012), p. 1
168. S.Y. Buhmann, *Dispersion Forces I* (Springer, Berlin, 2012)
169. S.Y. Buhmann, *Dispersion Forces II* (Springer, Berlin, 2012)
170. G. Baffou, *Thermoplasmonics* (Cambridge University Press, Cambridge, 2018)
171. W. Eckhardt, Macroscopic theory of electromagnetic fluctuations and stationary radiative heat transfer. Phys. Rev. A **29**, 1991 (1984)
172. H. Carmichael, *An Open Systems Approach to Quantum Optics*. Lecture Notes in Physics, vol. 18 (Springer, Berlin, 1991)
173. A.I. Volokitin, B.N.J. Persson, Near-field radiative heat transfer and noncontact friction. Rev. Mod. Phys. **79**, 1291 (2007)
174. H. Casimir, On the attraction between two perfectly conducting plates. Proc. K. Ned. Akad. Wet. **51**, 793 (1948)
175. M. Bordaga, U. Mohideenb, V.M. Mostepanenkoc, New developments in the Casimir effect. Phys. Rep. **353**(1), 1–205 (2001)
176. G.L. Klimchitskaya, U. Mohideen, V.M. Mostepanenko, The Casimir force between real materials: experiment and theory. Rev. Mod. Phys. **81**, 1827–1885 (2009)
177. E.M. Lifshitz, The theory of molecular attractive forces between solids. Soc. Phys. JETP **2**, 73 (1956)
178. J.L. Garrett, D.A. Somers, J.N. Munday, Measurement of the Casimir force between two spheres. Phys. Rev. Lett. **120**, 040401 (2018)
179. J.-J. Greffet, R. Carminati, K. Joulain, J.-P. Mulet, S. Mainguy, Y. Chen, Coherent emission of light by thermal sources. Nature **416**, 61 (2002)
180. J.R. Howell, M.P. Menguc, R. Siegel, *Thermal Radiation Heat Transfer* (CRC Press, Boca Raton, 2015)
181. W. Lei, Z. Shan, S. Fan, Photonic thermal management of coloured objects. Nature Commun. **9**, 4240 (2018)
182. M. Krüger, G. Bimonte, T. Emig, M. Kardar, Trace formulas for nonequilibrium Casimir interactions, heat radiation, and heat transfer for arbitrary objects. Phys. Rev. B **86**, 115423 (2012)
183. A. Narayanaswamy, Y. Zheng, A Green's function formalism of energy and momentum transfer in fluctuational electrodynamics. J. Quant. Spectrosc. Radiat. Transf. **132**, 12 (2014)

184. E.T. Jaynes, F.W. Cummings, Comparison of quantum and semiclassical radiation theories with application to the beam maser. Proc. IEEE **51**, 89 (1963)
185. S.M. Barnett, P.M. Radmore, *Methods in Theoretical Quantum Optics* (Clarendon, Oxford, 1997)
186. M.B. Plenio, P.L. Knight, The quantum-jump approach to dissipative dynamics in quantum optics. Rev. Mod. Phys. **70**, 101 (1998)
187. H.-P. Breuer, F. Petruccione, *Open Quantum Systems* (Oxford University Press, New York, 2002)
188. R.D. Artuso, G.W. Bryant, Strongly coupled quantum dot-metal nanoparticle systems: exciton-induced transparency, discontinuous response, and suppression as driven quantum oscillator effects. Phys. Rev. B **82**, 195419 (2010)
189. C.H. Townes, *How the Laser Happened* (Oxford University Press, Oxford, 1999)
190. D.J. Bergman, M.I. Stockman, Surface plasmon amplification by stimulated emission of radiation: quantum generation of coherent surface plasmons in nanosystems. Phys. Rev. Lett. **90**, 027402 (2003)
191. M.A. Noginov, G. Zhu, A.M. Belgrave, R. Bakker, V.M. Shalaev, E.E. Narimanov, S. Stout, E. Herz, T. Suteewong, U. Wiesner, Demonstration of a spaser-based nanolaser. Nature **460**, 1110 (2009)
192. M. Premaratne, M. I. Stockman, Theory and technology of SPASERs. Adv. Opt. Photonics **9**, 79 (2017)
193. R.F. Oulton, V.J. Sorger, T. Zentgraf, R.M. Ma, C. Gladden, L. Dai, G. Bartal, X. Zhang, Plasmon laser at deep subwavelength scale. Nature **461**, 629 (2009)
194. M.I. Stockman, The spaser as a nanoscale quantum generator and ultrafast amplifier. J. Opt. **12**, 024004 (2010)
195. E. Fick, G. Sauermann, *The Quantum Statistics of Dynamic Processes* (Springer, Berlin, 1990)
196. A.J. Leggett, S. Chakravarty, A.T. Dorsey, M.P. A. Fisher, A. Garg, W. Zwerger, Dynamics of the dissipative two-state system. Rev. Mod. Phys. **59**, 1 (1987).
197. S. Grandi, K.D. Major, C. Polisseni, S. Boissier, A.S. Clark, E.A. Hinds, Quantum dynamics of a driven two-level molecule with variable dephasing. Phys. Rev. A **94**, 063839 (2016)
198. T. Heindel, A. Thoma, M. von Helversen, M. Schmidt, A. Schlehahn, M. Gschrey, P. Schnauber, J. H. Schulze, A. Strittmatter, J. Beyer, S. Rodt, A. Carmele, A. Knorr, S. Reitzenstein, A bright triggered twin-photon source in the solid state. Nat. Commun. **8**, 14870 (2017)
199. P. Johansson, H. Xu, M. Käll, Surface-enhanced Raman scattering and fluorescence near metal nanoparticles. Phys. Rev. B **72**, 035427 (2005)
200. M.K. Schmidt, R. Esteban, A. Gonzalez-Tudela, G. Giedke, J. Aizpurua, Quantum mechanical description of raman scattering from molecules in plasmonic cavities. ACS Nano **10**, 6291 (2016)
201. M. Paulus, P. Gay-Balmaz, O.J.F. Martin, Accurate and efficient computation of the Green's tensor for stratified media. Phys. Rev. E **62**, 5797 (2000)
202. J.J. Sakurai, *Modern Quantum Mechanics* (Addison, Reading, 1994)
203. Y.S. Kim, P.T. Leung, T.F. George, Classical decay rates for molecules in the presence of a spherical surface: A complete treatment. Surf. Sci. **195**, 1 (1988)
204. J. Gersten, A. Nitzan, Radiative properties of solvated molecules in dielectric clusters and small particles. J. Chem. Phys. **95**, 686 (1991)

Index

U. Hohenester, *Nano and Quantum Optics*, Graduate Texts in Physics,
https://doi.org/10.1007/978-3-030-30504-8

N

O

P

Q

Printed in the United States
By Bookmasters